VOLUME ONE HUNDRED AND THIRTY ONE

INTERNATIONAL REVIEW OF NEUROBIOLOGY

Gut Microbiome and Behavior

INTERNATIONAL REVIEW OF NEUROBIOLOGY

VOLUME 131

SERIES EDITOR

PETER JENNER

Division of Pharmacology and Therapeutics
GKT School of Biomedical Sciences
King's College, London, UK

VOLUME ONE HUNDRED AND THIRTY ONE

INTERNATIONAL REVIEW OF NEUROBIOLOGY

Gut Microbiome and Behavior

Edited by

J. F. CRYAN

Department of Anatomy and Neuroscience, University College Cork, Cork, Ireland

G. CLARKE

Department of Psychiatry and Neurobehavioural Science, University College Cork, Cork, Ireland

ELSEVIER

AMSTERDAM • BOSTON • HEIDELBERG • LONDON
NEW YORK • OXFORD • PARIS • SAN DIEGO
SAN FRANCISCO • SINGAPORE • SYDNEY • TOKYO

Academic Press is an imprint of Elsevier

Academic Press is an imprint of Elsevier
50 Hampshire Street, 5th Floor, Cambridge, MA 02139, United States
525 B Street, Suite 1800, San Diego, CA 92101-4495, United States
The Boulevard, Langford Lane, Kidlington, Oxford OX5 1GB, United Kingdom
125 London Wall, London, EC2Y 5AS, United Kingdom

First edition 2016

ISBN: 978-0-12-803949-6
ISSN: 0074-7742

For information on all Academic Press publications
visit our website at https://www.elsevier.com/

Publisher: Zoe Kruze
Acquisition Editor: Kirsten Shankland
Editorial Project Manager: Hannah Colford
Production Project Manager: Surya Narayanan Jayachandran
Cover Designer: Miles Hitchens

CONTENTS

CONTRIBUTORS

M.T. Bailey
Center for Microbial Pathogenesis, The Research Institute at Nationwide Children's Hospital; The Institute for Behavioral Medicine Research (IBMR) at The Ohio State University; The Ohio State University College of Medicine, Columbus, OH, United States

K.L. Bates
US Air Force Academy, Colorado Springs, CO, United States

J.A. Bravo
Grupo de NeuroGastroBioquímica, Instituto de Química, Facultad de Ciencias, Pontificia Universidad Católica de, Valparaíso, Chile

L.A. Brenner
Military and Veteran Microbiome Consortium for Research and Education (MVM-CoRE); Rocky Mountain Mental Illness Research Education and Clinical Center, Denver; University of Colorado, Aurora, CO, United States

P.W.J. Burnet
University of Oxford, Oxford, United Kingdom

S.M. Collins
The Farncombe Family Digestive Health Research Centre, The Michael G DeGroote School of Medicine, McMaster University, Hamilton, ON, Canada

S.R. Dash
Food and Mood Centre, IMPACT SRC, Deakin University, School of Medicine, Geelong; Collaborative Research Centre for Mental Health, Carlton, VIC, Australia

S.L. Dawson
Food and Mood Centre, IMPACT SRC, Deakin University, School of Medicine, Geelong; Early Life Epigenetics Group, Murdoch Childrens Research Institute (MCRI), Royal Children's Hospital, Parkville, VIC, Australia

C.G.M. de Theije
Laboratory of Neuroimmunology and Developmental Origins of Disease, Academic Medical Centre, Utrecht University, Utrecht, The Netherlands

P.A. Engen
Rush University Medical Center, Chicago, IL, United States

S.E. Erdman
Division of Comparative Medicine, Massachusetts Institute of Technology, Cambridge, MA, United States; Aristotle University of Thessaloniki, Thessaloniki, Greece

M. Fleshner
Center for Neuroscience, University of Colorado, Boulder, CO, United States

C.B. Forsyth
Rush University Medical Center, Chicago, IL, United States

J.A. Foster
McMaster University, St. Joseph's Healthcare, Hamilton, ON, Canada

M.G. Gareau
School of Veterinary Medicine, University of California Davis, Davis, CA, United States

J. Garssen
Division of Pharmacology, Utrecht Institute for Pharmaceutical Sciences, Faculty of Science, Utrecht University; Nutricia Research, Utrecht, The Netherlands

O.M. Gomez
Center for Neuroscience, University of Colorado Boulder, Boulder, CO, United States

S.J. Green
DNA Services Facility, Research Resources Center, University of Illinois at Chicago, Chicago, IL, United States

T.L. Gur
Wexner Medical Center at The Ohio State University; The Institute for Behavioral Medicine Research (IBMR) at The Ohio State University, Columbus, OH, United States

A.L. Halweg-Edwards
University of Colorado Boulder, Boulder, CO, United States

S. Harty
University of Oxford, Oxford, United Kingdom

A.J. Hoisington
US Air Force Academy, Colorado Springs; Military and Veteran Microbiome Consortium for Research and Education (MVM-CoRE), Denver, CO, United States

P. Holzer
Research Unit of Translational Neurogastroenterology, Institute of Experimental and Clinical Pharmacology, Medical University of Graz; BioTechMed-Graz, Graz, Austria

F.N. Jacka
Food and Mood Centre, IMPACT SRC, Deakin University, School of Medicine, Geelong; Centre for Adolescent Health, Murdoch Children's Research Institute (MCRI), Royal Children's Hospital, Parkville; Royal Melbourne Hospital, University of Melbourne, Melbourne, VIC; Black Dog Institute, Prince of Wales Hospital, Randwick, NSW, Australia

L. Jones-Brando
Johns Hopkins School of Medicine, Baltimore, MD, United States

M. Julio-Pieper
Grupo de NeuroGastroBioquímica, Instituto de Química, Facultad de Ciencias, Pontificia Universidad Católica de, Valparaíso, Chile

A.C.C. Kao
University of Oxford, Oxford, United Kingdom

A. Keshavarzian
Rush University Medical Center, Chicago, IL, United States; Utrecht Institute for Pharmaceutical Sciences, Utrecht University, Utrecht, The Netherlands

K.A. Kinney
Military and Veteran Microbiome Consortium for Research and Education (MVM-CoRE), Denver, CO; University of Texas Austin, Austin, TX, United States

A.D. Kraneveld
Division of Pharmacology, Utrecht Institute for Pharmaceutical Sciences, Faculty of Science; Institute for Risk Assessment Sciences, Faculty of Veterinary Medicine, Utrecht University, Utrecht, The Netherlands

J.M. Krueger
Elson S. Floyd College of Medicine, Washington State University, Spokane, WA, United States

Y. Li
Johns Hopkins School of Medicine, Baltimore, MD, United States

A.B. Loughridge
Colorado State University, Fort Collins, CO, United States

C.A. Lowry
Center for Neuroscience, University of Colorado Boulder, Boulder; Military and Veteran Microbiome Consortium for Research and Education (MVM-CoRE); Rocky Mountain Mental Illness Research Education and Clinical Center, Denver; University of Colorado, Aurora, CO, United States

A.R. Mackos
Center for Microbial Pathogenesis, The Research Institute at Nationwide Children's Hospital, Columbus, OH, United States

R. Maltz
Center for Microbial Pathogenesis, The Research Institute at Nationwide Children's Hospital; Nationwide Children's Hospital, Columbus, OH, United States

A. Mika
Center for Neuroscience, University of Colorado, Boulder, CO, United States

M.R. Opp
University of Washington College of Medicine, Seattle, WA, United States

M. Pletnikov
Johns Hopkins School of Medicine, Baltimore, MD, United States

T.T. Postolache
Military and Veteran Microbiome Consortium for Research and Education (MVM-CoRE); University of Maryland School of Medicine; VISN 5 Mental Illness Research Education and Clinical Center (MIRECC); Rocky Mountain Mental Illness Research Education and Clinical Center, Denver, CO, United States

T. Poutahidis
Division of Comparative Medicine, Massachusetts Institute of Technology, Cambridge, MA, United States; Aristotle University of Thessaloniki, Thessaloniki, Greece

E. Prandovszky
Johns Hopkins School of Medicine, Baltimore, MD, United States

G.A.W. Rook
Center for Clinical Microbiology, UCL (University College London), London, United Kingdom

N. Rumian
Center for Neuroscience, University of Colorado, Boulder, CO, United States

S. Sabunciyan
Johns Hopkins School of Medicine, Baltimore, MD, United States

E.G. Severance
Johns Hopkins School of Medicine, Baltimore, MD, United States

D.G. Smith
Center for Neuroscience, University of Colorado Boulder, Boulder, CO, United States

C.E. Stamper
Center for Neuroscience, University of Colorado Boulder, Boulder, CO, United States

K. Szklany
Division of Pharmacology, Utrecht Institute for Pharmaceutical Sciences, Faculty of Science, Utrecht University, Utrecht, The Netherlands

V.A. Varaljay
Center for Microbial Pathogenesis, The Research Institute at Nationwide Children's Hospital, Columbus, OH, United States

R.M. Voigt
Rush University Medical Center, Chicago, IL, United States

J. Xiao
Johns Hopkins School of Medicine, Baltimore, MD, United States

R. Yolken
Johns Hopkins School of Medicine, Baltimore, MD, United States

PREFACE

The Gut Microbiome and Behavior under the microscope: Where to focus?

G. Clarke*[,‡,1], J.F. Cryan[†,‡]
*Department of Psychiatry and Neurobehavioural Science, University College Cork, Cork, Ireland
[†]Department of Anatomy and Neuroscience, University College Cork, Cork, Ireland
[‡]APC Microbiome Institute, University College Cork, Cork, Ireland
[1]Corresponding author: e-mail address: g.clarke@ucc.ie

SETTING THE STAGE

The impact of the gut microbiome on brain and behavior is now a frontier research area with considerable momentum generated by a wealth of preclinical data and supported by a growing clinical literature. The wide variety of topics covered in this volume of "*International Review of Neurobiology*" by experts across multiple disciplines highlights the broad scope of influence now under consideration. It is now clear that virtually no aspect of mammalian physiology and behavior is left untouched by the community of microorganisms resident in our gastrointestinal tract and that the resultant host–microbe interactions are critical for typical brain function and development (De Palma, Collins, Bercik, & Verdu, 2014; Dinan & Cryan, 2016; Mayer, Tillisch, & Gupta, 2015). Keeping abreast of these exciting developments requires diligence and careful consideration of the implications arising from each new observation begs the question of where the focus of our research efforts should lie. We offer some suggestions below, many of which are critically evaluated by the experts assembled for this volume.

BRAIN REGIONS OF INTEREST

From a classical neurobiological perspective, a reasonable appreciation exists of the brain regions and neurocircuitry underpinning mood disorders (Schloesser, Martinowich, & Manji, 2012). Although an oversimplification in some respects, it is nevertheless useful then to consider brain region-specific effects of the gut microbiome. The hippocampus, often deemed the seat of learning and memory, appears particularly receptive to microbiome–gut–brain axis signals. Thus, alterations have been noted in this brain region in germ-free animals related to serotonergic neurotransmission (Clarke et al., 2013), transcriptional regulation (Diaz Heijtz et al., 2011), neurogenesis (Ogbonnaya et al., 2015), and neuronal morphology (Luczynski et al., 2016b). It is perhaps not surprising then to see the association between the cognitive performance and the gut microbiome as documented by Gareau (Chapter 11) or that microbiota-directed interventions such as prebiotics impact on, for example, hippocampal BDNF expression (Chapter 2, Burnet). Beyond the hippocampus, studies have also revealed a role for the gut microbiome in the regulation of prefrontal cortex myelination (Gacias et al., 2016; Hoban et al., 2016), amygdalar transcription patterns (Stilling et al., 2015), and neuronal morphology (Luczynski et al., 2016b). Taken together, these neurobiological features of microbiome–gut–brain axis signaling can readily be aligned to the altered behavioral profile of germ-free animals including not just cognitive performance but also anxiety- and depressive-like behaviors, and sociability (Dinan, Stilling, Stanton, & Cryan, 2015). It is likely then that a continued emphasis on the brain region-specific effects of the gut microbiome within the context of known and emerging behavioral alterations can yield important insights. This approach also needs to take into account the more general impact of the gut microbiome on CNS features such as the blood–brain barrier integrity (Braniste et al., 2014) and microglial maturation and function (Erny et al., 2015).

ZOOMING IN ON THE GUT MICROBIOME AND BEHAVIOR AT THE EXTREMES OF LIFE

The roots of adult disease may take hold following developmental and biological disruption during early life (Shonkoff, Boyce, & McEwen, 2009). Conceptually, the overlaps during critical time windows in the assembly of the gut microbiome postweaning and the various phases of brain development in early life implicate microbiome–gut–brain axis signaling in neurodevelopment (Borre et al., 2014; Diaz Heijtz et al., 2011). A logical extension of this line of thought suggests consideration of the gut microbiome in neurodevelopmental disorders such as autism spectrum disorder (ASD) as put forward by Kraneveld and colleagues (Chapter 13). Although not without controversy given the complexity of ASD (Mayer, Padua, & Tillisch, 2014), this is particularly salient in the context of the social behavioral deficits noted in microbiota-deficient animals (Desbonnet, Clarke, Shanahan, Dinan, & Cryan, 2014; Stilling et al., 2015) and has seen interest in evaluating gut microbes as drivers of brain evolution and development (Stilling, Bordenstein, Dinan, & Cryan, 2014).

Associations between the gut microbiome and schizophrenia are also being considered (Dinan, Borre, & Cryan, 2014) with many lessons to be learned from the approaches used in the study of *Toxoplasma gondii* as outlined by Yolken and colleagues (Chapter 7). Research in this area has also reflected on the links between nutrition, the gut microbiota, and brain development (Goyal, Venkatesh, Milbrandt, Gordon, & Raichle, 2015). Interestingly, recent work has shown an association between the neurobehavioral changes induced by altering omega-3 polyunsaturated fatty acids status in early life and alterations in gut microbiota composition (Robertson et al., 2016). Meanwhile, dietary supplementation with an eicosapentaenoic acid/docosahexaenoic acid mixture can restore the disturbed gut microbiota composition of maternally separated female rats (Pusceddu et al., 2015). Disruption of the gut microbiome in early life also takes in the impact of transient disruptions on the subsequent expression of pain behavior in adulthood (O'Mahony et al., 2014) as well as the long-lasting metabolic consequences that might ensue (Cox et al., 2014). The adolescent brain is also distinguished by a vulnerability to gut microbiota alterations (Desbonnet et al., 2015; McVey Neufeld, Luczynski, Dinan, & Cryan, 2016). This theme is elaborated on more generally by Jacka and colleagues in their discussion of the clinical implications of diet and gut microbiome during various stages of life (Chapter 15).

There is also now a real possibility that healthy aging and gut microbiome status go hand in hand. There are certainly distinct microbiota profiles associated with increasing age (O'Toole & Jeffery, 2015) and extreme longevity (Biagi et al., 2016) and associated with frailty (Jackson et al., 2016). Narrowing of gut microbiota diversity in aging is likely linked to diet (Claesson et al., 2012). This research supports Metchnikoff's ideas on the role of bacteria in prolongation of life (Cryan & Dinan, 2015). Research in invertebrate models such as *Caenorhabditis elegans* also supports a role for the microbiota in the physiology of aging (Heintz & Mair, 2014). Clearly, the extremes of life are associated with marked changes in gut microbiota composition with profound implications for host health and well-being.

THE GUT MICROBIOME THOUGH THE APERTURE OF STRESS

The neurobiological consequences of stress on the brain are seen across the lifespan (Koenig, Walker, Romeo, & Lupien, 2011; Lupien, McEwen, Gunnar, & Heim, 2009). Regulation of the stress response and HPA axis programming by the gut microbiome as well as the reciprocal impact of stress on the gut microbiome is discussed in detail by Bailey and colleagues (Chapter 1). The clinical implications of this are also considered in the context of irritable bowel syndrome (Collins, Chapter 12), a stress-related microbiome–gut–brain axis disorder. Microbiota alterations also manifest in other stress-related disorders as recently demonstrated in depressed cohorts (Kelly et al., 2016; Zheng et al., 2016). Moreover, it has also been demonstrated, using a microbiota-deficient animal model, that many of the prominent features of depression could be transferred via the gut microbiota (Kelly et al., 2016). Links between stress, the vaginal microbiome, and infant health outcomes are also being investigated (Jasarevic, Howerton, Howard, & Bale, 2015; Jasarevic, Rodgers, & Bale, 2015). Other recent clinical and preclinical observations support the concept that prenatal stress might result in a suboptimal transmission of the microbiota from mother to infant at birth (Golubeva et al., 2015; Zijlmans, Korpela, Riksen-Walraven, de Vos, & de Weerth, 2015). Our knowledge of stress-microbiome interactions is probably at a more advanced stage than most avenues of research in this field and may well be the mostly likely to yield meaningful dividends in the short to medium term.

RESOLVING THE QUESTION OF MECHANISMS

Despite all the advances outlined earlier, one of the main limitations holding back this field relates to our rudimentary understanding of the mechanisms through which the gut microbiome influences brain and behavior. There are many possibilities in this regard including a role for neuroimmune interactions (Foster, Chapter 3) or the role of the gut microbiome in regulating intestinal permeability (Julio-Pieper and Bravo, Chapter 6). Other important factors include links to oxytocin (Erdman, Chapter 5) or other neuropeptides (Holzer, Chapter 4). Microbial regulation of tryptophan availability and onward metabolism into a variety of neuroactives has also been implicated and is reviewed in detail elsewhere (Kennedy, Cryan, Dinan, & Clarke, 2016; O'Mahony, Clarke, Borre, Dinan, & Cryan, 2015). Moreover, microbial metabolites such as short-chain fatty acids like butyrate are also considered important mediators of the observed impact on host physiology and behavior (Stilling et al., 2016). There remain few proven mechanisms outside of the demonstration that the beneficial effects of a putative psychobiotic (*Lactobacillus rhamnosus* JB-1) were mediated via the vagus nerve (Bravo et al., 2011). Even here, high-level detail is currently missing and the search continues for the microbial mediators that initiate the microbial dialogue with the vagus nerve. We note as well that such mechanisms are likely strain specific and will need to be defined on a case by case basis.

VISION FOR THE FUTURE

Research in this field continues to expand into new and interesting areas. This now includes sleep (Kreuger, Chapter 10), circadian rhythm

(Keshavarzian, Chapter 9), and exercise (Fleshner, Chapter 8). The intriguing possibility of links between the microbiome of the built environment, the human microbiome, and mental health is discussed by Lowry and colleagues (Chapter 14). More established subjects outside the scope of this volume such as obesity continue to be explored (Nehra, Allen, Mailing, Kashyap, & Woods, 2016), with open questions regarding the capacity of the microbiota to influence eating behaviors via the brain–gut axis (Alcock, Maley, & Aktipis, 2014). We can expect growth as well in the area of pharmacomicrobiomics (Carmody & Turnbaugh, 2014; Nayak & Turnbaugh, 2016).

Although all these avenues of research bring with them many possibilities, research in this area has also come in for some criticism. In particular, responsible researchers active in the field need to be cognizant of the fact that microbiome research has caught the public attention to a degree rarely seen in other areas of science. As each advance materializes, it comes with often exaggerated press coverage that can dangerously raise lay expectations. This has seen concern about overselling of the microbiome (https://phylogenomics.blogspot.ie/) and led to commentaries on the misleading "hyperbolome" that often accompanies gut microbiome research (Shanahan, 2015). This is also associated with unscrupulous pseudoscientific microbiome-based claims that surf the wave of legitimate scientific developments. The flip side of this coin is that this bubble of exaggerated expectations can too easily be punctured by minor setbacks. An example of this was seen recently with the doomsday reporting following the interim failure at phase 2 stage of an investigational oral microbiome therapeutic produced by Seres Therapeutics for the prevention of recurrent *Clostridium difficile* infection (http://www.businesswire.com/news/home/20160729005385/en/Seres-Therapeutics-Announces-Interim-Results-SER-109-Phase). We can expect the road ahead for microbiome-based interventions to be just as bumpy as for more traditional pharmacological agents but that should not be interpreted as the failure of the field in general. Perhaps the real message to be taken from these unexpectedly poor results is the need to redouble efforts toward understanding the mechanisms underpinning the benefits of fecal microbiota transplantation (Bojanova & Bordenstein, 2016; Khoruts & Sadowsky, 2016).

As more and more research groups look to the gut microbiome for answers, it is important as well that the information obtained from commonly applied experimental approaches are interpreted appropriately to maximize their translational potential. This includes the need for greater consideration of how we use germ-free animals and the need for complementary sources of information. Clearly, the data thus far illustrate that these animals are profoundly abnormal (Luczynski et al., 2016a). Nevertheless, they do provide valuable proof-of-principle insights regarding the aspects of brain function, development, and behavior that *could be* under the influence of the gut microbiome. We can take confidence from the fact that many of the initial observations regarding myelination, neurogenesis, and social behavior in germ-free studies were subsequently confirmed with complimentary approaches using alternative microbiota manipulation strategies (Buffington et al., 2016; Gacias et al., 2016; Mohle et al., 2016). As we continue to explore the microbial impact across the lifespan, it will be important to use models that allow for the initial assembly of the gut microbiome to eliminate the confounding neurodevelopmental abnormalities that arise from growing up germ free. We should also not forget the important insights that might be gained from studying the primate microbiome (Bailey & Coe, 2002; Bailey, Lubach, & Coe, 2004) and other model organisms (Borrelli et al., 2016; Heintz & Mair, 2014). While we are just starting to get to grips with the bacterial contribution to brain function, gut virome research waits on the horizon (Columpsi et al., 2016).

This volume provides the opportunity for reflection on these and other matters in the company of experts in this field. In charting the way forward, they discuss how to overcome the pitfalls and challenges that go hand in hand with these opportunities. As they survey the important neurobiological implications for many stress-related neuropsychiatric disorders, we can say we are just at the beginning of an exciting new research paradigm. With the right focus and direction, the outcomes of this research can inform future healthcare policy and yield novel medicinal strategies based on therapeutic targeting of the gut microbiome.

REFERENCES

Alcock, J., Maley, C. C., & Aktipis, C. A. (2014). Is eating behavior manipulated by the gastrointestinal microbiota? Evolutionary pressures and potential mechanisms. *BioEssays: News and Reviews in Molecular, Cellular and Developmental Biology*, *36*, 940–949.

Bailey, M. T., & Coe, C. L. (2002). Intestinal microbial patterns of the common marmoset and rhesus macaque. *Comparative Biochemistry and Physiology. Part A, Molecular & Integrative Physiology*, *133*, 379–388.

Bailey, M. T., Lubach, G. R., & Coe, C. L. (2004). Prenatal stress alters bacterial colonization of the gut in infant monkeys. *Journal of Pediatric Gastroenterology and Nutrition*, *38*, 414–421.

Biagi, E., et al. (2016). Gut microbiota and extreme longevity. *Current Biology: CB*, *26*, 1480–1485.

Bojanova, D. P., & Bordenstein, S. R. (2016). Fecal transplants: What is being transferred? *PLoS Biology, 14*, e1002503.

Borre, Y. E., et al. (2014). Microbiota and neurodevelopmental windows: Implications for brain disorders. *Trends in Molecular Medicine, 20*, 509–518.

Borrelli, L., et al. (2016). Probiotic modulation of the microbiota-gut-brain axis and behaviour in zebrafish. *Scientific Reports, 6*, 30046.

Braniste, V., et al. (2014). The gut microbiota influences blood–brain barrier permeability in mice. *Science Translational Medicine, 6*, 263ra158.

Bravo, J. A., et al. (2011). Ingestion of Lactobacillus strain regulates emotional behavior and central GABA receptor expression in a mouse via the vagus nerve. *Proceedings of the National Academy of Sciences of the United States of America, 108*, 16050–16055.

Buffington, S. A., et al. (2016). Microbial reconstitution reverses maternal diet-induced social and synaptic deficits in offspring. *Cell, 165*, 1762–1775.

Carmody, R. N., & Turnbaugh, P. J. (2014). Host-microbial interactions in the metabolism of therapeutic and diet-derived xenobiotics. *The Journal of Clinical Investigation, 124*, 4173–4181.

Claesson, M. J., et al. (2012). Gut microbiota composition correlates with diet and health in the elderly. *Nature, 488*, 178–184.

Clarke, G., et al. (2013). The microbiome-gut-brain axis during early life regulates the hippocampal serotonergic system in a sex-dependent manner. *Molecular Psychiatry, 18*, 666–673.

Columpsi, P., et al. (2016). Beyond the gut bacterial microbiota: The gut virome. *Journal of Medical Virology, 88*, 1467–1472.

Cox, L. M., et al. (2014). Altering the intestinal microbiota during a critical developmental window has lasting metabolic consequences. *Cell, 158*, 705–721.

Cryan, J. F., & Dinan, T. G. (2015). Gut microbiota: Microbiota and neuroimmune signalling—Metchnikoff to microglia. *Nature Reviews. Gastroenterology & Hepatology, 12*, 494–496.

De Palma, G., Collins, S. M., Bercik, P., & Verdu, E. F. (2014). The microbiota-gut-brain axis in gastrointestinal disorders: Stressed bugs, stressed brain or both? *The Journal of Physiology, 592*, 2989–2997.

Desbonnet, L., Clarke, G., Shanahan, F., Dinan, T. G., & Cryan, J. F. (2014). Microbiota is essential for social development in the mouse. *Molecular Psychiatry, 19*, 146–148.

Desbonnet, L., et al. (2015). Gut microbiota depletion from early adolescence in mice: Implications for brain and behaviour. *Brain, Behavior, and Immunity, 48*, 165–173.

Diaz Heijtz, R., et al. (2011). Normal gut microbiota modulates brain development and behavior. *Proceedings of the National Academy of Sciences of the United States of America, 108*, 3047–3052.

Dinan, T. G., Borre, Y. E., & Cryan, J. F. (2014). Genomics of schizophrenia: Time to consider the gut microbiome? *Molecular Psychiatry, 19*, 1252–1257.

Dinan, T. G., & Cryan, J. F. (2016). Gut instincts: Microbiota as a key regulator of brain development, ageing and neurodegeneration. *The Journal of Physiology*. http://dx.doi.org/10.1113/JP273106. [Epub ahead of print] PMID: 27641441.

Dinan, T. G., Stilling, R. M., Stanton, C., & Cryan, J. F. (2015). Collective unconscious: How gut microbes shape human behavior. *Journal of Psychiatric Research, 63*, 1–9.

Erny, D., et al. (2015). Host microbiota constantly control maturation and function of microglia in the CNS. *Nature Neuroscience, 18*, 965–977.

Gacias, M., et al. (2016). Microbiota-driven transcriptional changes in prefrontal cortex override genetic differences in social behavior. *eLife, 5*, e13442.

Golubeva, A. V., et al. (2015). Prenatal stress-induced alterations in major physiological systems correlate with gut microbiota composition in adulthood. *Psychoneuroendocrinology, 60*, 58–74.

Goyal, M. S., Venkatesh, S., Milbrandt, J., Gordon, J. I., & Raichle, M. E. (2015). Feeding the brain and nurturing the mind: Linking nutrition and the gut microbiota to brain development. *Proceedings of the National Academy of Sciences of the United States of America*, *112*, 14105–14112.

Heintz, C., & Mair, W. (2014). You are what you host: Microbiome modulation of the aging process. *Cell*, *156*, 408–411.

Hoban, A. E., et al. (2016). Regulation of prefrontal cortex myelination by the microbiota. *Translational Psychiatry*, *6*, e774.

Jackson, M. A., et al. (2016). Signatures of early frailty in the gut microbiota. *Genome Medicine*, *8*, 8.

Jasarevic, E., Howerton, C. L., Howard, C. D., & Bale, T. L. (2015). Alterations in the vaginal microbiome by maternal stress are associated with metabolic reprogramming of the offspring gut and brain. *Endocrinology*, *156*, 3265–3276.

Jasarevic, E., Rodgers, A. B., & Bale, T. L. (2015). A novel role for maternal stress and microbial transmission in early life programming and neurodevelopment. *Neurobiology of Stress*, *1*, 81–88.

Kelly, J. R., et al. (2016). Transferring the blues: Depression-associated gut microbiota induces neurobehavioural changes in the rat. *Journal of Psychiatric Research*, *82*, 109–118.

Kennedy, P. J., Cryan, J. F., Dinan, T. G., & Clarke, G. (2016). Kynurenine pathway metabolism and the microbiota-gut-brain axis. *Neuropharmacology*. http://dx.doi.org/10.1016/j.neuropharm.2016.07.002. [Epub ahead of print].

Khoruts, A., & Sadowsky, M. J. (2016). Understanding the mechanisms of faecal microbiota transplantation. *Nature Reviews. Gastroenterology & Hepatology*, *13*, 508–516.

Koenig, J. I., Walker, C. D., Romeo, R. D., & Lupien, S. J. (2011). Effects of stress across the lifespan. *Stress*, *14*, 475–480.

Luczynski, P., et al. (2016a). Growing up in a bubble: Using germ-free animals to assess the influence of the gut microbiota on brain and behavior. *The International Journal of Neuropsychopharmacology/Official Scientific Journal of the Collegium Internationale Neuropsychopharmacologicum*. *19*(8). http://dx.doi.org/10.1093/ijnp/pyw020.

Luczynski, P., et al. (2016b). Adult microbiota-deficient mice have distinct dendritic morphological changes: Differential effects in the amygdala and hippocampus. *The European Journal of Neuroscience*. http://dx.doi.org/10.1111/ejn.13291. [Epub ahead of print].

Lupien, S. J., McEwen, B. S., Gunnar, M. R., & Heim, C. (2009). Effects of stress throughout the lifespan on the brain, behaviour and cognition. *Nature Reviews. Neuroscience*, *10*, 434–445.

Mayer, E. A., Padua, D., & Tillisch, K. (2014). Altered brain-gut axis in autism: Comorbidity or causative mechanisms? *BioEssays: News and Reviews in Molecular, Cellular and Developmental Biology*, *36*, 933–939.

Mayer, E. A., Tillisch, K., & Gupta, A. (2015). Gut/brain axis and the microbiota. *The Journal of Clinical Investigation*, *125*, 926–938.

McVey Neufeld, K. A., Luczynski, P., Dinan, T. G., & Cryan, J. F. (2016). Reframing the teenage wasteland: Adolescent microbiota-gut-brain axis. *Canadian Journal of Psychiatry. Revue Canadienne de Psychiatrie*, *61*, 214–221.

Mohle, L., et al. (2016). Ly6C(hi) monocytes provide a link between antibiotic-induced changes in gut microbiota and adult hippocampal neurogenesis. *Cell Reports*, *15*, 1945–1956.

Nayak, R. R., & Turnbaugh, P. J. (2016). Mirror, mirror on the wall: Which microbiomes will help heal them all? *BMC Medicine*, *14*, 72.

Nehra, V., Allen, J. M., Mailing, L. J., Kashyap, P. C., & Woods, J. A. (2016). Gut microbiota: Modulation of host physiology in obesity. *Physiology*, *31*, 327–335.

Ogbonnaya, E. S., et al. (2015). Adult hippocampal neurogenesis is regulated by the microbiome. *Biological Psychiatry*, *78*, e7–e9.

O'Mahony, S. M., Clarke, G., Borre, Y. E., Dinan, T. G., & Cryan, J. F. (2015). Serotonin, tryptophan metabolism and the brain-gut-microbiome axis. *Behavioural Brain Research*, *277*, 32–48.

O'Mahony, S. M., et al. (2014). Disturbance of the gut microbiota in early-life selectively affects visceral pain in adulthood without impacting cognitive or anxiety-related behaviors in male rats. *Neuroscience*, *277*, 885–901.

O'Toole, P. W., & Jeffery, I. B. (2015). Gut microbiota and aging. *Science*, *350*, 1214–1215.

Pusceddu, M. M., et al. (2015). N-3 polyunsaturated fatty acids (PUFAs) reverse the impact of early-life stress on the gut microbiota. *PLoS One*, *10*, e0139721.

Robertson, R. C., et al. (2016). Omega-3 polyunsaturated fatty acids critically regulate behaviour and gut microbiota development in adolescence and adulthood. *Brain, Behavior, and Immunity*. http://dx.doi.org/10.1016/j.bbi.2016.07.145. [Epub ahead of print].

Schloesser, R. J., Martinowich, K., & Manji, H. K. (2012). Mood-stabilizing drugs: Mechanisms of action. *Trends in Neurosciences*, *35*, 36–46.

Shanahan, F. (2015). Separating the microbiome from the hyperbolome. *Genome Medicine*, *7*, 17.

Shonkoff, J. P., Boyce, W. T., & McEwen, B. S. (2009). Neuroscience, molecular biology, and the childhood roots of health disparities: Building a new framework for health promotion and disease prevention. *JAMA*, *301*, 2252–2259.

Stilling, R. M., Bordenstein, S. R., Dinan, T. G., & Cryan, J. F. (2014). Friends with social benefits: Host-microbe interactions as a driver of brain evolution and development? *Frontiers in Cellular and Infection Microbiology*, *4*, 147.

Stilling, R. M., et al. (2015). Microbes & neurodevelopment—Absence of microbiota during early life increases activity-related transcriptional pathways in the amygdala. *Brain, Behavior, and Immunity*, *50*, 209–220.

Stilling, R. M., et al. (2016). The neuropharmacology of butyrate: The bread and butter of the microbiota-gut-brain axis? *Neurochemistry International*, *99*, 110–132.

Zheng, P., et al. (2016). Gut microbiome remodeling induces depressive-like behaviors through a pathway mediated by the host's metabolism. *Molecular Psychiatry*, *21*, 786–796.

Zijlmans, M. A., Korpela, K., Riksen-Walraven, J. M., de Vos, W. M., & de Weerth, C. (2015). Maternal prenatal stress is associated with the infant intestinal microbiota. *Psychoneuroendocrinology*, *53*, 233–245.

CHAPTER ONE

Role of the Intestinal Microbiota in Host Responses to Stressor Exposure

A.R. Mackos*, V.A. Varaljay*, R. Maltz*[,†], T.L. Gur[‡,§], M.T. Bailey*[,§,¶,1]

*Center for Microbial Pathogenesis, The Research Institute at Nationwide Children's Hospital, Columbus, OH, United States
[†]Nationwide Children's Hospital, Columbus, OH, United States
[‡]Wexner Medical Center at The Ohio State University, Columbus, OH, United States
[§]The Institute for Behavioral Medicine Research (IBMR) at The Ohio State University, Columbus, OH, United States
[¶]The Ohio State University College of Medicine, Columbus, OH, United States
[1]Corresponding author: e-mail address: michael.bailey2@nationwidechildrens.org

Contents

Abstract

Humans have coevolved over time to not only tolerate but also rely on trillions of microbes that aid in the development of our immune system, provide nutrients, break down potentially noxious substances, and act as a barrier against potentially pathogenic organisms. These microbes, collectively known as the microbiota, live in relatively stable communities on mucosal surfaces such as the respiratory tract and gastrointestinal tract. Changes to the microbiota are often transient, due to changes in diet, antibiotic exposure, and psychological stressor exposure. This chapter will discuss how psychological stressors can shape the intestinal microbial community and how these perturbations can contribute to stressor-induced changes in immune function, neurodevelopment, and behavioral deficits.

International Review of Neurobiology, Volume 131
ISSN 0074-7742
http://dx.doi.org/10.1016/bs.irn.2016.08.002

1. INTRODUCTION

The mammalian body is colonized by a rich consortium of microbes, which consists of bacteria, archaea, fungi, and viruses. These microbes, more commonly called the microbiota, along with their gene products are collectively defined as the microbiome, a term first coined by Nobel laureate Joseph Lederberg in 2001 as an "ecological community of commensal, symbiotic, and pathogenic microorganisms that literally share our body space and have been all but ignored as determinants of health and disease" (Hooper & Gordon, 2001; Lederberg & McCray, 2001). Indeed, the significance and complexity of the human microbiome has been vastly underexplored. However, with new technologies (such as next-generation, high-throughput sequencing) and new initiatives (such as the NIH Human Microbiome Project), there has been an explosion of interest in the microbiome and the importance for health and disease.

Every surface of the body contains its own, distinct microbiomes (Costello et al., 2009). The most abundant and genetically diverse microbiomes are found within the gastrointestinal tract, which is home to approximately 10 trillion bacteria (Sender, Fuchs, & Milo, 2016). Proximal sections of the gastrointestinal tract, including the stomach and the duodenum, harbor low levels of microorganisms (typically between 100 and 1000 bacteria per mL of contents), but distal sections of the gastrointestinal tract, including the distal small intestine and the large intestine, contain higher levels of bacteria (as many as 10^6 to 10^{11} bacteria per mL of contents) (Bailey, 2014; Donaldson, Lee, & Mazmanian, 2016; Rastall, 2004; Sender et al., 2016). The majority of these microbes transiently pass through the gastrointestinal tract and thus are only found in the intestinal lumen, but some of these microbes can adhere to the intestinal mucous layer to create well-formed biofilms (de Vos, 2015; Donaldson et al., 2016). Although luminal and mucosa-associated microbial communities have different community profiles, there is substantial overlap between these communities (Galley, Yu, et al., 2014). Luminal bacteria can be trapped within the mucous layer, and muco-adherent bacteria can be shed into the lumen as the mucous layer is constantly breaking down and reforming. Despite the overlap, luminal and mucosa-associated bacteria are thought to have different effects on host physiology (Duerkop, Vaishnava, & Hooper, 2009; Galley, Yu, et al., 2014; Van den Abbeele, Van de Wiele, Verstraete, & Possemiers, 2011).

The human body is not always colonized by vast numbers of microbes, but colonization begins very early in life. It was once thought that the developing fetus is completely sterile and devoid of any microbes, but quickly becomes colonized by bacteria from the mother and from the environment during birth. However, there is accumulating evidence that colonization may actually begin in utero (Neu & Rushing, 2011). This is based on studies demonstrating that the human placenta contains bacterial DNA (Aagaard et al., 2014). Next-generation sequencing indicated that the DNA profiles in the placenta were more similar to profiles found in the oral cavity, suggesting that bacteria from the mouth translocate and seed the developing placenta (Aagaard et al., 2014). It is still to be determined whether there are live bacteria in the placenta, but findings of bacterial DNA in the placenta are consistent with findings of bacterial DNA in the neonatal meconium (Ardissone et al., 2014; Moles et al., 2013). The meconium is comprised of partially digested amniotic fluid, and thus the presence of bacteria in the meconium supports the contention that bacterial colonization begins in utero.

Birth is a period of rapid colonization. Newly colonizing microbes are initially derived from the mother depending on the type of delivery (e.g., maternal vaginal, skin, or gastrointestinal microbes), but eventually the microbiota diversifies and settles into a relatively stable microbiota. As infants begin to regularly consume an adult-like diet (typically by 2–3 years of age), the microbiome begins to reflect the adult microbiome that remains relatively stable throughout healthy adulthood (Bergström et al., 2014; Koenig et al., 2011; Kostic, Howitt, & Garrett, 2013; Spor, Koren, & Ley, 2011; Voreades, Kozil, & Weir, 2014). In the colon, the microbiota resides as a stable climax community due to the selection of microbes that are best adapted for life in the colon (Bailey, 2014). This climax community is relatively resilient and resistant to dramatic changes (Bailey, 2014). But, it is known that factors such as diet and antibiotics can cause alterations in microbial community composition (David et al., 2014; Francino, 2015). Importantly, these changes in microbial community composition can have significant effects on the function (or biological activities) of the microbial communities (Morgan, Segata, & Huttenhower, 2013). Changes in microbial community function have been linked to a myriad of diseases, including inflammatory bowel disease, irritable bowel syndrome, obesity, cancer, diarrhea, and infections (Chang et al., 2008; Sartor & Mazmanian, 2012; Sekirov, Russell, Antunes, & Finlay, 2010; Turnbaugh et al., 2006). Thus, understanding the complete set of factors that can change the composition of the gut microbiota has important implications for understanding health and disease. This chapter will outline the evidence that the physiological stress response can impact microbial community structure and function, and will discuss the potential impact on host health.

2. STRESS AND THE STRESS RESPONSE

Exposure to life stressors evokes a response that has been conserved over time and across species to prepare the organism to confront or flee from potentially threatening stimuli in the so-called fight-or-flight response. This stress response includes the rapid activation of multiple neuroendocrine pathways, including the hypothalamic–pituitary–adrenal (HPA) axis and the sympathetic nervous system (SNS) which culminate with the production of glucocorticoids (such as corticosterone in laboratory mice and cortisol in humans) and catecholamines (such as norepinephrine and epinephrine), respectively. Psychological stressor exposure has long been known to leave the host vulnerable to infectious diseases, which, until recently, has been solely attributed to stressor-induced alterations of immune functioning. However, it is now realized that in addition to impacting the immune system, stressor exposure can have a significant impact on the commensal microbiota. These effects on the microbiota can, in turn, impact host immunity and behavioral responses.

One of the first indications that the stress response might impact microbes came from pioneering studies by Dr. Mark Lyte, who demonstrated that both pathogenic and commensal microbes can recognize and respond to mammalian hormones. This was most clearly demonstrated with the catecholamine hormones that when given to bacteria have the capacity to increase bacterial growth by over 10,000-fold. In addition to bacterial growth, the expression of virulence factors in enterohemorrhagic *Escherichia coli*, such as toxins and adhesins, was increased after exposure to catecholamines (Chen, Lyte, Stevens, Vulchanova, & Brown, 2006; Freestone, Haigh, & Lyte, 2007a, 2007b; Green et al., 2004). These effects are not limited to pathogens, as other microbes, such as commensal *E. coli*, were found to respond to catecholamines (Freestone et al., 2002; Lyte & Ernst, 1992). The mechanisms by which this occurs are not yet completely understood, but it is thought that some microbes, such as pathogenic *Salmonella enterica* Serovar Typhimurium, contain a two-component sensor kinase that recognizes catecholamines in the environment, thus leading to more rapid replication and enhanced expression of virulence factors (Bailey, Karaszewski,

Lubach, Coe, & Lyte, 1999; Bearson & Bearson, 2008). In addition to the sensor kinases, catecholamines are thought to bind to iron and enhance its uptake into the microbe (Freestone et al., 2000; Sandrini et al., 2010). Because the increased bioavailability of iron helps with bacterial replication, microbes (either pathogens or commensals) found in the oral cavity, respiratory tract, or gastrointestinal tract all show growth increases when exposed to catecholamines in a low-iron/blood serum-based medium (Lyte, Vulchanova, & Brown, 2011). It is now recognized that a myriad of host-derived neuroendocrine mediators and other factors can influence bacterial growth and activity, thus providing ample rationale to suggest that the physiological stress response influences the gut microbiota.

3. STRESSOR EXPOSURE AND THE INTESTINAL MICROBIOTA

Studies by Tannock and Savage from the early 1970s were the first well-documented experimental studies to assess whether exposure to stressful stimuli could impact the composition of the gut microbiota. In these studies, mice that were deprived of bedding, in addition to food and water, were found to have lower levels of lactobacilli that could be cultured from all segments of the gastrointestinal tract, including the stomach, small and large intestines (Tannock & Savage, 1974). Interestingly, changes in lactobacilli are commonly found in stressor-exposed animals, including mice, nonhuman primates, and, in some studies, humans. For example, studies involving coprocultures from rhesus monkeys (*Macaca mulatta*) separated from their mothers found lower levels of lactobacilli in the stool the week following the separation (Bailey & Coe, 1999). This effect lasted for 3 days after the separation and correlated with stress-indicative behaviors. The more behavioral signs of distress the monkeys displayed, the lower their lactobacilli levels were. Interestingly, as the monkeys formed more stable social groups, the levels of lactobacilli returned back to baseline levels.

Stressor-induced reductions in lactobacilli have also been observed in the colon of stressor-exposed mice, but only in mucosa-associated microbial populations (not in luminal populations). For example, mice exposed to an overnight restraint stressor on seven consecutive nights were found to have significant reductions in the relative abundance of lactobacilli (based on 454 pyrosequencing of the V1–V3 region of the gene encoding the microbial 16S rRNA) in mucosa-associated microbial populations (Galley, Yu, et al., 2014). However, there was not a statistically significant reduction of *Lactobacillus* spp. in the lumen of the colon. This lack of a stressor-induced reduction in *Lactobacillus* abundance was also evident in a separate study involving an assessment of the luminal contents of the cecums from mice exposed to prolonged restraint stress vs nonstress controls (Bailey et al., 2010). Interestingly, when a different stressor was employed, namely social disruption for a 2-h period, reductions in mucosa-associated lactobacilli were again evident (Galley, Nelson, et al., 2014). However, when the luminal contents of the cecum were assessed in mice exposed to the same stressor, there was a smaller reduction in *Lactobacillus* abundance that approached, but did not quite reach statistical significance (with $p=0.08$). Thus, in the murine stressors currently used in our laboratory (namely prolonged restraint and social disruption), mucosa-associated *Lactobacillus* abundance decreases as a consequence of the stressor exposure. In contrast, the abundance of *Lactobacillus* found in the luminal contents of the cecum and of the colon (which is most closely reflective of stool) is less consistently affected by stressor exposure.

Stressor-induced reductions in the lactobacilli are not limited to prolonged restraint and social disruption stressors. Mice housed without bedding or in cages that are exposed to horizontal shaking have been found to lower levels of lactobacilli shed in the feces (Sakuma, Funabashi, Matsuoka, & Saito, 2013). Because the reduction in lactobacilli was consistent between the different stressors, the authors argued that stressor-induced reduction in lactobacilli could be a useful marker of for environmental stressor exposure in mice. However, not all studies find that stressor exposure results in reductions to the lactobacilli. For example, exposure to a water avoidance stressor has been observed to increase *Lactobacillus* levels (Aguilera, Vergara, & Martinez, 2013).

It is not yet known whether these changes in the gut microbiota are due to direct effects of stress-responsive hormones and neurotransmitters on the gut microbiota, but such a hypothesis should not be ruled out. In early studies, it was demonstrated that lysing sympathetic neurons, to induce a bolus release of norepinephrine, resulted in a significant bloom in Gram-negative bacteria (primarily *E. coli*) (Lyte & Bailey, 1997). Given the impact that norepinephrine can have on bacterial growth, it is possible that the large bolus release of norepinephrine led to the Gram-negative bacterial overgrowth, but further experiments are needed to confirm this conclusion. Although it is known from in vitro assays that catecholamine hormones can increase bacterial growth, relatively little is known about the impact of glucocorticoid hormones on bacterial growth. Stressor-induced HPA axis activity is associated with differences in microbial diversity in the gut (Stothart et al., 2016), but it is not yet known whether this association is due to a direct effect of glucocorticoids on gut microbes, due to HPA-induced effects on the immune system that then impacts the gut microbiota, or due to effects of HPA activation of gastrointestinal physiology (that then impacts the gut microbiota). More research is needed to begin to understand how the stress response can impact the composition of the gut microbiota.

4. ROLE OF THE MICROBIOTA IN THE BODY'S RESPONSE TO STRESS

4.1 Neuroendocrine and Behavioral Responses

There is now substantial evidence that stressor exposure, and the resultant stress response, can impact gut microbes. In 2004, Sudo et al. provided the first experimental evidence that stress physiology–microbe interactions are bidirectional. This group demonstrated that germfree mice have different patterns of HPA axis activity in response to stress in comparison to specific pathogen-free mice. Upon exposure to a restraint stressor, germfree mice had higher levels of corticosterone than did their colonized, specific pathogen-free counterparts. Colonizing germfree mice with microbes prevented this increased reactivity of the HPA axis to stress, demonstrating that differences between germfree and specific pathogen-free mice were due to the absence of bacteria. Colonization of the germfree mice with *Bifidobacterium infantis* (a beneficial microbe often found in neonatal animals and humans), but not enteropathogenic *E. coli*, was sufficient to prevent the increased reactivity of the HPA axis to stress. In addition, HPA axis hyperactivity could be prevented by colonizing germfree mice with the commensal microbiota from conventional specific pathogen-free mice. Interestingly, this effect only occurred if the germfree mice were colonized early in life, and had no effect if the mice were colonized in adulthood (Sudo et al., 2004).

The findings by Sudo et al. stimulated additional studies to assess microbiome–nervous system interactions in early life. The maternal separation model in rats or mice is one of the most well-characterized animal models for studying the effects of the early-life environment on offspring development. Separation of the offspring from their mothers results in significant changes to the composition of the gut microbiota in the developing offspring (Barouei, Moussavi, & Hodgson, 2012; De Palma et al., 2015; O'Mahony et al., 2014, 2009). Interestingly, these differences persist into adulthood and are associated with changes in HPA axis activity (i.e., increased corticosterone levels) and increased acetylcholine release in the intestines (De Palma et al., 2015). In addition to differences in markers of stress physiology, animals exposed to the maternal separation stressor also show different behavioral responses during development and also as adults. For example, mice show differences in tests of anxiety-like behavior (such as the light:dark preference and the step-down tests) as well as tests of depressive-like behavior (such as the tail suspension test). Interestingly, the stressor appears to impact the interactions between the microbiota and the host, because these behavioral differences were not evident in germfree mice exposed to the maternal separation. And, transplanting microbiota into adult germfree mice did not induce behavioral changes, but transplanting microbiota into adult germfree mice that were separated from their mothers during development resulted in significant changes in anxiety-like and depressive-like behavior (De Palma et al., 2015).

The importance of the microbiota for behavior and nervous system development may even extend into the prenatal period. Studies by Bale et al. highlight the consequences of maternal stressor exposure on the neurodevelopment of the fetus and subsequent development of behavioral disorders in the postnatal period (as reviewed in Howerton & Bale, 2012). It is known that exposure to stress during pregnancy can have negative effects on fetal development and has been thought to contribute to the development of schizophrenia, anxiety, depression, and autism through undefined mechanisms. Because gut bacteria have been implicated in numerous behavioral disorders, and it is known that the gut microbiota can be altered by stress, and offspring acquire their own microbiota during parturition, it is possible that stressor-induced changes within the maternal microbiota may contribute to the development of behavioral disorders. Early pregnancy stress (EPS) reduced the abundance of vaginal *Lactobacillus* in EPS-exposed dams on postnatal day 2 which translated to a reduced transmission of *Lactobacillus* to the gut of EPS-exposed offspring as compared to nonstressed controls (Jasarevic, Howerton, Howard, & Bale, 2015). The acquisition of an altered microbiota from EPS-exposed dams likely led to the alterations in host and bacterial metabolism in the colons and plasma of EPS-exposed offspring. One such alteration is the significant increase of colonic hippuric acid, a metabolite that has been associated with neuropsychiatric disorders including depression and autism (Jasarevic et al., 2015; Yap et al., 2010; Zheng et al., 2013). Sex differences, in which male offspring are more likely to develop behavioral disorders following EPS exposure, have been reported with this paradigm. In addition to the reduction of colonic *Lactobacillus*, there was also a significant increase in colonic *Clostridium* and *Bacteroides* bacteria in male, EPS-exposed offspring only compared to gender-matched controls. Metabolomic analysis of brain regions within offspring from EPS-exposed and control dams revealed a reduction in numerous amino acids in the paraventricular nucleus (PVN) in EPS-exposed male offspring only (Jasarevic et al., 2015). Because the PVN is highly influenced by circulating factors, it is possible that metabolic changes due to shifts in the maternally acquired microbiota can contribute to neurological development and behavioral deficits that occur in the male offspring of EPS-exposed dams.

4.2 Immune Responses

Stressor exposure is well known to impact immune system activity, and some of the mechanisms by which this occurs have been well defined. For example, stressor-induced increases in glucocorticoid hormones are well recognized to suppress immune system activity in part through repression of key transcription factors, such as NF-kB. However, with the increasing awareness that commensal microbes influence immune system activity, both within their local niche (such as the gut or vagina) as well as at distant sites (such as the spleen or the lung), we began to question whether the microbiota might be involved in stressor-induced immunomodulation.

As would be expected, microbes in the colon, particularly microbes associated with the mucosal tissue, can influence mucosal immune responses. And, diseases involving chronic inflammation in the intestines, such as the inflammatory bowel diseases, are thought to be due to disrupted homeostatic interactions between the gut microbiota and the host mucosal immune response as a result of multiple genetic, dietary, and environmental factors. Interestingly, psychological stress is one of those environmental factors that are thought to exacerbate the disease, but it is not known how. Thus, we questioned whether changes in the composition of the gut microbiota might be involved with dysregulation of mucosal immune responses. As predicted, mice exposed to either a prolonged restraint stressor or the social disruption stressor (both of which change the composition of the gut microbiota) show significant increases in colonic inflammation when challenged with the colonic murine pathogen *Citrobacter rodentium* (Mackos, Eubank, Parry, & Bailey, 2013; Mackos et al., 2016). Because previous results had indicated that the stressors reduced colonic mucosa-associated lactobacilli, and many lactobacilli, such as *Lactobacillus reuteri*, are able to decrease inflammatory responses, it was tested whether *L. reuteri* could prevent the deleterious effects of the stressor on colonic inflammation. As predicted, mice treated with an antiinflammatory strain of probiotic *L. reuteri* were resistant to stressor-induced increases in colonic inflammation upon oral challenge with *C. rodentium* (Mackos et al., 2016). Interestingly, mice treated with *L. reuteri* still showed stressor-induced disruptions to the structure of the commensal microbiota (Galley et al., under review); thus, it was not clear from these studies whether stressor-induced changes in the composition of the gut microbiota were associated with dysregulated mucosal immunity.

To get a better understanding of whether stressor-induced alterations of the gut microbiota influence mucosal immune responses, naïve germfree mice were colonized with microbiota from conventional, nonstressed control mice or from conventional mice exposed to the prolonged restraint stressor. Importantly, germfree mice colonized with microbiota from stressor-exposed donors showed significantly higher levels of inflammatory cytokines and colonic histopathology upon challenge with *C. rodentium* than did germfree mice colonized with microbiota from nonstressed, control donors (Galley, Parry, Ahmer, Fox, & Bailey, under review). These data demonstrate that stressor-induced changes in the composition of the gut microbiota contribute to exacerbated colonic inflammatory responses.

Often microbial shifts in the intestine are to blame for altering intestinal immune function to permit a hyperinflammatory state. However, a recent study from Reber et al. demonstrated the importance of the development of regulatory T cells (Treg) by repeated immunizations with heat-killed *Mycobacterium vaccae*, a naturally occurring soil microbe with immunoregulatory properties (Reber et al., 2016). Mice exposed to chronic subordinate colony housing (CSC), a psychosocial stressor, develop spontaneous colitis and also experience an exacerbation of symptoms following chemically induced infectious colitis when compared to nonstressed controls. Mice exposed to CSC following heat-killed *M. vaccae* immunization were protected from the development of spontaneous colitis and chemically induced colitis due to the development of IL-10^{+} Tregs.

While effects on mucosal immunity are somewhat intuitive, it is less intuitive that gut microbes also affect immune responses outside of the gut. However, there are now several studies demonstrating that immune responses in the lung and spleen are dependent upon the gut microbiota. Thus, studies using the psychosocial stressors social disruption and chronic social defeat were conducted to test the hypothesis that stressor-induced immunomodulation outside of the gut may also be influenced by the gut microbiota.

Chronic social defeat induces social and exploratory behavioral deficits in mice when compared to nonstressed control counterparts, which may be attributed to stressor-induced changes in reduced intestinal richness and diversity as assessed by fecal pellets (Bharwani et al., 2016). Splenic immune function was also impacted by chronic social defeat. Spleens from mice exposed to chronic social defeat had an increased population of activated dendritic cells for as long as 17 days postcompletion of the stressor. At this same time point, 17 days poststressor completion, the immunoregulatory IL-10$^+$CD4$^+$CD25$^+$ T cells were reduced. Based on the chronic social defeat stress-induced microbial changes, in silico metagenomic analysis predicted that defeated mice would have reduced short-chain fatty acid synthesis as well as a reduction in the production of the neurotransmitter precursors tyrosine and tryptophan. Neurotransmitters and short-chain fatty acids have the ability to modulate the central nervous system, and thereby behavior which makes these predictive findings an exciting avenue for new discoveries linking stressor-induced behavioral deficits with shifts in the intestinal microbiome (Bharwani et al., 2016). One such study involving prebiotic supplementation during stressor exposure was performed in which mice were fed a diet rich in human milk oligosaccharides, i.e., 3′- and 6′sialyllactose (SL), or a control diet for 2 weeks prior to exposure to social disruption. Prebiotic substances can be fermented to short-chain fatty acids and can aid the growth of beneficial bacteria. Mice fed the control diet had significant increases in anxiety-like behavior which was associated with reductions of immature neurons of the dentate gyrus as evidenced by doublecortin immunostaining following social disruption exposure (Tarr et al., 2015). In contrast, mice fed either 3′SL or 6′SL prior to social disruption had no changes in anxiety-like behavior or the neuronal effects observed in mice fed a control diet.

Exposure to the social disruption stressor has long been recognized to result in increases in circulating cytokines, such as IL-6 and IL-1α/β. Earlier studies demonstrated that stressor-induced increases in these cytokines are associated with changes in the relative abundance of commensal microbes in the gut, such as *Coprococcus*, *Pseudobutyrivibrio*, and *Dorea* (Bailey et al., 2011). Although it is not known whether these bacteria directly participate in stressor-induced increases in circulating cytokines, treating mice with broad spectrum antibiotics prevented the stressor-induced increase in cytokines, suggesting that the gut microbiota is necessary for the stressor-induced immune activation to occur. This finding is consistent with others demonstrating that rats exposed to a repeated tail shock stressor have higher levels of cytokines that can be abrogated through the use of broad spectrum antibiotics (Maslanik et al., 2013, 2012). Other aspects of immune system activity, such as macrophage functioning, that can be enhanced during stressor exposure are also dependent upon commensal microbes, thus demonstrating that the gut microbiota are an essential component of stressor-induced immunopotentiation.

The mechanisms by which the gut microbiota can enhance immune system activity during stressor exposure are not yet completely known. However, previous studies suggest that stressor exposure results in the translocation of microbes from the gut (as well as from the skin) to regional lymph nodes (Bailey, Engler, & Sheridan, 2006). In mice exposed to the SDR stressor, IL-1-producing neutrophils and Ly6C^{hi} monocytes within the spleens of stressor-exposed mice contain higher levels of bacterial RNA than do cells in the spleens of nonstressed control mice (Lafuse, Gearinger, Fisher, Nealer, & Bailey, under review). Moreover, treating mice with antibiotics to affect the microbiota prevents stressor-induced increases in IL-1β gene expression in the spleen (Bailey, unpublished results). These are important observations, because signaling through the IL-1 receptor type 1 is necessary for enhanced macrophage activity and the development of anxiety-like behavior in mice exposed to the social disruption stressor (Engler et al., 2008). The anxiety-like behavior is also associated with trafficking of Ly6C^{hi} inflammatory monocytes into the brain (Wohleb, Powell, Godbout, & Sheridan, 2013). With recent studies suggesting that trafficking of Ly6C^{hi} cells to the brain is dependent upon gut microbes, it is worth testing whether stressor-induced anxiety-like behavior is dependent upon the microbiota through effects on IL-1-producing Ly6C^{hi} inflammatory monocytes.

5. CONCLUSIONS

The bidirectional brain–gut–microbiota axis (Fig. 1) is emerging as an important component of the body's physiological response to stressor exposure. The microbiota and its host have coevolved, and it is evident that during stressor exposure, brain–gut microbiota interactions can be adaptive and help the host respond to stressor exposure. Gut microbes play an important role in regulating the HPA axis response to stressor exposure. Activation of the HPA axis is an essential component of the physiological stress response, because it helps to increase glucose levels that are needed for physical and

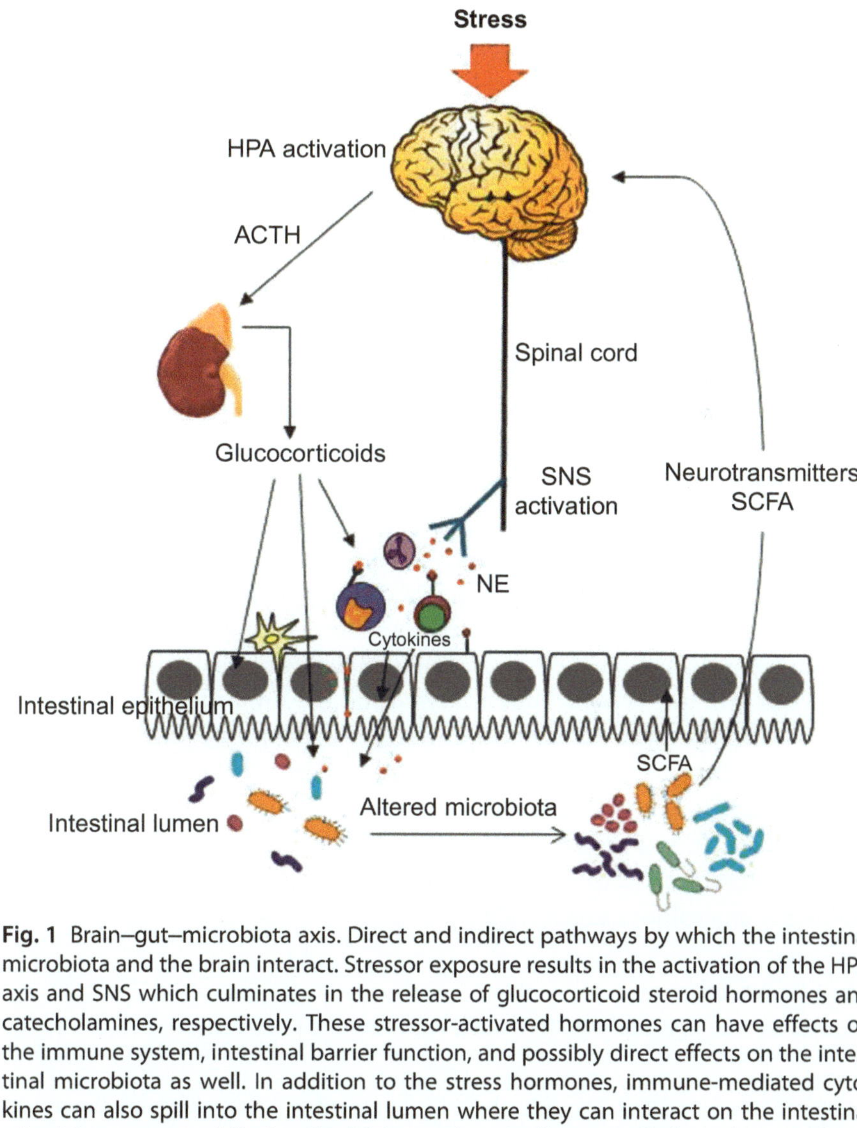

Fig. 1 Brain–gut–microbiota axis. Direct and indirect pathways by which the intestinal microbiota and the brain interact. Stressor exposure results in the activation of the HPA axis and SNS which culminates in the release of glucocorticoid steroid hormones and catecholamines, respectively. These stressor-activated hormones can have effects on the immune system, intestinal barrier function, and possibly direct effects on the intestinal microbiota as well. In addition to the stress hormones, immune-mediated cytokines can also spill into the intestinal lumen where they can interact on the intestinal microbiota to cause shifts in microbial growth and adherence. Shifts in the microbiota can lead to changes in neurotransmitter and SCFA production which can enter the circulation and enter the brain leading to neurological changes. *ACTH*, adrenocorticotropic hormone; *HPA*, hypothalamic–pituitary–adrenal; *NE*, norepinephrine; *SCFA*, short-chain fatty acids; *SNS*, sympathetic nervous system.

behavioral responses to stress. Indeed, studies now demonstrate that behavioral responses to stress, such as anxiety-like behavior, are influenced by gut microbes. Such effects on behavior are adaptive in the short term, because they can help the host limit their exposure to potentially threatening stimuli.

Although most stressors in human society today are psychological or psychosocial and not overtly physical, throughout evolution stressors were primarily physical (for example, predator–prey interactions). Wounding and exposure to contagion are more likely during physical stressors, and as a result, enhanced activation of the immune system would favor survival after aggressive confrontations, demonstrating that the effects of the microbiota on stressor-induced immunopotentiation are also adaptive.

Prolonged or dysregulated brain–gut microbiota interactions, however, can predispose to disease. Dysregulated homeostatic interactions between the gut microbiota and the mucosal, as well as systemic, immune system can lead to diseases involving enhanced inflammatory responses, such as the inflammatory bowel diseases and irritable bowel syndrome. It is likely that the brain–gut–microbiota axis is one reason that the symptoms of these diseases tend to be more severe during stressful periods. There is also increasing evidence that mental illness, which is strongly influenced by stressor exposure, involves dysregulation of the brain–gut–microbiota axis (Cryan & Dinan, 2012). This dysregulation may be particularly important early in life, where even small effects on the brain–gut–microbiota axis may severely change the developmental trajectories of the brain, gut microbiota, and immune system, thus underscoring the importance of the early-life environment for disease development later in life (Gur, Worly, & Bailey, 2015; O'Mahony, Clarke, Dinan, & Cryan, 2015). As a result, the brain–gut–microbiota axis is an attractive therapeutic target for the treatment of a broad spectrum of diseases. However, understanding how the brain and the gut microbiota interact is crucial to make therapeutic targeting of the brain–gut–microbiota axis a reality.

REFERENCES

Aagaard, K., Ma, J., Antony, K. M., Ganu, R., Petrosino, J., & Versalovic, J. (2014). The placenta harbors a unique microbiome. *Science Translational Medicine*, *6*(237), 237ra265. http://dx.doi.org/10.1126/scitranslmed.3008599.

Aguilera, M., Vergara, P., & Martinez, V. (2013). Stress and antibiotics alter luminal and wall-adhered microbiota and enhance the local expression of visceral sensory-related systems in mice. *Neurogastroenterology and Motility*, *25*(8), e515–e529. http://dx.doi.org/10.1111/nmo.12154.

Ardissone, A. N., de la Cruz, D. M., Davis-Richardson, A. G., Rechcigl, K. T., Li, N., Drew, J. C., et al. (2014). Meconium microbiome analysis identifies bacteria correlated with premature birth. *PLoS One*, *9*(3), e90784. http://dx.doi.org/10.1371/journal.pone.0090784.

Bailey, M. T. (2014). Influence of stressor-induced nervous system activation on the intestinal microbiota and the importance for immunomodulation. In M. Lyte & F. J. Cryan (Eds.), *Microbial endocrinology: The microbiota-gut-brain axis in health and disease* (pp. 255–276). New York, NY: Springer New York.

Bailey, M. T., & Coe, C. L. (1999). Maternal separation disrupts the integrity of the intestinal microflora in infant rhesus monkeys. *Developmental Psychobiology*, *35*(2), 146–155.

Bailey, M. T., Dowd, S. E., Galley, J. D., Hufnagle, A. R., Allen, R. G., & Lyte, M. (2011). Exposure to a social stressor alters the structure of the intestinal microbiota: Implications for stressor-induced immunomodulation. *Brain, Behavior, and Immunity*, *25*(3), 397–407. http://dx.doi.org/10.1016/j.bbi.2010.10.023.

Bailey, M. T., Dowd, S. E., Parry, N. M., Galley, J. D., Schauer, D. B., & Lyte, M. (2010). Stressor exposure disrupts commensal microbial populations in the intestines and leads to increased colonization by Citrobacter rodentium. *Infection and Immunity*, *78*(4), 1509–1519. http://dx.doi.org/10.1128/iai.00862-09.

Bailey, M. T., Engler, H., & Sheridan, J. F. (2006). Stress induces the translocation of cutaneous and gastrointestinal microflora to secondary lymphoid organs of C57BL/6 mice. *Journal of Neuroimmunology*, *171*(1–2), 29–37. http://dx.doi.org/10.1016/j.jneuroim.2005.09.008.

Bailey, M. T., Karaszewski, J. W., Lubach, G. R., Coe, C. L., & Lyte, M. (1999). In vivo adaptation of attenuated Salmonella typhimurium results in increased growth upon exposure to norepinephrine. *Physiology & Behavior*, *67*(3), 359–364.

Barouei, J., Moussavi, M., & Hodgson, D. M. (2012). Effect of maternal probiotic intervention on HPA axis, immunity and gut microbiota in a rat model of irritable bowel syndrome. *PLoS One*, 7(10), e46051. http://dx.doi.org/10.1371/journal.pone.0046051.

Bearson, B. L., & Bearson, S. M. (2008). The role of the QseC quorum-sensing sensor kinase in colonization and norepinephrine-enhanced motility of Salmonella enterica serovar Typhimurium. *Microbial Pathogenesis*, *44*(4), 271–278. http://dx.doi.org/10.1016/j.micpath.2007.10.001.

Bergström, A., Skov, T. H., Bahl, M. I., Roager, H. M., Christensen, L. B., Ejlerskov, K. T., et al. (2014). Establishment of intestinal microbiota during early life: A longitudinal, explorative study of a large cohort of Danish infants. *Applied and Environmental Microbiology*, *80*(9), 2889–2900. http://dx.doi.org/10.1128/aem.00342-14.

Bharwani, A., Mian, M. F., Foster, J. A., Surette, M. G., Bienenstock, J., & Forsythe, P. (2016). Structural & functional consequences of chronic psychosocial stress on the microbiome & host. *Psychoneuroendocrinology*, *63*, 217–227. http://dx.doi.org/10.1016/j.psyneuen.2015.10.001.

Chang, J. Y., Antonopoulos, D. A., Kalra, A., Tonelli, A., Khalife, W. T., Schmidt, T. M., et al. (2008). Decreased diversity of the fecal microbiome in recurrent Clostridium difficile-associated diarrhea. *Journal of Infectious Diseases*, *197*(3), 435–438. http://dx.doi.org/10.1086/525047.

Chen, C., Lyte, M., Stevens, M. P., Vulchanova, L., & Brown, D. R. (2006). Mucosally-directed adrenergic nerves and sympathomimetic drugs enhance non-intimate adherence of Escherichia coli O157:H7 to porcine cecum and colon. *European Journal of Pharmacology*, *539*(1–2), 116–124. http://dx.doi.org/10.1016/j.ejphar.2006.03.081.

Costello, E. K., Lauber, C. L., Hamady, M., Fierer, N., Gordon, J. I., & Knight, R. (2009). Bacterial community variation in human body habitats across space and time. *Science*, *326*(5960), 1694–1697. http://dx.doi.org/10.1126/science.1177486.

Cryan, J. F., & Dinan, T. G. (2012). Mind-altering microorganisms: The impact of the gut microbiota on brain and behaviour. *Nature Reviews. Neuroscience*, *13*(10), 701–712. http://dx.doi.org/10.1038/nrn3346.

David, L. A., Maurice, C. F., Carmody, R. N., Gootenberg, D. B., Button, J. E., Wolfe, B. E., et al. (2014). Diet rapidly and reproducibly alters the human gut microbiome. *Nature*, *505*(7484), 559–563. http://dx.doi.org/10.1038/nature12820.

De Palma, G., Blennerhassett, P., Lu, J., Deng, Y., Park, A. J., Green, W., et al. (2015). Microbiota and host determinants of behavioural phenotype in maternally separated mice. *Nature Communications*, *6*, 7735. http://dx.doi.org/10.1038/ncomms8735.

de Vos, W. M. (2015). Microbial biofilms and the human intestinal microbiome. *Npj Biofilms and Microbiomes*, *1*, 15005. http://dx.doi.org/10.1038/npjbiofilms.2015.5.

Donaldson, G. P., Lee, S. M., & Mazmanian, S. K. (2016). Gut biogeography of the bacterial microbiota. *Nature Reviews. Microbiology*, *14*(1), 20–32. http://dx.doi.org/10.1038/nrmicro3552.

Duerkop, B. A., Vaishnava, S., & Hooper, L. V. (2009). Immune responses to the microbiota at the intestinal mucosal surface. *Immunity*, *31*(3), 368–376.

Engler, H., Bailey, M. T., Engler, A., Stiner-Jones, L. M., Quan, N., & Sheridan, J. F. (2008). Interleukin-1 receptor type 1-deficient mice fail to develop social stress-associated glucocorticoid resistance in the spleen. *Psychoneuroendocrinology*, *33*(1), 108–117. http://dx.doi.org/10.1016/j.psyneuen.2007.10.007.

Francino, M. P. (2015). Antibiotics and the human gut microbiome: Dysbioses and accumulation of resistances. *Frontiers in Microbiology*, *6* 1543. http://dx.doi.org/10.3389/fmicb.2015.01543.

Freestone, P. P., Haigh, R. D., & Lyte, M. (2007a). Blockade of catecholamine-induced growth by adrenergic and dopaminergic receptor antagonists in Escherichia coli O157:H7, Salmonella enterica and Yersinia enterocolitica. *BMC Microbiology*, *7*, 8. http://dx.doi.org/10.1186/1471-2180-7-8.

Freestone, P. P., Haigh, R. D., & Lyte, M. (2007b). Specificity of catecholamine-induced growth in Escherichia coli O157:H7, Salmonella enterica and Yersinia enterocolitica. *FEMS Microbiology Letters*, *269*(2), 221–228. http://dx.doi.org/10.1111/j.1574-6968.2006.00619.x.

Freestone, P. P., Lyte, M., Neal, C. P., Maggs, A. F., Haigh, R. D., & Williams, P. H. (2000). The mammalian neuroendocrine hormone norepinephrine supplies iron for bacterial growth in the presence of transferrin or lactoferrin. *Journal of Bacteriology*, *182*(21), 6091–6098.

Freestone, P. P., Williams, P. H., Haigh, R. D., Maggs, A. F., Neal, C. P., & Lyte, M. (2002). Growth stimulation of intestinal commensal Escherichia coli by catecholamines: A possible contributory factor in trauma-induced sepsis. *Shock*, *18*(5), 465–470.

Galley, J. D., Nelson, M. C., Yu, Z., Dowd, S. E., Walter, J., Kumar, P. S., et al. (2014). Exposure to a social stressor disrupts the community structure of the colonic mucosa-associated microbiota. *BMC Microbiology*, *14*, 189. http://dx.doi.org/10.1186/1471-2180-14-189.

Galley, J.D., Parry, N.M., Ahmer, B.M.M., Fox, J.G., & Bailey, M.T. The commensal microbiota exacerbate infectious colitis in stressor-exposed mice. *Brain, Behavior, and Immunity*, under review.

Galley, J. D., Yu, Z., Kumar, P., Dowd, S. E., Lyte, M., & Bailey, M. T. (2014a). The structures of the colonic mucosa-associated and luminal microbial communities are distinct and differentially affected by a prolonged murine stressor. *Gut Microbes*, *5*(6), 748–760. http://dx.doi.org/10.4161/19490976.2014.972241.

Green, B. T., Lyte, M., Chen, C., Xie, Y., Casey, M. A., Kulkarni-Narla, A., et al. (2004). Adrenergic modulation of Escherichia coli O157:H7 adherence to the colonic mucosa. *American Journal of Physiology. Gastrointestinal and Liver Physiology*, *287*(6), G1238–G1246. http://dx.doi.org/10.1152/ajpgi.00471.2003.

Gur, T. L., Worly, B. L., & Bailey, M. T. (2015). Stress and the commensal microbiota: Importance in parturition and infant neurodevelopment. *Front Psychiatry*, *6*, 5. http://dx.doi.org/10.3389/fpsyt.2015.00005.

Hooper, L. V., & Gordon, J. I. (2001). Commensal host-bacterial relationships in the gut. *Science*, *292*(5519), 1115–1118. http://dx.doi.org/10.1126/science.1058709.

Howerton, C. L., & Bale, T. L. (2012). Prenatal programing: At the intersection of maternal stress and immune activation. *Hormones and Behavior*, *62*(3), 237–242. http://dx.doi.org/10.1016/j.yhbeh.2012.03.007.

Jasarevic, E., Howerton, C. L., Howard, C. D., & Bale, T. L. (2015). Alterations in the vaginal microbiome by maternal stress are associated with metabolic reprogramming of the offspring gut and brain. *Endocrinology, 156*(9), 3265–3276. http://dx.doi.org/10.1210/en.2015-1177.

Koenig, J. E., Spor, A., Scalfone, N., Fricker, A. D., Stombaugh, J., Knight, R., et al. (2011). Succession of microbial consortia in the developing infant gut microbiome. *Proceedings of the National Academy of Sciences of the United States of America, 108*(Suppl 1), 4578–4585. http://dx.doi.org/10.1073/pnas.1000081107.

Kostic, A. D., Howitt, M. R., & Garrett, W. S. (2013). Exploring host-microbiota interactions in animal models and humans. *Genes & Development, 27*(7), 701–718. http://dx.doi.org/10.1101/gad.212522.112.

Lafuse, W.P., Gearinger, R., Fisher, S., Nealer, C., & Bailey, M.T. Exposure to a social stressor induces translocation of commensal lactobacilli to the spleen and priming of the innate immune system. *Journal of Immunology*, under review.

Lederberg, J., & McCray, A. T. (2001). 'Ome sweet 'omics: A genealogical treasury of words. *The Scientist, 15*, 8.

Lyte, M., & Bailey, M. T. (1997). Neuroendocrine-bacterial interactions in a neurotoxin-induced model of trauma. *The Journal of Surgical Research, 70*(2), 195–201. http://dx.doi.org/10.1006/jsre.1997.5130.

Lyte, M., & Ernst, S. (1992). Catecholamine induced growth of gram negative bacteria. *Life Sciences, 50*(3), 203–212.

Lyte, M., Vulchanova, L., & Brown, D. R. (2011). Stress at the intestinal surface: Catecholamines and mucosa-bacteria interactions. *Cell and Tissue Research, 343*(1), 23–32. http://dx.doi.org/10.1007/s00441-010-1050-0.

Mackos, A. R., Eubank, T. D., Parry, N. M., & Bailey, M. T. (2013). Probiotic Lactobacillus reuteri attenuates the stressor-enhanced severity of Citrobacter rodentium infection. *Infection and Immunity, 81*(9), 3253–3263. http://dx.doi.org/10.1128/iai.00278-13.

Mackos, A. R., Galley, J. D., Eubank, T. D., Easterling, R. S., Parry, N. M., Fox, J. G., et al. (2016). Social stress-enhanced severity of Citrobacter rodentium-induced colitis is CCL2-dependent and attenuated by probiotic Lactobacillus reuteri. *Mucosal Immunology, 9*(2), 515–526. http://dx.doi.org/10.1038/mi.2015.81.

Maslanik, T., Mahaffey, L., Tannura, K., Beninson, L., Greenwood, B. N., & Fleshner, M. (2013). The inflammasome and danger associated molecular patterns (DAMPs) are implicated in cytokine and chemokine responses following stressor exposure. *Brain, Behavior, and Immunity, 28*, 54–62. http://dx.doi.org/10.1016/j.bbi.2012.10.014.

Maslanik, T., Tannura, K., Mahaffey, L., Loughridge, A. B., Beninson, L., Ursell, L., et al. (2012). Commensal bacteria and MAMPs are necessary for stress-induced increases in IL-1beta and IL-18 but not IL-6, IL-10 or MCP-1. *PLoS One*, 7(12), e50636. http://dx.doi.org/10.1371/journal.pone.0050636.

Moles, L., Gómez, M., Heilig, H., Bustos, G., Fuentes, S., de Vos, W., et al. (2013). Bacterial diversity in meconium of preterm neonates and evolution of their fecal microbiota during the first month of life. *PLoS One, 8*(6), e66986. http://dx.doi.org/10.1371/journal.pone.0066986.

Morgan, X. C., Segata, N., & Huttenhower, C. (2013). Biodiversity and functional genomics in the human microbiome. *Trends in Genetics, 29*(1), 51–58. http://dx.doi.org/10.1016/j.tig.2012.09.005.

Neu, J., & Rushing, J. (2011). Cesarean versus vaginal delivery: Long-term infant outcomes and the hygiene hypothesis. *Clinics in Perinatology, 38*(2), 321–331. http://dx.doi.org/10.1016/j.clp.2011.03.008.

O'Mahony, S. M., Clarke, G., Dinan, T. G., & Cryan, J. F. (2015). Early-life adversity and brain development: Is the microbiome a missing piece of the puzzle? *Neuroscience*. http://dx.doi.org/10.1016/j.neuroscience.2015.09.068. Epub ahead of print.

O'Mahony, S. M., Felice, V. D., Nally, K., Savignac, H. M., Claesson, M. J., Scully, P., et al. (2014). Disturbance of the gut microbiota in early-life selectively affects visceral pain in adulthood without impacting cognitive or anxiety-related behaviors in male rats. *Neuroscience, 277*, 885–901. http://dx.doi.org/10.1016/j.neuroscience.2014.07.054.

O'Mahony, S. M., Marchesi, J. R., Scully, P., Codling, C., Ceolho, A. M., Quigley, E. M., et al. (2009). Early life stress alters behavior, immunity, and microbiota in rats: Implications for irritable bowel syndrome and psychiatric illnesses. *Biological Psychiatry, 65*(3), 263–267. http://dx.doi.org/10.1016/j.biopsych.2008.06.026.

Rastall, R. A. (2004). Bacteria in the gut: Friends and Foes and how to alter the balance. *The Journal of Nutrition, 134*(8), 2022S–2026S.

Reber, S. O., Siebler, P. H., Donner, N. C., Morton, J. T., Smith, D. G., Kopelman, J. M., et al. (2016). Immunization with a heat-killed preparation of the environmental bacterium Mycobacterium vaccae promotes stress resilience in mice. *Proceedings of the National Academy of Sciences of the United States of America, 113*(22), E3130–E3139. http://dx.doi.org/10.1073/pnas.1600324113.

Sakuma, K., Funabashi, H., Matsuoka, H., & Saito, M. (2013). Potential use of Lactobacillus cell density in feces as a non-invasive bio-indicator for evaluating environmental stress during mouse breeding. *Biocontrol Science, 18*(2), 101–104.

Sandrini, S. M., Shergill, R., Woodward, J., Muralikuttan, R., Haigh, R. D., Lyte, M., et al. (2010). Elucidation of the mechanism by which catecholamine stress hormones liberate iron from the innate immune defense proteins transferrin and lactoferrin. *Journal of Bacteriology, 192*(2), 587–594. http://dx.doi.org/10.1128/jb.01028-09.

Sartor, R. B., & Mazmanian, S. K. (2012). Intestinal microbes in inflammatory bowel diseases. *The American Journal of Gastroenterology Supplements, 1*(1), 15–21.

Sekirov, I., Russell, S. L., Antunes, L. C., & Finlay, B. B. (2010). Gut microbiota in health and disease. *Physiological Reviews, 90*(3), 859–904. http://dx.doi.org/10.1152/physrev.00045.2009.

Sender, R., Fuchs, S., & Milo, R. (2016). Are we really vastly outnumbered? Revisiting the ratio of bacterial to host cells in humans. *Cell, 164*(3), 337–340. http://dx.doi.org/10.1016/j.cell.2016.01.013.

Spor, A., Koren, O., & Ley, R. (2011). Unravelling the effects of the environment and host genotype on the gut microbiome. *Nature Reviews. Microbiology, 9*(4), 279–290. http://dx.doi.org/10.1038/nrmicro2540.

Stothart, M. R., Bobbie, C. B., Schulte-Hostedde, A. I., Boonstra, R., Palme, R., Mykytczuk, N. C., et al. (2016). Stress and the microbiome: Linking glucocorticoids to bacterial community dynamics in wild red squirrels. *Biology Letters, 12*(1), 20150875. http://dx.doi.org/10.1098/rsbl.2015.0875.

Sudo, N., Chida, Y., Aiba, Y., Sonoda, J., Oyama, N., Yu, X. N., et al. (2004). Postnatal microbial colonization programs the hypothalamic-pituitary-adrenal system for stress response in mice. *The Journal of Physiology, 558*(Pt 1), 263–275. http://dx.doi.org/10.1113/jphysiol.2004.063388.

Tannock, G. W., & Savage, D. C. (1974). Influences of dietary and environmental stress on microbial populations in the murine gastrointestinal tract. *Infection and Immunity, 9*(3), 591–598.

Tarr, A. J., Galley, J. D., Fisher, S. E., Chichlowski, M., Berg, B. M., & Bailey, M. T. (2015). The prebiotics 3′sialyllactose and 6′sialyllactose diminish stressor-induced anxiety-like behavior and colonic microbiota alterations: Evidence for effects on the gut-brain axis. *Brain, Behavior, and Immunity, 50*, 166–177. http://dx.doi.org/10.1016/j.bbi.2015.06.025.

Turnbaugh, P. J., Ley, R. E., Mahowald, M. A., Magrini, V., Mardis, E. R., & Gordon, J. I. (2006). An obesity-associated gut microbiome with increased capacity for energy harvest. *Nature, 444*(7122), 1027–1131. http://dx.doi.org/10.1038/nature05414. http://www.nature.com/nature/journal/v444/n7122/suppinfo/nature05414_S1.html.

Van den Abbeele, P., Van de Wiele, T., Verstraete, W., & Possemiers, S. (2011). The host selects mucosal and luminal associations of coevolved gut microorganisms: A novel concept. *FEMS Microbiology Reviews*, *35*(4), 681–704. http://dx.doi.org/10.1111/j.1574-6976.2011.00270.x.

Voreades, N., Kozil, A., & Weir, T. (2014). Diet and the development of the human intestinal microbiome. *Frontiers in Microbiology*, *5*, 494.

Wohleb, E. S., Powell, N. D., Godbout, J. P., & Sheridan, J. F. (2013). Stress-induced recruitment of bone marrow-derived monocytes to the brain promotes anxiety-like behavior. *The Journal of Neuroscience*, *33*(34), 13820–13833. http://dx.doi.org/10.1523/jneurosci.1671-13.2013.

Yap, I. K., Angley, M., Veselkov, K. A., Holmes, E., Lindon, J. C., & Nicholson, J. K. (2010). Urinary metabolic phenotyping differentiates children with autism from their unaffected siblings and age-matched controls. *Journal of Proteome Research*, *9*(6), 2996–3004. http://dx.doi.org/10.1021/pr901188e.

Zheng, P., Chen, J. J., Huang, T., Wang, M. J., Wang, Y., Dong, M. X., et al. (2013). A novel urinary metabolite signature for diagnosing major depressive disorder. *Journal of Proteome Research*, *12*(12), 5904–5911. http://dx.doi.org/10.1021/pr400939q.

CHAPTER TWO

The Influence of Prebiotics on Neurobiology and Behavior

A.C.C. Kao, S. Harty, P.W.J. Burnet[1]
University of Oxford, Oxford, United Kingdom
[1]Corresponding author: e-mail address: phil.burnet@psych.ox.ac.uk

Contents

Abstract

Manipulating the intestinal microbiota for the benefit of the brain is a concept that has become widely acknowledged. Prebiotics are nondigestible nutrients (i.e., fibers, carbohydrates, or various saccharides) that proliferate intrinsic, beneficial gut bacteria, and so provide an alternative strategy for effectively altering the enteric ecosystem, and thence brain function. Rodent studies demonstrating neurobiological changes following prebiotic intake are slowly emerging, and have thus far revealed significant benefits in disease models, including antiinflammatory and neuroprotective actions. There are also compelling data showing the robust and favorable effects of prebiotics on several behavioral paradigms including, anxiety, learning, and memory. At present, studies in humans are limited, though there is strong evidence for prebiotics modulating emotional processes and the neuroendocrine stress response that may underlie the pathophysiology of anxiety. While the mechanistic details linking the enteric microbiota to the central nervous system remain to be elucidated, there are a number of considerations that can guide future studies. These include the modulation of intestinal endocrine systems and inflammatory cascades, as well as direct interaction with the enteric

International Review of Neurobiology, Volume 131
ISSN 0074-7742
http://dx.doi.org/10.1016/bs.irn.2016.08.007

nervous system and gut mucosa. Our knowledge of gut microbiome–brain communication is steadily progressing, and thorough investigations validating the use of prebiotics in the treatment of neuropsychiatric disorders would be highly valued and are encouraged.

1. INTRODUCTION

Compelling evidence for the role of the gut microbiota as a potential therapeutic target for a broad range of neuropsychological disorders has been accumulating over the past decade. Preclinical and clinical studies have suggested that altered composition of enteric microbial communities is fundamental to the modulation of the microbiome–gut–brain axis. For example, in a behavioral investigation with mice, greater bacterial diversity in the colon was associated with improved working and reference memory when fed with standard rodent chow containing 50% lean ground beef (Li, Dowd, Scurlock, Acosta-Martinez, & Lyte, 2009). Similarly, oral administration of antibiotics has been shown to reverse overt hepatic encephalopathy, possibly through acidification of the gastrointestinal tract (Schiano, 2010).

In addition to diet and antibiotics, there are several additional approaches to manipulate the intestinal microbiome including exposure to activated carbon (Khoder, Tsapis, Domergue-Dupont, Gueutin, & Fattal, 2010), fecal microbiota transplantation (Collins, Kassam, & Bercik, 2013), or ingestion of natural dietary compounds. This latter group, which exhibits a high safety profile and is therefore preferred, consists of such as probiotics, prebiotics, polyphenols, amino acids, and dietary fibers. In efforts to manipulate the gut microbiota for the benefit of the brain, the effect of consuming specific microbial strains as live cultures (i.e., probiotics) on the psychological state of healthy volunteers have been explored. Reinforcing bacterial populations such as *lactobacillus* and *bifidobacterium* resulted in significantly improved psychological well-being in healthy volunteers after 30-days of ingestion (Messaoudi et al., 2011) as well as in patients with chronic fatigue syndrome after 60-days of ingestion (Rao et al., 2009).

The composition of the gut microbiota can also be altered by prebiotic consumption. These are nondigestible compounds, comprising primarily of carbohydrates or short chains of saccharide molecules. These molecules are naturally found within the mammalian diet and are also commercially available as dietary supplements in their purified form. Prebiotics enhance the simultaneous proliferation of specific indigenous microbiota by providing them with a source of energy. It is conceivable, therefore, that the ingestion of prebiotics might impart greater health benefits than probiotics; they have the potential to augment the growth of many beneficial microbial species, which would ultimately lead to greater bacterial diversity in the gut.

2. PREBIOTICS

Over the past decade, the definition of prebiotics has developed and matured into a very specific set of requirements. When the concept was first proposed over 20 years ago, prebiotics were described as any "nondigestible food ingredient that beneficially affects the host by selectively stimulating the growth and/or activity of one or a limited number of bacteria already resident in the colon" (Gibson & Roberfroid, 1995). In hindsight this definition is narrow and dismisses two major properties. First, microflora in other parts of the gastrointestinal tract may also impart health benefits to the host; second, the definition lacks the metabolic details (i.e., fermentation) of converting indigestible food ingredients into health-promoting molecules. The latter is an important distinction as it identifies between molecules that require microbial metabolism to exert its actions vs molecules that alter the composition of the gut microbiota (e.g., antibiotics and bacteriocins). Therefore, current dietary prebiotics must be a selectively fermentable ingredient that results in specific changes in the composition and/or activity of microbiota within the gastrointestinal tract (Gibson et al., 2010). In this context, "selectivity" describes the specificity of a compound to proliferate a health-promoting taxonomic group such as *bifidobacteria* and *lactobacillus*, whereas fermentation produces metabolites (e.g., short-chain fatty acids [SCFAs]) that can be absorbed by the mammalian gut and influence host physiology.

Although prebiotics and dietary fibers share several characteristics (e.g., partial or total resistance to digestion, fermented by gut microbiota), the specificity condition excludes dietary fibers from having a "prebiotic" effect. This highlights the key condition of prebiotics which needs to be demonstrated in an in vivo experiment (e.g., complex human or animal gut microbiota) using relevant and validated methodologies to quantify a wide variety of genera/species composing the gut microbiota (Roberfroid et al., 2010). As a result, the majority of scientific data on prebiotics has focused on compounds belonging to two major chemical groups: fructans and oligosaccharides. In general, oligosaccharides and polysaccharides are composed of 3–9 and 10 or more saccharide units, respectively. Chemical nomenclature of carbohydrates is constructed upon the type of monosaccharide unit (i.e., glucose, fructose, galactose) and the type of linkage within the molecule. The most commonly studied prebiotics are discussed later.

2.1 Inulins and Fruto-Oligosaccharide

Inulins are a heterogeneous group of fructose-based polymers used as storage carbohydrates by numerous plant species including wheat, onion, bananas, garlic, and chicory (Niness, 1999). Chemically, inulin is a linear molecule composed of a core fructose and a terminal glucose moiety. The former moiety is the result of repeating D-fructofuranose units linked by a β-(2, 1)-fructosyl–fructose linkage, whereas the latter is composed of a single D-glucopyranose unit linked by an α-(1,2) linkage (Mutanda, Mokoena, Olaniran, Wilhelmi, & Whiteley, 2014). In general, inulins have a degree of polymerization (DF) between 10 and 65 and, therefore, can be hydrolyzed into smaller oligosaccharides termed oligofrutose and fruto-oligosaccharides (FOSs) depending on the synthesis process (Roberfroid et al., 2010). More specifically, an inulinases catalyzed hydrolysis reaction of the β-(2,1)-fructosyl–fructose bonds creates short-chain oligomers, often referred to as oligofructose; whereas fructosyltransferases are able to synthesize short-chain fructo-oligosaccharides (scFOS) from sucrose molecules by attaching additional fructose molecules (Niness, 1999). Both enzymatic processes result in FOS that contains similar β-linkages as inulin, rendering both molecules indigestible by human enzymes. This ensures their passage into the colon, satisfying one of the necessary prebiotic conditions.

Several studies have validated the effects of inulin in defined pure and mixed cultures, in vivo animal models, as well as in humans. In pure cultures, several species of bacteria were adept at utilizing inulin-type fructans including *bifidobacteria* and *Lactobacillus acidophilus* (Roberfroid, Van Loo, & Gibson, 1998). In mixed batch and continuous cultures containing both beneficial and pathogenic bacteria such as *Escherichia coli* and *Clostridium*, the addition of oligofructose and inulin proliferated *bifidobacterium* (i.e., the health-promoting genus), as well as limiting the levels of pathogenic species (Wang & Gibson, 1993). Similar beneficial effects have also been observed in preclinical studies that administered inulin-type fructans to laboratory (Campbell, Fahey, & Wolf, 1997) and germ-free (GF) rats (Kleessen, Hartmann, & Blaut, 2001). In these studies, health-promoting effects have been described as decreased fecal pH, increased production of SCFAs, including butyrate, as well as significantly increased number of *bifidobacteria* species. More interestingly, despite the heterogeneity in clinical trials involving dosing regimens and age of subjects, there is consensus on the bifidogenic nature of ingesting either scFOS, oligofructose, or inulin. This highlights the robustness in the inulins specificity, satisfying yet another one of the necessary prebiotic conditions.

2.2 Galacto-Oligosaccharides

The galacto-oligosaccharides (GOSs) naturally occur in legumes such as lentils, chickpeas, and beans. Chemically, GOS are the elongation products of lactose transgalactosylase which is catalyzed by β-galactosidase. The resulting molecule consists of a varying degree of β-glycosidic linkages due to the multiple number of positions galactose chains can be hydrolyzed as well as branching glucose residues (Otieno, 2010). For example, a commercially available formulation of GOS, Bimuno® (B-GOS), comprises of two structurally different β-galacto-oligosaccharides: β-1,3 galacto-oligosaccharides and β-1,4 and/or 1,6 galacto-oligosaccharides. Among the studies that investigate the effects of GOS, it is important to take note of the form, structure, and preparation of the prebiotic. Differences at the physiological level may originate from the high variability of chemical linkages, the DF of each molecule, as well as the relative abundance of these molecules in individual preparations.

3. NEUROBIOLOGICAL CHANGES ASSOCIATED WITH PREBIOTIC INTAKE

Prebiotic administration in preclinical models has shown significant peripheral and central changes. These results include altered expression of proteins associated with neuroprotection including antiinflammatory or antioxidant systems. Collectively, convincing data are available to support the use of prebiotics as a therapeutic agent in neurodegenerative disorders, although the robustness and replicability of these studies remain to be demonstrated. Prebiotics may also play a role in neurodevelopment in early life as it influences brain-derived neurotrophic factor (BDNF) levels in both adults and infants. Additional studies that examine the longevity of the prebiotic effects are necessary. Lastly, clinical studies that investigate central effects of prebiotics are currently lacking possibly due to the novelty of the field. However, these studies are highly encouraged as they will provide unequivocal validity to use prebiotics in the treatment of several neuropsychiatric disorders.

3.1 Prebiotics and Neuroinflammation

The neuroprotective properties of oligosaccharides have been demonstrated in animal models of neurodegenerative disorders such as amyotrophic lateral sclerosis (ALS) or Alzheimer's disease (AD). In $SOD1^{G93A}$ transgenic mouse model of ALS, 10-week oral administration of 2% GOS significantly delayed disease progression and increased survival by almost 2 weeks (Song, Gao, Zhang, & Le, 2013). This ground-breaking finding was further supported at the physiological level, wherein GOS administration significantly attenuated motor neuron degeneration in the anterior horn of the lumbar spinal cord, improved muscle atrophy, and significantly attenuated oxidative stress in skeletal muscles. The authors further reported antiinflammatory actions of GOS by examining the levels of glial fibrillary acidic protein (GFAP) and ionized calcium-binding adapter molecule (Iba-1), both of which are widely accepted cellular markers of activated astrocytes and microglia, respectively, the latter being a hallmark of an inflammatory response in the brain. Immunohistochemical analysis revealed significantly less GFAP and Iba-1 immunostaining in lumbar spinal cord sections from ALS animals that had received a GOS-supplemented diet, compared to those on a normal diet. As expected, decreased iNOS and TNFα levels in the spinal cord tissues were also reported. In addition, a significant reduction in the levels of proapoptotic factors (e.g., cleaved caspase-3 and Bax), together with an increase in antiapoptotic factors (i.e., Bcl-2) were observed in the spinal cord of GOS-fed mice (Song et al., 2013). This study provides unequivocal evidence for the potent antiinflammatory, and subsequent neuroprotective effects, of gut microbiota nurtured with GOS.

Though mechanistic details that link the cell and molecular changes are lacking, it is possible that the serotonin 5-HT2 receptor is involved. In spite of an ongoing debate on whether central levels of 5-HT receptors are affected by systemic inflammation, some groups have shown LPS-induced systemic inflammation results in elevated cortical 5-HT_{2A} receptor transcripts (Bouhnik et al., 2004) and hippocampal 5-HT release (Linthorst, Flachskamm, Muller-Preuss, Holsboer, & Reul, 1995). Pertinent to prebiotic administration, it has been shown recently that LPS-induced increase of 5-HT_{2A} receptor expression, as well as accompanying IL-1β and TNFα levels, can be normalized upon B-GOS administration (Savignac et al., 2016). However, functional data suggests that selective activation of 5-HT_{2A} results in a potent and specific blockade of TNFα-mediated inflammatory response (Nau, Yu, Martin, & Nichols, 2013). Since expression levels do not always correlate with functionality, it would be interesting to explore the downstream mechanisms that allow prebiotics to confer its antiinflammatory effects.

The chitosan oligosaccharides (COSs) have been shown to be effective at alleviating the pathological effects of an amyloid-β_{1-42} induced AD rat model (Jia et al., 2016). A 2-week treatment with COS, partially reversed hippocampal pyramidal neuronal damage and mitigated neuronal loss in the hippocampal CA1 subfield. In addition, rodents that received COS had a significantly decreased number of TUNEL positive staining in the hippocampus, suggesting antiapoptotic properties. The intake of COS has also been shown to increase the activity of antioxidant enzymes including, glutathione peroxidase, superoxide dismutase, and malondialdehyde in the hippocampus. At the molecular level, hippocampal expression of DNA oxidation marker 8-oxo-2′-deoxyguanosine as well as proinflammatory markers (i.e., TNFα and IL-1β) were significantly decreased upon COS administration. Together, these findings highlight the therapeutic potential of GOS and COS in neurodegenerative disorders, and the involvement of cellular pathways associated with inflammation and apoptosis.

3.2 Receptors and Signaling Molecules

BDNF is a member of the neurotrophin family, a heterogeneous class of proteins shown to control various aspects of survival, development, and function of neurons in both the peripheral and the central nervous systems (CNS). In cells, BDNF is synthesized as a proprotein and is secreted as a homodimeric protein into the extracellular space where it has autocrine and/or paracrine actions. The pioneering study that examined the influence of prebiotics on central BNDF expression was conducted by Savignac et al. (2013). In this study, relative to controls, a significant elevation of BDNF expression was observed in the dentate gyrus of adult male Sprague-Dawley (SD) rats that had received a daily administration of GOS or FOS, for 5 weeks (Savignac et al., 2013). Similarly, after 5 weeks of feeding with 2′-fucosyllactose, the most abundant glycan in human milk, a significant increase in BDNF expression was reported in the rat hippocampus (Vazquez et al., 2015). However, in a preclinical piglet model, a 30-day dietary supplementation containing prebiotics did not result in significant changes in BDNF gene expression in neither the hippocampus nor prefrontal cortex (Mudd et al., 2016). Notably, the dietary supplementation fed to piglets consisted of a blend of polydextrose, GOSs, bovine lactoferrin, and milk fat globule membrane, possibly adding noise to the data. The role of BDNF in the adult brain, particularly in the hippocampus, have been shown to be pivotal for learning and memory as well as protection from anxiety and depressive disorders (Heldt, Stanek, Chhatwal, & Ressler, 2007; Monteggia et al., 2004). In an adult mouse model with a site-specific deletion of BNDF, significant impairment of several behavioral paradigms has been observed including, novel object recognition, spatial learning, and reduced extinction of conditioned fear (Heldt et al., 2007). The augmentation of BDNF function via the administration of prebiotics, therefore, presents as a simple and natural strategy to ameliorate these types of cognitive/emotional deficits that manifest in several brain disorders.

Prebiotic feeding has also been shown to result in higher expression levels of *N*-methyl-D-aspartate receptor (NMDAR) subunits in the hippocampus and frontal cortex. The NMDAR is a type of ionotropic glutamate receptor and is composed of two obligatory GluN1 subunits and two regulatory GluN2A and/or GluN2B subunits. The activation of these receptors requires the concomitant binding of glutamate, the major excitatory neurotransmitter in the brain, and an NMDAR coagonist, D-Serine or glycine, and their optimal function is crucial for healthy learning and memory (Fedder & Sabo, 2015). B-GOS feeding resulted in a significant increase in the concentrations of cortical D-serine, and the expression of cortical GluN1 and hippocampal GluN2A subunits (Savignac et al., 2013). Although mechanistic details are lacking to explain why FOS administration did not result in these changes, it is possible that it was associated with the less potent *bifidogenic* properties of this prebiotic, compared to B-GOS.

The prebiotic B-GOS has also been shown to influence the levels of brain proteins in early life. Its administration to suckling neonatal rats prior to weaning, resulted in significant elevations of BDNF, synaptophysin, and the GluN2A subunit in the hippocampus, 1 and 26 days after discontinuing the prebiotic feed (Williams et al., 2016). This study not only highlights the consistent effects of B-GOS on central BDNF and NMDARs, but further illustrates the sustained influence of the gut microbiome on brain development, which is in keeping with studies in GF mice. These animals, which lack commensal bacteria, have provided the initial, and predominant, source of evidence for the importance of the microbiota on neurodevelopment. The GF mice exhibit a unique set of abnormal behavioral phenotypes including lapses in learning and memory (Gareau et al., 2011), decreases in locomotor activity (Diaz Heijtz et al., 2011), decreases in social cognition and preference (Arentsen, Raith, Qian, Forssberg, & Diaz Heijtz, 2015;

Desbonnet, Clarke, Shanahan, Dinan, & Cryan, 2014), and abnormal anxiety profiles (Arentsen et al., 2015; Clarke et al., 2013; Diaz Heijtz et al., 2011; Neufeld, Kang, Bienenstock, & Foster, 2011). Importantly, some of these studies have shown that colonization of the intestines with strain-matched microbiota can normalize abnormal behaviors, an effect that is only observed when colonization occurs in early life, and not in adulthood. This observation alone unreservedly supports a role of gut bacteria in neurodevelopment.

A recent randomized control trial in 2-year-old infants has shown that a 1-month dietary supplementation with a blend of prebiotics (i.e., scGOS, long chain FOS, pAOS) did not lead to significant improvements in neurodevelopmental outcomes (i.e., cognitive, social, language, gross, and fine motor skills) assessed by The Bayley Scales of Infant and Toddler Development (van den Berg, Westerbeek, Broring-Starre, Garssen, & van Elburg, 2016). Notably, this study was designed with a short intervention period and examined a smaller experimental window. In preclinical studies, however, it appears that early-life prebiotic intake may enrich neurodevelopment (Keunen, van Elburg, van Bel, & Benders, 2015; Krishna, Divyashri, Prapulla, & Muralidhara, 2015). For example, in piglets receiving dietary supplements which included GOS, significantly smaller cortical gray and white matter volumes were observed (Mudd et al., 2016). Although this may seem detrimental, the authors proposed that to ensure only vital neuronal connections are retained, the developing brain normally undergoes axon pruning, and that in prebiotic-fed animals synapse formation, followed by pruning, occurred earlier than in control piglets. Future studies, therefore, should evaluate whether improvements in cognitive performance in adulthood follow early-life prebiotic feeding.

4. PREBIOTIC-MEDIATED CHANGES IN BEHAVIOR

While 3–4 weeks of probiotic-based manipulations of the intestinal microbial population have been shown to significantly elevate mood (Benton, Williams, & Brown, 2007), or alleviate psychological distress (Messaoudi et al., 2011) in healthy volunteers, research on the behavioral effects of prebiotics remains limited (Tables 1 and 2). In one study, 5.5 g/day of either B-GOS, FOS, or a placebo was ingested for 3 weeks by healthy volunteers, and the psychological mechanisms that underlie anxiety, stress, and depression risk were evaluated by measuring the levels of waking cortisol response and administering a battery of neuropsychological tests (i.e.,

Table 1 Summary of Behavioral Changes as a Result of Prebiotic Administration

Prebiotic Compound	Dosing Regimen	Subjects (n)	Results	Reference
B-GOS	5.5 g/day for 3 weeks	22♂ 23♀ Healthy volunteers (n=45)	Decreased levels of waking cortisol; increased attentional vigilance to positive vs. negative stimuli	Schmidt et al. (2015)
GOS/polydextrose	3.5 g/100 g of milk for 4 weeks	♂ Piglets (n=12)	N.S. in T-maze	Mudd et al. (2016)
B-GOS	1.3% (w/v) in water bottles for 3 weeks	♂ CD-1 mice (n=15/group)	Decrease anxiety-like behavior	Savignac et al. (2016)
FOS	25–50 g/kg basal diet for 49 days	♂ BALB/cJ mice (n=7/group)	Improved spatial learning and memory as evaluated using the MWM	Yen, Wang, Wu, and Chen (2015)
COS	200–400 mg/kg for 14 days	♂ SD Rats (n=12)	Improved spatial learning and memory as evaluated using the MWM	Jia et al. (2016)
2′FL	350 mg/kg via diet (0.312% wt/wt) for 12 weeks	♂ C57/BL6 Mice (n=28)	Enhanced acquisition of conditioned task	Vazquez et al. (2015)
2′FL	0.625% (wt/wt) oral gavage of 2′FL via diet for 5 weeks (i.e., 350 mg/kg)	♂ SD Rats (n=10)	Enhanced associative learning (increased working memory)	Vazquez et al. (2015)

Abbreviations: *2′FL*, 2′-fucosyllactose; *B-GOS*, Bimuno galacto-oligosaccharide; *COS*, chitosan oligosaccharide; *FOS*, fructo-oligosaccharide; *GOS*, galacto-oligosaccharide; *MWM*, Morris Water Maze; *N.S.*, no significance.

Table 2 Central Cell and Molecular Changes in Prebiotic Studies

	Prebiotic Compound	Dose and Duration	Subjects (*n*)	Results	Reference
Cellular changes	GOS/ polydextrose	3.5 g/100 g for 4 weeks	♂ Piglets (n=12)	Decrease in cortical gray/white matter	Mudd et al. (2016)
	COS	200–400 mg/kg for 14 days	♂ Sprague-Dawley rats (n=12)	Decreased TUNEL positive staining; reversed hippocampal neuronal damage; increased activities of glutathione peroxidase and super oxide dismutase	Jia et al. (2016)
	GOS	Orally fed with 2% GOS at 8 mg/kg weight/day	♂ Transgenic ALZ mice (n=22)	Attenuated motor neuron degeneration; Improved skeletal muscle atrophy	Song et al. (2013)
Molecular changes	COS	200–400 mg/kg for 14 days	♂ Sprague-Dawley rats (n=12)	Reduced levels of 8-OHdG; reduced hippocampal expression of TNFα and IL-1β	Jia et al. (2016)
	GOS	Orally fed with 2% GOS at 8 mg/kg weight/day	♂ Transgenic ALZ mice (n=22)	Decreased expression of GFAP, Iba-1 in spinal cord tissue; reduced protein expression of iNOS, TNFα; increased protein expression of Bcl-2; decreased protein expression of caspase-3 and Bax	Song et al. (2013)

Continued

Table 2 Central Cell and Molecular Changes in Prebiotic Studies—cont'd

Prebiotic Compound	Dose and Duration	Subjects (*n*)	Results	Reference
B-GOS or FOS	3–4 g/kg, oral gavage for 5 weeks	♂ Sprague-Dawley rats ($n=8$/group)	FOS-fed rats: Elevated Hippocampal BDNF and GluN1 subunit. GOS-fed rats: Elevated hippocampal GLUN2A subunit and frontal cortex GLUN1 and D-serine. Elevated plasma D-alanine, peptide YY	Savignac et al. (2013)
B-GOS	1.3% (w/v) in drinking water for 3 weeks	♂ CD-1 mice ($n=15$/group)	Elevated cortical IL-1β concentrations in control mice 28 h after LPS were not observed in B-GOS-fed animals; significant B-GOS LPS interaction was also observed for 5HT2A receptors	Savignac et al. (2016)
B-GOS	4 g/kg/day, oral gavage for 3 weeks	♂ Sprague-Dawley pups	Elevated NMDAR GluN2A subunit, SYN and BDNF protein levels in the hippocampus which persists into adulthood (gender independent)	Williams et al. (2016)
2′FL	0.625% (wt/wt) oral gavage of 2′FL via diet for 5 weeks (i.e., 350 mg/kg)	♂ Sprague-Dawley rats ($n=10$)	Increased BDNF protein expression in cortical and subcortical regions	Vazquez et al. (2015)
GOS/polydextrose	3.5 g/100 g for 4 weeks	♂ Piglets ($n=12$)	N.S. in BDNF gene expression	Mudd et al. (2016)

2′FL, 2′-fucosyllactose; *ALZ*, Alzheimer's disease; *B-GOS*, Bimuno galacto-oligosaccharide; *COS*, chitosan oligosaccharide; *FOS*, fructo-oligosaccharide; *GOS*, g-alacto-oligosaccharide; *MWM*, Morris Water Maze; *N.S.*, no significance.

emotional processing, attentional dot-probe task, etc.). Participants who ingested B-GOS had significantly lower levels of waking cortisol and higher attentional vigilance to positive vs negative stimuli (Schmidt et al., 2015). These results were similar to the effects of various antidepressant medications (e.g., SSRIs, benzodiazepine, and diazepam) wherein threatening stimuli were less likely to be attended to, which has been suggested to underlie a anxiolytic-like profile. Additional studies involving candidate prebiotics such as plant-derived polysaccharides (Best, Kemps, & Bryan, 2010) or beta-glucans (Talbott & Talbott, 2009) have also shown significant improvements in cognitive function and mood in healthy subjects. Although these compounds have only been shown to satisfy the prebiotic definition (i.e., to influence the intestinal microbial population) in cultures of anaerobic microbial species (Lam & Chi-Keung Cheung, 2013; Marzorati et al., 2010), their beneficial effects on behavior are promising, and beg the need for further investigation.

Preclinical studies demonstrate similar behavioral effects following prebiotics ingestion. Human milk contains a diverse range of oligosaccharides that have *bifidogenic* properties. The most abundant oligosaccharide, 2′-fucosyllactose, was shown to confer enhanced associative learning and working memory in rodents (Vazquez et al., 2015). In this study, SD rats and C57BL/6 mice were fed 2′-fucosyllactose for 5 and 12 weeks, respectively. The effect of 2′-fucosyllactose in rats was evaluated using a basic Skinner box that exposed rats to an operant conditioning task (i.e., a food pellet following each lever press). Although rats receiving the prebiotic-infused diet acquired the task in slightly fewer sessions compared to control groups, the effect did not reach statistical significance, likely as the result of the study being under-powered ($n=7$; $p=0.053$). Nonetheless, 2′-fucosyllactose fed rats demonstrated a higher performance across 10 sessions compared to control groups. In mice, the 12-week administration of 2′-fucosyllactose was evaluated in an IntelliCage system that examined several behavioral paradigms, particularly working memory tasks. Compared to the control group, a higher percentage of prebiotics-fed mice were able to correctly identify the water containing corner (i.e., the rewarding corner). Since the rewarding corner changed dynamically clockwise, this meant that upon a correct visit mice were required to learn that the next correct visit would be in the adjacent corner. Using this behavioral task in various schedules of reinforcement, it appears that a 2′-fucosyllactose supplemented diet may result in enhanced spatial learning, operant conditioning, as well as working memory (Vazquez et al., 2015).

Like, 2′-fucosyllactose, 6′-sialyllactose, and 3″-sialyllactose are common oligosaccharides found in human milk and may confer anxiolytic effects. Significantly decreased anxiety-like behavior was reported in both light–dark preference task as well as the open field task in stressed C57/BL6 male mice fed with 6′-sialyllactose or 3′-sialyllactose 3 weeks (Tarr et al., 2015). Similarly, after a 3-week prebiotic feeding with B-GOS, CD-1 male mice were less anxious in the light–dark box compared to controls, but no significant differences were observed in marble burying behavior and locomotor activity (Savignac et al., 2016). Results obtained using the light–dark preference task appear to support an anxiolytic-effect of prebiotics. However, it is interesting that no significant differences were observed in the marble burying task as this is a widely used paradigm for examining anxiety (Deacon, 2006; Njung'e & Handley, 1991). However, behavioral tasks in rodents are highly sensitive to strain differences, animal-experimental interactions, as well as the possibility of producing a ceiling effect (Hanell & Marklund, 2014). This highlights the need for future investigations to employ additional behavioral tasks that examine anxiety in order to verify the anxiolytic potential of prebiotic feeding.

The cognitive effect of other prebiotics such as COS and FOS has been evaluated in preclinical models. In an AD model, the Morris Water Maze was used to evaluate the effect of 200–800 mg/kg of COS (Jia et al., 2016). It was reported that cognitive deficits induced by $A\beta_{1-42}$ administeration in SD rats were significantly improved after 2-week oral feeding of COS. Similarly, in a D-galactose rat model of AD, FOS administration improved spatial learning and memory in the Morris Water Maze (Yen et al., 2015). However, using a T-maze assessment, piglets receiving dietary supplementation that included GOS did not exhibit significantly different latency nor correct choices (Mudd et al., 2016).

5. MECHANISTIC CONSIDERATIONS

Both direct and indirect mechanisms have been proposed to mediate the central effects of the gut microbiome. However, for oligosaccharide prebiotics, the significant production of SCFAs that results from their fermentation, may be the predominant mediator that relays changes in the enteric environment to the brain, either directly or via the gut endocrine and immune systems. In the case of gut hormones, SCFA production may elevate levels of circulating satiety peptides which are able to penetrate the blood–brain barrier and initiate neurochemical and signaling cascades that underlie the observed behavioral changes. It is also possible that the antiinflammatory effects of prebiotic administration, which affect brain function (see earlier), are conferred by SCFAs interacting with enteric immunomodulatory cells. However, neurobiological changes resulting from the direct stimulation of enteric neurons or vagal afferent terminals by bacterial species or metabolites (e.g., neurotransmitters), or even physical microbial/prebiotic interactions with the gut lumen, should also be considered.

5.1 SCFAs and Gut Hormones

In healthy individuals, the gut microbiota maintains the proper functioning of several vital physiological functions such as synthesizing vitamins, maintaining the integrity of the intestinal barrier, and more importantly, extracting energy from indigestible carbohydrates (Jandhyala et al., 2015). This latter function is made possible by catabolic enzymes, not found in the human genome. These enzymes ferment the indigestible macromolecules into SCFAs, which can then be absorbed by the host providing energy and nutrients. Once SCFAs are produced in the intestinal tract, they can enter the circulatory system via active transport systems across the apical and basolateral membranes of colonocytes. The colonocytes metabolize SCFAs for energy, particularly butyrate, but the remainder SCFAs that are not consumed are transported across the basolateral membrane by similar active transport mechanisms (Wong, de Souza, Kendall, Emam, & Jenkins, 2006). Once the SCFAs are within the hepatic-portal system, a large portion of propionate is absorbed by the liver as a substrate for hepatic gluconeogenesis (Cummings, Pomare, Branch, Naylor, & Macfarlane, 1987). While it appears that majority of the SCFAs produced in the intestines are absorbed in the periphery, there is some evidence to suggest that a small fraction is able to cross into the CNS, where they might convey psychotropic properties. For example, a 2-week intragastric injection of sodium butyrate to male C57BL/B6 mice resulted in antidepressive effects, as well as increased central neurotransmission and BDNF expression (Sun et al., 2016). However, since free fatty acids interact with G-coupled protein receptors (GPRs) which are expressed at low levels in the brain (Erny et al., 2015; Khan & He, 2015), a direct agonistic action of SCFAs is unlikely. Rather, SCFAs may influence central neurotransmission by modulating the mucosal immune system (see later), exerting an effect on the hypothalamic–pituitary–adrenal axis, or by acting as an epigenetic modifier (e.g., histone deacetylase inhibitor) that translate environmental signals into differential expression of certain genes (Stilling, Dinan, & Cryan, 2014). Another important function of SCFAs is the stimulation of gut hormone release from enteroendocrine cells.

The production of SCFAs through prebiotic ingestion influences the secretion of peptide tyrosine tyrosine (PYY) and other satiety peptides such as cholecystokinin (CCK) and glucagon-like peptide-1 (GLP-1), from enteric L-cells. In rodents, propionic acid mediates the release of GLP-1 and PYY through GPR40 and GPR41 (Nohr et al., 2013; Psichas et al., 2015). The SCFA-induced secretion of PYY may play a significant role in central effects. Functional autoradiography studies have revealed abundant Y1 and Y2 receptor activity (Shaw, Gackenheimer, & Gehlert, 2003) and the blood–brain barrier has been shown to readily allow passage of PYY into and out of the mouse brain (Nonaka, Shioda, Niehoff, & Banks, 2003). Similarly, since GLP-1 receptors have also been identified in the brain modulating a spectrum of feeding responses, it is likely that circulating GLP-1 may elicit central effects via GLP-1 receptor signaling (Lockie, 2013; Trapp & Richards, 2013). Together, these data suggest a possible direct mechanism by which circulating PYY and GLP-1 modulates brain chemistry. Furthermore, the administration of PYY and GLP-1 to rodents has resulted in significant central changes such as the inhibition of neurotransmitter release, influencing learning and memory processes (Stadlbauer, Woods, Langhans, & Meyer, 2015), as well as mediating anxiogenic and antidepressant behavior (Anderberg et al., 2016). Alternatively, enteric PYY secretion may influence vagal nerve activity through local BNDF signaling pathways in the myenteric neurons.

Pertinent to prebiotics, the effects of a 3-week dietary enrichment (i.e., 6% GOS) on the expression of circulating PYY have been shown in Wistar rats (Overduin, Schoterman, Calame, Schonewille, & Ten Bruggencate, 2013). GOS-fed rats had significantly elevated gene expression of PYY and GLP-1 precursor proglucagon in the colonic mucosal. This result is accompanied by higher levels of circulating PYY in premeal plasma. Notably, plasma levels of GLP-1 were not elevated in spite of increased gene expression of proglucagon, possibly due to mRNA stability and cellular translation/secretion rates. GOS-fed rats were also reported to show significantly reduced energy intake and feeding, possibly relating to an anorexigenic mechanism involving PYY. GOS-associated increase in plasma PYY levels may result in changes to BNDF expression. In cultures of human

SH-SY5Y neuroblastoma cells, plasma collected from rats that received a 5-week administration of GOS, but not FOS, induced higher expression of BDNF. This observation was further confirmed by treating cultures in the presence of PYY antisera and GLP-1 antisera, which normalized BDNF secretion (Savignac et al., 2013).

5.2 Gut Microbiota and the Immune Response

Augmenting the population of *bifidobacteria* has been shown to exert direct antiinflammatory effects in the periphery. In a double-blind, placebo-controlled, crossover study, elderly volunteers were supplemented with B-GOS or maltodextrin for 10 weeks each, separated by a 4-week washout period (Vulevic, Drakoularakou, Yaqoob, Tzortzis, & Gibson, 2008; Vulevic et al., 2015). In this study, circulating proinflammatory cytokine levels were detected in both treatment periods wherein B-GOS consumption led to significantly decreased levels of IL-6, IL-1β, and TNFα. The therapeutic potential of prebiotics may extend beyond suppressing circulating levels of proinflammatory cytokines to increasing levels of antiinflammatory compounds such as IL-10. These properties are also displayed by specific probiotics (Davey et al., 2013; Klementowicz, Travis, & Grencis, 2012), wherein all likelihood, the antiinflammatory effects of prebiotics may be largely mediated by the SCFAs, and to a lesser extent though the direct interaction with the enteric mucosa.

The SCFAs have potent immune-related effects that result from their binding to GPR41, GPR43, and GPR109A (Louis, Hold, & Flint, 2014). GPR43 recognizes acetate, propionate and butyrate, and is highly expressed on neutrophils, macrophages, and monocytes; whereas, GPR41 expression is low or undetectable in the same cells (Brestoff & Artis, 2013). Immune-related effects of SCFA recognition include the modulation of antiinflammatory responses, intracellular cyclic adenosine monophosphate levels, calcium levels, and ERK1/2 activation. For example, Maslowski et al. found that GPR43 stimulation by SCFAs was necessary for the normal resolution of inflammatory responses in mice(Maslowski et al., 2009). Other immunoregulatory activities of SCFAs include the regulation of autophagy (Donohoe et al., 2011), T-cell differentiation (Kim, Park, & Kim, 2014), and stimulation of heat shock protein production (Ren et al., 2001). Although the full spectrum of molecular mechanisms by which SCFA regulate the development and functioning of immune cells remains far from known, it is clear that the production of SCFAs by the gut microbiota, through prebiotic feeding, would play a central role in host immunity. Of course, the physical impact of prebiotic ingestion and increased bacterial numbers on gut immunity cannot be ignored.

All microbes possess microbe-associated molecular patterns (MAMPs, formally known as pathogen-associated molecular patterns, PAMPs (Mackey & McFall, 2006)) which can be molecular components of their cell wall (e.g., LPS on *E. coli*), bacterial flagella, and/or microbial nucleic acids. The activation of pattern recognition receptors by MAMPs initiates the innate immune response which, in the case of the beneficial gut bacteria, may result in the secretion of antiinflammatory cytokines, such as IL-10 (Chu & Mazmanian, 2013; O'Mahony et al., 2005). Although the mechanistic pathways of attenuating an inflammatory response are uncertain, it is possible that *bifidobacteria*, and other commensal microbes, might act as a physical barrier by reducing pathogenic MAMPs (i.e., LPS) from binding to host enteric toll-like receptors (TLRs), such as TLR2 and TLR4, that mediate a proinflammatory response (Zhou et al., 2015). Notably, a direct interaction between prebiotic oligosaccharides and the gut mucosa, independent of gut bacteria, has been shown to influence the response of the immune system (Bode et al., 2004; Eiwegger et al., 2010). This suggests an avenue of research worth pursuing. Nonetheless, irrespective of the mechanisms of action, it is clear that prebiotic intake influences the immune system.

There is currently a growing body of evidence illustrating the impact of immunological changes on neuropsychiatric disorders. In preclinical models, systemic inflammation has been shown to lead to neurotoxicity and aberrant behaviors in models using intraperitoneal injection of LPS (Zarifkar et al., 2010) and parasitic models (Klementowicz et al., 2012). In humans, immunological dysfunction has been identified in schizophrenia (Khandaker et al., 2015), and the concurrent administration of antiinflammatory (e.g., nonsteroidal antiinflammatory drugs) and antipsychotic medication has shown some promising results including improving negative symptoms and cognitive functioning (Miyaoka et al., 2007; Muller et al., 2002; Muller, Riedel, Schwarz, & Engel, 2005) as well as reducing symptom severity (Sommer, de Witte, Begemann, & Kahn, 2012). In addition, minocycline, a broad-spectrum antibiotic with anti-inflammatory properties, may also be a valuable adjunctive therapy for patients with schizophrenia. A systematic review and meta-analysis of randomized controlled trials of minocycline has reported effective reduction of total symptom severity but not on PANSS subscale scores or depressive symptoms (Oya, Kishi, & Iwata, 2014). Inflammation may also be a potential etiological factor for mood disorders as worsening depressive symptoms are correlated with higher levels of inflammatory markers including C-reactive protein, IL-6, and IL-1 (Howren, Lamkin, & Suls, 2009). Current evidence for the influence of the immune system on the brain is compelling and supports the proposition that communication between the gut bacteria and host CNS may have a strong immunological component.

5.3 Gut Microbiome and the Enteric Nervous System

There is evidence that neuronal excitability can be altered by exposure to specific species of bacteria, or the lack thereof. The latter, as in the case of GF mice, normal excitability of myenteric neurons (i.e., enteric sensory neurons) was only observed after intestinal microbiota colonization (McVey Neufeld, Mao, Bienenstock, Foster, & Kunze, 2013). While this suggests that bacterial presence is necessary for normal activation of the enteric nervous system, the specific species, or collection of species, responsible remains to be identified. In rodents, enteric neuronal excitability was enhanced when animals were fed (gavage) with *Lactobacillus reuteri* for 9 days (Kunze et al., 2009). However, it was shown in one study that the dorsal root ganglion in the colon did not show hyperexcitabilty in animals treated with *Lactobacillus rhamnosus* (Ma et al., 2009). More convincingly, a 14-day oral administration of *bifidobacterium longum* NCC3001 resulted in greatly decreased action potentials of enteric nerves as well as anxiolytic effects mediated by the vagal nerve (Bercik et al., 2011). The central effects of probiotics, and potentially prebiotics, require afferent vagal signaling wherein vagotomy eliminates behavioral and neurobiological (e.g., gamma-aminobutyric acid (GABA) gene expression) changes (Bercik et al., 2011; Bravo et al., 2011). Notably, due to its *bifidogenic* properties, prebiotics may confer greater therapeutic effects due to an increased amount of *bifidobacteria* interactions with enteric neurons or vagal afferent terminals. It would be interesting to compare enteric nerve innervations between probiotics and prebiotics. Nonetheless, enhancing *bifidobacterium* is an encouraging therapeutic avenue.

Neurotransmitters produced by enteric bacteria may further stimulate the enteric nervous system. Different species in the gut microbiota are responsible for the production of different neurotransmitters through the catabolism of indigestible compounds. For example, GABA and acetylcholine are produced by members of the *lactobacilli* and *bifidobacteria* families whereas serotonin is produced by members of the *Enterococcus*, *Escherichia*, and *Streptococcus* families (Wall et al., 2014). Pertinent to the effect of prebiotics, a recent study in piglets fed with GOS containing formula for 31 days did not report significantly different levels of enteric serotonin (5-HT) compared to control subjects (Berding et al., 2016). However, additional replications are required to verify whether enteric serotonin levels are truly not affected by prebiotic administration. Further investigations on the effect of prebiotics on other neurotransmitters such as dopamine, noradrenalin, and GABA are warranted.

6. CONCLUSION

Specific prebiotics, through their proliferative action on indigenous beneficial gut bacteria, influence host neurobiology and behavior. Numerous well-designed preclinical studies that demonstrate the downstream central molecular effects of prebiotics are slowly accumulating. Changes in the expression of brain receptors and/or the levels of circulating hormones and immune molecules may be crucial for attaining healthy microbiome–gut–brain communication. Based on current prebiotic studies, it would be particularly worthwhile to extend investigations on the interactions between NMDARs and SCFAs and/or PYY; there are suggestions that these may be the predominant mechanisms underlying microbiome–brain communication and function (Fig. 1). However, inspection of other potential mechanisms including the inflammatory response and direct microbial-enteric interactions is strongly encouraged. Finally, clinical studies to support the validity of using prebiotics in the treatment of brain disorders, or even in the alleviation of the unwanted side-effects incurred with conventional medication, are urgently needed. In the latter instance, the chronic use of the antipsychotic olanzapine leads to metabolic syndrome and weight gain in schizophrenia patients wherein the drug has been shown to change the relative abundance of gut bacterial populations in animals (Davey et al., 2013; Morgan et al., 2014). Future pharmacological studies that demonstrate the potential interactions of relevant medications and prebiotics would be highly beneficial to support prebiotics as a cost-effective and safe adjunctive therapy for neuropsychiatric disorders.

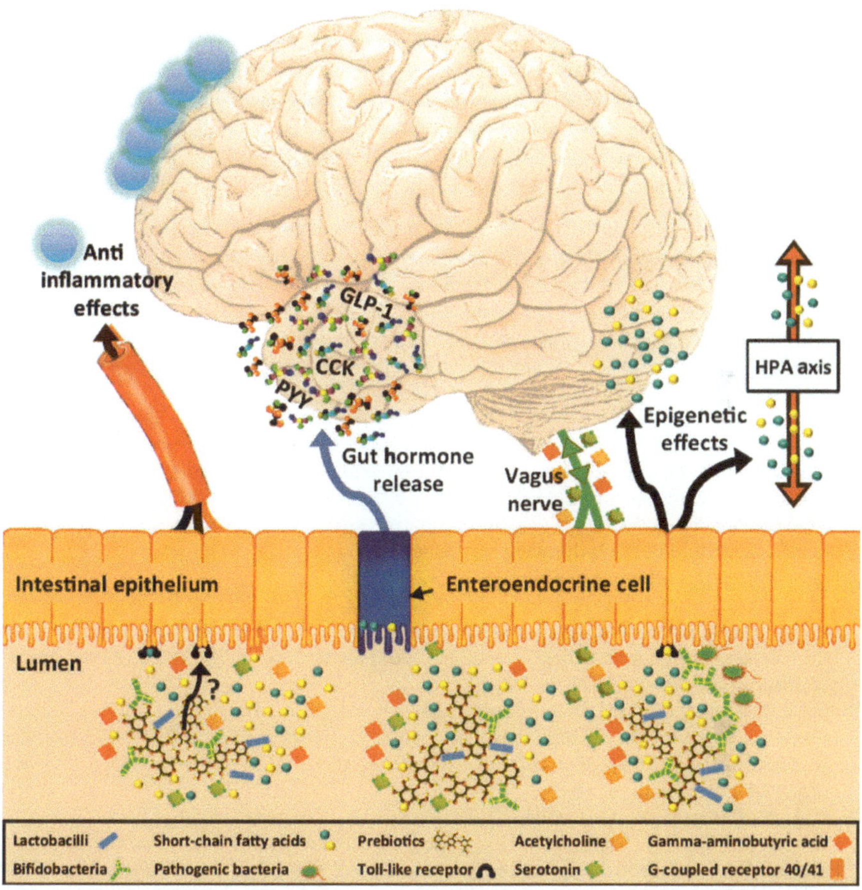

Fig. 1 Mechanisms underpinning gut microbiome–brain communication. Prebiotic ingestion introduces nondigestible food ingredients (i.e., fibers, carbohydrates, or various saccharides), which support the growth of health-promoting bacteria such as *lactobacilli* and *bifidobacteria*. These gut bacteria contribute to host physiology by producing metabolites such as short-chain fatty acids (SCFAs). SCFAs interact with enteroendocrine cells and stimulate the release of gut hormones such as peptide tyrosine tyrosine (PYY), cholecystokinin (CCK), and glucagon-like peptide-1 (GLP-1). These gut hormones enter the circulation and can migrate into the central nervous system, wherein they may contribute to the observed neuromodulatory and behavioral changes. By binding with host enteric receptors such as toll-like receptors and G-coupled protein receptors, the SCFA may also modulate central neurotransmission by triggering antiinflammatory responses. They may additionally exert an effect on the hypothalamic–pituitary–adrenal (HPA) axis or act as an epigenetic modifier directly on the brain. A direct interaction between prebiotics and host enteric receptors, independent of gut bacteria, has also been found to positively influence the immune system response. Commensal microbes such as *bifidobacteria* may also help to attenuate inflammatory responses by acting as a physical barrier against pathogenic bacteria, which may otherwise bind to toll-like receptors and trigger proinflammatory responses. Prebiotics can also increase the volume of neurotransmitters such as gamma-aminobutyric acid, serotonin, and acetylcholine generated by the health-promoting gut bacteria, which in turn, can modulate brain activity via the vagus nerve.

REFERENCES

Anderberg, R. H., Richard, J. E., Hansson, C., Nissbrandt, H., Bergquist, F., & Skibicka, K. P. (2016). GLP-1 is both anxiogenic and antidepressant; divergent effects of acute and chronic GLP-1 on emotionality. *Psychoneuroendocrinology*, *65*, 54–66. http://dx.doi.org/10.1016/j.psyneuen.2015.11.021.

Arentsen, T., Raith, H., Qian, Y., Forssberg, H., & Diaz Heijtz, R. (2015). Host microbiota modulates development of social preference in mice. *Microbial Ecology in Health and Disease*, *26*, 29719. http://dx.doi.org/10.3402/mehd.v26.29719.

Benton, D., Williams, C., & Brown, A. (2007). Impact of consuming a milk drink containing a probiotic on mood and cognition. *European Journal of Clinical Nutrition*, *61*(3), 355–361. http://dx.doi.org/10.1038/sj.ejcn.1602546.

Bercik, P., Park, A. J., Sinclair, D., Khoshdel, A., Lu, J., Huang, X., … Verdu, E. F. (2011). The anxiolytic effect of Bifidobacterium longum NCC3001 involves vagal pathways for gut-brain communication. *Neurogastroenterology and Motility*, *23*(12), 1132–1139. http://dx.doi.org/10.1111/j.1365-2982.2011.01796.x.

Berding, K., Wang, M., Monaco, M. H., Alexander, L. S., Mudd, A. T., Chichlowski, M., … Donovan, S. M. (2016). Prebiotics and bioactive milk fractions affect gut development, microbiota and neurotransmitter expression in piglets. *Journal of Pediatric Gastroenterology and Nutrition*. http://dx.doi.org/10.1097/MPG.0000000000001200.

Best, T., Kemps, E., & Bryan, J. (2010). Saccharide effects on cognition and well-being in middle-aged adults: A randomized controlled trial. *Developmental Neuropsychology*, *35*(1), 66–80. http://dx.doi.org/10.1080/87565640903325709.

Bode, L., Kunz, C., Muhly-Reinholz, M., Mayer, K., Seeger, W., & Rudloff, S. (2004). Inhibition of monocyte, lymphocyte, and neutrophil adhesion to endothelial cells by human milk oligosaccharides. *Thrombosis and Haemostasis*, *92*(6), 1402–1410. http://dx.doi.org/10.1160/TH04-01-0055.

Bouhnik, Y., Attar, A., Joly, F. A., Riottot, M., Dyard, F., & Flourie, B. (2004). Lactulose ingestion increases faecal bifidobacterial counts: A randomised double-blind study in healthy humans. *European Journal of Clinical Nutrition*, *58*(3), 462–466. http://dx.doi.org/10.1038/sj.ejcn.1601829.

Bravo, J. A., Forsythe, P., Chew, M. V., Escaravage, E., Savignac, H. M., Dinan, T. G., … Cryan, J. F. (2011). Ingestion of Lactobacillus strain regulates emotional behavior and central GABA receptor expression in a mouse via the vagus nerve. *Proceedings of the National Academy of Sciences of the United States of America*, *108*(38), 16050–16055. http://dx.doi.org/10.1073/pnas.1102999108.

Brestoff, J. R., & Artis, D. (2013). Commensal bacteria at the interface of host metabolism and the immune system. *Nature Immunology*, *14*(7), 676–684. http://dx.doi.org/10.1038/ni.2640.

Campbell, J. M., Fahey, G. C., Jr., & Wolf, B. W. (1997). Selected indigestible oligosaccharides affect large bowel mass, cecal and fecal short-chain fatty acids, pH and microflora in rats. *The Journal of Nutrition*, *127*(1), 130–136.

Chu, H., & Mazmanian, S. K. (2013). Innate immune recognition of the microbiota promotes host-microbial symbiosis. *Nature Immunology*, *14*(7), 668–675. http://dx.doi.org/10.1038/ni.2635.

Clarke, G., Grenham, S., Scully, P., Fitzgerald, P., Moloney, R. D., Shanahan, F., … Cryan, J. F. (2013). The microbiome-gut-brain axis during early life regulates the hippocampal serotonergic system in a sex-dependent manner. *Molecular Psychiatry*, *18*(6), 666–673. http://dx.doi.org/10.1038/mp.2012.77.

Collins, S. M., Kassam, Z., & Bercik, P. (2013). The adoptive transfer of behavioral phenotype via the intestinal microbiota: Experimental evidence and clinical implications. *Current Opinion in Microbiology*, *16*(3), 240–245. http://dx.doi.org/10.1016/j.mib.2013.06.004.

Cummings, J. H., Pomare, E. W., Branch, W. J., Naylor, C. P., & Macfarlane, G. T. (1987). Short chain fatty acids in human large intestine, portal, hepatic and venous blood. *Gut, 28*(10), 1221–1227.

Davey, K. J., Cotter, P. D., O'Sullivan, O., Crispie, F., Dinan, T. G., Cryan, J. F., & O'Mahony, S. M. (2013). Antipsychotics and the gut microbiome: Olanzapine-induced metabolic dysfunction is attenuated by antibiotic administration in the rat. *Translational Psychiatry, 3*, e309. http://dx.doi.org/10.1038/tp.2013.83.

Deacon, R. M. (2006). Digging and marble burying in mice: Simple methods for in vivo identification of biological impacts. *Nature Protocols, 1*(1), 122–124. http://dx.doi.org/10.1038/nprot.2006.20.

Desbonnet, L., Clarke, G., Shanahan, F., Dinan, T. G., & Cryan, J. F. (2014). Microbiota is essential for social development in the mouse. *Molecular Psychiatry, 19*(2), 146–148. http://dx.doi.org/10.1038/mp.2013.65.

Diaz Heijtz, R., Wang, S., Anuar, F., Qian, Y., Bjorkholm, B., Samuelsson, A., … Pettersson, S. (2011). Normal gut microbiota modulates brain development and behavior. *Proceedings of the National Academy of Sciences of the United States of America, 108*(7), 3047–3052. http://dx.doi.org/10.1073/pnas.1010529108.

Donohoe, D. R., Garge, N., Zhang, X., Sun, W., O'Connell, T. M., Bunger, M. K., & Bultman, S. J. (2011). The microbiome and butyrate regulate energy metabolism and autophagy in the mammalian colon. *Cell Metabolism, 13*(5), 517–526. http://dx.doi.org/10.1016/j.cmet.2011.02.018.

Eiwegger, T., Stahl, B., Haidl, P., Schmitt, J., Boehm, G., Dehlink, E., … Szepfalusi, Z. (2010). Prebiotic oligosaccharides: In vitro evidence for gastrointestinal epithelial transfer and immunomodulatory properties. *Pediatric Allergy and Immunology, 21*(8), 1179–1188. http://dx.doi.org/10.1111/j.1399-3038.2010.01062.x.

Erny, D., Hrabe de Angelis, A. L., Jaitin, D., Wieghofer, P., Staszewski, O., David, E., … Prinz, M. (2015). Host microbiota constantly control maturation and function of microglia in the CNS. *Nature Neuroscience, 18*(7), 965–977. http://dx.doi.org/10.1038/nn.4030.

Fedder, K. N., & Sabo, S. L. (2015). On the role of glutamate in presynaptic development: Possible contributions of presynaptic NMDA receptors. *Biomolecules, 5*(4), 3448–3466. http://dx.doi.org/10.3390/biom5043448.

Gareau, M. G., Wine, E., Rodrigues, D. M., Cho, J. H., Whary, M. T., Philpott, D. J., … Sherman, P. M. (2011). Bacterial infection causes stress-induced memory dysfunction in mice. *Gut, 60*(3), 307–317. http://dx.doi.org/10.1136/gut.2009.202515.

Gibson, G. R., & Roberfroid, M. B. (1995). Dietary modulation of the human colonic microbiota: Introducing the concept of prebiotics. *The Journal of Nutrition, 125*(6), 1401–1412.

Gibson, G. R., Scott, K. P., Rastall, R. A., Tuohy, K. M., Hotchkiss, A., Dubert-Ferrandon, A., … Buddington, R. (2010). Dietary prebiotics: Current status and new definition. *Food Science and Technology Bulletin: Functional Foods*, 7, 1–19. http://dx.doi.org/10.1616/1476-2137.15880.

Hanell, A., & Marklund, N. (2014). Structured evaluation of rodent behavioral tests used in drug discovery research. *Frontiers in Behavioral Neuroscience, 8*, 252. http://dx.doi.org/10.3389/fnbeh.2014.00252.

Heldt, S. A., Stanek, L., Chhatwal, J. P., & Ressler, K. J. (2007). Hippocampus-specific deletion of BDNF in adult mice impairs spatial memory and extinction of aversive memories. *Molecular Psychiatry, 12*(7), 656–670. http://dx.doi.org/10.1038/sj.mp.4001957.

Howren, M. B., Lamkin, D. M., & Suls, J. (2009). Associations of depression with C-reactive protein, IL-1, and IL-6: A meta-analysis. *Psychosomatic Medicine, 71*(2), 171–186. http://dx.doi.org/10.1097/PSY.0b013e3181907c1b.

Jandhyala, S. M., Talukdar, R., Subramanyam, C., Vuyyuru, H., Sasikala, M., & Nageshwar Reddy, D. (2015). Role of the normal gut microbiota. *World Journal of Gastroenterology*, *21*(29), 8787–8803. http://dx.doi.org/10.3748/wjg.v21.i29.8787.

Jia, S., Lu, Z., Gao, Z., An, J., Wu, X., Li, X., … Sun, Y. (2016). Chitosan oligosaccharides alleviate cognitive deficits in an amyloid-beta1-42-induced rat model of Alzheimer's disease. *International Journal of Biological Macromolecules*, *83*, 416–425. http://dx.doi.org/10.1016/j.ijbiomac.2015.11.011.

Keunen, K., van Elburg, R. M., van Bel, F., & Benders, M. J. (2015). Impact of nutrition on brain development and its neuroprotective implications following preterm birth. *Pediatric Research*, 77(1–2), 148–155. http://dx.doi.org/10.1038/pr.2014.171.

Khan, M. Z., & He, L. (2015). The role of polyunsaturated fatty acids and GPR40 receptor in brain. *Neuropharmacology*. http://dx.doi.org/10.1016/j.neuropharm.2015.05.013.

Khandaker, G. M., Cousins, L., Deakin, J., Lennox, B. R., Yolken, R., & Jones, P. B. (2015). Inflammation and immunity in schizophrenia: Implications for pathophysiology and treatment. *Lancet Psychiatry*, *2*(3), 258–270. http://dx.doi.org/10.1016/S2215-0366(14)00122-9.

Khoder, M., Tsapis, N., Domergue-Dupont, V., Gueutin, C., & Fattal, E. (2010). Removal of residual colonic ciprofloxacin in the rat by activated charcoal entrapped within zinc-pectinate beads. *European Journal of Pharmaceutical Sciences*, *41*(2), 281–288. http://dx.doi.org/10.1016/j.ejps.2010.06.018.

Kim, C. H., Park, J., & Kim, M. (2014). Gut microbiota-derived short-chain fatty acids, T cells, and inflammation. *Immune Network*, *14*(6), 277–288. http://dx.doi.org/10.4110/in.2014.14.6.277.

Kleessen, B., Hartmann, L., & Blaut, M. (2001). Oligofructose and long-chain inulin: Influence on the gut microbial ecology of rats associated with a human faecal flora. *The British Journal of Nutrition*, *86*(2), 291–300.

Klementowicz, J. E., Travis, M. A., & Grencis, R. K. (2012). Trichuris muris: A model of gastrointestinal parasite infection. *Seminars in Immunopathology*, *34*(6), 815–828. http://dx.doi.org/10.1007/s00281-012-0348-2.

Krishna, G., Divyashri, G., Prapulla, S. G., & Muralidhara (2015). A combination supplement of fructo- and xylo-oligosaccharides significantly abrogates oxidative impairments and neurotoxicity in maternal/fetal milieu following gestational exposure to acrylamide in rat. *Neurochemical Research*, *40*(9), 1904–1918. http://dx.doi.org/10.1007/s11064-015-1687-x.

Kunze, W. A., Mao, Y. K., Wang, B., Huizinga, J. D., Ma, X., Forsythe, P., & Bienenstock, J. (2009). Lactobacillus reuteri enhances excitability of colonic AH neurons by inhibiting calcium-dependent potassium channel opening. *Journal of Cellular and Molecular Medicine*, *13*(8B), 2261–2270. http://dx.doi.org/10.1111/j.1582-4934.2009.00686.x.

Lam, K.-L., & Chi-Keung Cheung, P. (2013). Non-digestible long chain beta-glucans as novel prebiotics. *Bioactive Carbohydrates and Dietary Fibre*, *2*(1), 45–64. http://dx.doi.org/10.1016/j.bcdf.2013.09.001.

Li, W., Dowd, S. E., Scurlock, B., Acosta-Martinez, V., & Lyte, M. (2009). Memory and learning behavior in mice is temporally associated with diet-induced alterations in gut bacteria. *Physiology & Behavior*, *96*(4–5), 557–567. http://dx.doi.org/10.1016/j.physbeh.2008.12.004.

Linthorst, A. C., Flachskamm, C., Muller-Preuss, P., Holsboer, F., & Reul, J. M. (1995). Effect of bacterial endotoxin and interleukin-1 beta on hippocampal serotonergic neurotransmission, behavioral activity, and free corticosterone levels: An in vivo microdialysis study. *The Journal of Neuroscience*, *15*(4), 2920–2934.

Lockie, S. H. (2013). Glucagon-like peptide-1 receptor in the brain: Role in neuroendocrine control of energy metabolism and treatment target for obesity. *Journal of Neuroendocrinology*, *25*(7), 597–604. http://dx.doi.org/10.1111/jne.12039.

Louis, P., Hold, G. L., & Flint, H. J. (2014). The gut microbiota, bacterial metabolites and colorectal cancer. *Nature Reviews. Microbiology, 12*(10), 661–672. http://dx.doi.org/10.1038/nrmicro3344.

Ma, X., Mao, Y. K., Wang, B., Huizinga, J. D., Bienenstock, J., & Kunze, W. (2009). Lactobacillus reuteri ingestion prevents hyperexcitability of colonic DRG neurons induced by noxious stimuli. *American Journal of Physiology. Gastrointestinal and Liver Physiology, 296*(4), G868–G875. http://dx.doi.org/10.1152/ajpgi.90511.2008.

Mackey, D., & McFall, A. J. (2006). MAMPs and MIMPs: Proposed classifications for inducers of innate immunity. *Molecular Microbiology, 61*(6), 1365–1371. http://dx.doi.org/10.1111/j.1365-2958.2006.05311.x.

Marzorati, M., Verhelst, A., Luta, G., Sinnott, R., Verstraete, W., Van de Wiele, T., & Possemiers, S. (2010). In vitro modulation of the human gastrointestinal microbial community by plant-derived polysaccharide-rich dietary supplements. *International Journal of Food Microbiology, 139*(3), 168–176. http://dx.doi.org/10.1016/j.ijfoodmicro.2010.02.030.

Maslowski, K. M., Vieira, A. T., Ng, A., Kranich, J., Sierro, F., Yu, D., … Mackay, C. R. (2009). Regulation of inflammatory responses by gut microbiota and chemoattractant receptor GPR43. *Nature, 461*(7268), 1282–1286. http://dx.doi.org/10.1038/nature08530.

McVey Neufeld, K. A., Mao, Y. K., Bienenstock, J., Foster, J. A., & Kunze, W. A. (2013). The microbiome is essential for normal gut intrinsic primary afferent neuron excitability in the mouse. *Neurogastroenterology and Motility, 25*(2), e183–e188. http://dx.doi.org/10.1111/nmo.12049.

Messaoudi, M., Lalonde, R., Violle, N., Javelot, H., Desor, D., Nejdi, A., … Cazaubiel, J. M. (2011). Assessment of psychotropic-like properties of a probiotic formulation (Lactobacillus helveticus R0052 and Bifidobacterium longum R0175) in rats and human subjects. *The British Journal of Nutrition, 105*(5), 755–764. http://dx.doi.org/10.1017/S0007114510004319.

Messaoudi, M., Violle, N., Bisson, J. F., Desor, D., Javelot, H., & Rougeot, C. (2011). Beneficial psychological effects of a probiotic formulation (Lactobacillus helveticus R0052 and Bifidobacterium longum R0175) in healthy human volunteers. *Gut Microbes, 2*(4), 256–261. http://dx.doi.org/10.4161/gmic.2.4.16108.

Miyaoka, T., Yasukawa, R., Yasuda, H., Hayashida, M., Inagaki, T., & Horiguchi, J. (2007). Possible antipsychotic effects of minocycline in patients with schizophrenia. *Progress in Neuro-Psychopharmacology & Biological Psychiatry, 31*(1), 304–307. http://dx.doi.org/10.1016/j.pnpbp.2006.08.013.

Monteggia, L. M., Barrot, M., Powell, C. M., Berton, O., Galanis, V., Gemelli, T., … Nestler, E. J. (2004). Essential role of brain-derived neurotrophic factor in adult hippocampal function. *Proceedings of the National Academy of Sciences of the United States of America, 101*(29), 10827–10832. http://dx.doi.org/10.1073/pnas.0402141101.

Morgan, A. P., Crowley, J. J., Nonneman, R. J., Quackenbush, C. R., Miller, C. N., Ryan, A. K., … Sullivan, P. F. (2014). The antipsychotic olanzapine interacts with the gut microbiome to cause weight gain in mouse. *PloS One, 9*(12), e115225. http://dx.doi.org/10.1371/journal.pone.0115225.

Mudd, A. T., Alexander, L. S., Berding, K., Waworuntu, R. V., Berg, B. M., Donovan, S. M., & Dilger, R. N. (2016). Dietary prebiotics, milk fat globule membrane, and lactoferrin affects structural neurodevelopment in the young piglet. *Frontiers in Pediatrics, 4*, 4. http://dx.doi.org/10.3389/fped.2016.00004.

Muller, N., Riedel, M., Scheppach, C., Brandstatter, B., Sokullu, S., Krampe, K., … Schwarz, M. J. (2002). Beneficial antipsychotic effects of celecoxib add-on therapy compared to risperidone alone in schizophrenia. *The American Journal of Psychiatry, 159*(6), 1029–1034. http://dx.doi.org/10.1176/appi.ajp.159.6.1029.

Muller, N., Riedel, M., Schwarz, M. J., & Engel, R. R. (2005). Clinical effects of COX-2 inhibitors on cognition in schizophrenia. *European Archives of Psychiatry and Clinical Neuroscience*, *255*(2), 149–151. http://dx.doi.org/10.1007/s00406-004-0548-4.

Mutanda, T., Mokoena, M. P., Olaniran, A. O., Wilhelmi, B. S., & Whiteley, C. G. (2014). Microbial enzymatic production and applications of short-chain fructooligosaccharides and inulooligosaccharides: Recent advances and current perspectives. *Journal of Industrial Microbiology & Biotechnology*, *41*(6), 893–906. http://dx.doi.org/10.1007/s10295-014-1452-1.

Nau, F., Jr., Yu, B., Martin, D., & Nichols, C. D. (2013). Serotonin 5-HT2A receptor activation blocks TNF-alpha mediated inflammation in vivo. *PLoS One*, *8*(10), e75426. http://dx.doi.org/10.1371/journal.pone.0075426.

Neufeld, K. M., Kang, N., Bienenstock, J., & Foster, J. A. (2011). Reduced anxiety-like behavior and central neurochemical change in germ-free mice. *Neurogastroenterology and Motility*, *23*(3), 255–264. http://dx.doi.org/10.1111/j.1365-2982.2010.01620.x. e119.

Niness, K. R. (1999). Inulin and oligofructose: What are they? *The Journal of Nutrition*, *129*(7 Suppl.), 1402S–1406S.

Njung'e, K., & Handley, S. L. (1991). Evaluation of marble-burying behavior as a model of anxiety. *Pharmacology, Biochemistry, and Behavior*, *38*(1), 63–67.

Nohr, M. K., Pedersen, M. H., Gille, A., Egerod, K. L., Engelstoft, M. S., Husted, A. S., … Schwartz, T. W. (2013). GPR41/FFAR3 and GPR43/FFAR2 as cosensors for short-chain fatty acids in enteroendocrine cells vs FFAR3 in enteric neurons and FFAR2 in enteric leukocytes. *Endocrinology*, *154*(10), 3552–3564. http://dx.doi.org/10.1210/en.2013-1142.

Nonaka, N., Shioda, S., Niehoff, M. L., & Banks, W. A. (2003). Characterization of blood–brain barrier permeability to PYY3-36 in the mouse. *The Journal of Pharmacology and Experimental Therapeutics*, *306*(3), 948–953. http://dx.doi.org/10.1124/jpet.103.051821.

O'Mahony, L., McCarthy, J., Kelly, P., Hurley, G., Luo, F., Chen, K., … Quigley, E. M. (2005). Lactobacillus and bifidobacterium in irritable bowel syndrome: Symptom responses and relationship to cytokine profiles. *Gastroenterology*, *128*(3), 541–551.

Otieno, D. O. (2010). Synthesis of β-galactooligosaccharides from lactose using microbial β-galactosidases. *Comprehensive Reviews in Food Science and Food Safety*, *9*(5), 471–482. http://dx.doi.org/10.1111/j.1541-4337.2010.00121.x.

Overduin, J., Schoterman, M. H., Calame, W., Schonewille, A. J., & Ten Bruggencate, S. J. (2013). Dietary galacto-oligosaccharides and calcium: Effects on energy intake, fat-pad weight and satiety-related, gastrointestinal hormones in rats. *The British Journal of Nutrition*, *109*(7), 1338–1348. http://dx.doi.org/10.1017/S0007114512003066.

Oya, K., Kishi, T., & Iwata, N. (2014). Efficacy and tolerability of minocycline augmentation therapy in schizophrenia: A systematic review and meta-analysis of randomized controlled trials. *Human Psychopharmacology*, *29*(5), 483–491. http://dx.doi.org/10.1002/hup.2426.

Psichas, A., Sleeth, M. L., Murphy, K. G., Brooks, L., Bewick, G. A., Hanyaloglu, A. C., … Frost, G. (2015). The short chain fatty acid propionate stimulates GLP-1 and PYY secretion via free fatty acid receptor 2 in rodents. *International Journal of Obesity*, *39*(3), 424–429. http://dx.doi.org/10.1038/ijo.2014.153.

Rao, A. V., Bested, A. C., Beaulne, T. M., Katzman, M. A., Iorio, C., Berardi, J. M., & Logan, A. C. (2009). A randomized, double-blind, placebo-controlled pilot study of a probiotic in emotional symptoms of chronic fatigue syndrome. *Gut Pathogens*, *1*(1), 6. http://dx.doi.org/10.1186/1757-4749-1-6.

Ren, H., Musch, M. W., Kojima, K., Boone, D., Ma, A., & Chang, E. B. (2001). Short-chain fatty acids induce intestinal epithelial heat shock protein 25 expression in rats and IEC 18 cells. *Gastroenterology*, *121*(3), 631–639.

Roberfroid, M., Gibson, G. R., Hoyles, L., McCartney, A. L., Rastall, R., Rowland, I., … Meheust, A. (2010). Prebiotic effects: Metabolic and health benefits. *The British Journal of Nutrition, 104*(Suppl. 2), S1–S63. http://dx.doi.org/10.1017/S0007114510003363.

Roberfroid, M. B., Van Loo, J. A., & Gibson, G. R. (1998). The bifidogenic nature of chicory inulin and its hydrolysis products. *The Journal of Nutrition, 128*(1), 11–19.

Savignac, H. M., Corona, G., Mills, H., Chen, L., Spencer, J. P., Tzortzis, G., & Burnet, P. W. (2013). Prebiotic feeding elevates central brain derived neurotrophic factor, N-methyl-D-aspartate receptor subunits and D-serine. *Neurochemistry International, 63*(8), 756–764. http://dx.doi.org/10.1016/j.neuint.2013.10.006.

Savignac, H. M., Couch, Y., Stratford, M., Bannerman, D. M., Tzortzis, G., Anthony, D. C., & Burnet, P. W. (2016). Prebiotic administration normalizes lipopolysaccharide (LPS)-induced anxiety and cortical 5-HT2A receptor and IL1-beta levels in male mice. *Brain, Behavior, and Immunity, 52*, 120–131. http://dx.doi.org/10.1016/j.bbi.2015.10.007.

Schiano, T. D. (2010). Treatment options for hepatic encephalopathy. *Pharmacotherapy, 30*(5 Pt. 2), 16S–21S. http://dx.doi.org/10.1592/phco.30.pt2.16S.

Schmidt, K., Cowen, P. J., Harmer, C. J., Tzortzis, G., Errington, S., & Burnet, P. W. (2015). Prebiotic intake reduces the waking cortisol response and alters emotional bias in healthy volunteers. *Psychopharmacology, 232*(10), 1793–1801. http://dx.doi.org/10.1007/s00213-014-3810-0.

Shaw, J. L., Gackenheimer, S. L., & Gehlert, D. R. (2003). Functional autoradiography of neuropeptide Y Y1 and Y2 receptor subtypes in rat brain using agonist stimulated [35S] GTPgammaS binding. *Journal of Chemical Neuroanatomy, 26*(3), 179–193.

Sommer, I. E., de Witte, L., Begemann, M., & Kahn, R. S. (2012). Nonsteroidal anti-inflammatory drugs in schizophrenia: Ready for practice or a good start? A meta-analysis. *The Journal of Clinical Psychiatry, 73*(4), 414–419. http://dx.doi.org/10.4088/JCP.10r06823.

Song, L., Gao, Y., Zhang, X., & Le, W. (2013). Galactooligosaccharide improves the animal survival and alleviates motor neuron death in SOD1G93A mouse model of amyotrophic lateral sclerosis. *Neuroscience, 246*, 281–290. http://dx.doi.org/10.1016/j.neuroscience.2013.05.002.

Stadlbauer, U., Woods, S. C., Langhans, W., & Meyer, U. (2015). PYY3-36: Beyond food intake. *Frontiers in Neuroendocrinology, 38*, 1–11. http://dx.doi.org/10.1016/j.yfrne.2014.12.003.

Stilling, R. M., Dinan, T. G., & Cryan, J. F. (2014). Microbial genes, brain & behaviour—Epigenetic regulation of the gut-brain axis. *Genes, Brain, and Behavior, 13*(1), 69–86. http://dx.doi.org/10.1111/gbb.12109.

Sun, J., Wang, F., Hong, G., Pang, M., Xu, H., Li, H., … Liu, J. (2016). Antidepressant-like effects of sodium butyrate and its possible mechanisms of action in mice exposed to chronic unpredictable mild stress. *Neuroscience Letters, 618*, 159–166. http://dx.doi.org/10.1016/j.neulet.2016.03.003.

Talbott, S., & Talbott, J. (2009). Effect of BETA 1, 3/1, 6 GLUCAN on upper respiratory tract infection symptoms and mood state in marathon athletes. *Journal of Sports Science & Medicine, 8*(4), 509–515.

Tarr, A. J., Galley, J. D., Fisher, S. E., Chichlowski, M., Berg, B. M., & Bailey, M. T. (2015). The prebiotics 3'Sialyllactose and 6'Sialyllactose diminish stressor-induced anxiety-like behavior and colonic microbiota alterations: Evidence for effects on the gut-brain axis. *Brain, Behavior, and Immunity, 50*, 166–177. http://dx.doi.org/10.1016/j.bbi.2015.06.025.

Trapp, S., & Richards, J. E. (2013). The gut hormone glucagon-like peptide-1 produced in brain: Is this physiologically relevant? *Current Opinion in Pharmacology, 13*(6), 964–969. http://dx.doi.org/10.1016/j.coph.2013.09.006.

van den Berg, J. P., Westerbeek, E. A., Broring-Starre, T., Garssen, J., & van Elburg, R. M. (2016). Neurodevelopment of preterm infants at 24 month after neonatal supplementation of a prebiotic mix: A randomized trial. *Journal of Pediatric Gastroenterology and Nutrition*, *63*, 270–276. http://dx.doi.org/10.1097/MPG.0000000000001148.

Vazquez, E., Barranco, A., Ramirez, M., Gruart, A., Delgado-Garcia, J. M., Martinez-Lara, E., ... Rueda, R. (2015). Effects of a human milk oligosaccharide, 2'-fucosyllactose, on hippocampal long-term potentiation and learning capabilities in rodents. *The Journal of Nutritional Biochemistry*, *26*(5), 455–465. http://dx.doi.org/10.1016/j.jnutbio.2014.11.016.

Vulevic, J., Drakoularakou, A., Yaqoob, P., Tzortzis, G., & Gibson, G. R. (2008). Modulation of the fecal microflora profile and immune function by a novel trans-galactooligosaccharide mixture (B-GOS) in healthy elderly volunteers. *The American Journal of Clinical Nutrition*, *88*(5), 1438–1446.

Vulevic, J., Juric, A., Walton, G. E., Claus, S. P., Tzortzis, G., Toward, R. E., & Gibson, G. R. (2015). Influence of galacto-oligosaccharide mixture (B-GOS) on gut microbiota, immune parameters and metabonomics in elderly persons. *The British Journal of Nutrition*, *114*(4), 586–595. http://dx.doi.org/10.1017/S0007114515001889.

Wall, R., Cryan, J. F., Ross, R. P., Fitzgerald, G. F., Dinan, T. G., & Stanton, C. (2014). Bacterial neuroactive compounds produced by psychobiotics. *Advances in Experimental Medicine and Biology*, *817*, 221–239. http://dx.doi.org/10.1007/978-1-4939-0897-4_10.

Wang, X., & Gibson, G. R. (1993). Effects of the in vitro fermentation of oligofructose and inulin by bacteria growing in the human large intestine. *The Journal of Applied Bacteriology*, *75*(4), 373–380.

Williams, S., Chen, L., Savignac, H. M., Tzortzis, G., Anthony, D. C., & Burnet, P. W. (2016). Neonatal prebiotic (BGOS) supplementation increases the levels of synaptophysin, GluN2A-subunits and BDNF proteins in the adult rat hippocampus. *Synapse*, *70*(3), 121–124. http://dx.doi.org/10.1002/syn.21880.

Wong, J. M., de Souza, R., Kendall, C. W., Emam, A., & Jenkins, D. J. (2006). Colonic health: Fermentation and short chain fatty acids. *Journal of Clinical Gastroenterology*, *40*(3), 235–243.

Yen, C. H., Wang, C. H., Wu, W. T., & Chen, H. L. (2015). Fructo-oligosaccharide improved brain beta-amyloid, beta-secretase, cognitive function, and plasma antioxidant levels in d-galactose-treated Balb/cJ mice. *Nutritional Neuroscience*. http://dx.doi.org/10.1080/1028415X.2015.1110952.

Zarifkar, A., Choopani, S., Ghasemi, R., Naghdi, N., Maghsoudi, A. H., Maghsoudi, N., ... Moosavi, M. (2010). Agmatine prevents LPS-induced spatial memory impairment and hippocampal apoptosis. *European Journal of Pharmacology*, *634*(1–3), 84–88. http://dx.doi.org/10.1016/j.ejphar.2010.02.029.

Zhou, W., Lv, H., Li, M. X., Su, H., Huang, L. G., Li, J., & Yuan, W. M. (2015). Protective effects of bifidobacteria on intestines in newborn rats with necrotizing enterocolitis and its regulation on TLR2 and TLR4. *Genetics and Molecular Research*, *14*(3), 11505–11514. http://dx.doi.org/10.4238/2015.September.28.2.

CHAPTER THREE

Gut Microbiome and Behavior: Focus on Neuroimmune Interactions

J.A. Foster[1]
McMaster University, St. Joseph's Healthcare, Hamilton, ON, Canada
[1]Corresponding author: e-mail address: jfoster@mcmaster.ca

Contents

Abstract

As neuroscientists, psychologists, and psychiatrists are starting to appreciate the importance of the gut microbiota to mental health, it is critical to determine the mechanisms of microbiota to brain communication and thereby provide a better understanding of the aspects that may be modifiable with proper intervention in individuals with mental illness. Microbiota–brain communication is emerging as an important factor in brain development and function. Further, immune dysfunction is clearly established to play a role in mental illness. Investigators in the field have established expertise in studying the microbiota, the immune system, brain, and behavior and are poised to contribute significant novel findings to our understanding of microbiota–immune–brain communication in mental illness. This chapter provides a review of the literature related to the influence of microbiota–immune–brain communication to behavior. This research has a clear translational relevance for mental health, contributing to extant findings that indicate a role for the microbiome in brain development and behavior.

1. MICROBIOTA–BRAIN AXIS

Excitement has been generated in mental health research by recent findings from animal and clinical studies demonstrating an important role for the gut microbiota in brain function and behavior (Cryan & Dinan, 2012; Foster & McVey Neufeld, 2013; Luna & Foster, 2014; Mayer,

International Review of Neurobiology, Volume 131
ISSN 0074-7742
http://dx.doi.org/10.1016/bs.irn.2016.07.005

Tillisch, & Gupta, 2015). The microbiota and its human host interact in a mutualistic relationship. The host provides bacteria with a rich environment to grow. In parallel, the microbiota contributes to healthy metabolism and is critical to the normal development of the immune, endocrine, and nervous system (Hooper et al., 2001; Macpherson & Harris, 2004; Macpherson, Martinic, & Harris, 2002; Macpherson & Uhr, 2004; Tlaskalova-Hogenova et al., 2004). The public and the scientific community is engaged in the topic of the microbiota leading to unprecedented attention to this field of research—this is very exciting and provides an opportunity for research results to have a broader impact across many disciplines. This chapter will provide an overview of advances in the field of microbiota–brain research that relate to immune mechanisms, microbiota, and behavior. The human microbiome includes bacteria, viruses, fungi, and other microorganisms that live in close association with us, and although a microbiota is found on multiple sites on the body, much of the information reviewed here is specifically related to commensal bacteria that are resident in the gastrointestinal tract.

Early postnatal life in mammals represents a period of bacteria colonization. Normal microbiota, referred to as commensal microbiota, colonize the mammalian gastrointestinal tract shortly after birth and remain there throughout life. Recently developed molecular and metagenomic tools have allowed researchers to better understand the structure and function of the human microbial gut community. Several bacterial phyla are represented in the gut (reviewed by Diamant, Blaak, & de Vos, 2011) and commensals exhibit considerable diversity, with more than 1000 distinct bacterial species involved (Qin et al., 2010). Further, an individual's profile of microbiota is continually influenced by genetics, age, sex, and diet (Jumpertz et al., 2011; Kau, Ahern, Griffin, Goodman, & Gordon, 2011). While metagenomic population approaches have shown that, in general, certain bacterial populations are shared among groups of people (Arumugam et al., 2011), it is important to note that detailed analyses demonstrate considerable variability in bacterial content between related and unrelated individuals (Costello et al., 2009; Gill et al., 2006). As such, the microbiota profile may be a good representation of the environmental history of the individual. This dynamic nature and the diversity of the microbiome determined to date extend far beyond what researchers expected. Overall, molecular and metagenomic studies emphasize that microbiota colonies are dynamic in structure and function.

2. MICROBIOTA AND IMMUNE SIGNALING INFLUENCE BEHAVIOR

In the past few years, a link between the gut microbiota and anxiety-like behavior has emerged in animal studies. The germ-free (GF) mouse model was established in 1957 where GF mice are raised in sterile/gnotobiotic environments and have no commensal bacteria (Gustafsson, 1959; Gustafsson, Kahlson, & Rosengren, 1957). Both the mucosal and the systemic immune systems of GF mice are immature with reduced numbers of B lymphocytes and T lymphocytes (Macpherson & Harris, 2004). A landmark study that showed GF mice have exaggerated stress reactivity in response to restraint stress was really the stimulus for neuroscientists to consider the GF mouse model (Sudo et al., 2004). Specifically, these investigators showed exaggerated levels of plasma corticosterone levels (CORT) and plasma adrenocorticotrophic hormone levels in stressed GF mice compared to specific pathogen-free mice (Sudo et al., 2004). This work sparked an interest for neuroscientists and work that followed demonstrated that GF mice showed reduced anxiety-like behavior (Neufeld, Kang, Bienenstock, & Foster, 2011a, 2011b). Further, this low anxiety-like phenotype was accompanied by long-term changes in plasticity-related genes in the hippocampus and amygdala (Neufeld et al., 2011a, 2011b). Two additional groups have confirmed the reduced anxiety-like behavioral phenotype in GF mice (Clarke et al., 2013; Heijtz et al., 2011). Interestingly, one can argue that the behavioral phenotype is linked directly to the microbiota as the transfer of stress-prone Balb/C microbiota to GF Swiss Webster (SW) mice reduced exploratory behavior compared to normal SW mice, while the transfer of SW microbiota to GF Balb/C mice increased exploratory behavior compared to normal Balb/C mice (Bercik et al., 2011). Certainly, strain can influence the relationship between the microbiota and anxiety-like behavior. For example, one study has shown increased anxiety-like behavior in Balb/C GF in the open field (OF) compared to Balb/C GF mice following 24 h exposure to normal housing conditions mice (EX-GF) (Nishino et al., 2013). Similarly, stress-sensitive F344 rats showed reduced center entries in the OF compared to conventionally housed rats (Crumeyrolle-Arias et al., 2014). Consideration of the contribution of strain to the composition and diversity of the gut microbiota is needed in ongoing studies to better determine the contribution of host genetics compared to environmental factors to behavior.

In addition to the microbiota, researchers in psychiatry and behavioral neuroscience are increasingly recognizing the importance of the immune system in behavior. Several groups have examined the role of the immune system in behavior using immunocompromised mice including GF mice (Foster & McVey Neufeld, 2013; McVey Neufeld, Mao, Bienenstock, Foster, & Kunze, 2013; Neufeld et al., 2011a, 2011b), mice lacking T cell receptor β and δ chains (*TCRβ−/−δ−/−*) (Rilett et al., 2015), mice lacking both β-2 microglobulin and transporter associated with antigen-processing genes (*β2M−/−TAP−/−*) resulting in the loss of functional class I MHC molecules and depleted CD8 T cells (Sankar, Mackenzie, & Foster, 2012), and B cell-deficient mice (Rilett et al., 2015). Recently, one study showed that mice deficient of T cells (*TCRβ−/−δ−/−*) showed reduced anxiety-like behavior in the elevated plus maze, light/dark test, and open field, whereas these behavioral differences are not observed in B cell-deficient mice (Rilett et al., 2015). In contrast, mice lacking functional class I MHC molecules and depleted CD8 T cells spent similar amounts of time in the open arm of the elevated plus maze (EPM) as WT mice; however, they did show increased risk assessment behaviors (Sankar et al., 2012), suggesting that subsets of T cells might have distinct roles in modulating behavior. Other groups have shown reduced anxiety-like behavior in mice with adaptive immune deficits such as *RAG1−/−* mice (Cushman, Lo, Huang, Wasserfall, & Petitto, 2003). Deletion of RAG-1 results in an ability of lymphocytes to execute VDJ recombination, a mechanism of genetic recombination that rearranges variable (V), joining (J), and diversity (D) gene segments to create diversity in the variable chain of the T cell receptor. This deletion generates mice that lack mature T and B cells and thus silencing the adaptive immune system (Mombaerts et al., 1992). Together, this work by our group and others demonstrates that the adaptive immune system, and in particular T cells, influences behavior. An interesting observation in our study with *β2M−/− TAP−/−* mice was that we observed a loss of sexual dimorphism in activity, exploratory, and anxiety-like behaviors compared to wild-type mice (Sankar et al., 2012). This was also observed in *TCRβ−/−δ−/−* mice. Considering the evidence of sexually dimorphisms in immune functioning (Da Silva, 1999; De Leon-Nava et al., 2009; Weinstein, Ran, & Segal, 1984), it seems reasonable and necessary to further examine a role for immune phenotype in sex differences in behavior.

3. PROBIOTICS ATTENUATE STRESS- AND IMMUNE-RELATED CHANGES IN BEHAVIOR

Probiotics are defined as "live microorganisms that, when administered in adequate amounts, confer a health benefit on the host" (Food and Agriculture Organization of the United Nations & World Health Organization, 2006; Hill et al., 2014). Animal studies have demonstrated that administration or feeding of probiotics to rodents has beneficial effects (Table 1). Effects of probiotic feeding on brain function and behavior have included experiments using healthy rodents. In one study, administration of *Bifidobacterium infantis* to Sprague–Dawley adult male rats for 14 days showed no effect on depressive-like behavior but was associated with altered plasma levels of tryptophan and metabolites, as well as a few small differences in monoamine levels in brain tissue from probiotic-treated rats (Desbonnet, Garrett, Clarke, Bienenstock, & Dinan, 2008). Male Wistar adult male rats fed a probiotic mixture of *Lactobacillus helveticus* and *B. longum* for 14 days showed reduced conditioned defensive burying compared to control rats (Messaoudi et al., 2011).

In mice, 28 days administration of *Lactobacillus rhamnosus* (JB-1) in male Balb/C mice showed reduced number of open arm entries with no difference in open arm time in the EPM and reduced immobility time in the forced swim test (FST) compared to broth-fed control mice (Bravo et al., 2011). These behavioral differences in JB-1 fed mice were accompanied by changes in GABA receptor mRNA expression in several stress-related brain regions and were absent in vagotomized mice (Bravo et al., 2011). In a follow-up study, Savignac et al. (2014) showed that two different *Bifidobacteria* strains had distinct behavioral effects when administered to male Balb/C adult mice. *B. longum 1714* and *B. breve 1205* were fed to mice for 6 weeks and behavior compared to mice that were administered escitalopram or vehicle. Both probiotic treatments and escitalopram treatment results in reduced marble burying, whereas only mice fed *B. breve 1205* showed reduced anxiety-like behavior in the EPM. Mice treated with *B. longum 1714* showed reduced depressive-like behavior in the tail suspension test, but no difference in immobility was observed in the FST for any treatment group (Savignac et al., 2014). Interestingly, in the previous study, feeding with *L. rhamnosus* reduced stress-induced plasma CORT in male

Table 1 Beneficial Probiotics

Bacteria	Benefits Observed	Animal/ Clinical	References
Bifidobacterium infantis	Reduced inflammation	Sprague–Dawley rats	Desbonnet et al. (2008)
B. longum	Reduced anxiety-like behavior	Balb/C mice	Savignac, Kiely, Dinan, and Cryan (2014)
	Reduced depressive-like behavior		
B. breve	Reduced anxiety-like behavior	Balb/C mice	Savignac et al. (2014)
B. longum	Improved recognition memory	Balb/C mice	Savignac, Tramullas, Kiely, Dinan, and Cryan (2015)
	Improved spatial learning		
	Reduced fear behavior		
B. breve	Improved recognition memory	Balb/C mice	Savignac et al. (2015)
B. longum	Reduced infection-related anxiety-like behavior	AKR mice	Bercik et al. (2010)
Lactobacillus rhamnosus	Reduced anxiety-like behavior	Balb/C mice	Bravo et al. (2011)
	Reduced depressive-like behavior		
	Reduced stress hormones		
L. helveticus	Reduced anxiety-like behavior	Sprague–Dawley rats	Liang et al. (2015)
	Reduced stress hormones		
L. reuteri	Reduced stress-related infectious colonization	CD1 mice	Mackos, Eubank, Parry, & Bailey (2013)
L. reuteri	Reduced stress/infection-related inflammatory mediators	CD1 mice	Mackos et al. (2016)

Table 1 Beneficial Probiotics—cont'd

Bacteria	Benefits Observed	Animal/ Clinical	References
L. pentosus	Improved age-related deficits in spatial memory	Fischer rats	Jeong, Woo, Kim, Han, and Kim (2015)
	Reduced inflammation		
L. pentosus	Protected against stress-induced deficits in memory	C57Bl/6 mice	Woo et al. (2014)
	Reduced inflammation		
Mycobacterium vaccae	Improved spatial learning	Balb/C mice	Matthews and Jenks (2013)
M. vaccae	Reduced stress-related infectious colonization	C57Bl/6 mice	Reber et al. (2016)
	Reduced stress-related anxiety-like behavior		
Combination: *L. helveticus, B. longum*	Reduced stress hormones	C57Bl/6 mice	Ait-Belgnaoui et al. (2012)
	Prevented stress-related decrease in neurogenesis		
	Protected gut barrier integrity		
Combination: *L. helveticus, L. rhamnosus*	Protected against stress-induced deficits in recognition memory	C57Bl/6 mice	Gareau et al. (2011)
Combination: *L. acidophilus, B. lactis, L. fermentum*	Improved spatial learning	Wistar rats	Davari, Talaei, Alaei, and Salami (2013)
Combination: *L. acidophilus, B. lactis, L. fermentum*	Synaptic transmission	Wistar rats	Davari et al. (2013)
Combination: *Streptococcus salivarius, B. breve, B. infanti, B. longum, L. acidophilus, L. plantarum, L. casei, L. delbrueckii*	Reduced inflammation	C57Bl/6 mice	D'Mello et al. (2015)

Balb/C mice (Bravo et al., 2011), but neither *Bifidobacteria* strain nor escitalopram treatment effected CORT following stress in this study (Savignac et al., 2014). In a related study, daily feeding for several weeks with *B. longum 1714* showed subtle positive effects on healthy male adult Balb/C mice in novel object recognition, in the Barnes maze, and in fear conditioning, whereas feeding with *B. breve 1205* improved novel object recognition but did not have an effect on other cognitive tests (Savignac et al., 2015). *Mycobacterium vaccae* is not a commensal bacteria; however, it has been shown to have immunomodulatory effects (Lowry et al., 2007) and was recently shown to have a positive effect on cognitive behavior in mice (Matthews & Jenks, 2013). *M. vaccae* was fed to mice 3 weeks and 1 week prior to behavioral testing and incorporated into a food reward following testing. *M. vaccae* treatment improved performance on a complex maze task that measures spatial learning (Matthews & Jenks, 2013).

The above noted studies examined probiotic feeding to healthy rodents; however, a key question is whether probiotics can be beneficial in animal models of disease. To date, beneficial effects of probiotics have been observed in stress, inflammatory, and disease models. Administration of a commercially available probiotic mixture that contained both *L. helveticus* R0052 and *B. longum* R0175 to C57Bl6 male mice for 2 weeks prior to water avoidance stress exposure significantly reduced the impact of the stressor (Ait-Belgnaoui et al., 2012). In particular, probiotic-treated mice showed reduced plasma CORT and catecholamine levels following chronic stress compared to vehicle-treated mice. In addition, probiotic pretreatment prevented stress-related decreases in hippocampal neurogenesis and prevented stress-related increases in gut barrier integrity (Ait-Belgnaoui et al., 2012). Exposure to stress has an impact on brain and behavior and these changes may be mediated by stress-induced changes in microbiota composition (Bailey & Coe, 1999; Bailey et al., 2011; Bharwani et al., 2016; Galley, Nelson, et al., 2014; Galley, Yu, et al., 2014). Microbiota changes also impact the sensitivity of the GI tract to infection by pathogenic bacteria. A comparison of GI colonization in unstressed and previously stressed CD1 male mice showed that exposure to prolonged restraint stress prior to infection a significant increase in *Citrobacter rodentium* colonization, suggesting that stress-mediated changes in microbiota may influence sensitivity to infectious pathogens (Bailey et al., 2010). Further these investigators showed that increased sensitivity to *C. rodentium* was associated with increased levels of tumor necrosis factor-α (TNF-α) in colonic tissue, supporting a role for microbiota–immune signaling as one signaling system that might mediate stress-related outcomes (Bailey et al., 2010). Coincident exposure to prolonged restraint stress and *C. rodentium* enhanced infection-induced colitis measured as increased colonic pathology, increased expression of inflammatory mediators, and increased pathogen translocation to the spleen (Mackos et al., 2013); whereas pretreatment with probiotic *L. reuteri* resulted in reduced expression of inflammatory mediators, there were still significant colitus-related symptoms in the probiotic-treated mice (Mackos et al., 2013). Exposure to social defeat stress also alters microbiota composition and results in reduced levels of *L. reuteri* (Bailey et al., 2011). To better understand the beneficial effects of *L. reuteri* as a probiotic, CD1 male mice were exposed to social defeat stress and infection with *C. rodentium* with either daily probiotic or vehicle treatment. *L. reuteri*-treated and vehicle-treated mice showed similar levels of fecal *C. rodentium*; however, inflammation-related stress outcomes were improved by probiotic treatment (Mackos et al., 2016). *M. vaccae* has also been shown to have beneficial effects in a stress model (Reber et al., 2016). Exposure to *M. vaccae* prior to chronic subordinate colony housing stress protected mice from behavioral effects of this stressor, including reduced submissive behaviors in response to an aggressor and reduced anxiety-like behavior on the EPM (Reber et al., 2016). These behavioral effects were associated with changes in serotonergic gene expression. A beneficial effect on stress-related colitis was also observed (Reber et al., 2016). Interestingly, some of the beneficial effects were linked to immunoregulation by *M. vaccae*, as depletion of regulatory T lymphocytes eliminated the beneficial effects on stress-related colitis and anxiety-like behavior in the EPM, whereas mice remained resilient in response to an aggressor, suggesting that multiple signaling pathways may be involved. Further, analysis of microbiota changes in this model suggested that these beneficial effects were not directly associated with changes in the composition of the gut microbiota (Reber et al., 2016).

In male Sprague–Dawley rats, repeated restraint stress was shown to increase anxiety-like behavior and memory deficits (Liang et al., 2015). In parallel, repeated restraint resulted in higher levels of brain monoamine and increased peripheral cytokine levels (Liang et al., 2015). Daily administration of *L. helveticus* NS8 strain during stress reduced the behavioral effects of chronic stress and normalized most monoamine brain levels. *L. helveticus* administration did reduce peripheral stress hormones and increased peripheral levels of antiinflammatory cytokine interleukin 10 (IL10), but did not reduce stress-induced peripheral proinflammatory cytokines, interferon-γ (IFN-γ) and TNF-α (Liang et al., 2015). Changes in stress hormones and brain serotonin levels were associated with probiotic-related changes in behavior (Liang et al., 2015).

The above studies link stress to changes in the microbiota. Further, this work demonstrates that microbiota–immune interactions that influence behavior can benefit from probiotic administration. It is important to note that the beneficial effects of probiotic administration vary across experiments. Differential effects of administration of *L. rhamnosus* compared to *B. longum* was reported in mice infected with the noninvasive parasite *Trichuris muris* (Bercik et al., 2010). *T. muris* infection is resulted in increased anxiety-like behavior and related decreased expression of brain-derived neurotrophic factor in the hippocampus (Bercik et al., 2010). Infected mice treated with *B. longum* showed normalization of behavior and hippocampal BDNF mRNA expression, whereas infected mice treated with *L. rhamnosus* showed no improvement in behavior or BDNF expression (Bercik et al., 2010). Interestingly, *B. longum* administration did not reduce peripheral cytokine levels in *T. muris*-infected mice, suggesting that CNS benefits of probiotics are possible in certain situations independent of changes in peripheral inflammation.

Several studies to date, including those cited earlier, have linked microbiota–brain signaling to anxiety- and depressive-like behaviors, and yet only a few studies have considered a role for microbiota–brain signaling in cognitive processes and memory. The few studies that have been conducted have demonstrated a beneficial role for probiotics in improving memory deficits (Gareau et al., 2011; Jeong et al., 2015; Woo et al., 2014). An indication that the microbiota influences memory was suggested by behavioral testing in GF mice that showed deficits in object recognition memory and working memory (Gareau et al., 2011). In addition, exposure to *C. rodentium* infection in combination with water avoidance stress exposure resulted in deficits in object recognition memory that were prevented by pretreatment with a combination of *L. helveticus* and *L. rhamnosus* (Gareau et al., 2011). Aging-related deficits in spatial memory were shown to be improved in Fischer 344 rats by 8-week oral administration of *L. pentosus* var. *plantarum* C29, a probiotic that is thought to have antiinflammatory action (Jeong et al., 2015). In parallel, probiotic treatment restored hippocampal levels of plasticity-related signaling molecules in old rats to levels observed in young rats (Jeong et al., 2015). Moreover, *L. pentosus* administration had an antiinflammatory effect in aged rats (Jeong et al., 2015). As with many of the stress experiments noted earlier, researchers have observed similar findings in both mice and rats related to microbiota and memory deficits. In an accelerated aging mouse models using C57Bl/6 mice, administration of *L. pentosus* var. *plantarum* C29 for 5 weeks prevented memory impairment, plasticity-related changes in brain tissue (Woo et al., 2014). This study also demonstrated reduced inflammatory mediators in probiotic-treated mice (Woo et al., 2014). These animal studies are impressive but clearly more studies, particularly in clinical populations, are needed to better understand how probiotics may be used to reduce inflammation associated with aging.

Depression or depressive symptoms are often comorbid with medical conditions. This observation is also apparent in animal models. For example, in mice, liver inflammation is accompanied by the presentation of sickness behavior, measured as reduced social exploration and immobility in the social exploration task (D'Mello, Le, & Swain, 2009). Peripheral immune signaling cascades involving TNFα-TNF receptor 1 and monocytes were shown to be associated with microglia activation in this inflammation model (D'Mello et al., 2013). Interestingly, administration of a probiotic mixture (VSL#3 containing *Streptococcus salivarius* subsp. *thermophilus, B. breve, B. infanti, B. longum, L. acidophilus, L. plantarum, L. casei*, and *L. delbrueckii* subsp. *bulgaricus*) reduced peripheral inflammation, reduced microglial activation, and attenuated the magnitude of sickness behavior, suggesting that probiotic administration has antiinflammatory effects that have benefits on reducing inflammatory-mediated sickness behaviors (D'Mello et al., 2015). The benefit on behavior in these studies was partial and probiotic treatment did not improve barrier function, microbiota composition, or disease severity (D'Mello et al., 2015).

Exposure of Sprague–Dawley rats to repeated injections of ammonium acetate results in elevated blood levels of ammonium (hyperammonemia—HA) and was associated with increased anxiety-like behaviors measured as reduced time spent in the open arms of the EPM and fewer open arm entries (Luo et al., 2014). HA treatment was also associated with a significant learning deficit in the Morris water maze and deficits in memory retention (Luo et al., 2014). Supplementation with *L. helveticus* strain NS8 in drinking water was effective in reversing learning deficits in HA rats and increasing the number of open arm entries in the EPM but did not significantly effect time spent in the open arm of the EPM. Probiotic treatment lasted 14 days and perhaps a longer treatment time may have improved this outcome measure. Both serotonergic and immune signaling pathways were implicated in HA-induced behavioral effects, and probiotic administration was shown to have antiinflammatory effects and reversed some of the HA-induced alterations in central serotonergic signaling and peripheral tryptophan metabolism (Luo et al., 2014).

Metabolic disorders including diabetes are also accompanied by deficits in brain function including learning and memory deficits. Diabetic rats, generated by administration of streptozocin, show elevation of serum glucose and decreased serum insulin (Davari et al., 2013). Diabetic rats showed spatial learning deficits in the Morris water maze and deficits in hippocampal long-term potentiation, effects that were reversed by 8-week administration of a probiotic mixture containing *L. acidophilus, B. lactis, L. fermentum* (Davari et al., 2013). However, it should be noted that while the probiotic treatment did reduce serum glucose levels compared to nonprobiotic diabetic rat group, these levels remained significantly higher than control rats (Davari et al., 2013). Interestingly, probiotic treatment further enhanced the spatial learning ability of control rats, suggesting a possible beneficial effect in normal, healthy conditions.

Overall, the animal studies demonstrate beneficial effects of probiotic treatment in both healthy rodents and animal models of disease (Ait-Belgnaoui et al., 2012; Bercik et al., 2010; Bravo et al., 2011; Davari et al., 2013; Desbonnet et al., 2008; D'Mello et al., 2015; Gareau et al., 2011; Hsiao et al., 2013; Jeong et al., 2015; Liang et al., 2015; Luo et al., 2014; Mackos et al., 2013; Matthews & Jenks, 2013; Savignac et al., 2014, 2015; Woo et al., 2014). A summary of the benefits suggested by animal work to date would include:

1. reduced anxiety and depressive-like behavior,
2. improved cognitive and spatial learning,
3. reduced stress hormone levels,
4. antiinflammatory action in peripheral and central immune signaling.

These results are promising and need to be considered in the context of the work to date that has been conducted in healthy individuals and clinical populations.

4. FUTURE DIRECTIONS

Overall, the evidence provided to date in the literature links the microbiota to behavior and suggests that immune signaling may be a key pathway involved. As neuroscientists, psychologists, and psychiatrists are starting to appreciate the importance of the gut microbiota to mental health, it is critical to determine the mechanisms of microbiota to brain communication so that we have a better understanding of the aspects that may be modifiable with proper intervention in individual with mental illness. Microbiota–brain communication is emerging as an important factor in brain development and function. Further, immune dysfunction is clearly established to play a role in mental illness. We have established expertise in studying microbiota, the immune system, brain, and behavior and poised to contribute significant novel findings to our understanding of microbiota–brain communication in mental illness. This research has a clear translational relevance for mental health, contributing to extant findings that indicate a role for the microbiome in brain development and behavior.

More research looking into mood and probiotics is of interest as mood disorders and anxiety disorders are very heterogeneous and the challenge is to identify subtypes of patients and match them to the best treatment from onset of disease. Gut-brain biomarkers may be helpful in predicting response to existing treatment and probiotics may be a good approach to treat some individuals. Importantly, there are many paths between gut and brain and we are only just starting to figure out which aspects are connected to the brain. More systematic work is needed in order for us to understand how and when probiotics may be beneficial to mental health.

Conflicts of interest: The author confirms that this chapter contents have no conflict of interest.

REFERENCES

Ait-Belgnaoui, A., Durand, H., Cartier, C., Chaumaz, G., Eutamene, H., Ferrier, L., … Theodorou, V. (2012). Prevention of gut leakiness by a probiotic treatment leads to attenuated HPA response to an acute psychological stress in rats. *Psychoneuroendocrinology*, *37*(11), 1885–1895. http://dx.doi.org/10.1016/j.psyneuen.2012.03.024.

Arumugam, M., Raes, J., Pelletier, E., Le Paslier, D., Yamada, T., Mende, D. R., … Bork, P. (2011). Enterotypes of the human gut microbiome. *Nature*, *473*(7346), 174–180. http://dx.doi.org/10.1038/nature09944. nature09944 [pii].

Bailey, M. T., & Coe, C. L. (1999). Maternal separation disrupts the integrity of the intestinal microflora in infant rhesus monkeys. *Developmental Psychobiology*, *35*(2), 146–155.

Bailey, M. T., Dowd, S. E., Galley, J. D., Hufnagle, A. R., Allen, R. G., & Lyte, M. (2011). Exposure to a social stressor alters the structure of the intestinal microbiota: Implications for stressor-induced immunomodulation. *Brain, Behavior, and Immunity*, *25*(3), 397–407. http://dx.doi.org/10.1016/j.bbi.2010.10.023.

Bailey, M. T., Dowd, S. E., Parry, N. M., Galley, J. D., Schauer, D. B., & Lyte, M. (2010). Stressor exposure disrupts commensal microbial populations in the intestines and leads to increased colonization by Citrobacter rodentium. *Infection and Immunity*, *78*(4), 1509–1519. http://dx.doi.org/10.1128/IAI.00862-09. IAI.00862-09 [pii].

Bercik, P., Denou, E., Collins, J., Jackson, W., Lu, J., Jury, J., … Collins, S. M. (2011). The intestinal microbiota affect central levels of brain-derived neurotropic factor and behavior in mice. *Gastroenterology*, *141*(2), 599–609. http://dx.doi.org/10.1053/j.gastro.2011.04.052. 609 e591–593.

Bercik, P., Verdu, E. F., Foster, J. A., Macri, J., Potter, M., Huang, X., … Collins, S. M. (2010). Chronic gastrointestinal inflammation induces anxiety-like behavior and alters central nervous system biochemistry in mice. *Gastroenterology*, *139*(6), 2102–2112. http://dx.doi.org/10.1053/j.gastro.2010.06.063. e2101, S0016-5085(10)01006-1 [pii].

Bharwani, A., Mian, M. F., Foster, J. A., Surette, M. G., Bienenstock, J., & Forsythe, P. (2016). Structural & functional consequences of chronic psychosocial stress on the microbiome & host. *Psychoneuroendocrinology*, *63*, 217–227. http://dx.doi.org/10.1016/j.psyneuen.2015.10.001.

Bravo, J. A., Forsythe, P., Chew, M. V., Escaravage, E., Savignac, H. M., Dinan, T. G., … Cryan, J. F. (2011). Ingestion of Lactobacillus strain regulates emotional behavior and central GABA receptor expression in a mouse via the vagus nerve. *Proceedings of the National Academy of Sciences of the United States of America*, *108*(38), 16050–16055. http://dx.doi.org/10.1073/pnas.1102999108.

Clarke, G., Grenham, S., Scully, P., Fitzgerald, P., Moloney, R. D., Shanahan, F., … Cryan, J. F. (2013). The microbiome-gut-brain axis during early life regulates the hippocampal serotonergic system in a sex-dependent manner. *Molecular Psychiatry*, *18*(6), 666–673. http://dx.doi.org/10.1038/mp.2012.77.

Costello, E. K., Lauber, C. L., Hamady, M., Fierer, N., Gordon, J. I., & Knight, R. (2009). Bacterial community variation in human body habitats across space and time. *Science*, *326*(5960), 1694–1697. http://dx.doi.org/10.1126/science.1177486. 1177486 [pii].

Crumeyrolle-Arias, M., Jaglin, M., Bruneau, A., Vancassel, S., Cardona, A., Dauge, V., … Rabot, S. (2014). Absence of the gut microbiota enhances anxiety-like behavior and neuroendocrine response to acute stress in rats. *Psychoneuroendocrinology*, *42*, 207–217. http://dx.doi.org/10.1016/j.psyneuen.2014.01.014.

Cryan, J. F., & Dinan, T. G. (2012). Mind-altering microorganisms: The impact of the gut microbiota on brain and behaviour. *Nature Reviews Neuroscience*, *13*(10), 701–712. http://dx.doi.org/10.1038/nrn3346.

Cushman, J., Lo, J., Huang, Z., Wasserfall, C., & Petitto, J. M. (2003). Neurobehavioral changes resulting from recombinase activation gene 1 deletion. *Clinical and Diagnostic Laboratory Immunology*, *10*(1), 13–18.

Da Silva, J. A. (1999). Sex hormones and glucocorticoids: Interactions with the immune system. *The Annals of the New York Academy of Sciences*, *876*, 102–117. discussion 117–108.

Davari, S., Talaei, S. A., Alaei, H., & Salami, M. (2013). Probiotics treatment improves diabetes-induced impairment of synaptic activity and cognitive function: Behavioral and electrophysiological proofs for microbiome-gut-brain axis. *Neuroscience*, *240*, 287–296. http://dx.doi.org/10.1016/j.neuroscience.2013.02.055.

De Leon-Nava, M. A., Nava, K., Soldevila, G., Lopez-Griego, L., Chavez-Rios, J. R., Vargas-Villavicencio, J. A., & Morales-Montor, J. (2009). Immune sexual dimorphism: Effect of gonadal steroids on the expression of cytokines, sex steroid receptors, and lymphocyte proliferation. *The Journal of Steroid Biochemistry and Molecular Biology*, *113*(1–2), 57–64. http://dx.doi.org/10.1016/j.jsbmb.2008.11.003. S0960-0760(08)00264-1 [pii].

Desbonnet, L., Garrett, L., Clarke, G., Bienenstock, J., & Dinan, T. G. (2008). The probiotic Bifidobacteria infantis: An assessment of potential antidepressant properties in the rat. *Journal of Psychiatric Research*, *43*(2), 164–174. http://dx.doi.org/10.1016/j.jpsychires.2008.03.009.

Diamant, M., Blaak, E. E., & de Vos, W. M. (2011). Do nutrient-gut-microbiota interactions play a role in human obesity, insulin resistance and type 2 diabetes? *Obesity Reviews*, *12*(4), 272–281. http://dx.doi.org/10.1111/j.1467-789X.2010.00797.x. OBR797 [pii].

D'Mello, C., Le, T., & Swain, M. G. (2009). Cerebral microglia recruit monocytes into the brain in response to tumor necrosis factoralpha signaling during peripheral organ inflammation. *The Journal of Neuroscience*, *29*(7), 2089–2102. http://dx.doi.org/10.1523/JNEUROSCI.3567-08.2009.

D'Mello, C., Riazi, K., Le, T., Stevens, K. M., Wang, A., McKay, D. M., ... Swain, M. G. (2013). P-selectin-mediated monocyte-cerebral endothelium adhesive interactions link peripheral organ inflammation to sickness behaviors. *The Journal of Neuroscience*, *33*(37), 14878–14888. http://dx.doi.org/10.1523/JNEUROSCI.1329-13.2013.

D'Mello, C., Ronaghan, N., Zaheer, R., Dicay, M., Le, T., MacNaughton, W. K., ... Swain, M. G. (2015). Probiotics improve inflammation-associated sickness behavior by altering communication between the peripheral immune system and the brain. *The Journal of Neuroscience*, *35*(30), 10821–10830. http://dx.doi.org/10.1523/JNEUROSCI.0575-15.2015.

Food and Agriculture Organization of the United Nations, & World Health Organization. (2006). Health and nutrition properties of probiotics in food including powder milk with live lactic acid bacteria. Guidelines for the evaluation of probiotics in food. *Probiotics in food: Health and nutritional properties and guidelines for evaluation*. Italy, Rome: Food and Agriculture Organization of the United Nations, World Health Organization.

Foster, J. A., & McVey Neufeld, K. A. (2013). Gut-brain axis: How the microbiome influences anxiety and depression. *Trends in Neurosciences*, *36*(5), 305–312. http://dx.doi.org/10.1016/j.tins.2013.01.005.

Galley, J. D., Nelson, M. C., Yu, Z., Dowd, S. E., Walter, J., Kumar, P. S., ... Bailey, M. T. (2014). Exposure to a social stressor disrupts the community structure of the colonic mucosa-associated microbiota. *BMC Microbiology*, *14*, 189. http://dx.doi.org/10.1186/1471-2180-14-189.

Galley, J. D., Yu, Z., Kumar, P., Dowd, S. E., Lyte, M., & Bailey, M. T. (2014). The structures of the colonic mucosa-associated and luminal microbial communities are distinct and differentially affected by a prolonged murine stressor. *Gut Microbes*, *5*(6), 748–760. http://dx.doi.org/10.4161/19490976.2014.972241.

Gareau, M. G., Wine, E., Rodrigues, D. M., Cho, J. H., Whary, M. T., Philpott, D. J., ... Sherman, P. M. (2011). Bacterial infection causes stress-induced memory dysfunction in mice. *Gut*, *60*(3), 307–317. http://dx.doi.org/10.1136/gut.2009.202515.

Gill, S. R., Pop, M., Deboy, R. T., Eckburg, P. B., Turnbaugh, P. J., Samuel, B. S., ... Nelson, K. E. (2006). Metagenomic analysis of the human distal gut microbiome. *Science*, *312*(5778), 1355–1359. http://dx.doi.org/10.1126/science.1124234. 312/5778/1355 [pii].

Gustafsson, B. E. (1959). Lightweight stainless steel systems for rearing germfree animals. *The Annals of the New York Academy of Sciences*, *78*, 17–28.

Gustafsson, B., Kahlson, G., & Rosengren, E. (1957). Biogenesis of histamine studied by its distribution and urinary excretion in germ free reared and not germ free rats fed a histamine free diet. *Acta Physiologica Scandinavica*, *41*(2–3), 217–228.

Heijtz, R. D., Wang, S., Anuar, F., Qian, Y., Bjorkholm, B., Samuelsson, A., ... Pettersson, S. (2011). Normal gut microbiota modulates brain development and behavior. *Proceedings of the National Academy of Sciences of the United States of America*, *108*(7), 3047–3052. http://dx.doi.org/10.1073/pnas.1010529108. 1010529108 [pii].

Hill, C., Guarner, F., Reid, G., Gibson, G. R., Merenstein, D. J., Pot, B., ... Sanders, M. E. (2014). Expert consensus document. The International Scientific Association for Probiotics and Prebiotics consensus statement on the scope and appropriate use of the term probiotic. *Nature Reviews. Gastroenterology & Hepatology*, *11*(8), 506–514. http://dx.doi.org/10.1038/nrgastro.2014.66.

Hooper, L. V., Wong, M. H., Thelin, A., Hansson, L., Falk, P. G., & Gordon, J. I. (2001). Molecular analysis of commensal host-microbial relationships in the intestine. *Science*, *291*(5505), 881–884.

Hsiao, E. Y., McBride, S. W., Hsien, S., Sharon, G., Hyde, E. R., McCue, T., ... Mazmanian, S. K. (2013). Microbiota modulate behavioral and physiological abnormalities associated with neurodevelopmental disorders. *Cell*, *155*(7), 1451–1463. http://dx.doi.org/10.1016/j.cell.2013.11.024.

Jeong, J. J., Woo, J. Y., Kim, K. A., Han, M. J., & Kim, D. H. (2015). Lactobacillus pentosus var. plantarum C29 ameliorates age-dependent memory impairment in Fischer 344 rats. *Letters in Applied Microbiology*, *60*(4), 307–314. http://dx.doi.org/10.1111/lam.12393.

Jumpertz, R., Le, D. S., Turnbaugh, P. J., Trinidad, C., Bogardus, C., Gordon, J. I., & Krakoff, J. (2011). Energy-balance studies reveal associations between gut microbes, caloric load, and nutrient absorption in humans. *The American Journal of Clinical Nutrition*, *94*(1), 58–65. http://dx.doi.org/10.3945/ajcn.110.010132. ajcn.110.010132 [pii].

Kau, A. L., Ahern, P. P., Griffin, N. W., Goodman, A. L., & Gordon, J. I. (2011). Human nutrition, the gut microbiome and the immune system. *Nature*, *474*(7351), 327–336. http://dx.doi.org/10.1038/nature10213. nature10213 [pii].

Liang, S., Wang, T., Hu, X., Luo, J., Li, W., Wu, X., … Jin, F. (2015). Administration of Lactobacillus helveticus NS8 improves behavioral, cognitive, and biochemical aberrations caused by chronic restraint stress. *Neuroscience*, *310*, 561–577. http://dx.doi.org/10.1016/j.neuroscience.2015.09.033.

Lowry, C. A., Hollis, J. H., de Vries, A., Pan, B., Brunet, L. R., Hunt, J. R., … Lightman, S. L. (2007). Identification of an immune-responsive mesolimbocortical serotonergic system: Potential role in regulation of emotional behavior. *Neuroscience*, *146*(2), 756–772.

Luna, R. A., & Foster, J. A. (2014). Gut brain axis: Diet microbiota interactions and implications for modulation of anxiety and depression. *Current Opinion in Biotechnology*, *32C*, 35–41. http://dx.doi.org/10.1016/j.copbio.2014.10.007.

Luo, J., Wang, T., Liang, S., Hu, X., Li, W., & Jin, F. (2014). Ingestion of Lactobacillus strain reduces anxiety and improves cognitive function in the hyperammonemia rat. *Science China. Life Sciences*, *57*(3), 327–335. http://dx.doi.org/10.1007/s11427-014-4615-4.

Mackos, A. R., Eubank, T. D., Parry, N. M., & Bailey, M. T. (2013). Probiotic Lactobacillus reuteri attenuates the stressor-enhanced severity of Citrobacter rodentium infection. *Infection and Immunity*, *81*(9), 3253–3263. http://dx.doi.org/10.1128/IAI.00278-13.

Mackos, A. R., Galley, J. D., Eubank, T. D., Easterling, R. S., Parry, N. M., Fox, J. G., … Bailey, M. T. (2016). Social stress-enhanced severity of Citrobacter rodentium-induced colitis is CCL2-dependent and attenuated by probiotic Lactobacillus reuteri. *Mucosal Immunology*, *9*(2), 515–526. http://dx.doi.org/10.1038/mi.2015.81.

Macpherson, A. J., & Harris, N. L. (2004). Interactions between commensal intestinal bacteria and the immune system. *Nature Reviews. Immunology*, *4*(6), 478–485.

Macpherson, A. J., Martinic, M. M., & Harris, N. (2002). The functions of mucosal T cells in containing the indigenous commensal flora of the intestine. *Cellular and Molecular Life Sciences*, *59*(12), 2088–2096.

Macpherson, A. J., & Uhr, T. (2004). Compartmentalization of the mucosal immune responses to commensal intestinal bacteria. *The Annals of the New York Academy of Sciences*, *1029*, 36–43.

Matthews, D. M., & Jenks, S. M. (2013). Ingestion of Mycobacterium vaccae decreases anxiety-related behavior and improves learning in mice. *Behavioural Processes*, *96*, 27–35. http://dx.doi.org/10.1016/j.beproc.2013.02.007.

Mayer, E. A., Tillisch, K., & Gupta, A. (2015). Gut/brain axis and the microbiota. *The Journal of Clinical Investigation*, *125*(3), 926–938. http://dx.doi.org/10.1172/JCI76304.

McVey Neufeld, K. A., Mao, Y. K., Bienenstock, J., Foster, J. A., & Kunze, W. A. (2013). The microbiome is essential for normal gut intrinsic primary afferent neuron excitability in the mouse. *Neurogastroenterology and Motility*, *25*(2), 183-e188. http://dx.doi.org/10.1111/nmo.12049.

Messaoudi, M., Lalonde, R., Violle, N., Javelot, H., Desor, D., Nejdi, A., … Cazaubiel, J. M. (2011). Assessment of psychotropic-like properties of a probiotic formulation (Lactobacillus helveticus R0052 and Bifidobacterium longum R0175) in rats and human

subjects. *The British Journal of Nutrition, 105*(5), 755–764. http://dx.doi.org/10.1017/S0007114510004319.
Mombaerts, P., Iacomini, J., Johnson, R. S., Herrup, K., Tonegawa, S., & Papaioannou, V. E. (1992). RAG-1-deficient mice have no mature B and T lymphocytes. *Cell, 68*(5), 869–877.
Neufeld, K. A., Kang, N., Bienenstock, J., & Foster, J. A. (2011a). Effects of intestinal microbiota on anxiety-like behavior. *Communicative & Integrative Biology, 4*(4), 492–494. http://dx.doi.org/10.4161/cib.4.4.15702.
Neufeld, K. M., Kang, N., Bienenstock, J., & Foster, J. A. (2011b). Reduced anxiety-like behavior and central neurochemical change in germ-free mice. *Neurogastroenterology and Motility, 23*(3), 255–264. http://dx.doi.org/10.1111/j.1365-2982.2010.01620.x. e119.
Nishino, R., Mikami, K., Takahashi, H., Tomonaga, S., Furuse, M., Hiramoto, T., … Sudo, N. (2013). Commensal microbiota modulate murine behaviors in a strictly contamination-free environment confirmed by culture-based methods. *Neurogastroenterology and Motility, 25*(6), 521–528. http://dx.doi.org/10.1111/nmo.12110.
Qin, J., Li, R., Raes, J., Arumugam, M., Burgdorf, K. S., Manichanh, C., … Ehrlich, S. D. (2010). A human gut microbial gene catalogue established by metagenomic sequencing. *Nature, 464*(7285), 59–65. http://dx.doi.org/10.1038/nature08821. nature08821 [pii].
Reber, S. O., Siebler, P. H., Donner, N. C., Morton, J. T., Smith, D. G., Kopelman, J. M., … Lowry, C. A. (2016). Immunization with a heat-killed preparation of the environmental bacterium Mycobacterium vaccae promotes stress resilience in mice. *Proceedings of the National Academy of Sciences of the United States of America, 113*(22), E3130–E3139. http://dx.doi.org/10.1073/pnas.1600324113.
Rilett, K. C., Friedel, M., Ellegood, J., MacKenzie, R. N., Lerch, J. P., & Foster, J. A. (2015). Loss of T cells influences sex differences in behavior and brain structure. *Brain, Behavior, and Immunity, 46*, 249–260. http://dx.doi.org/10.1016/j.bbi.2015.02.016.
Sankar, A., Mackenzie, R. N., & Foster, J. A. (2012). Loss of class I MHC function alters behavior and stress reactivity. *Journal of Neuroimmunology, 244*(1–2), 8–15. http://dx.doi.org/10.1016/j.jneuroim.2011.12.025. S0165-5728(11)00377-8 [pii].
Savignac, H. M., Kiely, B., Dinan, T. G., & Cryan, J. F. (2014). Bifidobacteria exert strain-specific effects on stress-related behavior and physiology in BALB/c mice. *Neurogastroenterology and Motility, 26*(11), 1615–1627. http://dx.doi.org/10.1111/nmo.12427.
Savignac, H. M., Tramullas, M., Kiely, B., Dinan, T. G., & Cryan, J. F. (2015). Bifidobacteria modulate cognitive processes in an anxious mouse strain. *Behavioural Brain Research, 287*, 59–72. http://dx.doi.org/10.1016/j.bbr.2015.02.044.
Sudo, N., Chida, Y., Aiba, Y., Sonoda, J., Oyama, N., Yu, X. N., … Koga, Y. (2004). Postnatal microbial colonization programs the hypothalamic-pituitary-adrenal system for stress response in mice. *The Journal of Physiology, 558*(Pt. 1), 263–275.
Tlaskalova-Hogenova, H., Stepankova, R., Hudcovic, T., Tuckova, L., Cukrowska, B., Lodinova-Zadnikova, R., … Kokesova, A. (2004). Commensal bacteria (normal microflora), mucosal immunity and chronic inflammatory and autoimmune diseases. *Immunology Letters, 93*(2–3), 97–108.
Weinstein, Y., Ran, S., & Segal, S. (1984). Sex-associated differences in the regulation of immune responses controlled by the MHC of the mouse. *The Journal of Immunology, 132*(2), 656–661.
Woo, J. Y., Gu, W., Kim, K. A., Jang, S. E., Han, M. J., & Kim, D. H. (2014). Lactobacillus pentosus var. plantarum C29 ameliorates memory impairment and inflammaging in a D-galactose-induced accelerated aging mouse model. *Anaerobe, 27*, 22–26. http://dx.doi.org/10.1016/j.anaerobe.2014.03.003.

CHAPTER FOUR

Neuropeptides, Microbiota, and Behavior

P. Holzer[1]
Research Unit of Translational Neurogastroenterology, Institute of Experimental and Clinical Pharmacology, Medical University of Graz, Graz, Austria
BioTechMed-Graz, Graz, Austria
[1]Corresponding author: e-mail address: peter.holzer@medunigraz.at

Contents

Abstract

The gut microbiota and the brain interact with each other through multiple bidirectional signaling pathways in which neuropeptides and neuroactive peptide messengers play potentially important mediator roles. Currently, six particular modes of a neuropeptide link are emerging. (i) Neuropeptides and neurotransmitters contribute to the mutual microbiota–host interaction. (ii) The synthesis of neuroactive peptides is influenced by microbial control of the availability of amino acids. (iii) The activity of neuropeptides is tempered by microbiota-dependent autoantibodies. (iv) Peptide signaling between periphery and brain is modified by a regulatory action of the gut microbiota on the blood–brain barrier. (v) Within the brain, gut hormones released under the influence of the gut microbiota turn into neuropeptides that regulate multiple aspects of brain

International Review of Neurobiology, Volume 131
ISSN 0074-7742
http://dx.doi.org/10.1016/bs.irn.2016.08.005

activity. (vi) Cerebral neuropeptides participate in the molecular, behavioral, and autonomic alterations which the brain undergoes in response to signals from the gut microbiota.

1. NEUROPEPTIDES TRANSCEND BOUNDARIES

The recognition of biologically active peptides as neuropeptides, being produced by neurons and/or having effects on both neurons and non-neuronal cells, is still relatively new and has evolved only in the past 60 years (Strand, 1999). Hypothalamic regulatory factors were among the first compounds discovered to be peptides, being subsequently joined by other peptidic compounds such as substance P, endorphin, and enkephalin and by an increasing number of gut hormones, some of which are also expressed in the central nervous system (CNS). Together with the identification of specific receptors (typically G protein coupled) for particular peptide families and the elaboration of distinct physiological functions, the neuropeptide concept gradually evolved. The recognition of neuropeptides as versatile messengers in the nervous, endocrine, and immune system also transcended several boundaries. The presence of neuropeptides along small-molecule transmitters such as acetylcholine, monoamines, and amino acids replaced the one neuron–one neurotransmitter idea by the cotransmission concept. The discovery that hormones such as adrenocorticotropic hormone (ACTH) not only acts on the adrenal cortex but also influences brain function and behavior disclosed that neuropeptides can subserve multiple functions, depending on the expression of specific receptors in diverse organs.

Neuropeptides and gut hormones and their mutual relationships have also been instrumental in the emerging view that brain and gut have much in common. When since the 1960s several biologically active peptides were discovered to occur both in the CNS and digestive tract, the term "brain–gut axis" was first coined. We now know that a vast number of neuropeptides are produced by central and peripheral neurons alongside with enteroendocrine cells in the gastrointestinal tract and other endocrinologically active cells (Burbach, 2010; Kastin, 2013; Strand, 1999). Gut hormones and neuropeptides form families of closely related peptides, exemplified, e.g., by peptide YY (PYY) and pancreatic polypeptide, on the one hand, and neuropeptide Y (NPY), on the other hand. With the genetic and molecular identification of a huge number of neuropeptide receptors and their transduction mechanisms, the diverse biological roles of neuropeptides in the bidirectional data highway between the gut and CNS are now well understood. Gut hormones and neuropeptides are not only relevant to the physiological regulation of digestion but are also involved in the control of food intake, energy homeostasis, autonomic regulation of gut function, and in the impact of gastrointestinal signals on sensation, emotion, affect, and cognition. It was already by the pioneering studies of De Wied (Kovács & De Wied, 1994) and Kastin (Sandman, Miller, & Kastin, 1977) with their associates in the early 1970s that an involvement of neuropeptides in regulating behavior was disclosed. Taken all these aspects together, it is still true what Strand (1999) stated almost 20 years ago, namely, "that one of the most important functions of neuropeptides is the integration of the functions of the brain and the systems of the body."

2. SIGNALING PATHWAYS IN GUT–BRAIN COMMUNICATION

As is discussed in detail in other chapters of this volume, gut and brain interact with each other in a bidirectional manner in which behavior is an important target, given that neurological and psychiatric disorders may have a root in the dysbiotic gut (Dinan & Cryan, 2015; Sampson & Mazmanian, 2015). Brain–gut communication involves the autonomic regulation of digestion by central, parasympathetic, sympathetic, and enteric neurons and neuroendocrine factors (derived from the adrenal medulla and cortex). Vice versa, there is ongoing signaling from the gut to the CNS, visceral information being continuously fed into subcortical regions of the brain including the limbic system and the autonomic and neuroendocrine centers (Mayer & Tillisch, 2011). This information is integrated with other interoceptive information from the body and with contextual information from the environment (Mayer & Tillisch, 2011). The input from the periphery to the brain may give rise to conscious sensations such as hunger, satiety, urgency, nausea, and pain but subconsciously also impacts on processes relevant to emotional, affective, and cognitive behavior. Gut–brain signaling involves four major pathways comprising vagal and spinal afferent neurons, circulating cytokines, circulating gut hormones, and microbial factors that may also reach the brain via the blood stream but in addition can interact directly with the other three communication pathways (Holzer & Farzi, 2014; Holzer, Hassan, Jain, Reichmann, & Farzi, 2015).

Each of the communication pathways between the gastrointestinal tract and CNS may involve neuropeptides and related molecules. Neuropeptides comprise a class of evolutionarily well-conserved molecules which, by definition, act as transmitters of enteric, peripheral, and central neurons. Apart from their origin, it is frequently difficult to distinguish between their function as neuropeptides or hormones because they operate often via the same receptors and cellular transduction systems. The microbiota residing in the gut is in immediate vicinity to the enteroendocrine cells of the gastrointestinal mucosa which produce more than 20 different gut hormones (Field, Chaudhri, & Bloom, 2010). It is emerging that these endocrine peptides operate as communicators between the gut microbiota and the host. Gut hormone signaling to the brain not only occurs by an endocrine route but may also involve activation of primary afferent neurons, especially in the vagus nerve, by cholecystokinin, ghrelin, and PYY, among others (Field et al., 2010; Holzer et al., 2015).

Although the specific involvement of neuropeptides in the gut microbiota–brain axis has not yet been systematically examined, I will consider six different scenarios that are emerging (Figs. 1 and 2). (i) Neuropeptides and neurotransmitters participate in the mutual communication between microbiota and host. (ii) The gut microbiota controls the availability of

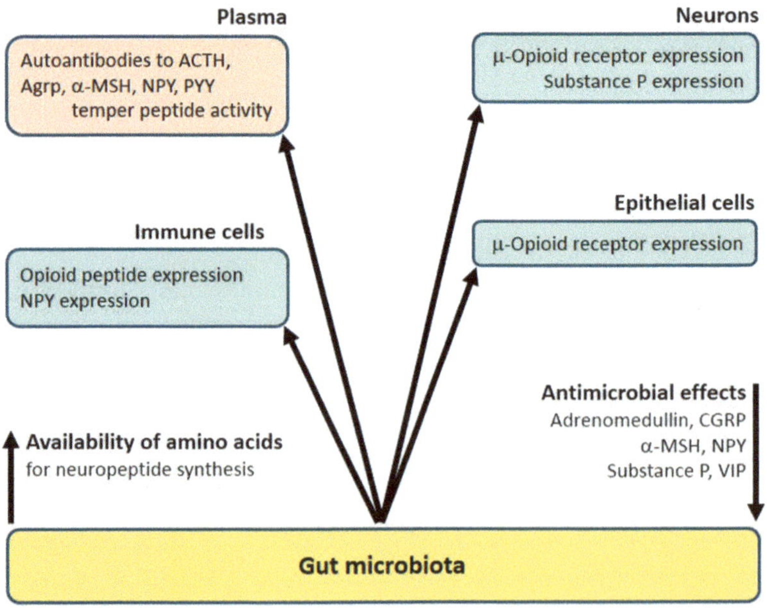

Fig. 1 Neuropeptides and neuroactive peptides in gut microbiota–host interaction in the periphery. Abbreviations: *ACTH*, adrenocorticotropic hormone; *Agrp*, agouti-related peptide; *CGRP*, calcitonin gene-related peptide; *α-MSH*, α-melanocyte-stimulating hormone; *NPY*, neuropeptide Y; *PYY*, peptide YY; *VIP*, vasoactive intestinal polypeptide.

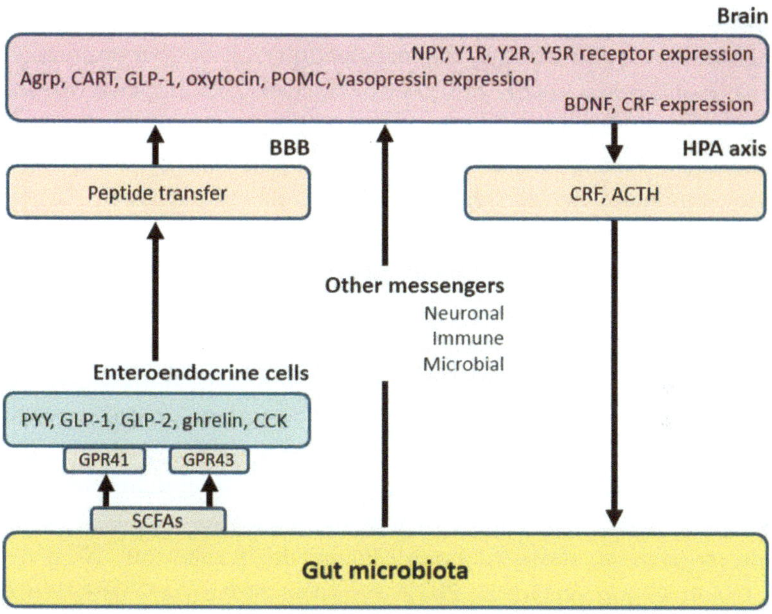

Fig. 2 Neuropeptides and neuroactive gut hormones in the interaction between gut microbiota and brain. Abbreviations: *ACTH*, adrenocorticotropic hormone; *Agrp*, agouti-related peptide; *BBB*, blood–brain barrier; *BDNF*, brain-derived neurotrophic factor; *CART*, cocaine- and amphetamine-regulated transcript; *CCK*, cholecystokinin; *CRF*, corticotropin-releasing factor; *GLP-1*, glucagon-like peptide-1; *GLP-2*, glucagon-like peptide-2; *GPR41*, *GPR43*, G protein-coupled receptors; *HPA*, hypothalamic–pituitary–adrenal; *NPY*, neuropeptide Y; *POMC*, proopiomelanocortin; *PYY*, peptide YY; *SCFAs*, short-chain fatty acids; *Y1R*, Y1 receptor; *Y2R*, Y2 receptor; *Y5R*, Y5 receptor.

amino acids required for the synthesis of neurotransmitters and neuroactive peptides. (iii) Neuropeptide activity is tempered via microbiota-dependent autoantibodies. (iv) The gut microbiota controls the function of the blood–brain barrier (BBB) and in this way affects peptide-mediated information transfer between periphery and brain. (v) Gut hormones are messengers between the gut microbiota and the brain where they act as neuropeptides to alter multiple aspects of brain function and behavior. (vi) Neuropeptides in the brain orchestrate the molecular, behavioral, and autonomic responses to the multiple inputs which the brain receives from the gut microbiota.

3. NEUROTRANSMITTERS AND NEUROPEPTIDES IN GUT MICROBIOTA–HOST COMMUNICATION

Through the sheer abundance of its cells, the gut microbiota acts as a huge metabolic organ which produces and releases compounds that target not only the immune but also the mucosal, endocrine, and neuronal cells of the digestive tract and through the circulation reach out beyond the gut. In view of its influence on the function of distal organs and systems, the gut microbiota resembles an endocrine gland (Clarke et al., 2014). There is reason to assume that a substantial portion of the metabolites present in the circulation originates from the gut microbiota or represents secondary metabolites that have been modified by the activity of the intestinal microbiota (Antunes et al., 2011; Fröhlich et al., 2016; Marcobal et al., 2013; Nicholson et al., 2012; Tremaroli & Bäckhed, 2012; Wikoff et al., 2009). The hormonally and neuronally active messengers derived from the gut microbiota are only beginning to be identified (Clarke et al., 2014). Specifically, the gut microbiota is capable of generating a number of neurotransmitters and neuromodulators including 5-hydroxytryptamine (5-HT, serotonin), noradrenaline, dopamine, histamine, acetylcholine, gamma-aminobutyric acid, and compounds with benzodiazepine-like structures and effects (Barrett, Ross, O'Toole, Fitzgerald, & Stanton, 2012; Clarke et al., 2014; Cryan & Dinan, 2012; Forsythe & Kunze, 2013; Holzer & Farzi, 2014; Nicholson et al., 2012; Yurdaydin et al., 1995). Through these neuroactive mediators the gut microbiota is capable of affecting neuronal systems in the periphery and brain.

Some of the microbial factors derived from the colon have particular effects on visceral nociception and opioid signaling. The probiotic *Lactobacillus acidophilus* NCFM, for example, induces μ-opioid and cannabinoid CB_2 receptors in epithelial cells of the rat colon (Fig. 1), this change going hand in hand with a decrease of the pain reaction to colorectal distension (Rousseaux et al., 2007). A similar effect is seen in patients with functional abdominal pain in whom the induction of μ-opioid receptors in the colon is associated with a trend toward symptom improvement (Ringel-Kulka et al., 2014). In a similar line, local release of opioids from T-helper 1 and 17 cells, which accumulate in response to microbe-derived antigens, ameliorates inflammation-induced visceral hypersensitivity in the mouse (Boué et al., 2014). The observations relating to the opioid system indicate that the gut microbiota has the potential to modulate neuropeptide-mediated transmission.

Although it remains to be established whether the gut microbiota itself can produce neuropeptide-like compounds, this possibility is not unlikely because certain neuropeptide families represent evolutionarily highly conserved messenger systems. It also need be taken into account that the host may control the activity of the gut microbiota with the help of neuropeptides (Fig. 1), a relationship that can be envisaged from in vitro studies of the effects of various biologically active peptides. For instance, the motility and epithelial pathogenicity of the commensal *Pseudomonas fluorescens* is reduced by substance P, serotonin, and adrenaline (Biaggini et al., 2015). Furthermore, substance P, calcitonin gene-related peptide, adrenomedullin, vasoactive intestinal polypeptide, NPY, and α-melanocyte-stimulating hormone (α-MSH) have been reported to exert species- and strain-related antimicrobial effects against various gut bacteria including *Escherichia coli*, *Enterococcus faecalis*, and *L. acidophilus* (Augustyniak, Nowak, & Lundy, 2012; El Karim, Linden, Orr, & Lundy, 2008). It would seem, therefore, that neuropeptides may play a relevant role in mutual microbiota–host homeostasis. This contention is to some extent supported by in vivo studies in which antibiotic-induced gut dysbiosis has been found to be associated with an increase in intestinal substance P expression along with a small increment in inflammatory activity and pain perception (Collins, Verdu, Denou, & Bercik, 2009). A delayed increase in colonic substance P expression has also been reported following colonization of the germ-free mouse intestine (El Aidy, Kunze, Bienenstock, & Kleerebezem, 2012).

There is circumstantial evidence that biologically active peptides also play a role in gut microbiota–immune interaction. This applies, for instance, to NPY which has a distinct impact on immune function, within and outside the gastrointestinal tract (Dimitrijević & Stanojević, 2013; Farzi, Reichmann, & Holzer, 2015; Holzer & Farzi, 2014). NPY released from sympathetic nerve fibers acts on Y receptors expressed by distinct classes of immune cells to modify their activity. In addition, NPY acts as a paracrine and/or autocrine immune mediator (Fig. 1), because immune cells themselves can express and release NPY (Dimitrijević & Stanojević, 2013; Farzi et al., 2015; Holzer & Farzi, 2014). With this activity profile, NPY regulates inflammatory processes in the gut, given that NPY-containing nerve fibers are in close contact with immune cells in the intestinal mucosa (Shibata, Hisajima, Nakano, Goris, & Funakoshi, 2008). Specifically, NPY is able to promote colonic inflammation via activation of Y1 receptors, an implication that is supported by distinct alterations of the gastrointestinal NPY and PYY system in experimentally induced colitis and inflammatory bowel disease (Farzi et al., 2015; Holzer & Farzi, 2014). The proinflammatory effect of NPY could in part be counterregulated by the vasoconstrictor effect of the peptide (Holzer, 2012).

4. MICROBIOTA CONTROLS OF THE AVAILABILITY OF AMINO ACIDS REQUIRED FOR NEUROPEPTIDE SYNTHESIS

Apart from producing and releasing neuroactive factors, the gut microbiota modifies the levels of components that are necessary for the synthesis of transmitters in the nervous system (Fig. 1). This impact is exemplified by studies in germ-free mice in which the plasma concentrations of tryptophan (the precursor of 5-HT), 5-HT, and tyrosine have been found to be elevated (Clarke et al., 2013; Wikoff et al., 2009), the levels of tryptophan and 5-HT being normalized after recolonization of the intestine with murine microbiota (Clarke et al., 2013; El Aidy et al., 2012). The mechanisms whereby the peripheral availability of tryptophan is regulated by gut bacteria are not yet fully understood, but it is likely that utilization and degradation of tryptophan by the microbiota and a redirection of the metabolic pathways of tryptophan in the host play a role (Clarke et al., 2013, 2014; El Aidy et al., 2012; Wikoff et al., 2009). On the one hand, the elevation of circulating tryptophan levels in germ-free mice has been related to the absence of bacterial tryptophanase and the use of tryptophan for microbial indole formation (Clarke et al., 2014; Wikoff et al., 2009). Similarly, the increase in plasma tyrosine seen in germ-free mice is paralleled by reduced metabolism to *p*-cresyl sulfate, this metabolite being absent from germ-free mice (Wikoff et al., 2009) and markedly reduced in antibiotic-treated mice (Fröhlich et al., 2016). On the other hand, the gut microbiota is able to redirect the metabolism of tryptophan toward the production of kynurenine, instead of 5-HT, by enhancing the activity of indoleamine-2,3-dioxygenase which catalyzes the initial step in this pathway. In line with this, the relative plasma levels of kynurenine are low in germ-free mice and increase following recolonization along with a rise of indoleamine-2,3-dioxygenase activity (Clarke et al., 2013; El Aidy et al., 2012).

How the gut microbiota controls the peripheral availability of amino acids other than tryptophan and tyrosine has not systematically been studied, nor is it known whether this modulatory influence of intestinal bacteria on amino acid sources translates to alterations of neuropeptide synthesis in neurons, especially in the CNS. This possibility may be envisaged from the finding that the concentrations of tryptophan, tyrosine (the precursor of dopamine and noradrenaline), and glutamine in the total brain of germ-free mice are lower than in mice that have been recolonized with gut microbiota

(Matsumoto et al., 2013). If particular brain nuclei are analyzed, the turnover of 5-HT, dopamine, and noradrenaline is found elevated in the striatum of germ-free mice (Diaz Heijtz et al., 2011) as are the concentrations of 5-HT and its main metabolite 5-hydroxyindoleacetic acid in the hippocampus (Clarke et al., 2013). Gut dysbiosis and the total absence of gut microbes do alter neuropeptide levels in the CNS as will be discussed later, but it remains unknown how these changes remote from the gut are in fact brought about.

5. INTERACTION OF THE GUT MICROBIOTA WITH NEUROACTIVE GUT HORMONES

Due to its proximity to the gastrointestinal epithelium, the gut microbiota is in a prime position to interact with the enteroendocrine cells of the mucosa. By releasing gut hormones locally and into the blood stream, the enteroendocrine cells convey messages within the digestive system as well as to distant organs including the brain (Fig. 2). The enteroendocrine L cells in the distal ileum and colon represent a particular example of these relationships. These cells are activated by nutrients and other products of digestion, which results in the release of PYY, glucagon-like peptide-1 (GLP-1), and GLP-2 (Holst, 2007; Holzer, Reichmann, & Farzi, 2012; Samuel et al., 2008). L cells are also stimulated by short-chain fatty acids (SCFAs, e.g., acetic, *n*-butyric, and propionic acid) which are generated by microbial fermentation of otherwise indigestible carbohydrate fibers. The interaction between SCFAs and L cells is mediated by G protein-coupled receptors such as GPR41 and GPR43 (Bindels, Dewulf, & Delzenne, 2013; Holst, 2007; Holzer et al., 2012; Samuel et al., 2008). This SCFA-mediated microbiota–L cell communication is confirmed by the finding that colonization of the mouse colon with a fermentative human microbial community increases the plasma level of PYY, an effect that is blunted by knockout of GPR41 (Samuel et al., 2008). GPR41 deficiency is associated with a reduced expression of PYY, an increase in intestinal transit rate, and an attenuation of energy harvest from food (Samuel et al., 2008). The close interaction between gut microbiota and gut hormone-producing cells is further underscored by the observation that germ-free mice have a smaller number of enteroendocrine cells than conventionally colonized animals (Duca, Swartz, Sakar, & Covasa, 2012).

Circulating SCFAs, particularly butyrate and propionate, travel to sites remote from their site of production and can also enter the brain through uptake by monocarboxylate transporters at the BBB (Clarke et al., 2014). Injected systemically to mice, sodium butyrate induces an antidepressant-like behavioral response which is associated with a transient acetylation of histones in the frontal cortex along with an increase in the expression of brain-derived neurotrophic factor (BDNF) (Schroeder, Lin, Crusio, & Akbarian, 2007). Butyrate is also able to ameliorate aging-related memory decline in rats (Reolon et al., 2011), while propionate has been shown to evoke autism spectrum disorder-related behaviors in rodents (MacFabe, Cain, Boon, Ossenkopp, & Cain, 2011; Thomas et al., 2012).

Apart from a direct action in the CNS and other organ systems, SCFAs are likely to utilize gut hormones released from enteroendocrine cells as secondary endocrine messengers (Fig. 2). Following their release from L cells, PYY and GLP-1 inhibit gastric motility, improve glucose homeostasis, induce satiety, and alter behavior (Holst, 2007; Holzer et al., 2012). These effects involve changes in the activity of peripheral and central neurons, which characterizes gut hormones as neuroactive peptides and places them in close vicinity to proper neuropeptides. The interaction between intestinal microbes and gut hormone-releasing cells can be mimicked by prebiotics (fermentable carbohydrates) which in humans increase the plasma concentrations of GLP-1 and PYY, cause satiety, and decrease postprandial glucose levels (Cani et al., 2009). It is likely that appetite-regulating hormones other than PYY, GLP-1, and GLP-2 are also under the influence of the gut microbiota. Emerging evidence indicates that this applies to ghrelin, cholecystokinin, as well as leptin (Duca et al., 2012; Queipo-Ortuño et al., 2013; Schellekens, Finger, Dinan, & Cryan, 2012). Ghrelin which is released from the upper gastrointestinal tract under conditions of hunger reduces both anxiety-like and depression-related behavior (Lutter et al., 2008). Under fed conditions, behavior is changed to a hedonic state as observed when PYY_{3-36} is administered to humans at a dose resulting in postprandial plasma concentrations of the peptide (Batterham et al., 2007). The ability of PYY to promote hedonic behavior is supported by the finding that knockout of PYY increases depression-like behavior but does not alter anxiety (Painsipp, Herzog, Sperk, & Holzer, 2011). Physiologically, however, emotion and mood under fed conditions will be shaped by the integrated effect of all gut hormones that are released postprandially. Pancreatic polypeptide, a hormone structurally related to PYY and NPY, likewise suppresses food intake and through activation of cerebral Y4 receptors promotes extinction and suppression of fear (Verma et al., 2016). Like PYY (Painsipp et al., 2011), GLP-2 attenuates depression-like behavior (Iwai et al., 2009), whereas GLP-1 enhances anxiety-related behavior (Kinzig et al., 2003; Möller et al., 2002).

6. CONTROL OF NEUROPEPTIDE ACTIVITY VIA GUT MICROBIOTA-DEPENDENT AUTOANTIBODIES

There is mounting evidence that autoantibodies against several neuropeptides, generated under the influence of the gut microbiota, contribute to disturbances of appetite, emotion, mood, and conduct. This indirect microbe–neuropeptide link seems to be a reflection of the important role of the gut microbiota in educating the immune system to recognize foreign antigens and tolerate commensal microbes (Lathrop et al., 2011). The gut microbiota–autoantibody link was first envisaged from studies in which IgG and IgA autoantibodies against α-MSH, NPY, PYY, agouti-related peptide (Agrp), ghrelin, leptin, and other neuropeptides/peptides involved in appetite control have been found to occur in the human blood (Fetissov, Hamze Sinno, Coëffier, et al., 2008; Fetissov et al., 2005). Several intestinal microbes including *Lactobacillus*, *Bacteroides*, *Helicobacter pylori*, *E. coli*, and *Candida* species contain proteins that have amino acid sequences identical with those of many appetite-regulating peptides (Fetissov, Hamze Sinno, Coquerel, et al., 2008). The circulating levels of autoantibodies against α-MSH, which are elevated in anorexia nervosa and bulimia nervosa, correlate with the psychobehavioral abnormalities accompanying these eating disorders (Fetissov et al., 2005). A role of the gut microbiota in the generation of peptide autoantibodies is supported by the findings that germ-free rats have decreased levels of circulating IgA autoantibodies against several appetite-regulating peptides, while the levels of antighrelin IgG are increased (Fetissov, Hamze Sinno, Coquerel, et al., 2008). Further analyses in rats have confirmed that α-MSH autoantibodies are involved in the regulation of feeding and anxiety (Fetissov, Hamze Sinno, Coquerel, et al., 2008). These observations indicate that the gut microbiota controls appetite and emotional behavior indirectly by provoking the formation of autoantibodies against neuropeptides involved in food intake control (Fig. 1).

Neuropeptide autoantibodies may in addition be implicated in psychopathologies such as depression and conduct disorder. NPY autoantibodies, for instance, have been proposed to contribute to alterations of appetite in depressive disorder (Garcia et al., 2012). Although the levels of anti-NPY IgG do not significantly differ between patients with depression and controls, the NPY autoantibodies found in depressed patients exhibit enhanced affinity for NPY, which is associated with a decreased body mass index (Garcia et al., 2012). Transfer of these high-affinity NPY autoantibodies to the mouse brain blunts the orexigenic response to NPY (Garcia et al., 2012). Elevated levels of ACTH-reactive immunoglobulins have been found in males with conduct disorder (Fetissov et al., 2006). This finding has been further elaborated in male adolescents in whom higher anti-ACTH IgG levels are associated with higher antisocial behavior scores (Schaefer et al., 2013). In addition, the cortisol release evoked by a social stress test is related to the anti-ACTH IgG levels. Taken together, there is reason to assume that ACTH autoantibodies play a role in the biology of antisocial behavior and conduct disorder (Schaefer et al., 2013).

7. CONTROL OF PEPTIDE SIGNALING THROUGH A GUT MICROBIOTA–BBB INTERACTION

Although the effect of neuroactive gut hormones on food intake control, energy homeostasis, and emotional-affective behavior indicates that these peptides can gain access to the brain, the BBB represents a significant barrier that limits the cerebral in- and outflow of peptides. The transport of peptides into, and out of, the brain is a multifactorial process (Banks & Kastin, 1992; Pan & Kastin, 2004), changes of which are likely to affect gut–brain as well as brain–gut signaling via peptidic and nonpeptidic messengers. However, the BBB is no longer considered a simple barrier but a regulatory interface that is essential for normal brain functioning and for that reason is in cross talk with the brain and other systems in the body (Banks, 2016). There is now evidence that the gut microbiota is among those systems that control the development and function of the BBB and that hence BBB permeability represents another potential interface for microbe–neuropeptide interactions (Fig. 2).

The impact of gut microbes on BBB functioning has been disclosed in germ-free mice which, beginning with intrauterine life, exhibit enhanced BBB permeability that is maintained into adulthood (Braniste et al., 2014). Increased BBB permeability is associated with reduced expression of the tight junction proteins occludin and claudin-5 in frontal cortex, striatum, and hippocampus (Braniste et al., 2014). A decrease in the expression of these tight junction proteins in the murine hippocampus, but not amygdala, prefrontal cortex, and hypothalamus, has likewise been found after antibiotic-induced gut dysbiosis (Fröhlich et al., 2016). Recolonization of the intestine of germ-free adult mice with a normal gut microbiota attenuates BBB permeability and upregulates the expression of tight junction proteins in the brain. This effect of recolonization is reproduced by administration of the SCFA butyrate (Braniste et al., 2014). The implications which the microbial control of BBB development and function has in health and disease have not yet been systematically explored, but they have very likely an impact on gut–brain and brain–gut communication.

Cerebral neuropeptides may themselves be involved in the regulation of the BBB. For instance, corticotropin-releasing factor (CRF) appears to be involved in the increase in BBB permeability evoked by stress (Theoharides & Konstantinidou, 2007). Stress causes a local release of CRF which, in turn, activates brain mast cells to release proinflammatory interleukins and vascular endothelial growth factor, this signaling cascade enhancing BBB permeability (Theoharides & Konstantinidou, 2007).

8. CEREBRAL NEUROPEPTIDES MEDIATING THE IMPACT OF THE GUT MICROBIOTA ON BRAIN FUNCTION AND BEHAVIOR

Accumulating evidence shows that the absence or disturbance of the gut microbiota has a significant impact on brain function and behavior including neurogenesis, stress responsiveness, locomotion, anxiety, sociability, learning, and memory (Fröhlich et al., 2016; Luczynski et al., 2016). The neurochemical correlates of these alterations in the brain are also beginning to unfold, but there is still limited information as to which neuropeptides are involved. Nevertheless, there is information that BDNF, CRF, GLP-1, the NPY system, Agrp, proopiomelanocortin, cocaine- and amphetamine-regulated transcript, vasopressin, and oxytocin play a role (Fig. 2).

8.1 Brain-Derived Neurotrophic Factor

BDNF is not a neuropeptide in its strict sense but a neurotrophic factor which through promoting the growth, development, and survival of neurons impacts on neuron-to-neuron signaling (Benarroch, 2015; Park & Poo, 2013). Several studies have shown that the gut microbiota affects the expression of BDNF in distinct brain nuclei, although the results are inconsistent. On the one hand, BDNF expression in the cortex, hippocampus, amygdala, and hypothalamus has been found decreased in germ-free mice (Clarke et al., 2013; Diaz Heijtz et al., 2011; Gareau et al., 2011; Sudo et al., 2004) while, on the other hand, an upregulation of BDNF in the hippocampus, hypothalamus, and brain stem of germ-free mice has also been reported (Neufeld, Kang, Bienenstock, & Foster, 2011; Schéle et al., 2013). Antibiotic-induced gut dysbiosis is likewise associated with a decrease of BDNF expression in the medial prefrontal cortex, hippocampus, and hypothalamus (Desbonnet et al., 2015; Fröhlich et al., 2016), although another study found BDNF upregulated in the hippocampus and down-regulated in the amygdala of antibiotic-treated mice (Bercik et al., 2011). To date, there has been no attempt, genetically or pharmacologically, to establish a causal relationship between the changes in BDNF expression and particular functional and behavioral alterations seen in germ-free and antibiotic-treated animals.

8.2 NPY System

NPY is one of the most abundant neuropeptides in the brain (Morales-Medina, Dumont, & Quirion, 2010), and there is ample evidence that this neuropeptide family plays a role at several levels of the gut–brain axis (Holzer et al., 2012). This implication also extends to gut microbiota–brain communication, given that the expression of NPY and some of its receptors is significantly altered by antibiotic-induced gut dysbiosis (Fröhlich et al., 2016). NPY, PYY, and pancreatic polypeptide form a family of peptides, the biological actions of which are mediated by multiple G protein-coupled Y receptors, the most relevant in mammals being the Y1, Y2, Y4, and Y5 receptors (Holzer & Farzi, 2014). This activity profile explains why PYY acts both as a gut hormone and as a neuroactive peptide in the brain, in this way subserving a number of functions beyond the induction of satiety (Painsipp et al., 2011; Stadlbauer, Woods, Langhans, & Meyer, 2015). The involvement of the NPY system in the control of many CNS functions including food intake, anxiety, mood, stress resilience, and cognition (Farzi et al., 2015; Fröhlich et al., 2016; Morales-Medina et al., 2010; Reichmann & Holzer, 2016) implies that dysbiosis-evoked alterations in the expression of NPY and its receptors could have multiple functional consequences in the CNS, although any causal relationship awaits to be proved.

Antibiotic-induced disruption of the intestinal microbial community leads to a specific increment of NPY mRNA expression in the amygdala and hypothalamus (Fröhlich et al., 2016). This finding is in keeping with data from germ-free mice in which an increase in the expression of NPY and Agrp in the hypothalamus has been observed (Schéle et al., 2013). On the one hand, hypothalamic upregulation of NPY could be considered as a measure counterbalancing the decrease of food intake caused by antibiotic-induced gut dysbiosis. On the other hand, the increased NPY expression in the hypothalamus and amygdala could also be thought of as a reaction to the gut dysbiosis-evoked rise of plasma corticosterone (Fröhlich et al., 2016), which in other studies has been found to amplify anxiety- and depression-like behavior (Kutiyanawalla, Terry, & Pillai, 2011; Lee, Shim, Lee, Yang, & Hahm, 2009). Given that NPY is able to suppress anxiety- and depression-related behavior (Morales-Medina et al., 2010; Painsipp et al., 2011; Sajdyk, Vandergriff, & Gehlert, 1999; Stogner & Holmes, 2000; Tasan et al., 2016; Thorsell et al., 2000), the enhanced expression of NPY mRNA in the hypothalamus and amygdala of antibiotic-treated mice may prevent any increase in anxiety- and depression-like behavior, which otherwise would result from the increase in corticosterone levels. In these implications, NPY is likely to interact with other neuronal signaling molecules in the brain (Fröhlich et al., 2016). For instance, the gut dysbiosis-induced decrease of the hypothalamic BDNF expression may be related to the enhanced NPY expression in this area (Gelfo et al., 2012).

Antibiotic-induced gut dysbiosis impacts not only on the cerebral transcription of NPY but also on that of Y1, Y2, and Y5 receptors in distinct areas of the mouse brain. The enhanced expression of NPY in the amygdala goes in parallel with a diminished expression of Y5 receptors, a change that it is not yet understood in functional terms (Fröhlich et al., 2016). In contrast, in the hippocampus of antibiotic-treated mice, in which there is no change in NPY expression, a specific downregulation of Y1 and Y2 receptors has been observed (Fröhlich et al., 2016). Although the functional relevance awaits to be disclosed, these changes attest to a dynamic regulation of the cerebral NPY system in response to gut dysbiosis. Both Y1 receptors, mostly located postsynaptically, and Y2 receptors, preferentially located presynaptically, are known to play an important role in the maintenance of emotional-affective homeostasis, stress resilience, and cognition (Farzi et al., 2015; Morales-Medina et al., 2010; Reichmann & Holzer, 2016). It should not go unnoticed in this context that a similar deficit in object recognition memory, as found in antibiotic-treated mice (Fröhlich et al., 2016), has been observed in Y2 receptor knockout mice (Painsipp et al., 2008; Redrobe, Dumont, Herzog, & Quirion, 2004). Hence, hippocampal downregulation of both BDNF and Y2 receptors may contribute to the cognitive impairment induced by gut dysbiosis.

8.3 Corticotropin-Releasing Factor

A further neuropeptide responding to alterations of the gut microbiota is CRF, which among other functions is an essential mediator of the hypothalamic–pituitary–adrenal (HPA) axis and the stress response mediated by this system. This contention is in keeping with the involvement of the gut microbiota in stress-mediated alterations of behavior as well as gastrointestinal function. The absence of bacteria in germ-free mice increases the neuroendocrine response (circulating ACTH and corticosterone) to acute restraint stress in adulthood (Sudo et al., 2004). Both in adult germ-free mice (Sudo et al., 2004) and germ-free F344 rats (Crumeyrolle-Arias et al., 2014), the expression of CRF mRNA in the hypothalamus is elevated when compared with control animals. The exaggerated stress response of germ-free mice is partly corrected by recolonization of the gut with fecal microbiota at an early stage of development, which suggests that the presence of the gut microbiota is required for the HPA system to become fully susceptible to inhibitory neural regulation (Sudo et al., 2004). Prenatal stress which leads to a long-lasting alteration of the intestinal microbiota composition also causes an exaggeration of the HPA axis response to stress, along with an impairment of cognitive function (Golubeva et al., 2015). In contrast, gut dysbiosis induced by antibiotic treatment of mice from weaning onward does not alter CRF expression in the hypothalamus of the adult brain, nor is the expression of CRF, oxytocin, and vasopressin altered by acute restraint stress (Desbonnet et al., 2015).

A link between the gut microbiota and cerebral CRF expression can in addition be deduced from a murine model of chronic depression induced by bilateral olfactory bulbectomy (Park et al., 2013). This model of sustained depression- and anxiety-related behavior is associated with a change in the intestinal microbial profile and an increase in hypothalamic CRF expression (Park et al., 2013). Intracerebral administration of CRF is able to reproduce the same changes in behavior and intestinal microbial community as bilateral olfactory bulbectomy, which suggests that the HPA system provides an interface between the behavioral and intestinal disturbances (Park et al., 2013). This relationship is further supported by the findings that stress-induced changes in gut microbiota composition and inflammasome inhibition are mediated by CRF (Sun et al., 2013) and that chronic peripheral administration of CRF causes colonic barrier dysfunction (increase in mucosal permeability and secretion) as does psychological stress (Teitelbaum, Gareau, Jury, Yang, & Perdue, 2008).

8.4 Other Neuropeptides

There is some evidence that neuropeptides other than BDNF, NPY, and CRF likewise play a role in gut microbiota–brain communication as deduced from comparative peptide expression studies in control and germ-free mice. While the expression of NPY and Agrp in the hypothalamus and that of the GLP-1 precursor proglucagon in the brain stem is enhanced in germ-free mice, the expression of proopiomelanocortin and cocaine- and amphetamine-regulated transcript in the hypothalamus is attenuated in the absence of the gut microbiota (Schéle et al., 2013). These data have been related to the metabolic disturbances observed in germ-free mice which have a reduced fat mass and respond to leptin with a more pronounced reduction of weight and NPY/Agrp expression (Schéle et al., 2013). Studies of antibiotic-treated mice extend the spectrum of cerebral neuropeptides that are potentially sensitive to gut dysbiosis. Thus, antibiotic treatment of mice from weaning onward induces cognitive deficits, reduces anxiety, and attenuates the expression of oxytocin and vasopressin in the hypothalamus and that of BDNF in the hippocampus of the adult brain (Desbonnet et al., 2015).

9. CONCLUSION: THE GUT MICROBIOTA–NEUROPEPTIDE NETWORK

As is emerging from this overview, neuropeptides contribute to gut microbiota–brain communication at several levels of the multiple signaling pathways between the two systems. Relative to fast-acting small-molecule transmitters, neuropeptides are specialized in the long-lasting fine-tuning of regulatory interactions (Strand, 1999), a mode of action that is characteristic of the gut microbiota–brain interaction. While increasing evidence shows that the expression of neuropeptides, notably BDNF, NPY, and CRF, in distinct brain areas is altered in relation to a change or the absence of the gut microbiota, studies confirming a causal implication of neuropeptides in the concomitant functional and behavioral alterations are still lacking. Nevertheless, the dynamic changes in neuropeptide and neuropeptide receptor expression in the brain of dysbiotic and germ-free animals attest to a profound impact of the gut microbiota on this type of messenger molecules. Neuropeptides in the brain may be envisaged to orchestrate the molecular, functional, behavioral, and autonomic reactions that take place in response to alterations of the gut microbial community (Collins,

Surette, & Bercik, 2012; Dinan & Cryan, 2015; Forsythe & Kunze, 2013; Sampson & Mazmanian, 2015). In this role, cerebral neuropeptides are complemented by gut hormones, the synthesis and release of which is directly or indirectly controlled by the gut microbiota. In their action as messengers of the gut microbiota, gut hormones subserve both an endocrine and neuropeptide function as they target receptors that are also operated by brain neuropeptides, an example being the Y receptor types that are activated by NPY, PYY, and pancreatic polypeptide with differential selectivity. In this perspective, neuroactive gut hormones and cerebral neuropeptides are of particular relevance to the analysis of how the gut microbiota and brain interact with each other in health and disease.

ACKNOWLEDGMENTS

This work was supported by the Austrian Science Fund (FWF grant P25912-B23) and BioTechMed-Graz.

REFERENCES

Antunes, L. C., Han, J., Ferreira, R. B., Lolić, P., Borchers, C. H., & Finlay, B. B. (2011). Effect of antibiotic treatment on the intestinal metabolome. *Antimicrobial Agents and Chemotherapy*, *55*, 1494–1503.

Augustyniak, D., Nowak, J., & Lundy, F. T. (2012). Direct and indirect antimicrobial activities of neuropeptides and their therapeutic potential. *Current Protein & Peptide Science*, *13*, 723–738.

Banks, W. A. (2016). From blood-brain barrier to blood-brain interface: New opportunities for CNS drug delivery. *Nature Reviews. Drug Discovery*, *15*, 275–292.

Banks, W. A., & Kastin, A. J. (1992). Bidirectional passage of peptides across the blood-brain barrier. *Progress in Brain Research*, *91*, 139–148.

Barrett, E., Ross, R. P., O'Toole, P. W., Fitzgerald, G. F., & Stanton, C. (2012). gamma-Aminobutyric acid production by culturable bacteria from the human intestine. *Journal of Applied Microbiology*, *113*, 411–417.

Batterham, R. L., ffytche, D. H., Rosenthal, J. M., Zelaya, F. O., Barker, G. J., Withers, D. J., et al. (2007). PYY modulation of cortical and hypothalamic brain areas predicts feeding behaviour in humans. *Nature*, *450*, 106–109.

Benarroch, E. E. (2015). Brain-derived neurotrophic factor: Regulation, effects, and potential clinical relevance. *Neurology*, *84*, 1693–1704.

Bercik, P., Denou, E., Collins, J., Jackson, W., Lu, J., Jury, J., et al. (2011). The intestinal microbiota affect central levels of brain-derived neurotrophic factor and behavior in mice. *Gastroenterology*, *141*, 599–609.

Biaggini, K., Barbey, C., Borrel, V., Feuilloley, M., Déchelotte, P., & Connil, N. (2015). The pathogenic potential of Pseudomonas fluorescens MFN1032 on enterocytes can be modulated by serotonin, substance P and epinephrine. *Archives of Microbiology*, *197*, 983–990.

Bindels, L. B., Dewulf, E. M., & Delzenne, N. M. (2013). GPR43/FFA2: Physiopathological relevance and therapeutic prospects. *Trends in Pharmacological Sciences*, *34*, 226–232.

Boué, J., Basso, L., Cenac, N., Blanpied, C., Rolli-Derkinderen, M., Neunlist, M., et al. (2014). Endogenous regulation of visceral pain via production of opioids by colitogenic $CD4^+$ T cells in mice. *Gastroenterology, 146*, 166–175.
Braniste, V., Al-Asmakh, M., Kowal, C., Anuar, F., Abbaspour, A., Tóth, M., et al. (2014). The gut microbiota influences blood-brain barrier permeability in mice. *Science Translational Medicine, 6*, 263ra158.
Burbach, J. P. (2010). Neuropeptides from concept to online database www.neuropeptides.nl. *European Journal of Pharmacology, 626*, 27–48.
Cani, P. D., Lecourt, E., Dewulf, E. M., Sohet, F. M., Pachikian, B. D., Naslain, D., et al. (2009). Gut microbiota fermentation of prebiotics increases satietogenic and incretin gut peptide production with consequences for appetite sensation and glucose response after a meal. *The American Journal of Clinical Nutrition, 90*, 1236–1243.
Clarke, G., Grenham, S., Scully, P., Fitzgerald, P., Moloney, R. D., Shanahan, F., et al. (2013). The microbiome-gut-brain axis during early-life regulates the hippocampal serotonergic system in a gender-dependent manner. *Molecular Psychiatry, 18*, 666–673.
Clarke, G., Stilling, R. M., Kennedy, P. J., Stanton, C., Cryan, J. F., & Dinan, T. G. (2014). Minireview: Gut microbiota: The neglected endocrine organ. *Molecular Endocrinology, 28*, 1221–1238.
Collins, S. M., Surette, M., & Bercik, P. (2012). The interplay between the intestinal microbiota and the brain. *Nature Reviews. Microbiology, 10*, 735–742.
Collins, S., Verdu, E., Denou, E., & Bercik, P. (2009). The role of pathogenic microbes and commensal bacteria in irritable bowel syndrome. *Digestive Diseases, 27*(Suppl. 1), 85–89.
Crumeyrolle-Arias, M., Jaglin, M., Bruneau, A., Vancassel, S., Cardona, A., Daugé, V., et al. (2014). Absence of the gut microbiota enhances anxiety-like behavior and neuroendocrine response to acute stress in rats. *Psychoneuroendocrinology, 42*, 207–217.
Cryan, J. F., & Dinan, T. G. (2012). Mind-altering microorganisms: The impact of the gut microbiota on brain and behaviour. *Nature Reviews. Neuroscience, 13*, 701–712.
Desbonnet, L., Clarke, G., Traplin, A., O'Sullivan, O., Crispie, F., Moloney, R. D., et al. (2015). Gut microbiota depletion from early adolescence in mice: Implications for brain and behaviour. *Brain, Behavior, and Immunity, 48*, 165–173.
Diaz Heijtz, R., Wang, S., Anuar, F., Qian, Y., Björkholm, B., Samuelsson, A., et al. (2011). Normal gut microbiota modulates brain development and behavior. *Proceedings of the National Academy of Sciences of the United States of America, 108*, 3047–3052.
Dimitrijević, M., & Stanojević, S. (2013). The intriguing mission of neuropeptide Y in the immune system. *Amino Acids, 45*, 41–53.
Dinan, T. G., & Cryan, J. F. (2015). The impact of gut microbiota on brain and behaviour: Implications for psychiatry. *Current Opinion in Clinical Nutrition and Metabolic Care, 18*, 552–558.
Duca, F. A., Swartz, T. D., Sakar, Y., & Covasa, M. (2012). Increased oral detection, but decreased intestinal signaling for fats in mice lacking gut microbiota. *PLoS One*, 7, e39748.
El Aidy, S., Kunze, W., Bienenstock, J., & Kleerebezem, M. (2012). The microbiota and the gut-brain axis: Insights from the temporal and spatial mucosal alterations during colonisation of the germfree mouse intestine. *Beneficial Microbes, 3*, 251–259.
El Karim, I. A., Linden, G. J., Orr, D. F., & Lundy, F. T. (2008). Antimicrobial activity of neuropeptides against a range of micro-organisms from skin, oral, respiratory and gastrointestinal tract sites. *Journal of Neuroimmunology, 200*, 11–16.
Farzi, A., Reichmann, F., & Holzer, P. (2015). The homeostatic role of neuropeptide Y in immune function and its impact on mood and behaviour. *Acta Physiologica (Oxford, England), 213*, 603–627.
Fetissov, S. O., Hallman, J., Nilsson, I., Lefvert, A. K., Oreland, L., & Hökfelt, T. (2006). Aggressive behavior linked to corticotropin-reactive autoantibodies. *Biological Psychiatry, 60*, 799–802.

Fetissov, S. O., Hamze Sinno, M., Coëffier, M., Bole-Feysot, C., Ducrotté, P., Hökfelt, T., et al. (2008). Autoantibodies against appetite-regulating peptide hormones and neuropeptides: Putative modulation by gut microflora. *Nutrition, 24*(4), 348–359.

Fetissov, S. O., Hamze Sinno, M., Coquerel, Q., Do Rego, J. C., Coëffier, M., Gilbert, D., et al. (2008). Emerging role of autoantibodies against appetite-regulating neuropeptides in eating disorders. *Nutrition, 24*, 854–859.

Fetissov, S. O., Harro, J., Jaanisk, M., Järv, A., Podar, I., Allik, J., et al. (2005). Autoantibodies against neuropeptides are associated with psychological traits in eating disorders. *Proceedings of the National Academy of Sciences of the United States of America, 102*, 14865–14870.

Field, B. C., Chaudhri, O. B., & Bloom, S. R. (2010). Bowels control brain: Gut hormones and obesity. *Nature Reviews Endocrinology, 6*, 444–453.

Forsythe, P., & Kunze, W. A. (2013). Voices from within: Gut microbes and the CNS. *Cellular and Molecular Life Sciences, 70*, 55–69.

Fröhlich, E. E., Farzi, A., Mayerhofer, R., Reichmann, F., Jačan, A., Wagner, B., et al. (2016). Cognitive impairment by antibiotic-induced gut dysbiosis: Analysis of gut microbiota-brain communication. *Brain, Behavior, and Immunity, 56*, 140–155.

Garcia, F. D., Coquerel, Q., do Rego, J. C., Cravezic, A., Bole-Feysot, C., Kiive, E., et al. (2012). Anti-neuropeptide Y plasma immunoglobulins in relation to mood and appetite in depressive disorder. *Psychoneuroendocrinology, 37*, 1457–1467.

Gareau, M. G., Wine, E., Rodrigues, D. M., Cho, J. H., Whary, M. T., Philpott, D. J., et al. (2011). Bacterial infection causes stress-induced memory dysfunction in mice. *Gut, 60*, 307–317.

Gelfo, F., Tirassa, P., De Bartolo, P., Croce, N., Bernardini, S., Caltagirone, C., et al. (2012). NPY intraperitoneal injections produce antidepressant-like effects and downregulate BDNF in the rat hypothalamus. *CNS Neuroscience & Therapeutics, 18*, 487–492.

Golubeva, A. V., Crampton, S., Desbonnet, L., Edge, D., O'Sullivan, O., Lomasney, K. W., et al. (2015). Prenatal stress-induced alterations in major physiological systems correlate with gut microbiota composition in adulthood. *Psychoneuroendocrinology, 60*, 58–74.

Holst, J. J. (2007). The physiology of glucagon-like peptide 1. *Physiological Reviews, 87*, 1409–1439.

Holzer, P. (2012). Neural regulation of gastrointestinal blood flow. In L. R. Johnson (Ed.), *Physiology of the gastrointestinal tract* (5th ed.). Oxford: Academic Press.

Holzer, P., & Farzi, A. (2014). Neuropeptides and the microbiota-gut-brain axis. *Adv. Exp. Med. Biol., 817*, 195–219.

Holzer, P., Hassan, A. M., Jain, P., Reichmann, F., & Farzi, A. (2015). Neuroimmune pharmacological approaches. *Current Opinion in Pharmacology, 25*, 13–22.

Holzer, P., Reichmann, F., & Farzi, A. (2012). Neuropeptide Y, peptide YY and pancreatic polypeptide in the gut-brain axis. *Neuropeptides, 46*, 261–274.

Iwai, T., Hayashi, Y., Narita, S., Kasuya, Y., Jin, K., Tsugane, M., et al. (2009). Antidepressant-like effects of glucagon-like peptide-2 in mice occur via monoamine pathways. *Behavioural Brain Research, 204*, 235–240.

Kastin, A. J. (2013). *Handbook of biologically active peptides* (2nd ed.). . San Diego: Academic Press.

Kinzig, K. P., D'Alessio, D. A., Herman, J. P., Sakai, R. R., Vahl, T. P., Figueiredo, H. F., et al. (2003). CNS glucagon-like peptide-1 receptors mediate endocrine and anxiety responses to interoceptive and psychogenic stressors. *The Journal of Neuroscience, 23*, 6163–6170.

Kovács, G. L., & De Wied, D. (1994). Peptidergic modulation of learning and memory processes. *Pharmacological Reviews, 46*, 269–291.

Kutiyanawalla, A., Terry, A. V., & Pillai, A. (2011). Cysteamine attenuates the decreases in TrkB protein levels and the anxiety/depression-like behaviors in mice induced by corticosterone treatment. *PLoS One*, *6*, e26153.

Lathrop, S. K., Bloom, S. M., Rao, S. M., Nutsch, K., Lio, C. W., Santacruz, N., et al. (2011). Peripheral education of the immune system by colonic commensal microbiota. *Nature*, *478*, 250–254.

Lee, B., Shim, I., Lee, H. J., Yang, Y., & Hahm, D. H. (2009). Effects of acupuncture on chronic corticosterone-induced depression-like behavior and expression of neuropeptide Y in the rats. *Neuroscience Letters*, *453*, 151–156.

Luczynski, P., McVey Neufeld, K. A., Oriach, C. S., Clarke, G., Dinan, T. G., & Cryan, J. F. (2016). Growing up in a bubble: Using germ-free animals to assess the influence of the gut microbiota on brain and behavior. *The International Journal of Neuropsychopharmacology*, *19*, 1–17.

Lutter, M., Sakata, I., Osborne-Lawrence, S., Rovinsky, S. A., Anderson, J. G., Jung, S., et al. (2008). The orexigenic hormone ghrelin defends against depressive symptoms of chronic stress. *Nature Neuroscience*, *11*, 752–753.

MacFabe, D. F., Cain, N. E., Boon, F., Ossenkopp, K. P., & Cain, D. P. (2011). Effects of the enteric bacterial metabolic product propionic acid on object-directed behavior, social behavior, cognition, and neuroinflammation in adolescent rats: Relevance to autism spectrum disorder. *Behavioural Brain Research*, *217*, 47–54.

Marcobal, A., Kashyap, P. C., Nelson, T. A., Aronov, P. A., Donia, M. S., Spormann, A., et al. (2013). A metabolomic view of how the human gut microbiota impacts the host metabolome using humanized and gnotobiotic mice. *The ISME Journal*, *7*, 1933–1943.

Matsumoto, M., Kibe, R., Ooga, T., Aiba, Y., Sawaki, E., Koga, Y., et al. (2013). Cerebral low-molecular metabolites influenced by intestinal microbiota: A pilot study. *Frontiers in Systems Neuroscience*, *7*, 9.

Mayer, E. A., & Tillisch, K. (2011). The brain-gut axis in abdominal pain syndromes. *Annual Review of Medicine*, *62*, 381–396.

Möller, C., Sommer, W., Thorsell, A., Rimondini, R., & Heilig, M. (2002). Anxiogenic-like action of centrally administered glucagon-like peptide-1 in a punished drinking test. *Progress in Neuro-Psychopharmacology & Biological Psychiatry*, *26*, 119–122.

Morales-Medina, J. C., Dumont, Y., & Quirion, R. (2010). A possible role of neuropeptide Y in depression and stress. *Brain Research*, *1314*, 194–205.

Neufeld, K. M., Kang, N., Bienenstock, J., & Foster, J. A. (2011). Reduced anxiety-like behavior and central neurochemical change in germ-free mice. *Neurogastroenterology and Motility*, *23*, 255–264. e119.

Nicholson, J. K., Holmes, E., Kinross, J., Burcelin, R., Gibson, G., Jia, W., et al. (2012). Host-gut microbiota metabolic interactions. *Science*, *336*, 1262–1267.

Painsipp, E., Herzog, H., Sperk, G., & Holzer, P. (2011). Sex-dependent control of murine emotional-affective behaviour in health and colitis by peptide YY and neuropeptide Y. *British Journal of Pharmacology*, *163*, 1302–1314.

Painsipp, E., Wultsch, T., Edelsbrunner, M. E., Tasan, R. O., Singewald, N., Herzog, H., et al. (2008). Reduced anxiety-like and depression-related behavior in neuropeptide Y Y4 receptor knockout mice. *Genes, Brain, and Behavior*, *7*, 532–542.

Pan, W., & Kastin, A. J. (2004). Polypeptide delivery across the blood-brain barrier. *Current Drug Targets CNS and Neurological Disorders*, *3*, 131–136.

Park, A. J., Collins, J., Blennerhassett, P. A., Ghia, J. E., Verdu, E. F., Bercik, P., et al. (2013). Altered colonic function and microbiota profile in a mouse model of chronic depression. *Neurogastroenterology and Motility*, *25*. 733-e575.

Park, H., & Poo, M. (2013). Neurotrophin regulation of neural circuit development and function. *Nature Reviews. Neuroscience*, *14*, 7–23.

Queipo-Ortuño, M. I., Seoane, L. M., Murri, M., Pardo, M., Gomez-Zumaquero, J. M., Cardona, F., et al. (2013). Gut microbiota composition in male rat models under different nutritional status and physical activity and its association with serum leptin and ghrelin levels. *PLoS One*, *8*, e65465.

Redrobe, J. P., Dumont, Y., Herzog, H., & Quirion, R. (2004). Characterization of neuropeptide Y, Y_2 receptor knockout mice in two animal models of learning and memory processing. *Journal of Molecular Neuroscience*, *22*, 159–166.

Reichmann, F., & Holzer, P. (2016). Neuropeptide Y: A stressful review. *Neuropeptides*, *55*, 99–109.

Reolon, G. K., Maurmann, N., Werenicz, A., Garcia, V. A., Schröder, N., Wood, M. A., et al. (2011). Posttraining systemic administration of the histone deacetylase inhibitor sodium butyrate ameliorates aging-related memory decline in rats. *Behavioural Brain Research*, *221*, 329–332.

Ringel-Kulka, T., Goldsmith, J. R., Carroll, I. M., Barros, S. P., Palsson, O., Jobin, C., et al. (2014). *Lactobacillus acidophilus* NCFM affects colonic mucosal opioid receptor expression in patients with functional abdominal pain—A randomised clinical study. *Alimentary Pharmacology & Therapeutics*, *40*, 200–207.

Rousseaux, C., Thuru, X., Gelot, A., Barnich, N., Neut, C., Dubuquoy, L., et al. (2007). *Lactobacillus acidophilus* modulates intestinal pain and induces opioid and cannabinoid receptors. *Nature Medicine*, *13*, 35–37.

Sajdyk, T. J., Vandergriff, M. G., & Gehlert, D. R. (1999). Amygdalar neuropeptide Y Y1 receptors mediate the anxiolytic-like actions of neuropeptide Y in the social interaction test. *European Journal of Pharmacology*, *368*, 143–147.

Sampson, T. R., & Mazmanian, S. K. (2015). Control of brain development, function, and behavior by the microbiome. *Cell Host & Microbe*, *17*, 565–576.

Samuel, B. S., Shaito, A., Motoike, T., Rey, F. E., Backhed, F., Manchester, J. K., et al. (2008). Effects of the gut microbiota on host adiposity are modulated by the short-chain fatty-acid binding G protein-coupled receptor, Gpr41. *Proceedings of the National Academy of Sciences of the United States of America*, *105*, 16767–16772.

Sandman, C. A., Miller, L. H., & Kastin, A. J. (1977). Neuropeptide influences on the brain and behavior: Introduction: Perspectives on the behavioral effects of the neuropeptides. *Advances in Biochemical Psychopharmacology*, *17*, 1–9.

Schaefer, J. M., Fetissov, S. O., Legrand, R., Claeyssens, S., Hoekstra, P. J., Verhulst, F. C., et al. (2013). Corticotropin (ACTH)-reactive immunoglobulins in adolescents in relation to antisocial behavior and stress-induced cortisol response. The TRAILS study. *Psychoneuroendocrinology*, *38*, 3039–3047.

Schéle, E., Grahnemo, L., Anesten, F., Hallén, A., Bäckhed, F., & Jansson, J. O. (2013). The gut microbiota reduces leptin sensitivity and the expression of the obesity-suppressing neuropeptides proglucagon (Gcg) and brain-derived neurotrophic factor (Bdnf) in the central nervous system. *Endocrinology*, *154*, 3643–3651.

Schellekens, H., Finger, B. C., Dinan, T. G., & Cryan, J. F. (2012). Ghrelin signalling and obesity: At the interface of stress, mood and food reward. *Pharmacology & Therapeutics*, *135*, 316–326.

Schroeder, F. A., Lin, C. L., Crusio, W. E., & Akbarian, S. (2007). Antidepressant-like effects of the histone deacetylase inhibitor, sodium butyrate, in the mouse. *Biological Psychiatry*, *62*, 55–64.

Shibata, M., Hisajima, T., Nakano, M., Goris, R. C., & Funakoshi, K. (2008). Morphological relationships between peptidergic nerve fibers and immunoglobulin A-producing lymphocytes in the mouse intestine. *Brain, Behavior, and Immunity*, *22*, 158–166.

Stadlbauer, U., Woods, S. C., Langhans, W., & Meyer, U. (2015). PYY3-36: Beyond food intake. *Frontiers in Neuroendocrinology*, *38*, 1–11.

Stogner, K. A., & Holmes, P. V. (2000). Neuropeptide-Y exerts antidepressant-like effects in the forced swim test in rats. *European Journal of Pharmacology*, *387*, R9–R10.

Strand, F. L. (1999). *Neuropeptides: Regulators of physiological processes*. Cambridge: MIT Press.

Sudo, N., Chida, Y., Aiba, Y., Sonoda, J., Oyama, N., Yu, X. N., et al. (2004). Postnatal microbial colonization programs the hypothalamic–pituitary–adrenal system for stress response in mice. *The Journal of Physiology*, *558*, 263–275.

Sun, Y., Zhang, M., Chen, C. C., Gillilland, M., Sun, X., El-Zaatari, M., et al. (2013). Stress-induced corticotropin-releasing hormone-mediated NLRP6 inflammasome inhibition and transmissible enteritis in mice. *Gastroenterology*, *144*, 1478–1487. 1487.e1-8.

Tasan, R. O., Verma, D., Wood, J., Lach, G., Hörmer, B., de Lima, T. C., et al. (2016). The role of neuropeptide Y in fear conditioning and extinction. *Neuropeptides*, *55*, 111–126.

Teitelbaum, A. A., Gareau, M. G., Jury, J., Yang, P. C., & Perdue, M. H. (2008). Chronic peripheral administration of corticotropin-releasing factor causes colonic barrier dysfunction similar to psychological stress. *The American Journal of Physiology*, *295*, G452–G459.

Theoharides, T. C., & Konstantinidou, A. D. (2007). Corticotropin-releasing hormone and the blood-brain-barrier. *Frontiers in Bioscience*, *12*, 1615–1628.

Thomas, R. H., Meeking, M. M., Mepham, J. R., Tichenoff, L., Possmayer, F., Liu, S., et al. (2012). The enteric bacterial metabolite propionic acid alters brain and plasma phospholipid molecular species: Further development of a rodent model of autism spectrum disorders. *Journal of Neuroinflammation*, *9*, 153.

Thorsell, A., Michalkiewicz, M., Dumont, Y., Quirion, R., Caberlotto, L., Rimondini, R., et al. (2000). Behavioral insensitivity to restraint stress, absent fear suppression of behavior and impaired spatial learning in transgenic rats with hippocampal neuropeptide Y overexpression. *Proceedings of the National Academy of Sciences of the United States of America*, *97*, 12852–12857.

Tremaroli, V., & Bäckhed, F. (2012). Functional interactions between the gut microbiota and host metabolism. *Nature*, *489*, 242–249.

Verma, D., Hörmer, B., Bellmann-Sickert, K., Thieme, V., Beck-Sickinger, A. G., Herzog, H., et al. (2016). Pancreatic polypeptide and its central Y4 receptors are essential for cued fear extinction and permanent suppression of fear. *British Journal of Pharmacology*, *173*, 1925–1938.

Wikoff, W. R., Anfora, A. T., Liu, J., Schultz, P. G., Lesley, S. A., Peters, E. C., et al. (2009). Metabolomics analysis reveals large effects of gut microflora on mammalian blood metabolites. *Proceedings of the National Academy of Sciences of the United States of America*, *106*, 3698–3703.

Yurdaydin, C., Walsh, T. J., Engler, H. D., Ha, J. H., Li, Y., Jones, E. A., et al. (1995). Gut bacteria provide precursors of benzodiazepine receptor ligands in a rat model of hepatic encephalopathy. *Brain Research*, *679*, 42–48.

CHAPTER FIVE

Microbes and Oxytocin: Benefits for Host Physiology and Behavior

S.E. Erdman*[,†,1], T. Poutahidis*[,†]
*Division of Comparative Medicine, Massachusetts Institute of Technology, Cambridge, MA, United States
†Aristotle University of Thessaloniki, Thessaloniki, Greece
[1]Corresponding author: e-mail address: serdman@mit.edu

Contents

Abstract

It is now understood that gut bacteria exert effects beyond the local boundaries of the gastrointestinal tract to include distant tissues and overall health. Prototype probiotic bacterium *Lactobacillus reuteri* has been found to upregulate hormone oxytocin and systemic immune responses to achieve a wide array of health benefits involving wound healing, mental health, metabolism, and myoskeletal maintenance. Together these display that the gut microbiome and host animal interact via immune–endocrine–brain signaling networks. Such findings provide novel therapeutic strategies to stimulate powerful homeostatic pathways and genetic programs, stemming from the coevolution of mammals and their microbiome.

1. INTRODUCTION

The gastrointestinal (GI) tract mucosa is a major host–microbe anatomical interface. Disturbances of the equilibrium between gut mucosa

International Review of Neurobiology, Volume 131
ISSN 0074-7742
http://dx.doi.org/10.1016/bs.irn.2016.07.004

and its bacterial flora have been convincingly linked with intestinal disease (Bull & Plummer, 2014; Chang & Lin, 2016; Fujimura, Slusher, Cabana, & Lynch, 2010; Manichanh, Borruel, Casellas, & Guarner, 2012; Nagao-Kitamoto, Kitamoto, Kuffa, & Kamada, 2016; Neish, 2009). There are numerous reports describing beneficial effects of probiotics on the gut health of both human and animals (Bull & Plummer, 2015; Ducatelle, Eeckhaut, Haesebrouck, & Van Immerseel, 2015; Floch et al., 2015; Fujimura et al., 2010; Marchesi et al., 2016). In the recent years, however, several studies have shown that the effects of gut bacteria expand beyond the local boundaries of the GI tract to include distant tissues and overall health (Belkaid & Hand, 2014; Clemente, Ursell, Parfrey, & Knight, 2012; Erdman & Poutahidis, 2010, 2015; Ho, Chan, & Li, 2015; Maynard, Elson, Hatton, & Weaver, 2012; Noverr & Huffnagle, 2004; Rao, Poutahidis, Fox, & Erdman, 2007; Rook, 2010; Round & Mazmanian, 2009).

This finding, although surprising on the surface, is in line with fundamental biological concepts. In particular, when viewed in the context of a multicellular organism as a holobiont, i.e., an ecosystem comprising a host organism along with its myriad of symbiotic microorganisms (Mcfall-Ngai et al., 2013; Rosenberg & Zilber-Rosenberg, 2011). Mammalian species, including humans, have interacted with bacteria and coevolved with their commensal microbiota for millions of years. It follows that mammals should have developed machinery to balance changes in their physiological status with matching bacterial flora compositions (Dethlefsen, Mcfall-Ngai, & Relman, 2007; Koren, Whiteside, et al., 2012; Mcfall-Ngai et al., 2013; Rosenberg & Zilber-Rosenberg, 2011; Sommer & Backhed, 2013; Sommer et al., 2016; Walter, Britton, & Roos, 2011).

Within the limits of homeostasis the physiological profile of a mammal undergoes changes to address not only special biological functions but environmental challenges as well. Developmental stage and age, reproductive cycles, social relationships, metabolic needs, and changing climatic conditions impact the physiology of a mammal during its lifetime. The major players involved in the induction and the orchestration of the complex physiological alterations belong to the immune, endocrine, and central nervous systems.

This line of reasoning, supported by accumulating experimental data, led to the notion that the gut microbiome and host animal interact via immune–endocrine–brain signaling networks. In that way, a dynamic, bidirectional regulation, controlled by genetic programs shaped over millenia ensures that the physiological status of an animal is in harmony with its gut bacteria community structure (Belkaid & Hand, 2014; Clemente et al., 2012; Cryan & Dinan, 2012; Lee & Mazmanian, 2010; Pittman, 2011; Sandrini, Aldriwesh, Alruways, & Freestone, 2015; Sherwin, Rea, Dinan, & Cryan, 2016).

Could this novel biological concept, however, be exploited to achieve health benefits for animals? Could we stimulate inherent homeostatic properties and introduce palpable changes in mammalian physiology simply by manipulating its gut microbiota? For example, is it possible to achieve youthful phenotypes in an aged animal by simply enriching its gut flora with key bacterial elements characteristic of reproductive fitness and youth? Recent findings in mice suggest that this is doable (Erdman & Poutahidis, 2014b; Ibrahim et al., 2014; Levkovich et al., 2013; Poutahidis & Erdman, 2016; Poutahidis, Kearney, et al., 2013; Poutahidis, Kleinewietfeld, et al., 2013; Poutahidis, Springer, et al., 2014; Varian et al., 2014).

As the understanding for the role of gut microbiota in health and disease increases rapidly (Bull & Plummer, 2014; Clemente et al., 2012; Erdman & Poutahidis, 2015; Fujimura et al., 2010; Marchesi et al., 2016), a translational biomedical research approach should focus on exploiting gut bacteria-induced signaling in humans. Dietary interventions together with edible bacterial cocktails may be used to activate quiescent host gene expression programs and impart systemic effects with healthful immune, hormonal, and neuroendocrine profiles.

Identifying useful gut bacteria and the key host factors may be the basis for disease preventing strategies and therapeutic modalities for an array of ailments including immune dysfunction-associated diseases, metabolic diseases, senility-associated disorders, psychiatric illness, and cancer (Clemente et al., 2012; Erdman & Poutahidis, 2015; Fujimura et al., 2010; Kelly, Clarke, Cryan, & Dinan, 2016; Marchesi et al., 2016; O'toole & Jeffery, 2015; Tremaroli & Backhed, 2012).

Along these lines, in a series of studies using mouse models, we have documented beneficial effects of the prototype probiotic gut bacterium *Lactobacillus reuteri* (*L. reuteri*). Daily consumption of *L. reuteri* alters the immunological and hormonal profile of mice and induces healthful phenotypes with luxuriant fur and enhanced reproductive behaviors in both sexes (see Fig. 1) (Erdman & Poutahidis, 2014b; Ibrahim et al., 2014; Lakritz et al., 2014; Levkovich et al., 2013). Mice fed with *L. reuteri* have increased lifespan with accelerated skin wound healing and resistance to obesity (Erdman & Poutahidis, 2014b; Ibrahim et al., 2014; Lakritz et al., 2014;

Poutahidis, Kearney, et al., 2013; Poutahidis, Kleinewietfeld, et al., 2013; Poutahidis, Springer, et al., 2014). In addition, they have less age-associated atrophic changes in skeletal muscles, testes, thymus, and thyroid glands (Poutahidis, Springer, et al., 2014; Varian, Goureshetti, et al., 2016; Varian, Levkovich, et al., 2016; Varian et al., 2014). The probiotic also counteracts carcinogenesis not only in the intestine but also in distant tissues such as mammary gland, liver, and lung (Erdman & Poutahidis, 2015; Lakritz et al., 2014; Poutahidis, Kleinewietfeld, & Erdman, 2014; Poutahidis et al., 2015).

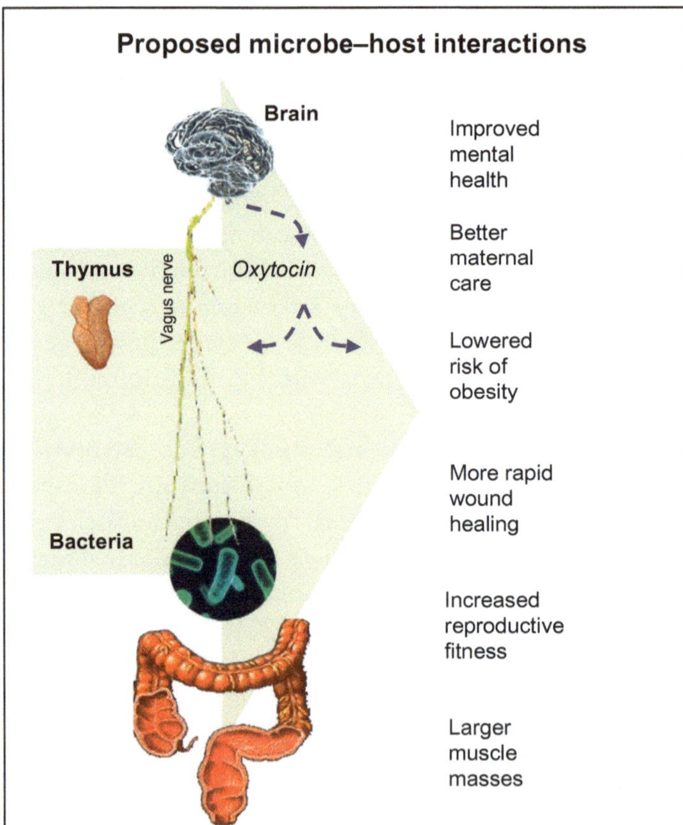

Fig. 1 Oral supplementation with probiotic microbes in drinking water conveys a wide range of health benefits to the mammalian host. Bacteria therapy upregulates endogenous levels of oxytocin leading to improved mental health, a more balanced immune system, a leaner physique, and a longer lifespan.

Interestingly, during our experiments using mice and *L. reuteri*, we consistently find that the probiotic supplement significantly increases blood plasma levels of the hormone oxytocin (Erdman & Poutahidis, 2014b; Ibrahim et al., 2014; Poutahidis, Kearney, et al., 2013; Poutahidis, Kleinewietfeld, et al., 2013; Varian, Goureshetti, et al., 2016; Varian, Levkovich, et al., 2016). This surprising result supports the presence of a gut-brain axis (Cryan & Dinan, 2012; Sherwin et al., 2016) and highlights the importance of gut bacteria in systemic health (Albenberg & Wu, 2014; Clemente et al., 2012; Fujimura et al., 2010; Ho et al., 2015; Marchesi et al., 2016). Using oxytocin-deficient mice we tested selected beneficial effects of *L. reuteri* and found them to be depending upon this pleiotropic hormone (Erdman & Poutahidis, 2014b; Ibrahim et al., 2014; Poutahidis, Kearney, et al., 2013; Varian, Goureshetti, et al., 2016). This suggests that oxytocin may be a key mediator of gut microbiota–host interactions.

2. OXYTOCIN: A MULTIFUNCTIONAL NEUROPEPTIDE

Oxytocin is a neuroendocrine oligopeptide that is primarily produced by the magnocellular neurosecretory system comprising large neurons of the paraventricular and supraoptic hypothalamic nuclei. Along with its carrier protein neurophysin I, oxytocin is transported via neuronal axons in the neurohypophysis where it is stored and secreted into the circulation. Smaller amounts of oxytocin that remain in the dendrites of magnocellular cells or originate from other smaller neurons of the parvocellular secretory system and other accessory hypothalamic nuclei used for local CNS signaling. Extra-CNS organs, such as the thymus, GI tract, pancreas, skin, testis, ovary, uterus, placenta, kidney, and heart also synthesize small quantities of oxytocin (Knobloch & Grinevich, 2014; Lee, Macbeth, Pagani, & Young, 2009; Young & Gainer, 2009).

The widespread expression of the oxytocin receptor in many different tissues throughout the body of mammals is indicative of the pleiotropic actions of oxytocin. Thus far the oxytocin receptor has been detected in the central and peripheral neurons, uterus, mammary glands, placenta, ovary, testis, pancreas, adipose tissue, skeletal and smooth muscles, thymus, adrenal glands, intestine, kidney, heart, bone, and immune cells (Baribeau & Anagnostou, 2015; Gimpl, Fahrenholz, & Gene, 2001; Wang et al., 2015; Zingg & Laporte, 2003).

The first recognized role of oxytocin in mammalian physiology was the induction of uterine contractions to facilitate labor. Sir Henry Dale who made the discovery in 1906 named the hormone using two ancient Greek words meaning "swift labor." Soon after that, oxytocin was shown to be integral for milk secretion by mammary glands during lactation of suckling mammals. To this end numerous studies have profoundly expanded the list of the documented actions of oxytocin. These include effects on renal sodium excretion, sperm transport in ejaculation, GI motility, cardiovascular function, thermoregulation, pancreatic glucagon and insulin production, bone remodeling, and muscle mass maintenance and regeneration (Gimpl et al., 2001; Lee et al., 2009; Viero et al., 2010; Wang et al., 2015; Yang, Wang, Han, & Wang, 2013; Zingg & Laporte, 2003).

Oxytocin, however, gained most of its recent popularity for another reason. It has been dubbed molecule of "life," "happiness," "love," "trust," and "cuddle." All this based in its psychotropic properties and central role in the neuroendocrine pathways that regulate behavior. The hormone affects parental behavior and especially maternal care, and promotes social bonding in many different contexts of relationship including pairing, falling in love, sexual receptivity, friendship, partnership, and cooperation. It reduces social stress and also regulates social memory and recognition, empathy, aggressiveness, and trust. Interestingly, oxytocin also affects the general sense of well-being and is involved in nonsocial cognitive functions related to learning, memory, pain perception, grooming, anxiety, and depression (Baribeau & Anagnostou, 2015; Carter, 2014; Donaldson & Young, 2008; Feldman, Monakhov, Pratt, & Ebstein, 2016; Gimpl et al., 2001; Knobloch & Grinevich, 2014; Kosfeld, Heinrichs, Zak, Fischbacher, & Fehr, 2005; Lee et al., 2009; Shen, 2015).

The additional findings that oxytocin influences energy balance and metabolism (Barengolts, 2016; Blevins & Ho, 2013; Blevins et al., 2016) and the immune system responses (Blevins & Ho, 2013), have added more to its reputation as an overall health and well-being promoter. The hormone has been shown to modulate feeding behavior toward smaller meals with accelerated satiety, decrease of appetite, and avoidance of carbohydrate-rich foods (Barengolts, 2016; Blevins & Ho, 2013; Blevins et al., 2015; Gimpl et al., 2001; Lee et al., 2009). The contributions of oxytocin in weight gain control and leaner physique do not depend solely on the decrease of caloric intake. Oxytocin boosts energy expenditure by contributing to a high metabolic rate with increased heart rate, body temperature, and oxygen consumption (Barengolts, 2016; Blevins & Ho, 2013; Blevins et al., 2015).

It also improves the secretion of and systemic resistance to insulin, induces glucose tolerance, and acts directly on adipose tissue to inhibit lipogenesis and promote lipolysis with increased lipid catabolism (Barengolts, 2016; Blevins & Ho, 2013; Blevins et al., 2015; Colaianni, Sun, Zaidi, & Zallone, 2015; Viero et al., 2010).

Oxytocin has a fundamental connection with the immune system, since it is an integral element of thymus and bone marrow, which are the major immune cell sources of mammals (Gimpl et al., 2001; Wang et al., 2015). It contributes to the maturation and selection of T-lymphocytes in the thymus and promotes the production and transport of hematopoietic progenitor cells from the bone marrow to blood and peripheral lymphoid tissues (Elabd et al., 2014; Gimpl et al., 2001; Hansenne et al., 2009; Wang et al., 2015). Most of oxytocin's reported actions in inflammatory processes are regulatory and suppressive (Wang et al., 2015). It has been shown to ameliorate GI, urinary tract, adipose and neuronal tissue, and joint inflammation. It also suppresses inflammation associated with atheromatous plaques, myocardial infarcts, sepsis, ischemia/reperfusion, and endotoxicosis. The hormone primarily downregulates neutrophils, oxidative stress and proinflammatory cytokines including TNF-α, IL-4, IL-6, macrophage inflammatory protein-1α and 1β, monocyte chemoattractant protein-1, and IL-1β. Importantly, it also connects with efficient regulatory T-cell functions (Hamasaki et al., 2016; Matsuura et al., 2016; Wang et al., 2015; Yuan et al., 2016).

3. PARALLELS BETWEEN GUT BACTERIA AND OXYTOCIN EFFECTS

The perinatal period including pregnancy and embryo development, labor, weaning, and early stages of newborn growth is when both oxytocin (Farshim et al., 2016; Gimpl et al., 2001; Lee et al., 2009) and gut microbiota (Clemente et al., 2012; Erdman & Poutahidis, 2010; Koren, Goodrich, et al., 2012; Poutahidis et al., 2015; Rook, 2013; Round & Mazmanian, 2009) are of peak physiological importance. This concurrence takes place in the context of mother and child both undergoing profound changes involving tissue architecture reconstruction at the whole organism level. At that time, genetic programs in cells switch-on and switch-off in a concerted fashion to achieve the "miracle of life." Under these conditions, it would be reasonable to hypothesize that two major players of perinatal events, oxytocin, and microbial flora, influence each other to bring out the best in both mother and infant. If this hypothesis is true then the reciprocal interactions of oxytocin and gut bacteria should be of great biomedical importance. First of all, because perinatal events are among the most fundamental pathways of living beings. Second, because newborn effects should logically be largely positive and health-promoting. Third, because early-life influences shaping the newborn are likely to affect their health later in life (Collado, Cernada, Bauerl, Vento, & Perez-Martinez, 2012; Erdman & Poutahidis, 2010; Poutahidis et al., 2015; Rautava, Luoto, Salminen, & Isolauri, 2012; Rook, 2013).

Although the basis of the hypothesis appears attractive, there are questions to be answered. Is there some evidence, at least indirect, to support this line of reasoning? As a general principle the effects of oxytocin in health and disease should parallel the effects of beneficial bacteria (probiotics) and vice versa. Is this, however, true? For the purposes of this review we will first attempt to draw parallels between beneficial effects of oxytocin and probiotics, with special reference to the prototype probiotic bacterium *L. reuteri*, which is contained in human milk and is the only microorganism thus far reported to induce upregulation of oxytocin upon ingestion. Next, we will provide more direct experimental evidence that health-promoting effects of *L. reuteri* rely on oxytocin.

3.1 *L. reuteri* and Oxytocin Promote Skin Wound Healing

Wound healing is an essential biological process that involves the timely implementation of basic physiological phenomena such as hemostasis, inflammation, extracellular matrix, and connective tissue formation and angiogenesis. Successful wound healing demands the orchestrated collaboration of many different cell lineages that proliferate, differentiate, migrate, and undergo apoptosis in the context of tissue remodeling. It's no wonder that optimal wound healing is considered to reflect overall good health and fitness, and connects with healthful longevity (Eming, Martin, & Tomic-Canic, 2014; Gurtner, Werner, Barrandon, & Longaker, 2008; Yanai, Budovsky, Tacutu, & Fraifeld, 2011).

The direct application of probiotics and their products in skin wounds has been shown to promote healing. This is a local effect that mostly relies on the antibiotic properties of probiotics and their interaction with local inflammatory cells and proliferating epidermis (Gueniche et al., 2010; Huseini, Rahimzadeh, Fazeli, Mehrazma, & Salehi, 2012; Peral, Martinez, & Valdez, 2009; Wong, Martindale, Longaker, & Gurtner,

2013). Can, however, consumed bacteria residing in the gut exert effects that control wound healing processes in distant tissue such as skin?

In a recent study we have used a well-characterized mouse model of skin wound healing to show that this is possible (Poutahidis, Kearney, et al., 2013) (Fig. 1). For that, excisional wounds were inflicted in the dorsal skin of C57BL/6 mice. Mice receiving orally *L. reuteri* probiotic bacteria healed the wounds in half the time required for matched control animals, comprising both untreated and *Escherichia coli* strain K12-treated mice. The wounds of mice with *L. reuteri*-enriched gut flora showed accelerated epidermal closure, maturation of the granulation tissue and collagen deposition in the wound bed. The central cell player in this phenomenon was the $CD4^+Foxp3^+CD25^+$ regulatory-T cell (Treg). Indeed, the depletion of $CD25^+$ cells negated the *L. reuteri*-induced hastening of the classical wound repair process. To further determine the importance of this inflammation-suppressive lymphocyte type, we then transferred highly purified $CD4^+Foxp3^+$ Treg cells originating from *L. reuteri*-treated donor mice into *Recombination activating gene 2* (*Rag2*)-deficient mice, otherwise lacking T and B lymphocytes. Surprisingly, the transferred Treg cells were sufficient to recapitulate the rapid wound closure in the hosts, which were not themselves receiving probiotic bacteria. The early influx of Tregs in the wound bed corresponded with a rapid clearance of neutrophils, and suppressed IL-17-driven proinflammatory responses. These IL-17-mediated host responses were detrimental to repair, since the depletion of IL-17A alone was sufficient enough to benefit the skin wound healing closure (Poutahidis, Kearney, et al., 2013).

By using outbred Swiss mice we found that the rapid wound healing effect of oral *L. reuteri* treatment does not restrict to the C57BL/6 strain of mice in the original study. Also, in a pilot double-blind placebo-controlled study involving a small cohort of healthy female volunteers, we find that edible *L. reuteri* holds promise for promoting skin wound healing in human beings as well (unpublished data). Likewise, in a recent clinical study, another probiotic organism *L. rhamnosus GG*, when orally introduced to acutely burn pediatric patients reduced the time required to complete wound healing. A trend toward decreased requirement for systemic antifungal therapy was also noted in probiotic-treated patients compared to controls (Mayes et al., 2015).

One of the classical actions of oxytocin relates to the promotion of hemostasis and healing of the postpartum uterine wounds (Dept. of Reproductive Health and Research, W, 2012). Accumulating evidence, however, links oxytocin with faster healing of skin wounds as well (Gouin et al., 2010; Poutahidis, Kearney, et al., 2013). Several studies have shown that oxytocin administration accelerates the healing of burn and irradiation injuries in the skin of rats (Iseri et al., 2010, 2008; Vitalo et al., 2009), while oxytocin-deficient mice show retarded healing of skin wounds (Poutahidis, Kearney, et al., 2013). In humans, exogenous oxytocin had a positive effect in boosting the healing of diabetic suppurative necrotic foot lesions; clearance of the necrotic tissues by macrophages, angiogenesis, and optimal granulation tissue formation was accelerated after oxytocin (Gavrilenko, Esipov, & Sivozhelezov, 2003). Likewise, high levels of plasma oxytocin correlated with faster healing of experimentally induced arm skin blister wounds in healthy individuals (Gouin et al., 2010). The positive effects of oxytocin in wound healing are attributed to its stress-reducing psychological effects, as well as to its antiinflammatory and antioxidant properties (Carter, 2014; Gouin et al., 2010; Poutahidis, Kearney, et al., 2013; Uvnas-Moberg, 1998; Viero et al., 2010; Wang et al., 2015).

In conclusion, studies in both rodents and humans show that bacterial elements of the gut flora and oxytocin exert systemic effects that contribute to accelerate healing of skin wounds.

3.2 *L. reuteri* and Oxytocin Counteract Obesity

Obesity becomes more and more prevalent due to modern lifestyle and dietary habits. It is now viewed as a world-wide epidemic that needs to be controlled. Obesity increases mortality by predisposing to serious pathological conditions including diabetes mellitus type 2, cardiovascular disease, nonalcoholic hepatic steatosis, arthritis, asthma, and cancer. Obesity coexists with a systemic smoldering subclinical inflammatory state and altered endocrine system status that create a vicious-circle of perturbed homeostasis contributing to the various adverse health effects (Gregor & Hotamisligil, 2011; Mozaffarian, Hao, Rimm, Willett, & Hu, 2011; Shoelson, Herrero, & Naaz, 2007).

In the recent years an overwhelming amount of data retrieved from both animal and human studies highlights the important role of the intestinal bacterial flora in the pathogenesis of obesity (Kallus & Brandt, 2012; Kobyliak, Virchenko, & Falalyeyeva, 2016; Nieuwdorp, Gilijamse, Pai, & Kaplan, 2014; Poutahidis, Kleinewietfeld, et al., 2013; Shoelson et al., 2007; Tremaroli & Backhed, 2012). Diet affects the composition of the gut microbiota (Albenberg & Wu, 2014; Clemente et al., 2012; Sommer & Backhed,

2013). Western-type, energy-rich diets have been linked with a decreased diversity and alterations in the balance between major bacterial phyla or even species of the gut microbial communities (Clemente et al., 2012; Kallus & Brandt, 2012; Kobyliak et al., 2016; Lau, Carvalho, Pina-Vaz, Barbosa, & Freitas, 2015, #9792; Nieuwdorp et al., 2014). The most convincing studies are those showing clear correlation of certain microbiota elements with weight loss or gain and obese status in humans (Clemente et al., 2012; Kallus & Brandt, 2012; Kobyliak et al., 2016; Nieuwdorp et al., 2014; Turnbaugh et al., 2009, #12904). Equally impressive are studies in mice where lean hosts eating normal diets become obese merely because of their experimental colonization with an obese-type mouse or human microbiota (Chassaing et al., 2015; Ridaura et al., 2013; Sommer et al., 2016; Turnbaugh et al., 2006; Vijay-Kumar et al., 2010).

Intestinal bacteria affect body weight by several, many times interrelated, mechanisms the most obvious and well studied of which relates to their role in harvesting energy from food. Other relevant mechanisms include effects on lipogenesis and fat storage, gluconeogenesis, bile acid metabolism, gut permeability, vitamin synthesis, and release of gut hormones involved in satiety and appetite signaling (Boulange, Neves, Chilloux, Nicholson, & Dumas, 2016; Drissi, Raoult, & Merhej, 2016; Kallus & Brandt, 2012; Kobyliak et al., 2016; Lau et al., 2015; Nieuwdorp et al., 2014; Tremaroli & Backhed, 2012; Woting & Blaut, 2016). Their interaction with the immune system, which is important in sustaining the bidirectional smoldering inflammation-obese metabolic status axis, involves not only local gut mucosa but also adipose tissue and systemic proinflammatory signaling as well signaling (Boulange et al., 2016; Gregor & Hotamisligil, 2011; Kallus & Brandt, 2012; Kobyliak et al., 2016; Poutahidis, Kleinewietfeld, et al., 2013; Shoelson et al., 2007). Bacterial LPS and flagellin have been shown to provide TLR-4- and TLR-5-mediated antigenic stimuli that may fuel the obesity-associated inflammation (Jia et al., 2014; Kobyliak et al., 2016; Poggi et al., 2007; Vijay-Kumar et al., 2010; Vila et al., 2014).

Studying the changes of the gut bacteria community in the context of obesity and metabolic syndrome as a whole many times yields results that are discrepant and difficult to explain (Drissi et al., 2016; Hildebrandt et al., 2009; Kobyliak et al., 2016; Murphy et al., 2010; Woting & Blaut, 2016). This is expected, given that gut microbiota is fluid and depended on a great number of host and environmental factors (Belkaid & Hand, 2014; Chassaing et al., 2015; Conlon & Bird, 2015; Cotillard et al., 2013; Maynard et al., 2012; Mozaffarian et al., 2011; Sonnenburg et al., 2016;

Tremaroli & Backhed, 2012). Nonetheless, the knowledge gained from these studies is essential for recognizing antiobesogenic gut bacteria targets for in vivo testing.

Following this line of thinking, a great number of investigators have tested probiotics for obesity control with promising results (Drissi et al., 2016; Mozaffarian et al., 2011; Woting & Blaut, 2016; Yoo & Kim, 2016). One such a probiotic is *L. reuteri*, a gut bacterium that upregulates oxytocin when fed to mice (Erdman & Poutahidis, 2014b; Ibrahim et al., 2014; Poutahidis, Kearney, et al., 2013; Poutahidis, Kleinewietfeld, et al., 2013; Varian, Goureshetti, et al., 2016; Varian, Levkovich, et al., 2016). Using outbred Swiss and inbred C57BL/6 mice fed with a Westernized-type diet as a model of obesity we found that dietary supplementation with *L. reuteri* ATCC 6475 organisms alone prevented weight gain without altering the existing GI flora in stool or amounts of calorie consumption. These slenderizing effects were irrespective of the baseline diet, since they were also observed in control mice fed with a normal chow. Both abdominal and subcutaneous fat depositions were significantly decreased in probiotic-treated animals compared to the control. *L. reuteri* suppressed obesity-associated inflammatory lesions, such as crown-like structures and pyogranuloma formation in the adipose tissue. Obese mice in this study were in a proinflammatory systemic status characterized by increased IL-17 and decreased Treg cells in the blood and peripheral lymphoid tissues, mimicking the immunological status of human beings consuming unhealthy "fast food" diets. The suppression of obesity using probiotic bacteria led to downregulated IL-17. Using knockout mice and adoptive cell transfer and cell depletion experiments we have shown that the slenderizing effect of *L. reuteri* depended on $CD4^+$ $CD25^+$ cells. The adoptive transfer of purified Foxp3 + Treg cells was sufficient to rescue fat pathology and decrease body fat in naive Rag2-deficient recipient animals. Importantly, the slenderizing effect occurred in host mice only when Tregs originated from *L. reuteri*-colonized donors (Poutahidis, Kleinewietfeld, et al., 2013).

Two related epidemiological studies reported that the presence of *L. reuteri* in human stool associates with obesity and a high body mass index (Million et al., 2013, 2012). Recent human clinical trial studies, however, yield conflicting results. In the first, glucose-tolerant patients were given *L. reuteri* SD5865 for 4 weeks. Anthropometric indices of obesity did not change, since the duration of the study was short. Nonetheless, *L. reuteri* improved insulin and incretin secretion, and the production of the intestinal integrity-protective glucagon-like peptide-2. These pilot results suggest that

L. reuteri should be further tested as a promising probiotic modality against obesity-associated metabolic syndrome (Simon et al., 2015). In a second study, the administration of free fatty acid-absorbing mutant strain of *L. reuteri* isolated from lean healthy individuals had a clear effect in counteracting obesity in both humans and mice (Chung et al., 2016). More recent work involving obesity in mouse models examines *L. reuteri* specifically in the context of oxytocin in mice, substantiating beneficial effects of consuming this organism (Buffington et al., 2016; Varian, Levkovich, et al., 2016).

Whether the comparison between epidemiological data and clinical trials underlies a discrepancy for the role of *L. reuteri* cannot be easily addressed. Should such a discrepancy exists though; it may be due to strain-specific attributes of *L. reuteri* (Fak & Backhed, 2012; Qiao et al., 2015). Several other studies in rodents, using various different strains of *L. reuteri*, however, consistently show antiobesogenic and hyperlipidemic effects (Hsieh et al., 2016; Huang et al., 2015; Poutahidis, Kleinewietfeld, et al., 2013; Singh, Malik, Katkamwar, & Kaur, 2015; Sun, Qiao, et al., 2016; Taranto, Medici, Perdigon, Ruiz Holgado, & Valdez, 1998; Ting et al., 2015). Our most recent mouse experiments using a novel *L. reuteri* isolate derived from canine saliva are in agreement with the majority of studies and to what we have reported earlier, in an oxytocin-dependent manner (Varian, Levkovich, et al., 2016).

How does oxytocin, however, affect weight gain and obesity? Oxytocin by large is connected with leaner physiques and weight gain control and has been shown to activate several different mechanisms toward that end (Barengolts, 2016; Blevins & Baskin, 2015; Ho & Blevins, 2013; Lee et al., 2009). Oxytocin acts as a satiety hormone and limits feed intake in rodents (Altirriba et al., 2014; Barengolts, 2016; Blevins & Baskin, 2015; Blevins et al., 2016; Iwasaki, 2015; Maejima et al., 2011; Morton et al., 2012; Zhang et al., 2013). It is involved in the regulation of gastric motility and distention (Blevins & Baskin, 2015; Ho & Blevins, 2013; Qin et al., 2009; Welch, Margolis, Li, & Gershon, 2014; Wu, Hung, Chang, Pau, & Wang, 2003). In the anorexigenic signaling cascade, it cross talks with other important satiety inducers including leptin, cholecystokinin, and nesfatin-1 (Blevins & Baskin, 2015; Blevins, Eakin, Murphy, Schwartz, & Baskin, 2003; Blevins et al., 2016; Ho & Blevins, 2013; Maejima et al., 2009; Yosten & Samson, 2010). Treatment with oxytocin rescued leptin-nonresponsive rats from diet-induced obesity, which suggests that oxytocin locates downstream from leptin signaling (Barengolts, 2016; Blevins &

Baskin, 2015; Iwasaki, 2015; Maejima et al., 2009). Also, the anorectic action of exogenously administered cholecystokinin-8 depends on intact brain oxytocin signaling (Blevins & Baskin, 2015; Blevins et al., 2003; Ho & Blevins, 2013).

Oxytocin's most powerful appetite modulating effect relates with suppressing carbohydrate predilection (Barengolts, 2016; Blevins & Baskin, 2015; Blevins et al., 2015; Ho & Blevins, 2013; Iwasaki, 2015; Lee et al., 2009; Maejima et al., 2009; Mullis, Kay, & Williams, 2013; Olszewski et al., 2010). Several studies show that the administration of oxytocin or oxytocin agonists decreased the consumption of feed in a dose-depended fashion (Arletti, Benelli, & Bertolini, 1990; Barengolts, 2016; Blevins & Baskin, 2015; Olson et al., 1991). Conversely, oxytocin antagonists stimulate increased chow consumption but also preference for glucose and sucrose (Blevins et al., 2015; Ho & Blevins, 2013; Lee et al., 2009; Lokrantz, Uvnas-Moberg, & Kaplan, 1997; Olszewski et al., 2010). Serum levels of oxytocin are decreased in most animal models of obesity (Barengolts, 2016; Blevins & Baskin, 2015), while oxytocin-deficient mice are prone to late life-onset obesity (Camerino, 2009; Varian, Goureshetti, et al., 2016). The same is also true for oxytocin receptor-deficient mice (Takayanagi et al., 2008), although the effect in this case depends on the background mouse strain (Lee, Caldwell, Macbeth, Tolu, & Young, 2008; Lee et al., 2009). Interestingly, oxytocin-deficient mice become progressively obese with age without meaningful changes to daily feed intake (Camerino, 2009; Poutahidis, Kearney, et al., 2013; Rinaman et al., 2005; Takayanagi et al., 2008; Varian, Goureshetti, et al., 2016). Likewise, rodents given exogenous oxytocin lose weight while showing only transient changes in their appetite (Blevins & Baskin, 2015; Deblon et al., 2011; Ho & Blevins, 2013; Maejima et al., 2011). These findings suggest that oxytocin's ability to counteract obesity does not rely solely on reducing intake of food. Indeed, oxytocin upregulates metabolic rate indices, such as heart beats, body temperature, and oxygen consumption (Barengolts, 2016; Blevins & Baskin, 2015; Ho & Blevins, 2013; Yoshida et al., 2009; Zhang & Cai, 2011; Zhang et al., 2011). Therefore, in many different animal models including primates, energy expenditure was increased with oxytocin treatment (Barengolts, 2016; Blevins & Baskin, 2015; Blevins et al., 2015; Deblon et al., 2011; Ho & Blevins, 2013). Moreover, oxytocin appears to target the adipose tissue directly. In rodents, oxytocin treatment decreases fat deposits, which show upregulated lipolysis and lipid utilization, and decreased lipogenesis, lipid uptake, and macrophage infiltration (Altirriba et al., 2014; Barengolts, 2016; Blevins & Baskin, 2015; Blevins et al., 2016; Deblon et al., 2011; Muchmore, Little, & De Haen, 1981; Varian, Goureshetti, et al., 2016). Depending on the oxytocin dose scheme, however, oxytocin may also induce increased lipogenesis along with the accelerated lipolysis of adipose tissue (Blevins & Baskin, 2015; Elabd et al., 2008; Ho & Blevins, 2013; Muchmore et al., 1981).

In humans, obese individuals have decreased serum oxytocin (Blevins & Baskin, 2015; Qian et al., 2014). Also, the increased appetite of pregnant women is attributed to the inhibition of oxytocin release from hypothalamic nuclei (Douglas, Johnstone, & Leng, 2007). Initial clinical studies, using exogenous administration of the hormone in humans report positive results in reduction of body weight and improving insulin abnormalities in diabetes type 2 (Barengolts, 2016; Blevins & Baskin, 2015; Zhang et al., 2013). Although oxytocin is emerging as a promising treatment for obesity, further tests are required to overcome concerns, which are always legitimate when hormones are being exogenously administered to human beings (Barengolts, 2016; Blevins & Baskin, 2015).

3.3 *L. reuteri* and Oxytocin Suppress Uncontrolled Inflammation

The importance of the gut microbiota in shaping the GI mucosal immune responses is now fully appreciated and highlighted in the recent literature (Belkaid & Hand, 2014; Lee & Mazmanian, 2010; Maynard et al., 2012; Round & Mazmanian, 2009). Emerging evidence, however, demonstrates that gut microbiota influences the immune system at the whole organism level and regulates its responses in tissues locating remotely from the gut (Belkaid & Hand, 2014; Clemente et al., 2012; Erdman & Poutahidis, 2010, 2015; Kamada & Nunez, 2013; Noverr & Huffnagle, 2004; Poutahidis, Kleinewietfeld, et al., 2014; Rao et al., 2007). This fact provides a physiological basis to support the "hygiene hypothesis," which attempts to explain why immune-dysregulation-associated disorders and cancer incidence has been increasing over the last decades in industrialized countries (Erdman & Poutahidis, 2010, 2015; Rook, 2010, 2013; Walter et al., 2011). According to this theory, the stringent hygiene conditions and the extensive use of antibiotics reduces the exposure of human beings in microbial antigens and undermines the efficient induction of immune tolerance during their early life. Consequently, their immune system has reduced regulatory competence and is unable to protect them from diseases that associate with uncontrolled and chronic inflammation later in life (Erdman & Poutahidis, 2010, 2015; Rook, 2010, 2013; Walter et al., 2011).

Going one step further, could it be that modern urban lifestyle deprives the adult human gut from exposures to transient or even resident gut bacteria, with which the mankind homeostasis balanced for millions of years? Could it be that many among us are in a subtle gut flora dysbiosis state that translates to an increased subclinical chronic inflammatory tone, which in turn increases the probability of developing certain diseases, including cancer over time? It is not just the rigorous hygiene practices to blame. According to this scenario, dietary practices causing temporal fluctuations in the intestinal flora (Albenberg & Wu, 2014; Milani et al., 2016; Sonnenburg et al., 2016) may play a role as well. For millenia the human diet depended on seasonal availability of food. Therefore, a degree of seasonal fluctuation of gut flora following seasonal diet transitions should be expected. Now that humans in most parts of the world can eat anything-anytime the gut flora cyclical changes should be more or less nonexisting. In the complex and competitive context of microbial symbiosis in the gut, this may also translate to reduced diversity of bacterial exposures.

Although *L. reuteri* has been isolated from human samples, including breast milk and feces, it is not a consistent colonizer of all humans (Walter et al., 2011). For example, one study found it in the feces of only 4% of the persons tested (Molin et al., 1993). However, experts in the field suspect that its prevalence used to be higher and declined dramatically over the last 50 years (Walter et al., 2011). This fact alone makes it an interesting research target for testing the hygiene hypothesis. *L. reuteri*, which is considered a classical probiotic bacterium, has been indeed tested and found to promote gut health in both animals and humans (Hou, Zeng, Yang, Liu, & Qiao, 2015; Salim et al., 2013; Szajewska, Urbanska, Chmielewska, Weizman, & Shamir, 2014; Urbanska, Gieruszczak-Bialek, & Szajewska, 2016; Walter et al., 2011). In mouse models *L. reuteri* has repeatedly shown to ameliorate inflammation and experimental colitis lesions (Ahl et al., 2016; Eaton, Honkala, Auchtung, & Britton, 2011; Fabia et al., 1993; Gao et al., 2015; Karimi, Inman, Bienenstock, & Forsythe, 2009; Liu, Fatheree, Mangalat, & Rhoads, 2010, 2012; Mechoud et al., 2012). Likewise, in humans, it alleviated intestinal inflammation in IBD, diarrhea, and irritable bowel syndrome (Lin, Thibodeaux, Pena, Ferry, & Versalovic, 2008; Lorea Baroja, Kirjavainen, Hekmat, & Reid, 2007; Niv, Naftali, Hallak, & Vaisman, 2005; Oliva et al., 2012; Szajewska et al., 2014; Urbanska et al., 2016).

We and others have also found that *L. reuteri* exerts systemic immunomodulation and suppresses deleterious inflammatory processes in tissues outside of the gut, such as skin, lungs, mammary gland, and adipose tissue (Lakritz et al., 2014; Poutahidis, Kearney, et al., 2013; Poutahidis, Kleinewietfeld, et al., 2014; Poutahidis et al., 2015; Varian, Goureshetti, et al., 2016). Collectively, findings in animal models and humans agree that *L. reuteri*-induced systemic effects on immune system include the upregulation of Tregs and downregulation of circulating neutrophils. TNF-α, IL-6, IL-17, IL-12, IL-1β, and other proinflammatory cytokines decrease, whereas the antiinflammatory cytokine IL-10 increases (Fabia et al., 1993; Gao et al., 2015; Karimi et al., 2009; Lakritz et al., 2014; Lin et al., 2008; Livingston, Loach, Wilson, Tannock, & Baird, 2010; Mechoud et al., 2012; Poutahidis, Kearney, et al., 2013; Poutahidis, Kleinewietfeld, et al., 2014; Poutahidis et al., 2015; Smits et al., 2005; Varian, Goureshetti, et al., 2016; Walters et al., 2011). In vitro assays suggest that *L. reuteri* primes dendritic cells via the C-type lectin DC-specific intercellular adhesion molecule 3-grabbing nonintegrin (DCSIGN), resulting in the induction of Tregs (Smits et al., 2005). Also, that it induces the suppression of *TNF-α* expression in antigen-presenting cells by inhibiting the activation of *c-Jun* and *AP-1* (Lin et al., 2008). In a rat model of necrotizing enterocolitis the antiinflammatory effect of the probiotic was connected with downregulation of mucosal TLR-4 and NF-κB signaling (Liu et al., 2012). In chemical colitis *L. reuteri* treatment reduced the expression of P-selectin and leucocyte- and platelet-endothelium adhesion (Schreiber et al., 2009).

The important role of Tregs in mediating the beneficial antiinflammatory effects of *L. reuteri* has been demonstrated with cell depletion and cell transfer experiments in mice. In the absence of Tregs the *L. reuteri* beneficial effects are negated. On the other hand, transfer of Tregs from *L. reuteri*-colonized donor mice is potent enough to counteract detrimental inflammatory processes in recipients that were noncolonized by the probiotic bacterium (Erdman & Poutahidis, 2014a; Lakritz et al., 2014; Poutahidis, Kleinewietfeld, et al., 2013).

During inflammatory insults the immune system cross talks with the central nervous and endocrine systems to coordinate metabolism and behavior and redirect them to meet the diseased organism needs (Pittman, 2011; Souza-Moreira, Campos-Salinas, Caro, & Gonzalez-Rey, 2011). Although the hypothalamus–pituitary–adrenal gland axis and glucocorticoid secretion has a central role in this process, the contributions of the oxytocin secreting system are also important (Wang et al., 2015). During infections, oxytocin downregulates proinflammatory responses and oxidative stress and has a protective role against immune-mediated damage (Wang et al., 2015).

Lymphocytes, monocytes, and macrophages have oxytocin receptors (Elands, Resink, & De Kloet, 1990; Hansenne et al., 2005; Maccio et al., 2010; Szeto et al., 2008; Wang et al., 2015). Oxytocin is important for T-lymphocyte differentiation and selection and blocking OXTRs inhibits the differentiation of T cells in the thymus (Hansenne et al., 2005). Interestingly, it has been recently shown that the autoimmune regulator gene/protein (Aire) which plays an important role in natural Treg differentiation in the thymus (Nomura & Sakaguchi, 2007) is induced by oxytocin in thymic epithelial cells (Hansenne et al., 2009). Oxytocin mimics IL-2 action and induces IFN-γ expression in lymphocytes (Johnson & Torres, 1985; Maccio et al., 2010; Wang et al., 2015). Also, it stimulates peripheral blood mononuclear cells to proliferate, promotes the expression of CD25 and CD95 ligands, and counteracts the suppressive effects of estrogens in the proliferation and differentiation of lymphocyte blood precursors (Johnson & Torres, 1985; Maccio et al., 2010; Wang et al., 2015).

Oxytocin is already being tested as an antiinflammatory therapy in experimental settings (Wang et al., 2015). In primary cultures of microglial cells and in a mouse model of LPS-induced microglial cell activation, oxytocin attenuated the expression of proinflammatory molecules and counteracted neuroinflammation (Yuan et al., 2016). Likewise, it ameliorated inflammation in rodent models of chemically induced colitis by decreasing oxidative stress and proinflammatory cytokine signaling (Balzola, Bernstein, Ho, & Lees, 2010; Iseri et al., 2008; Wang et al., 2015; Welch et al., 2014). In other in vivo models of inflammatory damage the major action of oxytocin related to the downregulation of neutrophils (Al-Amran & Shahkolahi, 2013; Biyikli et al., 2006; Iseri et al., 2005a, 2005b; Petersson, Wiberg, Lundeberg, & Uvnas-Moberg, 2001). This finding matches our observations in mice lacking oxytocin (Poutahidis, Kearney, et al., 2013; Poutahidis, Kleinewietfeld, et al., 2013). We also find that otherwise healthy oxytocin-deficient mice have significantly more circulating neutrophils by comparison to age-matched wild-type controls (Varian, Goureshetti, et al., 2016). In a detailed study oxytocin-receptor-deficient mice have been shown to be more susceptible to chemical colitis and cholera-toxin-induced epithelial permeability. In this study oxytocin protected wild-type but not oxytocin-receptor-deficient mice from colitis, whereas elegant experiments demonstrated the important role of oxytocin in regulating not only inflammation but also intestinal motility, permeability, and epithelial cell proliferation as well (Welch et al., 2014).

The protective effects of oxytocin against detrimental inflammatory responses have also been demonstrated in animal models of sepsis (Iseri et al., 2005a), urinary tract infections (Biyikli et al., 2006), ischemia/reperfusion injury (Erkanli Senturk et al., 2013; Houshmand, Faghihi, & Zahediasl, 2009; Moghimian, Faghihi, Karimian, & Imani, 2012), myocardial infarction (Jankowski et al., 2010), brain ischemia (Karelina et al., 2011), atherosclerosis (Nation et al., 2010), and adipose tissue inflammation (Nation et al., 2010; Varian, Goureshetti, et al., 2016). In a human study, intravenously administered oxytocin protected against LPS-induced systemic inflammation. Oxytocin downregulated plasma ACTH, cortisol, procalcitonin, and a number of proinflammatory cytokines. The effect of oxytocin was attributed to the activation of the cholinergic antiinflammatory pathway (Clodi et al., 2008).

3.4 *L. reuteri* and Oxytocin in Modulating Behavior

The recent evidence for the presence of a gut microbiota–brain axis has evoked a tremendous research interest on whether gut bacteria dysbiosis affects mental health. Data supporting a role for gut bacteria in emotional behavior, mood, anxiety, depression, and other neuropsychiatric disorders is mounting. Consequently, restoring gut microbiota balance with probiotics is coming forth as a novel preventive and therapeutic approach for managing stress-related disorders (Cryan & Dinan, 2012; Foster & Mcvey Neufeld, 2013; Kelly et al., 2016; Moloney, Desbonnet, Clarke, Dinan, & Cryan, 2014; Sherwin et al., 2016; Stilling, Dinan, & Cryan, 2016). These probiotics are now named psychobiotics (Dinan, Stanton, & Cryan, 2013); a term reflecting the justified enthusiasm for these novel discoveries, which reveal previously unknown aspects of mammalian physiology on the basis of the hologenome concept.

Interestingly, the most celebrated effects of oxytocin are those related with positive feelings and sense of well-being in the context of improved social and nonsocial behaviors and dampening of anxiety, stress, and depression (Baribeau & Anagnostou, 2015; Carter, 2014; Donaldson & Young, 2008; Feldman et al., 2016; Kosfeld et al., 2005; Lee et al., 2009). The parallels between psychobiotics and oxytocin are intriguing. Thus far, however, *L. reuteri* is the only bacteria known to upregulate endogenous oxytocin, and current literature contains few references other than

Ibrahim et al. (2014), suggesting this microbe may be a psychobiotic. A study in CD-1 mice, however, finds that experimentally introduced social disruption stress causes diminishing of colonic *L. reuteri* populations (Galley et al., 2014). A late-breaking recent study also found that consumption of *L. reuteri* lessened risk of autism spectrum disorder in infant mice (Buffington et al., 2016).

During our various studies involving feeding of *L. reuteri* in mice, we consistently observe that the probiotic relates with increased normal grooming behavior (Erdman & Poutahidis, 2014b; Ibrahim et al., 2014; Levkovich et al., 2013; Poutahidis, Kearney, et al., 2013). This behavior is an indicator of reduced stress, anxiety, and depression in mice (Smolinsky, Bergner, Laporte, & Kalueff, 2009) and is soundly connected with oxytocin (Amico, Mantella, Vollmer, & Li, 2004; Lee et al., 2009; Smolinsky et al., 2009). Also, *L. reuteri*-treated mother mice show less maternal neglect or cannibalism (Ibrahim et al., 2014), which also mimics increased maternal care conferred by oxytocin (Carter, 2014; Donaldson & Young, 2008; Farshim et al., 2016; Feldman et al., 2016; Lee et al., 2009).

Taken together these data, although preliminary, suggest that it may be worthwhile to further test dietary *L. reuteri* with its naturally induced endogenous oxytocin for psychobiotic properties.

3.5 *L. reuteri* and Oxytocin in Muscle Wasting and Bone Loss

We recently reported that *L. reuteri* counteracts age-associated sarcopenia- and cancer-induced cachexia (Varian, Goureshetti, et al., 2016). The beneficial effects on skeletal muscle coexisted with increased growth hormone levels, decreased systemic inflammatory tone, and increased Forkhead Box N1 (FoxN1) expression in the thymus gland. FoxN1-deficient mice (athymic nude) fail to inhibit sarcopenia after *L. reuteri* consumption, suggesting a FoxN1-mediated mechanism (Varian, Goureshetti, et al., 2016).

Oxytocin has also a trophic effect in skeletal muscle (Costa et al., 2014; Elabd et al., 2014). Experiments in mice show that plasma levels of oxytocin naturally decrease with age and that the hormone is necessary for muscle tissue regeneration. Oxytocin-deficient mice develop premature sarcopenia, while the administration of an oxytocin antagonist reduces muscle regeneration. By contrast, systemic administration of oxytocin activates the MAPK/ERK signaling pathway to promote aged muscle stem cell activation/proliferation and muscle regeneration (Elabd et al., 2014).

Several studies in mice demonstrate that *L. reuteri* increases bone density (Collins et al., 2016; Mccabe, Irwin, Schaefer, & Britton, 2013) and prevents bone loss in both menopausal ovariectomized and type 1 diabetes mouse models (Britton et al., 2014; Zhang et al., 2015). In the female-ovariectomized mice, the *L. reuteri* effects are mediated by the suppression of effector CD4+ T-cell-mediated osteoclastogenesis (Britton et al., 2014). In the male diabetic mice *L. reuteri* inhibited the TNF-α-induced reduction of osteoblast number and activity (Zhang et al., 2015).

Oxytocin has profound bone anabolic properties (Beranger et al., 2014; Colaianni et al., 2015; Sun, Tamma, et al., 2016; Tamma et al., 2009). Oxytocin- and oxytocin-receptor-deficient mice show osteoporosis (Tamma et al., 2009). Oxytocin is essential for balanced osteoblast and osteoclast functions. It promotes osteoblast bone formation. Although it induces osteoclast formation, at the same time inhibits their bone resorption function (Colaianni et al., 2015; Sun, Tamma, et al., 2016; Tamma et al., 2009).

4. DIRECT EVIDENCE FOR OXYTOCIN-DEPENDED GUT BACTERIA BENEFICIAL EFFECTS

Despite the compelling presumptive evidence that gut bacteria and oxytocin may interact as a part of an integral homeostatic mechanism, this hypothesis has only been recently addressed directly. In our studies we find that mice consuming *L. reuteri* have significantly increased levels of plasma oxytocin (Ibrahim et al., 2014; Poutahidis, Kearney, et al., 2013 #6581; Varian, Goureshetti, et al., 2016). In our mouse modes of skin wound healing the beneficial effect of oral *L. reuteri* treatment in wound closure and obesity is lost in the absence of oxytocin (Poutahidis, Kearney, et al., 2013 #6581; Varian, Goureshetti, et al., 2016). It is also lost after vagotomy (Poutahidis, Kearney, et al., 2013), which is consistent with dependence from oxytocin and other studies suggesting that one way of transduction of gut bacteria signals to the brain is via vagus nerve (Bravo et al., 2011).

By comparison with their wild-type counterparts, oxytocin-deficient mice, although treated with *L. reuteri*, have delayed reepithelialization, collagen maturation and fibrinogenesis, increased numbers of neutrophils, mast cells, and IL-17+ macrophages, along with decreased numbers of Tregs in their wounds at 6 days postwounding (Poutahidis, Kearney, et al., 2013).

Using adoptive cell transfer experiments, we find that the beneficial effects of edible *L. reuteri* in skin health, including accelerated wound healing and increased folliculogenesis and radiant fur, are transplantable to naïve mice receiving Tregs from *L. reuteri*-exposed donors. The ability of probiotic bacteria-primed Tregs to convey health benefits in recipient mice is lost when donor mice lack oxytocin (Erdman & Poutahidis, 2014b; Poutahidis, Kearney, et al., 2013). Likewise, the *L. reuteri*-induced protection from obesity and adipose tissue inflammation is lost when mice fed with the microbe are lacking oxytocin (Varian, Levkovich, et al., 2016).

Recently, we reported that maternal feeding with *L. reuteri* affects the sex ratio of offspring mice, i.e., mice give birth to more females. Again, this phenomenon is also not evident in oxytocin-deficient mice (Ibrahim et al., 2014).

5. OXYTOCIN AND GUT BACTERIA: AN ADVANCED QUORUM-SENSING MECHANISM OF MAMMALS?

Recent studies in free-ranging bears show that the seasonal fluctuation of their gut microbiota contributes to adjusting their energy metabolism during active or hibernating season. Germ-free mice transplanted with summer or winter bear microbiota recapitulated seasonal metabolic features of bears (Sommer et al., 2016). These facts taken together with numerous data supporting the presence of a gut microbiota-associated gut–immune–endocrine–CNS axis (Cryan & Dinan, 2012; Dinan et al., 2013; Foster & Mcvey Neufeld, 2013; Moloney et al., 2014; Pittman, 2011; Stilling et al., 2016), suggest that seasonal light cycles and seasonal foods may influence the host microbiome to communicate environmental conditions to the mammalian host.

Could, however, the same communication channels exist for mammals at a population level? If such a mechanism exists it would provide important clues for the evolutionary adjustment of mammalian populations in their environment. Our most recent studies examining connections between gut microbiota and cancer showed that dysbiosis of mothers profoundly affects physiology of F1 and F2 progeny risk for developing spontaneous cancers, obesity, reproductive and immune disorders, and a progeria-like syndrome (Poutahidis et al., 2015). Diseases were preventable by dietary supplementation with *L. reuteri*. Surely, such multigenerational phenomena of gut bacteria may affect mammalian population diversity as a whole.

In natural populations, empirical evidence indicates the maternal environment could influence offspring sex ratios. Environmental stress promotes birth of males. On the other hand, favorable environmental conditions impart a bias toward female births as a future population investment strategy during times of plenty (Kokko & Jennions, 2008; Nager, Monaghan, Griffiths, Houston, & Dawson, 1999; Pryke & Rollins, 2012; Pryke, Rollins, & Griffith, 2011; Trivers & Willard, 1973). Along these lines we find that feeding mother mice with *L. reuteri* increased infant survival and promoted a higher ratio of female:male offspring in an oxytocin-dependent manner. This unexpected transmission of environmental information by GI tract bacteria to the host population occurred via an oxytocin-dependent mechanism (Ibrahim et al., 2014). Hypothesizing a role for oxytocin in an advanced mammalian quorum-sensing mechanism would not be unsound, given the important role of this hormone in social and reproduction-associated behaviors, parental care as well as feeding, and metabolism (Baribeau & Anagnostou, 2015; Carter, 2014; Donaldson & Young, 2008; Feldman et al., 2016; Gimpl et al., 2001; Lee et al., 2009; Shen, 2015).

6. PROBIOTIC BACTERIA-INDUCED ENDOGENOUS OXYTOCIN FOR THERAPY

A major limitation of oxytocin therapy is difficulty in dosing. Oxytocin has a short life in the blood. In the cerebrospinal fluid it remains active much longer. Therefore it is usually administered to humans using a nasal spray (Lee et al., 2009; Shen, 2015; Spetter & Hallschmid, 2015; Taylor, Lee, & Buisman-Pijlman, 2014; Viero et al., 2010). This method bypasses the blood–brain barrier and allows oxytocin to reach the cerebrospinal fluid and consequently the brain. Recent evidence suggests that this route has satisfactory results in increasing both blood and cerebrospinal levels of oxytocin (Lefevre & Sirigu, 2016; Spetter & Hallschmid, 2015; Taylor et al., 2014; Viero et al., 2010). However, other authors believe that only a small proportion of intranasally delivered oxytocin finally reaches the brain (Leng & Ludwig, 2016; Viero et al., 2010). In the treatment of neuropsychiatric disorders several issues are being raised regarding effectiveness and safety of exogenous oxytocin chronic administration, especially in children and adolescents (Lefevre & Sirigu, 2016; Taylor et al., 2014). Also, the risk of desensitization of the endogenous oxytocinergic system following chronic intranasal administration is always an unwanted possibility (Lefevre & Sirigu, 2016).

While soundly designed trial studies are needed to access efficacy and safety of exogenous oxytocin, the possibility that certain probiotic bacteria could be a safe and natural modality to achieve high endogenous oxytocin levels should also be tested. Such probiotic bacteria would not simply raise oxytocin levels, as with exogenous administration strategy. They would rather stimulate a more complex homeostatic pathway involving interrelated gut, immune, endocrine, and brain functions. In that way, optimal therapeutic outcomes would not just be based in the chemistry of a single molecule. Instead, restoration of health will be based on stimulating powerful but latent genetic programs, stemming from the coevolution of mammals and their microbiome. These unified programs will together achieve peak reproductive fitness and youthfulness, and at the same time prevent or enhance treatments for important diseases including neuropsychiatric disorders, obesity and metabolic syndrome, uncontrolled and chronic inflammation, and cancer and senility-associated atrophic changes.

REFERENCES

Ahl, D., Liu, H., Schreiber, O., Roos, S., Phillipson, M., & Holm, L. (2016). Lactobacillus reuteri increases mucus thickness and ameliorates DSS-induced colitis in mice. *Acta Physiologica (Oxford, England)*, *217*(4), 300–310.

Al-Amran, F., & Shahkolahi, M. (2013). Oxytocin ameliorates the immediate myocardial injury in rat heart transplant through downregulation of neutrophil-dependent myocardial apoptosis. *Transplantation Proceedings*, *45*, 2506–2512.

Albenberg, L. G., & Wu, G. D. (2014). Diet and the intestinal microbiome: Associations, functions, and implications for health and disease. *Gastroenterology*, *146*, 1564–1572.

Altirriba, J., Poher, A. L., Caillon, A., Arsenijevic, D., Veyrat-Durebex, C., Lyautey, J., et al. (2014). Divergent effects of oxytocin treatment of obese diabetic mice on adiposity and diabetes. *Endocrinology*, *155*, 4189–4201.

Amico, J. A., Mantella, R. C., Vollmer, R. R., & Li, X. (2004). Anxiety and stress responses in female oxytocin deficient mice. *Journal of Neuroendocrinology*, *16*, 319–324.

Arletti, R., Benelli, A., & Bertolini, A. (1990). Oxytocin inhibits food and fluid intake in rats. *Physiology & Behavior*, *48*, 825–830.

Balzola, F., Bernstein, C., Ho, G. T., & Lees, C. (2010). Combined administration of secretin and oxytocin inhibits chronic colitis and associated activation of forebrain neurons: Commentary. *Inflammatory Bowel Disease Monitoring*, *22*(6)654-e202.

Barengolts, E. (2016). Oxytocin—An emerging treatment for obesity and dysglycemia: Review of randomized controlled trials and cohort studies. *Endocrine Practice*, *22*(7), 885–894.

Baribeau, D. A., & Anagnostou, E. (2015). Oxytocin and vasopressin: Linking pituitary neuropeptides and their receptors to social neurocircuits. *Frontiers in Neuroscience*, *9*, 335.

Belkaid, Y., & Hand, T. W. (2014). Role of the microbiota in immunity and inflammation. *Cell*, *157*, 121–141.

Beranger, G. E., Pisani, D. F., Castel, J., Djedaini, M., Battaglia, S., Amiaud, J., et al. (2014). Oxytocin reverses ovariectomy-induced osteopenia and body fat gain. *Endocrinology, 155*, 1340–1352.

Biyikli, N. K., Tugtepe, H., Sener, G., Velioglu-Ogunc, A., Cetinel, S., Midillioglu, S., et al. (2006). Oxytocin alleviates oxidative renal injury in pyelonephritic rats via a neutrophil-dependent mechanism. *Peptides, 27*, 2249–2257.

Blevins, J. E., & Baskin, D. G. (2015). Translational and therapeutic potential of oxytocin as an anti-obesity strategy: Insights from rodents, nonhuman primates and humans. *Physiology & Behavior, 152*, 438–449.

Blevins, J. E., Eakin, T. J., Murphy, J. A., Schwartz, M. W., & Baskin, D. G. (2003). Oxytocin innervation of caudal brainstem nuclei activated by cholecystokinin. *Brain Research, 993*, 30–41.

Blevins, J. E., Graham, J. L., Morton, G. J., Bales, K. L., Schwartz, M. W., Baskin, D. G., et al. (2015). Chronic oxytocin administration inhibits food intake, increases energy expenditure, and produces weight loss in fructose-fed obese rhesus monkeys. *American Journal of Physiology. Regulatory, Integrative and Comparative Physiology, 308*, R431–R438.

Blevins, J. E., & Ho, J. M. (2013). Role of oxytocin signaling in the regulation of body weight. *Reviews in Endocrine & Metabolic Disorders, 14*, 311–329.

Blevins, J. E., Thompson, B. W., Anekonda, V. T., Ho, J. M., Graham, J. L., Roberts, Z. S., et al. (2016). Chronic CNS oxytocin signaling preferentially induces fat loss in high-fat diet-fed rats by enhancing satiety responses and increasing lipid utilization. *American Journal of Physiology. Regulatory, Integrative and Comparative Physiology, 310*, R640–R658.

Boulange, C. L., Neves, A. L., Chilloux, J., Nicholson, J. K., & Dumas, M. E. (2016). Impact of the gut microbiota on inflammation, obesity, and metabolic disease. *Genome Medicine, 8*, 42.

Bravo, J. A., Forsythe, P., Chew, M. V., Escaravage, E., Savignac, H. M., Dinan, T. G., et al. (2011). Ingestion of Lactobacillus strain regulates emotional behavior and central GABA receptor expression in a mouse via the vagus nerve. *Proceedings of the National Academy of Sciences of the United States of America, 108*, 16050–16055.

Britton, R. A., Irwin, R., Quach, D., Schaefer, L., Zhang, J., Lee, T., et al. (2014). Probiotic L. reuteri treatment prevents bone loss in a menopausal ovariectomized mouse model. *Journal of Cellular Physiology, 229*, 1822–1830.

Buffington, S. A., Di Prisco, G. V., Auchtung, T. A., Ajami, N. J., Petrosino, J. F., & Costa-Mattioli, M. (2016). Microbial reconstitution reverses maternal diet-induced social and synaptic deficits in offspring. *Cell, 165*, 1762–1775.

Bull, M. J., & Plummer, N. T. (2014). Part 1: The human gut microbiome in health and disease. *Integrative Medicine (Encinitas), 13*, 17–22.

Bull, M. J., & Plummer, N. T. (2015). Part 2: Treatments for chronic gastrointestinal disease and gut dysbiosis. *Integrative Medicine (Encinitas), 14*, 25–33.

Camerino, C. (2009). Low sympathetic tone and obese phenotype in oxytocin-deficient mice. *Obesity (Silver Spring), 17*, 980–984.

Carter, C. S. (2014). Oxytocin pathways and the evolution of human behavior. *Annual Review of Psychology, 65*, 17–39.

Chang, C., & Lin, H. (2016). Dysbiosis in gastrointestinal disorders. *Best Practice and Research in Clinical Gastroenterology, 30*, 3–15.

Chassaing, B., Koren, O., Goodrich, J. K., Poole, A. C., Srinivasan, S., Ley, R. E., et al. (2015). Dietary emulsifiers impact the mouse gut microbiota promoting colitis and metabolic syndrome. *Nature, 519*, 92–96.

Chung, H. J., Yu, J. G., Lee, I. A., Liu, M. J., Shen, Y. F., Sharma, S. P., et al. (2016). Intestinal removal of free fatty acids from hosts by Lactobacilli for the treatment of obesity. *FEBS Open Bio, 6*, 64–76.

Clemente, J. C., Ursell, L. K., Parfrey, L. W., & Knight, R. (2012). The impact of the gut microbiota on human health: An integrative view. *Cell*, *148*, 1258–1270.
Clodi, M., Vila, G., Geyeregger, R., Riedl, M., Stulnig, T. M., Struck, J., et al. (2008). Oxytocin alleviates the neuroendocrine and cytokine response to bacterial endotoxin in healthy men. *American Journal of Physiology. Endocrinology and Metabolism*, *295*, E686–E691.
Colaianni, G., Sun, L., Zaidi, M., & Zallone, A. (2015). The "love hormone" oxytocin regulates the loss and gain of the fat-bone relationship. *Frontiers in Endocrinology (Lausanne)*, *6*, 79.
Collado, M. C., Cernada, M., Bauerl, C., Vento, M., & Perez-Martinez, G. (2012). Microbial ecology and host-microbiota interactions during early life stages. *Gut Microbes*, *3*, 352–365.
Collins, F. L., Irwin, R., Bierhalter, H., Schepper, J., Britton, R. A., Parameswaran, N., et al. (2016). Lactobacillus reuteri 6475 increases bone density in intact females only under an inflammatory setting. *PloS One*, *11*, e0153180.
Conlon, M. A., & Bird, A. R. (2015). The impact of diet and lifestyle on gut microbiota and human health. *Nutrients*, *7*, 17–44.
Costa, A., Rossi, E., Scicchitano, B. M., Coletti, D., Moresi, V., & Adamo, S. (2014). Neurohypophyseal hormones: Novel actors of striated muscle development and homeostasis. *European Journal of Translational Myology*, *24*, 3790.
Cotillard, A., Kennedy, S. P., Kong, L. C., Prifti, E., Pons, N., Le Chatelier, E., et al. (2013). Dietary intervention impact on gut microbial gene richness. *Nature*, *500*, 585–588.
Cryan, J. F., & Dinan, T. G. (2012). Mind-altering microorganisms: The impact of the gut microbiota on brain and behaviour. *Nature Reviews. Neuroscience*, *13*, 701–712.
Deblon, N., Veyrat-Durebex, C., Bourgoin, L., Caillon, A., Bussier, A. L., Petrosino, S., et al. (2011). Mechanisms of the anti-obesity effects of oxytocin in diet-induced obese rats. *PloS One*, *6*, e25565.
Dept. of Reproductive Health and Research, W. (2012). *WHO recommendations for the prevention and treatment of postpartum haemorrhage*. Geneva: World Health Organization. Accessed June 2016.
Dethlefsen, L., Mcfall-Ngai, M., & Relman, D. A. (2007). An ecological and evolutionary perspective on human-microbe mutualism and disease. *Nature*, *449*, 811–818.
Dinan, T. G., Stanton, C., & Cryan, J. F. (2013). Psychobiotics: A novel class of psychotropic. *Biological Psychiatry*, *74*, 720–726.
Donaldson, Z. R., & Young, L. J. (2008). Oxytocin, vasopressin, and the neurogenetics of sociality. *Science*, *322*, 900–904.
Douglas, A. J., Johnstone, L. E., & Leng, G. (2007). Neuroendocrine mechanisms of change in food intake during pregnancy: A potential role for brain oxytocin. *Physiology & Behavior*, *91*, 352–365.
Drissi, F., Raoult, D., & Merhej, V. (2016). Metabolic role of lactobacilli in weight modification in humans and animals. *Microbial Pathogenesis*. http://dx.doi.org/10.1016/j.micpath.2016.03.006pii: S0882-4010(15)30152-2. [Epub ahead of print].
Ducatelle, R., Eeckhaut, V., Haesebrouck, F., & Van Immerseel, F. (2015). A review on prebiotics and probiotics for the control of dysbiosis: Present status and future perspectives. *Animal*, *9*, 43–48.
Eaton, K. A., Honkala, A., Auchtung, T. A., & Britton, R. A. (2011). Probiotic Lactobacillus reuteri ameliorates disease due to enterohemorrhagic Escherichia coli in germfree mice. *Infection and Immunity*, *79*, 185–191.
Elabd, C., Basillais, A., Beaupied, H., Breuil, V., Wagner, N., Scheideler, M., et al. (2008). Oxytocin controls differentiation of human mesenchymal stem cells and reverses osteoporosis. *Stem Cells*, *26*(9), 2399–2407.

Elabd, C., Cousin, W., Upadhyayula, P., Chen, R. Y., Chooljian, M. S., Li, J., et al. (2014). Oxytocin is an age-specific circulating hormone that is necessary for muscle maintenance and regeneration. *Nature Communications*, *5*, 4082.

Elands, J., Resink, A., & De Kloet, E. R. (1990). Neurohypophyseal hormone receptors in the rat thymus, spleen, and lymphocytes. *Endocrinology*, *126*, 2703–2710.

Eming, S. A., Martin, P., & Tomic-Canic, M. (2014). Wound repair and regeneration: Mechanisms, signaling, and translation. *Science Translational Medicine*, *6*, 265sr6.

Erdman, S. E., & Poutahidis, T. (2010). Cancer inflammation and regulatory T cells. *International Journal of Cancer*, *127*, 768–779.

Erdman, S. E., & Poutahidis, T. (2014a). The microbiome modulates the tumor macroenvironment. *Oncoimmunology*, *3*, e28271.

Erdman, S. E., & Poutahidis, T. (2014b). Probiotic 'glow of health': It's more than skin deep. *Beneficial Microbes*, *5*, 109–119.

Erdman, S. E., & Poutahidis, T. (2015). Gut bacteria and cancer. *Biochimica et Biophysica Acta*, *1856*, 86–90.

Erkanli Senturk, G., Erkanli, K., Aydin, U., Yucel, D., Isiksacan, N., Ercan, F., et al. (2013). The protective effect of oxytocin on ischemia/reperfusion injury in rat urinary bladder. *Peptides*, *40*, 82–88.

Fabia, R., Ar'rajab, A., Johansson, M. L., Willen, R., Andersson, R., Molin, G., et al. (1993). The effect of exogenous administration of Lactobacillus reuteri R2LC and oat fiber on acetic acid-induced colitis in the rat. *Scandinavian Journal of Gastroenterology*, *28*, 155–162.

Fak, F., & Backhed, F. (2012). Lactobacillus reuteri prevents diet-induced obesity, but not atherosclerosis, in a strain dependent fashion in Apoe −/− Mice. *PloS One*, *7*, e46837.

Farshim, P., Walton, G., Chakrabarti, B., Givens, I., Saddy, D., Kitchen, I., et al. (2016). Maternal weaning modulates emotional behavior and regulates the gut-brain axis. *Scientific Reports*, *6*, 21958.

Feldman, R., Monakhov, M., Pratt, M., & Ebstein, R. P. (2016). Oxytocin pathway genes: Evolutionary ancient system impacting on human affiliation, sociality, and psychopathology. *Biological Psychiatry*, *79*, 174–184.

Floch, M. H., Walker, W. A., Sanders, M. E., Nieuwdorp, M., Kim, A. S., Brenner, D. A., et al. (2015). Recommendations for probiotic use—2015 update: Proceedings and consensus opinion. *Journal of Clinical Gastroenterology*, *49*(Suppl. 1), S69–S73.

Foster, J. A., & Mcvey Neufeld, K. A. (2013). Gut-brain axis: How the microbiome influences anxiety and depression. *Trends in Neurosciences*, *36*(5), 305–312.

Fujimura, K. E., Slusher, N. A., Cabana, M. D., & Lynch, S. V. (2010). Role of the gut microbiota in defining human health. *Expert Review of Anti-Infective Therapy*, *8*, 435–454.

Galley, J. D., Nelson, M. C., Yu, Z., Dowd, S. E., Walter, J., Kumar, P. S., et al. (2014). Exposure to a social stressor disrupts the community structure of the colonic mucosa-associated microbiota. *BMC Microbiology*, *14*, 189.

Gao, C., Major, A., Rendon, D., Lugo, M., Jackson, V., Shi, Z., et al. (2015). Histamine H2 receptor-mediated suppression of intestinal inflammation by probiotic Lactobacillus reuteri. *MBio*, *6*, e01358-15.

Gavrilenko, V. G., Esipov, V. K., & Sivozhelezov, K. G. (2003). Morphological characteristic of wound healing process in patients with diabetic purulent-necrotic foot lesion treated with oxytocin. *Morfologiia*, *124*, 24–27.

Gimpl, G., Fahrenholz, F., & Gene, C. (2001). The oxytocin receptor system: Structure, function, and regulation. *Physiological Reviews*, *81*, 629–683.

Gouin, J. P., Carter, C. S., Pournajafi-Nazarloo, H., Glaser, R., Malarkey, W. B., Loving, T. J., et al. (2010). Marital behavior, oxytocin, vasopressin, and wound healing. *Psychoneuroendocrinology*, *35*, 1082–1090.

Gregor, M. F., & Hotamisligil, G. S. (2011). Inflammatory mechanisms in obesity. *Annual Review of Immunology, 29*, 415–445.
Gueniche, A., Bastien, P., Ovigne, J. M., Kermici, M., Courchay, G., Chevalier, V., et al. (2010). Bifidobacterium longum lysate, a new ingredient for reactive skin. *Experimental Dermatology, 16*, 511–517.
Gurtner, G. C., Werner, S., Barrandon, Y., & Longaker, M. T. (2008). Wound repair and regeneration. *Nature, 453*, 314–321.
Hamasaki, M. Y., Barbeiro, H. V., Barbeiro, D. F., Cunha, D. M., Koike, M. K., Machado, M. C., et al. (2016). Neuropeptides in the brain defense against distant organ damage. *Journal of Neuroimmunology, 290*, 33–35.
Hansenne, I., Louis, C., Martens, H., Dorban, G., Charlet-Renard, C., Peterson, P., et al. (2009). Aire and Foxp3 expression in a particular microenvironment for T cell differentiation. *Neuroimmunomodulation, 16*, 35–44.
Hansenne, I., Rasier, G., Pequeux, C., Brilot, F., Renard, C., Breton, C., et al. (2005). Ontogenesis and functional aspects of oxytocin and vasopressin gene expression in the thymus network. *Journal of Neuroimmunology, 158*, 67–75.
Hildebrandt, M. A., Hoffmann, C., Sherrill-Mix, S. A., Keilbaugh, S. A., Hamady, M., Chen, Y. Y., et al. (2009). High-fat diet determines the composition of the murine gut microbiome independently of obesity. *Gastroenterology, 137*. 1716-24.e1–2.
Ho, J. M., & Blevins, J. E. (2013). Coming full circle: Contributions of central and peripheral oxytocin actions to energy balance. *Endocrinology, 154*, 589–596.
Ho, J. T., Chan, G. C., & Li, J. C. (2015). Systemic effects of gut microbiota and its relationship with disease and modulation. *BMC Immunology, 16*, 21.
Hou, C., Zeng, X., Yang, F., Liu, H., & Qiao, S. (2015). Study and use of the probiotic Lactobacillus reuteri in pigs: A review. *Journal of Animal Science and Biotechnology, 6*, 14.
Houshmand, F., Faghihi, M., & Zahediasl, S. (2009). Biphasic protective effect of oxytocin on cardiac ischemia/reperfusion injury in anaesthetized rats. *Peptides, 30*, 2301–2308.
Hsieh, F. C., Lan, C. C., Huang, T. Y., Chen, K. W., Chai, C. Y., Chen, W. T., et al. (2016). Heat-killed and live Lactobacillus reuteri GMNL-263 exhibit similar effects on improving metabolic functions in high-fat diet-induced obese rats. *Food & Function, 7*, 2374–2388.
Huang, W. C., Chen, Y. M., Kan, N. W., Ho, C. S., Wei, L., Chan, C. H., et al. (2015). Hypolipidemic effects and safety of Lactobacillus reuteri 263 in a hamster model of hyperlipidemia. *Nutrients, 7*, 3767–3782.
Huseini, H. F., Rahimzadeh, G., Fazeli, M. R., Mehrazma, M., & Salehi, M. (2012). Evaluation of wound healing activities of kefir products. *Burns, 38*, 719–723.
Ibrahim, Y. M., Kearney, S. M., Levkovich, T., Springer, A., Mirabal, S., Poutahidis, T., et al. (2014). Maternal gut microbes control offspring sex and survival. *Journal of Probiotics and Health, 2*, 6. http://dx.doi.org/10.4172/2329-8901.1000120.
Iseri, S. O., Dusunceli, F., Erzik, C., Uslu, B., Arbak, S., & Yegen, B. C. (2010). Oxytocin or social housing alleviates local burn injury in rats. *Journal of Surgical Research, 162*, 122–131.
Iseri, S. O., Gedik, I. E., Erzik, C., Uslu, B., Arbak, S., Gedik, N., et al. (2008). Oxytocin ameliorates skin damage and oxidant gastric injury in rats with thermal trauma. *Burns, 34*, 361–369.
Iseri, S. O., Sener, G., Saglam, B., Gedik, N., Ercan, F., & Yegen, B. C. (2005a). Oxytocin ameliorates oxidative colonic inflammation by a neutrophil-dependent mechanism. *Peptides, 26*, 483–491.
Iseri, S. O., Sener, G., Saglam, B., Gedik, N., Ercan, F., & Yegen, B. C. (2005b). Oxytocin protects against sepsis-induced multiple organ damage: Role of neutrophils. *Journal of Surgical Research, 126*, 73–81.
Iwasaki, M. (2015). Chromatin resetting mechanisms preventing transgenerational inheritance of epigenetic states. *Frontiers in Plant Science, 6*, 380.

Jankowski, M., Bissonauth, V., Gao, L., Gangal, M., Wang, D., Danalache, B., et al. (2010). Anti-inflammatory effect of oxytocin in rat myocardial infarction. *Basic Research in Cardiology, 105*, 205–218.

Jia, H. Y., Liu, J. L., Zhou, C. J., Kong, F., Yuan, M. Z., Sun, W. D., et al. (2014). High expression of MICA in human kidney cancer tissue and renal cell carcinoma lines. *Asian Pacific Journal of Cancer Prevention, 15*, 1715–1717.

Johnson, H. M., & Torres, B. A. (1985). Regulation of lymphokine production by arginine vasopressin and oxytocin: Modulation of lymphocyte function by neurohypophyseal hormones. *Journal of Immunology, 135*, 773s–775s.

Kallus, S. J., & Brandt, L. J. (2012). The intestinal microbiota and obesity. *Journal of Clinical Gastroenterology, 46*, 16–24.

Kamada, N., & Nunez, G. (2013). Role of the gut microbiota in the development and function of lymphoid cells. *Journal of Immunology, 190*, 1389–1395.

Karelina, K., Stuller, K. A., Jarrett, B., Zhang, N., Wells, J., Norman, G. J., et al. (2011). Oxytocin mediates social neuroprotection after cerebral ischemia. *Stroke, 42*, 3606–3611.

Karimi, K., Inman, M. D., Bienenstock, J., & Forsythe, P. (2009). Lactobacillus reuteri-induced regulatory T cells protect against an allergic airway response in mice. *American Journal of Respiratory and Critical Care Medicine, 179*, 186–193.

Kelly, J. R., Clarke, G., Cryan, J. F., & Dinan, T. G. (2016). Brain-gut-microbiota axis: Challenges for translation in psychiatry. *Annals of Epidemiology, 26*, 366–372.

Knobloch, H. S., & Grinevich, V. (2014). Evolution of oxytocin pathways in the brain of vertebrates. *Frontiers in Behavioral Neuroscience, 8*, 31.

Kobyliak, N., Virchenko, O., & Falalyeyeva, T. (2016). Pathophysiological role of host microbiota in the development of obesity. *Nutrition Journal, 15*, 43.

Kokko, H., & Jennions, M. D. (2008). Parental investment, sexual selection and sex ratios. *Journal of Evolutionary Biology, 21*, 919–948.

Koren, O., Goodrich, J. K., Cullender, T. C., Spor, A., Laitinen, K., Backhed, H. K., et al. (2012). Host remodeling of the gut microbiome and metabolic changes during pregnancy. *Cell, 150*, 470–480.

Koren, L., Whiteside, D., Fahlman, S., Ruckstuhl, K., Kutz, S., Checkley, S., et al. (2012). Cortisol and corticosterone independence in cortisol-dominant wildlife. *General and Comparative Endocrinology, 177*, 113–119.

Kosfeld, M., Heinrichs, M., Zak, P. J., Fischbacher, U., & Fehr, E. (2005). Oxytocin increases trust in humans. *Nature, 435*, 673–676.

Lakritz, J. R., Poutahidis, T., Levkovich, T., Varian, B. J., Ibrahim, Y. M., Chatzigiagkos, A., et al. (2014). Beneficial bacteria stimulate host immune cells to counteract dietary and genetic predisposition to mammary cancer in mice. *International Journal of Cancer, 135*, 529–540.

Lau, E., Carvalho, D., Pina-Vaz, C., Barbosa, J. A., & Freitas, P. (2015). Beyond gut microbiota: Understanding obesity and type 2 diabetes. *Hormones (Athens, Greece), 14*(3), 358–369.

Lee, H. J., Caldwell, H. K., Macbeth, A. H., Tolu, S. G., & Young, W. S., 3rd. (2008). A conditional knockout mouse line of the oxytocin receptor. *Endocrinology, 149*, 3256–3263.

Lee, H. J., Macbeth, A. H., Pagani, J. H., & Young, W. S., 3rd. (2009). Oxytocin: The great facilitator of life. *Progress in Neurobiology, 88*, 127–151.

Lee, Y. K., & Mazmanian, S. K. (2010). Has the microbiota played a critical role in the evolution of the adaptive immune system? *Science, 330*, 1768–1773.

Lefevre, A., & Sirigu, A. (2016). The two fold role of oxytocin in social developmental disorders: A cause and a remedy? *Neuroscience and Biobehavioral Reviews, 63*, 168–176.

Leng, G., & Ludwig, M. (2016). Intranasal oxytocin: Myths and delusions. *Biological Psychiatry, 79*, 243–250.

Levkovich, T., Poutahidis, T., Smillie, C., Varian, B. J., Ibrahim, Y. M., Lakritz, J. R., et al. (2013). Probiotic bacteria induce a 'glow of health'. *PloS One, 8*, e53867.

Lin, Y. P., Thibodeaux, C. H., Pena, J. A., Ferry, G. D., & Versalovic, J. (2008). Probiotic Lactobacillus reuteri suppress proinflammatory cytokines via c-Jun. *Inflammatory Bowel Diseases, 14*, 1068–1083.

Liu, Y., Fatheree, N. Y., Mangalat, N., & Rhoads, J. M. (2010). Human-derived probiotic Lactobacillus reuteri strains differentially reduce intestinal inflammation. *American Journal of Physiology. Gastrointestinal and Liver Physiology, 299*, G1087–G1096.

Liu, Y., Fatheree, N. Y., Mangalat, N., & Rhoads, J. M. (2012). Lactobacillus reuteri strains reduce incidence and severity of experimental necrotizing enterocolitis via modulation of TLR4 and NF-kappaB signaling in the intestine. *American Journal of Physiology. Gastrointestinal and Liver Physiology, 302*, G608–G617.

Livingston, M., Loach, D., Wilson, M., Tannock, G. W., & Baird, M. (2010). Gut commensal Lactobacillus reuteri 100–23 stimulates an immunoregulatory response. *Immunology and Cell Biology, 88*, 99–102.

Lokrantz, C. M., Uvnas-Moberg, K., & Kaplan, J. M. (1997). Effects of central oxytocin administration on intraoral intake of glucose in deprived and nondeprived rats. *Physiology & Behavior, 62*, 347–352.

Lorea Baroja, M., Kirjavainen, P. V., Hekmat, S., & Reid, G. (2007). Anti-inflammatory effects of probiotic yogurt in inflammatory bowel disease patients. *Clinical and Experimental Immunology, 149*, 470–479.

Maccio, A., Madeddu, C., Chessa, P., Panzone, F., Lissoni, P., & Mantovani, G. (2010). Oxytocin both increases proliferative response of peripheral blood lymphomonocytes to phytohemagglutinin and reverses immunosuppressive estrogen activity. *In Vivo, 24*, 157–163.

Maejima, Y., Iwasaki, Y., Yamahara, Y., Kodaira, M., Sedbazar, U., & Yada, T. (2011). Peripheral oxytocin treatment ameliorates obesity by reducing food intake and visceral fat mass. *Aging (Albany NY), 3*, 1169–1177.

Maejima, Y., Sedbazar, U., Suyama, S., Kohno, D., Onaka, T., Takano, E., et al. (2009). Nesfatin-1-regulated oxytocinergic signaling in the paraventricular nucleus causes anorexia through a leptin-independent melanocortin pathway. *Cell Metabolism, 10*, 355–365.

Manichanh, C., Borruel, N., Casellas, F., & Guarner, F. (2012). The gut microbiota in IBD. *Nature Reviews. Gastroenterology & Hepatology, 9*, 599–608.

Marchesi, J. R., Adams, D. H., Fava, F., Hermes, G. D., Hirschfield, G. M., Hold, G., et al. (2016). The gut microbiota and host health: A new clinical frontier. *Gut, 65*, 330–339.

Matsuura, T., Kawasaki, M., Hashimoto, H., Yoshimura, M., Motojima, Y., Saito, R., et al. (2016). Effects of central administration of oxytocin-saporin cytotoxin on chronic inflammation and feeding/drinking behaviors in adjuvant arthritic rats. *Neuroscience Letters, 621*, 104–110.

Mayes, T., Gottschlich, M. M., James, L. E., Allgeier, C., Weitz, J., & Kagan, R. J. (2015). Clinical safety and efficacy of probiotic administration following burn injury. *Journal of Burn Care & Research, 36*, 92–99.

Maynard, C. L., Elson, C. O., Hatton, R. D., & Weaver, C. T. (2012). Reciprocal interactions of the intestinal microbiota and immune system. *Nature, 489*, 231–241.

Mccabe, L. R., Irwin, R., Schaefer, L., & Britton, R. A. (2013). Probiotic use decreases intestinal inflammation and increases bone density in healthy male but not female mice. *Journal of Cellular Physiology, 228*, 1793–1798.

Mcfall-Ngai, M., Hadfield, M. G., Bosch, T. C., Carey, H. V., Domazet-Loso, T., Douglas, A. E., et al. (2013). Animals in a bacterial world, a new imperative for the life sciences. *Proceedings of the National Academy of Sciences of the United States of America, 110*, 3229–3236.

Mechoud, M. A., Mateos, M. V., De Valdez, G. F., Villena, J., Salvador, G. A., & Rodriguez, A. V. (2012). Lactobacillus reuteri CRL1098 soluble factors modulate tumor necrosis factor alpha production in peripheral blood mononuclear cells: Involvement of lipid rafts. *International Immunopharmacology, 14*, 446–453.

Milani, C., Ferrario, C., Turroni, F., Duranti, S., Mangifesta, M., Van Sinderen, D., et al. (2016). The human gut microbiota and its interactive connections to diet. *Journal of Human Nutrition and Dietetics*. http://dx.doi.org/10.1111/jhn.12371[Epub ahead of print].

Million, M., Angelakis, E., Maraninchi, M., Henry, M., Giorgi, R., Valero, R., et al. (2013). Correlation between body mass index and gut concentrations of Lactobacillus reuteri, Bifidobacterium animalis, Methanobrevibacter smithii and Escherichia coli. *International Journal of Obesity (London), 37*(11), 1460–1466.

Million, M., Maraninchi, M., Henry, M., Armougom, F., Richet, H., Carrieri, P., et al. (2012). Obesity-associated gut microbiota is enriched in Lactobacillus reuteri and depleted in Bifidobacterium animalis and Methanobrevibacter smithii. *International Journal of Obesity, 36*, 817–825.

Moghimian, M., Faghihi, M., Karimian, S. M., & Imani, A. (2012). The effect of acute stress exposure on ischemia and reperfusion injury in rat heart: Role of oxytocin. *Stress, 15*, 385–392.

Molin, G., Jeppsson, B., Johansson, M. L., Ahrne, S., Nobaek, S., Stahl, M., et al. (1993). Numerical taxonomy of Lactobacillus spp. associated with healthy and diseased mucosa of the human intestines. *Journal of Applied Bacteriology, 74*, 314–323.

Moloney, R. D., Desbonnet, L., Clarke, G., Dinan, T. G., & Cryan, J. F. (2014). The microbiome: Stress, health and disease. *Mammalian Genome, 25*, 49–74.

Morton, G. J., Thatcher, B. S., Reidelberger, R. D., Ogimoto, K., Wolden-Hanson, T., Baskin, D. G., et al. (2012). Peripheral oxytocin suppresses food intake and causes weight loss in diet-induced obese rats. *American Journal of Physiology. Endocrinology and Metabolism, 302*, E134–E144.

Mozaffarian, D., Hao, T., Rimm, E. B., Willett, W. C., & Hu, F. B. (2011). Changes in diet and lifestyle and long-term weight gain in women and men. *New England Journal of Medicine, 364*, 2392–2404.

Muchmore, D. B., Little, S. A., & De Haen, C. (1981). A dual mechanism of action of oxytocin in rat epididymal fat cells. *Journal of Biological Chemistry, 256*, 365–372.

Mullis, K., Kay, K., & Williams, D. L. (2013). Oxytocin action in the ventral tegmental area affects sucrose intake. *Brain Research, 1513*, 85–91.

Murphy, E. F., Cotter, P. D., Healy, S., Marques, T. M., O'sullivan, O., Fouhy, F., et al. (2010). Composition and energy harvesting capacity of the gut microbiota: Relationship to diet, obesity and time in mouse models. *Gut, 59*, 1635–1642.

Nagao-Kitamoto, H., Kitamoto, S., Kuffa, P., & Kamada, N. (2016). Pathogenic role of the gut microbiota in gastrointestinal diseases. *Intestinal Research, 14*, 127–138.

Nager, R. G., Monaghan, P., Griffiths, R., Houston, D. C., & Dawson, R. (1999). Experimental demonstration that offspring sex ratio varies with maternal condition. *Proceedings of the National Academy of Sciences of the United States of America, 96*, 570–573.

Nation, D. A., Szeto, A., Mendez, A. J., Brooks, L. G., Zaias, J., Herderick, E. E., et al. (2010). Oxytocin attenuates atherosclerosis and adipose tissue inflammation in socially isolated ApoE −/− mice. *Psychosomatic Medicine, 72*, 376–382.

Neish, A. S. (2009). Microbes in gastrointestinal health and disease. *Gastroenterology, 136*, 65–80.

Nieuwdorp, M., Gilijamse, P. W., Pai, N., & Kaplan, L. M. (2014). Role of the microbiome in energy regulation and metabolism. *Gastroenterology, 146*, 1525–1533.

Niv, E., Naftali, T., Hallak, R., & Vaisman, N. (2005). The efficacy of Lactobacillus reuteri ATCC 55730 in the treatment of patients with irritable bowel syndrome—A double blind, placebo-controlled, randomized study. *Clinical Nutrition, 24*, 925–931.

Nomura, T., & Sakaguchi, S. (2007). Foxp3 and aire in thymus-generated Treg cells: A link in self-tolerance. *Nature Immunology*, *8*, 333–334.

Noverr, M. C., & Huffnagle, G. B. (2004). Does the microbiota regulate immune responses outside the gut? *Trends in Microbiology*, *12*, 562–568.

Oliva, S., Di Nardo, G., Ferrari, F., Mallardo, S., Rossi, P., Patrizi, G., et al. (2012). Randomised clinical trial: The effectiveness of Lactobacillus reuteri ATCC 55730 rectal enema in children with active distal ulcerative colitis. *Alimentary Pharmacology & Therapeutics*, *35*, 327–334.

Olson, B. R., Drutarosky, M. D., Chow, M. S., Hruby, V. J., Stricker, E. M., & Verbalis, J. G. (1991). Oxytocin and an oxytocin agonist administered centrally decrease food intake in rats. *Peptides*, *12*, 113–118.

Olszewski, P. K., Klockars, A., Olszewska, A. M., Fredriksson, R., Schioth, H. B., & Levine, A. S. (2010). Molecular, immunohistochemical, and pharmacological evidence of oxytocin's role as inhibitor of carbohydrate but not fat intake. *Endocrinology*, *151*, 4736–4744.

O'toole, P. W., & Jeffery, I. B. (2015). Gut microbiota and aging. *Science*, *350*, 1214–1215.

Peral, M. C., Martinez, M. A., & Valdez, J. C. (2009). Bacteriotherapy with Lactobacillus plantarum in burns. *International Wound Journal*, *6*, 73–81.

Petersson, M., Wiberg, U., Lundeberg, T., & Uvnas-Moberg, K. (2001). Oxytocin decreases carrageenan induced inflammation in rats. *Peptides*, *22*, 1479–1484.

Pittman, Q. J. (2011). A neuro-endocrine-immune symphony. *Journal of Neuroendocrinology*, *23*, 1296–1297.

Poggi, M., Bastelica, D., Gual, P., Iglesias, M. A., Gremeaux, T., Knauf, C., et al. (2007). C3H/HeJ mice carrying a toll-like receptor 4 mutation are protected against the development of insulin resistance in white adipose tissue in response to a high-fat diet. *Diabetologia*, *50*, 1267–1276.

Poutahidis, T., & Erdman, S. E. (2016). Commensal bacteria modulate the tumor microenvironment. *Cancer Letters*, *380*(1), 356–358.

Poutahidis, T., Kearney, S. M., Levkovich, T., Qi, P., Varian, B. J., Lakritz, J. R., et al. (2013). Microbial symbionts accelerate wound healing via the neuropeptide hormone oxytocin. *PloS One*, *8*(10), e78898.

Poutahidis, T., Kleinewietfeld, M., & Erdman, S. E. (2014). Gut microbiota and the paradox of cancer immunotherapy. *Frontiers in Immunology*, *5*, 157.

Poutahidis, T., Kleinewietfeld, M., Smillie, C., Levkovich, T., Perrotta, A., Bhela, S., et al. (2013). Microbial reprogramming inhibits Western diet-associated obesity. *PloS One*, *8*, e68596.

Poutahidis, T., Springer, A., Levkovich, T., Qi, P., Varian, B. J., Lakritz, J. R., et al. (2014). Probiotic microbes sustain youthful serum testosterone levels and testicular size in aging mice. *PloS One*, *9*, e84877.

Poutahidis, T., Varian, B. J., Levkovich, T., Lakritz, J. R., Mirabal, S., Kwok, C., et al. (2015). Dietary microbes modulate transgenerational cancer risk. *Cancer Research*, *75*, 1197–1204.

Pryke, S. R., & Rollins, L. A. (2012). Mothers adjust offspring sex to match the quality of the rearing environment. *Proceedings of the Biological Sciences*, *279*, 4051–4057.

Pryke, S. R., Rollins, L. A., & Griffith, S. C. (2011). Context-dependent sex allocation: Constraints on the expression and evolution of maternal effects. *Evolution*, *65*, 2792–2799.

Qian, W., Zhu, T., Tang, B., Yu, S., Hu, H., Sun, W., et al. (2014). Decreased circulating levels of oxytocin in obesity and newly diagnosed type 2 diabetic patients. *Journal of Clinical Endocrinology and Metabolism*, *99*, 4683–4689.

Qiao, Y., Sun, J., Xia, S., Li, L., Li, Y., Wang, P., et al. (2015). Effects of different lactobacillus reuteri on inflammatory and fat storage in high-fat diet-induced obesity mice model. *Journal of Functional Foods*, *14*, 424–434.

Qin, J., Feng, M., Wang, C., Ye, Y., Wang, P. S., & Liu, C. (2009). Oxytocin receptor expressed on the smooth muscle mediates the excitatory effect of oxytocin on gastric motility in rats. *Neurogastroenterology and Motility, 21*, 430–438.

Rao, V. P., Poutahidis, T., Fox, J. G., & Erdman, S. E. (2007). Breast cancer: Should gastrointestinal bacteria be on our radar screen? *Cancer Research, 67*, 847–850.

Rautava, S., Luoto, R., Salminen, S., & Isolauri, E. (2012). Microbial contact during pregnancy, intestinal colonization and human disease. *Nature Reviews. Gastroenterology & Hepatology, 9*, 565–576.

Ridaura, V. K., Faith, J. J., Rey, F. E., Cheng, J., Duncan, A. E., Kau, A. L., et al. (2013). Gut microbiota from twins discordant for obesity modulate metabolism in mice. *Science, 341*, 1241214.

Rinaman, L., Vollmer, R. R., Karam, J., Phillips, D., Li, X., & Amico, J. A. (2005). Dehydration anorexia is attenuated in oxytocin-deficient mice. *American Journal of Physiology. Regulatory, Integrative and Comparative Physiology, 288*, R1791–R1799.

Rook, G. A. (2010). 99th Dahlem conference on infection, inflammation and chronic inflammatory disorders: Darwinian medicine and the 'hygiene' or 'old friends' hypothesis. *Clinical and Experimental Immunology, 160*, 70–79.

Rook, G. A. (2013). Regulation of the immune system by biodiversity from the natural environment: An ecosystem service essential to health. *Proceedings of the National Academy of Sciences of the United States of America, 110*, 18360–18367.

Rosenberg, E., & Zilber-Rosenberg, I. (2011). Symbiosis and development: The hologenome concept. *Birth Defects Research. Part C, Embryo Today, 93*, 56–66.

Round, J. L., & Mazmanian, S. K. (2009). The gut microbiota shapes intestinal immune responses during health and disease. *Nature Reviews. Immunology, 9*, 313–323.

Salim, H. M., Kang, H. K., Akter, N., Kim, D. W., Kim, J. H., Kim, M. J., et al. (2013). Supplementation of direct-fed microbials as an alternative to antibiotic on growth performance, immune response, cecal microbial population, and ileal morphology of broiler chickens. *Poultry Science, 92*, 2084–2090.

Sandrini, S., Aldriwesh, M., Alruways, M., & Freestone, P. (2015). Microbial endocrinology: Host-bacteria communication within the gut microbiome. *Journal of Endocrinology, 225*, R21–R34.

Schreiber, O., Petersson, J., Phillipson, M., Perry, M., Roos, S., & Holm, L. (2009). Lactobacillus reuteri prevents colitis by reducing P-selectin-associated leukocyte- and platelet-endothelial cell interactions. *American Journal of Physiology. Gastrointestinal and Liver Physiology, 296*, G534–G542.

Shen, H. (2015). Neuroscience: The hard science of oxytocin. *Nature, 522*, 410–412.

Sherwin, E., Rea, K., Dinan, T. G., & Cryan, J. F. (2016). A gut (microbiome) feeling about the brain. *Current Opinion in Gastroenterology, 32*, 96–102.

Shoelson, S. E., Herrero, L., & Naaz, A. (2007). Obesity, inflammation, and insulin resistance. *Gastroenterology, 132*, 2169–2180.

Simon, M. C., Strassburger, K., Nowotny, B., Kolb, H., Nowotny, P., Burkart, V., et al. (2015). Intake of Lactobacillus reuteri improves incretin and insulin secretion in glucose-tolerant humans: A proof of concept. *Diabetes Care, 38*, 1827–1834.

Singh, T. P., Malik, R. K., Katkamwar, S. G., & Kaur, G. (2015). Hypocholesterolemic effects of Lactobacillus reuteri LR6 in rats fed on high-cholesterol diet. *International Journal of Food Science and Nutrition, 66*, 71–75.

Smits, H. H., Engering, A., Van Der Kleij, D., De Jong, E. C., Schipper, K., Van Capel, T. M., et al. (2005). Selective probiotic bacteria induce IL-10-producing regulatory T cells in vitro by modulating dendritic cell function through dendritic cell-specific intercellular adhesion molecule 3-grabbing nonintegrin. *Journal of Allergy and Clinical Immunology, 115*, 1260–1267.

Smolinsky, A. N., Bergner, C. L., Laporte, J. L., & Kalueff, A. V. (2009). Analysis of grooming behavior and its utility in studying animal stress, anxiety, and depression. *Neuromethods*, *42*, 21–36.

Sommer, F., & Backhed, F. (2013). The gut microbiota—Masters of host development and physiology. *Nature Reviews. Microbiology*, *11*, 227–238.

Sommer, F., Stahlman, M., Ilkayeva, O., Arnemo, J. M., Kindberg, J., Josefsson, J., et al. (2016). The gut microbiota modulates energy metabolism in the hibernating brown bear Ursus arctos. *Cell Reports*, *14*, 1655–1661.

Sonnenburg, E. D., Smits, S. A., Tikhonov, M., Higginbottom, S. K., Wingreen, N. S., & Sonnenburg, J. L. (2016). Diet-induced extinctions in the gut microbiota compound over generations. *Nature*, *529*, 212–215.

Souza-Moreira, L., Campos-Salinas, J., Caro, M., & Gonzalez-Rey, E. (2011). Neuropeptides as pleiotropic modulators of the immune response. *Neuroendocrinology*, *94*, 89–100.

Spetter, M. S., & Hallschmid, M. (2015). Intranasal neuropeptide administration to target the human brain in health and disease. *Molecular Pharmaceutics*, *12*, 2767–2780.

Stilling, R. M., Dinan, T. G., & Cryan, J. F. (2016). The brain's Geppetto-microbes as puppeteers of neural function and behaviour? *Journal of Neurovirology*, *22*, 14–21.

Sun, J., Qiao, Y., Qi, C., Jiang, W., Xiao, H., Shi, Y., et al. (2016). High-fat-diet-induced obesity is associated with decreased antiinflammatory Lactobacillus reuteri sensitive to oxidative stress in mouse Peyer's patches. *Nutrition*, *32*, 265–272.

Sun, L., Tamma, R., Yuen, T., Colaianni, G., Ji, Y., Cuscito, C., et al. (2016). Functions of vasopressin and oxytocin in bone mass regulation. *Proceedings of the National Academy of Sciences of the United States of America*, *113*, 164–169.

Szajewska, H., Urbanska, M., Chmielewska, A., Weizman, Z., & Shamir, R. (2014). Meta-analysis: Lactobacillus reuteri strain DSM 17938 (and the original strain ATCC 55730) for treating acute gastroenteritis in children. *Beneficial Microbes*, *5*, 285–293.

Szeto, A., Nation, D. A., Mendez, A. J., Dominguez-Bendala, J., Brooks, L. G., Schneiderman, N., et al. (2008). Oxytocin attenuates NADPH-dependent superoxide activity and IL-6 secretion in macrophages and vascular cells. *American Journal of Physiology. Endocrinology and Metabolism*, *295*, E1495–E1501.

Takayanagi, Y., Kasahara, Y., Onaka, T., Takahashi, N., Kawada, T., & Nishimori, K. (2008). Oxytocin receptor-deficient mice developed late-onset obesity. *Neuroreport*, *19*, 951–955.

Tamma, R., Colaianni, G., Zhu, L. L., Dibenedetto, A., Greco, G., Montemurro, G., et al. (2009). Oxytocin is an anabolic bone hormone. *Proceedings of the National Academy of Sciences of the United States of America*, *106*, 7149–7154.

Taranto, M. P., Medici, M., Perdigon, G., Ruiz Holgado, A. P., & Valdez, G. F. (1998). Evidence for hypocholesterolemic effect of Lactobacillus reuteri in hypercholesterolemic mice. *Journal of Dairy Science*, *81*, 2336–2340.

Taylor, A. E., Lee, H. E., & Buisman-Pijlman, F. T. (2014). Oxytocin treatment in pediatric populations. *Frontiers in Behavioral Neuroscience*, *8*, 360.

Ting, W. J., Kuo, W. W., Kuo, C. H., Yeh, Y. L., Shen, C. Y., Chen, Y. H., et al. (2015). Supplementary heat-killed Lactobacillus reuteri GMNL-263 ameliorates hyperlipidaemic and cardiac apoptosis in high-fat diet-fed hamsters to maintain cardiovascular function. *British Journal of Nutrition*, *114*, 706–712.

Tremaroli, V., & Backhed, F. (2012). Functional interactions between the gut microbiota and host metabolism. *Nature*, *489*, 242–249.

Trivers, R. L., & Willard, D. E. (1973). Natural selection of parental ability to vary the sex ratio of offspring. *Science*, *179*, 90–92.

Turnbaugh, P. J., Hamady, M., Yatsunenko, T., Cantarel, B. L., Duncan, A., Ley, R. E., et al. (2009). A core gut microbiome in obese and lean twins. *Nature*, *457*(7228), 480–484.

Turnbaugh, P. J., Ley, R. E., Mahowald, M. A., Magrini, V., Mardis, E. R., & Gordon, J. I. (2006). An obesity-associated gut microbiome with increased capacity for energy harvest. *Nature, 444*, 1027–1031.

Urbanska, M., Gieruszczak-Bialek, D., & Szajewska, H. (2016). Systematic review with meta-analysis: Lactobacillus reuteri DSM 17938 for diarrhoeal diseases in children. *Alimentary Pharmacology & Therapeutics, 43*, 1025–1034.

Uvnas-Moberg, K. (1998). Oxytocin may mediate the benefits of positive social interaction and emotions. *Psychoneuroendocrinology, 23*, 819–835.

Varian, B. J., Goureshetti, S., Poutahidis, T., Lakritz, J. R., Levkovich, T., Kwok, C., et al. (2016). Beneficial bacteria inhibit cachexia. *Oncotarget, 7*, 11803–11816.

Varian, B. J., Levkovich, T., Poutahidis, T., Ibrahim, Y. M., Perrotta, A., Alm, E. J., et al. (2016). Beneficial dog bacteria up-regulate oxytocin and lower risk of obesity. *Journal of Probiotics & Health, 4*, 1–9.

Varian, B. J., Poutahidis, T., Levkovich, T., Ibrahim, Y. M., Lakritz, J. R., Chatzigiagkos, A., et al. (2014). Beneficial bacteria stimulate youthful thyroid gland activity. *Journal of Obesity and Weight Loss Therapy, 4*, 1–8.

Viero, C., Shibuya, I., Kitamura, N., Verkhratsky, A., Fujihara, H., Katoh, A., et al. (2010). REVIEW: Oxytocin: Crossing the bridge between basic science and pharmacotherapy. *CNS Neuroscience & Therapeutics, 16*, e138–e156.

Vijay-Kumar, M., Aitken, J. D., Carvalho, F. A., Cullender, T. C., Mwangi, S., Srinivasan, S., et al. (2010). Metabolic syndrome and altered gut microbiota in mice lacking toll-like receptor 5. *Science, 328*, 228–231.

Vila, I. K., Badin, P. M., Marques, M. A., Monbrun, L., Lefort, C., Mir, L., et al. (2014). Immune cell toll-like receptor 4 mediates the development of obesity- and endotoxemia-associated adipose tissue fibrosis. *Cell Reports, 7*, 1116–1129.

Vitalo, A., Fricchione, J., Casali, M., Berdichevsky, Y., Hoge, E. A., Rauch, S. L., et al. (2009). Nest making and oxytocin comparably promote wound healing in isolation reared rats. *PloS One, 4*, e5523.

Walter, J., Britton, R. A., & Roos, S. (2011). Host-microbial symbiosis in the vertebrate gastrointestinal tract and the Lactobacillus reuteri paradigm. *Proceedings of the National Academy of Sciences of the United States of America, 108*(Suppl. 1), 4645–4652.

Walters, W. A., Caporaso, J. G., Lauber, C. L., Berg-Lyons, D., Fierer, N., & Knight, R. (2011). PrimerProspector: De novo design and taxonomic analysis of barcoded polymerase chain reaction primers. *Bioinformatics, 27*, 1159–1161.

Wang, P., Yang, H. P., Tian, S., Wang, L., Wang, S. C., Zhang, F., et al. (2015). Oxytocin-secreting system: A major part of the neuroendocrine center regulating immunologic activity. *Journal of Neuroimmunology, 289*, 152–161.

Welch, M. G., Margolis, K. G., Li, Z., & Gershon, M. D. (2014). Oxytocin regulates gastrointestinal motility, inflammation, macromolecular permeability, and mucosal maintenance in mice. *American Journal of Physiology. Gastrointestinal and Liver Physiology, 307*, G848–G862.

Wong, V. W., Martindale, R. G., Longaker, M. T., & Gurtner, G. C. (2013). From germ theory to germ therapy: Skin microbiota, chronic wounds, and probiotics. *Plastic and Reconstructive Surgery, 132*, 854e–861e.

Woting, A., & Blaut, M. (2016). The intestinal microbiota in metabolic disease. *Nutrients, 8*(4), 202.

Wu, C. L., Hung, C. R., Chang, F. Y., Pau, K. Y., & Wang, P. S. (2003). Pharmacological effects of oxytocin on gastric emptying and intestinal transit of a non-nutritive liquid meal in female rats. *Naunyn-Schmiedeberg's Archives of Pharmacology, 367*, 406–413.

Yanai, H., Budovsky, A., Tacutu, R., & Fraifeld, V. E. (2011). Is rate of skin wound healing associated with aging or longevity phenotype? *Biogerontology, 12*, 591–597.

Yang, H. P., Wang, L., Han, L., & Wang, S. C. (2013). Nonsocial functions of hypothalamic oxytocin. *ISRN Neuroscience, 2013*, 179272.

Yoo, J. Y., & Kim, S. S. (2016). Probiotics and prebiotics: Present status and future perspectives on metabolic disorders. *Nutrients, 8*, 173–193.

Yoshida, M., Takayanagi, Y., Inoue, K., Kimura, T., Young, L. J., Onaka, T., et al. (2009). Evidence that oxytocin exerts anxiolytic effects via oxytocin receptor expressed in serotonergic neurons in mice. *Journal of Neuroscience, 29*, 2259–2271.

Yosten, G. L., & Samson, W. K. (2010). The anorexigenic and hypertensive effects of nesfatin-1 are reversed by pretreatment with an oxytocin receptor antagonist. *American Journal of Physiology. Regulatory, Integrative and Comparative Physiology, 298*, R1642–R1647.

Young, W. S., III, & Gainer, H. (2009). Vasopressin/oxytocin and receptors. In L. R. Squire (Ed.), *Encyclopedia of neuroscience.* (1st ed.). Oxford: Academic Press.

Yuan, L., Liu, S., Bai, X., Gao, Y., Liu, G., Wang, X., et al. (2016). Oxytocin inhibits lipopolysaccharide-induced inflammation in microglial cells and attenuates microglial activation in lipopolysaccharide-treated mice. *Journal of Neuroinflammation, 13*, 77.

Zhang, G., Bai, H., Zhang, H., Dean, C., Wu, Q., Li, J., et al. (2011). Neuropeptide exocytosis involving synaptotagmin-4 and oxytocin in hypothalamic programming of body weight and energy balance. *Neuron, 69*, 523–535.

Zhang, G., & Cai, D. (2011). Circadian intervention of obesity development via resting-stage feeding manipulation or oxytocin treatment. *American Journal of Physiology. Endocrinology and Metabolism, 301*, E1004–E1012.

Zhang, J., Motyl, K. J., Irwin, R., Macdougald, O. A., Britton, R. A., & Mccabe, L. R. (2015). Loss of bone and Wnt10b expression in male type 1 diabetic mice is blocked by the probiotic Lactobacillus reuteri. *Endocrinology, 156*, 3169–3182.

Zhang, H., Wu, C., Chen, Q., Chen, X., Xu, Z., Wu, J., et al. (2013). Treatment of obesity and diabetes using oxytocin or analogs in patients and mouse models. *PloS One, 8*, 1–11.

Zingg, H. H., & Laporte, S. A. (2003). The oxytocin receptor. *Trends in Endocrinology and Metabolism, 14*, 222–227.

CHAPTER SIX

Intestinal Barrier and Behavior

M. Julio-Pieper[1], J.A. Bravo[1]
Grupo de NeuroGastroBioquímica, Instituto de Química, Facultad de Ciencias, Pontificia Universidad Católica de, Valparaíso, Chile
[1]Corresponding authors: e-mail address: marcela.julio@pucv.cl; javier.bravo@pucv.cl

Contents

Abstract

The intestinal barrier function contributes to gut homeostasis by modulating absorption of water, electrolytes, and nutrients from the lumen into the circulation while restricting the passage of noxious luminal substances and microorganisms. Chronic conditions such as rheumatoid arthritis, inflammatory bowel disease, and celiac disease are associated to intestinal barrier dysfunction. Here, the hypothesis is that a leaky intestinal wall allowing for indiscriminate passage of intraluminal compounds to the vascular compartment could in turn lead to systemic inflammation. An increasing number of studies are now investigating the association between gut permeability and CNS disorders, under the premise that translocation of intestinal luminal contents could affect CNS function, either directly or indirectly. Still, it is unknown whether disruption of intestinal barrier is a causative agent or a consequence in these situations. Here, we discuss the latest evidence pointing to an association between increased gut permeability and disrupted behavioral responses.

1. THE INTESTINAL BARRIER: AN OVERVIEW

In superior animals, the intestinal mucosa is considered the largest surface of interaction between the internal milieu and the external environment (Barrett, 2008). It has a dual role involving both the absorption of water, electrolytes, and nutrients from the lumen into the circulation and restriction of permeation of noxious luminal substances and microorganisms (Farhadi, Banan, Fields, & Keshavarzian, 2003). This selective, highly

International Review of Neurobiology, Volume 131
ISSN 0074-7742
http://dx.doi.org/10.1016/bs.irn.2016.08.006

regulated passage of intestinal luminal contents is also known as the intestinal barrier function.

In the gut, barrier function is dynamic and depends on many cell types and cell-derived substances, including the intestinal microbiota, the epithelial cells which are apically sealed through tight junctions, secreted products of epithelial origin, the intestinal endothelium, the mucosal resident immune cells, and the enteric nervous system (ENS) (Neunlist et al., 2013). As shown in Fig. 1, the first obstacle faced by luminal substances is a mucus gel coat produced by the epithelial goblet cells. The mucus, composed of highly glycosylated proteins known as mucins, forms a viscoelastic

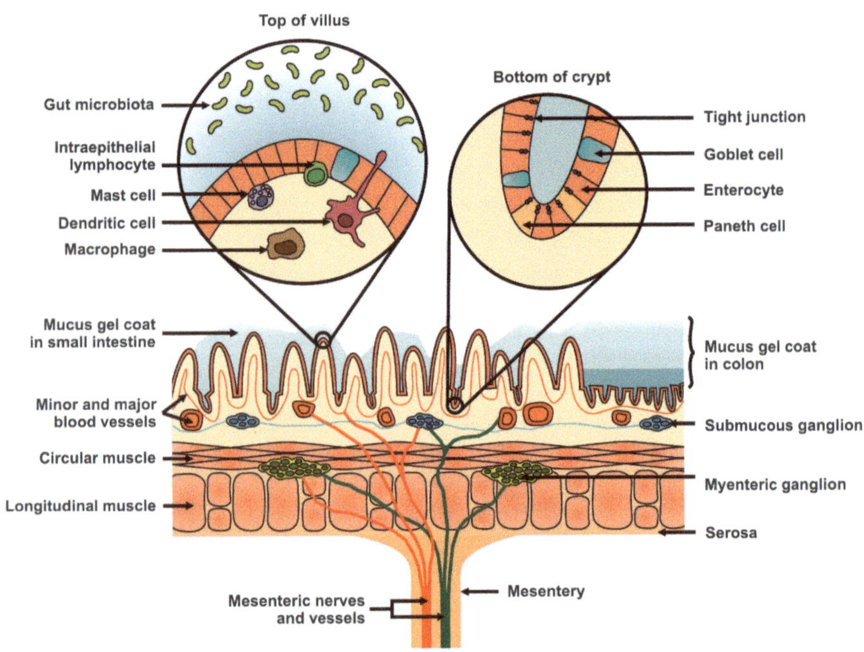

Fig. 1 Schematic view of the intestinal barrier. Commensal bacteria in the lumen together with the mucus layer inhibit pathogen colonization of the mucosa. Enterocytes, the main absorptive epithelial cell type, connected by junctional complexes, regulate the transport of luminal content. Immune cells, some of which are immersed in the epithelium, can detect changes in luminal microbial composition and respond by secreting immunoglobulins and cytokines. Enteric neurons and glial cells organized in submucous and myenteric ganglia are also able to produce soluble mediators that affect intestinal epithelial cell function. *Adapted from Díaz-Zepeda, C., Escobar-Luna, J., González-Arancibia, C., González-Toro, M. P., Olavarría-Ramírez, L., Zanelli-Massai, F., … Julio-Pieper, M. (2015). Blancos farmacológicos en el eje intestino-cerebro.* Rev. Farmacol. Chile, *8(1), 12.*

network which acts as a physical diffusion barrier to pathogens. In addition, the mucus layer contains antimicrobial peptides mainly secreted by the Paneth cells and a high concentration of secretory IgA (sIgA) transported by the enterocytes from the mucosa where they are released by plasma B cells (Soderholm & Perdue, 2001). Mucus properties are not the same throughout the intestine. The small intestine has a single layer of mucus which is loose and discontinuous; the colon on the other hand, has two layers of mucus, the inner being more firmly attached, as opposed to the outer layer which is unattached and less dense (Johansson, Larsson, & Hansson, 2011; Johansson, Sjovall, & Hansson, 2013). In the healthy colon, commensal bacteria are restricted to the outer mucus layer. It is believed that defects in the inner layer of mucus produced by the colon can make it penetrable by bacteria, which might represent a pathophysiological mechanism for diseases such as ulcerative colitis (Johansson et al., 2013).

The mucus layer covers the intestinal epithelium, a monocellular layer which constitutes the second obstacle limiting the passage of luminal content into the systemic compartment. This stratum is formed by enterocytes (or colonocytes in the colon) which are connected with each other and with the basement membrane through structures such as desmosomes and adherens junctions, integrins, and tight junctions (Daneman & Rescigno, 2009).

Transport of substances across the epithelium can take place by two pathways: the transcellular one (across the cell) is involved in the absorption and transport of nutrients, and is mediated by specific transporters and channels located at the epithelial cell membrane. A second pathway of transport, the paracellular one, occurs between epithelial cells. Tight junctions seal the intercellular spaces and therefore determine the permeability of the paracellular pathway, which represents the main component of the overall epithelium permeability (Suzuki, 2013). Tight junctions are protein complexes composed of transmembrane proteins as well as cytosolic scaffold proteins. The complex is anchored to the actin cytoskeleton. The modification of paracellular permeability through tight junction modulation is a dynamic process involving intra- and extracellular stimuli (Suzuki, 2013).

Intestinal epithelial cells also play a role in local immunity in several ways: they express HLA class II molecules after internalization of luminal antigens (Hershberg et al., 1998), secrete antimicrobial peptides such as beta-defensins (O'Neil et al., 1999), and release cytokines upon infection in order to activate further responses (Perdue, 1999; Ramirez et al., 2005). Paneth cells, a specialized epithelial lineage located at the base of small intestinal crypts (as seen in Fig. 1), are also important for limiting intestinal barrier penetration by commensal and pathogenic bacteria and are able to secrete antimicrobial products such as the RegIII family of proteins (Vaishnava, Behrendt, Ismail, Eckmann, & Hooper, 2008). Some enteroendocrine cells (L cells) present in the epithelium of the distal ileon and proximal colon have been shown to release GLP-2, which has trophic activity on the intestinal and colonic mucosa (Drucker, Erlich, Asa, & Brubaker, 1996). GLP-2 increases the expression of tight junction proteins in vitro and induces strengthening of the intestinal barrier (Moran, O'Neill, & McLaughlin, 2012).

Underneath and immersed in the intestinal epithelium are various types of more specialized immune cells. For example, intraepithelial lymphocytes can also initiate responses upon contact with foreign material in order to prevent microbial invasion (Li et al., 2012; Perdue, 1999). Antigen-presenting cells (APCs) can induce the differentiation of T helper cells which in turn, release cytokines capable of modifying flux across either tight junctions or cation pores (Turner, 2009). However, APCs also contribute to dampen down an otherwise exaggerated immune response to luminal agents. For example, APCs promote the differentiation of T regulatory cells which helps counteracting the effects of T helper cells; the balance between both routes may determine whether homeostasis is rapidly reestablished, or a transient barrier breach is perpetuated into a pathological condition (Fasano & Shea-Donohue, 2005; Turner, 2009). Dendritic cells are a subtype of APCs which can protrude transepithelial dendrites to sample luminal antigens (Niess et al., 2005). Dendritic cells are able to respond to constituents of intestinal mucus by increasing the expression of cytokines such as TGF-beta1, which again promotes the differentiation of T regulatory cells and a more tolerant phenotype toward food and commensal antigens (Shan et al., 2013).

Microorganisms that infringe the intestinal epithelial barrier are rapidly eliminated by lamina propia macrophages (Kelsall, 2008), which however do not induce a strong proinflammatory response (Smythies et al., 2005). Macrophages play a second role relevant to intestinal barrier maintenance: upon damage, they are recruited into the injured area to support and promote epithelial progenitor cell proliferation and tissue repair (Pull, Doherty, Mills, Gordon, & Stappenbeck, 2005). Toll-like receptors (TLRs) are a family of transmembrane receptors which play an important role in innate immunity as they recognize conserved molecular patterns found in microorganisms from the intestinal microbiota or in pathogens (Liew, Xu, Brint, & O'Neill, 2005). It has been shown that TLRs, signaling through the adaptor Myd88 in macrophages are required for proper regeneration of colon epithelium (Pull et al., 2005). Interestingly, Paneth cells sense and respond to bacteria also through a mechanism dependent on Myd88 pathway (Vaishnava et al., 2008) suggesting that intestinal cell types might have similar or complementary pathways activated upon exposure to microbes.

Peyer's patches are part of the gut-associated lymphoid tissue and are composed by aggregated lymphoid follicles surrounded by a specialized epithelium. This epithelium produces less mucus and has a more porous basement membrane than the regular one; it is also enriched in a particular type of enterocytes called M cells (Jung, Hugot, & Barreau, 2010). One main function of M cells is to transport luminal antigens and bacteria toward the underlying immune cells, which in turn modulate further responses leading to either tolerance or systemic immune cell response (Jung et al., 2010). M cells are largely devoid of lysosomes, and for that reason endocytosed products are less likely to undergo intracellular proteolysis. Also, when compared to regular enterocytes, M cells have a shorter distance between the apical and basolateral surfaces (Brayden & Baird, 1994; Jung et al., 2010). Therefore, Peyer's patches, and particularly M cells represent a site where permeability via the transcellular pathway is facilitated, a feature that has been exploited for the administration of diverse drugs and for vaccine delivery strategies (Shakweh, Ponchel, & Fattal, 2004).

The intestinal mucosa is a highly vascularized tissue (Kachlik, Baca, & Stingl, 2010). Being the last anatomic barrier limiting the access of luminal microbes to the circulation, intestinal endothelium plays an important role in preventing translocation of microbial material to other organs. In the piglet for example, it has been shown that *Clostridium perfringens* type C enteritis is associated with the adhesion of B-toxin to small intestinal endothelial cells throughout the lamina propia (Schumacher, Martel, Pasmans, Van Immerseel, & Posthaus, 2013). In vitro experiments show that intestinal microvascular endothelial cells respond to several microbial products, namely peptidoglycan, Poly (I:C) and flagellin (which are agonists to TLRs 2, 3, and 5, respectively) by increasing the expression of EBI3, a member of IL-12 family (Heidemann et al., 2007). Poly (I:C) also enhances leukocyte adhesion to intestinal microvascular endothelial cells in vitro (Heidemann et al., 2007). Therefore, it is possible to suggest that the intestinal endothelium is able to sense microbial antigens and elicit an array of innate immune responses in vivo.

The ENS is another important factor in the modulation of gut permeability. Intestinal mucosa is innervated by both submucosal and myenteric neurons able to release mediators that can affect wound healing, epithelial proliferation/differentiation, and paracellular permeability as well as hydroelectrolitic transport and nutrient absorption (Neunlist et al., 2013). Electrical activation of enteric neurons exerts a protective effect by promoting strengthening of the intestinal barrier in vitro (Neunlist et al., 2003). Acetylcholine and VIP, two important neurotransmitters of the ENS have also been shown to modulate intestinal paracellular permeability (Neunlist et al., 2003; Saunders, Hanssen, & Perdue, 1997): the addition of VIP to a coculture of submucosal neurons and colonic epithelial cells induced an increase in the expression of the tight junction protein ZO-1 (Neunlist et al., 2003), and atropine pretreatment prevented the increased permeability induced by stress in the rat (Saunders et al., 1997).

The establishment of a healthy intestinal barrier depends on adequate colonization by luminal microbiota at key developmental stages (Sommer & Backhed, 2013; Wagner, Taylor, & Johnson, 2008). When compared to conventional counterparts, animals reared in a germ-free environment display a dysfunctional intestinal barrier, with morphological, immune, biochemical, and biophysical alterations. To name a few, germ-free mice have fewer intestinal lymphoid follicles, a thinner mucus gel layer, lower sIgA concentrations, and a less complex intestinal vascular network than control mice (Sommer & Backhed, 2013). The ENS is also affected in germ-free mice, which show decreased excitability in a subgroup of myenteric neurons (McVey Neufeld, Mao, Bienenstock, Foster, & Kunze, 2013).

Diverse pathologies have been associated with disruption of the gut barrier, including frequent reports of critically ill patients (affected, for instance, with extensive burn, multiple trauma, hemorrhagic shock, or postoperative complications) displaying increased intestinal permeability (Aranow & Fink, 1996; Derikx et al., 2008; Doig et al., 1998). A leaky intestinal wall allowing for indiscriminate passage of intraluminal compounds to the vascular compartment could in turn lead to systemic inflammation (Luyer et al., 2004; Plotz, Slutsky, van Vught, & Heijnen, 2004). This hypothesis seems to explain the pathophysiology of multiple organ failure syndrome, which is an important cause of death in intensive care medical units and correlates to the extent of intestinal permeability displayed by patients (Doig et al., 1998). Moreover, chronic conditions such as rheumatoid arthritis and diabetes type 1, in addition to inflammatory bowel disease and celiac disease, also present with intestinal barrier dysfunction (Carratu et al., 1999; Fasano & Shea-Donohue, 2005; Jenkins, Rooney, Jones, Bienenstock, & Goodacre, 1987). An increasing number of studies are now investigating the association between gut permeability and CNS disorders (Maes, Kubera, Leunis, & Berk, 2012), under the premise that translocation of intestinal luminal contents could affect CNS function, either directly or indirectly. Still, it is unknown whether disruption of intestinal barrier is a causative agent or a consequence in these situations. However, restoration of gut barrier integrity might improve the quality of life of these patients by relieving symptoms that are worsened by inflammation.

2. CAN AN ALTERED BARRIER FUNCTION DISRUPT BEHAVIORAL RESPONSES?

Investigating whether intestinal barrier changes arise as consequence of CNS's response to stress (top-down point of view), or the altered stress-related behavioral traits seen in patients suffering from inflammatory bowel diseases are a consequence of increased intestinal permeability (bottom-up point of view) is a challenging task. There is preclinical evidence supporting both positions; for example, disruption of the intestinal epithelium with dextran sodium sulfate (DSS) in rodents increases anxiety- and depression-like behaviors (Bercik et al., 2011; Chen et al., 2015; Emge et al., 2016) and enhances corticosterone release (Reichmann et al., 2015) in comparison to control subjects. However, it is difficult to dissect if this artificially induced alteration in intestinal permeability is the sole culprit of an altered behavioral response to stress. This is mainly because such interventions affect visceral proprioception, as shown through colorectal distention (CRD) experiments, where DSS-treated animals have reduced pain thresholds (Chen et al., 2015). This finding suggests that behavioral alterations may not arise as a direct consequence of loss in intestinal permeability, but most likely due to a secondary cause: increased visceral pain perception. Moreover, Chen and colleagues (2015) demonstrate that desensitization of transient receptor potential vanilloid 1 from colonic afferents in DSS-exposed animals, prevents the anxiogenic effects of DSS, and increases the pain threshold to CRD in this model of colitis (Chen et al., 2015). Thus, a highly sensitive gastrointestinal tract could activate CNS structures such as the amygdala, where pain processing is mediated through direct pain-related inputs (Bourgeais, Gauriau, & Bernard, 2001) that modulate the emotional-affective dimension of pain (Ashorn et al., 2009), which in turn could lead to activation of the hypothalamus–pituitary–adrenal (HPA) axis. Therefore, the final outcome of increased visceral hypersensitivity as a result of chronic DSS exposure in rodents is a chronically activated HPA axis (Reichmann et al., 2015) that impacts the immune system (Oppong & Cato, 2015), affects neural plasticity (Madalena & Lerch, 2016), and further increases gut permeability (Bhatia & Tandon, 2005) generating a vicious cycle that could favor the occurrence of intestinal disorders and stress-related psychiatric illnesses.

Exposure to *Citrobacter rodentium*, a gram-negative bacterium, is commonly used as a model of biologically induced transient colitis in mice. The infection proceeds with intestinal inflammation (Rodrigues, Sousa, Johnson-Henry, Sherman, & Gareau, 2012) and increased paracellular permeability at the colon also involving disruption of tight junctions at 10 days postexposure, when the peak of infection takes place (Conlin et al., 2009; Rodrigues et al., 2012). Lyte et al. reports that 8 h after being exposed to *C. rodentium*, mice display increased anxiety-like behavior in the hole-board open field apparatus, with avoidance of the central area, and increased risk assessment behavior. However, plasma levels of IFN-gamma, TNF-alpha, and IL-12 were similar to noninfected control mice. Histological analysis of colonic tissue indicated a lack of obvious inflammation at the 8 h postchallenge time point, indicating that it was unlikely that behavioral alterations were consequence of stress due to an inflammatory status. Although pain perception was not assessed in this study, vagal sensory ganglia from infected animals displayed a higher number of neurons that were positive for c-Fos protein, suggesting vagal afferent transmission of signals associated to *C. rodentium* in a gut-to-brain fashion (Lyte, Li, Opitz, Gaykema, & Goehler, 2006). A second study did not find behavioral changes either at 10 days postchallenge (the peak of infection) or at 30 days postchallenge (clearance stage). The authors did report alterations in memory at both time points when infected mice were subjected to water avoidance stress (Gareau et al., 2011). Notably, when *C. rodentium*-infected mice received a daily administration of probiotic bacteria (a combination of *Lactobacillus rhamnosus* and *Lactobacillus helveticus*) both the increase in intestinal permeability (Rodrigues et al., 2012) and the memory dysfunction (Gareau et al., 2011) were prevented. These and the above studies underscore the need of animal models of increased intestinal permeability, where the contribution of stress is either dampened or completely absent, in order to effectively assess the effect of a disrupted gut barrier function on behavior.

3. STRESSORS AFFECTING BOTH BEHAVIOR AND GUT BARRIER FUNCTION

Preclinical evidence demonstrates that stress can contribute to gut alterations, especially in relation to barrier function. For instance, after active

DSS-induced inflammation in rodents, colitis can be reactivated by central depletion of catecholamines through icv. administration of reserpine (Ghia et al., 2009). This drug depletes catecholamines from nerve endings and is used to model the monoamine hypothesis of depression (Blasco-Serra et al., 2015; Ghia et al., 2009; Rojas-Corrales, Berrocoso, Gibert-Rahola, & Mico, 2004). The finding was confirmed when DSS animals were subjected to olfactory bulbectomy (Obx), which is another animal model that induces depression-like features (Ghia et al., 2009). Both results led Ghia and colleagues (2009) to propose that depression reactivates alterations in gut barrier function. However, reserpine as a depressogenic compound is still controversial, as reduction in brain monoamines as a main etiological factor for depression continues to be debated (for a review, see Baumeister, Hawkins, & Uzelac, 2003). On the other hand, Obx does seem to suggest that depression might be involved in colitis reactivation. However, depression-like features of Obx rats appear from a physical lesion that does not necessarily resemble the etiology of most cases of depression, so data interpretation for these results must be dealt with caution. Despite this, there is preclinical and clinical evidence suggesting that glucocorticoids are capable of altering intestinal permeability. Meddings and Swain (2000) show that 24 h after rats were subjected to a 20-min swim stress, they had higher urinary excretion of sucrose, lactulose/mannitol, and sucralose than nonstressed control rats, which suggests an increased gastrointestinal permeability. Interestingly, these effects were absent in adrenalectomized rats subjected to the same stress procedure. This was further confirmed by pharmacological antagonism of glucocorticoid receptors by RU-486, which also prevented the effects of the swim stress on gastrointestinal permeability (Meddings & Swain, 2000). This evidence supports the idea that heightened HPA axis activation promotes an increase in intestinal permeability, potentially contributing to the transit of luminal content through the intestinal wall, and an enhanced immune activation which could in turn perpetuate the inflammatory bowel condition. Interestingly, this hypothesis has been challenged in human subjects. For instance, healthy volunteers exposed to different types of stress show higher intestinal permeability than in a resting nonstressful situation (Alonso et al., 2012; Li et al., 2013). Moreover, there seems to be a gender bias when it comes to the effect of stress in gut permeability: when women are exposed to a cold painful stressor for 15 min, their duodenal water flux is higher than in stressed men (Alonso et al., 2012). However, in this case the changes in intestinal permeability cannot be entirely explained by changes in glucocorticoids, mostly because men produced higher levels or cortisol and ACTH in comparison to the age-matched women from the same study. One explanation could be that women's intestinal barrier has a higher sensitivity to glucocorticoids, which in part explains the gender bias observed by Alonso and colleagues (2012), being this the only experimental outcome that is still in line with preclinical findings suggesting a link between glucocorticoids and gut barrier permeability. However, when healthy young men are exposed to high levels of stress (i.e., 6 weeks of military combat training), not only there was an increase in anxiety and depression scores during combat training, but also 5 h sucrose urinary excretion was higher, as well as sucralose excretion at 5 and 24 h in comparison to their own baseline levels before combat training (Li et al., 2013). Moreover, the lactulose/mannitol ratio was also higher during training, but only in volunteers that had higher levels of IBS-symptoms severity scores (Li et al., 2013). These data suggest that stress affects intestinal permeability; however, these observations were made under very extreme conditions and only male volunteers were evaluated, with no indication at what could happen to women, which have the highest prevalence of functional–gastrointestinal disorders (Koloski, Jones, Young, & Talley, 2015). Nonetheless, it is an interesting finding that relates stress, altered behavior, and gut barrier dysfunction, which strongly supports the bidirectional communication between the CNS and the gastrointestinal tract.

Therefore, when new studies are proposed in the field of stress research, this brain–gut interaction cannot be ruled out, as the effects of environmental stressors will not only generate changes in brain neurochemistry and endocrine functions such as hyperactivation of the HPA axis, but will also affect gut barrier function, thus allowing for the translocation of intestinal contents (i.e., microorganisms or pathogen-associated molecular patterns [PAMPs]). The above could enhance the immune system and activate the antiinflammatory cholinergic reflex (Pavlov, Wang, Czura, Friedman, & Tracey, 2003). The increase in vagal activation could affect intestinal motility, which might contribute to the overall pathophysiology of the condition. On the other hand, there is evidence that glucocorticoid secretion is stimulated independently of ACTH. For example, intraperitoneal exposure to live bacteria or bacterial PAMPs (LPS) leads to an enhanced glucocorticoid secretion that is mediated by bacterially stimulated prostaglandin secretion and not by ACTH (Vakharia & Hinson, 2005; Zimomra, Porterfield, Camp, & Johnson, 2011). Although this effect may arise from bacteria and/or bacterial cell wall components coming from any source (i.e., acquired infections that reach the blood stream, or bacteria from skin, urogenital, or pulmonary microbiota), it does suggest that enhanced permeability, leading to an increased transit from the intestinal lumen might stimulate the secretion of cortical steroids, which could contribute to alterations in neuronal plasticity within the CNS and therefore generate alterations in behavioral responses to stressful situations.

4. CONCLUDING REMARKS

Most cellular components of the intestinal barrier are able to produce signals that can reach the periphery, including the CNS, either indirectly via activation of cells within the intestinal wall (e.g., immune cells, neurons) or directly by using humoral pathways. The nature or extent of these signals may be altered as result of perturbations in the luminal environment or in the inflammatory status of the gut, initiating alterations in the brain that may translate into altered behaviors. Therefore, we must consider that gut alterations may precede some psychiatric alterations associated with gastrointestinal disorders. However, the evidence supporting this hypothesis needs further testing, as stress itself increases gut permeability; for this reason, the interpretation of results on preclinical models and patients where alterations of the intestinal barrier function is indicated as the cause of behavioral alterations is intricate and needs to be supported by more adequate animal models.

REFERENCES

Alonso, C., Guilarte, M., Vicario, M., Ramos, L., Rezzi, S., Martinez, C., ... Santos, J. (2012). Acute experimental stress evokes a differential gender-determined increase in human intestinal macromolecular permeability. *Neurogastroenterology and Motility*, *24*(8), 740–746. e348–749.

Aranow, J. S., & Fink, M. P. (1996). Determinants of intestinal barrier failure in critical illness. *British Journal of Anaesthesia*, 77(1), 71–81.

Ashorn, S., Valineva, T., Kaukinen, K., Ashorn, M., Braun, J., Raukola, H., ... Iltanen, S. (2009). Serological responses to microbial antigens in celiac disease patients during a gluten-free diet. *Journal of Clinical Immunology*, *29*(2), 190–195.

Barrett, K. E. (2008). New ways of thinking about (and teaching about) intestinal epithelial function. *Advances in Physiology Education*, *32*(1), 25–34.

Baumeister, A. A., Hawkins, M. F., & Uzelac, S. M. (2003). The myth of reserpine-induced depression: Role in the historical development of the monoamine hypothesis. *Journal of the History of the Neurosciences*, *12*(2), 207–220.

Bercik, P., Park, A. J., Sinclair, D., Khoshdel, A., Lu, J., Huang, X., ... Verdu, E. F. (2011). The anxiolytic effect of Bifidobacterium longum NCC3001 involves vagal pathways for gut-brain communication. *Neurogastroenterology and Motility*, *23*(12), 1132–1139.

Bhatia, V., & Tandon, R. K. (2005). Stress and the gastrointestinal tract. *Journal of Gastroenterology and Hepatology*, *20*(3), 332–339.

Blasco-Serra, A., Escrihuela-Vidal, F., Gonzalez-Soler, E. M., Martinez-Exposito, F., Blasco-Ausina, M. C., Martinez-Bellver, S., ... Valverde-Navarro, A. A. (2015).

Depressive-like symptoms in a reserpine-induced model of fibromyalgia in rats. *Physiology & Behavior, 151*, 456–462.
Bourgeais, L., Gauriau, C., & Bernard, J. F. (2001). Projections from the nociceptive area of the central nucleus of the amygdala to the forebrain: A PHA-L study in the rat. *The European Journal of Neuroscience, 14*(2), 229–255.
Brayden, D. J., & Baird, A. W. (1994). A distinctive electrophysiological signature from the Peyer's patches of rabbit intestine. *British Journal of Pharmacology, 113*(2), 593–599.
Carratu, R., Secondulfo, M., de Magistris, L., Iafusco, D., Urio, A., Carbone, M. G., ... Prisco, F. (1999). Altered intestinal permeability to mannitol in diabetes mellitus type I. *Journal of Pediatric Gastroenterology and Nutrition, 28*(3), 264–269.
Chen, J., Winston, J. H., Fu, Y., Guptarak, J., Jensen, K. L., Shi, X. Z., ... Sarna, S. K. (2015). Genesis of anxiety, depression, and ongoing abdominal discomfort in ulcerative colitis-like colon inflammation. *American Journal of Physiology. Regulatory, Integrative and Comparative Physiology, 308*(1), R18–R27.
Conlin, V. S., Wu, X., Nguyen, C., Dai, C., Vallance, B. A., Buchan, A. M., ... Jacobson, K. (2009). Vasoactive intestinal peptide ameliorates intestinal barrier disruption associated with Citrobacter rodentium-induced colitis. *American Journal of Physiology. Gastrointestinal and Liver Physiology, 297*(4), G735–G750.
Daneman, R., & Rescigno, M. (2009). The gut immune barrier and the blood-brain barrier: Are they so different? *Immunity, 31*(5), 722–735.
Derikx, J. P., van Waardenburg, D. A., Thuijls, G., Willigers, H. M., Koenraads, M., van Bijnen, A. A., ... Buurman, W. A. (2008). New insight in loss of gut barrier during major non-abdominal surgery. *PLoS One, 3*(12), e3954.
Doig, C. J., Sutherland, L. R., Sandham, J. D., Fick, G. H., Verhoef, M., & Meddings, J. B. (1998). Increased intestinal permeability is associated with the development of multiple organ dysfunction syndrome in critically ill ICU patients. *American Journal of Respiratory and Critical Care Medicine, 158*(2), 444–451.
Drucker, D. J., Erlich, P., Asa, S. L., & Brubaker, P. L. (1996). Induction of intestinal epithelial proliferation by glucagon-like peptide 2. *Proceedings of the National Academy of Sciences of the United States of America, 93*(15), 7911–7916.
Emge, J. R., Huynh, K., Miller, E. N., Kaur, M., Reardon, C., Barrett, K. E., & Gareau, M. G. (2016). Modulation of the microbiota-gut-brain axis by probiotics in a murine model of inflammatory bowel disease. *American Journal of Physiology. Gastrointestinal and Liver Physiology, 310*, G989–G998.
Farhadi, A., Banan, A., Fields, J., & Keshavarzian, A. (2003). Intestinal barrier: An interface between health and disease. *Journal of Gastroenterology and Hepatology, 18*(5), 479–497.
Fasano, A., & Shea-Donohue, T. (2005). Mechanisms of disease: The role of intestinal barrier function in the pathogenesis of gastrointestinal autoimmune diseases. *Nature Clinical Practice. Gastroenterology & Hepatology, 2*(9), 416–422.
Gareau, M. G., Wine, E., Rodrigues, D. M., Cho, J. H., Whary, M. T., Philpott, D. J., ... Sherman, P. M. (2011). Bacterial infection causes stress-induced memory dysfunction in mice. *Gut, 60*(3), 307–317.
Ghia, J. E., Blennerhassett, P., Deng, Y., Verdu, E. F., Khan, W. I., & Collins, S. M. (2009). Reactivation of inflammatory bowel disease in a mouse model of depression. *Gastroenterology, 136*(7), 2280–2288. e2281–2284.
Heidemann, J., Ruther, C., Kebschull, M., Domschke, W., Bruwer, M., Koch, S., ... Maaser, C. (2007). Expression of IL-12-related molecules in human intestinal microvascular endothelial cells is regulated by TLR3. *American Journal of Physiology. Gastrointestinal and Liver Physiology, 293*(6), G1315–G1324.
Hershberg, R. M., Cho, D. H., Youakim, A., Bradley, M. B., Lee, J. S., Framson, P. E., & Nepom, G. T. (1998). Highly polarized HLA class II antigen processing and presentation by human intestinal epithelial cells. *The Journal of Clinical Investigation, 102*(4), 792–803.

Jenkins, R. T., Rooney, P. J., Jones, D. B., Bienenstock, J., & Goodacre, R. L. (1987). Increased intestinal permeability in patients with rheumatoid arthritis: A side-effect of oral nonsteroidal anti-inflammatory drug therapy? *British Journal of Rheumatology*, *26*(2), 103–107.

Johansson, M. E., Larsson, J. M., & Hansson, G. C. (2011). The two mucus layers of colon are organized by the MUC2 mucin, whereas the outer layer is a legislator of host-microbial interactions. *Proceedings of the National Academy of Sciences of the United States of America*, *108*(Suppl. 1), 4659–4665.

Johansson, M. E., Sjovall, H., & Hansson, G. C. (2013). The gastrointestinal mucus system in health and disease. *Nature Reviews. Gastroenterology & Hepatology*, *10*(6), 352–361.

Jung, C., Hugot, J. P., & Barreau, F. (2010). Peyer's patches: The immune sensors of the intestine. *International Journal of Inflammation*, *2010*, 823710.

Kachlik, D., Baca, V., & Stingl, J. (2010). The spatial arrangement of the human large intestinal wall blood circulation. *Journal of Anatomy*, *216*(3), 335–343.

Kelsall, B. (2008). Recent progress in understanding the phenotype and function of intestinal dendritic cells and macrophages. *Mucosal Immunology*, *1*(6), 460–469.

Koloski, N. A., Jones, M., Young, M., & Talley, N. J. (2015). Differentiation of functional constipation and constipation predominant irritable bowel syndrome based on Rome III criteria: A population-based study. *Alimentary Pharmacology & Therapeutics*, *41*(9), 856–866.

Li, X., Kan, E. M., Lu, J., Cao, Y., Wong, R. K., Keshavarzian, A., & Wilder-Smith, C. H. (2013). Combat-training increases intestinal permeability, immune activation and gastrointestinal symptoms in soldiers. *Alimentary Pharmacology & Therapeutics*, *37*(8), 799–809.

Li, Z., Zhang, C., Zhou, Z., Zhang, J., Zhang, J., & Tian, Z. (2012). Small intestinal intraepithelial lymphocytes expressing CD8 and T cell receptor gammadelta are involved in bacterial clearance during Salmonella enterica serovar Typhimurium infection. *Infection and Immunity*, *80*(2), 565–574.

Liew, F. Y., Xu, D., Brint, E. K., & O'Neill, L. A. (2005). Negative regulation of toll-like receptor-mediated immune responses. *Nature Reviews. Immunology*, *5*(6), 446–458.

Luyer, M. D., Buurman, W. A., Hadfoune, M., Jacobs, J. A., Konstantinov, S. R., Dejong, C. H., & Greve, J. W. (2004). Pretreatment with high-fat enteral nutrition reduces endotoxin and tumor necrosis factor-alpha and preserves gut barrier function early after hemorrhagic shock. *Shock*, *21*(1), 65–71.

Lyte, M., Li, W., Opitz, N., Gaykema, R. P., & Goehler, L. E. (2006). Induction of anxiety-like behavior in mice during the initial stages of infection with the agent of murine colonic hyperplasia Citrobacter rodentium. *Physiology & Behavior*, *89*(3), 350–357.

Madalena, K. M., & Lerch, J. K. (2016). Glucocorticoids and nervous system plasticity. *Neural Regeneration Research*, *11*(1), 37–41.

Maes, M., Kubera, M., Leunis, J. C., & Berk, M. (2012). Increased IgA and IgM responses against gut commensals in chronic depression: Further evidence for increased bacterial translocation or leaky gut. *Journal of Affective Disorders*, *141*(1), 55–62.

McVey Neufeld, K. A., Mao, Y. K., Bienenstock, J., Foster, J. A., & Kunze, W. A. (2013). The microbiome is essential for normal gut intrinsic primary afferent neuron excitability in the mouse. *Neurogastroenterology and Motility*, *25*(2), 183. e188.

Meddings, J. B., & Swain, M. G. (2000). Environmental stress-induced gastrointestinal permeability is mediated by endogenous glucocorticoids in the rat. *Gastroenterology*, *119*(4), 1019–1028.

Moran, G. W., O'Neill, C., & McLaughlin, J. T. (2012). GLP-2 enhances barrier formation and attenuates TNFalpha-induced changes in a Caco-2 cell model of the intestinal barrier. *Regulatory Peptides*, *178*(1–3), 95–101.

Neunlist, M., Toumi, F., Oreschkova, T., Denis, M., Leborgne, J., Laboisse, C. L., … Jarry, A. (2003). Human ENS regulates the intestinal epithelial barrier permeability and a tight junction-associated protein ZO-1 via VIPergic pathways. *American Journal of Physiology. Gastrointestinal and Liver Physiology*, *285*(5), G1028–G1036.

Neunlist, M., Van Landeghem, L., Mahe, M. M., Derkinderen, P., des Varannes, S. B., & Rolli-Derkinderen, M. (2013). The digestive neuronal-glial-epithelial unit: A new actor in gut health and disease. *Nature Reviews. Gastroenterology & Hepatology*, *10*(2), 90–100.

Niess, J. H., Brand, S., Gu, X., Landsman, L., Jung, S., McCormick, B. A., … Reinecker, H. C. (2005). CX3CR1-mediated dendritic cell access to the intestinal lumen and bacterial clearance. *Science*, *307*(5707), 254–258.

O'Neil, D. A., Porter, E. M., Elewaut, D., Anderson, G. M., Eckmann, L., Ganz, T., & Kagnoff, M. F. (1999). Expression and regulation of the human beta-defensins hBD-1 and hBD-2 in intestinal epithelium. *Journal of Immunology*, *163*(12), 6718–6724.

Oppong, E., & Cato, A. C. (2015). Effects of glucocorticoids in the immune system. *Adv. Exp. Med. Biol.*, *872*, 217–233.

Pavlov, V. A., Wang, H., Czura, C. J., Friedman, S. G., & Tracey, K. J. (2003). The cholinergic anti-inflammatory pathway: A missing link in neuroimmunomodulation. *Molecular Medicine*, *9*(5–8), 125–134.

Perdue, M. H. (1999). Mucosal immunity and inflammation. III. The mucosal antigen barrier: Cross talk with mucosal cytokines. *The American Journal of Physiology*, *277*(1 Pt. 1), G1–G5.

Plotz, F. B., Slutsky, A. S., van Vught, A. J., & Heijnen, C. J. (2004). Ventilator-induced lung injury and multiple system organ failure: A critical review of facts and hypotheses. *Intensive Care Medicine*, *30*(10), 1865–1872.

Pull, S. L., Doherty, J. M., Mills, J. C., Gordon, J. I., & Stappenbeck, T. S. (2005). Activated macrophages are an adaptive element of the colonic epithelial progenitor niche necessary for regenerative responses to injury. *Proceedings of the National Academy of Sciences of the United States of America*, *102*(1), 99–104.

Ramirez, K., Huerta, R., Oswald, E., Garcia-Tovar, C., Hernandez, J. M., & Navarro-Garcia, F. (2005). Role of EspA and intimin in expression of proinflammatory cytokines from enterocytes and lymphocytes by rabbit enteropathogenic Escherichia coli-infected rabbits. *Infection and Immunity*, *73*(1), 103–113.

Reichmann, F., Hassan, A. M., Farzi, A., Jain, P., Schuligoi, R., & Holzer, P. (2015). Dextran sulfate sodium-induced colitis alters stress-associated behaviour and neuropeptide gene expression in the amygdala-hippocampus network of mice. *Scientific Reports*, *5*, 9970.

Rodrigues, D. M., Sousa, A. J., Johnson-Henry, K. C., Sherman, P. M., & Gareau, M. G. (2012). Probiotics are effective for the prevention and treatment of Citrobacter rodentium-induced colitis in mice. *The Journal of Infectious Diseases*, *206*(1), 99–109.

Rojas-Corrales, M. O., Berrocoso, E., Gibert-Rahola, J., & Mico, J. A. (2004). Antidepressant-like effect of tramadol and its enantiomers in reserpinized mice: Comparative study with desipramine, fluvoxamine, venlafaxine and opiates. *Journal of Psychopharmacology*, *18*(3), 404–411.

Saunders, P. R., Hanssen, N. P., & Perdue, M. H. (1997). Cholinergic nerves mediate stress-induced intestinal transport abnormalities in Wistar–Kyoto rats. *The American Journal of Physiology*, *273*(2 Pt. 1), G486–G490.

Schumacher, V. L., Martel, A., Pasmans, F., Van Immerseel, F., & Posthaus, H. (2013). Endothelial binding of beta toxin to small intestinal mucosal endothelial cells in early stages of experimentally induced Clostridium perfringens type C enteritis in pigs. *Veterinary Pathology*, *50*(4), 626–629.

Shakweh, M., Ponchel, G., & Fattal, E. (2004). Particle uptake by Peyer's patches: A pathway for drug and vaccine delivery. *Expert Opinion on Drug Delivery*, *1*(1), 141–163.

Shan, M., Gentile, M., Yeiser, J. R., Walland, A. C., Bornstein, V. U., Chen, K., … Cerutti, A. (2013). Mucus enhances gut homeostasis and oral tolerance by delivering immunoregulatory signals. *Science*, *342*(6157), 447–453.

Smythies, L. E., Sellers, M., Clements, R. H., Mosteller-Barnum, M., Meng, G., Benjamin, W. H., … Smith, P. D. (2005). Human intestinal macrophages display profound inflammatory anergy despite avid phagocytic and bacteriocidal activity. *The Journal of Clinical Investigation*, *115*(1), 66–75.

Soderholm, J. D., & Perdue, M. H. (2001). Stress and gastrointestinal tract. II. Stress and intestinal barrier function. *American Journal of Physiology. Gastrointestinal and Liver Physiology*, *280*(1), G7–G13.

Sommer, F., & Backhed, F. (2013). The gut microbiota—Masters of host development and physiology. *Nature Reviews. Microbiology*, *11*(4), 227–238.

Suzuki, T. (2013). Regulation of intestinal epithelial permeability by tight junctions. *Cellular and Molecular Life Sciences*, *70*(4), 631–659.

Turner, J. R. (2009). Intestinal mucosal barrier function in health and disease. *Nature Reviews. Immunology*, *9*(11), 799–809.

Vaishnava, S., Behrendt, C. L., Ismail, A. S., Eckmann, L., & Hooper, L. V. (2008). Paneth cells directly sense gut commensals and maintain homeostasis at the intestinal host-microbial interface. *Proceedings of the National Academy of Sciences of the United States of America*, *105*(52), 20858–20863.

Vakharia, K., & Hinson, J. P. (2005). Lipopolysaccharide directly stimulates cortisol secretion by human adrenal cells by a cyclooxygenase-dependent mechanism. *Endocrinology*, *146*(3), 1398–1402.

Wagner, C. L., Taylor, S. N., & Johnson, D. (2008). Host factors in amniotic fluid and breast milk that contribute to gut maturation. *Clinical Reviews in Allergy & Immunology*, *34*(2), 191–204.

Zimomra, Z. R., Porterfield, V. M., Camp, R. M., & Johnson, J. D. (2011). Time-dependent mediators of HPA axis activation following live Escherichia coli. *American Journal of Physiology. Regulatory, Integrative and Comparative Physiology*, *301*(6), R1648–R1657.

CHAPTER SEVEN

Toxoplasma gondii—A Gastrointestinal Pathogen Associated with Human Brain Diseases

E.G. Severance, J. Xiao, L. Jones-Brando, S. Sabunciyan, Y. Li, M. Pletnikov, E. Prandovszky, R. Yolken[1]
Johns Hopkins School of Medicine, Baltimore, MD, United States
[1]Corresponding author: e-mail address: rhyolken@gmail.com

Contents

Abstract

Serious psychiatric disorders such as schizophrenia, bipolar disorder, and major depression are important causes of mortality and morbidity worldwide. While these are primarily diseases involving altered brain functioning, numerous studies have documented increased rates of gastrointestinal inflammation and dysfunction in many individuals with these disorders.

Toxoplasma gondii is an apicomplexan protozoan intracellular parasite with a widespread distribution in both developed and developing countries. *Toxoplasma* organisms enter the ecosystem through the shedding of oocysts by *Toxoplasma*-infected felines. In almost all cases of postnatal human infection, *Toxoplasma* enters its hosts through the intestinal tract either by the ingestion of oocysts or by the consumption of meat from food animals which themselves were infected by *Toxoplasma* oocysts.

It had previously been thought that most cases of *Toxoplasma* infection in immune competent children and adults were inapparent and asymptomatic. However, recent studies cast doubt on this concept as exposure to *Toxoplasma* has been associated with a range of acute and chronic symptoms. Of particular note has been the finding of an

International Review of Neurobiology, Volume 131
ISSN 0074-7742
http://dx.doi.org/10.1016/bs.irn.2016.08.008

increased rate of a range of neurological and psychiatric disorders associated with serological evidence of *Toxoplasma* exposure. A role of *Toxoplasma* infection in brain diseases is also supported by the consistent finding of altered cognition and behavior in animal models of infections. Much of the attention relating to the role of *Toxoplasma* infection in neuropsychiatric disorders has focused on the brain, where *Toxoplasma* tissue cysts can persist for extended periods of time. However, recent discoveries relating to the role of the gastrointestinal tract in cognition and behavior suggest that *Toxoplasma* may also increase susceptibility to human brain diseases through immune activation, particularly involving the gastrointestinal mucosa.

The study of the pathways relating to the pathobiology and immunology of *Toxoplasma* infection may provide insights into the pathogenesis of a range of human neuropsychiatric disorders as well as into cognitive functioning in otherwise healthy individuals.

1. THE BIOLOGY OF *TOXOPLASMA* INFECTION

Toxoplasma gondii is an apicomplexan protozoan with a worldwide distribution. *Toxoplasma* organisms can undergo a complete cycle of replication in feline species, which thus serve as complete hosts for this organism. However, *Toxoplasma* can also undergo incomplete replication in virtually any warm-blooded animal, with those animals constituting intermediate hosts (Fig. 1). In order to adapt to these multiple environments, *Toxoplasma* has developed an intricate and complex genome and system of gene expression capable of encoding enzymes and other proteins required for intracellular replication and maintenance in different host species. It has also developed a complex set of regulatory molecules and pathways allowing for persistence in tissues, particularly within the central nervous system (White, Radke, & Radke, 2014).

Initial infection of a host is usually accomplished by the rapidly replicating form of the organism called tachyzoites (Fig. 2). However, in an immune competent host following activation of the immune system, these tachyzoites undergo a conversion to the slowly replicating forms of the organism called bradyzoites (Fig. 3) which cluster to form tissue cysts. These tissue cysts can persist in the brain and other organs for extended periods of time without the generation of an apparent immune response. At various times, a portion of these tissue cysts can reactivate into tachyzoites and, if the immune response is adequate, with subsequent reconversion into bradyzoite-containing tissue cysts. Through these interconversions *Toxoplasma* can establish lifelong persistence in immune competent hosts.

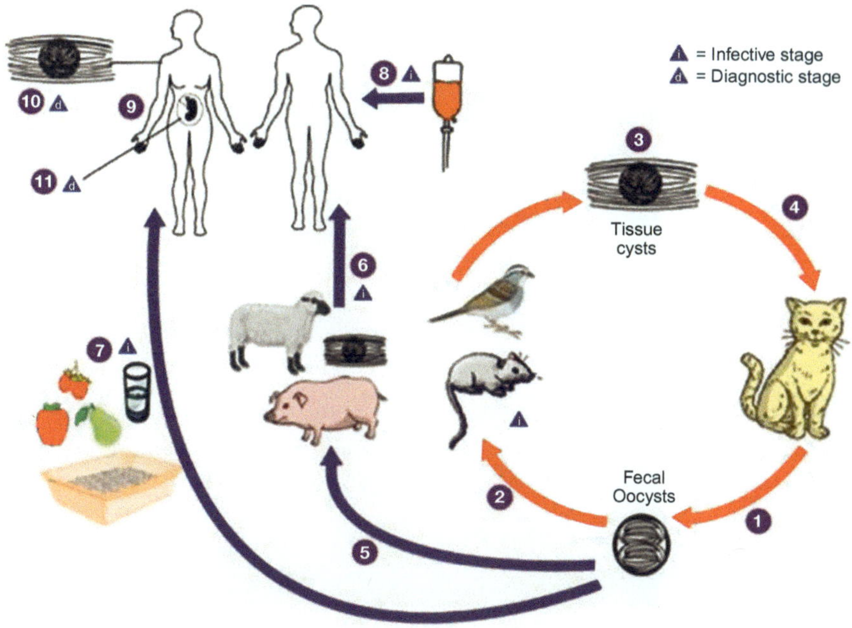

Fig. 1 Life cycle of *Toxoplasma gondii*. The only known definitive hosts for *Toxoplasma gondii* are members of family Felidae (domestic cats and their relatives). Unsporulated oocysts are shed in the cat's feces (1). Although oocysts are usually only shed for 1–2 weeks, large numbers may be shed especially in kittens of your cats undergoing their first infection. Oocysts take 1–5 days to sporulate in the environment and become infective. Intermediate hosts in nature (including birds, rodents, and farm animals) become infected after ingesting soil, water, or plant material contaminated with oocysts (2). Oocysts transform into tachyzoites shortly after ingestion. These tachyzoites localize in neural and muscle tissue and develop into tissue cyst bradyzoites (3). Cats become infected after consuming intermediate hosts harboring tissue cysts (4). Cats may also become infected directly by ingestion of sporulated oocysts although this is a less common form of infection. Animals bred for human consumption and wild game may also become infected with tissue cysts after ingestion of sporulated oocysts in the environment (5). Humans can become infected by any of several routes:eating undercooked meat of animals harboring tissue cysts (6) consuming food or water contaminated with cat feces or by contaminated environmental samples such as fecal-contaminated soil or changing the litter box of a pet cat (7). More rarely humans can become infected through blood transfusion or organ transplantation (8) transplacentally from mother to fetus (9). In the human host, the parasites form tissue cysts, most commonly in skeletal muscle, myocardium, brain, and eyes; these cysts may remain throughout the life of the host. Diagnosis is usually achieved by serology, although tissue cysts may be observed in stained biopsy specimens, particularly in immune-compromised individuals (10). Diagnosis of congenital infections can be achieved by detecting *T. gondii* DNA in amniotic fluid using molecular methods such as PCR (11). *Adapted from http://www.cdc.gov/parasites/toxoplasmosis/biology.html.*

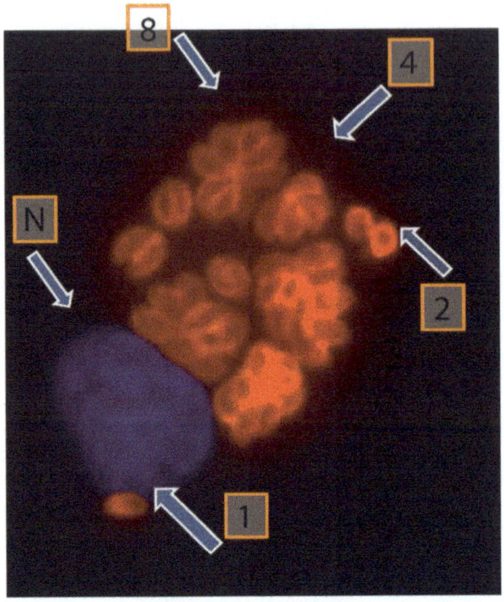

Fig. 2 Immunofluorescent visualization of intracellular tachyzoites of *T. gondii* strain RH shown 26 h after infection of human fibroblasts. The tachyzoites are reacted with rabbit antibody to the *Toxoplasma* p30 (SAG1) protein and then with antirabbit antibodies labeled with Alexa Fluor 594. Host cell nuclei are visualized using 4′,6-diamidino-2-phenylindole (DAPI). The tachyzoites thus are stained *red*, and the fibroblast nucleus (N) is stained *blue*. Tachyzoites typically replicate by endodyogeny. Thus, a cluster of 2 tachyzoites (2) is indicative of one cycle of replication, 4 tachyzoites of two cycles (4), and cluster of 8 tachyzoites (8) representing three cycles of replication. A single tachyzoite (1) represents one that has not yet undergone a cycle of replication. *Figure courtesy of Claudia Bordón, Johns Hopkins School of Medicine.*

2. EPIDEMIOLOGY OF *TOXOPLASMA* INFECTION

This complex life cycle has facilitated the widespread prevalence of *Toxoplasma* in animal populations, as the organism has been found to infect a wide range of warm-blooded animals living in different environments (Pittman & Knoll, 2015). In the case of humans, *Toxoplasma* infection exists in essentially every human population (Pappas, Roussos, & Falagas, 2009). As in the case of most other protozoa, the prevalence of *Toxoplasma* is higher in developing areas of the world. However, *Toxoplasma* is one of the few protozoans which have maintained a significant prevalence in developed countries such as the United States (Jones, Kruszon-Moran, Rivera, Price, & Wilkins, 2014). This high level of prevalence in human populations

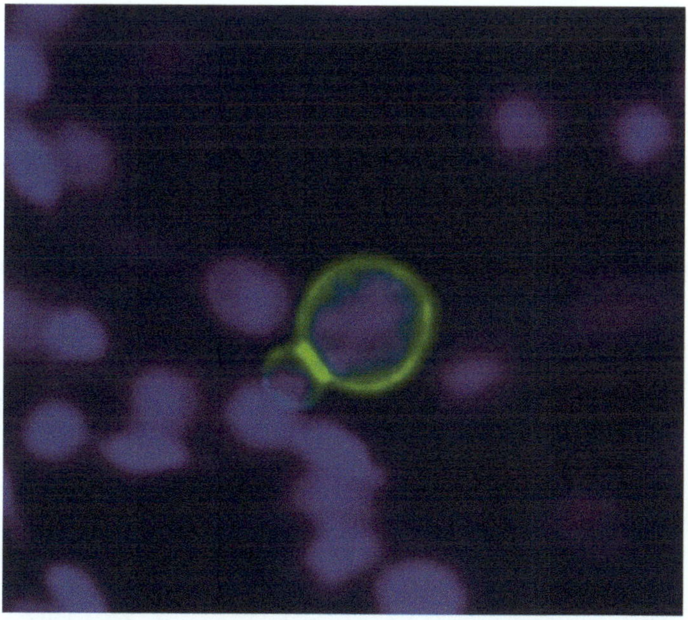

Fig. 3 A section of brain from chronically infected mice showing two *T. gondii* bradyzoite tissue cysts (*green*), at 400 × magnification. Note that the tissue cyst on the *left* is younger than the one on the *right* because of the difference in cyst size.

is likely due to the fact that individuals can become infected with *Toxoplasma* by a number of routes. For example, humans can become infected following the ingestion of oocysts shed in the feces of infected cats which reside in soil and other environmental niches. In many areas of the world, water contaminated with oocysts is a major source of environmental infections (Jones & Dubey, 2010). Since water systems which employ filtration and chlorination destroy most oocysts, this type of infection is less common in countries with well-functioning water purification infrastructure. Differential exposure to oocyst-contaminated water is thus likely to be the environmental factor which is largely responsible for the increased rates of prevalence of *Toxoplasma* infection in many areas of the developing world (Flegr, Prandota, Sovickova, & Israili, 2014).

Humans can also become infected through ingestion of meat from animals harboring tissue cysts. Since the level of heat used in cooking disrupts tissue cysts, most infections by this route occur through the ingestion of uncooked or undercooked meat. Also food animals vary in terms of resistance to tissue cysts and thus the ability to infect humans. This is the reason why the consumption of meat from some species, such as bovines, which are relatively resistant to *Toxoplasma*, is a less common source of transmission as compared to the ingestion of meat from other animal food sources such as ovine or porcine species (Dubey et al., 2005; Jones & Dubey, 2012).

Humans and other mammals can also become infected vertically by the passage of tachyzoites from mother to fetus. While the effects of fetal infection can be devastating, active fetal infection is a relatively rare event (Evengård et al., 2001; Yamada et al., 2015). This fact, in addition to the fact that *Toxoplasma* prevalence increased with age (Wilking, Thamm, Stark, Aebischer, & Seeber, 2016), suggests that most cases of *Toxoplasma* in humans are acquired after birth. Risk factors associated with the prevalence of schizophrenia in adults are largely based on age, geographic location, and household and occupational environmental exposures. Household environmental exposures include lack of access to clean water, eating vegetables washed with contaminated water, eating undercooked meat, and soil floors (Alvarado-Esquivel et al., 2006). Occupational risks associated with increased *Toxoplasma* exposure include gardening, working with, and raising farm animals (Alvarado-Esquivel, Campillo-Ruiz, & Liesenfeld, 2013; Alvarado-Esquivel, Sánchez-Anguiano, et al., 2013). Several studies have also found household cats to be a risk factor (Chiang et al., 2014), particularly if the exposure is to multiple kittens, although some studies have not found increased risks of *Toxoplasma* in households with only one adult cat (Esch & Peterson, 2013). Nonetheless, careful handling of cat feces and strategies to minimize *Toxoplasma* infection in cats is recommended in terms of the prevention of household transmission of infection (Opsteegh, Kortbeek, Havelaar, & van der Giessen, 2015).

It is of note that most available data relating to *Toxoplasma* exposure have been collected in adults. Data relating to the age of acquisition of *Toxoplasma* in children living under differing environmental conditions are currently lacking and are needed to better understand the epidemiology of *Toxoplasma* infection in childhood and adolescence. Also, the proportion of individuals infected from the ingestion of oocysts shed from infected cats as compared to tissue cysts from consumed meat is difficult to determine since, as discussed earlier, both forms of the organism are often generated in the host following infection despite source of the initial infection. Recent studies suggest that antibodies to sporozoites are present only when humans or other animals have been infected with *T. gondii* oocysts (Munoz-Zanzi, Fry, Lesina, & Hill, 2010). While sporozoite antibodies have the potential to differentiate oocyst- vs tissue cyst-induced infection, such antibodies are only detectable in humans within 6–8 months of initial oocyst-acquired infection. This time window considerably limits the utility of the sporozoite antibody, given the majority of human infections are chronic and thus would exceed the detectable period. The availability of assays capable of distinguishing the life cycle form of the initial infecting organisms in a large population over an extended period of time would be an important step in terms of developing efficient methods for the control of infection within a population (Hill et al., 2011).

3. CHRONIC *TOXOPLASMA* INFECTION OF HUMANS AND EXPERIMENTAL ANIMALS

Toxoplasma is well recognized as a cause of serious central nervous system infections in neonates and in individuals with depressed T-cell immunity as can occur in HIV infection, congenital immunodeficiency diseases, hematological malignancies, and during the course of immunosuppressive chemotherapy (Robert-Gangneux & Dardé, 2012). Considerably less attention has been given to the consequences of *Toxoplasma* infection in immune competent individuals. It has been previously thought that *Toxoplasma* infection in immune competent individuals was inapparent and not associated with measurable health consequences. However, there are a number of recent observations which seriously challenge this supposition. For example, there have been a number of waterborne outbreaks which have been described to be associated with acute fever, lymphadenopathy, retinopathy, and altered mental state (Bowie et al., 1997). The acute symptoms generally resolve but are sometimes followed by long-term sequellae such as retinal infection and decreased visual acuity (Burnett et al., 1998).

In addition, a number of animal models of chronic *Toxoplasma* infection have been developed, particularly in rodents such as mice and rats. While chronically infected animals gain weight normally and appear healthy, they often have measurable changes in behavior and cognition (Xiao et al., 2012). One of the interesting findings is that chronic *T. gondii* infection triggers abnormal response to dopamine, as evidenced by *T. gondii*-infected mice displayed a striking behavioral deficit in amphetamine-trigged locomotor response (Xiao et al., 2016). These models indicate that *Toxoplasma* infection can have lifelong effects on the brain functioning of intermediate hosts.

4. *TOXOPLASMA* EXPOSURE AND NEUROPSYCHIATRIC DISORDERS

An increased understanding of the role of persistent *Toxoplasma* infection in humans and animals has led to a reconsideration of the pathogenic effect of acquired *Toxoplasma* infections in immune competent individuals. Of particular importance are studies investigating the potential role of *Toxoplasma* infection in human neuropsychiatric disorders.

The neuropsychiatric disorder which has been studied in most detail is schizophrenia. Schizophrenia is a severe brain disorder involving altered perception and cognition. The etiology of schizophrenia is uncertain but is likely to involve both genetic and environmental factors (Børglum et al., 2014). A possible role of infectious agents in some cases of schizophrenia has been suspected since the disease was first characterized in the early part of the 20th century (Torrey, Bartko, Lun, & Yolken, 2007). The first published association between *Toxoplasma* exposure and schizophrenia was in 1952, and there have been many subsequent studies (Yolken & Torrey, 2008). A meta-analysis published in 2012 calculated a pooled odds ratio relating seropositivity to *Toxoplasma* and risk of schizophrenia of 2.71 (95% CI 1.93–3.80) based on 38 published studies (Fig. 4; Torrey et al.,

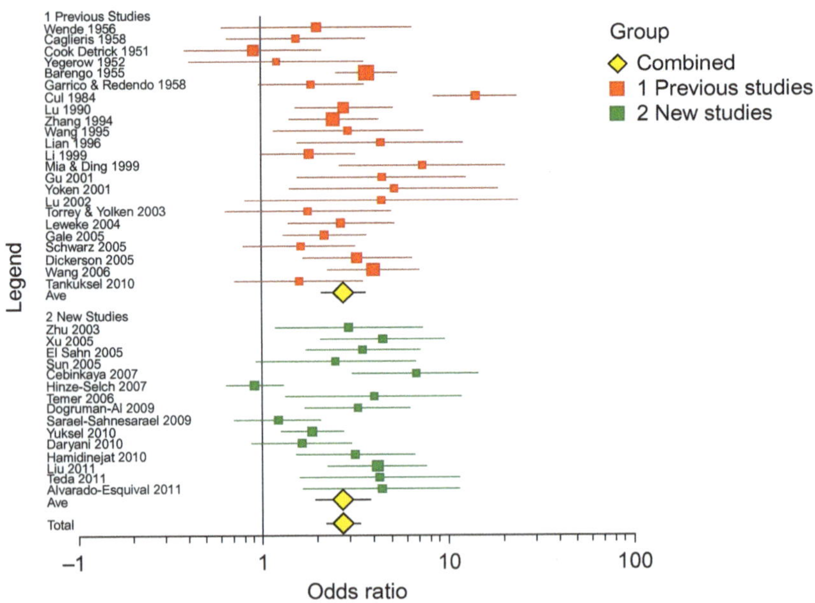

Fig. 4 Meta-analysis of published studies describing associations between *Toxoplasma* exposure and schizophrenia or related disorders. *Red boxes* indicate studies published before 2007 and summarized in Torrey et al. (2007). *Green boxes* indicate studies published between 2007 and 2012 and summarized in Torrey, Bartko, and Yolken (2012). *Yellow diamonds* indicate pooled odds ratios. *Reprinted from Torrey, E. F., Bartko, J. J., & Yolken, R. H. (2012). Toxoplasma gondii and other risk factors for schizophrenia: An update.* Schizophrenia Bulletin, 38*(3), 642–647.*

2012). Several additional studies reporting an association between exposure to *Toxoplasma* and either increased risk of schizophrenia or increased symptoms have been published since 2012. It is of note that some studies report an association between risk of schizophrenia and *Toxoplasma* prevalence as defined by detectable levels of antibodies, while other studies report a quantitative association between the risk of schizophrenia and the level of antibodies (Hinze-Selch et al., 2007). It should also be noted that several studies have been published which have failed to find a statistically significant association between exposure to *Toxoplasma* and risk of schizophrenia (Avramopoulos et al., 2015; Sugden et al., 2016). Reasons for this variation are not known with certainty but might include differences in the clinical status of the participants (recent onset vs chronic), methodological differences in the antibody measurement systems, differences in the prevalence of *Toxoplasma* infection, the timing of infection, the form of the infecting organism (tissue cysts from infected meat vs oocysts from cat feces), and the genetic background of the host. It is also possible that differences in the strain of *Toxoplasma* contribute to differences in psychiatric manifestations of infection (Xiao & Yolken, 2015). In animal models, behavioral abnormalities are associated with increased levels of antibodies to the *Toxoplasma* cyst protein MAG1, which is a serological marker of parasite burden (Xiao et al., 2016). It will be of great interest to determine if the measurement of specific antibodies to cyst proteins will correlate better with psychiatric outcomes than the measurement of antibodies to whole organisms or tachyzoite proteins as provided by currently available assay systems.

The mechanisms by which *Toxoplasma* exposure might be related to the risk of schizophrenia in humans are not known with certainty. However, the facts that *Toxoplasma* infection alters dopamine levels in the brain of some experimentally infected animals and the levels of dopamine are abnormal in schizophrenia suggest that alterations in this neurotransmitter may be a common link between *Toxoplasma* infection and schizophrenia (Flegr, 2015). A possible association between *Toxoplasma* infection, dopamine metabolism, and schizophrenia is also suggested by the finding that a number of pharmacological inhibitors of dopamine receptor binding, which are used for the treatment of schizophrenia, are also inhibitors of *Toxoplasma* replication in cell culture and animals (Dittmar, Drozda, & Blader, 2016). It is of interest that valproic acid, a medication used for the treatment of schizophrenia, the mechanisms of action of which are unknown, can also inhibit the in vitro replication of *Toxoplasma* (Jones-Brando, Torrey, & Yolken, 2003). The source of increased dopamine is not known with certainty but may be related to both the generation of dopamine by the *Toxoplasma* organisms or by the host immune response (Martin et al., 2015). It is of note, however, that not all investigators have found increased levels of dopamine in *Toxoplasma*-infected mice (Wang, Harmon, O'Malley, & Sibley, 2015), suggesting that other neurophysiological mechanisms may be involved as well. Additional mechanisms which might be operant in *Toxoplasma* and which might be relevant to human psychiatric disorders include the generation of microRNAs (Li, Kannan, Pletnikov, Yolken, & Xiao, 2015) and alterations in the metabolism of kynurenine. It is also of note that *Toxoplasma* encodes a protein which is annotated as an NMDA receptor (*T. gondii GT1* protein-coding gene on TGGT1_chrXI), since the NMDA receptor is also an important component of human psychiatric disorders (Balu, 2016). In particular, experimental infection of mice with *Toxoplasma* results in the generation of antibodies to the NMDA receptor (Kannan et al., 2016). Antibodies to the NMDA receptor have been noted to be associated with a number of human brain diseases including schizophrenia, mania (Dickerson et al., 2013), and other forms of acute psychosis.

Exposure to *Toxoplasma* has also been associated with other psychiatric conditions, albeit less consistently than with schizophrenia. Psychiatric conditions which have been associated with either increased prevalence of *Toxoplasma* infection or increased levels of antibodies include bipolar disorder (Pearce, Kruszon-Moran, & Jones, 2012), general anxiety disorder (Markovitz et al., 2015), mixed anxiety and depressive disorder (Alvarado-Esquivel et al., 2016), aggressive behavior (Cook et al., 2015), and acute convulsive epilepsy (Ae-Ngibise et al., 2015). Exposure to *Toxoplasma* has also been associated with increased rates of suicidality as measured by the number of actual suicide attempts (Alvarado-Esquivel, Campillo-Ruiz, et al., 2013; Alvarado-Esquivel, Sánchez-Anguiano, et al., 2013) or episodes of self-directed violence (Zhang et al., 2012). The association with suicide attempts is of interest in light of experimental models in which infected rodents lose their natural fear of feline predators, a process which has been termed "fatal attraction" (Berdoy, Webster, & Macdonald, 2000). This alteration in behavior is believed to be evolutionarily favorable to the *Toxoplasma* organism by facilitating its transmission from intermediate hosts to the complete feline host where it can undergo sexual reproduction and complete its life cycle. Some studies indicate that the *Toxoplasma* organism achieves the manipulation of host behavior through alterations in brain dopamine levels as described earlier. Suicidality and *Toxoplasma* infection have also been postulated to be linked by other pathways such as ones which metabolize kynurenine (Okusaga et al., 2016). Regardless of mechanism, it remains an intriguing possibility that suicidal behavior in humans represents a vestigial effect of this behavior, despite the fact that humans have not been common prey for carnivorous felines for many thousands of years. Consistent with this hypothesis are studies indicating an association between the exposure to *Toxoplasma* and other risk-taking behaviors such as automobile accidents (Alvarado-Esquivel, Torres-Castorena, Liesenfeld, Estrada-Martinez, & Urbina-Ålvarez, 2012) and impulsivity (Cook et al., 2015). Additional studies are needed to confirm the extent of these associations and mechanisms by which *Toxoplasma* infection might elicit these behavioral changes. In addition, it will be of interest to determine if the effects of *Toxoplasma* infection require the continued presence of organisms or if it can persist following organism clearance, as has been documented in animal models (Ingram, Goodrich, Robey, & Eisen, 2013).

In addition to being associated with altered behavior, *Toxoplasma* infections have shown consistent effects on cognitive functioning in experimental animals, particularly in the domains measuring memory (Wang et al., 2013). Similar alterations have also been noted in some populations of humans. For example, exposure to *Toxoplasma* has been associated with lower levels of cognitive functioning in children 12–16 years of age (Mendy, Vieira, Albatineh, & Gasana, 2015). Exposure to *Toxoplasma* has also been associated with decreased memory functioning in individuals who are more than 64 years (Gajewski, Falkenstein, Hengstler, & Golka, 2014) of age as well as a decline in memory and other cognitive functioning after that age (Nimgaonkar et al., 2016). On the other hand, exposure to *Toxoplasma* was not associated with altered levels of memory in unselected adults of unknown age (Gale, Brown, Erickson, Berret, & Hedges, 2015) or adults with schizophrenia of unknown age (Yolken, Torrey, Lieberman, Yang, & Dickerson, 2011). Results of studies examining the possible association between exposure to *Toxoplasma* and Alzheimer's disease have shown mixed results (Perry et al., 2016). It is of note that *Toxoplasma* infection of cells can actually decrease the formation of beta-amyloid plaques (Möhle et al., 2016), suggesting that *Toxoplasma* may be involved in other forms of cognitive impairment and dementia in the elderly through an alternative mechanism.

5. *TOXOPLASMA* AND INTESTINAL INFLAMMATION

While most of the interest in mechanisms relating *Toxoplasma* infection and altered cognition or behavior has focused on the brain, the fact that *Toxoplasma* enters most hosts through the intestinal tract has also led to investigation of the possibility that *Toxoplasma* might also affect behavior through its effects on the intestinal tract. Numerous studies have documented alterations in intestinal functioning and inflammation in individuals with psychiatric disorders (reviewed in Severance, Gressitt, et al., 2015; Severance et al., 2014). In addition, antibodies to *Toxoplasma* have been associated with markers of intestinal inflammation in individuals with psychiatric disorders (Severance et al., 2012, 2014). Recent studies linking changes in the intestinal microbiome to altered behavior in humans and experimental animals (Borre, Moloney, Clarke, Dinan, & Cryan, 2014) suggest that some of the effect of *Toxoplasma* on behavior and cognition may be related to changes at the level of the gastrointestinal tract.

Following oral infection of experimental animals, *T. gondii* parasites can be found within hours in the surface epithelium and lamina propria of the small intestine, and particularly in the ileum (Dubey, 1997). Within days of entry to the intestinal tract, parasites can migrate into systemic circulation from the lamina propria where they then have access to host organs (Lisenfeld, 2002). Presumably this translocation of *T. gondii* into the blood stream is a consequence of localized intestinal inflammation and enteropathy generated by the parasite that collectively results in impaired integrity of the intestinal mucosa and gut–blood barrier. Indeed, the gut-targeted inflammatory state elicited by *T. gondii* has been adapted for experimental animal models of inflammatory bowel diseases and of ileitis in particular, although evidence in support of cellular pathologies in the duodenum and jejunum is surfacing as well (Araújo et al., 2015; Trevizan et al., 2016). The intra- and paracellular mechanisms for parasite invasion of gut epithelial or Peyer's patch-associated cells are not known with certainty but are actively investigated topics (Briceño et al., 2016; Gregg et al., 2013).

Loss of cellular barrier integrity at the gut–vasculature interface has implications for the blood–brain barrier in psychiatric disorders. In light of the similarities of the of gut–blood and brain–blood barriers, this cellular permeability offers a route by which products of gut-based processes may impact the brain. A permeabilized gut–blood barrier in psychiatric disorders has been inferred from studies of microbial translocation and microbiome sequencing which indicate an actively dysbiotic environment in subsets of individuals with psychiatric disorders (Castro-Nallar et al., 2015; Severance et al., 2013; Severance, Gressitt, et al., 2016; Severance, Yolken, & Eaton, 2016; Yolken et al., 2015). A functional pathological outcome of barrier permeability is the translocation of gut-dwelling microbes including bacteria and yeast into systemic circulation, and these translocation rates are increased in people with psychiatric disorders (Severance et al., 2013). Studies of gut-derived markers in the CNS also point toward barrier permeability issues of the CSF–blood interface including the choroid plexus in individuals with psychiatric disorders (Severance, Prandovszky, Castiglione, & Yolken, 2015). In the context of the present review, microbial dysbiosis is an effective perpetrator of intestinal inflammation and subsequent permeability of the gut barrier and importantly has distal consequences on the blood–brain barrier (Braniste et al., 2014). Thus, a parasite-mediated endothelial barrier compromise in psychiatric disorders could be a function of intestinal inflammation produced directly by invasion or indirectly by the parasite effects on the gut microbiome. Oral infection has been demonstrated to perpetrate changes in the dynamics of the gut microbiome that are sometimes immunopathogenic and sometimes immunoprotective (Bereswill et al., 2014; Craven et al., 2012; Egan, Cohen, & Denkers, 2012; Haag et al., 2012).

Thus, the immune system and its response to *T. gondii* both systemically and locally in the gut mediate the degree of eventual neuropathy of the parasite. The gut derived immune response which even when operating functionally is complicated and dependent on a stable and homeostatically balanced gut microbiome. *Toxoplasma* infection in experimental animals has been shown to alter several aspects of intestinal immunity (Cohen & Denkers, 2015a, 2015b). Conversely, the intestinal microflora may be one factor controlling the immune response to *Toxoplasma* in the intestinal tract and hence host resistance to *Toxoplasma* infection (Ribeiro, Zorgi, Meireles, Garcia, & de Andrade Junior, 2016). Acute *Toxoplasma* infection has also been shown to change the microbiome of experimentally infected mice. In these studies, the relative abundance of Gram-negative bacteria such as Enterobacteria and Prevotella increases, while relative abundances of Gram-positive bacteria such as Clostridia and other Firmicutes are decreased. Despite the increase in total eubacteria load, acute *T. gondii* infection was accompanied by loss of microbial diversity (Craven et al., 2012; Molloy et al., 2013). Initial studies also indicate that chronic *Toxoplasma* infection of mice is also associated with changes in the intestinal microbiome, suggesting that alterations in the microbiome may play a role in the altered behavior noted in such animals (E. Prandovszky et al., unpublished). Studies of the effect of *Toxoplasma* infection on the microbiome of humans are currently lacking.

6. CURRENT STATUS OF ANTI-*TOXOPLASMA* MEDICATIONS

One limitation in terms of the study and management of *Toxoplasma* infections relates to the paucity of effective medications. The ideal treatment against *Toxoplasma* would be effective at inhibiting the different life stages of the parasite. However, current treatments are only capable of suppressing the rapidly dividing tachyzoite stage of the organism. The folate synthesis pathway is the best-known therapeutic target against *Toxoplasma.* Compared to mammalian cells that make extensive use of exogenous folate transport pathways to obtain folate, *Toxoplasma* mainly relies on folate synthesis pathways to produce most of the folate it metabolizes (Allegra et al., 1987; Kovacs et al., 1989; Massimine et al., 2005). The main enzymes targeted by pharmaceuticals in the folate synthesis pathway are the dihydropteroate synthase and dihydrofolate reductase. Since the dihydropteroate enzyme is not found in mammalian cells, it provides a specific target against *Toxoplasma.* Pyrimethamine and trimethoprim are the most widely used dihydrofolate reductase inhibitors. Pyrimethamine is a known terattogen but is more effective then trimethoprim, which is usually administered in conjunction with a dihydropteroate synthase inhibitor (sulfamethoxazole). The sulfonamide compounds sulfadiazine and sulfamethoxazole, which inhibit the dihydropteroate synthase, are highly effective against *Toxoplasma.* However, the cessation of the use of these compounds is frequently accompanied by relapse. It is also worth noting that congenital *Toxoplasma* infections in the first trimester of pregnancy are commonly treated with spiramycin, a macrolide antibiotic that has low toxicity but limited potency against the parasite.

In immune competent individuals, the most widely accepted regimen involves the use of the folic acid antagonist pyrimethamine with or without added sulfadiazine with the main application being the treatment of *Toxoplasma*-associated eye disease (Pradhan, Bhandari, Gilbert, & Stanford, 2016). In addition to having toxicities related to its mechanism of action as noted earlier, this mechanism is generally ineffective against the bradyzoite form of the organism and thus cannot be used to treat tissue cysts, the form of the organism resident in the central nervous system of immune competent individuals. Additional medications, largely antibacterials and antiparasitics, have been employed to treat *Toxoplasma* infections in immune-compromised individuals who do not respond to, or cannot tolerate, treatment with folate antagonists (Wei, Wei, Lindsay, & Peng, 2015). However, the efficacy of these regimens is difficult to evaluate and, as in the case of the folate antagonists, these medications are not effective against tissue cysts. Recently a number of pharmacological approaches have been developed for the treatment of *Toxoplasma* tissue cysts in the brains of experimentally infected mice (Doggett et al., 2012). The development of these compounds as human medications would represent an important step in the ability to control *Toxoplasma* infections and to evaluate the role of *Toxoplasma* brain cysts in human diseases. Similarly, the application to feline and human populations of immunization regimens for the prevention of *Toxoplasma* infections, which are currently in the experimental stage (Opsteegh et al., 2015), would represent another important tool both the prevention of human *Toxoplasma* infections and the study of the role of these infections in human pathobiology.

7. ONGOING RESEARCH NEEDS

During the past decade a great deal has been learned regarding the biology of chronic *Toxoplasma* infection in experimental animals, particularly in rodents. However, an understanding of the role of *Toxoplasma* in human biology and pathology has proceeded more slowly. The main limitation of human studies is that, due to the encysted nature of the organism, it is very difficult to obtain *Toxoplasma* organisms or DNA from accessible body fluids such as blood or urine. Hence most studies have relied on serological methods to define exposure and the immune response to infection. While accurate and reproducible in terms of differentiating exposed from unexposed populations, currently available serological methods have a number of limitations. Of particular importance, assays which are currently in widespread usage cannot easily determine the timing of infection earlier in life, the source of infection (tissue cysts vs oocysts), and the biotype of the infecting organism. The more widespread development of assays which have been reported to accomplish some of these goals would represent a major step forward in terms of the ability to study *Toxoplasma* infection in humans. Similarly, infection and inflammation within the body organs infected with *Toxoplasma*, such as the brain and the gastrointestinal tract, can be difficult to study in a noninvasive manner. The further development of diffusion-weighted imaging (Maschke, Kastrup, Forsting, & Diener, 2004) and other modalities capable of measuring infection and within these and other body organs would also represent an important step in terms of identifying individuals with clinically significant *Toxoplasma* infections and guiding therapeutic interventions. Finally, the role of *Toxoplasma* in cognition and behavior suggests that population with higher rates of *Toxoplasma* infections based on differential exposures may suffer societal consequences based on this exposure. The study of *Toxoplasma* exposure as a health disparity which can potentially be corrected through preventative measures such as improved water purification remains an important goal of Public Health research.

8. CONCLUSIONS

T. gondii is an organism associated with infection of the gastrointestinal tract, local and systemic inflammation, and alterations in brain functioning. The study of the pathways relating to the gastrointestinal biology and immunology of *Toxoplasma* infection may provide novel insights into the pathogenesis of a range of human neuropsychiatric disorders. The development of effective means for the prevention of *Toxoplasma* infections and for the control of immune activation may lead to new methods for the prevention and treatment of these devastating disorders as well as an overall improvement in the physical and mental health of exposed individuals.

REFERENCES

Ae-Ngibise, K. A., Akpalu, B., Ngugi, A., Akpalu, A., Agbokey, F., Adjei, P., et al. (2015). Prevalence and risk factors for active convulsive epilepsy in Kintampo, Ghana. *The Pan African Medical Journal*, *21*, 29.

Allegra, C. J., Kovacs, J. A., Drake, J. C., Swan, J. C., Chabner, B. A., & Masur, H. (1987). Potent in vitro and in vivo antitoxoplasma activity of the lipid-soluble antifolate trimetrexate. *Journal of Clinical Investigation*, *79*, 478–482.

Alvarado-Esquivel, C., Campillo-Ruiz, F., & Liesenfeld, O. (2013). Seroepidemiology of infection with Toxoplasma gondii in migrant agricultural workers living in poverty in Durango, Mexico. *Parasites & Vectors*, *6*, 113.

Alvarado-Esquivel, C., Sánchez-Anguiano, L. F., Arnaud-Gil, C. A., López-Longoria, J. C., Molina-Espinoza, L. F., Estrada-Martinez, S., et al. (2013). Toxoplasma gondii infection and suicide attempts: A case-control study in psychiatric outpatients. *The Journal of Nervous and Mental Diseases*, *201*(11), 948–952.

Alvarado-Esquivel, C., Sanchez-Anguiano, L. F., Hernandez-Tinoco, J., Berumen-Segovia, L. O., Torres-Prieto, Y. E., Estrada-Martinez, S., et al. (2016). Toxoplasma gondii infection and mixed anxiety and depressive disorder: A case-control seroprevalence study in Durango, Mexico. *Journal of Clinical Medicine Research*, *8*(7), 519–523.

Alvarado-Esquivel, C., Sifuentes-Alvarez, A., Narro-Duarte, S. G., Estrada-Martinez, S., Diaz-Garcia, J. H., Liesenfeld, O., et al. (2006). Seroepidemiology of Toxoplasma gondii infection in pregnant women in a public hospital in northern Mexico. *BMC Infectious Diseases*, *6*, 113.

Alvarado-Esquivel, C., Torres-Castorena, A., Liesenfeld, O., Estrada-Martinez, S., & Urbina-Álvarez, J. D. (2012). High seroprevalence of Toxoplasma gondii infection in a subset of Mexican patients with work accidents and low socioeconomic status. *Parasites & Vectors*, *5*, 13.

Araújo, E. J., Zaniolo, L. M., Vincentino, S. L., Góis, M. B., Zanoni, J. N., da Silva, A. V., et al. (2015). Toxoplasma gondii causes death and plastic alteration in the jejunal myenteric plexus. *World Journal of Gastroenterology*, *21*(16), 4829–4839.

Avramopoulos, D., Pearce, B. D., McGrath, J., Wolyniec, P., Wang, R., Eckart, N., et al. (2015). Infection and inflammation in schizophrenia and bipolar disorder: A genome wide study for interactions with genetic variation. *PloS One*, *10*(3), e016696.

Balu, D. T. (2016). The NMDA receptor and schizophrenia: From pathophysiology to treatment. *Advanced in Pharmacology*, *76*, 351–382.

Berdoy, M., Webster, J. P., & Macdonald, D. W. (2000). Fatal attraction in rats infected with Toxoplasma gondii. *Proceedings of the Biological Sciences*, *267*(1452), 1591–1594.

Bereswill, S., Kühl, A. A., Alutis, M., Fischer, A., Möhle, L., Struck, D., et al. (2014). The impact of Toll-like-receptor-9 on intestinal microbiota composition and extra-intestinal sequellae in experimental Toxoplasma gondii induced ileitis. *Gut Pathogens*, *6*, 19.

Børglum, A. D., Demontis, D., Grove, J., Pallesen, J., Hollegaard, M. V., Pedersen, C. B., et al. (2014). Genome-wide study of association and interaction with maternal cytomegalovirus infection suggests new schizophrenia loci. *Molecular Psychiatry*, *19*(3), 325–333.

Borre, Y. E., Moloney, R. D., Clarke, G., Dinan, T. G., & Cryan, J. F. (2014). The impact of microbiota on brain and behavior: Mechanisms and therapeutic potential. *Advances in Experimental Medicine and Biology*, *817*, 373–403.

Bowie, W. R., King, A. S., Werker, D. H., Isaac-Renton, J. L., Bell, A., Eng, S. B., et al. (1997). Outbreak of toxoplasmosis associated with municipal drinking water. The BC Toxoplasma Investigation Team. *Lancet*, *19*(9072), 173–177.

Braniste, V., Al-Asmakh, M., Kowal, C., Anuar, F., Abbaspour, A., Tóth, M., et al. (2014). The gut microbiota influences blood-brain barrier permeability in mice. *Science Translational Medicine*, *6*(263), 263ral58.

Briceño, M. P., Nascimento, L. A., Nogueira, N. P., Barenco, P. V., Ferro, E. A., Rezende-Oliveira, K., et al. (2016). Toxoplasma gondii infection promotes epithelial barrier dysfunction of Caco-2 cells. *The Journal of Histochemistry and Cytochemistry*, *64*(8), 459–469.

Burnett, A. J., Shortt, S. G., Isaac-Renton, J., King, A., Werker, D., & Bowie, W. R. (1998). Multiple cases of acquired toxoplasmosis retinitis presenting in an outbreak. *Ophthalmology*, *105*(6), 1032–1037.

Castro-Nallar, E., Bendall, M. L., Pérez-Losada, M., Sabunciyan, S., Severance, E. G., Dickerson, F., et al. (2015). Composition, taxonomy and functional diversity of the oropharynx microbiome in individuals with schizophrenia and controls. *PeerJ*, *3*, e1140.

Chiang, T. Y., Kuo, M. C., Chen, C. H., Yang, J. Y., Kao, C. F., Ji, D. D., et al. (2014). Risk factors for acute Toxoplasma gondii diseases in Taiwan: A population-based case-control study. *PloS One*, *9*(3), e90880.

Cohen, S. B., & Denkers, E. Y. (2015a). The gut mucosal immune response to Toxoplasma gondii. *Parasite Immunology*, *37*(3), 108–117.

Cohen, S. B., & Denkers, E. Y. (2015b). Impact of Toxoplasma gondii on dendritic cell subset function in the intestinal mucosa. *Journal of Immunology*, *195*(6), 2754–2762.

Cook, T. B., Brenner, L. A., Cloninger, C. R., Langenberg, P., Igbide, A., Giegling, I., et al. (2015). "Latent" infection with Toxoplasma gondii: Association with trait aggression and impulsivity in healthy adults. *Journal of Psychiatric Research*, *60*, 87–94.

Craven, M., Egan, C. E., Dowd, S. E., McDonough, S. P., Dogan, B., Denkers, E. Y., et al. (2012). Inflammation drives dysbiosis and bacterial invasion in murine models of ileal Crohn's disease. *PloS One*, *7*(7), e41594.

Dickerson, F., Stallings, C., Origoni, A., Vaughan, C., Katsafanas, E., Khushalani, S., et al. (2013). A combined marker of inflammation in individuals with mania. *PloS One*, *8*(9), e73520.

Dittmar, A. J., Drozda, A. A., & Blader, I. J. (2016). Drug repurposing screening identifies novel compounds that effectively inhibit Toxoplasma gondii growth. *mSphere*, *1*(2). e00042-15.

Doggett, J. S., Nilsen, A., Forquer, I., Wegman, K. W., Jones-Brando, L., Yolken, R. H., et al. (2012). Endochin-like quinolones are highly efficacious against acute and latent experimental toxoplasmosis. *Proceedings of the National Academy of Sciences of the United States of America*, *109*(39), 15936–15941.

Dubey, J. P. (1997). Bradyzoite-induced murine toxoplasmosis: Stage conversion, pathogenesis, and tissue formation in mice fed bradyzoites of different strains of Toxoplasma gondii. *Journal of Eukaryotic Microbiology*, *44*, 592–602.

Dubey, J. P., Hill, D. E., Jones, J. L., Hightower, A. W., Kirkland, E., Roberts, J. M., et al. (2005). Prevalence of viable Toxoplasma gondii in beef, chicken, and pork from retail meat stores in the United States: Risk assessment to consumers. *Journal of Parasitology*, *91*(5), 1082–1093.

Egan, C. E., Cohen, S. B., & Denkers, E. Y. (2012). Insights into inflammatory bowel disease using Toxoplasma gondii as an infectious trigger. *Immunology and Cell Biology*, *90*(7), 668–675.

Esch, K., & Peterson, C. A. (2013). Transmission and epidemiology of zoonotic protozoal diseases of companion animals. *Clinical Microbiology Reviews*, *26*(1), 58–85.

Evengård, B., Petersson, K., Engman, M. L., Wiklund, S., Ivarsson, S. A., Teär-Fahnehjelm, K., et al. (2001). Low incidence of Toxoplasma infection during pregnancy and in newborns in Sweden. *Epidemiology and Infection*, *127*(1), 121–127.

Flegr, J. (2015). Schizophrenia and Toxoplasma gondii: An undervalued association? *Expert Review of Anti-Infective Therapy*, *13*(7), 817–820.

Flegr, J., Prandota, J., Sovickova, M., & Israili, Z. H. (2014). Toxoplasmosis—A global treat. Correlation of latent toxoplasmosis with specific disease burden in a set of 88. *PloS One*, *9*(3), e90203.

Gajewski, P. D., Falkenstein, M., Hengstler, J. P., & Golka, K. (2014). Toxoplasma gondii impairs memory in infected seniors. *Brain, Behavior and Immunity*, *36*, 193–199.

Gale, S. D., Brown, B. L., Erickson, L. D., Berret, A., & Hedges, D. W. (2015). Association between latent toxoplasmosis and cognition in adults: A cross-sectional study. *Parasitology*, *142*(4), 557–565.

Gregg, B., Taylor, B. C., John, B., Tait-Wojno, E. D., Girgis, N. M., Miller, N., et al. (2013). Replication and distribution of Toxoplasma gondii in the small intestine after oral infection with tissue cysts. *Infection and Immunity*, *81*(5), 1635–1643.

Haag, L. M., Fischer, A., Otto, B., Plickert, R., Kühl, A. A., Göbel, U. B., et al. (2012). Intestinal microbiota shifts towards elevated commensal Escherichia coli loads abrogate colonization resistance against Campylobacter jejuni in mice. *PloS One*, *7*(5), e35988.

Hill, D., Coss, C., Dubey, J. P., Wroblewski, K., Sautter, M., Hosen, T., et al. (2011). Identification of a sporozoite-specific antigen from Toxoplasma gondii. *Journal of Parasitology*, *97*(2), 328–337.

Hinze-Selch, D., Däubener, W., Eggert, L., Erdag, S., Stoltenberg, R., & Wilms, S. (2007). A controlled prospective study of Toxoplasma gondii infection in individuals with schizophrenia: Beyond seroprevalence. *Schizophrenia Bulletin*, *33*(3), 782–788.

Ingram, W. M., Goodrich, L. M., Robey, E. A., & Eisen, M. B. (2013). Mice infected with low-virulence strains of Toxoplasma gondii lose their innate aversion to cat urine, even after extensive parasite clearance. *PloS One*, *8*(9), e75246.

Jones, J. L., & Dubey, J. P. (2010). Waterborne toxoplasmosis—Recent developments. *Experimental Parasitology*, *124*(1), 10–25.

Jones, J. L., & Dubey, J. P. (2012). Foodborne toxoplasmosis. *Clinical Infectious Diseases, 55*(6), 845–851.

Jones, J. L., Kruszon-Moran, D., Rivera, H. N., Price, C., & Wilkins, P. P. (2014). Toxoplasma gondii seroprevalence in the United States 2009–2010 and comparison with the past two decades. *American Journal of Tropical Medicine and Hygiene, 90*(6), 1135–1139.

Jones-Brando, L., Torrey, E. F., & Yolken, R. (2003). Drugs used in the treatment of schizophrenia and bipolar disorder inhibit the replication of Toxoplasma gondii. *Schizophrenia Research, 62*(3), 237–244.

Kannan, G., Crawford, J. A., Yang, C., Gressitt, K. L., Ihenatu, C., Krasnova, I. N., et al. (2016). Anti-NMDA receptor autoantibodies and associated neurobehavioral pathology in mice are dependent on age of first exposure to Toxoplasma gondii. *Neurobiology of Disease, 91*, 307–314.

Kovacs, J. A., Allegra, C. J., Beaver, J., Boarman, D., Lewis, M., Parrillo, J. E., et al. (1989). Characterization of de novo folate synthesis in Pneumocystis carinii and Toxoplasma gondii: Potential for screening therapeutic agents. *Journal of Infectious Diseases, 160*, 312–320.

Li, Y., Kannan, G., Pletnikov, M. V., Yolken, R. H., & Xiao, J. (2015). Chronic infection of Toxoplasma gondii downregulates miR-132 expression in multiple brain regions in a sex-dependent manner. *Parasitology, 142*(4), 623–632.

Lisenfeld, O. (2002). Oral infection of C57BL/6 mice with Toxoplasma gondii: A new model of inflammatory bowel disease? *Journal of Infectious Diseases, 185*(Suppl.), S96–S101.

Markovitz, A. A., Simanek, A. M., Yolken, R. H., Galea, S., Koenen, K. C., Chen, S., et al. (2015). Toxoplasma gondii and anxiety disorders in a community-based sample. *Brain, Behavior, and Immunity, 43*, 192–197.

Martin, H. L., Alsasdy, I., Howell, G., Prandovszky, E., Peers, C., Robinson, P., et al. (2015). Effects of parasitic infection on dopamine biosynthesis in dopaminergic cells. *Neuroscience, 306*, 50–62.

Maschke, M., Kastrup, O., Forsting, M., & Diener, H. C. (2004). Update on neuroimaging in infectious central nervous system disease. *Current Opinion in Neurobiology, 17*(4), 475–480.

Massimine, K. M., Doan, L. T., Atreya, C. A., Stedman, T. T., Anderson, K. S., Joiner, K. A., et al. (2005). Toxoplasma gondii is capable of exogenous folate transport. A likely expansion of the BT1 family of transmembrane proteins. *Molecular and Biochemical Parasitology, 144*, 44–54.

Mendy, A., Vieira, E. R., Albatineh, A. N., & Gasana, J. (2015). Toxoplasma gondii seropositivity and cognitive functions in school-aged children. *Parasitology, 142*(9), 1221–1227.

Möhle, L., Israel, N., Paarmann, K., Krohn, M., Pietkiewicz, S., Müller, A., et al. (2016). Chronic Toxoplasma gondii infection enhances β-amyloid phagocytosis and clearance by recruited monocytes. *Acta Neuropathologica Communications, 4*, 25.

Molloy, M. J., Grainger, J. R., Bouladoux, N., Hand, T. W., Koo, L. Y., Naik, S., et al. (2013). Intraluminal containment of commensal outgrowth in the gut during infection-induced dysbiosis. *Cell Host & Microbe, 14*(3), 318–328.

Munoz-Zanzi, C. A., Fry, P., Lesina, B., & Hill, D. (2010). Toxoplasma gondii oocyst-specific antibodies and source of infection. *Emerging Infectious Diseases, 16*(10), 1591–1593.

Nimgaonkar, V. L., Yolken, R. H., Wang, T., Chung-Chou, H. C., McClain, L., McDade, E., et al. (2016). Temporal cognitive decline associated with exposure to infectious agents in a population-based, aging cohort. *Alzheimer Disease and Associated Disorders, 30*, 216–222. (Epub ahead of print).

Okusaga, O., Duncan, E., Langenberg, P., Brundin, L., Fuchs, D., Groer, M. W., et al. (2016). Combined Toxoplasma gondii seropositivity and high blood kynurenine—

Linked with nonfatal suicidal self-directed violence in patients with schizophrenia. *Journal of Psychiatric Research*, *72*, 74–81.
Opsteegh, M., Kortbeek, T. M., Havelaar, A. H., & van der Giessen, J. W. (2015). Intervention strategies to reduce human Toxoplasma gondii disease burden. *Clinical Infectious Diseases*, *60*(1), 101–107.
Pappas, G., Roussos, N., & Falagas, M. E. (2009). Toxoplasmosis snapshots: Global status of Toxoplasma gondii seroprevalence and implications for pregnancy and congenital toxoplasmosis. *International Journal for Parasitology*, *39*(12), 1385–1394.
Pearce, B. D., Kruszon-Moran, D., & Jones, J. L. (2012). The relationship between Toxoplasma gondii infection and mood disorders in the third National Health and Nutrition Survey. *Biological Psychiatry*, *72*(4), 290–295.
Perry, C. E., Gale, S. D., Erickson, L., Wilson, E., Nielsen, B., et al. (2016). Seroprevalence and serointensity of latent Toxoplasma gondii in a sample of elderly adults with and without Alzheimer disease. *Alzheimer Disease and Associated Disorders*, *30*(2), 123–126.
Pittman, K. J., & Knoll, L. J. (2015). Long-term relationships: The complicated interplay between the host and the developmental stages of Toxoplasma gondii during acute and chronic infections. *Microbiology and Molecular Biology Reviews*, *79*(4), 387–401.
Pradhan, E., Bhandari, S., Gilbert, R. E., & Stanford, M. (2016). Antibiotics versus no treatment for Toxoplasma retinochoroiditis. *Cochrane Database of Systematic Reviews*, *5*, CD002218.
Ribeiro, Cde M., Zorgi, N. E., Meireles, L. R., Garcia, J. L., & de Andrade Junior, H. F. (2016). CD19 lymphocyte proliferation induced by Bifidobacterium animalis subsp. Lactis in C57Bl/6 mice experimentally infected with Toxoplasma gondii. *Revista do Instituto de Medicina Tropical de São Paulo*, *58*, 26.
Robert-Gangneux, F., & Dardé, M. L. (2012). Epidemiology of and diagnostic strategies for toxoplasmosis. *Clinical Microbiology Reviews*, *25*(2), 264–296.
Severance, E. G., Alaedini, A., Yang, S., Halling, M., Gressitt, K. L., Stallings, C. R., et al. (2012). Gastrointestinal inflammation and associated immune activation in schizophrenia. *Schizophrenia Research*, *138*(1), 48–53.
Severance, E. G., Gressitt, K. L., Alaedini, A., Rohleder, C., Enning, F., Bumb, J. M., et al. (2015). IgG dynamics of dietary antigens point to cerebrospinal fluid barrier or flow dysfunction in first-episode schizophrenia. *Brain, Behavior, and Immunity*, *44*, 148–158.
Severance, E. G., Gressitt, K. L., Stallings, C. R., Origoni, A. E., Khushalani, S., Leweke, F. M., et al. (2013). Discordant patterns of bacterial translocation markers and implications for innate immune imbalances in schizophrenia. *Schizophrenia Research*, *148*(1–3), 130–137.
Severance, E. G., Gressitt, K. L., Yang, S., Katsafanas, E., Schweinfurth, L., Savage, C., et al. (2016). Candida albicans exposures, sex-specificity, and cognitive deficits in schizophrenia and bipolar disorder. *npj Schizophrenia*, *2*, 16018.
Severance, E. G., Gressitt, K. L., Yang, S., Stallings, C. R., Origoni, A. E., Vaughn, C., et al. (2014). Seroreactive marker for inflammatory bowel disease and associations with antibodies to dietary proteins in bipolar disorder. *Bipolar Disorders*, *16*(3), 230–240.
Severance, E. G., Prandovszky, E., Castiglione, J., & Yolken, R. H. (2015). Gastroenterology issues in schizophrenia: Why the gut matters. *Current Psychiatry Reports*, *17*(5), 574.
Severance, E. G., Yolken, R. H., & Eaton, W. W. (2016). Autoimmune diseases, gastrointestinal disorders and the microbiome in schizophrenia: More than a gut feeling. *Schizophrenia Research*, *176*, 23–25. (Epub ahead of print).
Sugden, K., Moffitt, T. E., Pinto, L., Poulton, R., Williams, B. S., & Caspi, A. (2016). Is Toxoplasma gondii infection related to brain and behavior impairments in humans? Evidence from a population-representative birth cohort. *PloS One*, *11*(2), e0148435.

Torrey, E. F., Bartko, J. J., Lun, Z. R., & Yolken, R. H. (2007). Antibodies to Toxoplasma gondii in patients with schizophrenia: A meta-analysis. *Schizophrenia Bulletin, 33*(3), 729–736.

Torrey, E. F., Bartko, J. J., & Yolken, R. H. (2012). Toxoplasma gondii and other risk factors for schizophrenia: An update. *Schizophrenia Bulletin, 38*(3), 642–647.

Trevizan, A. R., Vicentino-Vieira, S. L., da Silva Watanabe, P., Góis, M. B., de Melo Gde, A., Garcia, J. L., et al. (2016). Kinetics of acute infection with Toxoplasma gondii and histopathological changes in the duodenum of rats. *Experimental Parasitology, 165*, 22–29.

Wang, Z. T., Harmon, S., O'Malley, K. L., & Sibley, L. D. (2015). Reassessment of the role of aromatic amino acid hydroxylases and the effect of infection by Toxoplasma gondii on host dopamine. *Infection and Immunity, 83*(3), 1039–1047.

Wang, T., Tang, Z. H., Li, J. F., Li, X. N., Wang, X., & Xhao, Z. J. (2013). A potential association between Toxoplasma gondii infection and schizophrenia in mouse models. *Experimental Parasitology, 135*(3), 497–502.

Wei, H. X., Wei, S. S., Lindsay, D. S., & Peng, H. J. (2015). A systematic review and meta-analysis of the efficacy of anti-Toxoplasma gondii medicines in humans. *PloS One, 10*(9), e0128204.

White, M. W., Radke, J. R., & Radke, J. B. (2014). Toxoplasma development—Turn the switch on or off? *Cellular Microbiology, 16*(4), 466–472.

Wilking, H., Thamm, M., Stark, K., Aebischer, T., & Seeber, F. (2016). Prevalence, incidence estimations, and risk factors of Toxoplasma gondii infection in Germany: A representative, cross-sectional, serological study. *Scientific Reports, 6*, 22551.

Xiao, J., Kannan, G., Jones-Brando, L., Brannock, C., Krasnova, I. N., Cadet, J. L., et al. (2012). Sex-specific changes in gene expression and behavior induced by chronic Toxoplasma infection in mice. *Neuroscience, 206*, 39–48.

Xiao, J., Li, Y., Prandovszky, E., Kannan, G., Viscidi, R. P., Pletnikov, M. V., et al. (2016). Behavioral abnormalities in a mouse model of chronic toxoplasmosis are associated with MAG1 antibody levels and cyst burden. *PLoS Neglected Tropical Diseases, 10*(4), e0004674.

Xiao, J., & Yolken, R. H. (2015). Strain hypothesis of Toxoplasma gondii infection on the outcome of human diseases. *Acta Physiologica (Oxford), 213*(4), 828–845.

Yamada, H., Tairaku, S., Morioka, I., Sonoyama, A., Tanimura, K., Deguchi, M., et al. (2015). Nationwide survey of mother-to-child infections in Japan. *Journal of Infection and Chemotherapy, 21*(3), 161–164.

Yolken, R. H., Severance, E. G., Sabunciyan, S., Gressitt, K., Chen, O., Origoni, A., et al. (2015). Metagenomic sequencing indicates that the oropharyngeal virome of individuals with schizophrenia differs from that of controls. *Schizophrenia Bulletin, 41*(5), 1153–1161.

Yolken, R. H., & Torrey, E. F. (2008). Are some cases of psychosis caused by microbial agents? A review of the evidence. *Molecular Psychiatry, 13*(5), 470–479.

Yolken, R. H., Torrey, E. F., Lieberman, J. A., Yang, S., & Dickerson, F. B. (2011). Serological evidence of exposure to Herpes Simplex Virus type 1 is associated with cognitive deficits in the CATIE schizophrenia sample. *Schizophrenia Research, 128*(1–3), 61–65.

Zhang, Y., Träskman-Bendz, L., Janelidze, S., Langenberg, P., Saleh, A., Constantine, N., et al. (2012). Toxoplasma gondii immunoglobulin G antibodies and nonfatal suicidal self-directed violence. *Journal of Clinical Psychiatry, 73*(8), 1069–1076.

CHAPTER EIGHT

Exercise and Prebiotics Produce Stress Resistance: Converging Impacts on Stress-Protective and Butyrate-Producing Gut Bacteria

A. Mika*, N. Rumian*, A.B. Loughridge†, M. Fleshner*,[1]
*Center for Neuroscience, University of Colorado, Boulder, CO, United States
†Colorado State University, Fort Collins, CO, United States
[1]Corresponding author: e-mail address: fleshner@colorado.edu

Contents

International Review of Neurobiology, Volume 131
ISSN 0074-7742
http://dx.doi.org/10.1016/bs.irn.2016.08.004

Abstract

The gut microbial ecosystem can mediate the negative health impacts of stress on the host. Stressor-induced disruptions in microbial ecology (dysbiosis) can lead to maladaptive health effects, while certain probiotic organisms and their metabolites can protect against these negative impacts. Prebiotic diets and exercise are feasible and cost-effective strategies that can increase stress-protective bacteria and produce resistance against the detrimental behavioral and neurobiological impacts of stress. The goal of this review is to describe research demonstrating that both prebiotic diets and exercise produce adaptations in gut ecology and the brain that arm the organism against inescapable stress-induced learned helplessness. The results of this research support the novel hypothesis that some of the stress-protective effects of prebiotics and exercise are due to increases in stress-protective gut microbial species and their metabolites. In addition, new evidence also suggests that prebiotic diet or exercise interventions are most effective if given early in life (juvenile–adolescence) when both the gut microbial ecosystem and the brain are plastic. Based on our new understanding of the mechanistic convergence of these interventions, it is feasible to propose that in adults, both interventions delivered in combination may elevate their efficacy to promote a stress-resistant phenotype.

1. STRESS AND HEALTH

1.1 Introduction

The majority of health-care professionals and stress physiologists agree that repeated, continuous, and excessive stressor exposure adversely affects both mental and physical health. The biological response initiated by stressor exposure is a highly conserved cascade of events orchestrated by the sympathetic nervous system and hypothalamic–pituitary–adrenal (HPA) axis in efforts to arm the organism with the metabolic substrates needed for survival; coordinated increases in energy mobilization, heart rate, respiration, and immunity serve to prepare the organism to either successfully fight or flee an impeding threat. While these survival-promoting aspects of the stress response are considered adaptive, excessive or prolonged stimulation of the stress response produces maladaptive health consequences that persist long after the stressor subsides. These widespread negative health impacts are well documented and include increased vulnerability to mental health disorders, such as anxiety disorders and depression (Cohen, Janicki-Deverts, & Miller, 2007; Grippo & Johnson, 2009), and increased susceptibility to illness due to pathogens (Padgett & Glaser, 2003; Sheridan et al., 1998), neoplastic cellular replication (e.g., cancer; Moreno-Smith, Lutgendorf, & Sood, 2010), cardiovascular disease (Iso et al., 2002), as well as inflammatory disease (Reber et al., 2011). Given that exposure to stressors is often an inevitable part of life, avoiding stress is not an effective means to maintain health. Identifying factors that can mitigate the undesirable consequences of stress and promote a stress-resistant phenotype is therefore an important goal for stress researchers.

The degree to which stressor exposure impacts our health depends on a variety of factors. Some of these factors are outside of our control; these may include the nature of the stressor (such as intensity, controllability, and chronicity) as well as predetermined biological features (such as age, gender, and genes). For instance, the mental health impacts of stressors experienced earlier in life can be longer lasting and more severe compared with the effects of stressors experienced in adulthood. Indeed, epidemiological and clinical literature in humans demonstrates that adverse experiences occurring earlier in life are major risk factors for psychiatric disorders in adulthood (Culpin, Stapinski, Miles, Araya, & Joinson, 2015; Heim & Nemeroff, 2001; Heim, Plotsky, & Nemeroff, 2004; Whitesell et al., 2009), including depression, anxiety, illicit drug use, stunted learning, and attention disorders. Environmental factors we are capable of controlling, such as physical fitness, social support, and diet, can also regulate the degree to which stressor exposure impacts health. These variables are of special interest to researchers, as they represent feasible, cost-effective strategies that can promote stress robustness. Identifying such strategies, and exploring the mechanism by which they mitigate the health consequences of stress can lead to effective treatments and interventions for stress-related disorders.

2. GUT MICROBIAL ORGANISMS AND THEIR METABOLITES ARE EMERGING MEDIATORS OF THE HEALTH IMPACTS OF STRESS

2.1 Stress Disrupts Health by Disturbing Gut Microbes

Commensal gut bacterial symbiosis (composition and balance) plays an important role in many aspects of physiology and health. It is now well understood that gut microbes maintain and influence key aspects of host physiology, including immune development and function (Hrncir,

Stepankova, Kozakova, Hudcovic, & Tlaskalova-Hogenova, 2008; Tlaskalova-Hogenova et al., 2004), energy metabolism (Geurts, Neyrinck, Delzenne, Knauf, & Cani, 2014; Ridaura et al., 2013; Turnbaugh et al., 2006), nutrient production (Conly, Stein, Worobetz, & Rutledge-Harding, 1994; Hill, 1997; Sommer & Backhed, 2013), gastrointestinal barrier integrity (Berg & Garlington, 1979), circadian rhythms (Leone et al., 2015), as well as brain function and behavior (Clarke et al., 2013; Desbonnet, Clarke, Shanahan, Dinan, & Cryan, 2014; Diaz Heijtz et al., 2011; Naseribafrouei et al., 2014; Tarr et al., 2015). Stressor exposure can damage this dynamic ecology of intestinal microbes by producing dysbiosis. Dysbiosis is implicated in the pathogenesis of a wide variety of disease states, including obesity (Ley et al., 2005; Ridaura et al., 2013), kidney disease (Nallu, Sharma, Ramezani, Muralidharan, & Raj, 2016), cancer (Moore & Moore, 1995), and psychiatric disorders (Desbonnet et al., 2014; Yarandi, Peterson, Treisman, Moran, & Pasricha, 2016), to name a few. Given that the adverse health consequences of stress overlap with those that are produced by gut microbial dysbiosis, stress-induced disruptions in gut ecology constitute a promising potential mechanism by which stress disrupts health. Indeed, gut microbes contribute to many of the physiological and behavioral consequences of stressor exposure, including immune modulation (Bailey, 2012; Bailey et al., 2011; Maslanik et al., 2012), disruptions in intestinal permeability and bacterial translocation (Ait-Belgnaoui et al., 2012; Eutamene et al., 2007; Zareie et al., 2006), altered social and emotional behavior (Bailey & Coe, 1999), impaired memory (Gareau et al., 2011), and HPA axis dysregulation (Gareau, Jury, MacQueen, Sherman, & Perdue, 2007; Sudo et al., 2004). Stressor-induced disruptions in the gut microbial ecosystem have thus emerged as a means by which stress disrupts health.

2.2 Stress-Protective Microbes: Probiotic Bacteria

Recognizing the significant role of the gut microbial ecology as a mediator of stress-evoked adverse health consequences has exposed its promise as a target for intervention. If stress-induced disruptions in gut microbial ecology can produce and potentiate disease, then perhaps maintaining the integrity of this ecosystem can protect host health in the face of stress. Indeed, certain probiotic bacterial species, largely consisting of lactic acid-producing bacteria (Lutgendorff, Akkermans, & Soderholm, 2008), have been shown to mitigate some of the maladaptive health consequences of stressor exposure. For instance, *Lactobacillus farciminis* strengthened intestinal barrier integrity in the face of water-avoidance stress (Da Silva et al., 2014), while combined administration of *Lactobacillus acidophilus* and *Bifidobacterium longum* attenuated stress-induced gastrointestinal symptoms in humans (Diop, Guillou, & Durand, 2008). A large extent of this literature also demonstrates the effectiveness of probiotic bacteria in lessening the impact of stress on the central nervous system. Treatment with *Bifidobacteria infantis* normalized stress-evoked behavioral deficits in the forced swim test as well as accompanying immune and neurotransmitter perturbations (Desbonnet et al., 2010); *B. longum* and *Bifidobacterium breve* differentially attenuated stress-induced anxiety (Savignac, Kiely, Dinan, & Cryan, 2014). *Lactobacillus helveticus* and *B. longum*, in combination, prevented stress-induced decreases in hippocampal neurogenesis (Ait-Belgnaoui et al., 2014). *L. farciminis* was capable of normalizing HPA responses (Ait-Belgnaoui et al., 2012), and comparably, *Lactobacillus rhamnosus* diminished stress-evoked anxiety- and depressive-like symptoms and HPA responses (Bravo et al., 2011). Along the same lines, *Lactobacillus plantarum* diminished depressive-like behavior, HPA responses, and proinflammatory cytokine profiles following early-life stressor exposure (Liu et al., 2016).

2.3 Stress-Protective Microbes: Butyrate-Producing Bacteria

The fermentation products of certain microorganisms produce a wide variety of health benefits. Selective bacterial species are capable of producing short-chain fatty acids (SCFAs) through fermentation of complex carbohydrates. SCFAs, consisting of acetate, propionate, and butyrate produced in a ratio of 60:20:20, respectively, have been implicated as a primary mechanism by which certain microbial organisms affect host physiology.

Notably, butyrate was recently shown to produce beneficial adaptations in brain plasticity and function. Butyrate is a potent histone deacetylase (HDAC) inhibitor both in vitro and in vivo (Boffa, Vidali, Mann, & Allfrey, 1978; Candido, Reeves, & Davie, 1978; Riggs, Whittaker, Neumann, & Ingram, 1977; Sealy & Chalkley, 1978), meaning that it can facilitate acetylation of histone proteins and promote gene transcription by attenuating the acetyl group turnover rate (Reeves & Candido, 1978; Sealy & Chalkley, 1978). One study demonstrated that an IP dose of butyrate can increase brain-derived neurotrophic factor (BDNF) promoter acetylation, enhance BDNF transcription within the hippocampus, and promote hippocampal-dependent learning (Intlekofer et al., 2013). Beyond hippocampal function, other work demonstrated that butyrate can produce antidepressant-like effects (Schroeder, Lin, Crusio, & Akbarian, 2007). Recently, in a rodent model, butyrate administered IP attenuated chronic stress-induced depression, as measured by sucrose preference, light/dark test, forced swim test, and tail suspension, as well as attenuated stress-induced decreases in hippocampal immediate-early gene expression, histone acetylation, and BDNF protein expression (Han, Sung, Chung, & Kwon, 2014). Although it is still unclear whether endogenous butyrate, produced by intestinal microbes, is capable of interacting with the brain, this SCFA serves as a notable potential mechanism by which certain microbes alter brain and behavior. This body of work collectively demonstrates that promoting gut microbial ecology by increasing probiotic and butyrate-producing microorganisms may protect the host from many the negative impacts of stress on neurobiology and behavior.

3. PREBIOTIC DIETS AND EXERCISE CAN PROMOTE STRESS-PROTECTIVE PROBIOTIC BACTERIA

3.1 Prebiotics Promote the Growth and Function of Probiotic and Butyrate-Producing Bacteria

Ingesting prebiotic dietary fiber is a feasible and effective means by which to increase probiotic bacterial species. Prebiotic dietary ingredients constitute a class of compounds that are inadequately metabolized by the host, and instead travel to the lower GI tract where they undergo fermentation by gut microbes that are armed with the enzymatic machinery to utilize prebiotic fibers as sources of fuel. Interactions between prebiotics and gut microbes support the growth and function of existing probiotic bacterial species residing within the colon (Bouhnik et al., 1997; Cardelle-Cobas et al., 2011; Herfel et al., 2011; Moro et al., 2002; Schwab & Ganzle, 2011), often resulting in the proliferation of probiotic bacterial species and enhancements in host health (Roberfroid, 2007; Roberfroid et al., 2010). Indeed, a variety of synthetic and naturally occurring compounds have been described as having prebiotic capabilities, ranging from polysaccharides, oligosaccharides, and polyols (Ouwehand, Derrien, de Vos, Tiihonen, & Rautonen, 2005; Roberfroid et al., 2010). Importantly, prebiotics may offer several benefits to probiotics, including the ability to produce broader and potentially more stable changes in microbial ecology through the expansion of numerous probiotic species.

A variety of prebiotics have been shown to enhance the growth and activity of stress-protective microbial organisms. For instance, the synthetic soluble fiber polydextrose (PDX) successfully increased stress-protective lactic acid-producing bacteria *Lactobacillus* spp. in piglets (Herfel et al., 2011), and the polyol disaccharide lactitol and xylooligosaccharides promoted the growth of *Lactobacillus* and *Bifidobacteria* spp. in culture (Makelainen, Saarinen, Stowell, Rautonen, & Ouwehand, 2010). PDX and soluble corn fiber favorably shifted gut microbial composition in a group of healthy adults; soluble corn fiber increased *Lactobacillus* spp., and both successfully increased bacteria with antiinflammatory properties (Hooda et al., 2012). Similarly, healthy adults fed the synthetic soluble fiber galactooligosaccharide (GOS) exhibited increases in *Bifidobacteria* spp. (Davis, Martinez, Walter, & Hutkins, 2010).

Due to their resistance to digestion in upper GI tract, prebiotics also play a significant role in increasing SCFA production by serving as substrates for colonic fermentation (Macfarlane, Macfarlane, & Cummings, 2006). For instance, chemically modified resistance starches were capable of modulating concentrations of all three SCFAs. Fructooligosaccharides have been shown to increase butyrate in humans (Vitali et al., 2012). Similarly, arabinoxylans and arabinoxylans oligosaccharides, types of dietary fibers found in wheat (Damen et al., 2011), increased butyrate and butyrate-producing bacteria, while inulin (Jung, Jeon, & Han, 2015) was also capable of increasing butyrate levels. Prebiotic dietary ingredients therefore present a promising tool by which to expand stress-protective microbial species (lactic acid- and butyrate-producing bacteria) and protect the host from the behavioral, neurobiological, and physiological effects of stress.

3.2 Exercise Promotes the Growth and Function of Probiotic and Butyrate-Producing Bacteria

Researchers are also realizing that environmental and lifestyle factors known to promote stress resistance may do so by adaptively modulating the gut microbiota. Exercise is one such factor that can both produce stress resistance and modulate the gut microbiota. Recently, it has become clear that exercise is also capable of producing adaptations within gut microbial ecology in both rodents and humans (Allen et al., 2015; Campbell et al., 2016; Clarke et al., 2014; Denou, Marcinko, Surette, Steinberg, & Schertzer, 2016; Evans et al., 2014; Kang et al., 2014; Matsumoto et al., 2008; Petriz et al., 2014; Welly et al., 2016), including promoting the growth of probiotic bacterial species such as *Bifidobacteria* (Lambert et al., 2015) and *Lactobacillus* spp. (Mika &

Fleshner, 2016; Queipo-Ortuno et al., 2013) as well as butyrate-producing bacteria (Matsumoto et al., 2008; Mika & Fleshner, 2016). Given that exercise produces increases in stress-protective bacteria, it is possible that exercise-induced increases in these bacteria are mechanistically involved in the stress-protective effects of exercise. Indeed, we as well as others have postulated that exercise-induced changes in gut bacterial species contribute to the health-enhancing effects of exercise, including the ability of exercise to exert positive effects on mood and behavior (Cerda et al., 2016; Kang et al., 2014; Mika & Fleshner, 2016; Yuan et al., 2015).

4. PREBIOTIC DIETS AND EXERCISE PROMOTE RESISTANCE AGAINST THE BEHAVIORAL AND NEUROBIOLOGICAL CONSEQUENCES OF INESCAPABLE STRESS THROUGH UNIQUE AND OVERLAPPING MECHANISMS

In the remainder of this chapter, we discuss our research demonstrating that both prebiotic diets and exercise are capable of promoting resistance against the anxiety- and depressive-like behaviors produced by inescapable stress (IS). Given that the neurobiological mechanisms by which IS produces anxiety- and depressive-like behaviors are well known, this stress paradigm can be very useful for investigating the mechanisms by which novel treatments, like prebiotics, can protect against the negative mental health effects of excessive stress. Using this paradigm, we discuss our findings demonstrating that both exercise and prebiotic diet can alter gene expression within the brain's serotonergic circuits that subserve IS-induced anxiety- and depressive-like behaviors. We will discuss the potential mechanisms by which gut bacteria and their metabolites influence these neurobiological adaptations. We will discuss the importance of age, specifically how these interventions may be more successful earlier in life due to the inherent plasticity of the central nervous system as well as gut microbial ecology and lastly, the possibility of utilizing these interventions in tandem to promote more robust and lasting adaptive changes in adult organisms.

4.1 Prebiotics Positively Impact Brain and Behavior

There has recently been increasing interest in investigating the impact of prebiotics on the CNS and behavior. Though this literature is still in its infancy, early evidence suggests that prebiotic dietary ingredients are indeed capable of modulating central BDNF among other plasticity-related proteins, neurotransmitters, cytokines, as well as anxiety, and emotional processing (Savignac et al., 2013, 2016; Schmidt et al., 2015; Williams et al., 2016). Preliminary evidence also suggests that prebiotic diets are capable of protecting the host against the negative mental health effects of stress. Prebiotics derived from human milk were recently shown to successfully attenuate stress-induced anxiety-like symptoms in a battery of tests (Tarr et al., 2015). There is thus compelling evidence that prebiotics are capable of protecting against the negative mental health consequences of stress, though the neural mechanisms by which this occurs are unclear.

4.2 Prebiotics Protect Against IS-Induced Learned Helplessness

We have recently demonstrated that ingestion of diets containing the synthetic, soluble prebiotics GOS and PDX as well as the naturally occurring iron-binding glycoprotein lactoferrin (LAC) for 4 weeks attenuated the behavioral and neurological consequences of IS (Mika et al., 2016, in review; Rumian et al., 2014). IS, consisting of a series of uncontrollable and unpredictable local shocks over a 2-h period, leads to the development of a series of behaviors that arguably represent symptoms of human stress-related psychiatric disorders, such as depression and anxiety (Greenwood & Fleshner, 2008). The depressive- and anxiety-like behaviors produced by exposure to IS are collectively termed learned helplessness (LH), and include exaggerated shock-elicited fear, a measure of anxiety-like behavior, and instrumental escape learning deficits, a measure of a cognitive deficit associated with depression. Three dietary formulations, consisting of GOS and PDX, LAC alone, and GOS, PDX with LAC, all attenuated IS-induced LH.

Furthermore, each diet produced distinct adaptations in gene expression within central stress circuits regulating the development of these LH behaviors. The neurobiological mechanisms by which IS alters 5-HT circuits and produces LH behaviors have been demonstrated by Maier and Watkins (2005). Exposure to IS hyperactivates dorsal raphe nucleus (DRN) 5-HT neurons and desensitizes the inhibitory 5-HT1A autoreceptor (5-HT1AR) on 5-HT neurons within the DRN. Since activation of this presynaptic G_i/G_o-coupled autoreceptor reduces 5-HT synthesis and neuronal firing (Chen & Penington, 1996; Valdizan, Castro, & Pazos, 2010), 5-HT1AR serves as an important negative feedback mechanism that guards against excessive 5-HT release. However, when IS-induced activation of DRN neurons desensitizes 5-HT1AR (Rozeske et al., 2011), this negative feedback signal is rendered temporarily inactive, and DRN

5-HT neurons become temporarily sensitized. Sensitized DRN neurons then release copious amounts of 5-HT in response to even mild stimuli, such as those experienced during subsequent behavior testing. Excessive release of 5-HT in downstream brain structures (for review, see Abrams, Johnson, Hollis, & Lowry, 2004), such as the amygdala and the dorsal striatum (for reviews, see Graeff, Guimaraes, De Andrade, & Deakin, 1996; Lowry, Johnson, Hay-Schmidt, Mikkelsen, & Shekhar, 2005), produces deficits in shock-elicited fear and instrumental escapes learning, respectively (Amat, Matus-Amat, Watkins, & Maier, 1998a, 1998b).

Although each diet was capable of protecting against IS-induced LH behaviors, they produced both overlapping and unique neurobiological adaptations within 5-HT circuits. All diets decreased IS-induced cfos mRNA within DRN neurons, indicating that these diets may be capable of producing neurobiological adaptations within DRN 5-HT circuits that constrained DRN activity during stress. Interestingly, we also observed an increase in cfos mRNA expression in home cage rats (rats not exposed to IS) following diets containing GOS and PDX, perhaps indicating that diets containing this particular blend of prebiotics are capable of impacting baseline DRN neuronal activity. In addition, diets containing all three ingredients (GOS, PDX, and LAC) attenuated stress-evoked decreases in 5-HT1AR mRNA, and increased BDNF mRNA within the prefrontal cortex. These results demonstrate that these blends of prebiotics can protect against IS-induced LH, perhaps by producing neuroplastic adaptations that constrain 5-HT DRN neurons in the face of IS. The mechanisms by which this occurs may depend on the structure and function of each unique prebiotic ingredient.

4.3 Exercise Positively Impacts Brain and Behavior

In contrast to the impact of prebiotics on neurobiology and behavior, the stress-protective effects of exercise have been well established. Human and rodent studies demonstrate that regular, daily physical activity can protect against the development of stress-related psychiatric disorders, such as anxiety and depression. Clinical and epidemiological studies report that regular exercise can improve symptoms associated with depression and anxiety (Carek, Laibstain, & Carek, 2011; Mason & Powell, 1985; Mortazavi et al., 2012; Paluska & Schwenk, 2000), whereas physical inactivity is associated with increased incidence of psychological disorders (Hamer & Stamatakis, 2014).

4.4 Exercise Protects Against IS-Induced LH

Using rodent models, we have repeatedly demonstrated that 6 weeks of exercise in adulthood can prevent LH behaviors after IS exposure, demonstrating that physical activity is a powerful tool that can protect the organism against the debilitating effects of stress (Greenwood & Fleshner, 2008; Greenwood, Foley, Burhans, Maier, & Fleshner, 2005; Greenwood et al., 2003; Greenwood, Loughridge, Sadaoui, Christianson, & Fleshner, 2012).

We have also previously characterized a number of neurobiological mechanisms by which exercise can protect against the behavioral consequences of IS. Our lab has demonstrated that 6 weeks of exercise in adult rats can increase 5-HT1AR mRNA in the dorsal aspect of the rostral and mid DRN (Greenwood et al., 2003; Loughridge, Greenwood, Day, McQueen, & Fleshner, 2013). Exercise-induced increases in 5-HT1AR mRNA can protect against IS-induced 5-HT sensitization through a variety of mechanisms (reviewed in Greenwood & Fleshner, 2011). Evidence from prior studies demonstrates that 6 weeks of physical activity can attenuate the stress-induced activation of 5-HT DRN neurons (Greenwood et al., 2003), suggesting that an increase in 5-HT1AR number may perhaps maintain negative feedback in the face of stress. Recent work supports this hypothesis; using in vivo microdialysis, Clark et al. (2015) reported that 6 weeks of wheel running prior to IS prevented the exaggerated release of 5-HT in the dorsal striatum evoked by mild stress 24 h later. Collectively, this work has shown that exercise can produce adaptations that protect against the negative behavioral and neurobiological consequences of IS-induced exaggerated 5-HT release.

4.5 Prebiotics and Exercise Produce Stress Resistance via Unique Neuroplastic Changes

As we have just begun to investigate the neurobiological adaptations produced by prebiotics, the neural mechanisms by which exercise protects against LH are better characterized in comparison to the mechanisms by which prebiotics protect against LH. However, from our work to date, we can conclude that exercise and prebiotic diets both produce unique, neuroplastic changes within 5-HT circuits that result in constrained DRN activity during stress. For instance, exercise has repeatedly produced an increase in 5-HT1AR mRNA, whereas prebiotic diet did not. Exercise produces a wide variety of neurobiological adaptations that are likely independent of exercise-induced adaptations within gut microbial organisms. Thus, one would expect the neurobiological adaptations produced by exercise to be more extensive. Nonetheless, despite these differential neuroplastic changes, both manipulations alter gene expression within serotonergic circuits and produce similar behavioral effects as well as similar adaptations in microbial ecology.

5. THE STRESS-PROTECTIVE EFFECTS OF PREBIOTICS AND EXERCISE MAY BE AGE DEPENDENT

5.1 Age-Dependent Effects of Exercise and Prebiotic Diet on Stress-Protective Bacteria and Butyrate

Prebiotic diets and exercise can improve health throughout the life span, and the adaptations in gut bacteria as well as the central nervous system and behavior produced by exercise training and prebiotics are well documented at all ages. However, it is important to note that our observations demonstrating that prebiotic diets containing GOS, PDX, and LAC can increase stress-protective bacteria were completed in young rats; diets were initiated at postnatal day 24 (Mika et al., 2016, in review). Although others have demonstrated that increases in endogenous probiotic bacteria are possible in adults following GOS and PDX, it is unclear whether the tested prebiotic dietary formulations would have produced similar changes within the same time course.

On the other hand, we have demonstrated that exercise-induced changes in gut microbiota are more robust in early life. We have recently compared the influence of early vs late-life exercise on gut microbial ecology and demonstrated that 6 weeks of exercise initiated in early life have a greater impact on the gut microbiota than exercise initiated in adulthood (Mika et al., 2015). Importantly, exercise begun earlier in life increased *Lactobacillus* spp. and bacterial genera that are capable of producing butyrate (Mika & Fleshner, 2016), while on the other hand, adult-onset exercise modestly impacted the gut microbiota and failed to produce alterations in stress-protective bacteria. Similarly, 3 weeks of exercise initiated earlier in life, but not exercise initiated in adulthood, are capable of increasing fecal butyrate concentrations (Fig. 1). Thus, exercise earlier in life more effectively produces a stress robust microbial ecosystem by increasing probiotic bacteria, butyrate-producing bacteria, as well as butyrate production.

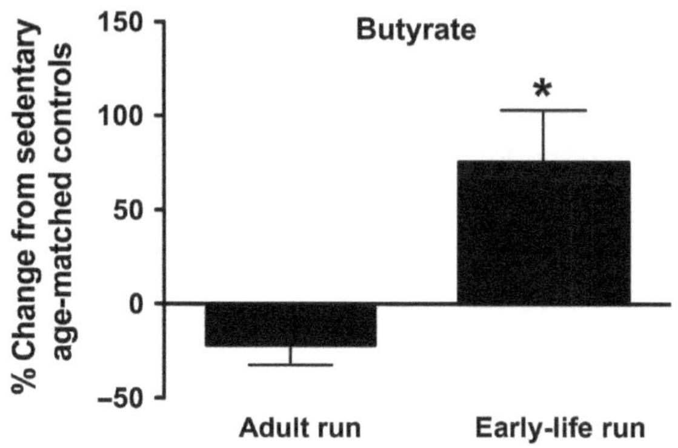

Fig. 1 Percent change in butyrate levels of runners from their respective age-matched sedentary controls. Briefly, rats were given access to voluntary running wheels or remained sedentary for 3 weeks. Fecal samples were collected at the end of the 3 weeks and sent to the Metabolomics Core of the University of Michigan for short-chain fatty acid analysis, and processed in accordance with previous protocols (Chassaing et al., 2015). ANOVA shows there is a significant increase in percent change of butyrate levels in the rats that began running in early life (postnatal day 24 at the start of exercise) as compared to adult runners (postnatal day 70 at the start of exercise ($F_{1,8}=0.0064$, $p=0.0064$). $^*p<0.05$.

5.2 Age-Dependent Effects of Exercise and Prebiotic Diet on Neurobiology and Behavior

Because our observations demonstrating that prebiotic diets can protect against IS-induced LH behaviors were completed in young rats, we cannot be sure if similar effects are possible in adults. We have, however, compared the neurobiological and behavioral effects of exercise initiated in early life with those produced by adult-onset exercise. We have demonstrated that exercise initiated during early, critical periods in development can produce behavioral protection against LH behaviors that is more persistent following cessation of exercise, compared with exercise initiated in adulthood (Mika, Bouchet, Spence, Greenwood, & Fleshner, 2013). Preliminary data also demonstrate that exercise in early life can produce long-lasting increases in 5-HT1AR mRNA. In comparison, increases in 5-HT1AR mRNA were transient in adults.

Interestingly, early-onset exercise can also impact a wide variety of genes within DRN 5-HT neurons that may be necessary for conferring stress resistance. Using laser capture microdissection of the DRN and subsequent Affymetrix microarray for the investigation of gene expression within the DRN neurons, we have also demonstrated that exercise initiated earlier in life (during adolescence, prior to adulthood) can produce unique changes in gene expression for genes involved in regulating 5-HT1A as well as those involved in producing epigenetic changes within the DRN. As demonstrated in Loughridge et al. (2013), mRNA expression levels of HTR1A, the gene responsible for 5-HT1AR expression, increased in adolescent runners compared to their sedentary counterparts (Fig. 2A). On the other hand, exercise decreased mRNA expression level of both FEV and CC2D1b, known transcriptional regulators capable of repressing HTR1A (Albert, Le Francois, & Millar, 2011; Le Francois, Czesak, Steubl, & Albert, 2008).

Fig. 3 depicts early-life exercise-induced changes in gene expression for genes that are implicated in epigenetic regulation; exercise altered gene expression for a variety of these epigenetic regulators. Specifically, exercise significantly decreased mRNA expression levels of HDAC3 and produced trends toward decreases in HDAC5 and HDAC11. HDACs remove an acetyl group from acetylated histone proteins and typically lead to a decrease in transcription of the associated DNA. Studies show these HDACs are each capable of reducing transcription of various genes (Chen, Sahakian, et al., 2016; Lewandowski, Janardhan, & Trivedi, 2015; Wein et al., 2015; Zhao et al., 2016). In contrast to the mechanism by which HDACs regulate gene expression, histone acetyltransferases (HATs) add an acetyl group to histone proteins, which typically results in enhanced gene transcription of associated genes. Fig. 3D–F consists of HATs; exercise increased mRNA expression levels of Myst1, Myst2, and Myst3, respectively. Myst1, Myst2, and Myst3 have been shown to acetylate certain lysine residues on various histone proteins (Chen et al., 2014; Kim et al., 2015; Mishima et al., 2011; Sheikh et al., 2015; Zhu et al., 2015) and are capable of increasing expression of their associated genes. Collectively, these results demonstrate that exercise orchestrates epigenetic changes that lead to increases in gene expression within the DRN. This pattern is continued in Fig. 3G, where exercise produces a trend toward an increase in Kdm2a mRNA expression; Kdm2a acts as lysine demethylase to enhance transcription (Chen, Li, et al., 2016). On the other hand, in Fig. 3H, exercise increases Kdm5b, which has been shown to repress gene transcription (Yamane et al., 2007).

Because we only measured global mRNA expression of DRN-specific genes involved in epigenetic modifications, we cannot say for certain what genes these epigenetic regulators are acting on, nor whether early-life exercise actually produced epigenetic modifications. However, these data imply that epigenetic modifications in the DRN following early-life exercise are

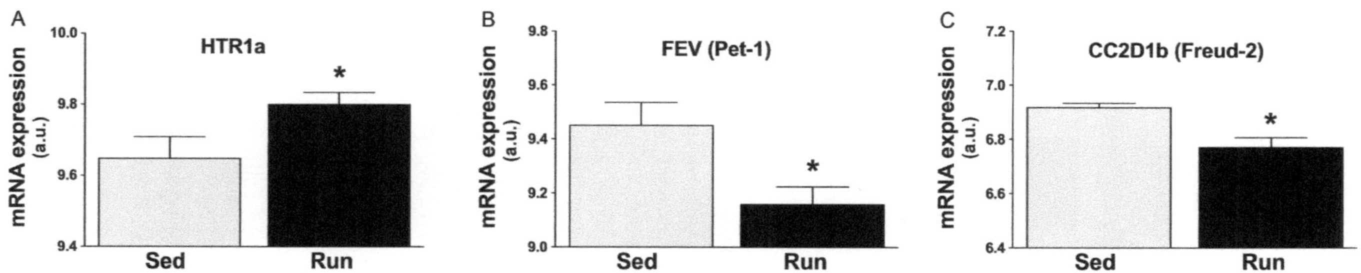

Fig. 2 mRNA expression levels of genes relevant to 5-hydroxytryptamine (serotonin) receptor 1A (HTR1A) gene expression within the DRN, determined by microarray analysis (Loughridge et al., 2013). Briefly, adolescent rats were given access to a voluntary running wheel or remained sedentary for 6 weeks, and were subsequently sacrificed and brains extracted for analysis of gene expression (Loughridge et al., 2013). Briefly, laser capture microdissection of the DRN allowed for the analysis of gene expression within DRN neurons only, and mRNA expression was measured using Affymetrix microarray. As depicted in Loughridge et al. (2013), mRNA expression levels of HTR1A increased ($F_{1,12} = 5.353$, $p = 0.0392$; A) in runners as compared to their sedentary counterparts, whereas there was a decrease in mRNA expression level of both FEV ($F_{1,12} = 7.649$, $p = 0.0171$; B) and CC2D1b ($F_{1,12} = 11.035$, $p = 0.0061$; C) in runners as compared to sedentary controls. All data were analyzed by ANOVA, $^*p < 0.05$ as compared to sedentary group.

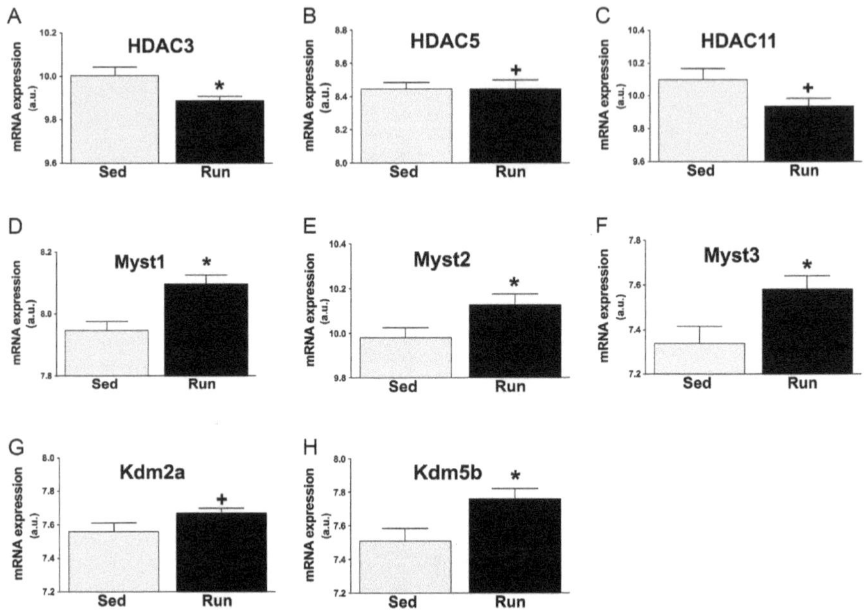

Fig. 3 mRNA expression levels of genes related to epigenetic processes within the DRN, determined by microarray analysis. Briefly, adolescent rats were given access to a voluntary running wheel or remained sedentary for 6 weeks, and were subsequently sacrificed and brains extracted for analysis of gene expression (Loughridge et al., 2013). Briefly, laser capture microdissection of the DRN allowed for the analysis of gene expression within DRN neurons only, and mRNA expression was measured using Affymetrix microarray. Data are grouped based upon function. Exercise decreased mRNA expression levels of HDAC3 ($F_{1,12}=8.087$, $p=0.0148$; A), HDAC5 ($F_{1,12}=4.178$, $p=0.0635$; B), and HDAC11 ($F_{1,12}=4.151$, $p=0.0643$; C). Exercise increased mRNA expression levels of Myst1 ($F_{1,12}=8.425$, $p=0.0133$; D), Myst2 ($F_{1,12}=4.682$, $p=0.0514$; E), and Myst3 ($F_{1,12}=6.567$, $p=0.0249$; F). Additionally, exercise increased mRNA expression levels of Kdm2a ($F_{1,12}=3.99$, $p=0.06990$; G) and Kdm5b ($F_{1,12}=7.096$, $p=0.0206$; H). All data were analyzed by ANOVA, $*p<0.05$ as compared to sedentary group; $+p<0.07$ as compared to sedentary group.

possible and could potentially explain the longer-lasting increases in 5-HT1AR mRNA that are also observed following exercise in early life.

5.3 Early-Life Increases in Stress-Protective Bacteria and Butyrate Can Promote Robust and Lasting Stress Resistance

We have discussed our observations demonstrating that exercise initiated during early, sensitive periods can produce persistent adaptations in brain function and gut microbial ecology, two seemingly separate physiological systems of the developing host. We have demonstrated that early-life exercise increases stress-protective and butyrate-producing microbes, and is capable of producing long-lasting stress resistance and changes in gene expression for 5-HT1AR and epigenetic regulators. Although we have not demonstrated that these observations are causally related, it is possible that early-life exercise-induced increases in butyrate production communicate with the brain to alter and maintain gene expression for genes that are necessary for producing longer-lasting stress resistance.

5.4 Potential Synergy Between Prebiotics and Exercise: Hope for Adults

These data beg the question of whether it is possible to achieve lasting and robust stress resistance in adult organisms. If some of the stress-protective effects of exercise are produced by early-life exercise-induced changes in gut microbes, then perhaps it would be possible to recapitulate some of these effects in adults by producing more robust increases in stress-protective bacteria. In adulthood, the microbial ecosystem is less receptive to change. Gut microbial ecology demonstrates a greater susceptibility for change in early life, perhaps partially because microbial diversity progressively increases with age (Yatsunenko et al., 2012). Furthermore, the gut ecology of younger children (Ringel-Kulka et al., 2013) and juvenile rats (Mika et al., 2015) is more plastic, variable, and less stable than adults. The plasticity and lack of diversity characteristic of an immature gut suggests that it may be more sensitive to manipulation (Ringel-Kulka et al., 2013), while the increase in diversity as organisms age eventually renders the adult gut less susceptible. We have previously shown that exercise alone in adults is not capable of increasing stress-protective probiotic and butyrate-producing bacteria, but perhaps exercise in tandem with a prebiotic diet will synergistically enhance these bacteria in adults (Fig. 4). Furthermore, if both manipulations could simultaneously evoke greater microbial change, it would be prudent to investigate whether these changes would lead to more robust and lasting protection against the negative mental health effects of stress. Future research should investigate this possibility, as it may provide a feasible and cost-effective means by which previously sedentary adults can increase stress robustness.

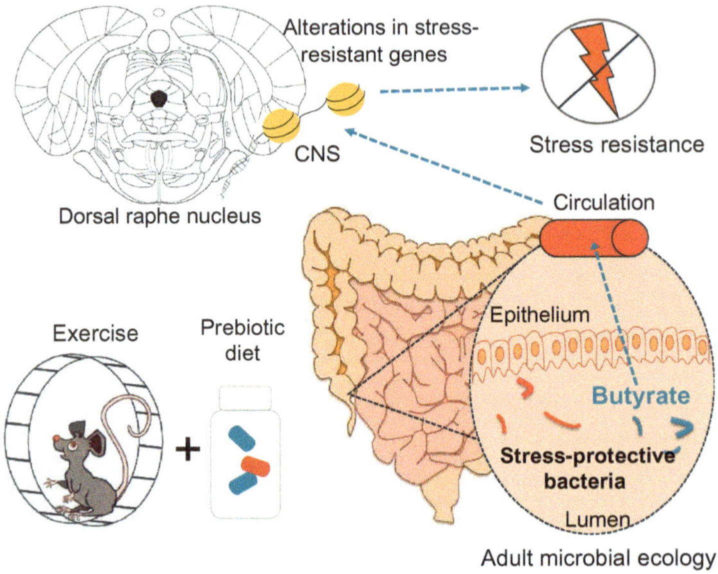

Fig. 4 Potential synergistic effects of exercise and prebiotic diet. Combining exercise and prebiotic diet can increase the amount of stress-protective bacteria within the gut. Butyrate produced by these bacteria can enter the circulation and cross the blood–brain barrier to induce epigenetic changes that modify and maintain genes involved in promoting stress resistance within the DRN, thus potentially helping to produce longer-lasting and more robust stress resistance.

REFERENCES

Abrams, J. K., Johnson, P. L., Hollis, J. H., & Lowry, C. A. (2004). Anatomic and functional topography of the dorsal raphe nucleus. *Annals of the New York Academy of Sciences*, *1018*, 46–57. http://dx.doi.org/10.1196/annals.1296.005.

Ait-Belgnaoui, A., Colom, A., Braniste, V., Ramalho, L., Marrot, A., Cartier, C., ... Tompkins, T. (2014). Probiotic gut effect prevents the chronic psychological stress-induced brain activity abnormality in mice. *Neurogastroenterology and Motility*, *26*(4), 510–520. http://dx.doi.org/10.1111/nmo.12295.

Ait-Belgnaoui, A., Durand, H., Cartier, C., Chaumaz, G., Eutamene, H., Ferrier, L., ... Theodorou, V. (2012). Prevention of gut leakiness by a probiotic treatment leads to attenuated HPA response to an acute psychological stress in rats. *Psychoneuroendocrinology*, *37*(11), 1885–1895. http://dx.doi.org/10.1016/j.psyneuen.2012.03.024.

Albert, P. R., Le Francois, B., & Millar, A. M. (2011). Transcriptional dysregulation of 5-HT1A autoreceptors in mental illness. *Molecular Brain*, *4*, 21. http://dx.doi.org/10.1186/1756-6606-4-21.

Allen, J. M., Berg Miller, M. E., Pence, B. D., Whitlock, K., Nehra, V., Gaskins, H. R., ... Woods, J. A. (2015). Voluntary and forced exercise differentially alters the gut microbiome in C57BL/6J mice. *Journal of Applied Physiology (Bethesda, Md.: 1985)*, *118*(8), 1059–1066. http://dx.doi.org/10.1152/japplphysiol.01077.2014.

Amat, J., Matus-Amat, P., Watkins, L. R., & Maier, S. F. (1998a). Escapable and inescapable stress differentially alter extracellular levels of 5-HT in the basolateral amygdala of the rat. *Brain Research*, *812*(1–2), 113–120.

Amat, J., Matus-Amat, P., Watkins, L. R., & Maier, S. F. (1998b). Escapable and inescapable stress differentially and selectively alter extracellular levels of 5-HT in the ventral hippocampus and dorsal periaqueductal gray of the rat. *Brain Research*, *797*(1), 12–22.

Bailey, M. T. (2012). The contributing role of the intestinal microbiota in stressor-induced increases in susceptibility to enteric infection and systemic immunomodulation. *Hormones and Behavior*, *62*, 286–294. http://dx.doi.org/10.1016/j.yhbeh.2012.02.006.

Bailey, M. T., & Coe, C. L. (1999). Maternal separation disrupts the integrity of the intestinal microflora in infant rhesus monkeys. *Developmental Psychobiology*, *35*(2), 146–155.

Bailey, M., Dowd, S. E., Galley, J. D., Hufnagle, A. R., Allen, R. G., & Lyte, M. (2011). Exposure to a social stressor alters the structure of the intestinal microbiota: Implications for stressor-induced immunomodulation. *Brain, Behavior and Immunity*, *25*, 397–407.

Berg, R. D., & Garlington, A. W. (1979). Translocation of certain indigenous bacteria from the gastrointestinal tract to the mesenteric lymph nodes and other organs in a gnotobiotic mouse model. *Infection and Immunity*, *23*(2), 403–411.

Boffa, L. C., Vidali, G., Mann, R. S., & Allfrey, V. G. (1978). Suppression of histone deacetylation in vivo and in vitro by sodium butyrate. *The Journal of Biological Chemistry*, *253*(10), 3364–3366.

Bouhnik, Y., Flourie, B., D'Agay-Abensour, L., Pochart, P., Gramet, G., Durand, M., & Rambaud, J. C. (1997). Administration of transgalacto-oligosaccharides increases fecal bifidobacteria and modifies colonic fermentation metabolism in healthy humans. *The Journal of Nutrition*, *127*(3), 444–448.

Bravo, J. A., Forsythe, P., Chew, M. V., Escaravage, E., Savignac, H. M., Dinan, T. G., … Cryan, J. F. (2011). Ingestion of Lactobacillus strain regulates emotional behavior and central GABA receptor expression in a mouse via the vagus nerve. *Proceedings of the National Academy of Sciences of the United States of America*, *108*(38), 16050–16055. http://dx.doi.org/10.1073/pnas.1102999108.

Campbell, S. C., Wisniewski, P. J., Noji, M., McGuinness, L. R., Haggblom, M. M., Lightfoot, S. A., … Kerkhof, L. J. (2016). The effect of diet and exercise on intestinal integrity and microbial diversity in mice. *PLoS One*, *11*(3), e0150502. http://dx.doi.org/10.1371/journal.pone.0150502.

Candido, E. P., Reeves, R., & Davie, J. R. (1978). Sodium butyrate inhibits histone deacetylation in cultured cells. *Cell*, *14*(1), 105–113.

Cardelle-Cobas, A., Corzo, N., Olano, A., Pelaez, C., Requena, T., & Avila, M. (2011). Galactooligosaccharides derived from lactose and lactulose: Influence of structure on Lactobacillus, Streptococcus and Bifidobacterium growth. *International Journal of Food Microbiology*, *149*(1), 81–87. http://dx.doi.org/10.1016/j.ijfoodmicro.2011.05.026.

Carek, P. J., Laibstain, S. E., & Carek, S. M. (2011). Exercise for the treatment of depression and anxiety. *International Journal of Psychiatry in Medicine*, *41*(1), 15–28.

Cerda, B., Perez, M., Perez-Santiago, J. D., Tornero-Aguilera, J. F., Gonzalez-Soltero, R., & Larrosa, M. (2016). Gut microbiota modification: Another piece in the puzzle of the benefits of physical exercise in health? *Frontiers in Physiology*, 7, 51. http://dx.doi.org/10.3389/fphys.2016.00051.

Chassaing, B., Koren, O., Goodrich, J. K., Poole, A. C., Srinivasan, S., Ley, R. E., & Gewirtz, A. T. (2015). Dietary emulsifiers impact the mouse gut microbiota promoting colitis and metabolic syndrome. *Nature*, *519*(7541), 92–96. http://dx.doi.org/10.1038/nature14232.

Chen, J. Y., Li, C. F., Chu, P. Y., Lai, Y. S., Chen, C. H., Jiang, S. S., … Hung, W. C. (2016). Lysine demethylase 2A promotes stemness and angiogenesis of breast cancer by upregulating Jagged1. *Oncotarget*, 7(19), 27689–27710. http://dx.doi.org/10.18632/oncotarget.8381.

Chen, Y., & Penington, N. J. (1996). Differential effects of protein kinase C activation on 5-HT1A receptor coupling to Ca2+ and K+ currents in rat serotonergic neurones. *The Journal of Physiology*, *496*(Pt. 1), 129–137.

Chen, J., Sahakian, E., Powers, J., Lienlaf, M., Perez-Villarroel, P., Knox, T., & Villagra, A. (2016). Functional analysis of histone deacetylase 11 (HDAC11). *Methods in Molecular Biology*, *1436*, 147–165. http://dx.doi.org/10.1007/978-1-4939-3667-0_11.

Chen, Z., Ye, X., Tang, N., Shen, S., Li, Z., Niu, X., … Xu, L. (2014). The histone acetylranseferase hMOF acetylates Nrf2 and regulates anti-drug responses in human non-small cell lung cancer. *British Journal of Pharmacology*, *171*(13), 3196–3211. http://dx.doi.org/10.1111/bph.12661.

Clark, P. J., Amat, J., McConnell, S. O., Ghasem, P. R., Greenwood, B. N., Maier, S. F., & Fleshner, M. (2015). Running reduces uncontrollable stress-evoked serotonin and potentiates stress-evoked dopamine concentrations in the rat dorsal striatum. *PLoS One*, *10*(11), e0141898. http://dx.doi.org/10.1371/journal.pone.0141898.

Clarke, G., Grenham, S., Scully, P., Fitzgerald, P., Moloney, R. D., Shanahan, F., … Cryan, J. F. (2013). The microbiome-gut-brain axis during early life regulates the hippocampal serotonergic system in a sex-dependent manner. *Molecular Psychiatry*, *18*(6), 666–673. http://dx.doi.org/10.1038/mp.2012.77.

Clarke, S. F., Murphy, E. F., O'Sullivan, O., Lucey, A. J., Humphreys, M., Hogan, A., … Cotter, P. D. (2014). Exercise and associated dietary extremes impact on gut microbial diversity. *Gut*, *63*, 1913–1920. http://dx.doi.org/10.1136/gutjnl-2013-306541.

Cohen, S., Janicki-Deverts, D., & Miller, G. E. (2007). Psychological stress and disease. *JAMA: The Journal of the American Medical Association*, *298*(14), 1685–1687. http://dx.doi.org/10.1001/jama.298.14.1685.

Conly, J. M., Stein, K., Worobetz, L., & Rutledge-Harding, S. (1994). The contribution of vitamin K2 (menaquinones) produced by the intestinal microflora to human nutritional requirements for vitamin K. *The American Journal of Gastroenterology*, *89*(6), 915–923.

Culpin, I., Stapinski, L., Miles, O. B., Araya, R., & Joinson, C. (2015). Exposure to socioeconomic adversity in early life and risk of depression at 18 years: The mediating role of locus of control. *Journal of Affective Disorders*, *183*, 269–278. http://dx.doi.org/10.1016/j.jad.2015.05.030.

Damen, B., Verspreet, J., Pollet, A., Broekaert, W. F., Delcour, J. A., & Courtin, C. M. (2011). Prebiotic effects and intestinal fermentation of cereal arabinoxylans and arabinoxylan oligosaccharides in rats depend strongly on their structural properties and joint presence. *Molecular Nutrition & Food Research*, *55*(12), 1862–1874. http://dx.doi.org/10.1002/mnfr.201100377.

Da Silva, S., Robbe-Masselot, C., Ait-Belgnaoui, A., Mancuso, A., Mercade-Loubiere, M., Salvador-Cartier, C., … Mercier-Bonin, M. (2014). Stress disrupts intestinal mucus barrier in rats via mucin O-glycosylation shift: Prevention by a probiotic treatment. *American Journal of Physiology. Gastrointestinal and Liver Physiology*, *307*(4), G420–G429. http://dx.doi.org/10.1152/ajpgi.00290.2013.

Davis, L. M., Martinez, I., Walter, J., & Hutkins, R. (2010). A dose dependent impact of prebiotic galactooligosaccharides on the intestinal microbiota of healthy adults. *International Journal of Food Microbiology*, *144*(2), 285–292. http://dx.doi.org/10.1016/j.ijfoodmicro.2010.10.007.

Denou, E., Marcinko, K., Surette, M. G., Steinberg, G. R., & Schertzer, J. D. (2016). High-intensity exercise training increases the diversity and metabolic capacity of the mouse distal gut microbiota during diet-induced obesity. *American Journal of Physiology. Endocrinology and Metabolism*, *310*(11), E982–E993. http://dx.doi.org/10.1152/ajpendo.00537.2015.

Desbonnet, L., Clarke, G., Shanahan, F., Dinan, T. G., & Cryan, J. F. (2014). Microbiota is essential for social development in the mouse. *Molecular Psychiatry*, *19*(2), 146–148. http://dx.doi.org/10.1038/mp.2013.65.

Desbonnet, L., Garrett, L., Clarke, G., Kiely, B., Cryan, J. F., & Dinan, T. G. (2010). Effects of the probiotic Bifidobacterium infantis in the maternal separation model of depression. *Neuroscience*, *170*(4), 1179–1188. http://dx.doi.org/10.1016/j.neuroscience.2010.08.005.

Diaz Heijtz, R., Wang, S., Anuar, F., Qian, Y., Bjorkholm, B., Samuelsson, A., … Pettersson, S. (2011). Normal gut microbiota modulates brain development and behavior. *Proceedings of the National Academy of Sciences of the United States of America*, *108*(7), 3047–3052. http://dx.doi.org/10.1073/pnas.1010529108.

Diop, L., Guillou, S., & Durand, H. (2008). Probiotic food supplement reduces stress-induced gastrointestinal symptoms in volunteers: A double-blind, placebo-controlled, randomized trial. *Nutrition Research*, *28*(1), 1–5. http://dx.doi.org/10.1016/j.nutres.2007.10.001.

Eutamene, H., Lamine, F., Chabo, C., Theodorou, V., Rochat, F., Bergonzelli, G. E., … Bueno, L. (2007). Synergy between Lactobacillus paracasei and its bacterial products to counteract stress-induced gut permeability and sensitivity increase in rats. *The Journal of Nutrition*, *137*(8), 1901–1907.

Evans, C. C., LePard, K. J., Kwak, J. W., Stancukas, M. C., Laskowski, S., Dougherty, J., … Ciancio, M. J. (2014). Exercise prevents weight gain and alters the gut microbiota in a mouse model of high fat diet-induced obesity. *PLoS One*, *9*(3), e92193. http://dx.doi.org/10.1371/journal.pone.0092193.

Gareau, M. G., Jury, J., MacQueen, G., Sherman, P. M., & Perdue, M. H. (2007). Probiotic treatment of rat pups normalises corticosterone release and ameliorates colonic dysfunction induced by maternal separation. *Gut*, *56*(11), 1522–1528. http://dx.doi.org/10.1136/gut.2006.117176.

Gareau, M. G., Wine, E., Rodrigues, D. M., Cho, J. H., Whary, M. T., Philpott, D. J., … Sherman, P. M. (2011). Bacterial infection causes stress-induced memory dysfunction in mice. *Gut*, *60*(3), 307–317. http://dx.doi.org/10.1136/gut.2009.202515.

Geurts, L., Neyrinck, A. M., Delzenne, N. M., Knauf, C., & Cani, P. D. (2014). Gut microbiota controls adipose tissue expansion, gut barrier and glucose metabolism: Novel insights into molecular targets and interventions using prebiotics. *Beneficial Microbes*, *5*(1), 3–17. http://dx.doi.org/10.3920/BM2012.0065.

Graeff, F. G., Guimaraes, F. S., De Andrade, T. G., & Deakin, J. F. (1996). Role of 5-HT in stress, anxiety, and depression. *Pharmacology, Biochemistry and Behavior*, *54*(1), 129–141.

Greenwood, B. N., & Fleshner, M. (2008). Exercise, learned helplessness, and the stress-resistant brain. *Neuromolecular Medicine*, *10*(2), 81–98. http://dx.doi.org/10.1007/s12017-008-8029-y.

Greenwood, B. N., & Fleshner, M. (2011). Exercise, stress resistance, and central serotonergic systems. *Exercise and Sport Sciences Reviews*, *39*(3), 140–149. http://dx.doi.org/10.1097/JES.0b013e31821f7e45.

Greenwood, B. N., Foley, T. E., Burhans, D., Maier, S. F., & Fleshner, M. (2005). The consequences of uncontrollable stress are sensitive to duration of prior wheel running. *Brain Research*, *1033*(2), 164–178. http://dx.doi.org/10.1016/j.brainres.2004.11.037.

Greenwood, B. N., Foley, T. E., Day, H. E., Campisi, J., Hammack, S. H., Campeau, S., … Fleshner, M. (2003). Freewheel running prevents learned helplessness/behavioral depression: Role of dorsal raphe serotonergic neurons. *The Journal of Neuroscience*, *23*(7), 2889–2898.

Greenwood, B. N., Loughridge, A. B., Sadaoui, N., Christianson, J. P., & Fleshner, M. (2012). The protective effects of voluntary exercise against the behavioral consequences of uncontrollable stress persist despite an increase in anxiety following forced cessation of exercise. *Behavioural Brain Research*, *233*(2), 314–321. http://dx.doi.org/10.1016/j.bbr.2012.05.017.

Grippo, A. J., & Johnson, A. K. (2009). Stress, depression and cardiovascular dysregulation: A review of neurobiological mechanisms and the integration of research from preclinical disease models. *Stress*, *12*(1), 1–21.

Hamer, M., & Stamatakis, E. (2014). Prospective study of sedentary behavior, risk of depression, and cognitive impairment. *Medicine and Science in Sports and Exercise*, *46*(4), 718–723. http://dx.doi.org/10.1249/MSS.0000000000000156.

Han, A., Sung, Y. B., Chung, S. Y., & Kwon, M. S. (2014). Possible additional antidepressant-like mechanism of sodium butyrate: Targeting the hippocampus. *Neuropharmacology*, *81*, 292–302. http://dx.doi.org/10.1016/j.neuropharm.2014.02.017.

Heim, C., & Nemeroff, C. B. (2001). The role of childhood trauma in the neurobiology of mood and anxiety disorders: Preclinical and clinical studies. *Biological Psychiatry*, *49*(12), 1023–1039.

Heim, C., Plotsky, P. M., & Nemeroff, C. B. (2004). Importance of studying the contributions of early adverse experience to neurobiological findings in depression. *Neuropsychopharmacology*, *29*(4), 641–648. http://dx.doi.org/10.1038/sj.npp.1300397.

Herfel, T. M., Jacobi, S. K., Lin, X., Fellner, V., Walker, D. C., Jouni, Z. E., & Odle, J. (2011). Polydextrose enrichment of infant formula demonstrates prebiotic characteristics by altering intestinal microbiota, organic acid concentrations, and cytokine expression in suckling piglets. *The Journal of Nutrition*, *141*(12), 2139–2145. http://dx.doi.org/10.3945/jn.111.143727.

Hill, M. J. (1997). Intestinal flora and endogenous vitamin synthesis. *European Journal of Cancer Prevention*, *6*(Suppl. 1), S43–S45.

Hooda, S., Boler, B. M., Serao, M. C., Brulc, J. M., Staeger, M. A., Boileau, T. W., … Swanson, K. S. (2012). 454 pyrosequencing reveals a shift in fecal microbiota of healthy adult men consuming polydextrose or soluble corn fiber. *The Journal of Nutrition*, *142*(7), 1259–1265. http://dx.doi.org/10.3945/jn.112.158766.

Hrncir, T., Stepankova, R., Kozakova, H., Hudcovic, T., & Tlaskalova-Hogenova, H. (2008). Gut microbiota and lipopolysaccharide content of the diet influence development of regulatory T cells: Studies in germ-free mice. *BMC Immunology*, *9*, 65. http://dx.doi.org/10.1186/1471-2172-9-65.

Intlekofer, K. A., Berchtold, N. C., Malvaez, M., Carlos, A. J., McQuown, S. C., Cunningham, M. J., … Cotman, C. W. (2013). Exercise and sodium butyrate transform a subthreshold learning event into long-term memory via a brain-derived neurotrophic factor-dependent mechanism. *Neuropsychopharmacology*, *38*(10), 2027–2034. http://dx.doi.org/10.1038/npp.2013.104.

Iso, H., Date, C., Yamamoto, A., Toyoshima, H., Tanabe, N., Kikuchi, S., … Ohno, Y. (2002). Perceived mental stress and mortality from cardiovascular disease among Japanese men and women: The Japan Collaborative Cohort Study for Evaluation of Cancer Risk Sponsored by Monbusho (JACC Study). *Circulation*, *106*(10), 1229–1236.

Jung, T. H., Jeon, W. M., & Han, K. S. (2015). In vitro effects of dietary inulin on human fecal microbiota and butyrate production. *Journal of Microbiology and Biotechnology*, *25*(9), 1555–1558. http://dx.doi.org/10.4014/jmb.1505.05078.

Kang, S. S., Jeraldo, P. R., Kurti, A., Miller, M. E., Cook, M. D., Whitlock, K., … Fryer, J. D. (2014). Diet and exercise orthogonally alter the gut microbiome and reveal independent associations with anxiety and cognition. *Molecular Neurodegeneration*, *9*, 36. http://dx.doi.org/10.1186/1750-1326-9-36.

Kim, M. S., Cho, H. I., Park, S. H., Kim, J. H., Chai, Y. G., & Jang, Y. K. (2015). The histone acetyltransferase Myst2 regulates Nanog expression, and is involved in maintaining pluripotency and self-renewal of embryonic stem cells. *FEBS Letters*, *589*(8), 941–950. http://dx.doi.org/10.1016/j.febslet.2015.02.029.

Lambert, J. E., Myslicki, J. P., Bomhof, M. R., Belke, D. D., Shearer, J., & Reimer, R. A. (2015). Exercise training modifies gut microbiota in normal and diabetic mice. *Applied*

Physiology, Nutrition and Metabolism, 40(7), 749–752. http://dx.doi.org/10.1139/apnm-2014-0452.

Le Francois, B., Czesak, M., Steubl, D., & Albert, P. R. (2008). Transcriptional regulation at a HTR1A polymorphism associated with mental illness. *Neuropharmacology, 55*(6), 977–985. http://dx.doi.org/10.1016/j.neuropharm.2008.06.046.

Leone, V., Gibbons, S. M., Martinez, K., Hutchison, A. L., Huang, E. Y., Cham, C. M., … Chang, E. B. (2015). Effects of diurnal variation of gut microbes and high-fat feeding on host circadian clock function and metabolism. *Cell Host & Microbe, 17*(5), 681–689. http://dx.doi.org/10.1016/j.chom.2015.03.006.

Lewandowski, S. L., Janardhan, H. P., & Trivedi, C. M. (2015). Histone deacetylase 3 coordinates deacetylase-independent epigenetic silencing of transforming growth factor-beta1 (TGF-beta1) to orchestrate second heart field development. *The Journal of Biological Chemistry, 290*(45), 27067–27089. http://dx.doi.org/10.1074/jbc.M115.684753.

Ley, R. E., Backhed, F., Turnbaugh, P., Lozupone, C. A., Knight, R. D., & Gordon, J. I. (2005). Obesity alters gut microbial ecology. *Proceedings of the National Academy of Sciences of the United States of America, 102*(31), 11070–11075.

Liu, Y. W., Liu, W. H., Wu, C. C., Juan, Y. C., Wu, Y. C., Tsai, H. P., … Tsai, Y. C. (2016). Psychotropic effects of Lactobacillus plantarum PS128 in early life-stressed and naive adult mice. *Brain Research, 1631*, 1–12. http://dx.doi.org/10.1016/j.brainres.2015.11.018.

Loughridge, A. B., Greenwood, B. N., Day, H. E., McQueen, M. B., & Fleshner, M. (2013). Microarray analyses reveal novel targets of exercise-induced stress resistance in the dorsal raphe nucleus. *Frontiers in Behavioral Neuroscience*, 7, 37. http://dx.doi.org/10.3389/fnbeh.2013.00037.

Lowry, C. A., Johnson, P. L., Hay-Schmidt, A., Mikkelsen, J., & Shekhar, A. (2005). Modulation of anxiety circuits by serotonergic systems. *Stress, 8*(4), 233–246. http://dx.doi.org/10.1080/10253890500492787.

Lutgendorff, F., Akkermans, L. M., & Soderholm, J. D. (2008). The role of microbiota and probiotics in stress-induced gastro-intestinal damage. *Current Molecular Medicine, 8*(4), 282–298.

Macfarlane, S., Macfarlane, G. T., & Cummings, J. H. (2006). Review article: Prebiotics in the gastrointestinal tract. *Alimentary Pharmacology and Therapeutics, 24*(5), 701–714. http://dx.doi.org/10.1111/j.1365-2036.2006.03042.x.

Maier, S. F., & Watkins, L. R. (2005). Stressor controllability and learned helplessness: The roles of the dorsal raphe nucleus, serotonin, and corticotropin-releasing factor. *Neuroscience and Biobehavioral Reviews, 29*(4–5), 829–841. http://dx.doi.org/10.1016/j.neubiorev.2005.03.021.

Makelainen, H., Saarinen, M., Stowell, J., Rautonen, N., & Ouwehand, A. C. (2010). Xylo-oligosaccharides and lactitol promote the growth of Bifidobacterium lactis and Lactobacillus species in pure cultures. *Beneficial Microbes, 1*(2), 139–148. http://dx.doi.org/10.3920/BM2009.0029.

Maslanik, T., Tannura, K., Mahaffey, L., Loughridge, A. B., Benninson, L., Ursell, L., … Fleshner, M. (2012). Commensal bacteria and MAMPs are necessary for stress-induced increases in IL-1beta and IL-18 but not IL-6, IL-10 or MCP-1. *PLoS One*, 7(12), e50636. http://dx.doi.org/10.1371/journal.pone.0050636.

Mason, J. O., & Powell, K. E. (1985). Physical activity, behavioral epidemiology, and public health. *Public Health Reports, 100*(2), 113–115.

Matsumoto, M., Inoue, R., Tsukahara, T., Ushida, K., Chiji, H., Matsubara, N., & Hara, H. (2008). Voluntary running exercise alters microbiota composition and increases n-butyrate concentration in the rat cecum. *Bioscience, Biotechnology, and Biochemistry, 72*(2), 572–576. http://dx.doi.org/10.1271/bbb.70474.

Mika, A., Bouchet, C. A., Spence, K. G., Greenwood, B. N., & Fleshner, M. (2013). The persistence of exercise-induced stress resistance depends on the developmental stage during which exercise is initiated. Abstract presented at the Society for Neuroscience, San Diego, CA.

Mika, A., Day, H. E., Martinez, A., Rumian, N. L., Greenwood, B. N., Chichlowski, M., Berg, B. M., & Fleshner, M. (2016). Early life diets with prebiotics and bioactive milk fractions attenuate the impact of stress on learned helplessness behaviors and alter gene expression within neural circuits important for stress resistance. *European Journal of Neuroscience*, in review.

Mika, A., & Fleshner, M. (2016). Early-life exercise may promote lasting brain and metabolic health through gut bacterial metabolites. *Immunology and Cell Biology*, *94*(2), 151–157. http://dx.doi.org/10.1038/icb.2015.113.

Mika, A., Van Treuren, W., Gonzalez, A., Herrera, J. J., Knight, R., & Fleshner, M. (2015). Exercise is more effective at altering gut microbial composition and producing stable changes in lean mass in juvenile versus adult male F344 rats. *PLoS One*, *10*(5), e0125889. http://dx.doi.org/10.1371/journal.pone.0125889.

Mishima, Y., Miyagi, S., Saraya, A., Negishi, M., Endoh, M., Endo, T. A., … Iwama, A. (2011). The Hbo1-Brd1/Brpf2 complex is responsible for global acetylation of H3K14 and required for fetal liver erythropoiesis. *Blood*, *118*(9), 2443–2453. http://dx.doi.org/10.1182/blood-2011-01-331892.

Moore, W. E., & Moore, L. H. (1995). Intestinal floras of populations that have a high risk of colon cancer. *Applied and Environmental Microbiology*, *61*(9), 3202–3207.

Moreno-Smith, M., Lutgendorf, S. K., & Sood, A. K. (2010). Impact of stress on cancer metastasis. *Future Oncology*, *6*(12), 1863–1881. http://dx.doi.org/10.2217/fon.10.142.

Moro, G., Minoli, I., Mosca, M., Fanaro, S., Jelinek, J., Stahl, B., & Boehm, G. (2002). Dosage-related bifidogenic effects of galacto- and fructooligosaccharides in formula-fed term infants. *Journal of Pediatric Gastroenterology and Nutrition*, *34*(3), 291–295.

Mortazavi, S. S., Mohammad, K., Ardebili, H. E., Beni, R. D., Mahmoodi, M., & Keshteli, A. H. (2012). Mental disorder prevention and physical activity in Iranian elderly. *International Journal of Preventive Medicine*, *3*(Suppl. 1), S64–S72.

Nallu, A., Sharma, S., Ramezani, A., Muralidharan, J., & Raj, D. (2016). Gut microbiome in chronic kidney disease: Challenges and opportunities. *Translational Research*. http://dx.doi.org/10.1016/j.trsl.2016.04.007.

Naseribafrouei, A., Hestad, K., Avershina, E., Sekelja, M., Linlokken, A., Wilson, R., & Rudi, K. (2014). Correlation between the human fecal microbiota and depression. *Neurogastroenterology and Motility*, *26*(8), 1155–1162. http://dx.doi.org/10.1111/nmo.12378.

Ouwehand, A. C., Derrien, M., de Vos, W., Tiihonen, K., & Rautonen, N. (2005). Prebiotics and other microbial substrates for gut functionality. *Current Opinion in Biotechnology*, *16*(2), 212–217. http://dx.doi.org/10.1016/j.copbio.2005.01.007.

Padgett, D. A., & Glaser, R. (2003). How stress influences the immune response. *Trends in Immunology*, *24*(8), 444–448.

Paluska, S. A., & Schwenk, T. L. (2000). Physical activity and mental health: Current concepts. *Sports Medicine*, *29*(3), 167–180.

Petriz, B. A., Castro, A. P., Almeida, J. A., Gomes, C. P., Fernandes, G. R., Kruger, R. H., … Franco, O. L. (2014). Exercise induction of gut microbiota modifications in obese, non-obese and hypertensive rats. *BMC Genomics*, *15*, 511. http://dx.doi.org/10.1186/1471-2164-15-511.

Queipo-Ortuno, M. I., Seoane, L. M., Murri, M., Pardo, M., Gomez-Zumaquero, J. M., Cardona, F., … Tinahones, F. J. (2013). Gut microbiota composition in male rat models under different nutritional status and physical activity and its association with serum leptin and ghrelin levels. *PLoS One*, *8*(5), e65465. http://dx.doi.org/10.1371/journal.pone.0065465.

Reber, S. O., Peters, S., Slattery, D. A., Hofmann, C., Scholmerich, J., Neumann, I. D., & Obermeier, F. (2011). Mucosal immunosuppression and epithelial barrier defects are key events in murine psychosocial stress-induced colitis. *Brain, Behavior, and Immunity*, *25*(6), 1153–1161. http://dx.doi.org/10.1016/j.bbi.2011.03.004.

Reeves, R., & Candido, E. P. (1978). Turnover of histone acetyl groups in cultured cells is inhibited by sodium butyrate. *FEBS Letters*, *91*(1), 117–120.

Ridaura, V. K., Faith, J. J., Rey, F. E., Cheng, J., Duncan, A. E., Kau, A. L., … Gordon, J. I. (2013). Gut microbiota from twins discordant for obesity modulate metabolism in mice. *Science*, *341*(6150), 1241214. http://dx.doi.org/10.1126/science.1241214.

Riggs, M. G., Whittaker, R. G., Neumann, J. R., & Ingram, V. M. (1977). n-Butyrate causes histone modification in HeLa and Friend erythroleukemia cells. *Nature*, *268*(5619), 462–464.

Ringel-Kulka, T., Cheng, J., Ringel, Y., Salojarvi, J., Carroll, I., Palva, A., … Satokari, R. (2013). Intestinal microbiota in healthy U.S. young children and adults—A high throughput microarray analysis. *PLoS One*, *8*(5), e64315. http://dx.doi.org/10.1371/journal.pone.0064315.

Roberfroid, M. (2007). Prebiotics: The concept revisited. *The Journal of Nutrition*, *137*(3 Suppl. 2), 830S–837S.

Roberfroid, M., Gibson, G. R., Hoyles, L., McCartney, A. L., Rastall, R., Rowland, I., … Meheust, A. (2010). Prebiotic effects: Metabolic and health benefits. *The British Journal of Nutrition*, *104*(Suppl. 2), S1–S63. http://dx.doi.org/10.1017/S0007114510003363.

Rozeske, R. R., Evans, A. K., Frank, M. G., Watkins, L. R., Lowry, C. A., & Maier, S. F. (2011). Uncontrollable, but not controllable, stress desensitizes 5-HT1A receptors in the dorsal raphe nucleus. *The Journal of Neuroscience*, *31*(40), 14107–14115. http://dx.doi.org/10.1523/JNEUROSCI.3095-11.2011.

Rumian, N. L., Mika, A., Greenwood, B. N., Day, H. E., Borchert, D., Paton, M. M., Chichlowski, M., Berg, B. M., & Fleshner, M. (2014). Dietary prebiotics increase *Bifidobacterium* spp. and *Lactobacillus* spp. in the gut and promote stress resistance. Abstract presented at the Society for Neuroscience, Washington, DC.

Savignac, H. M., Corona, G., Mills, H., Chen, L., Spencer, J. P., Tzortzis, G., & Burnet, P. W. (2013). Prebiotic feeding elevates central brain derived neurotrophic factor, N-methyl-D-aspartate receptor subunits and D-serine. *Neurochemistry International*, *63*(8), 756–764. http://dx.doi.org/10.1016/j.neuint.2013.10.006.

Savignac, H. M., Couch, Y., Stratford, M., Bannerman, D. M., Tzortzis, G., Anthony, D. C., & Burnet, P. W. (2016). Prebiotic administration normalizes lipopolysaccharide (LPS)-induced anxiety and cortical 5-HT2A receptor and IL1-beta levels in male mice. *Brain, Behavior, and Immunity*, *52*, 120–131. http://dx.doi.org/10.1016/j.bbi.2015.10.007.

Savignac, H. M., Kiely, B., Dinan, T. G., & Cryan, J. F. (2014). Bifidobacteria exert strain-specific effects on stress-related behavior and physiology in BALB/c mice. *Neurogastroenterology and Motility*, *26*(11), 1615–1627. http://dx.doi.org/10.1111/nmo.12427.

Schmidt, K., Cowen, P. J., Harmer, C. J., Tzortzis, G., Errington, S., & Burnet, P. W. (2015). Prebiotic intake reduces the waking cortisol response and alters emotional bias in healthy volunteers. *Psychopharmacology*, *232*(10), 1793–1801. http://dx.doi.org/10.1007/s00213-014-3810-0.

Schroeder, F. A., Lin, C. L., Crusio, W. E., & Akbarian, S. (2007). Antidepressant-like effects of the histone deacetylase inhibitor, sodium butyrate, in the mouse. *Biological Psychiatry*, *62*(1), 55–64. http://dx.doi.org/10.1016/j.biopsych.2006.06.036.

Schwab, C., & Ganzle, M. (2011). Lactic acid bacteria fermentation of human milk oligosaccharide components, human milk oligosaccharides and galactooligosaccharides. *FEMS Microbiology Letters*, *315*(2), 141–148. http://dx.doi.org/10.1111/j.1574-6968.2010.02185.x.

Sealy, L., & Chalkley, R. (1978). The effect of sodium butyrate on histone modification. *Cell*, *14*(1), 115–121.

Sheikh, B. N., Phipson, B., El-Saafin, F., Vanyai, H. K., Downer, N. L., Bird, M. J., … Thomas, T. (2015). MOZ (MYST3, KAT6A) inhibits senescence via the INK4A-ARF pathway. *Oncogene*, *34*(47), 5807–5820. http://dx.doi.org/10.1038/onc.2015.33.

Sheridan, J. F., Dobbs, C., Jung, J., Chu, X., Konstantinos, A., Padgett, D., & Glaser, R. (1998). Stress-induced neuroendocrine modulation of viral pathogenesis and immunity. *Annals of the New York Academy of Sciences*, *840*, 803–808.

Sommer, F., & Backhed, F. (2013). The gut microbiota—Masters of host development and physiology. *Nature Reviews Microbiology*, *11*(4), 227–238. http://dx.doi.org/10.1038/nrmicro2974.

Sudo, N., Chida, Y., Aiba, Y., Sonoda, J., Oyama, N., Yu, X. N., … Koga, Y. (2004). Postnatal microbial colonization programs the hypothalamic-pituitary-adrenal system for stress response in mice. *The Journal of Physiology*, *558*(Pt. 1), 263–275. http://dx.doi.org/10.1113/jphysiol.2004.063388.

Tarr, A. J., Galley, J. D., Fisher, S. E., Chichlowski, M., Berg, B. M., & Bailey, M. T. (2015). The prebiotics 3′Sialyllactose and 6′Sialyllactose diminish stressor-induced anxiety-like behavior and colonic microbiota alterations: Evidence for effects on the gut-brain axis. *Brain, Behavior, and Immunity*, *50*, 166–177. http://dx.doi.org/10.1016/j.bbi.2015.06.025.

Tlaskalova-Hogenova, H., Stepankova, R., Hudcovic, T., Tuckova, L., Cukrowska, B., Lodinova-Zadnikova, R., … Kokesova, A. (2004). Commensal bacteria (normal microflora), mucosal immunity and chronic inflammatory and autoimmune diseases. *Immunology Letters*, *93*(2–3), 97–108. http://dx.doi.org/10.1016/j.imlet.2004.02.005.

Turnbaugh, P. J., Ley, R. E., Mahowald, M. A., Magrini, V., Mardis, E. R., & Gordon, J. I. (2006). An obesity-associated gut microbiome with increased capacity for energy harvest. *Nature*, *444*(7122), 1027–1031.

Valdizan, E. M., Castro, E., & Pazos, A. (2010). Agonist-dependent modulation of G-protein coupling and transduction of 5-HT1A receptors in rat dorsal raphe nucleus. *The International Journal of Neuropsychopharmacology*, *13*(7), 835–843. http://dx.doi.org/10.1017/S1461145709990940.

Vitali, B., Ndagijimana, M., Maccaferri, S., Biagi, E., Guerzoni, M. E., & Brigidi, P. (2012). An in vitro evaluation of the effect of probiotics and prebiotics on the metabolic profile of human microbiota. *Anaerobe*, *18*(4), 386–391. http://dx.doi.org/10.1016/j.anaerobe.2012.04.014.

Wein, M. N., Spatz, J., Nishimori, S., Doench, J., Root, D., Babij, P., … Kronenberg, H. M. (2015). HDAC5 controls MEF2C-driven sclerostin expression in osteocytes. *Journal of Bone and Mineral Research*, *30*(3), 400–411. http://dx.doi.org/10.1002/jbmr.2381.

Welly, R. J., Liu, T. W., Zidon, T. M., Rowles, J. L., 3rd, Park, Y. M., Smith, T. N., … Vieira-Potter, V. J. (2016). Comparison of diet vs exercise on metabolic function and gut microbiota in obese rats. *Medicine and Science in Sports and Exercise*, *48*, 1688–1698. http://dx.doi.org/10.1249/MSS.0000000000000964.

Whitesell, N. R., Beals, J., Mitchell, C. M., Manson, S. M., Turner, R. J., & Team, A.-S. (2009). Childhood exposure to adversity and risk of substance-use disorder in two American Indian populations: The meditational role of early substance-use initiation. *Journal of Studies on Alcohol and Drugs*, *70*(6), 971–981.

Williams, S., Chen, L., Savignac, H. M., Tzortzis, G., Anthony, D. C., & Burnet, P. W. (2016). Neonatal prebiotic (BGOS) supplementation increases the levels of synaptophysin, GluN2A-subunits and BDNF proteins in the adult rat hippocampus. *Synapse*, *70*(3), 121–124. http://dx.doi.org/10.1002/syn.21880.

Yamane, K., Tateishi, K., Klose, R. J., Fang, J., Fabrizio, L. A., Erdjument-Bromage, H., … Zhang, Y. (2007). PLU-1 is an H3K4 demethylase involved in transcriptional repression and breast cancer cell proliferation. *Molecular Cell, 25*(6), 801–812. http://dx.doi.org/10.1016/j.molcel.2007.03.001.

Yarandi, S. S., Peterson, D. A., Treisman, G. J., Moran, T. H., & Pasricha, P. J. (2016). Modulatory effects of gut microbiota on the central nervous system: How gut could play a role in neuropsychiatric health and diseases. *Journal of Neurogastroenterology and Motility, 22*(2), 201–212. http://dx.doi.org/10.5056/jnm15146.

Yatsunenko, T., Rey, F. E., Manary, M. J., Trehan, I., Dominguez-Bello, M. G., Contreras, M., … Gordon, J. I. (2012). Human gut microbiome viewed across age and geography. *Nature, 486*(7402), 222–227. http://dx.doi.org/10.1038/nature11053.

Yuan, T. F., Ferreira Rocha, N. B., Paes, F., Arias-Carrion, O., Machado, S., & de Sa Filho, A. S. (2015). Neural mechanisms of exercise: Effects on gut microbiota and depression. *CNS & Neurological Disorders Drug Targets, 14*(10), 1312–1314.

Zareie, M., Johnson-Henry, K., Jury, J., Yang, P. C., Ngan, B. Y., McKay, D. M., … Sherman, P. M. (2006). Probiotics prevent bacterial translocation and improve intestinal barrier function in rats following chronic psychological stress. *Gut, 55*(11), 1553–1560. http://dx.doi.org/10.1136/gut.2005.080739.

Zhao, N., Li, S., Wang, R., Xiao, M., Meng, Y., Zeng, C., … Zhuang, S. M. (2016). Expression of microRNA-195 is transactivated by Sp1 but inhibited by histone deacetylase 3 in hepatocellular carcinoma cells. *Biochimica et Biophysica Acta, 1859*, 933–942. http://dx.doi.org/10.1016/j.bbagrm.2016.05.006.

Zhu, L., Yang, J., Zhao, L., Yu, X., Wang, L., Wang, F., … Jin, J. (2015). Expression of hMOF, but not HDAC4, is responsible for the global histone H4K16 acetylation in gastric carcinoma. *International Journal of Oncology, 46*(6), 2535–2545. http://dx.doi.org/10.3892/ijo.2015.2956.

CHAPTER NINE

Circadian Rhythm and the Gut Microbiome

R.M. Voigt*, C.B. Forsyth*, S.J. Green†, P.A. Engen*, A. Keshavarzian*,‡,1

*Rush University Medical Center, Chicago, IL, United States
†DNA Services Facility, Research Resources Center, University of Illinois at Chicago, Chicago, IL, United States
‡Utrecht Institute for Pharmaceutical Sciences, Utrecht University, Utrecht, The Netherlands
[1]Corresponding author: e-mail address: ali_keshavarzian@rush.edu

Contents

Abstract

Circadian rhythms are 24-h patterns regulating behavior, organs, and cells in living organisms. These rhythms align biological functions with regular and predictable environmental patterns to optimize function and health. Disruption of these rhythms can be detrimental resulting in metabolic syndrome, cancer, or cardiovascular disease, just to name a few. It is now becoming clear that the intestinal microbiome is also regulated by circadian rhythms via intrinsic circadian clocks as well as via the host organism. Microbiota rhythms are regulated by diet and time of feeding which can alter both microbial community structure and metabolic activity which can significantly impact host immune and metabolic function. In this review, we will cover how host circadian rhythms are generated and maintained, how host circadian rhythms can be disrupted, as well as the consequences of circadian rhythm disruption. We will further highlight the newly emerging literature indicating the importance of circadian rhythms of the intestinal microbiota.

International Review of Neurobiology, Volume 131
ISSN 0074-7742
http://dx.doi.org/10.1016/bs.irn.2016.07.002

1. CIRCADIAN RHYTHMS IN HEALTH

1.1 What Are Circadian Rhythms?

Circadian rhythms are rhythmic patterns of approximately 24 h that are exhibited by most organisms including bacteria, fungi, plants, and animals (Hastings, Reddy, et al., 2003). The circadian clock regulates and optimizes the function of cells, organs, systems, and behavior based on the 24-h day (Mohawk, Green, et al., 2012). Activity–rest cycles and feast–famine cycles that are features of our 24-h day drive physiological and cellular adaptations in a wide variety of processes including gastrointestinal function, metabolic processes, and cellular transcription and translation, just to name a few (Reddy & O'Neill, 2010). However, the ability to respond and adapt to changes in the environment is vital for survival; thus, physiological patterns controlled by circadian rhythms also can be influenced by external cues.

Circadian rhythms are endogenous which means that they are observed independent of external cues and this rhythmicity is based on the function of the molecular circadian clock (Reppert & Weaver, 2002). The molecular clock is a transcriptional–translational autoregulatory feedback loop that takes approximately 24 h to complete, and while the specifics of the molecular clock differ amongst various organisms, they all exhibit an endogenous 24 h pattern. The mammalian core circadian clock is comprised of the transcription factors "Clock" (circadian locomotor output cycles kaput) and "Bmal1" (brain and muscle aryl hydrocarbon receptor nuclear translocator-protein 1) which bind to the E-box promoter initiating transcription and subsequent translation of so-called clock-controlled genes (Bunger, Wilsbacher, et al., 2000; Mohawk et al., 2012; Schibler, 2005) (Fig. 1). Two clock-controlled genes are period (*Per*) and cryptochrome (*Cry*), and the accumulation and dimerization of PER and CRY proteins result in feedback inhibition whereby further CLOCK and BMAL1-mediated transcription is inhibited (van der Horst, Muijtjens, et al., 1999). Degradation of PER and CRY releases the feedback inhibition and the cycle begins anew. In addition to this core clock, there are other mechanisms of regulation including posttranscriptional and posttranslational modifications that result in "fine-tuning" of the clock, including nuclear receptors retinoic acid-related orphan receptor alpha (*Rora*), reverse erythroblastosis virus alpha (*Rev-erba*), and sirtuin 1 (*Sirt1*) (Bass & Takahashi, 2011; Grimaldi, Nakahata, et al., 2009).

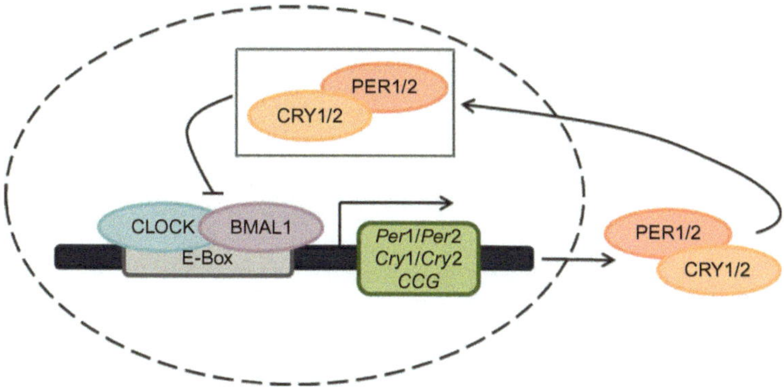

Fig. 1 The molecular circadian clock. The mammalian core circadian clock is comprised of the transcription factors "CLOCK" (circadian locomotor output cycles kaput) and "BMAL1" (brain and muscle aryl hydrocarbon receptor nuclear translocator-protein 1) which bind to the E-box promoter initiating transcription and subsequent translation of so-called clock-controlled genes (CCG). Two clock-controlled genes are period (*Per1/Per2*) and cryptochrome (*Cry1/Cry2*), and the accumulation and dimerization of PER and CRY proteins result in feedback inhibition whereby further CLOCK and BMAL1-mediated transcription is inhibited. Degradation of PER and CRY releases the feedback inhibition and the cycle begins anew.

Microorganisms colonize every accessible surface of the host organism. Thus, microorganisms are found on our skin, in our nasal passages, and in our gastrointestinal tract. Not surprisingly, given the importance and ubiquitous nature of circadian rhythms across the kingdoms, some bacteria also exhibit circadian rhythms. The bacterial clock has primarily been studied in Cyanobacteria (i.e., a bacterial phylum of photosynthetic microorganisms) which contain a molecular clock that is comprised of only three proteins KaiA, KaiB, and KaiC. These proteins function in a transcriptional feedback loop similar to that in mammals (Cohen & Golden, 2015). Also, one species of bacteria that is found in the human intestine, *Enterobacter aerogenes*, was recently shown to be responsive to the circadian hormone melatonin and exhibits an endogenous daily rhythm (Paulose, Wright, et al., 2016).

Circadian rhythms can be entrained or adjusted by cues in the environment in order to synchronize the organism with external cues which are known as zeitgebers (i.e., time givers) (Fig. 2). Light–dark cycles strongly regulate the mammalian central clock via inputs from the eye into the master circadian pacemaker located in the suprachiasmatic nucleus (SCN) of the hypothalamus. The ability of circadian rhythms to entrain to the environment is important for several reasons: (1) the ability to adapt to changes

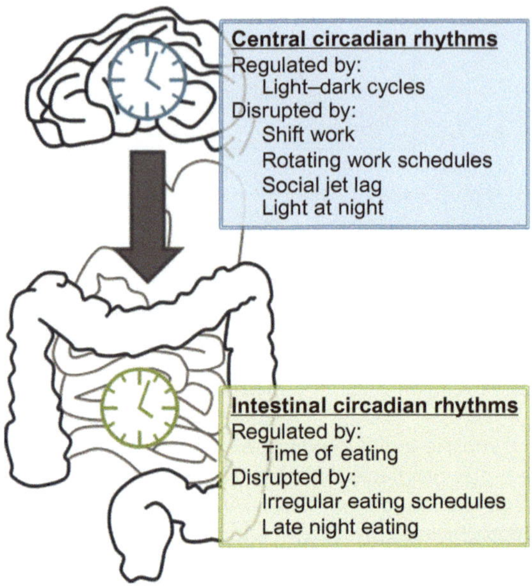

Fig. 2 Central and peripheral circadian clocks. Circadian rhythms and the molecular circadian clock are found in nearly every mammalian cell, with different molecular clocks regulated by different environmental cues. Light–dark cycles regulate the mammalian central circadian clock located in the suprachiasmatic nucleus (SCN) located in the hypothalamus. The SCN integrates inputs from the eye synchronizing circadian rhythms in the periphery via sympathetic and parasympathetic signals. Time of meal consumption (i.e., timing of nutrient availability) strongly regulates circadian clocks in the intestine and liver. Alterations in light:dark cycles or time of eating can disrupt central and peripheral rhythms, respectively, which can have detrimental health outcomes.

in the environment is critical for survival (e.g., seasonal changes in light–dark cycles and food availability) and (2) the molecular circadian clock is not exactly 24 h in duration and must constantly be readjusted to align with the environment. Human circadian rhythms average 24.2 h per day; however, variations among humans are observed, and this accounts for different chronotypes: night owls vs morning larks (Ehret, 1974). Without the ability of circadian rhythms to be adjusted by environmental cues, rhythms would gradually become desynchronized from the environment, with a resulting loss in the ability of the circadian rhythms to optimize biological functions.

1.2 Central vs Peripheral Circadian Clocks

Circadian rhythms and the molecular circadian clock are found in nearly every mammalian cell, with different molecular clocks regulated by different environmental cues (Yoo, Yamazaki, et al., 2004) (Fig. 2). Light–dark cycles are important regulators of the mammalian central circadian clock located in the SCN. The SCN has two essential functions: integrating inputs from the optic nerve and synchronizing circadian rhythms in the periphery through sympathetic and parasympathetic signals (Welsh, Takahashi, et al., 2010; West & Bechtold, 2015). For some time, it was believed that the ablation of the SCN would abolish circadian rhythms in the periphery. However, subsequent studies have shown that destruction or inactivation of the SCN does not prevent circadian rhythms from being observed in the periphery, but rather the rhythms become gradually desynchronized from one another without input from the SCN (Guo, Brewer, et al., 2006; Yoo et al., 2004). This observation is consistent with the idea that circadian rhythms are endogenous, but that cues from the environment help entrain the rhythms to synchronize them with the external environment. While light is a regulator of the central circadian clock, it is important to note that other external cues can regulate molecular clocks in peripheral tissues. For example, time of meal consumption (i.e., timing of nutrient availability) strongly regulates circadian clocks in the intestine and liver and these clocks are also impacted by diet composition (e.g., high-fat or high-sugar diets) (Mattson, Allison, et al., 2014; Mendoza, 2007). Exercise also can regulate circadian rhythms in muscle and lung tissue (Youngstedt, Kline, et al., 2016). Thus, regular and predictable patterns of eating or exercise will allow an organism to best adapt to and respond to stresses induced by these activities (e.g., oxidative stress).

2. CIRCADIAN RHYTHMS IN DISEASE

2.1 What Factors Disrupt Circadian Rhythms?

Disruption of circadian rhythms in humans can be the consequence of numerous lifestyle factors, with shift work and traveling between time zones being some of the most obvious causes. However, other factors contribute to disrupted circadian homeostasis. These include (but are not limited to): (1) light exposure at night (e.g., use of light emitting electronic devices), which can alter rhythmicity via inputs from the eye directly to the SCN (Burgess & Molina, 2014; Fonken, Weil, et al., 2013; Fonken, Workman, et al., 2010); (2) time of eating close to or during the rest period (i.e., typically within 2 h of the normal rest time) can effectively uncouple central circadian rhythms from those in the intestine and liver (Asher & Sassone-Corsi, 2015; Mattson et al., 2014); (3) social jet lag, wherein daily schedules are altered on work-free days compared to work days a leading to disruptions in central and peripheral circadian rhythmicity (Roenneberg, Allebrandt, et al., 2012; Wittmann, Dinich, et al., 2006); and (4) diet composition. For example, alcohol consumption disrupts central circadian rhythms (Forsyth, Voigt, et al., 2015; Swanson, Gorenz, et al., 2015) and diets high in fat can disrupt circadian rhythms in the intestine (Leone, Gibbons, et al., 2015; Zarrinpar, Chaix, et al., 2014).

2.2 Consequences of Circadian Rhythm Disruption

A number of studies have demonstrated higher rates of cancer (breast and prostate), cardiovascular disease, obesity, psychiatric, and neurodegenerative diseases in shift workers (Bechtold, Gibbs, et al., 2010; Golombek, Casiraghi, et al., 2013; Zelinski, Deibel, et al., 2014). Individuals with a late chronotype (i.e., night owl) are at a higher risk for developing poor health outcomes than individuals with an early chronotype, possibly because these individuals tend to eat close to the rest period, thereby uncoupling rhythms in the liver/intestine from the central pacemaker (Foster, Peirson, et al., 2013; Golombek et al., 2013; Roenneberg et al., 2012; Zelinski et al., 2014). But why do these detrimental consequences occur?

One common feature among the diseases associated with circadian rhythm disruption is that they appear to be triggered or promoted by inflammatory processes. One source of "sterile inflammation" is the intestine which is in constant contact with bacteria, fungi, and viruses in the intestinal tract (Clemente, Ursell, et al., 2012). There is an intimate relationship between the intestine and its contents (Bernardo, Sanchez, et al., 2012; Wells, Rossi, et al., 2011) and there are several factors that can have significant proinflammatory changes in the host, including alterations in the intestinal microbiota leading to dysfunction of intestinal barrier integrity (Caricilli, Castoldi, et al., 2014; Malago, 2015). Our group has demonstrated that both environmental circadian rhythm disruption (once weekly changes in the light–dark cycle) and genetic perturbation of the molecular clock (mutation in the core molecular clock, the Clock$^{\Delta 19}$ mouse) cause intestinal microbiota dysbiosis, especially when paired with a dietary stress such as a high-fat diet (HFD) or alcohol consumption (Voigt, Forsyth, et al., 2014; Voigt, Summa, et al., 2016). The changes in the microbiota are characterized by an increase in proinflammatory bacteria, a decrease in putative antiinflammatory, butyrate-producing bacteria, and a shift in the Bacteroidetes/Firmicutes ratio (Voigt et al., 2014, 2016). Our group has also demonstrated disrupted intestinal barrier function in mice exhibiting circadian rhythm disruption (both environmental and genetic manipulation) (Summa, Voigt, et al., 2013). The integrity of the intestinal barrier is critical for keeping the proinflammatory contents of the intestine separate from the intestinal mucosa and the systemic circulation (Farhadi, Banan, et al., 2003). Intestinal dysbiosis and intestinal hyperpermeability can have proinflammatory consequences in the intestinal mucosa, but additionally, these factors can alter immune function. Under normal circumstances, the immune system is tightly regulated by circadian rhythmicity and disrupted circadian rhythms can have devastating consequences (Cermakian, Westfall, et al., 2014; Curtis, Bellet, et al., 2014; Logan & Sarkar, 2012). Circadian disruption by shifts in light–dark cycles results in increased intestinal Th17 cells (Yu, Rollins, et al., 2013). Circadian rhythm disruption is also associated with higher rates of infection (Everson, 1993; Mohren, Jansen, et al., 2002) and a clear time of day effect exists in response to *Salmonella* infections (Bellet, Deriu, et al., 2013). Finally, mice with disrupted circadian homeostasis have exaggerated proinflammatory responses to lipopolysaccharide (i.e., a component in the outer membrane of Gram-negative bacteria) exposure and greater mortality (Castanon-Cervantes, Wu, et al., 2010). These data suggest that an increased abundance of proinflammatory bacteria in the intestine, coupled with intestinal barrier dysfunction, can promote inflammation-mediated diseases.

3. CIRCADIAN RHYTHMS AND THE INTESTINAL MICROBIOTA

It is well established that the intestinal microbiome changes according to dietary intake, and likewise, it is clear that different stages of life (birth to old age) are associated with shifts in bacterial populations (Bischoff, 2016; Clemente et al., 2012; Yatsunenko, Rey, et al., 2012). What was not known until recently was whether or not intestinal microbiota exhibit circadian patterns of composition or function (i.e., metabolism). While circadian fluctuations in metabolism have been observed in cyanobacteria (light-sensitive bacteria) (Cohen & Golden, 2015; Kondo, Mori, et al., 1997), it was not clear if microbes that are not light sensitive (or that are not exposed to sunlight) also have a genetically encoded endogenous circadian rhythm.

Our work has shown that disrupted circadian rhythmicity in the host can influence bacterial populations in the intestine (Voigt et al., 2014, 2016); however, it is becoming clear that intestinal microbiota also have circadian fluctuations (Fig. 3). Thaiss et al. subsequently demonstrated that up to 20% of intestinal bacteria exhibit diurnal fluctuations in relative abundance and

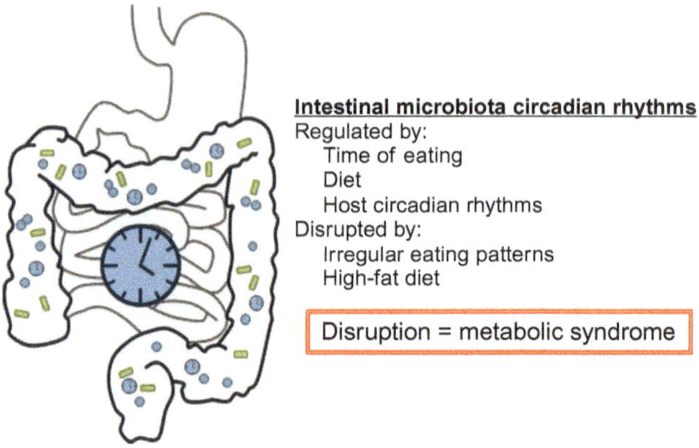

Fig. 3 Circadian rhythms in the intestinal microbiota. Intestinal microbiota exhibit diurnal fluctuations community populations (shifts in dominant bacteria) as well as fluctuations in bacterial function including metabolism. These rhythmic changes in the microbiota appear to be driven by the host via time of eating (i.e., nutrient availability), diet composition (i.e., high-fat diet), and circadian status of the host (i.e., genetic mutations in the host circadian clock, Clock, Per1/2, and Bmal1). Several studies demonstrate a clear link between disruption of microbiota rhythms with host metabolic syndrome and obesity.

activity (Thaiss, Zeevi, et al., 2014). Other groups have also observed circadian oscillations in bacterial abundance (Liang, Bushman, et al., 2015; Zarrinpar et al., 2014). Interestingly, one study has shown that bacterial rhythms in female mice are more robust than those observed in male mice (Liang et al., 2015). These rhythmic changes in the microbiota appear to be driven by the host. Time of eating, via changes in nutrient availability, regulates a variety of microbial functions (e.g., energy harvest, cell growth, and DNA repair are predominant during the periods of nutrient availability, while detoxification is predominant during periods of fasting) and overall community populations (shifts in dominant bacteria) also appear to be driven by time of eating (Thaiss et al., 2014). Other studies have also shown evidence of clear circadian patterns of metabolism in intestinal bacterial populations (Liang et al., 2015; Thaiss et al., 2014). Thus, while light/dark cycles are zeitgebers for the central circadian clock in mammals, time of eating is the zeitgeber for intestinal circadian rhythms in mammals as well as the zeitgeber for circadian rhythms in intestinal bacteria.

As previously mentioned, environmental factors can influence circadian rhythms and a recent study examined the impact of a (HFD) on circadian rhythmicity of intestinal microbiota (Leone et al., 2015). Circadian oscillations in bacterial abundance were dampened in mice fed a HFD. Affected bacteria included members of the family Lachnospiraceae (a short-chain fatty acid (SCFA)-producing bacteria). Conversely, bacteria that previously had no rhythm, such as H_2S-producing bacteria (i.e., sulfate-reducing bacteria), demonstrated a rhythm when the host mice were fed a HFD. However, circadian rhythms in intestinal bacteria can be partially restored when the HFD is fed only during the dark (i.e., time-restricted feeding). As might be expected based on these results, the HFD blunts the robust metabolic rhythms that are observed in chow-fed mice. SCFAs are a metabolic byproduct of fermentation by specific microbial taxa, including a number of members of the family Lachnospiraceae, and circadian oscillations in SCFA production are lost under certain conditions such as HFD consumption (Leone et al., 2015; Thaiss et al., 2014). Alterations in SCFA-producing bacteria and SCFA production could be important for several reasons: (1) SCFA (particularly butyrate) has beneficial effects on intestinal barrier function, (2) SCFA (particularly butyrate) is antiinflammatory via a histone deacetylase mechanism, and (3) SCFAs could be a feedback mechanism by which the intestinal bacteria communicate with the host (i.e., SCFAs may regulate the circadian metabolic clock in the brain and liver).

In addition to environmental perturbations of circadian rhythms (e.g., light–dark cycles, HFD consumption), genetic mutations of the core circadian clock also can result in intestinal dysbiosis and/or disrupted circadian rhythmicity of intestinal bacteria. The circadian gene mutations studied thus far include Clock (Voigt et al., 2016), Per1/2 (Thaiss et al., 2014), and Bmal1 (Liang et al., 2015). The $Clock^{\Delta19}$ mutation is associated with significant dysbiosis (an increase in proinflammatory bacteria), an effect that was exacerbated by alcohol consumption (Voigt et al., 2016). Per1/2 KO mice exhibit dysbiosis and lack of circadian rhythmicity in the intestinal tract (Thaiss et al., 2014) with similar results found in Bmal1 KO mice (Liang et al., 2015). It is interesting that these effects appear to be sex specific with female mice displaying greater microbiota rhythmicity as well as differing responses to the BMAL1 KO (Liang et al., 2015). Taken together, these studies support a role for the host genome in regulating the circadian pattern of the microbiota, which may, in part, be driven by circadian rhythm-induced differences in eating patterns or circadian-induced changes in host immune function.

3.1 How Do Bacterial Circadian Rhythms Impact Host Metabolism?

Host circadian rhythmicity can alter the composition and activity of the intestinal microbiota, and conversely, the intestinal microbiota can also influence the host. Several studies demonstrate a clear link between intestinal dysbiosis and disruption of microbiota rhythms with host metabolic syndrome and obesity. A compelling study by Thaiss et al. showed that when intestinal microbiota from jet-lagged humans (i.e., circadian disrupted) was transferred into germ-free mice, this resulted in obesity and glucose intolerance, an effect that was not observed when microbiota was transferred from nonjet-lagged humans (Thaiss et al., 2014). Consumption of a HFD blunts intestinal bacterial rhythms and results in obesity and signs of metabolic syndrome; however, time-restricted feeding of the HFD restores glucose tolerance and prevents obesity in mice (Leone et al., 2015). These effects could be attributed to time-restricted feeding-induced partial restoration of intestinal microbiota rhythms in HFD-fed mice, a reduction in obesogenic bacteria and an increase in bacterial taxa that promote healthy metabolism (e.g., organisms from the genera *Oscillibacter* and *Ruminococcus*), or changes in abundance or function of SCFA-producing bacteria (Zarrinpar et al., 2014). Leone et al. proposed that butyrate produced by specific populations of intestinal bacteria is a powerful signal for the liver that entrains the liver circadian clock; however, under conditions where SCFA production is altered, that signal is no longer present and this can result in metabolic syndrome in the host. Data from our studies show that the jet-lagged mice fed a HFD exhibit dysbiosis and increased gut leakiness and endotoxemia (Summa et al., 2013). Such increased intestinal permeability and endotoxemia have been associated with increased systemic inflammation and metabolic syndrome and obesity that was eliminated when normal gut permeability was restored (Cani, Bibiloni, et al., 2008). Thus, increased gut leakiness resulting from microbiota circadian disruption could be a key pathogenic factor and target for therapy.

4. CONCLUSION

Despite significant progress in intestinal microbiota research, the microbiota still seems to be characterized by more mystery than established principles and facts. The emerging consensus is of a dynamic system in which the host-associated microbiota is involved in a complex conversation with the host. Host-regulated conditions include those under endogenous circadian control, diet, time of feeding, and other factors can all lead to an altered microbial community structure and altered microbial activity. In turn, microbial metabolic activity, in part through the production of SCFAs, can serve as a key regulator of mammalian immune and metabolic function.

As such, it must now be considered in any model related to circadian homeostasis as a key component of circadian-related disease processes as well as potential therapies. The overall conclusion from these studies is that there is a clear circadian rhythm to the intestinal microbiota (i.e., largely regulated by food timing) and that this rhythm is closely tied to the microbiota function with substantial effects on the host in immunity and metabolism. We propose that this relationship holds the potential for microbiota-directed therapies such as probiotics and targeted prebiotics to ameliorate the effects of disrupted circadian homeostasis.

REFERENCES

Asher, G., & Sassone-Corsi, P. (2015). Time for food: The intimate interplay between nutrition, metabolism, and the circadian clock. *Cell*, *161*(1), 84–92.

Bass, J., & Takahashi, J. S. (2011). Circadian integration of metabolism and energetics. *Science*, *330*(6009), 1349–1354.

Bechtold, D. A., Gibbs, J. E., et al. (2010). Circadian dysfunction in disease. *Trends in Pharmacological Sciences*, *31*(5), 191–198.

Bellet, M. M., Deriu, E., et al. (2013). Circadian clock regulates the host response to Salmonella. *Proceedings of the National Academy of Sciences of the United States of America*, *110*(24), 9897–9902.

Bernardo, D., Sanchez, B., et al. (2012). Microbiota/host crosstalk biomarkers: Regulatory response of human intestinal dendritic cells exposed to Lactobacillus extracellular encrypted peptide. *PLoS One*, 7(5), e36262.

Bischoff, S. C. (2016). Microbiota and aging. *Current Opinion in Clinical Nutrition and Metabolic Care*, *19*(1), 26–30.

Bunger, M. K., Wilsbacher, L. D., et al. (2000). Mop3 is an essential component of the master circadian pacemaker in mammals. *Cell*, *103*(7), 1009–1017.

Burgess, H. J., & Molina, T. A. (2014). Home lighting before usual bedtime impacts circadian timing: A field study. *Photochemistry and Photobiology*, *90*(3), 723–726.

Cani, P. D., Bibiloni, R., et al. (2008). Changes in gut microbiota control metabolic endotoxemia-induced inflammation in high-fat diet-induced obesity and diabetes in mice. *Diabetes*, *57*(6), 1470–1481.

Caricilli, A. M., Castoldi, A., et al. (2014). Intestinal barrier: A gentlemen's agreement between microbiota and immunity. *World Journal of Gastrointestinal Pathophysiology*, *5*(1), 18–32.

Castanon-Cervantes, O., Wu, M., et al. (2010). Dysregulation of inflammatory responses by chronic circadian disruption. *The Journal of Immunology*, *185*(10), 5796–5805.

Cermakian, N., Westfall, S., et al. (2014). Circadian clocks and inflammation: Reciprocal regulation and shared mediators. *Archivum Immunologiae et Therapiae Experimentalis*, *62*(4), 303–318.

Clemente, J. C., Ursell, L. K., et al. (2012). The impact of the gut microbiota on human health: An integrative view. *Cell*, *148*(6), 1258–1270.

Cohen, S. E., & Golden, S. S. (2015). Circadian rhythms in cyanobacteria. *Microbiology and Molecular Biology Reviews*, *79*(4), 373–385.

Curtis, A. M., Bellet, M. M., et al. (2014). Circadian clock proteins and immunity. *Immunity*, *40*(2), 178–186.

Ehret, C. F. (1974). The sense of time: Evidence for its molecular basis in the eukaryotic gene-action system. *Advances in Biological and Medical Physics*, *15*, 47–77.

Everson, C. A. (1993). Sustained sleep deprivation impairs host defense. *The American Journal of Physiology*, *265*(5 Pt 2), R1148–R1154.
Farhadi, A., Banan, A., et al. (2003). Intestinal barrier: An interface between health and disease. *Journal of Gastroenterology and Hepatology*, *18*(5), 479–497.
Fonken, L. K., Weil, Z. M., et al. (2013). Mice exposed to dim light at night exaggerate inflammatory responses to lipopolysaccharide. *Brain, Behavior, and Immunity*, *34*, 159–163.
Fonken, L. K., Workman, J. L., et al. (2010). Light at night increases body mass by shifting the time of food intake. *Proceedings of the National Academy of Sciences of the United States of America*, *107*(43), 18664–18669.
Forsyth, C. B., Voigt, R. M., et al. (2015). Circadian rhythms, alcohol and gut interactions. *Alcohol*, *49*(4), 389–398.
Foster, R. G., Peirson, S. N., et al. (2013). Sleep and circadian rhythm disruption in social jetlag and mental illness. *Progress in Molecular Biology and Translational Science*, *119*, 325–346.
Golombek, D. A., Casiraghi, L. P., et al. (2013). The times they're a-changing: Effects of circadian desynchronization on physiology and disease. *Journal of Physiology (Paris)*, *107*(4), 310–322.
Grimaldi, B., Nakahata, Y., et al. (2009). Chromatin remodeling, metabolism and circadian clocks: The interplay of CLOCK and SIRT1. *The International Journal of Biochemistry & Cell Biology*, *41*(1), 81–86.
Guo, H., Brewer, J. M., et al. (2006). Suprachiasmatic regulation of circadian rhythms of gene expression in hamster peripheral organs: Effects of transplanting the pacemaker. *The Journal of Neuroscience*, *26*(24), 6406–6412.
Hastings, M. H., Reddy, A. B., et al. (2003). A clockwork web: Circadian timing in brain and periphery, in health and disease. *Nature Reviews. Neuroscience*, *4*(8), 649–661.
Kondo, T., Mori, T., et al. (1997). Circadian rhythms in rapidly dividing cyanobacteria. *Science*, *275*(5297), 224–227.
Leone, V., Gibbons, S. M., et al. (2015). Effects of diurnal variation of gut microbes and high-fat feeding on host circadian clock function and metabolism. *Cell Host & Microbe*, *17*(5), 681–689.
Liang, X., Bushman, F. D., et al. (2015). Rhythmicity of the intestinal microbiota is regulated by gender and the host circadian clock. *Proceedings of the National Academy of Sciences of the United States of America*, *112*(33), 10479–10484.
Logan, R. W., & Sarkar, D. K. (2012). Circadian nature of immune function. *Molecular and Cellular Endocrinology*, *349*(1), 82–90.
Malago, J. J. (2015). Contribution of microbiota to the intestinal physicochemical barrier. *Beneficial Microbes*, *6*(3), 295–311.
Mattson, M. P., Allison, D. B., et al. (2014). Meal frequency and timing in health and disease. *Proceedings of the National Academy of Sciences of the United States of America*, *111*(47), 16647–16653.
Mendoza, J. (2007). Circadian clocks: Setting time by food. *Journal of Neuroendocrinology*, *19*(2), 127–137.
Mohawk, J. A., Green, C. B., et al. (2012). Central and peripheral circadian clocks in mammals. *Annual Review of Neuroscience*, *35*, 445–462.
Mohren, D. C., Jansen, N. W., et al. (2002). Prevalence of common infections among employees in different work schedules. *Journal of Occupational and Environmental Medicine*, *44*(11), 1003–1011.
Paulose, J. K., Wright, J. M., et al. (2016). Human gut bacteria are sensitive to melatonin and express endogenous circadian rhythmicity. *PLoS One*, *11*(1), e0146643.
Reddy, A. B., & O'Neill, J. S. (2010). Healthy clocks, healthy body, healthy mind. *Trends in Cell Biology*, *20*(1), 36–44.

Reppert, S. M., & Weaver, D. R. (2002). Coordination of circadian timing in mammals. *Nature*, *418*(6901), 935–941.

Roenneberg, T., Allebrandt, K. V., et al. (2012). Social jetlag and obesity. *Current Biology*, *22*(10), 939–943.

Schibler, U. (2005). The daily rhythms of genes, cells and organs. Biological clocks and circadian timing in cells. *EMBO Reports*, *6 Spec No*, S9–S13.

Summa, K. C., Voigt, R. M., et al. (2013). Disruption of the circadian clock in mice increases intestinal permeability and promotes alcohol-induced hepatic pathology and inflammation. *PLoS One*, *8*(6), e67102.

Swanson, G. R., Gorenz, A., et al. (2015). Decreased melatonin secretion is associated with increased intestinal permeability and marker of endotoxemia in alcoholics. *American Journal of Physiology. Gastrointestinal and Liver Physiology*, *308*(12), G1004–G1011.

Thaiss, C. A., Zeevi, D., et al. (2014). Transkingdom control of microbiota diurnal oscillations promotes metabolic homeostasis. *Cell*, *159*(3), 514–529.

van der Horst, G. T., Muijtjens, M., et al. (1999). Mammalian Cry1 and Cry2 are essential for maintenance of circadian rhythms. *Nature*, *398*(6728), 627–630.

Voigt, R. M., Forsyth, C. B., et al. (2014). Circadian disorganization alters intestinal microbiota. *PLoS One*, *9*(5), e97500.

Voigt, R. M., Summa, K. C., et al. (2016). The circadian clock mutation promotes intestinal dysbiosis. *Alcoholism, Clinical and Experimental Research*, *40*(2), 335–347.

Wells, J. M., Rossi, O., et al. (2011). Epithelial crosstalk at the microbiota-mucosal interface. *Proceedings of the National Academy of Sciences of the United States of America*, *108*(Suppl. 1), 4607–4614.

Welsh, D. K., Takahashi, J. S., et al. (2010). Suprachiasmatic nucleus: Cell autonomy and network properties. *Annual Review of Physiology*, *72*, 551–577.

West, A. C., & Bechtold, D. A. (2015). The cost of circadian desynchrony: Evidence, insights and open questions. *Bioessays*, *37*(7), 777–788.

Wittmann, M., Dinich, J., et al. (2006). Social jetlag: Misalignment of biological and social time. *Chronobiology International*, *23*(1–2), 497–509.

Yatsunenko, T., Rey, F. E., et al. (2012). Human gut microbiome viewed across age and geography. *Nature*, *486*(7402), 222–227.

Yoo, S. H., Yamazaki, S., et al. (2004). PERIOD2::LUCIFERASE real-time reporting of circadian dynamics reveals persistent circadian oscillations in mouse peripheral tissues. *Proceedings of the National Academy of Sciences of the United States of America*, *101*(15), 5339–5346.

Youngstedt, S. D., Kline, C. E., et al. (2016). Circadian phase-shifting effects of bright light, exercise, and bright light + exercise. *Journal of Circadian Rhythms*, *14*, 2.

Yu, X., Rollins, D., et al. (2013). TH17 cell differentiation is regulated by the circadian clock. *Science*, *342*(6159), 727–730.

Zarrinpar, A., Chaix, A., et al. (2014). Diet and feeding pattern affect the diurnal dynamics of the gut microbiome. *Cell Metabolism*, *20*(6), 1006–1017.

Zelinski, E. L., Deibel, S. H., et al. (2014). The trouble with circadian clock dysfunction: Multiple deleterious effects on the brain and body. *Neuroscience and Biobehavioral Reviews*, *40C*, 80–101.

CHAPTER TEN

Sleep and Microbes

J.M. Krueger*,[1], M.R. Opp†
*Elson S. Floyd College of Medicine, Washington State University, Spokane, WA, United States
†University of Washington College of Medicine, Seattle, WA, United States
[1]Corresponding author: e-mail address: krueger@vetmed.wsu.edu

Contents

Abstract

Sleep is profoundly altered during the course of infectious diseases. The typical response to infection includes an initial increase in nonrapid eye movement sleep (NREMS) followed by an inhibition in NREMS. REMS is inhibited during infections. Bacterial cell wall components, such as peptidoglycan and lipopolysaccharide, macrophage digests of these components, such as muramyl peptides, and viral products, such as viral double-stranded RNA, trigger sleep responses. They do so via pathogen-associated molecular pattern recognition receptors that, in turn, enhance cytokine production. Altered sleep and associated sleep-facilitated fever responses are likely adaptive responses to infection. Normal sleep in physiological conditions may also be influenced by gut microbes because the microbiota is affected by circadian rhythms, stressors, diet, and exercise. Furthermore, sleep loss enhances translocation of viable bacteria from the intestine, which provides another means by which sleep–microbe interactions impact neurobiology.

International Review of Neurobiology, Volume 131
ISSN 0074-7742
http://dx.doi.org/10.1016/bs.irn.2016.07.003

1. INTRODUCTION: HISTORY

People have likely always been aware of the feelings of sleepiness and excess sleep that accompany many diseases including multiple microbial infections. However, it has only been over the past four decades that direct links between microbes and sleep have been established. This work had its origins in the laboratories of Pappenheimer and Karnovsky in the 1960s and 1970s. At the time they were seeking to isolate and identify a sleep-promoting substance, called Factor S, from cerebral spinal fluid and brains of sleep-deprived animals (Fencl, Koski, & Pappenheimer, 1971; Pappenheimer, Miller, & Goodrich, 1967). By the early 1980s, they had processed very large amounts of brain and, in separate experiments, urine with the goal of chemical identification of the somnogenic agent(s) responsible. By the early 1980s, they had characterized somnogenic muramyl peptides in both urine and brain (Krueger, Bacsik, & García-Arrarás, 1980; Krueger, Karnovsky, et al., 1984; Krueger, Pappenheimer, & Karnovsky, 1982). At the time it was recognized that the somnogenic muramyl peptides may have their origins from intestinal lumen bacterial cell wall peptidoglycan (Krueger et al., 1982; Fig. 1).

After the initial identification of somnogenic muramyl peptides, the field expanded rapidly. Muramyl peptides of bacterial origin were tested (Krueger et al., 1987), other bacterial cell wall products such as lipopolysaccharide (LPS) were shown to be somnogenic (Krueger, Kubillus, Shoham, & Davenne, 1986), multiple synthetic muramyl peptides (Krueger, Walter, et al., 1984), and LPS (Cady, Kotani, Shiba, Kusumoto, & Krueger, 1989) derivatives, developed as potential immune adjuvants, were used to define structure-somnogenic activities. Evidence, although remaining insufficient, was obtained showing muramyl peptide components of in normal mammalian tissues (Johannsen & Krueger, 1988; Krysciak, 1980; Zhai & Karnovsky, 1984). Further, the translocation across the intestinal wall of hydrophilic molecules, including muramyl peptides, was demonstrated (Pappenheimer & Zich, 1987). It was known that macrophages could digest peptidoglycan and release muramyl peptides (Vermeulen & Gray, 1984), including somnogenic muramyl peptides (Johannsen, Wecke, Obál, & Krueger, 1991), suggesting that muramyl peptides could be produced in the gut and other tissues. Further, the translocation of bacteria, including lactobacilli, was known in health and disease (Berg & Garlington, 1979). These findings led to the first quantification of sleep changes across the

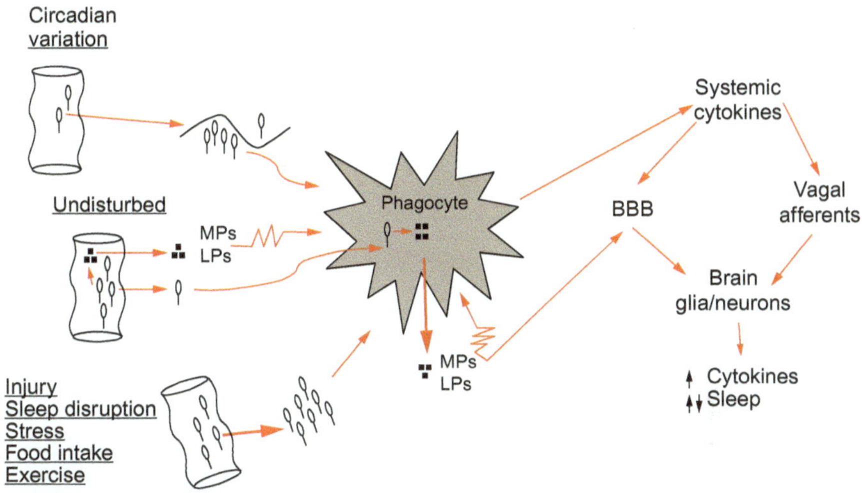

Fig. 1 Pathway for intestinal bacteria to affect sleep. From *left* to *right*: intestinal bacteria, and/or bacteria cell wall degradation products, such as muramyl peptides (MPs) or lipopolysaccharide (LPS), translocate across the intestinal epithelial barrier. Sleep loss and several conditions that affect sleep, e.g., injury, food intake, stress, circadian rhythm, and exercise, affect bacteria translocation. Bacteria are engulfed by phagocytes, such as macrophages or neutrophils, and digested; digest products (e.g., MPs, LPS) are released into the surrounding intercellular fluid. MPs and LPS in turn activate phagocytes (illustrated by the jagged cell membrane) that then release cytokines such as interleukin-1 and tumor necrosis factor. Systemic cytokines access the brain by at least two routes. Cytokines can signal the brain via vagus nerve afferents whose action potentials induce further cytokine production in the brain by glia and neurons. Cytokines can also cross the blood–brain barrier (BBB) to induce their own and other cytokine productions. Brain cytokines at low concentrations enhance sleep, while at high concentrations fragment sleep. Other microbes, e.g., viruses, and their components also enhance cytokine production via endogenous receptors that recognize pathogen-associated molecular patterns, e.g., Toll-like receptors, to affect sleep (not illustrated).

course of bacterial (Toth & Krueger, 1988) and viral (Kimura-Takeuchi, Majde, Toth, & Krueger, 1992a; Toth, Rehg, & Webster, 1995) infections. They also led to investigations of how sleep could affect bacterial translocation across the intestinal wall (Everson, 1993; Everson & Toth, 2000).

This review will focus on bacteria, although an interesting parallel story relating the influence of viruses on sleep is also developing. We will begin by describing how bacterial infections affect sleep and that sleep loss is associated with bacteremia. Then we will describe the profound prolonged effects of cecal ligation an experimental bacteremia model. We describe mechanisms by which bacteria initiate sleep responses including macrophage processing of bacterial cell walls and the subsequent products affect sleep. The effects of other microbes, including viruses, on sleep are described. We end with a summary of how the molecules derived from bacteria and viruses in turn induce cytokine production and the role of cytokines in sleep regulation in health and disease.

2. SLEEP PHYSIOLOGY

The reader is referred to Principles and Practices in Sleep Medicine (Kryger, Roth, & Dement, 2011) for an extensive discussion of sleep physiology and pathology; an abbreviated summary is presented. There are two sleep states, rapid eye movement sleep (REMS) and non-REMS (NREMS). In humans and experimental animals such as rats and mice, NREMS occupies most of the time asleep and there is an oscillation between NREMS and REMS with about a 90-min cycle in humans. In humans, most sleep occurs during the dark phase of the 24-h day. In contrast, mice and rats sleep mostly during the day. NREMS intensity is inferred from the amplitude of electroencephalogram (EEG) slow waves (SWs) (0.5–4 Hz). High-amplitude EEG SWs occur after sleep deprivation (Pappenheimer, Koski, Fencl, Karnovsky, & Krueger, 1975), and EEG SW activity is higher in humans during the initial bout of sleep at night, and in mice and rats during the initial bout of NREMS during the day. Different areas in brain are involved in NREMS and REMS, e.g., the anterior hypothalamus regulates, in part, NREMS, whereas more posterior brain stem nuclei regulate REMS. State oscillations occurring within small neuronal/glial circuits are posited to be a fundamental building block of organism sleep (Krueger & Roy, 2016); e.g., synchronization of cortical column sleep-like states emerges as organism sleep (Krueger, Huang, Rector, & Buysee, 2013; Krueger & Obál, 1993; Rector, Topchiy, Carter, & Rojas, 2005; Roy, Krueger, Rector, & Wan, 2008). The molecules involved in state regulation, whether within small circuits or the entire brain, include multiple cytokines (Churchill et al., 2008; Jewett et al., 2015), hormones (e.g., Chang & Opp, 2001; Obál, Fang, Payne, & Krueger, 1995), neurotransmitters (Hinard et al., 2012), and substances such as ATP (Krueger et al., 2010), adenosine, glutamic acid, GABA, and prostaglandins (reviewed Krueger et al., 2008; Imeri & Opp, 2009). All molecules involved in sleep regulation, including brain cytokines, show enhanced expression or release in response to cell activity, e.g., neuron action potentials, suggesting that, in brain, sleep serves a plasticity function (Krueger & Obál, 1993; Krueger & Tononi, 2011).

Mild infectious challenges or low doses of microbial products enhance duration and intensity of NREMS while simultaneously decreasing duration of REMS (Opp & Krueger, 2015). However, as severe infections progress, or after high doses of bacterial or viral components, sleep becomes fragmented often causing a reduction in duration of sleep. Microbes/microbe components promote sleep via their ability to enhance proinflammatory cytokines whether systemically or centrally. Cytokines in turn act on both individual small neuronal circuits and sleep regulatory centers to affect sleep via effector mechanisms that include NO, adenosine, and glutamate receptor trafficking (Imeri & Opp, 2009; Krueger et al., 2013). Systemic cytokines affect brain function either by stimulating vagal afferents that in turn enhance brain production of cytokines or are transported across the blood–brain barrier (BBB) (Dantzer, O'Connor, Freund, Johnson, & Kelley, 2008). Of intense interest is our hypothesis that under physiological conditions these mechanisms operate at a basal level contributing to normal sleep regulation, while the major changes in sleep occurring during pathology are an amplification of these physiological processes.

3. BACTERIAL CHALLENGE AFFECTS SLEEP

Although Hippocrates made the association between diseases and sleep, the first systematic studies of the effects of a bacterial infection on sleep were not done until about 30 years ago (Toth & Krueger, 1988). In that study, rabbits were inoculated with gram-positive bacteria, *Staphylococcus aureus*. Within a few hours after challenge, duration of NREMS increased and lasted for about 24 h. During the second day after challenge NREMS was reduced to below baseline control values. The intensity of NREMS, as determined from the amplitude of EEG delta wave (0.5–4 Hz) during NREMS, followed a similar time course, initially increasing followed by prolonged decreases. Initially average individual NREMS bout length increased, but within 20 h of challenge, NREMS bout length decreased to values below baseline. In subsequent studies of the effects on sleep of multiple bacteria and other microbes, a general feature found was that if the induced disease is severe, sleep becomes fragmented in the later stages of the disease. After *S. aureus* challenge, REMS was inhibited throughout the first 48 h after challenge.

If rabbits were inoculated with the same number of *S. aureus* that were first heat killed, there was no response to the challenge. However, if a 100-fold higher dose of heat-killed *S. aureus* was used, changes in sleep occurred, but they were attenuated. Thus, the initial increases in NREMS occurred and were more rapid in onset as occurred after viable *S. aureus* inoculations, although the subsequent inhibition of sleep was not observed. These results suggested that bacterial replication was required for the full course of the sleep responses. They also suggested that bacteria components could drive initial sleep responses; this is expanded upon below.

The time course of the fever responses to *S. aureus* challenge was distinct from the sleep response time course. Fevers were initiated during the first phase of the excessive sleep response but then persisted during the time when NREMS was lower than baseline values. These results clearly indicated that the sleep responses are independent of *S. aureus*-induced fever because temperatures were elevated during both periods of excess NREMS and during periods of NREMS inhibition.

In subsequent studies, other gram-positive (*Streptococcus pyogenes*) and gram-negative (*Escherichia coli, Pasteurella multocida*) were also tested with similar results. A notable exception was that after *E. coli* challenge the onset of NREMS was more rapid in onset and larger in magnitude, but these initial responses only lasted a few hours compared to responses induced by *S. aureus* (Krueger et al., 1994; Toth & Krueger, 1989). Heat-killed *E. coli* also elicited sleep responses, again suggesting that bacterial components could trigger sleep responses.

4. SLEEP LOSS PROMOTES INTESTINAL BACTERIAL TRANSLOCATION

Chronic sleep deprivation of rats leads to death after about 19 days (Everson, 1993). The experimental model used to reach this conclusion is called the disc over water model. Rats are placed one on each side of a rotating disc with a physical barrier separating the two rats and causing the rats to either wake up or be knocked into the water below the rotating disc. The disc rotation is turned on when the experimental rat enters sleep as determined from his EEG; the sleeping rat is then knocked into the water and wakes up. In this condition the experimental rat loses most of their sleep for days on end, while the control, yoked, rat maintains most of their sleep because the disc rotation is not timed to its sleep. Under these conditions, rats develop a hypercatabolic state with progressive enhanced food intake and body weight losses. As death approaches, the experimental rat develops a wasted appearance and hypothermia. Although extensive experimentation failed to reveal the cause of death (reviewed Rechtschaffen, Bergmann,

Everson, Kushida, & Gilliland, 1989), Everson determined that the experimental sleep-deprived rats became septicemic concluding that host defense failure was a result of sleep loss. In a subsequent study, Everson and Toth (2000) characterized live bacteria in normally sterile body tissues in sleep-deprived rats. Their major conclusion was that mesenteric lymph nodes contained viable bacteria as a result of bacterial translocation from the intestine. Although these studies suggested a relationship between sleep loss, intestinal bacterial translocation, and immune system failure, the disc over water model has been criticized. For example, the sleep-deprived rat lacks control over its environment because when the disc begins to rotate the rat is asleep, while the control rat maintains some degree of control because it knows when the disc will turn due to the inactivity of the experimental rat. Environmental control affects sleep deprivation outcomes (Oonk, Krueger, & Davis, 2016). Thus, the septicemia may result from learned helpless rather than of sleep loss per se.

Sleep is firmly linked to circadian rhythms, food intake, exercise, and stressors; these variables also affect each other compounding their actions on sleep. Feeding rhythms and disruption of circadian rhythms induce time-specific changes in intestinal bacteria (Thaiss et al., 2014; Voigt et al., 2014). Circadian clock disruption also increases permeability of the intestinal epithelial barrier (Summa et al., 2013). Similarly, exercise changes intestinal flora and the clearance of specific bacterial phyla from blood (Shukla et al., 2015; reviewed Bermon et al., 2015). Further, intestinal bacteria influence responses to stressors as they are necessary for stress-induced increases in cytokines such as interleukin-1 (IL1) (Maslanik et al., 2012). IL1, as discussed briefly later, is involved in physiological sleep regulation and in sleep responses to microbes (reviewed Imeri & Opp, 2009). Such findings could indicate a role for gut flora in sleep regulation in health and disease; however, much work is needed to verify and clarify this hypothesis.

5. CECAL LIGATION

Most experimental studies of bacterial infections and sleep have used inoculation of a single pathogen species as the infectious challenge. The gut microbiome, however, is polymicrobial and many infections result from invasion by multiple pathogen species. Such is the case in sepsis, during which polymicrobial infections are routinely the case. The preclinical sepsis model considered to be the gold standard is cecal ligation and puncture (CLP; Nemzek, Hugunin, & Opp, 2008). CLP produces a polymicrobial infection that is considered clinically relevant because of its time course, the dynamic changes in cardiac function, and because there is a progressive release of inflammatory mediators. Sleep is altered during the acute phase of CLP sepsis, which occurs from 1 to 4 days after sepsis induction (Baracchi, Ingiosi, Raymond, & Opp, 2011). During this period, NREMS and REMS of rats increase during the dark period, whereas these sleep phases are reduced during the light period. These changes in sleep coincide with increased cytokine mRNA and protein in brain (Granger, Ratti, Datta, Raymond, & Opp, 2013). Of interest, effects of sepsis on body temperature and activity rhythms of animals that survive persist long after recovery when the subject is no longer at risk of dying (Granger et al., 2013). These observations suggest that sepsis alters brain function, and are in agreement with observations that patients surviving sepsis often suffer severe and debilitating cognitive impairment. Aging and sleep status also impact BBB transport of tumor necrosis factor (TNF). BBB transport of TNF increases during sepsis in young mice, but not in aged (Opp & Krueger, 2015). Interactions among age, sleep status, and BBB function have received little attention, but are likely to be important determinants of outcomes in old and oldest old persons in response to inflammatory insult.

6. BACTERIAL COMPONENTS DRIVING SLEEP RESPONSES

Bacterial walls contain several layers (reviewed Pabst, Beranova, & Krueger, 1999). They have an inner membrane, also called the plasma membrane, which is similar in function and structure to those of other living organisms. Most bacteria also have a cell wall exterior to the plasma membrane that forms a flexible sturdy sheath surrounding the plasma membrane serving to protect the cell from osmotic shock. These cell walls are composed of peptidoglycan, a polymer that gives the bacteria their particular shape. Outside the peptidoglycan layer there is much diversity among bacterial species. For example, in gram-negative bacteria, there is an outer membrane of lipid bilayer of phospholipids and lipoproteins forming the inner leaflet and an outer leaflet composed of lipopolysaccharide (called endotoxin or LPS). Gram-positive bacteria lack an outer membrane but have other components such as teichoic acids anchored to the peptidoglycan layer.

As bacteria divide, grow, or die, the peptidoglycan, LPS, and other components are degraded or altered by bacterial enzymes. Host phagocytic cells like macrophages and neutrophils also can digest peptidoglycan producing muramyl peptides (small glycopeptides). Peptidoglycan, isolated from either gram-positive or gram-negative bacteria, induces sleep responses similar to those induced by whole dead bacteria described earlier. For example, NREMS duration and intensity is enhanced for several hours (Johannsen et al., 1990). If phagocytic mammalian cells are fed bacteria, they release small biologically active muramyl peptides (Johannsen, Wecke, et al., 1991). Some of these muramyl peptides induce sleep responses that mimic those induced by intact peptidoglycan and heat-killed whole bacteria. The somnogenic properties of muramyl peptides are dependent upon precise biochemical structure (reviewed Johannsen et al., 1989; Krueger & Johannsen, 1989). For example, β(1 → 4)-*N*-acetylglucosaminyl-β(1 → 4)-*N*-acetylmuramyl-L-alanyl-D-iso-glutaminyl-L-lysyl-D-ananyl-D-alanine is somnogenic. Somnogenic activity is retained if either the *N*-acetylglucosamine or terminal alanines are removed. Further, several naturally occurring muramyl peptides, containing diaminopimelic acid instead of lysine, are somnogenic. In fact, the sleep-promoting material isolated from human urine and rabbit brain contained diaminopimelic acid. It also contained 1,6-anhydro-*N*-acetylmuramic acid although the anhydro ring could be hydrated without loss of somnogenic activity. Muramyl peptides have multiple activities including induction of fever and are immune adjuvants. The structural requirements for fever induction or for adjuvant activity are distinct from those for sleep promotion, although some muramyl peptides possess all these activities, e.g., *N*-acetylmuramyl-L-alanyl-D-isoglutamine, also called muramyl dipeptide or MDP (Shoham & Krueger, 1988). MDP was isolated from the killed mycobacteria component of Freund's complete adjuvant; it is the component responsible for adjuvanticity.

LPS somnogenic activity–LPS structure relationships also were described although to a more limited extent. Thus, for example, lipid A is a biologically active component of LPS; both LPS and lipid A promote NREMS (Krueger et al., 1986). Somnogenic activity of synthetic lipid analogs is dependent on acylation or phosphorylation patterns and backbone structures of these molecules (Cady et al., 1989).

These findings led to the idea that specific muramyl peptides and lipid A molecules are tailor made from peptidoglycan and LPS by mammalian phagocyte enzymes to meet host innate immunity and sleep requirements in health and disease states (Krueger & Johannsen, 1989). Regardless, much investigative work is needed before acceptance of this speculation.

7. SLEEP RESPONSES TO VIRUS CHALLENGE

Viral infections are also associated with changes in sleep although the unfolding of this story is not as extensive as that described for the effects of bacterial infections. Thus, intravenous influenza virus inoculation induces large increases in rabbit sleep time as well as several other facets of the acute-phase response, e.g., fever. These responses occurred despite the inability of influenza virus to undergo complete replication in the rabbit (Kimura-Takeuchi et al., 1992a). Unlike bacteria, heat inactivation of the virus blocked the ability of the virus to induce acute-phase responses. Toth et al. (1995) extended the influenza—sleep work to mice, a species within which influenza can replicate, using a mouse-adapted human influenza virus. In this case, influenza challenge induces large increases in NREMS time in C57BL6 mice but not in Balb/c mice although both strains exhibited increases in NREMS intensity as determined from enhanced EEG delta wave (0.5–4 Hz) amplitudes. Changes in NREMS are most often accompanied by reductions in duration of REMS and hypothermia.

Influenza virus drives sleep responses via viral double-stranded (ds) RNA synthesized in lungs (Majde et al., 1991) and in the olfactory bulb after intranasal challenge in mice (Majde et al., 2007). dsRNA is made by all viruses and thus could serve as a common signal triggering the acute-phase response (Majde et al., 1991). Single-stranded poly I or poly C do not induce sleep responses, whereas if they are annealed together to form dsRNA, it is somnogenic (Kimura-Takeuchi, Majde, Toth, & Krueger, 1992b). Similarly, single-stranded 661 mer or 108 mer containing influenza sequences fail to induce acute-phase responses, whereas if the complement strand of either is annealed to form short dsRNA, the product is somnogenic (Fang, Bredow, Taishi, Majde, & Krueger, 1999). Viral dsRNA is recognized by Toll-like receptors (TLRs) 3 and 7 (Majde & Krueger, 2005) that in turn initiate a cytokine response.

Several mutant mouse models have been used to help decipher the pathways involved in virus-induced sleep responses; of great interest to sleep research are two models. Mice that lack a functional growth hormone-releasing hormone (GHRH) receptor, called *lit/lit* mice, have poor sleep responses to influenza and higher morbidity and mortality than corresponding wild-type (WT) mice (Alt et al., 2003). Similar weak sleep responses to influenza occur in mice lacking the neuron-specific IL1 receptor accessory protein (Davis et al., 2015). If WT mice are treated with anti-GHRH antibodies, it blocks the sleep responses induced by IL1 (Obál et al., 1995). Further, both GHRH and IL1 induce increases in intracellular Ca^{2+} hypothalamic neurons (De, Churchill, Obál, Simasko, & Krueger, 2002). IL1 induces GHRH receptor expression in brain (Taishi et al., 2004). These results are exciting because they link two well-known sleep regulatory substances, GHRH and IL1, to virus-induced changes in sleep. Perhaps more important, they demonstrate a brain mechanism involved in host responses to viral challenge that if disturbed leads to impaired outcomes. This mechanism is not yet fully investigated despite its potential for combating viral diseases.

Although much of the work relating viruses to sleep has used influenza virus, other viruses also affect sleep. Thus, humans infected with rhinovirus, or with influenza, exhibit altered sleep (Smith, 1992). Several disorders accompanied by excessive sleepiness and/or fatigue including mononucleosis, chronic fatigue syndrome, and sudden infant death syndrome (Guilleminault & Mondini, 1986; Hoffman, Damus, Hillman, & Krongrad, 1988; Holmes et al., 1988; Komaroff, 1988) are linked to viruses. In contrast, sleep intensity (reduced EEG SWA) and prolonged periods of little or no sleep are reported in mice with fatal experimental rabies infections (Gourmelon, Briet, Clarencom, Court, & Tsiang, 1991).

Humans seropositive for human immunodeficiency virus, but otherwise healthy, have excessive amounts of NREMS. As the disease progresses, sleep becomes disrupted (Norman, Chediak, Kiel, & Chon, 1990; Norman et al., 1988). Preclinical studies demonstrate that HIV envelope glycoproteins alter sleep (Gemma & Opp, 1999; Opp et al., 1996). The effects of HIV on sleep are likely mediated, in part, by cytokines as they upregulate cytokine expression in rat brain (Gemma, Smith, Hughes, & Opp, 2000) and in human plasma (Darko et al., 1995).

8. SLEEP RESPONSES TO OTHER MICROBES

The fungal organism, *Candida albicans*, live brewer's yeast, the protozoan *Trypanosoma brucei brucei* induce sleep responses (Kent, Price, & Satinoff, 1988; Toth & Krueger, 1989, 1995; Toth, Tolley, Broad, Blakelym, & Krueger, 1994). *T. brucei brucei*'s enhanced sleep is associated with the cyclic parasitemia occurring roughly every 3 weeks, and these increases are superimposed upon a longer-term reduction in sleep caused by the infection in rabbits. This suggests that the changes in sleep are induced by the accompanying immune response to the protozoan antigenic shift

(Toth et al., 1994). Finally, pseudomurein derived from *Methanobacterium thermoautotrophicum* is capable of inducing sleep responses in rabbits (Johannsen, Labischinski, & Krueger, 1991). Pseudomurein is a cell wall component but is chemically distinct from eubacterial peptidoglycan. Pseudomurein has been detected in the rabbit gut but does not induce any known disease.

9. MECHANISMS

There are a variety of pathogen-associated molecular patterns that are recognized by mammalian cells. The recognition receptors include peptidoglycan recognition proteins (PGRP), TLRs, and nucleotide-binding oligomerization domain receptors. Some of these have been linked to sleep (reviewed Majde & Krueger, 2005; Zielinski & Krueger, 2012). Thus, PGRP is constitutively expressed in brain and its hypothalamic and brainstem levels increase after sleep deprivation (Rehman, Taishi, Fang, Majde, & Krueger, 2001). Several TLRs are also linked to sleep (Chen et al., 2015; Hakim et al., 2014; Majde, Kapas, Bohnet, De, & Krueger, 2010; Sartorius et al., 2012; Wisor, Clegern, & Schmidt, 2011). For example, sleep responses to influenza virus are attenuated in TLR-3-deficient mice (Majde et al., 2010). Activation of these pathogen recognition receptors induces proinflammatory and antimicrobial responses by activation of many intracellular pathways (reviewed Zielinski & Krueger, 2012). These actions lead to central and systemic upregulation of many cytokines that are involved in sleep regulation, including IL1 and TNF (reviewed Besedovsky, Lange, & Born, 2012; Imeri & Opp, 2009; Krueger et al., 2008). The sleep-linked cytokines in turn act on brain sleep regulatory centers such as the hypothalamus (Alam et al., 2004; Kubota, Li, Guan, Brown, & Krueger, 2002) and brain stem nuclei (Brambilla, Barajon, Bianchi, Opp, & Imeri, 2010) to promote sleep. They also can act on local neuronal/glial circuits such as cortical columns to affect local state (Churchill et al., 2008; Jewett et al., 2015). Systemic cytokines also can alter sleep either by binding to vagal afferents that in turn upregulate brain cytokine expression (e.g., Zielinski, Dunbrasky, Taishi, Souza, & Krueger, 2013) or by crossing the BBB (reviewed Dantzer et al., 2008).

10. ARE SLEEP RESPONSES TO MICROBES ADAPTIVE?

Caring relatives have likely always advised rest and sleep for loved ones to help recuperate from diseases. Surprisingly there is relatively little evidence that sleep helps recuperate from microbial infections or other diseases. This is a consequence of the almost impossibility of isolating sleep as an independent variable. For instance, when you go to sleep, almost every physiological variable changes, e.g., body temperature, hormone levels, respiratory rate, kidney filtration rates, etc. Thus, it is not possible to know if improved morbidity or mortality rates are due to sleep per se, or to changes in one of the other physiological parameters that change with sleep. Nevertheless, some data suggest that sleep may contribute to, or facilitate, recuperation. First, there is a large and actively growing literature showing that sleep and sleep loss alter many facets of immune function (reviewed Besedovsky et al., 2012; Imeri & Opp, 2009; Majde & Krueger, 2005). There is also evidence that changes in sleep patterns during infection correlate with morbidity and mortality. For example, Toth, Tolley, and Krueger (1993) reported that during experimentally induced infections, long periods of enhanced sleep were associated with lower mortality, reduced morbidity, and less severe clinical symptoms. Conversely, animals that died, or had to be euthanized during the experimental infectious challenge, slept less and had poorer sleep quality than those that survived. These correlative data suggest that sleep may serve as an aid in recuperation.

Although little research has focused directly on the extent to which changes in sleep during microbial infection are adaptive, there is a large literature demonstrating the adaptive value of fever. A comprehensive review of the adaptive nature of fever is beyond the scope of this article, and the interested reader is referred to a classic literature on this topic (Kluger, 1979; Kluger, Kozak, Conn, Leon, & Soszynski, 1996; Kluger, Ringler, & Anver, 1975). Suffice it to say, moderate fevers are a double-edged host defense weapon in that they enhance immune function and make the environment less suitable for microbial replication. As briefly summarized, little evidence suggests that changes in sleep per se contribute to recuperative processes. However, changes in sleep during infection may be adaptive because they facilitate the generation of fever. The most efficient way to raise body temperature is to increase heat production and reduce heat loss. Endothermic animals increase heat production by shivering, yet there are sleep state-specific changes in thermoregulatory effector mechanisms such that during REM sleep shivering does not occur (Glotzbach & Heller, 1976; Parmeggiani, 2003); during infection, REM sleep is essentially abolished. It is unlikely that sleep evolved solely to support the generation of fever during infection, but the changes in sleep during microbial infection are exquisitely designed to fulfill this role (Imeri & Opp, 2009; Opp, 1999).

11. CONCLUSIONS

Thirty-five years ago it was scientific heresy when we first suggested that gut bacteria affect brain state. It should not have been given that it was known that some microbial infections induce coma and that bacteria could translocate across the gut epithelial barrier. Remarkably, a patent was issued 14 years ago entitled "Administering Bacteria to Improve Sleep" (#6.444.203B2). Regardless, in the intervening years, it has become accepted that microbes affect a variety of physiological functions and they are just part of a much larger symbiotic relationship between microbes and mammals. Microbiome analyses are accelerating the progress relating bacteria to physiological sleep, and we anticipate many important interesting findings emanating from that research endeavor.

ACKNOWLEDGMENTS

This work was supported by the National Institutes of Health Grant Numbers NS025378 and HD036520 to J.M.K. and AG041827 and AI115706 to M.R.O.

REFERENCES

Alam, M. N., McGinty, D., Bashir, T., Kumar, S., Imeri, L., Opp, M. R., et al. (2004). Interleukin-1beta modulates state-dependent discharge activity of preoptic area and basal forebrain neurons: Role in sleep regulation. *The European Journal of Neuroscience*, *20*, 207–216.

Alt, J., Obál, F., Jr., Traynor, T. R., Gard, J., Majde, J. A., & Krueger, J. M. (2003). Sleep responses to influenza viral infection in GHRH-receptor deficient mice. *Journal of Applied Physiology*, *95*, 460–468.

Baracchi, F., Ingiosi, A. M., Raymond, R. M., Jr., & Opp, M. R. (2011). Sepsis-induced alterations in sleep of rats. *The American Journal of Physiology*, *301*, R1467–R1478.

Berg, R. D., & Garlington, A. W. (1979). Translocation of certain indigenous bacteria from the gastrointestinal tract to the mesenteric lymph nodes and other organs in a gnotobiotic mouse model. *Infection and Immunity*, *23*, 403–411.

Bermon, S., Petriz, B., Kajeniene, A., Prestes, J., Castell, L., & Franco, O. L. (2015). The microbiota: An exercise immunology perspective. *Exercise Immunology Review*, *21*, 70–79.

Besedovsky, L., Lange, T., & Born, J. (2012). Sleep and immune function. *Pflügers Archiv*, *463*, 121–137.

Brambilla, D., Barajon, I., Bianchi, S., Opp, M. R., & Imeri, L. (2010). Interleukin-1 inhibits putative cholinergic neurons in vitro and REM sleep when microinjected into the rat laterodorsal tegmental nucleus. *Sleep*, *33*, 919–929.

Cady, A. B., Kotani, S., Shiba, T., Kusumoto, S., & Krueger, J. M. (1989). Somnogenic activities of synthetic lipid A. *Infection and Immunity*, *57*, 396–403.

Chang, F. C., & Opp, M. R. (2001). Corticotropin-releasing hormone (CRH) as a regulator of waking. *Neuroscience and Biobehavioral Reviews*, *25*, 445–453.

Chen, Y. C., Su, M. C., Liou, C. W., Liu, S. F., Chen, C. J., Lin, H. C., et al. (2015). Co-upregulation of Toll-like receptors 2 and 6 on peripheral blood cells in patients with obstructive sleep apnea. *Sleep & Breathing*, *19*, 873–882.

Churchill, L., Rector, D. M., Yasuda, K., Fix, C., Rojas, M. J., Yasuda, T., et al. (2008). Tumor necrosis factor α: Activity dependent expression and promotion of cortical column sleep in rats. *Neuroscience*, *156*, 71–80.

Dantzer, R., O'Connor, J. C., Freund, G. G., Johnson, R. W., & Kelley, K. W. (2008). From inflammation to sickness and depression: When the immune system subjugates the brain. *Nature Reviews. Neuroscience*, *9*, 46–56.

Darko, D. F., Miller, J. C., Gallen, C., White, J., Koziol, J., Brown, S. J., et al. (1995). Sleep electroencephalogram delta-frequency amplitude, night plasma levels of tumor necrosis factor a, and human immunodeficiency virus infection. *Proceedings of the National Academy of Sciences of the United States of America*, *92*, 12080–12084.

Davis, C. J., Dunbrasky, D., Oonk, M., Taishi, P., Opp, M. R., & Krueger, J. M. (2015). The neuron-specific interleukin-1 receptor accessory protein is required for homeostatic sleep and sleep responses to influenza viral challenge in mice. *Brain, Behavior, Immunity*, *47*, 35–43.

De, A., Churchill, L., Obál, F., Jr., Simasko, S. M., & Krueger, J. M. (2002). GHRH and IL1β increase cytoplasmic Ca^{2+} levels in cultured hypothalamic GABAergic neurons. *Brain Research*, *949*, 209–212.

Everson, C. A. (1993). Sustained sleep deprivation impairs host defense. *The American Journal of Physiology*, *34*, R1148–R1154.

Everson, C. A., & Toth, L. A. (2000). Systemic bacterial invasion induced by sleep deprivation. *The American Journal of Physiology*, *278*, R905–R916.

Fang, J., Bredow, S., Taishi, P., Majde, J. A., & Krueger, J. M. (1999). Synthetic influenza viral double-stranded RNA induces an acute phase response in rabbits. *Journal of Medical Virology*, *57*, 198–203.

Fencl, V., Koski, G., & Pappenheimer, J. R. (1971). Factors in cerebrospinal fluid from goats that affect sleep and activity in rats. *The Journal of Physiology*, *216*, 565–589.

Gemma, C., & Opp, M. R. (1999). Human immunodeficiency virus glycoproteins 160 and 41 alter sleep and brain temperature of rats. *Journal of Neuroimmunology*, *97*, 94–101.

Gemma, C., Smith, E. M., Hughes, T. K., Jr., & Opp, M. R. (2000). Human immunodeficiency virus glycoprotein 160 induces cytokine mRNA expression in the rat central nervous system. *Cellular and Molecular Neurobiology*, *20*, 419–431.

Glotzbach, S. F., & Heller, H. C. (1976). Central nervous regulation of body temperature during sleep. *Science*, *194*, 537–538.

Gourmelon, P., Briet, D., Clarencom, D., Court, L., & Tsiang, H. (1991). Sleep alterations in experimental street rabies virus infection occur in the absence of major EEG abnormalities. *Brain Research*, *554*, 159–165.

Granger, J. I., Ratti, P. L., Datta, S. C., Raymond, R. M., & Opp, M. R. (2013). Sepsis-induced morbidity in mice: Effects on body temperature, body weight, cage activity, social behavior and cytokines in brain. *Psychoneuroendocrinology*, *38*, 1047–1057.

Guilleminault, C., & Mondini, S. (1986). Mononucleosis and chronic daytime sleepiness: A long-term follow-up study. *Archives of Internal Medicine*, *146*, 1333–1335.

Hakim, F., Wang, Y., Zhang, S. X., Zheng, J., Yolcu, E. S., Carreras, A., et al. (2014). Fragmented sleep accelerates tumor growth and progression through recruitment of tumor-associated macrophages and TLR4 signaling. *Cancer Research*, *74*, 1329–1337.

Hinard, V., Mikhail, C., Pradervand, S., Curie, T., Houtkooper, R. H., Auwerx, J., et al. (2012). Key electrophysiological, molecular, and metabolic signatures of sleep and wakefulness revealed in primary cortical cultures. *The Journal of Neuroscience*, *32*, 12506–12517.

Hoffman, H. J., Damus, K., Hillman, L., & Krongrad, E. (1988). Risk factors for SIDS: Results of the National Instute of Child Health and Human Development SIDS cooperative epidemiological study. *Annals of the New York Academy of Sciences*, *533*, 13–30.
Holmes, G. P., Kaplan, J. E., Gantz, N. M., Komaroff, A. L., Schonberger, L. B., Straus, S. E., et al. (1988). Chronic fatigue syndrome: A working case definition. *Annals of Internal Medicine*, *108*, 387–389.
Imeri, L., & Opp, M. R. (2009). How (and why) the immune system makes us sleep. *Nature Reviews. Neuroscience*, *10*, 199–210.
Jewett, K. A., Taishi, P., Sengupta, P., Roy, S., Davis, C. J., & Krueger, J. M. (2015). Tumor necrosis factor enhances the sleep-like state and electrical stimulation induces a wake-like state in co-cultures of neurons and glia. *The European Journal of Neuroscience*, *42*, 2078–2090.
Johannsen, L., & Krueger, J. M. (1988). Quantitation of diaminopimelic acid in human urine. *Advances in the Biosciences*, *68*, 445–449.
Johannsen, L., Labischinski, H., & Krueger, J. M. (1991). Somnogenic activity of pseudomurein in rabbits. *Infection and Immunity*, *59*, 2502–2504.
Johannsen, L., Rosenthal, R. S., Martin, S. A., Cady, A. B., Obál, F., Jr., Guinand, M., et al. (1989). Somnogenic activity of O-acetylated and dimeric muramyl peptides. *Infection and Immunity*, *57*, 2726–2732.
Johannsen, L., Toth, L. A., Rosenthal, R. S., Opp, M. R., Obál, F., Jr., Cady, A. B., et al. (1990). Somnogenic, pyrogenic, and hematologic effects of bacterial peptidoglycan. *The American Journal of Physiology*, *259*, R182–R186.
Johannsen, L., Wecke, J., Obál, F., Jr., & Krueger, J. M. (1991). Macrophages produce somnogenic and pyrogenic muramyl peptides during the digestion of staphylococci. *The American Journal of Physiology*, *260*, R126–R133.
Kent, S., Price, M., & Satinoff, E. (1988). Fever alters characteristics of sleep in rats. *Physiology & Behavior*, *44*, 709–715.
Kimura-Takeuchi, M., Majde, J. A., Toth, L. A., & Krueger, J. M. (1992a). Influenza virus-induced changes in rabbit sleep and acute phase responses. *The American Journal of Physiology*, *263*, R1115–R1121.
Kimura-Takeuchi, M., Majde, J. A., Toth, L. A., & Krueger, J. M. (1992b). The role of double-stranded RNA in the induction of the acute-phase response in an abortive influenza virus infection model. *The Journal of Infectious Diseases*, *166*, 1266–1275.
Kluger, M. J. (1979). *Fever: It's biology, evolution and function*. Princeton: Princeton University Press.
Kluger, M. J., Kozak, W., Conn, C. A., Leon, L. R., & Soszynski, D. (1996). The adaptive value of fever. *Infectious Disease Clinics of North America*, *10*, 1–21.
Kluger, M. J., Ringler, D. H., & Anver, M. R. (1975). Fever and survival. *Science*, *188*, 166–168.
Komaroff, A. L. (1988). Chronic fatigue syndromes: Relationships to chronic viral infections. *Journal of Virological Methods*, *21*, 3–10.
Krueger, J. M., Bacsik, J., & García-Arrarás, J. (1980). Sleep-promoting material from human urine and its relation to factor S from brain. *The American Journal of Physiology*, *238*, E116–E123.
Krueger, J. M., Huang, Y., Rector, D. M., & Buysee, D. J. (2013). Sleep: A synchrony of cell activity-driven small network states. *The European Journal of Neuroscience*, *38*, 2199–2209.
Krueger, J. M., & Johannsen, L. (1989). Bacterial products, cytokines and sleep. *The Journal of Rheumatology. Supplement*, *16*, 52–57.
Krueger, J. M., Karnovsky, M. L., Martin, S. L., Pappenheimer, J. R., Walter, J., & Biemann, K. (1984). Peptidoglycans as promoters of slow-wave sleep. II. Somnogenic and pyrogenic activities of some naturally occurring muramyl peptides; correlations with

mass spectrometric structure determination. *The Journal of Biological Chemistry, 259*, 12659–12662.
Krueger, J. M., Kubillus, S., Shoham, S., & Davenne, D. (1986). Enhancement of slow-wave sleep by endotoxin and lipid A. *The American Journal of Physiology, 251*, R591–R597.
Krueger, J. M., & Obál, F., Jr. (1993). A neuronal group theory of sleep function. *Journal of Sleep Research, 2*, 63–69.
Krueger, J. M., Pappenheimer, J. R., & Karnovsky, M. L. (1982). The composition of sleep-promoting factor isolated from human urine. *The Journal of Biological Chemistry, 257*, 1664–1669.
Krueger, J. M., Rector, D. M., Roy, S., Van Dongen, H. P., Belenky, G., & Panksepp, J. (2008). Sleep as a fundamental property of neuronal assemblies. *Nature Reviews. Neuroscience, 9*, 910–919.
Krueger, J. M., Rosenthal, R. S., Martin, S. A., Walter, J., Davenne, D., Shoham, S., et al. (1987). Bacterial peptidoglycans as modulators of sleep. I. Anhydro forms of muramyl peptides enhance somnogenic potency. *Brain Research, 403*, 249–257.
Krueger, J. M., & Roy, S. (2016). Sleep's kernel. *The Scientist*, 36–41. March 1, 2016 issue.
Krueger, J. M., Taishi, P., De, A., Davis, C., Winters, B. D., Clinton, J., et al. (2010). ATP and the purine type 2 X7 receptor affect sleep. *Journal of Applied Physiology, 109*, 1318–1327.
Krueger, J. M., & Tononi, G. (2011). Local use-dependent sleep; synthesis of the new paradigm. *Current Topics in Medicinal Chemistry, 11*, 2490–2492.
Krueger, J. M., Toth, L. A., Floyd, R., Fang, J., Kapás, L., Bredow, S., et al. (1994). Sleep, microbes and cytokines. *Neuroimmunomodulation, 1*, 100–109.
Krueger, J. M., Walter, J., Karnovsky, M., Chedid, L., Choay, J. P., Lefrancier, P., et al. (1984). Muramyl peptides: Variation of somnogenic activity with structure. *The Journal of Experimental Medicine, 159*, 68–76.
Kryger, M. H., Roth, T., & Dement, W. C. (2011). *Principles and practice of sleep medicine* (5th ed.). Philadelphia, PA: Elsevier.
Krysciak, J. (1980). Diaminopimelate in mammalian urine. *Folia Biologica, 28*, 47–51.
Kubota, T., Li, N., Guan, Z., Brown, R., & Krueger, J. M. (2002). Intrapreoptic microinjection of TNFα enhances non-REM sleep in rats. *Brain Research, 932*, 37–44.
Majde, J. A., Bohnet, S. G., Ellis, G. A., Churchill, L., Leyva-Grado, V., Wu, M., et al. (2007). Detection of a mouse-adapted human influenza virus in the olfactory bulb of mice within hours after intranasal infection. *Journal of Neurovirology, 13*, 399–409.
Majde, J. A., Brown, R. K., Jones, M. W., Dieffenbach, C. W., Maitra, N., Krueger, J. M., et al. (1991). Detection of toxic viral-associated double-stranded RNA (dsRNA) in influenza-infected lung. *Microbial Pathogenesis, 10*, 105–115.
Majde, J. A., Kapas, L., Bohnet, S. G., De, A., & Krueger, J. M. (2010). Attenuation of the influenza virus sickness behavior in mice deficient in Toll-like receptor 3. *Brain, Behavior, and Immunity, 24*, 306–315.
Majde, J. A., & Krueger, J. M. (2005). Links between the innate immune system and sleep. *The Journal of Allergy and Clinical Immunology, 116*, 1188–1198.
Maslanik, T., Tannura, K., Mahaffey, L., Loughridge, A. B., Benninson, L., Ursell, L., et al. (2012). Commensal bacteria and MMAPs are necessary for stress-induced increases in IL-1β and IL-18 but not IL-6, IL-10 or MCP-1. *PLoS One*, 7, e50636.
Nemzek, J. A., Hugunin, K. M., & Opp, M. R. (2008). Modeling sepsis in the laboratory: Merging sound science with animal well-being. *Comparative Medicine, 58*, 120–128.
Norman, S. E., Chediak, H. D., Kiel, M., & Chon, M. A. (1990). Sleep disturbances in HIV-infected homosexual men. *AIDS, 4*, 775–781.
Norman, S. E., Resnik, L., Cohn, M. A., Duara, R., Herbst, J., & Berger, J. R. (1988). Sleep disturbances in HIV-seropositive patients. *JAMA, 260*, 922.

Obál, F., Jr., Fang, J., Payne, L. C., & Krueger, J. M. (1995). Growth hormone-releasing hormone (GHRH) mediates the sleep promoting activity of interleukin-1 (IL1) in rats. *Neuroendocrinology*, *61*, 559–565.

Oonk, M., Krueger, J. M., & Davis, C. J. (2016). Voluntary sleep loss in rats. *Sleep*, *39*, 1467–1479.

Opp, M. R. (1999). Fever, body temperature, and levels of arousal. In R. Lydic & H. A. Baghdoyan (Eds.), *Handbook of behavioral state control: Cellular and molecular mechanisms* (pp. 623–640). Boca Raton: CRC Press.

Opp, M. R., & Krueger, J. M. (2015). Sleep and host defense. In M. Kryger (Ed.), *Principals and practice of sleep medicine* (6th ed.). Philadelphia, PA: Elsevier Science.

Opp, M. R., Rady, P. L., Hughes, T. K., Jr., Cadet, P., Tyring, S. K., & Smith, E. M. (1996). Human immunodeficiency virus envelope glycoprotein 120 alters sleep and induces cytokine mRNA expression in rats. *The American Journal of Physiology*, *270*, R963–R970.

Pabst, M. J., Beranova, S., & Krueger, J. M. (1999). A review of the effects of muramyl peptides on macrophages, monokines and sleep. *Neuroimmunomodulation*, *6*, 261–283.

Pappenheimer, J. R., Koski, G., Fencl, V., Karnovsky, M. L., & Krueger, J. M. (1975). Extraction of sleep-promoting factor S from cerebrospinal fluid and from brains of sleep-deprived animals. *Journal of Neurophysiology*, *38*, 1299–1311.

Pappenheimer, J. R., Miller, T. B., & Goodrich, C. A. (1967). Sleep promoting effects of cerebrospinal fluid from sleep-deprived goats. *Proceedings of the National Academy of Sciences of the United States of America*, *58*, 513–517.

Pappenheimer, J. R., & Zich, K. E. (1987). Absorption of hydrophilic solutes from the small intestine. *The Journal of Physiology*, *371*, 138P.

Parmeggiani, P. L. (2003). Thermoregulation and sleep. *Frontiers in Bioscience*, *8*, s557–s567.

Rechtschaffen, A., Bergmann, B. M., Everson, C. A., Kushida, C. A., & Gilliland, M. A. (1989). Sleep deprivation in the rat: X. Integration and discussion of the findings. *Sleep*, *12*, 68–87.

Rector, D. M., Topchiy, I. A., Carter, K. M., & Rojas, M. J. (2005). Local functional state differences between rat cortical columns. *Brain Research*, *1047*, 45–55.

Rehman, A., Taishi, P., Fang, J., Majde, J. A., & Krueger, J. M. (2001). The cloning of a rat peptidoglycan recognition protein (PGRP) and its induction in brain by sleep deprivation. *Cytokine*, *13*, 8–17.

Roy, S., Krueger, J. M., Rector, D. M., & Wan, Y. (2008). Network models for activity-dependent sleep regulation. *Journal of Theoretical Biology*, *253*, 462–468.

Sartorius, T., Lutz, S. Z., Hoene, M., Waak, J., Peter, A., Weigert, C., et al. (2012). Toll-like receptors 2 and 4 impair insulin-mediated brain activity by interleukin-6 and osteopontin and alter sleep architecture. *The FASEB Journal*, *26*, 1799–1809.

Shoham, S., & Krueger, J. M. (1988). Muramyl dipeptide-induced sleep and fever: Effects of ambient temperature and time of injections. *The American Journal of Physiology*, *255*, R157–R165.

Shukla, S. K., Cook, D., Meyer, J., Vernon, S. D., Le, T., Clevidence, D., et al. (2015). Changes in gut and plasma microbiome following exercise challenge in myalgic encephalomyelitis/chronic fatigue syndrome (ME/CFS). *PLoS One*, *10*, e0145453.

Smith, A. (1992). Sleep, colds, and performance. In R. J. Broughton & R. D. Ogilvie (Eds.), *Sleep, arousal and performance* (pp. 233–242). Boston: Birkhauser.

Summa, K. C., Voight, R. M., Forsyth, C. B., Shaikh, M., Cavanaugh, K., Tang, Y., et al. (2013). Disruption of the circadian clock in mice increases intestinal permeability and promotes alcohol-induced hepatic pathology and inflammation. *PLoS One*, *8*, e67102.

Taishi, P., De, A., Alt, J., Gardi, J., Obál, F., Jr., & Krueger, J. M. (2004). Interleukin-1β stimulates GHRH receptor mRNA expression in the rat hypothalamus *in vitro* and *in vivo*. *Journal of Neuroendocrinology*, *16*, 113–118.

Thaiss, C. A., Zeevi, D., Levy, M., Zilberman-Schapira, G., Zuez, J., Tengeler, A. C., et al. (2014). Transkingdom control of microbiota diurnal oscillations promotes metabolic homeostasis. *Cell*, *159*, 514–529.

Toth, L. A., & Krueger, J. M. (1988). Alteration of sleep in rabbits by *Staphylococcus aureus* infection. *Infection and Immunity*, *56*, 1785–1791.

Toth, L. A., & Krueger, J. M. (1989). Effects of microbial challenge on sleep in rabbits. *The FASEB Journal*, *3*, 2062–2066.

Toth, L. A., & Krueger, J. M. (1995). Lighting conditions alter *Candida albicans*-induced sleep responses. *The American Journal of Physiology*, *269*, R1441–R1447.

Toth, l. A., Rehg, J. E., & Webster, R. G. (1995). Strain differences in sleep and other pathophysiological sequelae of influenza virus infection in naïve and immunized mice. *Journal of Neuroimmunology*, *58*, 89–99.

Toth, L. A., Tolley, E. A., Broad, Y. R., Blakelym, B., & Krueger, J. M. (1994). Sleep during experimental trypanosomiasis in rabbits. *Proceedings of the Society for Experimental Biology and Medicine*, *205*, 174–181.

Toth, L. A., Tolley, E. A., & Krueger, J. M. (1993). Sleep as a prognostic indicator during infectious disease in rabbits. *Proceedings of the Society for Experimental Biology and Medicine*, *203*, 179–192.

Vermeulen, M. W., & Gray, G. R. (1984). Processing of Bacillus subtilis peptidoglycan by a mouse macrophage cell line. *Infection and Immunity*, *46*, 476–483.

Voigt, R. M., Forsyth, C. B., Green, S. J., Mutlu, E., Engen, P., Vitaterna, M. H., et al. (2014). Circadian disorganization alters intestinal microbiota. *PLoS One*, *9*, e97500.

Wisor, J. P., Clegern, W. C., & Schmidt, M. A. (2011). Toll-like receptor 4 is a regulator of monocyte and electroencephalographic responses to sleep loss. *Sleep*, *34*, 1335–1345.

Zhai, S., & Karnovsky, M. L. (1984). Qualitative detection of muramic acid in normal mammalian tissues. *Infection and Immunity*, *43*, 937–941.

Zielinski, M. R., Dunbrasky, D. L., Taishi, P., Souza, G., & Krueger, J. M. (2013). Vagotomy attenuates brain cytokines and sleep induced by peripherally administered tumor necrosis factor alpha and lipopolysaccharide in mice. *Sleep*, *36*, 1227–1238.

Zielinski, M. R., & Krueger, J. M. (2012). Inflammation and sleep. In T. J. Barkoukis, J. K. Matheson, R. Ferber, & K. Doghramji (Eds.), *Therapy in sleep medicine* (pp. 607–616). Philadelphia, PA: Elsevier.

CHAPTER ELEVEN

Cognitive Function and the Microbiome

M.G. Gareau[1]
School of Veterinary Medicine, University of California Davis, Davis, CA, United States
[1]Corresponding author: e-mail address: mgareau@ucdavis.edu

Contents

Abstract

There is increasing evidence that the composition of the resident bacteria within the gastrointestinal tract can influence the brain and behavior, particularly with respect to cognitive function. Cognitive function encompasses the life-long process of learning, both long- and short-term processes. Cognition was originally thought to be exclusively regulated by the central nervous system, with long-term potentiation and neurogenesis contributing to the creation and storage of memories, but now other systems, including, for example, the immune system and the intestinal microbiome may also be involved. Cognitive impairment has been identified in numerous disease states, both gastrointestinal and extraintestinal in nature, many of which have also been characterized as having a role for dysbiosis in disease pathogenesis. This includes, but is not limited to, inflammatory bowel diseases, irritable bowel syndrome, type 1 diabetes, obesity, major depressive disorder, and autism spectrum disorder. The role of cognition and the microbiome will be discussed in this chapter for all these diseases, as well as evidence for a role in maintaining overall human health and well being. Finally, evidence for a role for probiotics in beneficially modulating the microbiota and leading to improved cognition will be discussed.

International Review of Neurobiology, Volume 131
ISSN 0074-7742
http://dx.doi.org/10.1016/bs.irn.2016.08.001

1. INTRODUCTION

Cognitive function encompasses the life-long process of learning, ranging from quantitative reasoning to memory formation—both long- and short-term processes. Cognition was originally thought to be exclusively regulated by the central nervous system (CNS), with long-term potentiation and neurogenesis contributing to the creation and storage of memories. Increasingly, it is becoming clear that other organ systems and processes including the immune system and more recently the resident bacteria of the gastrointestinal tract regulate how we form, process, and store memories, collectively forming cognitive function. Evidence for cognitive deficits have now been identified in numerous intestinal and extraintestinal diseases, highlighting the importance of characterizing these deficits and the mechanism by which they occur in patients. This chapter will explore the emerging evidence supporting how the intestinal microbiota can modulate cognitive function in both health and disease (Fig. 1).

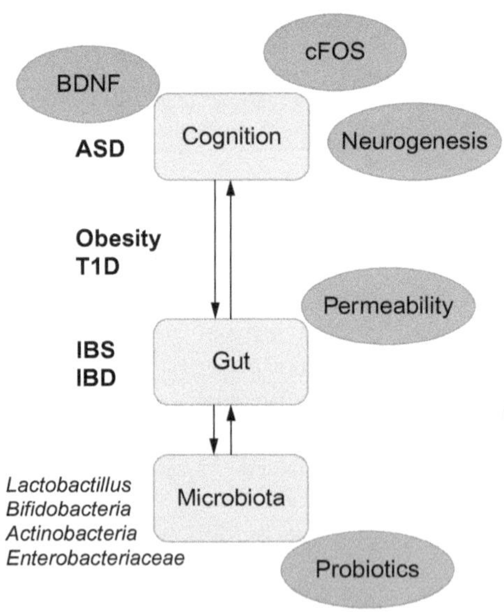

Fig. 1 Chapter summary. In this chapter, we will discuss how the microbiota–gut–brain axis affects cognitive function in health and disease, including irritable bowel syndrome (IBS), inflammatory bowel disease (IBD), autism spectrum disorder (ASD), obesity, and type 1 diabetes (T1D). Evidence exists that probiotics can beneficially modulate the microbiota, leading to improved gut physiology and cognitive function. Studies have identified changes in neurogenesis, expression of cFOS, and brain-derived neurotropic factor (BDNF) as possible mechanisms of communication between the microbiota and the brain. We will discuss these in detail.

2. DEVELOPMENT OF THE MICROBIOTA–GUT–BRAIN AXIS

Early life is an important period of rapid development of the host. Immediately following birth, the host is rapidly colonized with microbes, with changes in the composition of the microbiota occurring within the first few weeks of life and the establishment of unique site-specific microbial niches (Dominguez-Bello et al., 2016). It is a generally accepted concept that the composition of the microbiota is very plastic early in life and begins to mature with ongoing development, with the first major change occurring with the start of cessation of breast feeding in mammals (Backhed et al., 2015). This event triggers compositional changes in the microbiota population with the diversity and number of species beginning to resemble the microbial community in adults by 2–3 years of age in humans (Kostic et al., 2015).

During this colonization of the microbiota, the gastrointestinal tract is rapidly developing and maturing. Mucosal barrier function, including ion transport and macromolecular permeability in the colon, is greatly enhanced with the onset of weaning in rats (Gareau, Jury, Yang, MacQueen, & Perdue, 2006). Intestinal development can be greatly affected by exposure to early life stress, with neonatal maternal separation in rats delaying tightening of the barrier leading to increased macromolecular permeability and increasing baseline secretory state (Gareau et al., 2006). This defect in colonic mucosal barrier function is regulated in part by altered cholinergic signaling, with blocking of both muscarinic and nicotinic receptors in vitro preventing the increased macromolecular permeability associated with early life stress (Gareau, Jury, & Perdue, 2007). In addition, blocking the HPA-axis by daily administration of a corticotropin-releasing factor receptor antagonist in vivo could also prevent the early life stress-induced changes in mucosal barrier defects (Gareau et al., 2006). Whether cholinergic signaling is working in tandem with the HPA-axis in regulating intestinal barrier function remains to be determined. This maturation of the intestinal mucosal barrier is coordinated with the onset of consumption of complex foods, after cessation of maternal milk as the sole nutrient source. Interfering with maturation of intestinal physiology by exposure to early life stress could have downstream consequences.

The limbic system is important for regulating complex emotions and storing of memories, and while development is initiated early in utero, this process continues in postnatal life. Hippocampal volume, for example, continues to grow rapidly until age 2 in humans, after which growth decreases substantially (Utsunomiya, Takano, Okazaki, & Mitsudome, 1999). Neurogenesis, i.e., the formation of new neurons, is a process that occurs throughout life in specific regions of the brain, including the hippocampus. The level of neurogenesis and persistence of memories are causally related, with a reciprocal relationship between plasticity, and the ability to incorporate new information with stability, or ensuring that the incorporation of new information does not degrade stored information (Zhao, Deng, & Gage, 2008). Reducing neurogenesis during early postnatal development can increase the persistence of hippocampus-dependent memories; however, during adulthood, when neurogenesis levels are low, memories are more resistant to remodeling (Akers et al., 2014). Microglia are important in modulating neurogenesis, as they can phagocytose neural precursor cells (NPC) in the proliferative subventricular zone of the cortex, limiting the production of neurons (Cunningham, Martinez-Cerdeno, & Noctor, 2013). Therefore, activated microglia as expected to be present following an inflammatory insult, could decrease neurogenesis, leading to changes in neural development and consequently detrimental behavioral and cognitive outcomes. Microglial activation following neonatal lipopolysaccharide (LPS) administration is associated with learning impairments in adulthood (Williamson, Sholar, Mistry, Smith, & Bilbo, 2011). Therefore, changes in levels of neurogenesis during early life, for example, due to peripheral or neural inflammation, may have long-term impacts on cognitive function and memory formation.

A potential role for the microbiota in modulating neurogenesis is emerging. Evidence supporting this phenomenon was demonstrated recently in germ-free mice. Mice maintained under axenic or completely germ-free conditions can provide important information on the role of the microbiota by studying what occurs in its absence, even though humans are never in a germ-free state. Adult germ-free mice displayed increased neurogenesis compared to colonized controls, as measured by incorporation of BrdU in the adult hippocampus (Ogbonnaya et al., 2015). Altered neurogenesis could not be restored by postweaning recolonization (Ogbonnaya et al., 2015), suggesting perhaps that the microbial mechanisms for regulation of neurogenesis need to be established during early life, concurrent with establishment of the microbiota and/or development of intestinal physiology. Germ-free mice were also used to identify that the microbiota is required for the normal gross morphology and ultrastructure of the amygdala and hippocampus, as demonstrated by increased volume and altered dentritic morphology (Luczynski et al., 2016). Further, supporting a role for the microbiota in regulating neurogenesis can be assumed based on the finding that stress-induced decreases in hippocampal neurogenesis can be prevented by pretreatment with a probiotic combination of *Lactobacillus helveticus* strain R0052 and *Bifidobacterium longum* strain R0175 (Ait-Belgnaoui et al., 2014). Hippocampal neurogenesis in adults, as demonstrated in mice, is also controlled by tonic and stimuli-evoked toll-like receptor (TLR) signaling, suggesting a role for microbes or microbial components in regulating neurogenesis. While TLR2 deficiency reduced both neurogenesis and hippocampal volume, TLR4 KO mice exhibited increased neurogenesis (Rolls et al., 2007) and shaped spatial reference memory and fear learning (Okun et al., 2012). Mechanistically it appears as though binding of TLR4 on NPC by LPS inhibits proliferation and neural differentiation via MyD88 and PKC α/β-dependent NF-κB signaling (Rolls et al., 2007). While these experiments demonstrated a cell intrinsic effect of TLR agonists on NPC, and consequently physiology, treatment of NPC cultures during differentiation with the inflammatory cytokine TNFα reduced neurogenesis and increased astrocyte production (Keohane, Ryan, Maloney, Sullivan, & Nolan, 2010). These studies highlight the role of bacteria and bacterial products, and their interaction with cognate receptors within the CNS, on neural development and their potential impact on regulation of cognitive function.

Taken together, the simultaneous development of the microbiota, the maturation of the gastrointestinal tract and hippocampal neurogenesis during early postnatal life together forming the microbiota–gut–brain axis highlights the impact that dysbiosis, for example, may have on cognitive function. While these studies are still early in their respective fields, they point to a highly promising interaction that may shape postnatal development. Identifying the mechanisms and pathways through which this complex interaction is mediated may have significant therapeutic implications for numerous diseases.

3. COGNITION IN GASTROENTEROLOGY

Many of these initial gut–brain studies began with the observation that patients suffering from gastrointestinal diseases have a high prevalence of concomitant psychiatric comorbidities, including anxiety and depression. Patients with inflammatory bowel disease (IBD) and irritable bowel syndrome (IBS) often suffer from mood disorders and cognitive deficits in addition to their classical gastrointestinal symptoms. This association has a significantly detrimental impact on quality of life and complicate therapeutic treatment strategies. These patients often require multiple pharmacological modalities and an increased rate of patient noncompliance. The association between psychological comorbidity and gastrointestinal disease appears to be bidirectional in nature; however, the pathobiology underlying this interaction remains poorly understood (Bernstein, 2016).

IBS is classified as a functional gastrointestinal disorder involving a collection of intestinal symptoms such as bloating, gas, constipation, and/or diarrhea along with a high prevalence of extraintestinal manifestations of disease, which negatively impact patient quality of life. While anxiety and depression are well characterized to occur in IBS patients at a higher rate than healthy controls, negative cognitive function has also been observed in some patient populations. Cognitive response to recurrent gastrointestinal symptoms was found to negatively impact coping mechanisms for patients, with these dysfunctional cognitive processes leading to exacerbation of gastrointestinal symptoms (Hauser, Pletikosic, & Tkalcic, 2014). This included a negative bias in their central processing pathways to visceral pain, for example (Hauser et al., 2014). In a clinical study, altered cognitive function was demonstrated using a hippocampal-mediated test for visuospatial memory as assessed using the paired associates learning, intra-extradimensional set shift and spatial working memory tests from the Cambridge Neuropsychological Test Automated Battery (CANTAB) and a computerized Stroop test (Kennedy et al., 2014). This finding was related to patient cortisol levels, which were lower in IBS, but was independent of having been diagnosed with a psychiatric comorbidity (Kennedy et al., 2014). Exposure-based cognitive behavioral therapy in a small cohort of patient significantly improved gastrointestinal symptoms, pain catastrophizing, and quality of life while also displaying a modest effect on avoidance behavior (Boersma et al., 2016). Potential mechanisms of action for these cognitive effects in patients with IBS remain largely unknown, due in large part to a lack of preclinical mouse data. Despite the existence of murine models of IBS, most common for postinfectious IBS, little in the way of cognitive data has been published using these models, highlighting a need for additional studies to identify the impacts of IBS on cognitive function.

IBDs, consisting of Crohn's disease and ulcerative colitis, involve the presence of overt inflammation of the gastrointestinal tract of unknown etiology, likely caused by a combination of genetic, microbial, and unknown environmental insults. The association of extraintestinal manifestations of

IBD, particularly pertaining to mood and cognitive function, is now increasingly appreciated. While anxiety and depression are the primary behaviors demonstrated in patients, mild changes in cognition are seen in different subpopulations of patients, particularly in pediatric and adolescent patients. Cognitive functioning, in particular mild verbal memory problems, was observed in 13–19-year-old IBD patients, particularly those in acute phase of illness (Castaneda, Tuulio-Henriksson, Aronen, Marttunen, & Kolho, 2013). Again in children (age 8–17), administration of the corticosteroid prednisone, a common first-line treatment for IBD, led to cognitive deficits when compared to patients in remission and off steroids (Mrakotsky et al., 2013). Structural MRI and diffusion tensor imaging studies in children (10–14 years old) found that Crohn's disease patients showed reduced cortical thickness in posterior regions and middle frontal gyrus, along with reduced subcortical volume, altered fractional anisotropy in limbic tracts, and poorer verbal memory and cognitive function (Mrakotsky et al., 2016). The authors also identified an association between inflammation during active disease and cortical thinning along with poorer memory and cognition (Mrakotsky et al., 2016). In a study of adult patients, no overt cognitive deficits were observed between patients with IBD and healthy controls; however, the presence of concurrent mood disorders, particularly depression, was associated with impaired performance in specific tasks (Berrill et al., 2013). Despite these supportive studies others failed to demonstrate similar correlations. For example, an observational study performed in IBD and IBS patients did not identify altered cognitive function in either patient population following a series of computerized neuropsychological performance tests that assessed a range of cognitive function, including psychomotor speed, memory, and intelligence (Berrill et al., 2013). Based on these studies, it is clear that evidence exists supporting an impact of IBD on cognitive function, even though collectively these effects are considered mild in nature. Nevertheless, taken into account with mood disorders and intestinal symptoms more studies should be undertaken to better understand this association.

In a murine model of colitis, administration of dextran–sodium–sulfate (DSS) caused transient cognitive deficits, as demonstrated by the novel object recognition (NOR) task during the height of inflammation in the presence of intestinal dysbiosis (Emge et al., 2016). Cognitive deficit following acute colonic inflammation was transient, however, with recognition memory restored along with ameliorated inflammation of the colon and weight gain (Emge et al., 2016). Cognitive deficit was seen in conjunction with deceased hippocampal cFos expression, suggesting that neuronal activation in the CA1 region was modulating this behavioral deficit. These findings supported results demonstrating the presence of anxiety-like behavior in a chronic model of DSS-induced colitis in mice, which was demonstrated to be vagally mediated (Bercik et al., 2011). Together, along with the clinical evidence in patients with IBD, these findings in mice highlight the need to characterize and define the cognitive deficits associated with colonic inflammation and identify the mechanisms through which this communication is maintained.

4. COGNITION IN EXTRAINTESTINAL MANIFESTATIONS

Obesity rates are increasing in the United States, resulting in an increase need to identify potential therapeutic targets for patients. In addition to cardiovascular and diabetes risks associated with obesity, it is increasingly appreciated that this disease is characterized by deficits in the gut–brain axis (Ochoa-Reparaz & Kasper, 2016). While it is now clear that these gut–brain deficits include altered satiety signaling, in part via dysregulation of the vagus nerve (de La Serre, de Lartigue, & Raybould, 2015), it has now been identified that cognitive function, related to speed, attention, and cognitive flexibility, is affected in obese vs nonobese individuals (Fernandez-Real et al., 2015). Type 2 diabetes, commonly seen in obese individuals, is associated with cognitive impairment, increased cognitive decline, and increased risk of dementia compared to controls. In a clinical study, diabetic, prediabetic, and control subjects underwent detailed cognitive testing and 3-T resting-state functional MRI. Participants with type 2 diabetes have altered functional brain networks as measured by fMRI, which could be identified (at a lower rate) in the prediabetic stage (van Bussel et al., 2016). Diabetic participants also displayed altered network measures, characterized by a higher normalized cluster coefficient and higher local efficiency compared with controls, with the prediabetic participants falling in between the diabetics and the controls. Lower processing speed was associated with shorter path length and higher global efficiency, indicative of cognitive decline (van Bussel et al., 2016).

Obesity is often induced in rodents by exposure to a Western-style diet high in fat and/or sugar for multiple weeks. Dramatic effects can be observed after providing mice access to a Western diet for 9 weeks, with altered spatial recognition memory, as determined using the Y-maze (Andre, Dinel, Ferreira, Laye, & Castanon, 2014). Exposure to chronic Western diet increased peripheral inflammatory responses to LPS treatment, suggesting that obesity-associated inflammatory priming may be involved in modulating cognitive deficits seen in obese mice. Another study similarly used exposure to a Western diet, in this case either high fat or high sugar, which led to changes in the composition of the microbiota, with mice on a high-sucrose diet displaying significant deficits in cognitive flexibility, working memory, and evidence of a spatial bias during training for long-term memory formation (Magnusson et al., 2015), highlighting a potential role for the microbiota in modulating cognition; this concept will be pursued in greater detail in the next section of this chapter. Finally, a Western-style diet exposure in rats resulted in poor performance in the two-way active avoidance test during both the acquisition and retrieval phases as compared to controls (Noble et al., 2014). Exercise, either by voluntary running wheel access or forced treadmill, reversed high-fat diet-induced memory impairment, and increased BDNF expression in neurons of the hippocampal CA3 region (Noble et al., 2014). Together these findings suggest that diet and the resulting obesity have a negative impact on cognitive function. Given the strong association with obesity and alterations in the composition of the microbiota, this together suggests that diet, microbiota, and cognition are closely associated.

5. MICROBIOTA AND COGNITION

Changes in the composition of the microbiota in the presence of cognitive deficits are increasingly appreciated. Studies in germ-free mice are employed to demonstrate an association between microbes and a particular phenotype, given the complete absence of microorganisms. Numerous studies have identified behavioral deficits observed in germ-free mice, including cognitive deficits (Gareau et al., 2011), suggesting a role for microbes in regulating memory and cognition. This model provides tremendous power to studies, in that complete elimination of organisms leading to a change in physiology points to that absence as the inherent cause; however, due to the multitude of systems now shown to be regulated by microbes (i.e., mucosal immune system), it is becoming increasingly difficult to tease out the cause of these changes being either directly or indirectly associated with the microorganisms.

Antibiotics are known to significantly modify the composition of the microbiota and are increasingly being used as a reproducible model to induced intestinal dysbiosis. Administration of antibiotics from weaning onward caused significant reductions in total bacterial counts and decreased the overall fecal microbiota diversity (Desbonnet et al., 2015). This dysbiosis was accompanied by cognitive deficits, demonstrated by the NOR task, and decreased hippocampal BDNF expression in adult mice (Desbonnet et al., 2015). More recently, it was demonstrated that administration of a non-absorbable antibiotic cocktail starting in adulthood in mice caused cognitive deficits, including recognition memory but not spatial memory. These deficits were found to be in part mediated by altered BDNF expression, neuropeptide y receptor, and tight junction protein expression within the hippocampus, but not other brain regions such as the prefrontal cortex or hypothalamus (Frohlich et al., 2016). Whether the effects of antibiotics on cognitive deficits are transient and can be reversed following removal of the treatment remains unknown.

Despite the abundance of evidence from preclinical animal models supporting a role for the microbiota in modulating cognitive function, very little human clinical data exists. Obesity is well known to have a strong association with dysbiosis of the intestinal microbiota (Turnbaugh et al., 2009, 2006). A recent study involving obese and nonobese individuals compared the composition of the gut microbiota by 16S pyrosequencing, brain microstructure by diffusion tensor imaging white and gray matter, MRI, and performing behavioral tests including the Trail Making Test for motor speed, attention, and cognitive flexibility. The composition of the microbiota profile was demonstrated to cluster with cognitive function, for example, with the association between the relative abundance of actinobacteria and the brain confirmed at functional level. The relative abundance of actinobacteria correlated with the Trail Making Test scores, suggesting better motor speed and attention, which correlated with the better organization of amygdalar and thalamic microstructure seen on MRI. The relative abundance of actinobacteria linked not only to MRI diffusion tensor imaging in the thalamus, hypothalamus, and amygdala as well as to cognitive test scores (Fernandez-Real et al., 2015). The authors, therefore, demonstrated that obesity status affects microbiota–brain microstructure and function (i.e., cognition). A recent study demonstrated that increased presence of circulating progenitor cells (defined as being $CD34^{+}CD45^{dim}$) in peripheral blood from overweight children were positively correlated with cognitive performance using the Woodcock-Johnson III Tests of Cognitive Abilities (Niemiro et al., 2016). While circulating progenitor cell content and cytokine secretion are associated with increased response to acute inflammatory stimuli, it is not known whether they can enter the CNS in overweight children (Niemiro et al., 2016) and directly affect cognitive function. Together, these studies highlight an association between the microbiota, inflammation, and cognition in obese individuals; however, further studies are warranted to identify mechanistic pathways for this association.

6. PROBIOTICS AND COGNITION

The use of probiotics, or beneficial microbes, to modulate health and improve disease is rapidly increasing in the general population, in part due to increased awareness. Increasing evidence points toward a role for beneficial effects of probiotics that extend beyond the modification of the composition of the microbiota, including a beneficial impact on behavior, mood, and cognition. While the pathways for mediating this benefit remain largely unknown, a combination of changes to the composition of the intestinal microbiota, improving the physiology of the gastrointestinal tract leading to beneficially modulate the physiology of the brain and subsequently behavior, including cognitive function, are currently proposed.

6.1 Gut–Brain Axis and Probiotics

One of the initial studies identifying a potential role for probiotics in modulating the microbiota–gut–brain axis came from our group where we identified a combination of *Lactobacillus* containing probiotics (*Lactobacillus rhamnosus* R0011 + *L. helveticus* R0052) prevents stress-induced deficits in recognition memory via the NOR task following infection with the enteric bacterial pathogen *Citrobacter rodentium* (Gareau et al., 2011). We found that probiotics could not only dampen the HPA-axis but they could also restore neuronal activation indicated by nuclear cFos and BDNF expression in the CA1 region of the hippocampus which were decreased following exposure to acute stress in infected mice (Gareau et al., 2011). This study suggested that probiotics could influence expression of key mediators of cognitive behavior within structures of the limbic system. Follow-up studies revealed a role for the adaptive immune system in modulating the microbiota–gut–brain axis. Using $Rag1^{-/-}$ mice, we demonstrated cognitive deficits at baseline with decreased recognition memory, which could be restored by administration of probiotics (*L. rhamnosus* + *L. helveticus*) starting at weaning and continued until adulthood (Smith et al., 2014). We speculate that this is mediated in part by $CD4^+$ T cells, which have been demonstrated to play a key role in memory formation (Brynskikh, Warren, Zhu, & Kipnis, 2008; Kipnis, Cohen, Cardon, Ziv, & Schwartz, 2004). In healthy Balb/c mice, administration of *Bifidobacterium* (*B. longum* but not *B. breve*) could improve cognitive function as measured using the NOR task and the Barnes maze compared to vehicle, highlighting a strain-specific beneficial effect (Savignac, Kiely, Dinan, & Cryan, 2014). Feeding mice a Western-style diet can also induce weight gain and cognitive deficits, which can be ameliorated by administration of *L. helveticus* R0052 (Ohland et al., 2013). In a murine model of acute DSS-induced colitis, administration of probiotics starting 1 week prior to induction of colitis could restore the deficit in recognition memory induced by colonic inflammation (Emge et al., 2016). This restoration was correlated to changes in the composition of the microbiota and expression of cFos levels in the CA1 region of the hippocampus (Emge et al., 2016). The precise mechanism through which probiotics can regulate cognitive function in health and disease remains to be fully elucidated.

6.2 Healthy Human Population Studies

The beneficial impact of probiotic administration on cognitive function in humans is also being investigated. In a recent clinical study, consumption of multispecies probiotics (*Bifidobacterium bifidum* strain W23, *Bifidobacterium lactis* strain W52, *Lactobacillus acidophilus* strain W37, *Lactobacillus brevis* strain W63, *Lactobacillus casei* W56, *Lactobacillus salivarius* strain W24, and *L. lactis* [strains W19 and W58]) improved cognitive reactivity to sad mood in a healthy adult population as determined using the revised Leiden index of depression sensitivity scale (Steenbergen, Sellaro, van Hemert, Bosch, & Colzato, 2015). In a small cohort of healthy women with no history of gastrointestinal or psychiatric symptoms randomly assigned to consume a fermented milk product with probiotics, fMRI before and after intervention revealed the ability of the probiotic to modulate brain activity in distinct brain regions involved in mediating cognitive performance (Tillisch et al., 2013). In a population of healthy adults, administration of probiotics (*L. helveticus* strain R0052 and *B. longum* strain R0175) for 30 days led to a significant improvement in overall behavioral scores, including the hospital anxiety and depression scale and coping checklist, which is a questionnaire that measures problem solving strategies when faced with an adverse event (Messaoudi et al., 2011). These studies, although small in size, suggest that consumption of probiotic-containing products may have a beneficial impact on overall cognitive function in healthy adult populations.

6.3 Type 1 Diabetes and Probiotics

In addition to intestinal diseases, probiotics are being employed for extraintestinal disease, such as type 1 diabetes (T1D). T1D mellitus is a chronic autoimmune disease, which is characterized by a loss of insulin-producing β-cells in the pancreatic islets, and associated with intestinal dysbiosis (Knip & Siljander, 2016). In rats and mice, a single-intraperitoneal administration of streptozocin, a glucosamine–nitrosourea compound derived from *Streptomyces achromogenes* that is used clinically as a chemotherapeutic agent, damages pancreatic β cells, resulting in hypoinsulinemia and hyperglycemia, mimicking T1D (Lenzen, 2008). Administration of a combination of probiotic organisms (*L. acidophilus* [American type culture collection (ATCC) strain 4356], *B. lactis* [Dutch chemical company strain 10140]), and *L. fermentum* (ATCC 9338) dissolved in drinking water starting immediately following onset of diabetes prevented the development of deficits in spatial memory seen in diabetic rats (Davari, Talaei, Alaei, & Salami, 2013). While the composition of the microbiota was not characterized in this particular study, T1D is associated with dysbiosis in human clinical studies, in particular with a decrease in alpha diversity (Kostic et al., 2015), with Bacteroidetes dominating at the phylum level and reduced butyrate-producing bacteria (Knip & Siljander, 2016). Administration of probiotics in a small cohort of infants predisposed to developing T1D reduced the risk for islet autoimmunity (Uusitalo et al., 2016). Although this pilot study did not extend to assess any behavioral measurements, this warrants further study.

6.4 Autism Spectrum Disorders and Probiotics

Autism spectrum disorders (ASD) are etiologically heterogeneous and range from mild to very severe in symptomology. While exact cause of disease is unknown, it is generally believed that a combination of genetic and environmental factors influence disease development. Many of the behavioral symptoms of ASD, including social deficits, repetitive behaviors, and impaired cognition, are increasingly associated with the presence of intestinal dysbiosis, with the severity of gastrointestinal symptoms correlating with severity of behavioral deficits (Tomova et al., 2015). Subsets of children with ASD are associated with having significant gastrointestinal symptoms and those with gastrointestinal complications, primarily constipation and unwillingness to eat, and that these gastrointestinal symptoms correlate with increased severity of behavioral symptoms in a population of preschool children and compared to typical development controls (Fulceri et al., 2016). Numerous murine models of ASD exist, and a connection with the microbiota–gut–brain axis in disease development is evident in many of them. For example, in the maternal immune activation mouse model of ASD, intestinal barrier defects and dysbiosis were both observed in the presence of autistic-like behaviors, such as anxiety, sociability, and stereotypic behaviors (Hsiao et al., 2013). In addition to the presence of these altered behaviors, mice were observed to have intestinal permeability defects and dysbiosis (Hsiao et al., 2013). Interestingly, behavior, gastrointestinal physiology, and dysbiosis were ameliorated by administration of the human commensal, probiotic-like species *Bacteroides fragilis* (Hsiao et al., 2013). The inbred BTBR T+tf/J (BTBR) strain of mice, demonstrates ASD-like behaviors associated with decreased BDNF and tyrosine kinase B protein levels measured in the hippocampal region (Scattoni, Martire, Cartocci, Ferrante, & Ricceri, 2013). A randomized-controlled trial in children with ASD will seek to determine the effects of supplementation with a probiotic mixture (Vivomixx®) in ASD children to ameliorate not only their specific GI symptoms but also on the core deficits of the disorder, on cognitive and language development, and on brain function and connectivity (Santocchi et al., 2016). Given the current literature, it seems possible that administration of probiotics could ameliorate gastrointestinal symptoms and levels of BDNF in the brain to help improve the microbiota–gut–brain axis in patients.

7. FUTURE DIRECTIONS

It is increasingly appreciated that the composition of the intestinal microbiota is important for maintaining physiology of the host—including cognitive function. Additional preclinical mouse studies are warranted to identify new mechanisms of action by which the communication between the bacteria and the host occur, to determine how we can further manipulate this signaling pathway to treat disease. A possible candidate for modulating the microbiota–gut–brain axis that warrants further study is serotonin. Serotonin receptors are present within the limbic system and drugs targeting several serotonin-receptor subtypes have shown to be involved in cognition and memory (Jenkins, Nguyen, Polglaze, & Bertrand, 2016). Serotonin is well known to impact intestinal physiology, with the intestinal enterochromaffin cell the primary source for serotonin in the body, and important in regulation of intestinal permeability (Bischoff et al., 2014). Finally, the recent identification of indigenous spore-forming bacteria from the mouse and human microbiota that can promote 5-HT biosynthesis from colonic enterochromaffin cells and impact host serotonin biosynthesis (Yano et al., 2015) suggests that serotonin may play a key role in modulating the microbiota–gut–brain axis and specifically with respect to cognitive function. The complexity of the host-microbe interaction makes these studies more challenging to design and will likely yield in multiple mechanisms working in tandem or parallel to maintain host health. It is very unlikely that a single molecule or signaling pathway is responsible for modulating the host response to bacteria, even within a single readout, such as cognitive function.

These preclinical studies need to lead the way to support well-designed clinical trials to identify the beneficial effects of probiotics on cognitive function, for example, in ASD, IBD, IBS, T1D, and in obesity with multiple end points and readouts. While it is highly likely that certain probiotic strains will work while others will not, identifying those that demonstrate a clinical benefit and identifying the mechanisms through which this benefit is being translated are key to understanding the pathways by which the microbiota affects cognitive function. Of particular interest are studies in the pediatric population, whereby cognitive function is rapidly developing and where therapeutic intervention may have the greatest impact.

8. CONCLUSIONS

Based on the increasing evidence that the composition of the intestinal microbiota can influence cognitive function, in both mouse models and more recently from support from clinical studies, the future of the field lies in identifying the mechanisms through which this microbiota–gut–brain axis interaction is mediated. The complex nature of this interaction means that the mechanism(s) will not likely be a single molecule, or pathway, but until some putative process of communication is identified, the scientific sceptics will remain.

REFERENCES

Ait-Belgnaoui, A., Colom, A., Braniste, V., Ramalho, L., Marrot, A., Cartier, C., … Tompkins, T. (2014). Probiotic gut effect prevents the chronic psychological stress-induced brain activity abnormality in mice. *Neurogastroenterology and Motility*, *26*(4), 510–520. http://dx.doi.org/10.1111/nmo.12295.

Akers, K. G., Martinez-Canabal, A., Restivo, L., Yiu, A. P., De Cristofaro, A., Hsiang, H. L., … Frankland, P. W. (2014). Hippocampal neurogenesis regulates forgetting during adulthood and infancy. *Science*, *344*(6184), 598–602. http://dx.doi.org/10.1126/science.1248903.

Andre, C., Dinel, A. L., Ferreira, G., Laye, S., & Castanon, N. (2014). Diet-induced obesity progressively alters cognition, anxiety-like behavior and lipopolysaccharide-induced depressive-like behavior: Focus on brain indoleamine 2,3-dioxygenase activation. *Brain, Behavior, and Immunity*, *41*, 10–21. http://dx.doi.org/10.1016/j.bbi.2014.03.012.

Backhed, F., Roswall, J., Peng, Y., Feng, Q., Jia, H., Kovatcheva-Datchary, P., … Wang, J. (2015). Dynamics and stabilization of the human gut microbiome during the first year of life. *Cell Host & Microbe*, *17*(6), 852. http://dx.doi.org/10.1016/j.chom.2015.05.012.

Bercik, P., Park, A. J., Sinclair, D., Khoshdel, A., Lu, J., Huang, X., … Verdu, E. F. (2011). The anxiolytic effect of Bifidobacterium longum NCC3001 involves vagal pathways for gut-brain communication. *Neurogastroenterology and Motility*, *23*(12), 1132–1139. http://dx.doi.org/10.1111/j.1365-2982.2011.01796.x.

Bernstein, C. N. (2016). Psychological stress and depression: Risk factors for IBD? *Digestive Diseases*, *34*(1–2), 58–63. http://dx.doi.org/10.1159/000442929.

Berrill, J. W., Gallacher, J., Hood, K., Green, J. T., Matthews, S. B., Campbell, A. K., & Smith, A. (2013). An observational study of cognitive function in patients with irritable bowel syndrome and inflammatory bowel disease. *Neurogastroenterology and Motility*, *25*(11). http://dx.doi.org/10.1111/nmo.12219. 918–e704.

Bischoff, S. C., Barbara, G., Buurman, W., Ockhuizen, T., Schulzke, J. D., Serino, M., … Wells, J. M. (2014). Intestinal permeability—A new target for disease prevention and therapy. *BMC Gastroenterology*, *14*, 189. http://dx.doi.org/10.1186/s12876-014-0189-7.

Boersma, K., Ljotsson, B., Edebol-Carlman, H., Schrooten, M., Linton, S. J., & Brummer, R. J. (2016). Exposure-based cognitive behavioral therapy for irritable bowel syndrome. A single-case experimental design across 13 subjects. *Cognitive Behaviour Therapy*, 1–16. http://dx.doi.org/10.1080/16506073.2016.1194455.

Brynskikh, A., Warren, T., Zhu, J., & Kipnis, J. (2008). Adaptive immunity affects learning behavior in mice. *Brain, Behavior, and Immunity*, *22*(6), 861–869. http://dx.doi.org/10.1016/j.bbi.2007.12.008. S0889-1591(07)00341-8 [pii].

Castaneda, A. E., Tuulio-Henriksson, A., Aronen, E. T., Marttunen, M., & Kolho, K. L. (2013). Cognitive functioning and depressive symptoms in adolescents with inflammatory bowel disease. *World Journal of Gastroenterology*, *19*(10), 1611–1617. http://dx.doi.org/10.3748/wjg.v19.i10.1611.

Cunningham, C. L., Martinez-Cerdeno, V., & Noctor, S. C. (2013). Microglia regulate the number of neural precursor cells in the developing cerebral cortex. *Journal of Neuroscience*, *33*(10), 4216–4233. http://dx.doi.org/10.1523/JNEUROSCI.3441-12.2013.

Davari, S., Talaei, S. A., Alaei, H., & Salami, M. (2013). Probiotics treatment improves diabetes-induced impairment of synaptic activity and cognitive function: Behavioral and electrophysiological proofs for microbiome-gut-brain axis. *Neuroscience*, *240*, 287–296. http://dx.doi.org/10.1016/j.neuroscience.2013.02.055. S0306-4522(13)00196-6 [pii].

de La Serre, C. B., de Lartigue, G., & Raybould, H. E. (2015). Chronic exposure to low dose bacterial lipopolysaccharide inhibits leptin signaling in vagal afferent neurons. *Physiology and Behavior*, *139*, 188–194. http://dx.doi.org/10.1016/j.physbeh.2014.10.032.

Desbonnet, L., Clarke, G., Traplin, A., O'Sullivan, O., Crispie, F., Moloney, R. D., … Cryan, J. F. (2015). Gut microbiota depletion from early adolescence in mice: Implications for brain and behaviour. *Brain, Behavior, and Immunity*, *48*, 165–173. http://dx.doi.org/10.1016/j.bbi.2015.04.004.

Dominguez-Bello, M. G., De Jesus-Laboy, K. M., Shen, N., Cox, L. M., Amir, A., Gonzalez, A., … Clemente, J. C. (2016). Partial restoration of the microbiota of cesarean-born infants via vaginal microbial transfer. *Nature Medicine*, *22*(3), 250–253. http://dx.doi.org/10.1038/nm.4039.

Emge, J. R., Huynh, K., Miller, E. N., Kaur, M., Reardon, C., Barrett, K. E., & Gareau, M. G. (2016). Modulation of the microbiota-gut-brain axis by probiotics in a murine model of inflammatory bowel disease. *American Journal of Physiology. Gastrointestinal and Liver Physiology, 310*(11), G989–G998. http://dx.doi.org/10.1152/ajpgi.00086.2016.

Fernandez-Real, J. M., Serino, M., Blasco, G., Puig, J., Daunis-i-Estadella, J., Ricart, W., … Portero-Otin, M. (2015). Gut microbiota interacts with brain microstructure and function. *Journal of Clinical Endocrinology and Metabolism, 100*(12), 4505–4513. http://dx.doi.org/10.1210/jc.2015-3076.

Frohlich, E. E., Farzi, A., Mayerhofer, R., Reichmann, F., Jacan, A., Wagner, B., … Holzer, P. (2016). Cognitive impairment by antibiotic-induced gut dysbiosis: Analysis of gut microbiota-brain communication. *Brain, Behavior, and Immunity, 56*, 140–155. http://dx.doi.org/10.1016/j.bbi.2016.02.020.

Fulceri, F., Morelli, M., Santocchi, E., Cena, H., Del Bianco, T., Narzisi, A., … Muratori, F. (2016). Gastrointestinal symptoms and behavioral problems in preschoolers with autism spectrum disorder. *Digestive and Liver Disease, 48*(3), 248–254. http://dx.doi.org/10.1016/j.dld.2015.11.026.

Gareau, M. G., Jury, J., & Perdue, M. H. (2007). Neonatal maternal separation of rat pups results in abnormal cholinergic regulation of epithelial permeability. *American Journal of Physiology. Gastrointestinal and Liver Physiology, 293*(1), G198–G203. http://dx.doi.org/10.1152/ajpgi.00392.2006. 00392.2006 [pii].

Gareau, M. G., Jury, J., Yang, P. C., MacQueen, G., & Perdue, M. H. (2006). Neonatal maternal separation causes colonic dysfunction in rat pups including impaired host resistance. *Pediatric Research, 59*(1), 83–88. http://dx.doi.org/10.1203/01.pdr.0000190577.62426.45. 01.pdr.0000190577.62426.45 [pii].

Gareau, M. G., Wine, E., Rodrigues, D. M., Cho, J. H., Whary, M. T., Philpott, D. J., … Sherman, P. M. (2011). Bacterial infection causes stress-induced memory dysfunction in mice. *Gut, 60*(3), 307–317. http://dx.doi.org/10.1136/gut.2009.202515. gut.2009.202515 [pii].

Hauser, G., Pletikosic, S., & Tkalcic, M. (2014). Cognitive behavioral approach to understanding irritable bowel syndrome. *World Journal of Gastroenterology, 20*(22), 6744–6758. http://dx.doi.org/10.3748/wjg.v20.i22.6744.

Hsiao, E. Y., McBride, S. W., Hsien, S., Sharon, G., Hyde, E. R., McCue, T., … Mazmanian, S. K. (2013). Microbiota modulate behavioral and physiological abnormalities associated with neurodevelopmental disorders. *Cell, 155*(7), 1451–1463. http://dx.doi.org/10.1016/j.cell.2013.11.024.

Jenkins, T. A., Nguyen, J. C., Polglaze, K. E., & Bertrand, P. P. (2016). Influence of tryptophan and serotonin on mood and cognition with a possible role of the gut-brain axis. *Nutrients, 8*(1), pii: E56. http://dx.doi.org/10.3390/nu8010056.

Kennedy, P. J., Clarke, G., O'Neill, A., Groeger, J. A., Quigley, E. M., Shanahan, F., … Dinan, T. G. (2014). Cognitive performance in irritable bowel syndrome: Evidence of a stress-related impairment in visuospatial memory. *Psychological Medicine, 44*(7), 1553–1566. http://dx.doi.org/10.1017/S0033291713002171.

Keohane, A., Ryan, S., Maloney, E., Sullivan, A. M., & Nolan, Y. M. (2010). Tumour necrosis factor-alpha impairs neuronal differentiation but not proliferation of hippocampal neural precursor cells: Role of Hes1. *Molecular and Cellular Neurosciences, 43*(1), 127–135. http://dx.doi.org/10.1016/j.mcn.2009.10.003.

Kipnis, J., Cohen, H., Cardon, M., Ziv, Y., & Schwartz, M. (2004). T cell deficiency leads to cognitive dysfunction: Implications for therapeutic vaccination for schizophrenia and other psychiatric conditions. *Proceedings of the National Academy of Sciences of the United States of America, 101*(21), 8180–8185. http://dx.doi.org/10.1073/pnas.0402268101.0402268101 [pii].

Knip, M., & Siljander, H. (2016). The role of the intestinal microbiota in type 1 diabetes mellitus. *Nature Reviews Endocrinology, 12*(3), 154–167. http://dx.doi.org/10.1038/nrendo.2015.218.

Kostic, A. D., Gevers, D., Siljander, H., Vatanen, T., Hyotylainen, T., Hamalainen, A. M., … Xavier, R. J. (2015). The dynamics of the human infant gut microbiome in development and in progression toward type 1 diabetes. *Cell Host & Microbe, 17*(2), 260–273. http://dx.doi.org/10.1016/j.chom.2015.01.001.

Lenzen, S. (2008). The mechanisms of alloxan- and streptozotocin-induced diabetes. *Diabetologia, 51*(2), 216–226. http://dx.doi.org/10.1007/s00125-007-0886-7.

Luczynski, P., Whelan, S.O., O'Sullivan, C., Clarke, G., Shanahan, F., Dinan, T.G., & Cryan, J.F. (2016). Adult microbiota-deficient mice have distinct dendritic morphological changes: Differential effects in the amygdala and hippocampus. *European Journal of Neuroscience* (in press). http://dx.doi.org/10.1111/ejn.13291.

Magnusson, K. R., Hauck, L., Jeffrey, B. M., Elias, V., Humphrey, A., Nath, R., … Bermudez, L. E. (2015). Relationships between diet-related changes in the gut microbiome and cognitive flexibility. *Neuroscience, 300*, 128–140. http://dx.doi.org/10.1016/j.neuroscience.2015.05.016.

Messaoudi, M., Lalonde, R., Violle, N., Javelot, H., Desor, D., Nejdi, A., … Cazaubiel, J. M. (2011). Assessment of psychotropic-like properties of a probiotic formulation (Lactobacillus helveticus R0052 and Bifidobacterium longum R0175) in rats and human subjects. *British Journal of Nutrition, 105*(5), 755–764. http://dx.doi.org/10.1017/S0007114510004319. S0007114510004319 [pii].

Mrakotsky, C., Anand, R., Watson, C., Vu, C., Matos, A., Friel, S., … Snapper, S. (2016). O-018 New evidence for structural brain differences in pediatric Crohn's disease: Impact of underlying disease factors. *Inflammatory Bowel Diseases, 22*(Suppl. 1), S6–S7. http://dx.doi.org/10.1097/01.MIB.0000480115.50061.db.

Mrakotsky, C., Forbes, P. W., Bernstein, J. H., Grand, R. J., Bousvaros, A., Szigethy, E., & Waber, D. P. (2013). Acute cognitive and behavioral effects of systemic corticosteroids in children treated for inflammatory bowel disease. *Journal of the International Neuropsychological Society, 19*(1), 96–109. http://dx.doi.org/10.1017/S1355617712001014. S1355617712001014 [pii].

Niemiro, G.M., Raine, L.B., Khan, N.A., Emmons, R., Little, J., Kramer, A.F., … De Lisio, M. (2016). Circulating progenitor cells are positively associated with cognitive function among overweight/obese children. *Brain, Behavior, and Immunity* (in press). http://dx.doi.org/10.1016/j.bbi.2016.03.018.

Noble, E. E., Mavanji, V., Little, M. R., Billington, C. J., Kotz, C. M., & Wang, C. (2014). Exercise reduces diet-induced cognitive decline and increases hippocampal brain-derived neurotrophic factor in CA3 neurons. *Neurobiology of Learning and Memory, 114*, 40–50. http://dx.doi.org/10.1016/j.nlm.2014.04.006.

Ochoa-Reparaz, J., & Kasper, L. H. (2016). The second brain: Is the gut microbiota a link between obesity and central nervous system disorders? *Current Obesity Reports, 5*(1), 51–64. http://dx.doi.org/10.1007/s13679-016-0191-1.

Ogbonnaya, E. S., Clarke, G., Shanahan, F., Dinan, T. G., Cryan, J. F., & O'Leary, O. F. (2015). Adult hippocampal neurogenesis is regulated by the microbiome. *Biological Psychiatry, 78*(4), e7–e9. http://dx.doi.org/10.1016/j.biopsych.2014.12.023.

Ohland, C. L., Kish, L., Bell, H., Thiesen, A., Hotte, N., Pankiv, E., & Madsen, K. L. (2013). Effects of Lactobacillus helveticus on murine behavior are dependent on diet and genotype and correlate with alterations in the gut microbiome. *Psychoneuroendocrinology, 38*(9), 1738–1747. http://dx.doi.org/10.1016/j.psyneuen.2013.02.008. S0306-4530(13)00046-2 [pii].

Okun, E., Barak, B., Saada-Madar, R., Rothman, S. M., Griffioen, K. J., Roberts, N., … Mattson, M. P. (2012). Evidence for a developmental role for TLR4 in learning and memory. *PloS One*, 7(10), e47522. http://dx.doi.org/10.1371/journal.pone.0047522.

Rolls, A., Shechter, R., London, A., Ziv, Y., Ronen, A., Levy, R., & Schwartz, M. (2007). Toll-like receptors modulate adult hippocampal neurogenesis. *Nature Cell Biology*, *9*(9), 1081–1088. http://www.nature.com/ncb/journal/v9/n9/suppinfo/ncb1629_S1.html.

Santocchi, E., Guiducci, L., Fulceri, F., Billeci, L., Buzzigoli, E., Apicella, F., … Muratori, F. (2016). Gut to brain interaction in autism spectrum disorders: A randomized controlled trial on the role of probiotics on clinical, biochemical and neurophysiological parameters. *BMC Psychiatry*, *16*(1), 183. http://dx.doi.org/10.1186/s12888-016-0887-5.

Savignac, H. M., Kiely, B., Dinan, T. G., & Cryan, J. F. (2014). Bifidobacteria exert strain-specific effects on stress-related behavior and physiology in BALB/c mice. *Neurogastroenterology and Motility*, *26*(11), 1615–1627. http://dx.doi.org/10.1111/nmo.12427.

Scattoni, M. L., Martire, A., Cartocci, G., Ferrante, A., & Ricceri, L. (2013). Reduced social interaction, behavioural flexibility and BDNF signalling in the BTBR T+ tf/J strain, a mouse model of autism. *Behavioural Brain Research*, *251*, 35–40. http://dx.doi.org/10.1016/j.bbr.2012.12.028.

Smith, C. J., Emge, J. R., Berzins, K., Lung, L., Khamishon, R., Shah, P., … Gareau, M. G. (2014). Probiotics normalize the gut-brain-microbiota axis in immunodeficient mice. *American Journal of Physiology. Gastrointestinal and Liver Physiology*, *307*(8), G793–G802. http://dx.doi.org/10.1152/ajpgi.00238.2014.

Steenbergen, L., Sellaro, R., van Hemert, S., Bosch, J. A., & Colzato, L. S. (2015). A randomized controlled trial to test the effect of multispecies probiotics on cognitive reactivity to sad mood. *Brain, Behavior, and Immunity*, *48*, 258–264. http://dx.doi.org/10.1016/j.bbi.2015.04.003.

Tillisch, K., Labus, J., Kilpatrick, L., Jiang, Z., Stains, J., Ebrat, B., … Mayer, E. A. (2013). Consumption of fermented milk product with probiotic modulates brain activity. *Gastroenterology*, *144*(7), 1394–1401. http://dx.doi.org/10.1053/j.gastro.2013.02.043. e1394, S0016-5085(13)00292-8 [pii].

Tomova, A., Husarova, V., Lakatosova, S., Bakos, J., Vlkova, B., Babinska, K., & Ostatnikova, D. (2015). Gastrointestinal microbiota in children with autism in Slovakia. *Physiology and Behavior*, *138*, 179–187. http://dx.doi.org/10.1016/j.physbeh.2014.10.033.

Turnbaugh, P. J., Hamady, M., Yatsunenko, T., Cantarel, B. L., Duncan, A., Ley, R. E., … Gordon, J. I. (2009). A core gut microbiome in obese and lean twins. *Nature*, *457*(7228), 480–484.

Turnbaugh, P. J., Ley, R. E., Mahowald, M. A., Magrini, V., Mardis, E. R., & Gordon, J. I. (2006). An obesity-associated gut microbiome with increased capacity for energy harvest. *Nature*, *444*(7122), 1027–1031.

Utsunomiya, H., Takano, K., Okazaki, M., & Mitsudome, A. (1999). Development of the temporal lobe in infants and children: Analysis by MR-based volumetry. *AJNR. American Journal of Neuroradiology*, *20*(4), 717–723.

Uusitalo, U., Liu, X., Yang, J., Aronsson, C. A., Hummel, S., Butterworth, M., & T.S. Group. (2016). Association of early exposure of probiotics and islet autoimmunity in the TEDDY study. *JAMA Pediatrics*, *170*(1), 20–28. http://dx.doi.org/10.1001/jamapediatrics.2015.2757.

van Bussel, F. C., Backes, W. H., van Veenendaal, T. M., Hofman, P. A., van Boxtel, M. P., Schram, M. T., … Jansen, J. F. (2016). Functional brain networks are altered in type 2 diabetes and pre-diabetes: Signs for compensation of cognitive decrements?—The Maastricht Study. *Diabetes*, *65*(8), 2404–2413. http://dx.doi.org/10.2337/db16-0128.

Williamson, L. L., Sholar, P. W., Mistry, R. S., Smith, S. H., & Bilbo, S. D. (2011). Microglia and memory: Modulation by early-life infection. *Journal of Neuroscience*, *31*(43), 15511–15521. http://dx.doi.org/10.1523/JNEUROSCI.3688-11.2011.

Yano, J. M., Yu, K., Donaldson, G. P., Shastri, G. G., Ann, P., Ma, L., … Hsiao, E. Y. (2015). Indigenous bacteria from the gut microbiota regulate host serotonin biosynthesis. *Cell*, *161*(2), 264–276. http://dx.doi.org/10.1016/j.cell.2015.02.047.

Zhao, C., Deng, W., & Gage, F. H. (2008). Mechanisms and functional implications of adult neurogenesis. *Cell*, *132*(4), 645–660. http://dx.doi.org/10.1016/j.cell.2008.01.033.

CHAPTER TWELVE

The Intestinal Microbiota in the Irritable Bowel Syndrome

S.M. Collins[1]

The Farncombe Family Digestive Health Research Centre, The Michael G DeGroote School of Medicine, McMaster University, Hamilton, ON, Canada

[1]Corresponding author: e-mail address: scollins@mcmaster.ca

Contents

Abstract

The irritable bowel syndrome (IBS) is a chronic abdominal symptom complex occurring in a bowel devoid of discernible relevant pathology. There is growing interest in the role of the intestinal microbiota as a basis for the intestinal and possibly behavioral manifestations of this condition. Molecular-based microbial profiling has revealed compositional changes in the microbiota of at least a subset of IBS patients but the data are often conflicting and no microbial signature for this condition has yet been defined. Animal studies in which a previously stable intestinal microbiota is perturbed, by antibiotics or dietary change, results in alterations in intestinal function reminiscent of that seen in IBS patients. These include visceral sensitivity to painful stimuli, altered motility and intestinal barrier function as well as immune activation, and low-grade inflammation. More recent studies have shown that perturbation of the microbial composition of the gut alters brain chemistry and behavior. In a step toward establishing a causal link between an altar microbiota and gut–brain manifestations of IBS, colonization of germ-free mice with microbiota from IBS patients results in an IBS-like phenotype, including alterations and behavior if the donor exhibited psychiatric comorbidity, such as high levels of anxiety. This model provides an opportunity for exploring the mechanisms

International Review of Neurobiology, Volume 131
ISSN 0074-7742
http://dx.doi.org/10.1016/bs.irn.2016.08.003

underlying host–microbe interactions relevant to the pathogenesis of IBS and for developing novel therapeutic targets.

1. INTRODUCTION

The irritable bowel syndrome (IBS) is a chronic abdominal symptom complex in which pain or discomfort related to altered bowel habit are the clinical hallmarks. These symptoms arise in the absence of any demonstrable pathology and the disorder is considered to be one of function rather than structure. Psychiatric comorbidity is very common and there is a strong link between psychological stress and the expression of IBS. It is therefore considered to be a disorder of the gut–brain axis.

While, the IBS is the most common intestinal disorders in our society, our understanding of its underlying pathogenesis remains far from clear. IBS is heterogeneous in terms of its clinical manifestation and much effort has gone into defining clinically homogeneous IBS subgroups. However, there is unlikely to be homogeneity of the underlying pathogenesis even within these subgroups. IBS is simply a clinical descriptor for which there are likely to be several underlying etiological mechanisms—mechanisms that are not necessarily congruous with the clinical presentation. The appeal of a microbiota-based etiology lies in the fact that it is a dynamic system that can affect every aspect of intestinal function and can also impact on the brain and behavior. Despite this appeal, it is already evident that establishing a causal linkage between IBS and the intestinal microbiota represents a daunting and ongoing challenge.

2. THE GENERAL APPEAL OF THE MICROBIOTA AS PUTATIVE PATHOGENETIC FACTOR IN IBS

IBS is a chronic relapsing condition and its clinical expression may change over time within an individual sufferer. Thus, the underlying driving mechanism must have inherent flexibility in terms of its ability to alter various aspects of intestinal physiology, as well as to potentially influence the brain and behavior. The intestinal microbiota has these properties; it influences all aspects of gut function, can influence behavior and is susceptible to many of the environmental factors that are linked to the expression of IBS, including psychological stress and dietary factors. "Dysbiosis" is a term of convenience that is a widely used to describe a situation in which the composition or metabolism of the microbiota is disturbed; there is an implicit assumption that this is accompanied by changes in one or more host systems. Intestinal dysbiosis is an attractive mechanism to explain many aspects of the natural history of IBS and several lines of reasoning support this popular proposition.

3. FACTORS KNOWN TO PRECIPITATE OR EXACERBATE IBS ALSO INDUCE INTESTINAL DYSBIOSIS

3.1 Antibiotic Exposure

Antibiotic exposure clearly has the capacity to alter the intestinal microbiota at least on a temporary, if not a long-term basis in some individuals (Blaser, 2016; Rashid, Weintraub, & Nord, 2015). Several studies have shown an association between antibiotic exposure and a risk for developing chronic functional gastrointestinal (GI) symptoms. In a case–control study, there was a threefold increase in the development of IBS and those who had received antibiotics in the previous year (Maxwell, Rink, Kumar, & Mendall, 2002). In a recent nested case–control study, 83% of those reporting new functional GI symptoms had received antibiotics, generating an odds ratio of 1.95 (95% CI: 1.21–2.98; $p=0.005$) (Paula et al., 2015). A recent Swedish study showed that antibiotic exposure at a very young age (0–2 years) was associated with a higher prevalence of abdominal pain in young women at the age of 12 years. Interestingly, antibiotic exposure in years 9–12 did not have a similar effect. The difference in timing of exposure most likely relates to the impact of the antibiotic on the colonization process by intestinal bacteria (Uusijarvi et al., 2014). While acute enteric infection is a recognized risk factor for the development of IBS, Gwee et al. found that those who received antibiotics have a higher risk of developing functional symptoms but the relationship did not reach statistical significance (Gwee et al., 1999). While the apparent benefit of the nonabsorbable antibiotic rifamixin in some IBS patients may be construed as evidence supporting underlying dysbiosis in IBS, the interpretation is undermined by some uncertainty regarding the influences of the drug on the colonic microbiota vs those on host inflammatory responses (Pimentel, 2016).

3.2 Enteric Infection

Acute gastroenteritis, whether caused by bacterial, viral, or parasitic infection, is a recognized risk factor for the development or exacerbation of preexisting IBS (Spiller & Garsed, 2009). Studies in human subjects recovering from acute gastroenteritis have demonstrated, not surprisingly, changes in the intestinal microbial composition that usually recover within a few weeks. Longer-term studies reveal that those patients who go on to develop postinfectious IBS have compositional changes in their microbiota, compared to healthy subjects. For example, one study, using a combination of a phylogenetic microarray and (Spiller & Garsed, 2009) selected qPCR approach, characterized the microbiome in a total of 57 individuals and found that the microbiota of those patients with postinfectious IBS exhibited differences demonstrable at the phylum level compared to healthy controls but similar in profile to that seen in patients with diarrhea predominant IBS without a previous history of enteric infection. Differences between symptomatic and healthy groups were predominant in the Bacteroidetes phylum, where there was overexpression, and in the reduced expression of mainly uncultured Clostridia compared to the healthy group (Jalanka-Tuovinen et al., 2014), findings that were extended in a subsequent study (Jalanka, Salonen, Fuentes, & de Vos, 2015). Interestingly, this group found evidence of increased expression of host genes relevant to intestinal barrier function and inflammatory responses, findings previously observed in those patients who developed IBS following the outbreak of water poisoning in Walkerton Ontario in May 2000 (Villani et al., 2010). Thus, while enteric infection acts as the trigger for the onset of IBS, longer-term dysfunction may result from a combination of host factors and intestinal display analysis to produce the chronicity of postinfectious IBS (Marshall et al., 2010).

3.3 Psychological Stress Including Early Life Stress

There is a relationship between early life stress and the development of IBS later in life (Bradford et al., 2012) and recent studies in animals have shown that maternal separation produces long-lasting anxiety- and depression-like behavior in the offspring and this is associated with a compositional change in the microbiota (De Palma et al., 2015) and a susceptibility to inflammatory stimuli (Varghese et al., 2006). The changes in the microbiota, together with host factors that include altered colonic function, converge to induce the altered behavioral phenotype that characterizes this model (De Palma et al., 2015). A recent study showed that long-term supplementation of an eicosapentaenoic acid/docosahexaenoic acid (Pusceddu et al., 2015) *n*-3 PUFAs mixture not only improved behavior in maternal separated animals but also normalized the compositional changes in the microbiota.

Specifically, there was restoration of the relative abundance of the bacterium *Akkermansia* in the fatty acid-treated animals (Pusceddu et al., 2015). This bacterium is mucolytic and would promote translocation and a susceptibility to inflammation. By inference, these results describe a linkage between early life stress, dysbiosis, altered microbiome function, low-grade inflammation, and anxiety–depression-like behavior reminiscent of changes that have been described in some IBS patients. Stress is a well-accepted factor in the expression of IBS (Qin, Cheng, Tang, & Bian, 2014). Stress has also been shown to alter the composition of the microbiota in adult mice. Using a model of social stress, Bailley et al. demonstrated a shift in microbiota composition and this was accompanied by an increase in circulating interleukin-6 (IL-6), suggesting the induction of a proinflammatory microbiota (Bailey et al., 2011). Thus, stress, in early or later life, may be a factor contributing to dysbiosis in IBS.

3.4 Dietary Factors

Diet has always been implicated in the generation or exacerbation of symptoms such as abdominal pain or bloating in IBS. Erratic eating habits are common among IBS patients and this and fat are commonly reported to provoke symptoms such as bloating or abdominal pain. Recently, attention has focused on the role of wheat in symptom generation in IBS patients in whom there is an absence of features of classic celiac disease and this remains a work in progress (Pinto-Sanchez, Bercik, & Verdu, 2015). In addition, much attention has focused recently on the role of FODMAPS in symptoms generation in IBS but the data, that were initially promising, remain controversial (De Giorgio, Volta, & Gibson, 2016). For the purposes of this review, the question is whether these dietary components induce symptoms expression via changes in microbiota composition or metabolism or whether they reflect direct interactions between food components and host physiological or immunological systems. Studies that have attempted to link diet with altered microbiota profiles have often used comparisons of communities in which there are stable dietary differences among culturally or geographically distinct populations (De Filippo et al., 2010). It is difficult to reconcile these findings with the less stable and often erratic food intake patents characteristically seen in IBS patients. Intervention studies have added little to our understanding as they often involve large and dramatic changes in dietary composition that bare little relevance to the fluctuations in diet seen in IBS patients.

4. EVIDENCE OF DYSBIOSIS IN IBS PATIENTS

Despite a large number of studies using molecular-based approaches to profile the microbiome in IBS patients, the field remains unclear, and no microbial signature currently exists that distinguishes IBS or its subtypes from healthy controls. There are several factors the contribute to this and they include the inherent variability in the intestinal microbiota among healthy individuals, the confounding factor of diet, and differences in the technology used to profile the microbiome, rendering comparisons between studies difficult. In addition, and perhaps most important, the fact is that the majority of studies rely on a snapshot of the microbiome at a single point in time and therefore do not accommodate the many lifestyle factors that contribute to variability in both healthy subjects and controls. Commensal bacteria most relevant to the expression of IBS are likely those that reside in the distal small bowel, cecum, and colon. We are just beginning to characterize the small intestinal microbiome but it is clear that this is more likely to be subject to short-term influences of dietary variation than the colonic microbiome, due to the influence of simple dietary carbohydrates on bacterial species such as *Streptococcus* spp. (Zoetendal et al., 2012). The question as to whether bacterial overgrowth of the small intestine is a component of the pathophysiology of IBS remains controversial due to uncertainties regarding the methodologies used in many of these studies and the mechanism of action of the antibiotic rixaminin (Pimentel, 2016; Spiegel, 2011).

With respect to the colonic microbiome, there is general agreement there is a reduction in richness/diversity as well as some evidence of temporal instability. In addition, not all IBS patients exhibit discernible compositional differences in their microbiome compared to healthy controls, although this conclusion is subject to the caveats described earlier (Jeffery, Claesson, O'Toole, & Shanahan, 2012). It is beyond the scope of this review to detail the often conflicting microbiome profiles that have been associated with IBS patients and with symptom expression and IBS subgroups. This is referred to recent excellent reviews on this subject (Rajilic-Stojanovic et al., 2015; Simren et al., 2013).

Given the characteristic fluctuation in symptom type and severity seen in IBS patients, this author believes that documentation of temporal instability of the microbiota is more important than the compositional profiles reported to date in IBS subjects. In a study using culture techniques and denaturing gradient gel electrophoresis, Matto et al. found only small compositional differences between the IBS and control groups. However, significantly more temporal instability was evident in the IBS group, some of whom were followed up 6 months (Matto et al., 2005). Unfortunately, the impact of antibiotic usage in some patients undermined the impact of the study. Using an RNA-based approach, Maukonen et al. extended these findings, excluding patients that had received antibiotics 2 months prior to the study and confirmed greater instability of the microbiota in the IBS patients compared to healthy controls (Maukonen et al., 2006). A more recent study involving only two diarrhea predominant IBS patients and two healthy controls (including the husband of one of the patients) followed over 6–8 weeks used metagenomic and metatranscriptomic approaches. The IBS in these two cases was characterized by frequent relapses and episodes of remission and enabled the investigators to correlate changes in composition and activity with symptoms. They were able to identify a fraction of the microbiota whose transcriptomic based activity was related to IBS symptoms and showed considerable instability over time. The compositional shifts, particularly in the less active microbiota population of the IBS patients were less prominent. These findings provide encouragement in terms of providing a strategy to better understand host–microbial interactions in this condition. Clearly, while it is tempting to speculate that instability in the microbiota induced fluctuations in symptom profile, as has been implied in a probiotic study that improved both microbiota stability and IBS symptoms (Kajander et al., 2008), the possibility that the community instability simply reflects reactions to changes in intestinal physiology cannot be ruled out. In all likelihood, the findings represent a combination of factors of bacterial and host origin and constitute the mechanism for perpetuating intestinal dysfunction in this chronic condition (Durban et al., 2013).

There is growing interest in correlating the products of microbial metabolism with symptom generation in IBS. Increased fecal levels of acetic and propionic acid have been found in IBS patients compared to healthy controls in the findings correlated with the severity of IBS symptoms (Tana et al., 2010). Intestinal bacteria determine the bioavailability of dietary tryptophan to the host and consequently to the synthesis of serotonin. High levels of circulating serotonin have been found in patients with diarrhea predominant IBS and low levels in constipation predominant IBS (Dunlop et al., 2005) in these differences could reflect alterations in the intestinal microbiota across these IBS subgroups. The intestinal microbiota could also influence dietary tryptophan metabolism along a pathway that results in the formation of neurotoxic metabolites that may in turn contribute to behavioral changes in those patients with psychiatric comorbidity (Jenkins, Nguyen, Polglaze, & Bertrand, 2016).

5. PROOF OF PRINCIPLE THAT INTESTINAL DYSBIOSIS ALTERS FUNCTION IN THE GUT AND BRAIN

There is extensive evidence derived from animal studies showing that the intestinal microbiota can alter parameters of gut dysfunction that have been demonstrated in IBS patients. These include changes in intestinal motility, changes in visceral pain responses, alterations in epithelial secretory and barrier function, and in inflammatory and/or immunological function. Many of these studies are based on comparisons between germ-free and colonized animals and while they demonstrate the importance of the influence of the microbiota on these systems, they have limited relevance to the implied dysbiosis implicated in IBS. In addition, virtually all of the host systems relevant to IBS, including the intestinal physiological apparatus, the mucosal immune system including barrier function, and the brain are immature in germ-free animals. A strategy that is more relevant to the implied dysbiosis of IBS is to experimentally disrupt the intestinal microbiota in an otherwise healthy animal with a previously stable microbiota. This can be achieved through diet or through the use of nonabsorbable orally administered antimicrobial agents.

The first study to evaluate changes in host following experimental despite doses utilized a cocktail of antimicrobials that included bacitracin, neomycin, and pimaricin administered in the drinking water for a total of 10 days two specific pathogen free mice. The study was aimed at determining whether experimental dysbiosis could induce changes in visceral pain responses elicited by balloon distention of the colorectum. Antimicrobial treatment resulted in a significant increase in responses to colorectal distention and this was accompanied by an increase in immunoreactive substance P and CGP in the myenteric plexus (Verdu et al., 2006). Antimicrobial therapy was associated with depletion of lactobacilli, *Bacteroides* and enterococci but, in a separate experiment, administration of lactobacillus *paracaseii* prevented the dysbiosis-induced changes in visceral pain responses. The sensory changes induced by dysbiosis where accompanied by a small low-grade inflammatory response. Treatment with dexamethasone prevented the inflammatory response as well as the changes in visceral pain responses. Taken together, these findings provided proof of principle that experimental dysbiosis induces low-grade inflammation and increases visceral pain responses—findings reminiscent of IBS

(Verdu et al., 2006). A subsequent study extended these findings by incorporating a mild stressor with nonabsorbable antibiotics to induce dysbiosis, which was monitored in mucosal adherent bacterial populations rather than in the feces. While total bacterial counts were reduced, adherence was enhanced and there was a corresponding increase in luminal s-IgA. Stress alone increased visceral pain responses and upregulated expression of cannabinoid receptor 2 and this increase was abrogated by the concomitant use of antimicrobials (Aguilera, Vergara, & Martinez, 2013). In a later study, the same authors used intraperitoneal acetic acid to induce visceral hypersensitivity and used neomycin and bacitracin to induce dysbiosis. There was an increase in the expression of the cannabinoid receptor 2 and downregulation of the cannabinoid 1 and the μ-opioid receptors. Visceral pain responses were substantially reduced in the presence of dysbiosis, although colonic muscle contractility was enhanced (Aguilera, Cerda-Cuellar, & Martinez, 2015). The study showing the induction of visceral hypersensitivity in the presence of intestinal dysbiosis and the studies showing that dysbiosis was accompanied by an inhibition of enhanced visceral pain responses were performed in completely different laboratories in which the endogenous microbiota was likely different. This illustrates two points. First, it is difficult to extrapolate results involving experimental dysbiosis between laboratories. Second, they show that depending on the composition of the microbiota, experimental dysbiosis can produce diametrically opposite effects on a given host parameter. This may have bearing on the fluctuations often seen in the nature of symptoms reported by an individual IBS patient over time.

A recent study examined the consequences of antibiotic exposure early in life, a subject of considerable current interest in the natural history of IBS. Vancomycin was administered from the 4th to the 13th postnatal days in rat pups. This resulted in transient dysbiosis with restoration of the normal microbiome by 8 weeks. Nevertheless, these mice displayed visceral hypersensitivity later in life (O'Mahony et al., 2014). A similar phenomenon was observed when the combination of neomycin, bacitracin, and pimaricin was used in the postnatal period. Thus, disruption of the bacterial colonization process early in life impacts on enteric sensory function later in life. An interesting phenomenon observed in this study was that the effect on sensory function did not occur in female rats.

As mentioned earlier, experimental dysbiosis can lead to immune activation and low-grade inflammation, which in turn led to changes in gut physiology. A study examined the role of the innate immune system in a model of antibiotic induced dysbiosis. In this study, investigators used mice lacking critical components of the innate immune system. Antibiotic induced dysbiosis resulted in a significant delay of GI motility and this was accompanied by a reduction in the number of nitrergic neurons in the colon. Because a similar profile was observed in mice with neural crest-specific deletion of Myd88 (Wnt1Cre(+/−)/Myd88(fl/fl)) as well as TLR4, the authors concluded that the microbiota influence GI motor function via activation of the innate immune system via TLR4 receptors (Anitha, Vijay-Kumar, Sitaraman, Gewirtz, & Srinivasan, 2012).

There is increasing interest in the ability of the intestinal microbiota to influence the brain and behavior, prompting consideration of a microbiota–gut–brain axis (Collins, Surette, & Bercik, 2012) and this may be relevant to the psychiatric comorbidity that is common among IBS patients (Fond et al., 2014). Several strategies have been used to explore microbiota–brain interactions and they include comparisons between germ-free and colonized mice (Bercik et al., 2011; Diaz Heijtz et al., 2011; Neufeld, Kang, Bienenstock, & Foster, 2011; Sudo et al., 2004), the adoptive transfer of behavioral phenotype between murine strains via the intestinal microbiota (Bercik et al., 2011; Collins, Kassam, & Bercik, 2013), and the effect of experimental dysbiosis on brain and behavior (Bercik et al., 2011; Li, Dowd, Scurlock, Acosta-Martinez, & Lyte, 2009). For reasons already mentioned, the latter strategy is more relevant to IBS and will be addressed here. In one study, mice were fed on either a standard rodent chow or a chow containing 50% lean beef for 3 months. This resulted in a significant increase in diversity of the microbiome, as assessed by bacterial tag-encoded FLX amplicon pyrosequencing, in mice fed the beef enriched diet. Performance in memory and learning behavioral testing was enhanced in the mice fed the beef enriched diet and they also exhibited less anxiety like behavior (Li et al., 2009). While it was concluded that the diet-induced dysbiosis was responsible for the behavioral change, the authors acknowledge that direct contributions from the beef enriched diet could have contributed to the changes in brain function. In a more recent study involving antibiotic-induced dysbiosis, Bercik et al. found oral but not intraperitoneal administration of nonabsorbable antibiotics increased exploratory behavior as well as increased levels of brain-derived neurotropic factor in the hippocampus. The changes in neurochemistry and behavior were independent of the integrity of the vagus nerve and the sympathetic nervous system (Bercik et al., 2011). A recent study by Frohlich et al. showed that antibiotic induced dysbiosis in mice reduced microbiota-derived metabolites in the colon and their metabolic derivatives in the plasma. This was accompanied by cognitive impairment and changes in chemistry, indicating that intestinal dysbiosis may lead to behavioral changes by influencing host metabolism (http://www.sciencedirect.com/science/article/pii/S088915911630040X).

5.1 Is There a Causal Link Between Dysbiosis and Symptom Expression in IBS?

With limited therapeutic or investigative tools, coupled with the absence of confidently identified "bacteria of interest," it is almost impossible to establish causal links between the intestinal microbiota and symptom generation in IBS. The benefit of selected probiotics in improving symptoms and quality of life in IBS patients provides insufficient support for a causal role of the microbiota in the expression of this condition, as the underlying modes of action and impact on the microbiota community are poorly understood. Currently, our best strategy toward establishing the functional relevance of the IBS microbiota in producing gut dysfunctions is the use of human microbiota-associated animals. This strategy involves colonizing germ-free mice with microbiota taken from IBS patients or healthy controls. Crouzet et al. use this strategy to demonstrate the ability of the microbiota to induce visceral sensitivity in rats. The colonization of germ-free rats with microbiota from IBS patients with demonstrable visceral hypersensitivity, but not healthy controls, resulted in increased responses to colorectal distention in the recipient animals (Crouzet et al., 2013). This was not accompanied by changes in intestinal permeability or in mast cell density in the recipient mice. In a recent preliminary study (De Palma et al., 2014), germ-free mice were colonized with microbiota from diarrhea predominant IBS patients with or without anxiety, or from healthy controls and gut function and behavior were assessed. IBS microbiota-associated mice developed rapid transit, increased intestinal permeability in and secretion, and evidence of innate immune activation. Mice colonized with microbiota from patients with anxiety showed evidence of anxiety-like behavior whereas mice receiving microbiota from IBS patients without anxiety showed normal behavior (De Palma et al., 2014). These results provide strong support for the notion that the intestinal microbiota of IBS-d patients contributes to both the intestinal and behavioral manifestations of this condition. Zheng et al. recently provided further support for the notion that the intestinal microbiota contributes to psychopathology by showing that fecal microbiota from depressed patients altered brain chemistry and behavior following colonization of germ-free animals (Zheng et al., 2016).

6. FUTURE DIRECTIONS

Much evidence now points toward a role of the intestinal microbiota in the expression of IBS, the most common intestinal disorder in our society today. Clinical studies must move away from single time point comparisons of microbiome profiles between IBS patients and healthy controls. More emphasis should be placed on longitudinal studies in which patients serve as their own controls, thus minimizing the problem of inter-subject variation and microbiota composition. The field must now move forward toward establishing evidence of causality between the microbiota and symptom expression, including psychiatric comorbidity. Increased use of human microbiota-associated animals should generate important new information regarding a functional role of the microbiota in this condition. Our increasing ability to culture the "noncultivable" intestinal microbial population will enhance our ability to comprehensively profile the bacterial community, to understand community dynamics, and to adopt prescribed colonization strategies using germ-free mice in order to better understand microbe–host interactions. This in turn will help deliver novel therapeutic targets and the strategies for improving this common condition. Last, but by no means least, the influence of the intestinal virome on the intestinal microbiota must now be studied and exploited for therapeutic gain in IBS and other conditions.

REFERENCES

Aguilera, M., Cerda-Cuellar, M., & Martinez, V. (2015). Antibiotic-induced dysbiosis alters host–bacterial interactions and leads to colonic sensory and motor changes in mice. *Gut Microbes*, *6*(1), 10–23. http://dx.doi.org/10.4161/19490976.2014.990790.

Aguilera, M., Vergara, P., & Martinez, V. (2013). Stress and antibiotics alter luminal and wall-adhered microbiota and enhance the local expression of visceral sensory-related systems in mice. *Neurogastroenterology and Motility*, *25*(8), e515–e529. http://dx.doi.org/10.1111/nmo.12154.

Anitha, M., Vijay-Kumar, M., Sitaraman, S. V., Gewirtz, A. T., & Srinivasan, S. (2012). Gut microbial products regulate murine gastrointestinal motility via Toll-like receptor 4 signaling. *Gastroenterology*, *143*(4), 1006–1016. http://dx.doi.org/10.1053/j.gastro.2012.06.034. e1004.

Bailey, M. T., Dowd, S. E., Galley, J. D., Hufnagle, A. R., Allen, R. G., & Lyte, M. (2011). Exposure to a social stressor alters the structure of the intestinal microbiota: Implications for stressor-induced immunomodulation. *Brain, Behavior, and Immunity*, *25*(3), 397–407. http://dx.doi.org/10.1016/j.bbi.2010.10.023.

Bercik, P., Denou, E., Collins, J., Jackson, W., Lu, J., Jury, J., … Collins, S. M. (2011). The intestinal microbiota affect central levels of brain-derived neurotropic factor and behavior in mice. *Gastroenterology*, *141*(2), 599–609. http://dx.doi.org/10.1053/j.gastro.2011.04.052. e591–593.

Blaser, M. J. (2016). Antibiotic use and its consequences for the normal microbiome. *Science, 352*(6285), 544–545. http://dx.doi.org/10.1126/science.aad9358.

Bradford, K., Shih, W., Videlock, E. J., Presson, A. P., Naliboff, B. D., Mayer, E. A., & Chang, L. (2012). Association between early adverse life events and irritable bowel syndrome. *Clinical Gastroenterology and Hepatology, 10*(4), 385–390. http://dx.doi.org/10.1016/j.cgh.2011.12.018. e381–383.

Collins, S. M., Kassam, Z., & Bercik, P. (2013). The adoptive transfer of behavioral phenotype via the intestinal microbiota: Experimental evidence and clinical implications. *Current Opinion in Microbiology, 16*(3), 240–245. http://dx.doi.org/10.1016/j.mib.2013.06.004.

Collins, S. M., Surette, M., & Bercik, P. (2012). The interplay between the intestinal microbiota and the brain. *Nature Reviews Microbiology, 10*(11), 735–742. http://dx.doi.org/10.1038/nrmicro2876.

Crouzet, L., Gaultier, E., Del'Homme, C., Cartier, C., Delmas, E., Dapoigny, M., … Bernalier-Donadille, A. (2013). The hypersensitivity to colonic distension of IBS patients can be transferred to rats through their fecal microbiota. *Neurogastroenterology and Motility, 25*(4), e272–e282. http://dx.doi.org/10.1111/nmo.12103.

De Filippo, C., Cavalieri, D., Di Paola, M., Ramazzotti, M., Poullet, J. B., Massart, S., … Lionetti, P. (2010). Impact of diet in shaping gut microbiota revealed by a comparative study in children from Europe and rural Africa. *Proceedings of the National Academy of Sciences of the United States of America, 107*(33), 14691–14696. http://dx.doi.org/10.1073/pnas.1005963107.

De Giorgio, R., Volta, U., & Gibson, P. R. (2016). Sensitivity to wheat, gluten and FODMAPs in IBS: Facts or fiction? *Gut, 65*(1), 169–178. http://dx.doi.org/10.1136/gutjnl-2015-309757.

De Palma, G., Blennerhassett, P., Lu, J., Deng, Y., Park, A. J., Green, W., … Bercik, P. (2015). Microbiota and host determinants of behavioural phenotype in maternally separated mice. *Nature Communications, 6*, 7735. http://dx.doi.org/10.1038/ncomms8735.

De Palma, G., Lynch, M. D., Lu, J., Deng, Y., Jury, J., McLean, P., … Bercik, P. (2014). The adoptive transfer of anxiety and gut dysfunction from IBS patients to axenic mice through microbiota transplantation. *Gastroenterology, 146*(5), S845.

Diaz Heijtz, R., Wang, S., Anuar, F., Qian, Y., Bjorkholm, B., Samuelsson, A., … Pettersson, S. (2011). Normal gut microbiota modulates brain development and behavior. *Proceedings of the National Academy of Sciences of the United States of America, 108*(7), 3047–3052. http://dx.doi.org/10.1073/pnas.1010529108.

Dunlop, S. P., Coleman, N. S., Blackshaw, E., Perkins, A. C., Singh, G., Marsden, C. A., & Spiller, R. C. (2005). Abnormalities of 5-hydroxytryptamine metabolism in irritable bowel syndrome. *Clinical Gastroenterology and Hepatology, 3*(4), 349–357.

Durban, A., Abellan, J. J., Jimenez-Hernandez, N., Artacho, A., Garrigues, V., Ortiz, V., … Moya, A. (2013). Instability of the faecal microbiota in diarrhoea-predominant irritable bowel syndrome. *FEMS Microbiology Ecology, 86*(3), 581–589. http://dx.doi.org/10.1111/1574-6941.12184.

Fond, G., Loundou, A., Hamdani, N., Boukouaci, W., Dargel, A., Oliveira, J., … Boyer, L. (2014). Anxiety and depression comorbidities in irritable bowel syndrome (IBS): A systematic review and meta-analysis. *European Archives of Psychiatry and Clinical Neuroscience, 264*(8), 651–660. http://dx.doi.org/10.1007/s00406-014-0502-z.

Gwee, K. A., Leong, Y. L., Graham, C., McKendrick, M. W., Collins, S. M., Walters, S. J., … Read, N. W. (1999). The role of psychological and biological factors in postinfective gut dysfunction. *Gut, 44*(3), 400–406.

Jalanka, J., Salonen, A., Fuentes, S., & de Vos, W. M. (2015). Microbial signatures in post-infectious irritable bowel syndrome—Toward patient stratification for improved diagnostics and treatment. *Gut Microbes, 6*(6), 364–369. http://dx.doi.org/10.1080/19490976.2015.1096486.

Jalanka-Tuovinen, J., Salojarvi, J., Salonen, A., Immonen, O., Garsed, K., Kelly, F. M., … de Vos, W. M. (2014). Faecal microbiota composition and host-microbe cross-talk following gastroenteritis and in postinfectious irritable bowel syndrome. *Gut*, *63*(11), 1737–1745. http://dx.doi.org/10.1136/gutjnl-2013-305994.

Jeffery, I. B., Claesson, M. J., O'Toole, P. W., & Shanahan, F. (2012). Categorization of the gut microbiota: Enterotypes or gradients? *Nature Reviews. Microbiology*, *10*(9), 591–592.

Jenkins, T. A., Nguyen, J. C., Polglaze, K. E., & Bertrand, P. P. (2016). Influence of tryptophan and serotonin on mood and cognition with a possible role of the gut-brain axis. *Nutrients*, *8*(1), 56. http://dx.doi.org/10.3390/nu8010056. http://www.mdpi.com/2072-6643/8/1/56.

Kajander, K., Myllyluoma, E., Rajilic-Stojanovic, M., Kyronpalo, S., Rasmussen, M., Jarvenpaa, S., … Korpela, R. (2008). Clinical trial: Multispecies probiotic supplementation alleviates the symptoms of irritable bowel syndrome and stabilizes intestinal microbiota. *Alimentary Pharmacology & Therapeutics*, *27*(1), 48–57. http://dx.doi.org/10.1111/j.1365-2036.2007.03542.x.

Li, W., Dowd, S. E., Scurlock, B., Acosta-Martinez, V., & Lyte, M. (2009). Memory and learning behavior in mice is temporally associated with diet-induced alterations in gut bacteria. *Physiology & Behavior*, *96*(4–5), 557–567. http://dx.doi.org/10.1016/j.physbeh.2008.12.004.

Marshall, J. K., Thabane, M., Garg, A. X., Clark, W. F., Moayyedi, P., & Collins, S. M. (2010). Eight year prognosis of postinfectious irritable bowel syndrome following waterborne bacterial dysentery. *Gut*, *59*(5), 605–611. http://dx.doi.org/10.1136/gut.2009.202234.

Matto, J., Maunuksela, L., Kajander, K., Palva, A., Korpela, R., Kassinen, A., & Saarela, M. (2005). Composition and temporal stability of gastrointestinal microbiota in irritable bowel syndrome—A longitudinal study in IBS and control subjects. *FEMS Immunology and Medical Microbiology*, *43*(2), 213–222.

Maukonen, J., Satokari, R., Matto, J., Soderlund, H., Mattila-Sandholm, T., & Saarela, M. (2006). Prevalence and temporal stability of selected clostridial groups in irritable bowel syndrome in relation to predominant faecal bacteria. *Journal of Medical Microbiology*, *55*(Pt. 5), 625–633. http://dx.doi.org/10.1099/jmm.0.46134-0.

Maxwell, P. R., Rink, E., Kumar, D., & Mendall, M. A. (2002). Antibiotics increase functional abdominal symptoms. *The American Journal of Gastroenterology*, *97*(1), 104–108. http://dx.doi.org/10.1111/j.1572-0241.2002.05428.x.

Neufeld, K. M., Kang, N., Bienenstock, J., & Foster, J. A. (2011). Reduced anxiety-like behavior and central neurochemical change in germ-free mice. *Neurogastroenterology and Motility*, *23*(3), 255–264. http://dx.doi.org/10.1111/j.1365-2982.2010.01620.x. e119.

O'Mahony, S. M., Felice, V. D., Nally, K., Savignac, H. M., Claesson, M. J., Scully, P., … Cryan, J. F. (2014). Disturbance of the gut microbiota in early-life selectively affects visceral pain in adulthood without impacting cognitive or anxiety-related behaviors in male rats. *Neuroscience*, *277*, 885–901. http://dx.doi.org/10.1016/j.neuroscience.2014.07.054.

Paula, H., Grover, M., Halder, S. L., Locke, G. R., 3rd, Schleck, C. D., Zinsmeister, A. R., & Talley, N. J. (2015). Non-enteric infections, antibiotic use, and risk of development of functional gastrointestinal disorders. *Neurogastroenterology and Motility*, *27*(11), 1580–1586. http://dx.doi.org/10.1111/nmo.12655.

Pimentel, M. (2016). Review article: Potential mechanisms of action of rifaximin in the management of irritable bowel syndrome with diarrhoea. *Alimentary Pharmacology & Therapeutics*, *43*(Suppl. 1), 37–49. http://dx.doi.org/10.1111/apt.13437.

Pinto-Sanchez, M. I., Bercik, P., & Verdu, E. F. (2015). Motility alterations in celiac disease and non-celiac gluten sensitivity. *Digestive Diseases*, *33*(2), 200–207. http://dx.doi.org/10.1159/000371400.

Pusceddu, M. M., Kelly, P., Ariffin, N., Cryan, J. F., Clarke, G., & Dinan, T. G. (2015). n-3 PUFAs have beneficial effects on anxiety and cognition in female rats: Effects of early life stress. *Psychoneuroendocrinology*, *58*, 79–90. http://dx.doi.org/10.1016/j.psyneuen.2015.04.015.

Qin, H. Y., Cheng, C. W., Tang, X. D., & Bian, Z. X. (2014). Impact of psychological stress on irritable bowel syndrome. *World Journal of Gastroenterology*, *20*(39), 14126–14131. http://dx.doi.org/10.3748/wjg.v20.i39.14126.

Rajilic-Stojanovic, M., Jonkers, D. M., Salonen, A., Hanevik, K., Raes, J., Jalanka, J., … Penders, J. (2015). Intestinal microbiota and diet in IBS: Causes, consequences, or epiphenomena? *The American Journal of Gastroenterology*, *110*(2), 278–287. http://dx.doi.org/10.1038/ajg.2014.427.

Rashid, M. U., Weintraub, A., & Nord, C. E. (2015). Development of antimicrobial resistance in the normal anaerobic microbiota during one year after administration of clindamycin or ciprofloxacin. *Anaerobe*, *31*, 72–77. http://dx.doi.org/10.1016/j.anaerobe.2014.10.004.

Simren, M., Barbara, G., Flint, H. J., Spiegel, B. M., Spiller, R. C., Vanner, S., … Zoetendal, E. G. (2013). Intestinal microbiota in functional bowel disorders: A Rome foundation report. *Gut*, *62*(1), 159–176. http://dx.doi.org/10.1136/gutjnl-2012-302167.

Spiegel, B. M. (2011). Questioning the bacterial overgrowth hypothesis of irritable bowel syndrome: An epidemiologic and evolutionary perspective. *Clinical Gastroenterology and Hepatology*, *9*(6), 461–469. http://dx.doi.org/10.1016/j.cgh.2011.02.030. quiz e459.

Spiller, R., & Garsed, K. (2009). Postinfectious irritable bowel syndrome. *Gastroenterology*, *136*(6), 1979–1988. http://dx.doi.org/10.1053/j.gastro.2009.02.074.

Sudo, N., Chida, Y., Aiba, Y., Sonoda, J., Oyama, N., Yu, X. N., … Koga, Y. (2004). Postnatal microbial colonization programs the hypothalamic–pituitary–adrenal system for stress response in mice. *The Journal of Physiology*, *558*(Pt. 1), 263–275. http://dx.doi.org/10.1113/jphysiol.2004.063388.

Tana, C., Umesaki, Y., Imaoka, A., Handa, T., Kanazawa, M., & Fukudo, S. (2010). Altered profiles of intestinal microbiota and organic acids may be the origin of symptoms in irritable bowel syndrome. *Neurogastroenterology and Motility*, *22*(5), 512–519. http://dx.doi.org/10.1111/j.1365-2982.2009.01427.x. e114–515.

Uusijarvi, A., Bergstrom, A., Simren, M., Ludvigsson, J. F., Kull, I., Wickman, M., … Olen, O. (2014). Use of antibiotics in infancy and childhood and risk of recurrent abdominal pain—A Swedish birth cohort study. *Neurogastroenterology and Motility*, *26*(6), 841–850. http://dx.doi.org/10.1111/nmo.12340.

Varghese, A. K., Verdu, E. F., Bercik, P., Khan, W. I., Blennerhassett, P. A., Szechtman, H., & Collins, S. M. (2006). Antidepressants attenuate increased susceptibility to colitis in a murine model of depression. *Gastroenterology*, *130*(6), 1743–1753. http://dx.doi.org/10.1053/j.gastro.2006.02.007.

Verdu, E. F., Bercik, P., Verma-Gandhu, M., Huang, X. X., Blennerhassett, P., Jackson, W., … Collins, S. M. (2006). Specific probiotic therapy attenuates antibiotic induced visceral hypersensitivity in mice. *Gut*, *55*(2), 182–190. http://dx.doi.org/10.1136/gut.2005.066100.

Villani, A. C., Lemire, M., Thabane, M., Belisle, A., Geneau, G., Garg, A. X., … Marshall, J. K. (2010). Genetic risk factors for post-infectious irritable bowel syndrome following a waterborne outbreak of gastroenteritis. *Gastroenterology*, *138*(4), 1502–1513. http://dx.doi.org/10.1053/j.gastro.2009.12.049.

Zheng, P., Zeng, B., Zhou, C., Liu, M., Fang, Z., Xu, X., … Xie, P. (2016). Gut microbiome remodeling induces depressive-like behaviors through a pathway mediated by the host's metabolism. *Molecular Psychiatry*, *21*(6), 786–796. http://dx.doi.org/10.1038/mp.2016.44.

Zoetendal, E. G., Raes, J., van den Bogert, B., Arumugam, M., Booijink, C. C., Troost, F. J., … Kleerebezem, M. (2012). The human small intestinal microbiota is driven by rapid uptake and conversion of simple carbohydrates. *The ISME Journal*, *6*(7), 1415–1426. http://dx.doi.org/10.1038/ismej.2011.212.

CHAPTER THIRTEEN

Gut-to-Brain Axis in Autism Spectrum Disorders: Central Role for the Microbiome

A.D. Kraneveld*,†,1, K. Szklany*, C.G.M. de Theije‡, J. Garssen*,§
*Division of Pharmacology, Utrecht Institute for Pharmaceutical Sciences, Faculty of Science, Utrecht University, Utrecht, The Netherlands
†Institute for Risk Assessment Sciences, Faculty of Veterinary Medicine, Utrecht University, Utrecht, The Netherlands
‡Laboratory of Neuroimmunology and Developmental Origins of Disease, Academic Medical Centre, Utrecht University, Utrecht, The Netherlands
§Nutricia Research, Utrecht, The Netherlands
[1]Corresponding author: e-mail address: A.D.Kraneveld@uu.nl

Contents

Abstract

Autism spectrum disorders (ASDs) are neurodevelopmental disorders, which occur in early childhood and persist into adulthood. Although the etiology of these disorders is largely unknown, genetic and environmental factors are thought to interplay in the development of ASD. Intestinal microbial dysbiosis, in prenatal and postnatal phases, is an important example of these environmental factors, and gastrointestinal problems including adverse reactions to foods are often reported in these children.

International Review of Neurobiology, Volume 131
ISSN 0074-7742
http://dx.doi.org/10.1016/bs.irn.2016.09.001

In this review, we address the clinical and preclinical findings on the role of the intestinal microbiome in ASD and suggest possible underlying mechanisms. Furthermore, opportunities for (nutritional) interventions in ASD are provided.

1. INTRODUCTION

Austim spectrum disorders (ASDs) are pervasive neurodevelopmental disorders with a shared core of symptoms characterized by impairments in communication/speech, social interaction, presence of limited, stereotype, and repetitive behaviors and interests (Kohane, 2015; Matson & Goldin, 2014). Due to the lack of biomarkers, the diagnosis of ASD depends on behavioral observations according to the American Psychiatric Association (2016). The conventional view of ASD is that it is a genetic disorder involving a complex genetic background. Genome-wide association studies, copy number variation screening, and SNP analyses have identified several ASD candidate genes, and a large number of genetic mutations have been proposed to cause predisposition to ASD (Carter & Scherer, 2013; De Rubeis et al., 2014; Ehninger & Silva, 2009; Iossifov et al., 2014). Some of these genes may be involved in abnormal development of the nervous system, including the central nervous system (CNS) as well as the enteric nervous system (ENS) (Bernier et al., 2014; Kozol et al., 2015).

Since ASD is a spectrum of related disorders it is of particular interest to identify specific biomarkers for patient stratification and possible common pathways. In addition, there is an alarming rise of the number of cases diagnosed ASD in developed countries: increasing 35-fold since 1970s. Now ASD affect around 1 in 68 people (Christensen et al., 2016). The reasons for this tremendous increase of cases of ASD are not known; selected studies (Blumberg et al., 2013) suggest that part of the recent prevalence increase is attributable to environmental factors. Many ASD patients have comorbid medical conditions such as sleep problems, metabolic conditions, feeding difficulties, and gastrointestinal disorders, which have a significant impact on the quality of life of the patients and their caregivers (Frye & Rossignol, 2016; Kohane et al., 2012). The behavioral, neurological, and biochemical characteristics associated with ASD pathology might involve low-grade inflammatory processes (Rossignol & Frye, 2012). About 40–60% of ASD children suffer from gastrointestinal problems, ranging from frequent abdominal pain and bloating to diarrhea and constipation. A recent metaanalysis demonstrated children with ASD experienced significantly more general intestinal symptoms than comparison groups (OR: 4.42;

95% CI: 1.90–10.28) (McElhanon, McCracken, Karpen, & Sharp, 2014). It should be noted that due to social and communicative impairments, it is in particular challenging to determine the true prevalence of gastrointestinal problems in patients with ASD.

The intestinal tract regulates the immune system, in particular, during the first 1000 days of life (Kostic et al., 2015; Wopereis, Oozeer, Knipping, Belzer, & Knol, 2014), and disruption of immune regulation and inflammation are associated with brain disorders including ASD. In addition, psychological stress triggers inflammation further via the intestinal tract and its microbiome (Borre et al., 2014; Rook, Raison, & Lowry, 2014). There is increasing evidence indicating that intestinal immune dysfunction, including food allergy, in ASD patients may be caused by disturbances in the pathways underlying the so-called gut–immune–brain axis, with a central role of the intestinal microbiome (Croen et al., 2015; de Theije et al., 2011; Lyall, Van de Water, Ashwood, & Hertz-Picciotto, 2015; Onore, Careaga, & Ashwood, 2012; Zerbo et al., 2015). In the last decade, the role of the microbiome in ASD has received a lot of attention and many reviews on this topic have been published (a.o. Cao, Lin, Jiang, & Li, 2013; Carabotti, Scirocco, Maselli, & Severi, 2015; De Angelis, Francavilla, Piccolo, De Giacomo, & Gobbetti, 2015; Frye, Rose, Slattery, & MacFabe, 2015; Li & Zhou, 2016; Luna, Savidge, & Williams, 2016; Mulle, Sharp, & Cubells, 2013; Reddy & Saier, 2015; Rosenfeld, 2015).

2. THE RODAKIS CASE

At the age of 3 Tommy, son of John Rodakis, was diagnosed with ASD. Rodakis, a molecular biologist working as medical venture capitalist, kept a diary of his son's life tracking more than 20 different parameters of Tommy's ASD using a self-developed rating system. The parents became aware that their son's symptoms changed depending on environmental factors. The most striking effects on their son's behavior were observed after a 10-day course of the antibiotic, amoxicilline. A physician prescribed the antibiotic after conformation of streptococcus throat infection in Tommy and his sister. They had never received an antibiotic before and after 2 days they seemed to be recovered from their throat problems. It was on day 4 that Rodakis noticed changes in his son. From that day on the boy started to make eye contact, his speech improved, he became less rigid and repetitive, and he displaced more energy. Based on literature, Rodakis started his quest to find out more about the link between ASD and alteration in the microbiome (Rodakis, 2015).

Rodakis's observations are not standing alone and based on parental reports and limited medical research (Sandler et al., 2000), it is now postulated that the gut microbiome shapes the brain and might be a root cause of ASD. The link between ASD and the microbiome is hypothesized to be a new target for treatment of at least a substantial subpopulation of ASD patients.

3. MICROBIOME IN ASD: CORRELATION STUDIES

Dysbiosis has been demonstrated in ASD, however, studies on the intestinal microbiome composition in ASD are limited and it is not possible to pool results into a metaanalysis (Cao et al., 2013; Luna et al., 2016; Rosenfeld, 2015). Table 1 shows that in 12 (with a combined sample of 510 individuals: 296 ASD, 73 siblings, and 139 healthy controls) of the 14 studies described, significant differences were found in microbiome composition between patients suffering from ASD and (sibling) controls. Two studies with a combined sample of 205 individuals found no differences between ASD (108) and controls (97), however, in both studies the control groups consisted of sibling controls.

Based on bacterial cultures on stool samples, studies showed that Clostridial groups are significantly elevated in ASD (Finegold et al., 2002; Martirosian et al., 2011). PCR and 16S rRNA gene sequencing as well as FISH approaches led to more in depth investigations into microbiome in ASD. These techniques confirmed the association between toxin-producing *Clostridium* species and ASD (De Angelis et al., 2013; Song et al., 2004; Tomova et al., 2015; Wang et al., 2011, 2013). Siblings showed intermediate levels of changed *Clostridium* species suggesting that the living environment is of great influence (Parracho et al., 2005). In two studies, the anaerobic bacillus, *Desulfovibrio*, was found to be increased and associated with ASD severity (Finegold et al., 2010; Tomova et al., 2015) and non-ASD siblings contained intermediate amount of *Desulfovibrio*. Dysbiosis is also evident at phylum level in ASD where the fecal ratio of Bacteriodetes/Firmicutes is enhanced (De Angelis et al., 2013; Finegold et al., 2010) and on the other hand this ratio also is reported to be decreased in intestinal biopsy material and stool samples (Tomova et al., 2015; Williams et al., 2011, 2012). Possibly a decrease in beneficial bacteria (*Bifidobacterium* species, *Eubacteriaceae*, *Enterobactericeae*) is associated with gastrointestinal problems in ASD (De Angelis et al., 2013; Wang et al., 2011). Four studies reported on increases in *Sutterella* in the stool of ASD patients

Table 1 Recent Studies on Gut Microbiome in ASD

Study (Year)	Country	Study Group: ASD (GI+/GI−)	SIB (GI+/GI−)	CON (GI+GI−)	Sample Type	Method	Changes in Fecal Microbiome in ASD
Finegold et al. (2002)	USA	13	–	8	Stool	Bacterial cultures	↑ Nine species of *Clostridium*
Song, Liu, and Finegold (2004)	USA	15	–	8	Stool	16S rRNA gene sequencing	↑ *C. bolteae* and cluster I/IX
Parracho, Bingham, Gibson, and McCartney (2005)	United Kingdom	58	12	10	Stool	FISH analysis	↑ *C. histolyticum* and cluster I/II Siblings show intermediate levels
Finegold et al. (2010)	USA	33 (33/0)	7 (0/7)	8 (0/8)	Stool	16S rRNA gene sequencing	↑ Bacteroidetes and Proteobacteria: *Desulfovibrio*, *B. Alkaliflexus*, *Acetanaerobacterium*, *Parabacteroides* ↓ Firmicutes and Actinobacteria: *Clostridium*, *Weissella*, *Turicibacter*, *Anaerofilum*, *Pseudoramibacter*, *Ruminococcus*, *Streptococcus*

Continued

Table 1 Recent Studies on Gut Microbiome in ASD—cont'd

Study (Year)	Country	Study Group			Sample Type	Method	Changes in Fecal Microbiome in ASD
		ASD (GI+/GI−)	SIB (GI+/GI−)	CON (GI+GI−)			
Adams, Johansen, Powell, Quig, and Rubin (2011)	USA	58 (58/0)	–	39 (0/39)	Stool	Bacterial cultures	↓ *Bifidobacterium* and *Enterococcus* ↑ *Bacillus* spp. (*Lactobacillus*)
Martirosian et al. (2011)	Poland	41	–	10	Stool	Bacterial cultures	↑ *Clostridium perfringens*
Williams et al. (2011), Williams, Hornig, Parekh, and Lipkin (2012)	USA	23 (23/0)	–	9 (9/0)	Intestinal biopsies	16S rRNA gene sequencing	↓ Bacteroidetes ↑ Firmicutes, Proteobacteria, *Sutterella*
Wang et al. (2011, 2013)	Australia	23 (9/14)	22 (6/16)	9 (1/8)	Stool	Targeted qPCR	↓*Bifidobacterium* spp., *Akkermansia muciniphilia* ↑ *Sutterella* spp. (↑ *Clostridium diffile* ns.) ↑ *Ruminococcus torques* (only in ASD-GI+)
Gondalia et al. (2012)	Australia	51 (28/23)	53 (4/49)	–	Stool	16S rRNA gene sequencing	No differences
Kang et al. (2013)	USA	20 (20/0)	–	20 (0/20)	Stool	16S rRNA gene sequencing	↓ *Prevotella*, *Coprocuccus*, *Veillonellaceae*

De Angelis et al. (2013)	Italy	10	10	10	Stool	16S rRNA gene sequencing	↑*Caloramator*, *Sarcina*, *Clostridium*, *Sutterellaceae* ↓ *Eubacterium*, *Bifidobacterium*
Son et al. (2015)	USA	59 (25/34)	44 (13/31)	–	Stool	16S rRNA gene sequencing	No differences
Tomova et al. (2015)	Slovakia	10	10	10	Stool	Targeted qPCR	↓ *Bacteroidetes/Firmicutes* ↑ *Lactobacillus* spp. (↑*Clostridia* cluster I and *Desulfovibrio*, ns.) *Desulfovibrio* spp.: strong positive association with ASD severity

Based on Cao et al., 2013; Rosenfeld, 2015; Luna et al., 2016. *ASD*, autism spectrum disorder; *CON*, unrelated controls; *FISH*, fluorescence in sit hybridization; *GI+/GI–:* with/without gastrointestinal dysfunction; *ns.*, non significant; *SIB*, non-ASD siblings; *PCR*, polymerase chain reaction.

(De Angelis et al., 2013; Wang et al., 2011, 2013; Williams et al., 2011, 2012). Interestingly *Sutterella* spp. was predominantly found in ASD patients suffering from gastrointestinal problems and not in children with gastrointestinal symptoms only (Williams et al., 2012).

Taken together, the microbiome data suggests that in the stool of patients with ASD, elevated levels of *Clostridia*, *Desulfovibrio*, and *Sutterella* are evident. *Prevotella* and *Bifidobacter* seem to be reduced in ASD. At phylum levels the *Bacteroidetes/Firmicutus* ratio's in ASD did not show consistent results over different cohorts. In conclusion, significant, but not consistent, distinctive different microbiome compositions have been demonstrated in ASD patients with or without gastrointestinal problems compared to controls (non-ASD siblings and nonrelated healthy individuals or patients suffering from gastrointestinal problems). Larger well-designed studies are needed to confirm a distinctive ASD-subtype microbiome for better patient stratification as well for finding new targets for treatment.

4. FROM CORRELATION TO CAUSATION

The impact of microbiota in ASD has been further supported by studies in humans and rodent models for ASD through manipulation of microbiota composition and activity with antibiotics, pro- and prebiotics, and fecal microbiota transplantation (FMT).

4.1 Antibiotics

As described before parental reports suggest that the use of antibiotics improves ASD behaviors. It should be noted that some parents commented that their children's ASD symptoms became worse after receiving antibiotics (Rodakis, 2015). Two human studies report on the effects of antibiotic use on the (development) of ASD. Children with ASD have significantly more ear infections and use significantly more antibiotics (Fallon, 2005; Niehus & Lord, 2006). Moreover, the use of antibiotics during pregnancy was regarded as potential risk factors for ASD (Atladóttir, Henriksen, Schendel, & Parner, 2012). Contrary to this report, in a small clinical trial with the minimally absorbed antibiotic, vancomycin, 8 of the 10 ASD children showed improvement of behavior (Sandler et al., 2000). In addition in preclinical studies, treatment of valproic acid-induced autistic rats and Fragile X mice with the tetracycline derivate minocycline, an antibiotic with neuroprotective properties, resulted in behavioral benefits as well as in reduction of inflammation and blood–brain barrier (BBB) impairments (Dansie et al., 2013; Kumar & Sharma, 2016).

4.2 Probiotics

Limited animal studies are published on the effects of probiotic treatments on ASD behavioral deficits. Myocardial infarction in rats, associated with systemic inflammation and increased intestinal permeability, results in disturbed social behavior. *A Lactobacillus helveticus* and *Bifidobacterium longum* combination prevented and interfered with the development of post-myocardial infarction deficits in social behavior and protected intestinal barrier integrity (Arseneault-Bréard et al., 2012). A more ASD-related preclinical study was published in 2013. In a maternal immune activation (MIA) mouse model of ASD, the male offspring developed ASD-like behavior associated with an altered intestinal microbiota composition (Hsiao et al., 2013). Oral treatment of the MIA offspring with the human *Bacteroides fragilis* ameliorated deficits in communicative and stereotype behaviors as well as the associated intestinal permeability problems and changes in the intestinal microbial composition. However, no effects of this commensal were observed on the MIA-induced social behavior impairments. Of interest is the finding that behavioral deficits (e.g. social behavior) were caused by the *Clostridium*-associated metabolite 4-ethylphenyl sulfate, a molecule resembling the only identified (urinary) biomarker for ASD *p*-cresol, an environmental organic aromatic compound (Altieri et al., 2011; Gabriele et al., 2014).

Epidemiological studies suggest that maternal obesity during pregnancy is associated with a higher risk of ASD (Connolly et al., 2016; Sullivan, Riper, Lockard, & Valleau, 2015). A recent mouse study shows that maternal high-fat diet-induced social deficits and intestinal dysbiosis (marked reduction of *Lactobacillus reuteri*) via reduction of oxytocin immunoreactive neurons and impaired ventral tegmental area plasticity (Buffington et al., 2016). Colonization of the maternal high-fat diet-exposed offspring with *L. reuteri* restored the disturbed social behavior.

Only a limited amount of clinical studies have been performed to study the effects of probiotic treatment in ASD. In a follow-up study, 75 Finnish children were treated with *Lactobacillus rhamnosis GG* or placebo during the first 6 months of life and were followed-up for 13 years (Pärtty et al., 2015). At the age of 13 years ASD, specifically Asperger syndrome, or ADHD was diagnosed in 6/35 (17%) of the placebo-treated children whereas none in the probiotic group. A significantly lower number of *B. longum* was found in fecal samples of children with neuropsychiatric disorders compared to healthy children. These preliminary finding demonstrate a possible preventive effect of *L. rhamnosis* GG on later development of ADHD and ASD. Early in 2016, a randomized controlled clinical trial in ASD was approved to study the effect of a probiotic mixture containing *Streptococcus thermophilus, Bifidobacterium breve, B. longum, Bifidobacterium infantis, Lactobacillus acidophilus, Lactobacillus plantarum, Lactobacillus paracasei, and Lactobacillus delbrueckii* subsp. *bulgaricus* (Santocchi et al., 2016). One hundred children suffering from ASD with or without gastrointestinal symptoms will receive a placebo or probiotic supplemented diet for 6 months. At baseline, after 3 and 6 months the following will be evaluated: intestinal symptoms, ASD severity, affective and behavioral comorbid symptoms; plasma, urine and fecal biomarkers and neurophysiological patterns.

4.3 Fecal Microbiota Transplantation

One of the first studies showing that FMT can change behavior in mice was published in 2011 and clearly showed the impact of the host microbiota on behavior (Bercik et al., 2011). Balb/c mice when compared to NIH Swiss mice display more timid and anxious behavior, indicating influence of genetic background. However, when germ-free Balb/c mice are colonized with NIH microbiota they showed more exploratory behavior associated with higher levels of brain-derived neurotropic factor (BDNF) in the amygdala and hippocampus. In addition, another study demonstrates that social behavioral deficits of germ-free mice were normalized after conventional bacterial colonization showing that microbiota are crucial for the development of normal social behavior (Desbonnet, Clarke, Shanahan, Dinan, & Cryan, 2014). No other studies have been performed with FMT in rodent models for ASD, however, a human trial is on the way.

In 2014 a FDA-approved open-label clinical trial was initiated at the Arizona State University Autism/Asperger's Research Program examining the effect of FMT in children that have both ASD and gastrointestinal problems (Adams, 2015). The children aged 7–17 years old will receive a treatment with vancomycin to deplete resident bacteria followed by a FMT from a healthy donor. FMT is only successfully used for the treatment of *Clostridium difficile* infections (Vrieze et al., 2013). The ASD-BBT trial is still ongoing, but not recruiting participants. There is a case report of beneficial effects in two ASD children receiving FMT (Aroniadis & Brandt, 2013).

5. POSSIBLE MECHANISMS OF MICROBIOME–BRAIN AXIS IN AUTISM

Because of the fact that gastrointestinal disturbances correlate with ASD, it is hypothesized that the presence of an altered microbiome and associated gastrointestinal problems makes an individual with a genetic predisposition for ASD more prone to express or to increase severity of the autistic phenotype (de Theije et al., 2011). In Fig. 1 the possible pathways are shown by which the microbiome in the intestinal tract, possibly through immunological and neuronal pathways, can influence central neuronal functioning and behavior.

5.1 Barrier Pathways

The intestinal barrier is regulated via the microbiota and their metabolites. The dysbiosis in ASD is associated with increased permeability of the gastrointestinal tract referred to as "leaky gut" (De Magistris et al., 2010). Disruption of the intestinal barrier leads to the entry of endotoxins and other bacterial products into the bloodstream. For example, lipopolysaccharide (LPS) is a potent endotoxin of cell walls of gram-negative bacteria and it has been demonstrated that LPS can alter neuronal as well as microglial activity in regions involved in emotional control such as the amygdala (Audet, Jacobson-Pick, Wann, & Anisman, 2011; Haba et al., 2012; Qin et al., 2007; Van Heesch et al., 2013). In ASD, LPS serum levels were significantly higher compared to healthy individuals and correlated with impaired social behavioral scores (Emanuele et al., 2010). Besides transmission of bacterial products also antigen, virulence factors or other pathogens might get systemically in the case of disrupted intestinal barrier. These factors may also impact brain function and inflammatory processes. Improving the epithelial barrier in ASD may reduce the entrance of microbial products and thereby normalizing this gut–brain pathway.

The BBB protects the brain against the infiltration of pathogens and harmful particles into the brain parenchyma. The integrity of this barrier is essential for normal brain development and function. In germ-free mice, it has been demonstrated that the absence of intestinal microbiota leads to an increased permeability of the BBB due to a reduction of endothelial tight junction molecules expression (occludin and claudin-5) (Braniste et al., 2014). The reduced BBB of germ-free mice was already evident in utero. BBB dysfunction can be caused by multiple prenatal and postnatal risk factors that are also evident in ASD.

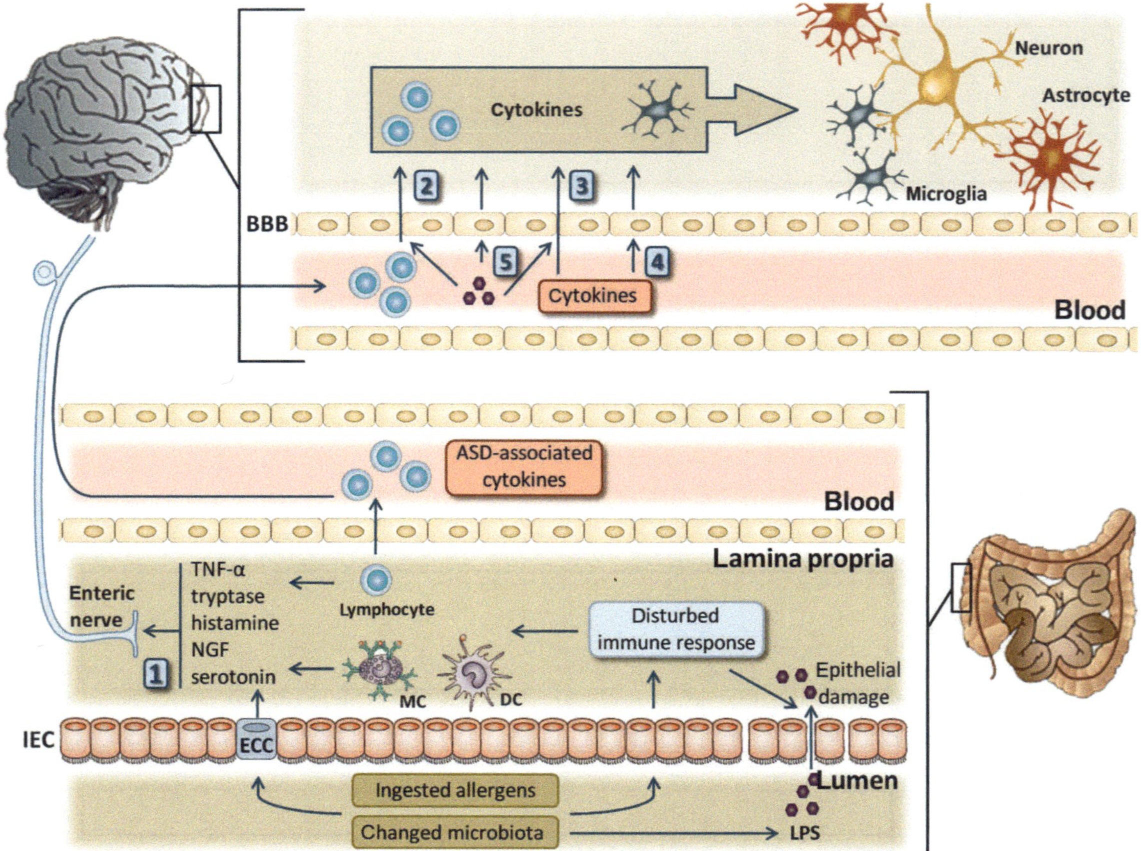

Fig. 1 Possible pathways involved in microbiome–gut–brain axis in ASD. *BBB*, blood brain barrier; *LPS*, lipopolysaccharide; *EEC*, enterochromaffin epithelial cell; *IEC*, intestinal epithelial cells; *MC*, mast cell; *DC*, dendritic cell; *TNF*, tumor necrosis factor; *NGF*, nerve growth factor. *Adapted from de Theije et al. (2011).*

5.2 Neuronal Pathways

The ENS is a complex network of neurons in the intestinal tract that independently regulate gastrointestinal functions and is often regarded as our "second" brain. Intestinal microbiota can signal via the ENS or via the afferent fibers of the vagus nerve to the CNS (Douglas-Escobar, Elliott, & Neu, 2013; Stilling, Bordenstein, Dinan, & Cryan, 2014). Using germ-free mice or studies with the probiotic *L. reuteri*, it has been demonstrated that the microbiota are important for a proper excitability of the ENS (Kunze et al., 2009; McVey Neufeld, Mao, Bienenstock, Foster, & Kunze, 2013). This might have consequences for brain functioning.

The vagus nerve is another important neuronal pathway in the microbiome–gut–brain axis. Studies in vagotomized mice have, for example, shown that the antidepressive and anxiolytic effect as well as altered gamma-aminobutyric acid (GABA) receptor expression of *L. rhamnosis* in mice was dependent on the vagus nerve (Bravo et al., 2011). A similar role for the vagus nerve has been demonstrated in bacterial infection-associated behavioral changes (Goehler et al., 2005; Wang et al., 2002). In contrast, antibiotic-induced microbiota alterations and associated behavioral deficits in mice did not depend on the vagus nerve (Bercik et al., 2011). In a study with ASD patients suffering from epilepsy, stimulation of the vagus nerve, besides reducing the seizure frequency, resulted in improved verbal skills, mood, and alertness (Park, 2003). In about 30% of ASD patients epilepsy is observed (Berney, 2000). These data suggest that vagus stimulation possibly through a "healthy" microbiome might be beneficial in ASD.

Several bacterial strains can produce monoamines, such as noradrenaline (*Escherichia* spp., *Bacillus* spp., and *Saccharomyces* spp.), dopamine (*Bacillus* spp.), and serotonin (*Candida* spp., *Streptococcus* spp., *Escherichia* spp., and *Enterococcus* spp.) as well as GABA (*Lactobacillus* spp. and *Bifidobacterium* spp.) and acetylcholine (*Lactobacillus* spp.) (Al Mardini, Al Jumaili, Record, & Burke, 1991; Barrett, Ross, O'Toole, Fitzgerald, & Stanton, 2012; Cryan & Dinan, 2012; Li & Cao, 2010). It might be that microbial neurotransmitters affect the ENS and afferent nerves function directly or via the intestinal epithelium. Based on the fact that stress-related host neurotransmitter release increases the proliferation rate as well as the activity of intestinal microbiota (Karavolos et al., 2011; Lyte, 2014; Pande, Suong, Bossier, & Defoirdt, 2014), it has been postulated that microbiota-derived neurotransmitters have a primary role in sustainability of the microbes themselves in the intestinal tract in stressful situations.

The prefrontal cortex is essential for social behavior and implicated in ASD (Amaral, Schumann, & Nordahl, 2008). Through a genome-wide transcriptome profiling approach, very recently it has been shown that in the prefrontal of germ-free mice axons are hypermyelinated. Colonization with conventional microbiota could reverse this hypermyelination (Hoban et al., 2016). These data demonstrate that the intestinal microbiome is necessary for appropriate regulation of myelin-related genes in the PFC. This is of particular interest since germ-free mice exhibit ASD-like deficits in social behavior. In addition, hypermyelination and accelerated brain growth are described in children with ASD (Ben Bashat et al., 2007). Mechanistic studies should be performed to examine whether specific bacterial metabolites, the ENS or the nervus vagus plays a role in the myelinating properties of the intestinal microbiota.

Taken together, the role of the ENS, the vagus nerve and bacterial neuroactive metabolites and molecular pathways in relation to the microbiome–gut–brain axis remains to be established in ASD.

5.3 The Serotonin Pathway: Neurotransmitter and Mediator of Inflammation

Serotonin (5-hydroxytryptamine) is a neurotransmitter important for regulating intestinal secretion, motility, and pain perception (Costedio, Hyman, & Mawe, 2007; McLean, Borman, & Lee, 2007). In the brain, serotonin is important for the regulation of mood and cognition (Cryan & Leonard, 2000). Serotonin synthesis in the intestines and brain depends on the availability of dietary tryptophan. Of special interest is a potential role for the serotonergic system in the pathogenesis of ASD, since already in the 1960s increased levels of blood serotonin are described in children with ASD (Anderson et al., 1987; Hanley, Stahl, & Freedman, 1977; Schain & Freedman, 1961). In contrast, serotonin synthesis in the brain is decreased in patients with ASD (Chugani et al., 1999). ASD-related hyperserotonemia is thought to be cased by genetic, gastrointestinal, or immune changes. Very recently, a significant correlation between whole-blood serotonin levels and intestinal symptoms in ASD was demonstrated (Marler et al., 2016). Regarding the low-grade intestinal inflammation, blood hyperserotonemia, and low serotonin synthesis in the brain observed in ASD, the following hypothesis was postulated (de Theije et al., 2011): Following inflammation in the intestinal tract, serotonin is produced by enterochromaffin epithelial cells and intestinal immune cells (mast cells and platelets), resulting in changes in motility, secretion, vasodilation, and increased (vascular) permeability. These serotonin-induced intestinal changes lead to functional intestinal dysmotility, problems in stool consistency, low-grade inflammation, and abdominal pain. Because of the increased use of dietary tryptophan in the intestinal tract during inflammation, less tryptophan will be available for the brain. Brain serotonin levels will be reduced resulting in mood and cognitive dysfunction in ASD. Indeed dietary tryptophan depletion in ASD patients resulted in increased autistic behavior (McDougle et al., 1996). In addition, intestinal dysbiosis in ASD might also affect tryptophan availability and metabolism. Indirect microbial regulation of tryptophan metabolism and serotonin synthesis is demonstrated in studies using germ-free mice. These mice have high levels of tryptophan and serotonin in their circulation, which was restored after recolonization (Wikoff et al., 2009). In a murine model of ASD induced by prenatal exposure to valproic acid, the impairments of social behavior were associated with intestinal inflammation and a disturbed serotonergic system in the brain and intestinal tract (de Theije, Koelink, et al., 2014). In the prefrontal cortex as well as in the amygdala, reduced levels of serotonin and increased turnover were found in VPA-exposed male offspring. The reduction in intestinal serotonin in VPA-exposed mice was attributable to reduced number of serotonin-positive cells (possibly enterochromaffin cells) in the small intestine. The latter observation was significantly correlated with VPA-induced changes in microbiota composition and activity (de Theije, Wopereis, et al., 2014).

Moreover, inflammatory mediators (specifically, interferon γ, INF γ) can activate indoleamine-2,3-dioxygenase and thereby skew tryptophan metabolism toward the kynurenine pathway and away from serotonin production. Several probiotic lactic acid-producing strains examined in rodents induced IDO inhibition and reduced levels of kynurenine in the plasma demonstrating that the gut microbiota can influence host tryptophan metabolism (Freewan et al., 2013; Valladares et al., 2013). Finally, the intestinal microbiota can directly affect the availability of tryptophan to the host through utilizing the dietary amino acid themselves (O'Mahony, Clarke, Borre, Dinan, & Cryan, 2015).

More studies are needed to establish the role of the microbiome in the disturbed serotonergic pathway in ASD.

5.4 Intestinal Immune System Pathway

The intestinal microbiota are very important for the development of the mucosal immune system in early life that, in turn, being the largest immune organ of the body, regulates the systemic immune system (Macpherson & Uhr, 2004; Vighi, Marcucci, Sensi, Di Cara, & Frati, 2008). Early life manifestation of allergic disease is associated with deficits in neurodevelopment and behavior such as enhanced internalizing behavior and low social emotional scores (Meldrum et al., 2012). Parents of ASD children report more often food allergies than parents of healthy children (Chandler et al., 2013; Gurney, McPheeters, & Davis, 2006). Serum levels of immunoglobulins IgA, IgG, and IgM specific for cow's milk-derived allergens and total IgE were shown to be increased in ASD children (Lucarelli et al., 1995) and peripheral blood mononuclear cells of ASD patients produce more proinflammatory cytokines upon stimulation cow's milk-derived allergens (Jyonouchi, Geng, Ruby, Reddy, & Zimmerman-Bier, 2005). The persistent default state of mucosal immune tolerance observed in food allergy is strongly associated with a changed microbiota composition such as enhanced *Bacteroidetes* and *Enterobacter* (Bunyavanich et al., 2016). The majority of allergies are characterized by a Thelper 2-type immune response with the characteristic cytokines interleukin (IL) 4, IL5, and IL13. Supporting the role of allergy in ASD, PBMCs of children produced significant higher levels of the mentioned cytokines (Gupta, Aggarwal, Rashanravan, & Lee, 1998; Molloy et al., 2006). In addition, less IL-10-producing T cell are present in the periphery and intestinal mucosa as well as reduced plasma levels of tumor necrosis factorβ in ASD patients suffering from intestinal problems (Ashwood & Wakefield, 2006; Okada et al., 2007). Taken together, there seems to be a disturbed T cell balance in the intestinal tract of ASD patients. It is hypothesized that during allergic exacerbation proinflammatory cytokines as well as mast cell mediators are able to trigger enteric neurons and signal through afferent pathway of vagal and spinal nerves to the CNS. Recently, ASD-like behavioral changes in cow's milk allergic Balb/c mice are demonstrated to be accompanied by reduced dopaminergic activity in the prefrontal cortex (de Theije, Wu, et al., 2014) and dietary intervention with *B. breve* and prebiotic fibers restored the disturbed social behavior, reduced BBB permeability, and reduced hippocampal BDNF (Y.E. Borre et al., unpublished data). Indeed, it has been demonstrated that a gluten and/or cow's milk-free diet, besides reducing the gastrointestinal problems, improve behavior in ASD patients as well (Elder et al., 2006; Rodakis, 2015). In conclusion, together with a genetic predisposition, food-related allergic immune activation may exacerbate behavioral abnormalities in ASD.

5.5 Bacterial Metabolites

Intestinal bacteria produce several volatile short-chain fatty acids (SCFAs) upon fermentation of nondigestible carbohydrates in the colon. Butyrate, acetate, and propionate are often regarded to have health benefits such as energy supply for epithelial cells, restoring epithelial barrier function, antiinflammatory, and immunomodulating activities (Richards, Yap, McLeod, Mackay, & Mariño, 2016). Human epidemiological studies as well as animal studies suggest that enteric SCFAs may also have a pathological role in ASD (MacFabe, 2015). Propionic acid, a major SCFA produced by *Clostridia*, *Bacteroides*, and *Desulfovibrio* that have been associated with ASD, can induce ASD-like behavioral deficits in rats (Foley, MacFabe, Kavaliers, & Ossenkopp, 2015; Foley, MacFabe, Vaz, Ossenkopp, & Kavaliers, 2014; Thomas et al., 2012). The detrimental effects of propionic acid are suggested to be through mitochondrial and epigenetic modulation of ASD-associated genes. Indeed, elevated levels of SCFAs are described in the stool of ASD children (Wang, Conlon, Christophersen, Sorich, & Angley, 2014, Wang et al., 2012). However, it needs to be established whether these elevated intestinal levels of SCFA are high enough to reach substantial levels in the brain.

6. CONCLUSION

Future studies on the role of the intestinal microbiota in the gut–brain axis in ASD could greatly enhance the understanding of the pathogenesis of ASD, lead to the identification of biomarkers for better patient's stratification, and realize the identification of new targets for therapy. All together, the management of the multifactorial ASD needs new and integrated therapeutic approaches. Pharmacological bioactive molecules as well as medical nutrition (multi-)targeting the microbiome–gut–brain axis could be such an approach.

REFERENCES

Adams, J. B. (2015). *Treating gastrointestinal problems in children with autism using Beneficial Bacteria Treatment (BBT)*. Arizona State University. https://clinicaltrials.gov/ct2/show/record/NCT02504554.

Adams, J. B., Johansen, L. J., Powell, L. D., Quig, D., & Rubin, R. A. (2011). Gastrointestinal flora and gastrointestinal status in children with autism—Comparisons to typical children and correlation with autism severity. *BMC Gastroenterology*, *11*, 22.

Al Mardini, H., Al Jumaili, B., Record, C. O., & Burke, D. (1991). Effect of protein and lactulose on the production of gamma-aminobutyric acid by faecal Escherichia coli. *Gut*, *32*(9), 1007–1010.

Altieri, L., Neri, C., Sacco, R., Curatolo, P., Benvenuto, A., Muratori, F., et al. (2011). Urinary p-cresol is elevated in small children with severe autism spectrum disorder. *Biomarkers*, *16*(3), 252–260.

Amaral, D. G., Schumann, C. M., & Nordahl, C. W. (2008). Neuroanatomy of autism. *Trends in Neurosciences*, *31*(3), 137–145.

American Psychiatric Association. (2016). *Diagnostic and statistical manual of mental disorders* (5th ed.). . Washington DC: American Psychiatric press.

Anderson, G. M., Freedman, D. X., Cohen, D. J., Volkmar, F. R., Hoder, E. L., McPhedran, P., et al. (1987). Whole blood serotonin in autistic and normal subjects. *Journal of Child Psychology and Psychiatry*, *28*(6), 885–900.

Aroniadis, O. C., & Brandt, L. J. (2013). Fecal microbiota transplantation: Past, present and future. *Current Opinion in Gastroenterology*, *29*(1), 79–84.

Arseneault-Bréard, J., Rondeau, I., Gilbert, K., Girard, S. A., Tompkins, T. A., Godbout, R., et al. (2012). Combination of Lactobacillus helveticus R0052 and Bifidobacterium longum R0175 reduces post-myocardial infarction depression symptoms and restores intestinal permeability in a rat model. *The British Journal of Nutrition*, *107*(12), 1793–1799.

Ashwood, P., & Wakefield, A. J. (2006). Immune activation of peripheral blood and mucosal CD3+ lymphocyte cytokine profiles in children with autism and gastrointestinal symptoms. *Journal of Neuroimmunology*, *173*(1–2), 126–134.

Atladóttir, H. Ó., Henriksen, T. B., Schendel, D. E., & Parner, E. T. (2012). Autism after infection, febrile episodes, and antibiotic use during pregnancy: An exploratory study. *Pediatrics*, *130*(6), e1447–e1454.

Audet, M. C., Jacobson-Pick, S., Wann, B. P., & Anisman, H. (2011). Social defeat promotes specific cytokine variations within the prefrontal cortex upon subsequent aggressive or endotoxin challenges. *Brain, Behavior, and Immunity*, *25*(6), 1197–1205.

Barrett, E., Ross, R. P., O'Toole, P. W., Fitzgerald, G. F., & Stanton, C. (2012). γ-Aminobutyric acid production by culturable bacteria from the human intestine. *Journal of Applied Microbiology*, *113*(2), 411–417.

Ben Bashat, D., Kronfeld-Duenias, V., Zachor, D. A., Ekstein, P. M., Hendler, T., Tarrasch, R., et al. (2007). Accelerated maturation of white matter in young children with autism: A high b value DWI study. *NeuroImage*, *37*(1), 40–47.

Bercik, P., Denou, E., Collins, J., Jackson, W., Lu, J., Jury, J., et al. (2011). The intestinal microbiota affect central levels of brain-derived neurotropic factor and behavior in mice. *Gastroenterology*, *141*(2), 599–609.

Berney, T. P. (2000). Autism: An evolving concept. *The British Journal of Psychiatry*, *176*, 20–25.

Bernier, R., Golzio, C., Xiong, B., Stessman, H. A., Coe, B. P., Penn, O., et al. (2014). Disruptive CHD8 mutations define a subtype of autism early in development. *Cell*, *158*(2), 263–276.

Blumberg, S. J., Bramlett, M. D., Kogan, M. D., Schieve, L. A., Jones, J. R., & Lu, M. C. (2013). Changes in prevalence of parent-reported autism spectrum disorder in school-aged U.S. children: 2007 to 2011-2012. *National Health Statistics Reports*, *65*, 1–11.

Borre, Y. E., O'Keeffe, G. W., Clarke, G., Stanton, C., Dinan, T. G., & Cryan, J. F. (2014). Microbiota and neurodevelopmental windows: Implications for brain disorders. *Trends in Molecular Medicine*, *20*(9), 509–518.

Braniste, V., Al-Asmakh, M., Kowal, C., Anuar, F., Abbaspour, A., Tóth, M., et al. (2014). The gut microbiota influences blood-brain barrier permeability in mice. *Science Translational Medicine, 6*(263), 263ra158.

Bravo, J. A., Forsythe, P., Chew, M. V., Escaravage, E., Savignac, H. M., Dinan, T. G., et al. (2011). Ingestion of Lactobacillus strain regulates emotional behavior and central GABA receptor expression in a mouse via the vagus nerve. *Proceedings of the National Academy of Sciences of the United States of America, 108*(38), 16050–16055.

Buffington, S. A., Di Prisco, G. V., Auchtung, T. A., Ajami, N. J., Petrosino, J. F., & Costa-Mattioli, M. (2016). Microbial reconstitution reverses maternal diet-induced social and synaptic deficits in offspring. *Cell, 165*(7), 1762–1775.

Bunyavanich, S., Shen, N., Grishin, A., Wood, R., Burks, W., Dawson, P., et al. (2016). Early-life gut microbiome composition and milk allergy resolution. *The Journal of Allergy and Clinical Immunology*.

Cao, X., Lin, P., Jiang, P., & Li, C. (2013). Characteristics of the gastrointestinal microbiome in children with autism spectrum disorder: A systematic review. *Shanghai Archives of Psychiatry, 25*(6), 342–353.

Carabotti, M., Scirocco, A., Maselli, M. A., & Severi, C. (2015). The gut-brain axis: Interactions between enteric microbiota, central and enteric nervous systems. *Annals of Gastroenterology, 28*(2), 203–209.

Carter, M. T., & Scherer, S. W. (2013). Autism spectrum disorder in the genetics clinic: A review. *Clinical Genetics, 83*(5), 399–407.

Chandler, S., Carcani-Rathwell, I., Charman, T., Pickles, A., Loucas, T., Meldrum, D., et al. (2013). Parent-reported gastro-intestinal symptoms in children with autism spectrum disorders. *Journal of Autism and Developmental Disorders, 43*(12), 2737–2747.

Christensen, D. L., Baio, J., Braun, K. V., Bilder, D., Charles, J., Constantino, J. N., et al. (2016). Prevalence and characteristics of autism spectrum disorder among children aged 8 years—Autism and developmental disabilities monitoring network, 11 sites, United States, 2012. *MMWR Surveillance Summaries, 65*(3), 1–23.

Chugani, D. C., Muzik, O., Behen, M., Rothermel, R., Janisse, J. J., Lee, J., et al. (1999). Developmental changes in brain serotonin synthesis capacity in autistic and nonautistic children. *Annals of Neurology, 45*(3), 287–295.

Connolly, N., Anixt, J., Manning, P., Ping-I Lin, D., Marsolo, K. A., & Bowers, K. (2016). Maternal metabolic risk factors for autism spectrum disorder—An analysis of electronic medical records and linked birth data. *Autism Research: Official Journal of the International Society for Autism Research, 9*, 829–837.

Costedio, M. M., Hyman, N., & Mawe, G. M. (2007). Serotonin and its role in colonic function and in gastrointestinal disorders. *Diseases of the Colon and Rectum, 50*(3), 376–388.

Croen, L. A., Zerbo, O., Qian, Y., Massolo, M. L., Rich, S., Sidney, S., et al. (2015). The health status of adults on the autism spectrum. *Autism, 19*(7), 814–823.

Cryan, J. F., & Dinan, T. G. (2012). Mind-altering microorganisms: The impact of the gut microbiota on brain and behaviour. *Nature Reviews. Neuroscience, 13*(10), 701–712.

Cryan, J. F., & Leonard, B. E. (2000). 5-HT1A and beyond: The role of serotonin and its receptors in depression and the antidepressant response. *Human Psychopharmacology, 15*(2), 113–135.

Dansie, L. E., Phommahaxay, K., Okusanya, A. G., Uwadia, J., Huang, M., Rotschafer, S. E., et al. (2013). Long-lasting effects of minocycline on behavior in young but not adult Fragile X mice. *Neuroscience, 246*, 186–198.

De Angelis, M., Francavilla, R., Piccolo, M., De Giacomo, A., & Gobbetti, M. (2015). Autism spectrum disorders and intestinal microbiota. *Gut Microbes, 6*(3), 207–213.

De Angelis, M., Piccolo, M., Vannini, L., Siragusa, S., De Giacomo, A., Serrazzanetti, D. I., et al. (2013). Fecal microbiota and metabolome of children with autism and pervasive developmental disorder not otherwise specified. *PLoS One, 8*(10), e76993.

De Magistris, L., Familiari, V., Pascotto, A., Sapone, A., Frolli, A., Iardino, P., et al. (2010). Alterations of the intestinal barrier in patients with autism spectrum disorders and in their first-degree relatives. *Journal of Pediatric Gastroenterology and Nutrition, 51*(4), 418–424.

De Rubeis, S., He, X., Goldberg, A. P., Poultney, C. S., Samocha, K., Cicek, A. E., et al. (2014). Synaptic, transcriptional and chromatin genes disrupted in autism. *Nature, 515*(7526), 209–215.

de Theije, C. G., Koelink, P. J., Korte-Bouws, G. A., Lopes da Silva, S., Korte, S. M., Olivier, B., et al. (2014-a). Intestinal inflammation in a murine model of autism spectrum disorders. *Brain, Behavior, and Immunity, 37*, 240–247.

de Theije, C. G., Wopereis, H., Ramadan, M., van Eijndthoven, T., Lambert, J., Knol, J., et al. (2014). Altered gut microbiota and activity in a murine model of autism spectrum disorders. *Brain, Behavior, and Immunity, 37*, 197–206.

de Theije, C. G., Wu, J., da Silva, S. L., Kamphuis, P. J., Garssen, J., Korte, S. M., et al. (2011). Pathways underlying the gut-to-brain connection in autism spectrum disorders as future targets for disease management. *European Journal of Pharmacology, 668*(Suppl. 1), S70–S80.

de Theije, C. G., Wu, J., Koelink, P. J., Korte-Bouws, G. A., Borre, Y., Kas, M. J., et al. (2014). Autistic-like behavioural and neurochemical changes in a mouse model of food allergy. *Behavioural Brain Research, 261*, 265–274.

Desbonnet, L., Clarke, G., Shanahan, F., Dinan, T. G., & Cryan, J. F. (2014). Microbiota is essential for social development in the mouse. *Molecular Psychiatry, 19*(2), 146–148.

Douglas-Escobar, M., Elliott, E., & Neu, J. (2013). Effect of intestinal microbial ecology on the developing brain. *JAMA Pediatrics, 167*(4), 374–379.

Ehninger, D., & Silva, A. J. (2009). Genetics and neuropsychiatric disorders: Treatment during adulthood. *Nature Medicine, 15*, 849–850.

Elder, J. H., Shankar, M., Shuster, J., Theriaque, D., Burns, S., & Sherrill, L. (2006). The gluten-free, casein-free diet in autism: Results of a preliminary double blind clinical trial. *Journal of Autism and Developmental Disorders, 36*(3), 413–420.

Emanuele, E., Orsi, P., Boso, M., Broglia, D., Brondino, N., Barale, F., et al. (2010). Low-grade endotoxemia in patients with severe autism. *Neuroscience Letters, 471*(3), 162–165.

Fallon, J. (2005). Could one of the most widely prescribed antibiotics amoxicillin/clavulanate "augmentin" be a risk factor for autism? *Medical Hypotheses, 64*(2), 312–315.

Finegold, S. M., Dowd, S. E., Gontcharova, V., Liu, C., Henley, K. E., Wolcott, R. D., et al. (2010). Pyrosequencing study of fecal microflora of autistic and control children. *Anaerobe, 16*(4), 444–453.

Finegold, S. M., Molitoris, D., Song, Y., Liu, C., Vaisanen, M. L., Bolte, E., et al. (2002). Gastrointestinal microflora studies in late-onset autism. *Clinical Infectious Diseases, 35*(Suppl. 1), S6–S16.

Foley, K. A., MacFabe, D. F., Kavaliers, M., & Ossenkopp, K. P. (2015). Sexually dimorphic effects of prenatal exposure to lipopolysaccharide, and prenatal and postnatal exposure to propionic acid, on acoustic startle response and prepulse inhibition in adolescent rats: Relevance to autism spectrum disorders. *Behavioural Brain Research, 278*, 244–256.

Foley, K. A., MacFabe, D. F., Vaz, A., Ossenkopp, K. P., & Kavaliers, M. (2014). Sexually dimorphic effects of prenatal exposure to propionic acid and lipopolysaccharide on social behavior in neonatal, adolescent, and adult rats: Implications for autism spectrum disorders. *International Journal of Developmental Neuroscience, 39*, 68–78.

Freewan, M., Rees, M. D., Plaza, T. S., Glaros, E., Lim, Y. J., Wang, X. S., et al. (2013). Human indoleamine 2,3-dioxygenase is a catalyst of physiological heme peroxidase reactions: Implications for the inhibition of dioxygenase activity by hydrogen peroxide. *The Journal of Biological Chemistry, 288*, 1548–1567.

Frye, R. E., Rose, S., Slattery, J., & MacFabe, D. F. (2015). Gastrointestinal dysfunction in autism spectrum disorder: The role of the mitochondria and the enteric microbiome. *Microbial Ecology in Health and Disease*, *26*, 27458.

Frye, R. E., & Rossignol, D. A. (2016). Identification and treatment of pathophysiological comorbidities of autism spectrum disorder to achieve optimal outcomes. *Clinical Medicine Insights. Pediatrics*, *10*, 43–56.

Gabriele, S., Sacco, R., Cerullo, S., Neri, C., Urbani, A., Tripi, G., et al. (2014). Urinary p-cresol is elevated in young French children with autism spectrum disorder: A replication study. *Biomarkers*, *19*(6), 463–470.

Goehler, L. E., Gaykema, R. P., Opitz, N., Reddaway, R., Badr, N., & Lyte, M. (2005). Activation in vagal afferents and central autonomic pathways: Early responses to intestinal infection with Campylobacter jejuni. *Brain, Behavior, and Immunity*, *19*(4), 334–344.

Gondalia, S. V., Palombo, E. A., Knowles, S. R., Cox, S. B., Meyer, D., & Austin, D. W. (2012). Molecular characterisation of gastrointestinal microbiota of children with autism (with and without gastrointestinal dysfunction) and their neurotypical siblings. *Autism Research*, *5*(6), 419–427.

Gupta, S., Aggarwal, S., Rashanravan, B., & Lee, T. (1998). Th1- and Th2-like cytokines in CD4+ and CD8+ T cells in autism. *Journal of Neuroimmunology*, *85*(1), 106–109.

Gurney, J. G., McPheeters, M. L., & Davis, M. M. (2006). Parental report of health conditions and health care use among children with and without autism: National Survey of Children's Health. *Archives of Pediatrics & Adolescent Medicine*, *160*(8), 825–830.

Haba, R., Shintani, N., Onaka, Y., Wang, H., Takenaga, R., Hayata, A., et al. (2012). Lipopolysaccharide affects exploratory behaviors toward novel objects by impairing cognition and/or motivation in mice: Possible role of activation of the central amygdala. *Behavioural Brain Research*, *228*(2), 423–431.

Hanley, H. G., Stahl, S. M., & Freedman, D. X. (1977). Hyperserotonemia and amine metabolites in autistic and retarded children. *Archives of General Psychiatry*, *34*(5), 521–531.

Hoban, A. E., Stilling, R. M., Ryan, F. J., Shanahan, F., Dinan, T. G., Claesson, M. J., et al. (2016). Regulation of prefrontal cortex myelination by the microbiota. *Translational Psychiatry*, *6*(4), e774.

Hsiao, E. Y., McBride, S. W., Hsien, S., Sharon, G., Hyde, E. R., McCue, T., et al. (2013). Microbiota modulate behavioral and physiological abnormalities associated with neurodevelopmental disorders. *Cell*, *155*(7), 1451–1463.

Iossifov, I., O'Roak, B. J., Sanders, S. J., Ronemus, M., Krumm, N., Levy, D., et al. (2014). The contribution of de novo coding mutations to autism spectrum disorder. *Nature*, *515*(7526), 216–221.

Jyonouchi, H., Geng, L., Ruby, A., Reddy, C., & Zimmerman-Bier, B. (2005). Evaluation of an association between gastrointestinal symptoms and cytokine production against common dietary proteins in children with autism spectrum disorders. *The Journal of Pediatrics*, *146*(5), 605–610.

Kang, D. W., Park, J. G., Ilhan, Z. E., Wallstrom, G., Labaer, J., Adams, J. B., et al. (2013). Reduced incidence of Prevotella and other fermenters in intestinal microflora of autistic children. *PLoS One*, *8*(7), e68322.

Karavolos, M. H., Bulmer, D. M., Spencer, H., Rampioni, G., Schmalen, I., Baker, S., et al. (2011). *Salmonella Typhi* sense host neuroendocrine stress hormones and release the toxin haemolysin E. *EMBO Reports*, *12*(3), 252–258.

Kohane, I. S. (2015). An autism case history to review the systematic analysis of large-scale data to refine the diagnosis and treatment of neuropsychiatric disorders. *Biological Psychiatry*, 77, 59–65.

Kohane, I. S., McMurry, A., Weber, G., MacFadden, D., Rappaport, L., Kunkel, L., et al. (2012). The co-morbidity burden of children and young adults with autism spectrum disorders. *PLoS One*, 7(4), e33224.

Kostic, A. D., Gevers, D., Siljander, H., Vatanen, T., Hyötyläinen, T., Hämäläinen, A. M., et al. (2015). The dynamics of the human infant gut microbiome in development and in progression toward type 1 diabetes. *Cell Host & Microbe*, *17*(2), 260–273.

Kozol, R. A., Cukier, H. N., Zou, B., Mayo, V., De Rubeis, S., Cai, G., et al. (2015). Two knockdown models of the autism genes SYNGAP1 and SHANK3 in zebrafish produce similar behavioral phenotypes associated with embryonic disruptions of brain morphogenesis. *Human Molecular Genetics*, *24*(14), 4006–4023.

Kumar, H., & Sharma, B. (2016). Minocycline ameliorates prenatal valproic acid induced autistic behaviour, biochemistry and blood brain barrier impairments in rats. *Brain Research*, *1630*, 83–97.

Kunze, W. A., Mao, Y. K., Wang, B., Huizinga, J. D., Ma, X., Forsythe, P., et al. (2009). Lactobacillus reuteri enhances excitability of colonic AH neurons by inhibiting calcium-dependent potassium channel opening. *Journal of Cellular and Molecular Medicine*, *13*(8B), 2261–2270.

Li, H., & Cao, Y. (2010). Lactic acid bacterial cell factories for gamma-aminobutyric acid. *Amino Acids*, *39*(5), 1107–1116.

Li, Q., & Zhou, J. M. (2016). The microbiota-gut-brain axis and its potential therapeutic role in autism spectrum disorder. *Neuroscience*, *324*, 131–139.

Lucarelli, S., Frediani, T., Zingoni, A. M., Ferruzzi, F., Giardini, O., Quintieri, F., et al. (1995). Food allergy and infantile autism. *Panminerva Medica*, *37*(3), 137–141.

Luna, R. A., Savidge, T. C., & Williams, K. C. (2016). The brain-gut-microbiome axis: What role does It play in autism spectrum disorder? *Current Developmental Disorders Reports*, *3*(1), 75–81.

Lyall, K., Van de Water, J., Ashwood, P., & Hertz-Picciotto, I. (2015). Asthma and allergies in children with autism spectrum disorders: Results from the CHARGE study. *Autism Research*, *8*(5), 567–574.

Lyte, M. (2014). The effect of stress on microbial growth. *Animal Health Research Reviews*, *15*(2), 172–174.

MacFabe, D. F. (2015). Enteric short-chain fatty acids: Microbial messengers of metabolism, mitochondria, and mind: Implications in autism spectrum disorders. *Microbial Ecology in Health and Disease*, *26*, 28177.

Macpherson, A. J., & Uhr, T. (2004). Compartmentalization of the mucosal immune responses to commensal intestinal bacteria. *Annals of the New York Academy of Sciences*, *1029*, 36–43.

Marler, S., Ferguson, B. J., Lee, E. B., Peters, B., Williams, K. C., McDonnell, E., et al. (2016). Brief report: Whole blood serotonin levels and gastrointestinal symptoms in autism spectrum disorder. *Journal of Autism and Developmental Disorders*, *46*(3), 1124–1130.

Martirosian, G., Ekiel, A., Aptekorz, M., Wiechuła, B., Kazek, B., Jankowska-Steifer, E., et al. (2011). Fecal lactoferrin and Clostridium spp. in stools of autistic children. *Anaerobe*, *17*(1), 43–45.

Matson, J. L., & Goldin, R. L. (2014). Diagnosing young children with autism. *International Journal of Developmental Neuroscience*, *39*, 44–48.

McDougle, C. J., Naylor, S. T., Cohen, D. J., Aghajanian, G. K., Heninger, G. R., & Price, L. H. (1996). Effects of tryptophan depletion in drug-free adults with autistic disorder. *Archives of General Psychiatry*, *53*(11), 993–1000.

McElhanon, B. O., McCracken, C., Karpen, S., & Sharp, W. G. (2014). Gastrointestinal symptoms in autism spectrum disorder: A meta-analysis. *Pediatrics*, *133*(5), 872–883.

McLean, P. G., Borman, R. A., & Lee, K. (2007). 5-HT in the enteric nervous system: Gut function and neuropharmacology. *Trends in Neurosciences*, *30*(1), 9–13.

McVey Neufeld, K. A., Mao, Y. K., Bienenstock, J., Foster, J. A., & Kunze, W. A. (2013). The microbiome is essential for normal gut intrinsic primary afferent neuron excitability in the mouse. *Neurogastroenterology and Motility*, *25*(2),183-e88.

Meldrum, S. J., D'Vaz, N., Dunstan, J. A., Mori, T. A., Hird, K., Simmer, K., et al. (2012). Allergic disease in the first year of liufe is associated with differences in subsequent neurodevelopment and behavior. *Early Human Development*, *88*(7), 567–573.

Molloy, C. A., Morrow, A. L., Meinzen-Derr, J., Schleifer, K., Dienger, K., Manning-Courtney, P., et al. (2006). Elevated cytokine levels in children with autism spectrum disorder. *Journal of Neuroimmunology*, *172*(1–2), 198–205.

Mulle, J. G., Sharp, W. G., & Cubells, J. F. (2013). The gut microbiome: A new frontier in autism research. *Current Psychiatry Reports*, *15*(2), 337.

Niehus, R., & Lord, C. (2006). Early medical history of children with autism spectrum disorders. *Journal of Developmental and Behavioral Pediatrics*, *27*(2 Suppl.), S120–S127.

Okada, K., Hashimoto, K., Iwata, Y., Nakamura, K., Tsujii, M., Tsuchiya, K. J., et al. (2007). Decreased serum levels of transforming growth factor-beta1 in patients with autism. *Progress in Neuro-Psychopharmacology & Biological Psychiatry*, *31*(1), 187–190.

O'Mahony, S. M., Clarke, G., Borre, Y. E., Dinan, T. G., & Cryan, J. F. (2015). Serotonin, tryptophan metabolism and the brain-gut-microbiome axis. *Behavioural Brain Research*, *277*, 32–48.

Onore, C., Careaga, M., & Ashwood, P. (2012). The role of immune dysfunction in the pathophysiology of autism. *Brain, Behavior, and Immunity*, *26*(3), 383–392.

Pande, G. S., Suong, N. T., Bossier, P., & Defoirdt, T. (2014). The catecholamine stress hormones norepinephrine and dopamine increase the virulence of pathogenic *Vibrio anguillarum* and *Vibrio campbellii*. *FEMS Microbiology Ecology*, *90*(3), 761–769.

Park, Y. D. (2003). The effects of vagus nerve stimulation therapy on patients with intractable seizures and either Landau-Kleffner syndrome or autism. *Epilepsy & Behavior*, *4*(3), 286–290.

Parracho, H. M., Bingham, M. O., Gibson, G. R., & McCartney, A. L. (2005). Differences between the gut microflora of children with autistic spectrum disorders and that of healthy children. *Journal of Medical Microbiology*, *54*(Pt. 10), 987–991.

Pärtty, A., Kalliomäki, M., Wacklin, P., Salminen, S., & Isolauri, E. (2015). A possible link between early probiotic intervention and the risk of neuropsychiatric disorders later in childhood: A randomized trial. *Pediatric Research*, 77(6), 823–828.

Qin, H., Roberts, K. L., Niyongere, S. A., Cong, Y., Elson, C. O., & Benveniste, E. N. (2007). Molecular mechanism of lipopolysaccharide-induced SOCS-3 gene expression in macrophages and microglia. *Journal of Immunology*, *179*(9), 5966–5976.

Reddy, B. L., & Saier, M. H. (2015). Autism and our intestinal microbiota. *Journal of Molecular Microbiology and Biotechnology*, *25*(1), 51–55.

Richards, J. L., Yap, Y. A., McLeod, K. H., Mackay, C. R., & Mariño, E. (2016). Dietary metabolites and the gut microbiota: An alternative approach to control inflammatory and autoimmune diseases. *Clinical & Translational Immunology*, *5*(5), e82.

Rodakis, J. (2015). An n = 1 case report of a child with autism improving on antibiotics and a father's quest to understand what it may mean. *Microbial Ecology in Health and Disease*, *26*, 26382.

Rook, G. A., Raison, C. L., & Lowry, C. A. (2014). Microbiota, immunoregulatory old friends and psychiatric disorders. *Advances in Experimental Medicine and Biology*, *817*, 319–356.

Rosenfeld, C. S. (2015). Microbiome disturbances and autism spectrum disorders. *Drug Metabolism and Disposition*, *43*(10), 1557–1571.

Rossignol, D. A., & Frye, R. E. (2012). A review of research trends in physiological abnormalities in autism spectrum disorders: Immune dysregulation, inflammation, oxidative stress, mitochondrial dysfunction and environmental toxicant exposures. *Molecular Psychiatry*, *17*(4), 389–401.

Sandler, R. H., Finegold, S. M., Bolte, E. R., Buchanan, C. P., Maxwell, A. P., Väisänen, M. L., et al. (2000). Short-term benefit from oral vancomycin treatment of regressive-onset autism. *Journal of Child Neurology*, *15*(7), 429–435.

Santocchi, E., Guiducci, L., Fulceri, F., Billeci, L., Buzzigoli, E., Apicella, F., et al. (2016). Gut to brain interaction in Autism Spectrum Disorders: A randomized controlled trial on the role of probiotics on clinical, biochemical and neurophysiological parameters. *BMC Psychiatry*, *16*, 183.

Schain, R. J., & Freedman, D. X. (1961). Studies on 5-hydroxyindole metabolism in autistic and other mentally retarded children. *The Journal of Pediatrics*, *58*, 315–320.

Son, J. S., Zheng, L. J., Rowehl, L. M., Tian, X., Zhang, Y., Zhu, W., et al. (2015). Comparison of fecal microbiota in children with autism spectrum disorders and neurotypical siblings in the Simons simplex collection. *PLoS One*, *10*(10), e0137725.

Song, Y., Liu, C., & Finegold, S. M. (2004). Real-time PCR quantitation of clostridia in feces of autistic children. *Applied and Environmental Microbiology*, *70*(11), 6459–6465.

Stilling, R. M., Bordenstein, S. R., Dinan, T. G., & Cryan, J. F. (2014). Friends with social benefits: Host-microbe interactions as a driver of brain evolution and development? *Frontiers in Cellular and Infection Microbiology*, *4*, 147.

Sullivan, E. L., Riper, K. M., Lockard, R., & Valleau, J. C. (2015). Maternal high-fat diet programming of the neuroendocrine system and behavior. *Hormones and Behavior*, *76*, 153–161.

Thomas, R. H., Meeking, M. M., Mepham, J. R., Tichenoff, L., Possmayer, F., Liu, S., et al. (2012). The enteric bacterial metabolite propionic acid alters brain and plasma phospholipid molecular species: Further development of a rodent model of autism spectrum disorders. *Journal of Neuroinflammation*, *9*, 153.

Tomova, A., Husarova, V., Lakatosova, S., Bakos, J., Vlkova, B., Babinska, K., et al. (2015). Gastrointestinal microbiota in children with autism in Slovakia. *Physiology & Behavior*, *138*, 179–187.

Valladares, R., Bojilova, L., Potts, A. H., Cameron, E., Gardner, C., Lorca, G., et al. (2013). Lactobacillus johnsonii inhibits indoleamine 2,3-dioxygenase and alters tryp- tophan metabolite levels in BioBreeding rats. *The FASEB Journal*, *27*, 1711–1720.

Van Heesch, F., Prins, J., Konsman, J. P., Westphal, K. G., Olivier, B., Kraneveld, A. D., et al. (2013). Lipopolysaccharide-induced anhedonia is abolished in male serotonin transporter knockout rats: An intracranial self-stimulation study. *Brain, Behavior, and Immunity*, *29*, 98–103.

Vighi, G., Marcucci, F., Sensi, L., Di Cara, G., & Frati, F. (2008). Allergy and the gastrointestinal system. *Clinical and Experimental Immunology*, *153*(Suppl. 1), 3–6.

Vrieze, A., de Groot, P. F., Kootte, R. S., Knaapen, M., van Nood, E., & Nieuwdorp, M. (2013). Fecal transplant: A safe and sustainable clinical therapy for restoring intestinal microbial balance in human disease? *Best Practice & Research. Clinical Gastroenterology*, *27*(1), 127–137.

Wang, L., Christophersen, C. T., Sorich, M. J., Gerber, J. P., Angley, M. T., & Conlon, M. A. (2011). Low relative abundances of the mucolytic bacterium Akkermansia muciniphila and Bifidobacterium spp. in feces of children with autism. *Applied and Environmental Microbiology*, 77(18), 6718–6721.

Wang, L., Christophersen, C. T., Sorich, M. J., Gerber, J. P., Angley, M. T., & Conlon, M. A. (2012). Elevated fecal short chain fatty acid and ammonia concentrations in children with autism spectrum disorder. *Digestive Diseases and Sciences*, *57*(8), 2096–2102.

Wang, L., Christophersen, C. T., Sorich, M. J., Gerber, J. P., Angley, M. T., & Conlon, M. A. (2013). Increased abundance of Sutterella spp. and Ruminococcus torques in feces of children with autism spectrum disorder. *Molecular Autism*, *4*(1), 42.

Wang, L., Conlon, M. A., Christophersen, C. T., Sorich, M. J., & Angley, M. T. (2014). Gastrointestinal microbiota and metabolite biomarkers in children with autism spectrum disorders. *Biomarkers in Medicine*, *8*(3), 331–344.

Wang, X., Wang, B. R., Zhang, X. J., Xu, Z., Ding, Y. Q., & Ju, G. (2002). Evidences for vagus nerve in maintenance of immune balance and transmission of immune information from gut to brain in STM-infected rats. *World Journal of Gastroenterology*, *8*(3), 540–545.

Wikoff, W. R., Anfora, A. T., Liu, J., Schultz, P. G., Lesley, S. A., Peters, E. C., et al. (2009). Metabolomics analysis reveals large effects of gut microflora on mammalian blood metabolites. *Proceedings of the National Academy of Sciences of the United States of America*, *106*(10), 3698–3703.

Williams, B. L., Hornig, M., Buie, T., Bauman, M. L., Cho Paik, M., Wick, I., et al. (2011). Impaired carbohydrate digestion and transport and mucosal dysbiosis in the intestines of children with autism and gastrointestinal disturbances. *PLoS One*, *6*, e24585.

Williams, B. L., Hornig, M., Parekh, T., & Lipkin, W. I. (2012). Application of novel PCR-based methods for detection, quantitation, and phylogenetic characterization of Sutterella species in intestinal biopsy samples from children with autism and gastrointestinal disturbances. *mBio*, *3*, 3.

Wopereis, H., Oozeer, R., Knipping, K., Belzer, C., & Knol, J. (2014). The first thousand days—Intestinal microbiology of early life: Establishing a symbiosis. *Pediatric Allergy and Immunology*, *25*(5), 428–438.

Zerbo, O., Leong, A., Barcellos, L., Bernal, P., Fireman, B., & Croen, L. A. (2015). Immune mediated conditions in autism spectrum disorders. *Brain, Behavior, and Immunity*, *46*, 232–236.

CHAPTER FOURTEEN

The Microbiome of the Built Environment and Human Behavior: Implications for Emotional Health and Well-Being in Postmodern Western Societies

C.E. Stamper*, A.J. Hoisington[†,‡], O.M. Gomez*, A. L. Halweg-Edwards[§], D.G. Smith*, K.L. Bates[†], K.A. Kinney[‡,¶], T.T. Postolache[‡,||,#,**], L.A. Brenner[‡,**,††], G.A.W. Rook[‡‡], C.A. Lowry[*,‡,**,††,1]

*Center for Neuroscience, University of Colorado Boulder, Boulder, CO, United States
†US Air Force Academy, Colorado Springs, CO, United States
‡Military and Veteran Microbiome Consortium for Research and Education (MVM-CoRE), Denver, CO, United States
§University of Colorado Boulder, Boulder, CO, United States
¶University of Texas Austin, Austin, TX, United States
||University of Maryland School of Medicine, Baltimore, MD, United States
#VISN 5 Mental Illness Research Education and Clinical Center (MIRECC), Baltimore, MD, United States
**Rocky Mountain Mental Illness Research Education and Clinical Center, Denver, CO, United States
††University of Colorado, Aurora, CO, United States
‡‡Center for Clinical Microbiology, UCL (University College London), London, United Kingdom
[1]Corresponding author: e-mail address: christopher.lowry@colorado.edu

Contents

International Review of Neurobiology, Volume 131
ISSN 0074-7742
http://dx.doi.org/10.1016/bs.irn.2016.07.006

Abstract

It is increasingly evident that inflammation is an important determinant of cognitive function and emotional behaviors that are dysregulated in stress-related psychiatric disorders, such as anxiety and affective disorders. Inflammatory responses to physical or psychological stressors are dependent on immunoregulation, which is indicated by a balanced expansion of effector T-cell populations and regulatory T cells. This balance is in part driven by microbial signals. The hygiene or "old friends" hypothesis posits that exposure to immunoregulation-inducing microorganisms is reduced in modern urban societies, leading to an epidemic of inflammatory disease and increased vulnerability to stress-related psychiatric disorders. With the global trend toward urbanization, humans are progressively spending more time in built environments, thereby, experiencing limited exposures to these immunoregulatory "old friends." Here, we evaluate the implications of the global trend toward urbanization, and how this transition may affect human microbial exposures and human behavior.

1. INTRODUCTION

It is well documented that exaggerated or inappropriate inflammation is strongly associated with numerous physical and mental health disorders that affect human behavior (for review, see Miller & Raison, 2016; Rook, 2010). Inflammatory disorders have increased in prevalence in Western societies (Graham-Rowe, 2011). In sample populations of children living in urban locations, over 40% report atopic sensitization (Downs et al., 2001; Zekveld et al., 2006). Researchers, including authors of this chapter, have suggested that increases in inflammation associated with urbanization may be due in part to decreased exposure to microorganisms that humans have coevolved with (Rook et al., 2004; Strachan, 1989). In this chapter, we evaluate the evidence that decreased exposure to these microorganisms is related to urbanization and the corresponding shift in human behavior from spending the majority of their time outdoors to spending the majority of their time in the built environment.

The biodiversity hypothesis states that current deficits in the exposure of individuals living in Westernized civilizations to diverse natural environments, and therefore microbial diversity (due to the majority of time spent in the built environment), has unintended adverse health consequences (von Hertzen et al., 2015). Lack of exposure to microbial biodiversity, especially during the crucial timeframe of prenatal development through the first years of life, may lead to the inadequate development of immunoregulatory circuits (Dietert & Dietert, 2008). Improper or inadequate development of immunoregulation in turn can lead to chronic low-grade inflammation. Chronic, inappropriate inflammation is known to have negative health repercussions such as atopic sensitization (asthma, pollen (Burr, Butland, King, & Vaughan-Williams, 1989; von Mutius et al., 1994; Yunginger et al., 1992), food allergies (Bartra, Garcia-Moral, & Enrique, 2016)), autoimmune disease (multiple sclerosis, type 1 diabetes (Bach, 2002)), and inflammatory bowel disease (IBD; Crohn's disease and ulcerative colitis (Bach, 2002)), which are all cited as increasingly prevalent in Westernized cultures.

The hygiene or "old friends" hypothesis is consistent with the biodiversity hypothesis in that Westernized humans have lost contact with microbes with which we have coevolved, and also points toward specific features of microbial ecosystems that contribute to immunoregulation. There are three broad categories of "old friends" that have coevolved with mammals and contribute to adequate immunoregulation. These include: (i) the commensal microbiota (Sonnenburg & Sonnenburg, 2014; von Hertzen et al., 2015); (ii) pathogens associated with the "old infections" that were present throughout life in evolving human hunter-gatherer populations (Atherton & Blaser, 2009); and (iii) organisms from the natural environment with which humans were inevitably in daily contact, including soil-derived bacteria, and human-adapted gut microbiota present in the environment as spores (Browne et al., 2016; Rook, Raison, & Lowry, 2014). Frequent contact with these "old friends" necessitated that these microorganisms were tolerated by the immune system. Conversely, in order for these organisms to persist for long periods of time in human hosts, they evolved the ability to suppress host inflammatory responses. As discussed in more detail later, this often involves strategies to mimic human antigens or produce metabolites that act as pharmacological agents to control host-signaling mechanisms and activate immunoregulatory pathways. Together, the biodiversity and "old friends" hypotheses provide a framework within which scientists, healthcare providers, public health workers, and policymakers can investigate the potential health impacts of transitioning Westernized lifestyles and increased time spent in the built environment.

Thus, both the biodiversity hypothesis and the "old friends" hypothesis suggest that reduced exposure to the diverse microbial ecosystems present in outdoor environments has adverse health consequences. In this review, we argue that reduced exposure to diverse microbial communities, as often occurs in urban living, has important implications, both in the context of inflammatory diseases and in the context of behaviors that are influenced by inflammation, such as cognitive function and anxiety-related behaviors. The specific types of microorganisms implicated in the "old friends" hypothesis and psychiatric disorders have been reviewed extensively (Rook et al., 2014). For a review of specific "old friends" that have been shown to increase immunoregulation, enhance cognitive function, and reduce anxiety-related behaviors, see review (Lowry et al., 2016). We also briefly cover this topic at the end of this review. First, we turn our attention to current global trends toward urbanization, increases in time spent in the built environment, implications for the human microbiome, and microbial exposures relevant to immunoregulation and human behavior.

2. GLOBAL TRENDS TOWARD URBANIZATION

There is an increasing global trend toward urbanization. As reported by the United Nations (2014) revision of "World Urbanization Prospects," in 1950 one-third of the earth's human inhabitants lived in an urban setting. The year 2007 was the first time in history that more people were living in urban than in rural environments (Fig. 1). Moreover, by 2050, two-thirds of the world's population will be living in cities (United Nations, 2014). Urbanization is associated with many benefits, including better access to health care, education, social services, and increased average lifespans (United Nations, 2014). However, there are multiple health benefits inherent to an agricultural lifestyle that may be lost or reduced in the transition to an urban lifestyle. Arguably, the largest negative health impacts resulting from urbanization and living in the built environment are in Westernized nations where 82% and 73% of the populations in North America and Europe live in urban environments, respectively. Importantly, urbanization in these Western countries is also linked to a desire to reduce energy

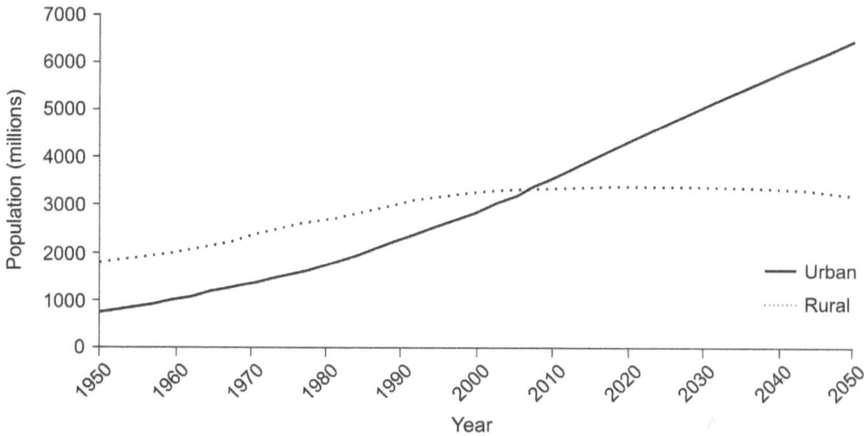

Fig. 1 Urbanization rates and projections from 1950 to 2050. *Solid* and *dotted lines* represent urban and rural populations, respectively. Worldwide population is represented in millions. Urban population surpassed rural population in 2007. *Figure adapted from "World Urbanization Prospects" 2014 revision, © (2014) United Nations. Reprinted with the permission of the United Nations.*

demands of heating and cooling loads, further separating individuals in the built environment from the outdoor environment. Dramatic rates of urbanization are increasing in many countries worldwide including China, where nearly half of China's current rural population (300 million people) is projected to become urbanized by 2050 (United Nations, 2014).

Given the rapid worldwide urbanization of human populations, it is important to consider the extent to which the microbial ecology of the built environment reflects the microbial ecosystems with which humans evolved. In addition, it is important to consider the increasing evidence supporting the hypothesis that these coevolved symbiotic microorganisms are essential for optimal health and well-being (Rook, 2013; Rook et al., 2004, 2014). Later, we outline the global trend for the movement of humans into the built environment, the effects of this transition on the human microbiome, and how this change may affect human behavior.

3. SHIFT IN THE HUMAN MICROBIOME DURING THE FIRST AND SECOND EPIDEMIOLOGICAL TRANSITIONS

In a simplistic model of the shift to urbanization, we compare the individual microbiomes at different points starting with hunter-gatherers and transitioning to inhabitants of Westernized urban societies. Although this is not a direct comparison, it does provide an indication of how human microbiomes may shift as societies become more urbanized. One major limitation of this approach is that individual microbiome differences are often linked to diet, activity level, and lifestyle. Some clues as to what factors really contribute to the variance seen among human microbiomes under these diverse conditions can be derived from an examination of postulated "enterotypes" or "biomarkers" of bacterial signatures associated with different societies that diverge with diet and lifestyle (Arumugam et al., 2011; Gorvitovskaia, Holmes, & Huse, 2016; Wu et al., 2011). Gorvitovskaia and colleagues combined many studies on the microbiomes collected in five continents from individuals living in a range of conditions, from primitive tribes to Westernized societies. Their analysis showed that, in contrast to the previously suggested clustering into distinct "enterotypes" (Arumugam et al., 2011), subjects' microbiomes fall into a continuum of dominant bacterial taxa (Gorvitovskaia et al., 2016). Individuals that consume Westernized diets, high in protein and animal fat, have gut microbiomes that are enriched in the genus *Bacteroides*, while the microbiomes of individuals that consume traditional diets, high in complex carbohydrates and fiber, are enriched in the genus *Prevotella*. It was also discovered that breast-fed children in both Westernized and traditional hunter-gatherer societies have microbiomes enriched in the genus *Bifidobacterium*, presumably because of the ability of these bacteria to efficiently metabolize the oligosaccharides found in human breast milk (De Filippo et al., 2010; Gorvitovskaia et al., 2016; Sela & Mills, 2010). However, it is not clear if collectively these shifts are due to differences in diet alone or if they may be linked to additional variables.

Distinct differences of microbiomes between humans living in Western, urban societies and those in rural hunter-gatherer societies have been reported in a number of studies. The microbiomes (gut, skin, and oral) of hunter-gatherers have significantly more biodiversity and stability than the microbiomes of those living in Westernized societies. Surprisingly, the microbiomes of unrelated tribes of hunter-gatherers in different regions of the world are more similar to each other than to the microbiomes of Westernized populations (Blaser et al., 2013; Clemente et al., 2015; Contreras et al., 2010; Obregon-Tito et al., 2015; Smith et al., 2013; Suzuki & Worobey, 2014; Yatsunenko et al., 2012). Even at an early age (<3 years), there are significant differences between the gut microbiomes of subjects in the United States and subjects in hunter-gatherer populations like the Malawian and Amerindian cultures (Yatsunenko et al., 2012).

Moreover, one study observed phylum level differences in Firmicutes/Bacteroidetes ratios between individuals with Westernized and traditional hunter-gatherer diets and lifestyles (Suzuki & Worobey, 2014). Interestingly, studies in obese and lean mice show an alignment of the Firmicutes/Bacteroidetes ratio of Westernized humans with that of obese mice, and of hunter-gatherer humans with that of lean mice (Ley et al., 2005).

Additionally, differences are encountered between the microbiomes of humans living in Western urban societies and those living in developing, rural communities. Studies comparing inhabitants of developing rural communities of Papua New Guinea and the Tunapuco tribe of the Peruvian Andes, where inhabitants practice subsistence farming and raise selective livestock, to inhabitants of the United States again demonstrated more biodiversity in the microbiome of the residents of the developing rural communities (Martinez et al., 2015; Obregon-Tito et al., 2015). As seen with the hunter-gatherer populations, the compositions of microbial communities of individuals living in the United States are distinct when compared to the developing rural societies. However, there is more intersubject variability between the microbial communities of United States residents than the Papua New Guineans (Martinez et al., 2015). This observation may be due to the standardization of the diet seen in Papua New Guinea, which is not the case for Westernized civilizations. Of potential importance, both hunter-gatherer and developing rural individuals have microbiomes enriched in Spirochaetes, specifically *Treponema*, as well as *Helicobacter* spp. (Martinez et al., 2015; Obregon-Tito et al., 2015), while the relative abundances of these bacterial species are greatly reduced in the microbiomes of individuals living in Westernized societies. *Helicobacter*, in particular, potently induces immunoregulation during early life (Arnold, Dehzad, et al., 2011; Arnold, Hitzler, & Muller, 2012; Arnold, Lee, et al., 2011), and the scarcity of *Helicobacter* in Westernized societies is an example of reduction of exposure to "old friends" (Rook et al., 2014).

4. DIFFERENCES BETWEEN THE MICROBIOMES OF THE OUTDOOR ENVIRONMENT AND THE BUILT ENVIRONMENT

Evidence suggests that the microbiomes of the built environments can be dramatically different from the microbiomes of outdoor environments in which humans evolved. As discussed here, a number of factors can contribute to the differences encountered between the microbial environment of the built environment and the outdoors. For example, microbial growth, accumulation, and diversity indoors can result from excess moisture and water damage in an indoor space (Emerson et al., 2015; Miller, Rand, & Jarvis, 2003; Szponar & Larsson, 2000), and can be further directed by type of building materials used (Green, 2014; Hoang, Kinney, Corsi, & Szaniszlo, 2010; Kembel et al., 2014). Water damage and moisture can promote the growth of potentially toxic mold (Korkalainen et al., 2016) and other microorganisms of concern to human health. Further, varying levels of ambient relative humidity can influence the extent of microbial growth found in the built environment (Frankel et al., 2012; Leung & Lee, 2016). The type and condition of ventilation present in the built environment has also been noted to significantly influence the microbiome of the built environment (Kembel et al., 2012; Meadow et al., 2014). Ventilation can be achieved in a variety of ways that typically depend on the type of space, regulatory measures, and energy preferences. Another main determinant of the microbial ecosystems of built environments is occupant-associated factors such as: the number of occupants in a space (Hospodsky et al., 2012), the gender of occupants (Flores et al., 2011; Luongo et al., 2016), household cleaning product type and frequency of use (Rusin, Orosz-Coughlin, & Gerba, 1998), and presence of domestic household animals (Fujimura et al., 2010; Ownby, Johnson, & Peterson, 2002). These drivers inform the focus of the microbiome of the built environment here to include: (i) contribution from, and interaction with, human occupants; (ii) contributions from the outdoor microbiome; (iii) influences of building design; (iv) influences of indoor water sources; and (v) urban design and health.

4.1 Contributions from, and Interactions with, Human Occupants and the Microbiome of the Built Environment

It has been argued that the occupants of a built environment are the most significant contributors to its microbial composition (Adams et al., 2015; Leung & Lee, 2016). Since humans shed between 10^6 and 10^7 skin-associated microbes per hour (Hospodsky et al., 2015; Qian, Hospodsky, Yamamoto, Nazaroff, & Peccia, 2012) and typically spend up to 90% of their time indoors (Evans & McCoy, 1998), human occupants deposit a significant amount of biomass in the built environment. When new occupants enter a space, those individuals can colonize the space with their microbial signature in a matter of hours or days (Lax, Nagler, & Gilbert, 2015; Lax et al., 2014). Further, the microbiomes of different rooms and surfaces in a built environment can actually be linked to different parts of the body (Flores et al., 2011; Lax et al., 2014). Both the presence of humans (occupancy) and the amount and frequency of humans (traffic) impact the diversity and distribution of the built environment microbiome (Adams et al., 2015; Hospodsky et al., 2012; Qian et al., 2012; Yamamoto, Hospodsky, Dannemiller, Nazaroff, & Peccia, 2015). Locations that experience more human traffic, such as hallways, have different microbial signatures than locations with lower traffic (Kembel et al., 2014). Even in relatively sparsely occupied retail stores, the human microbial signature can be detected and may be dependent on occupant density (Hoisington, Maestre, Kinney, & Siegel, 2015). Intuitively, the microbial signature of places with higher occupancy and traffic show more biomass and biodiversity than places of lower traffic and occupancy (Hospodsky et al., 2012; Meadow et al., 2014; Miletto & Lindow, 2015; Ross & Neufeld, 2015). Many of the studies focusing on human occupation and traffic indoors show that outdoor microbial "migrant" (i.e., microbes picked up in the outdoor environment and deposited in the built environment) can be a source of the human contribution to the microbial makeup of the built environment (Adams, Miletto, Lindow, Taylor, & Bruns, 2014; Hospodsky et al., 2012; Leung & Lee, 2016).

4.2 Contributions from the Outdoor Microbiome to the Microbiome of the Built Environment

A number of observations suggest that there are important differences between the microbiomes of the outdoor environment and the built environment. Soil, leaves, grasses, trees, and other organic and inorganic materials support miniature ecosystems with a seasonal cycle that promotes organic matter decay and therefore a regenerative food source. In contrast, the built environment lacks the decay of organic matter that encourages microbial growth. Yet microbes are still in our living spaces. These microbes can be brought into the built environment from the outside by the human occupants, pets, through open windows and doors, nonfiltered ventilation, and by other sources. Depending on these variables and others, the microbial composition of the built environment can be similar to the microbiome of the outdoor environment but with a prominent decrease in biomass (Adams et al., 2015; Hospodsky et al., 2012; Leung & Lee, 2016; Qian et al., 2012).

Converging lines of evidence show that the microbiome of the built environment is more composed of "migrant" microbes brought into the built environment from the outdoors, rather than "residential" microbes that originate from within the built environment (Adams et al., 2015;

Kelley & Gilbert, 2013; Leung & Lee, 2016). Different climates with variations in temperature, humidity, and vegetation, among other factors, play a central role in determining the microbial communities that are capable of flourishing in the outdoor environment (Bottos, Woo, Zawar-Reza, Pointing, & Cary, 2014; Bowers, McLetchie, Knight, & Fierer, 2011; DeLeon-Rodriguez et al., 2013; Seifried, Wichels, & Gerdts, 2015). The microbial communities associated with these different geographical regions are a product of dynamic ecosystems influenced by, and contributing to, the soil (Little, Robinson, Peterson, Raffa, & Handelsman, 2008), water (Kemp & Aller, 2004), vegetation (Lymperopoulou, Adams, & Lindow, 2016), and air masses interacting with these regions (Bowers et al., 2011). Though these ecological systems are dynamic in nature, a microbial make-up of a given microbiome can actually be used as a fingerprint of a certain geographical region (Grantham et al., 2015). Certain bacteria and fungi found in outdoor dust across the United States have been mapped and regions with similar climates show similar bacterial and fungal colonization patterns (Barberan et al., 2015). These same bacterial and fungal signatures are found between and within houses in the same region, with locations within the house more proximal to entry and exit points being more similar to dust found outside the house (Dunn, Fierer, Henley, Leff, & Menninger, 2013). The frequency and consistency with which humans pass into and out of the built environment, while carrying "migrant" microbes along the way, can over time shape the make-up of the microbiome of the built environment.

It is not clear if the biomass and diversity of outdoor microorganisms transported into the built environment in rural or urban environments are sufficient to influence immunoregulation. A recent study showed that urban built environment microbiomes were significantly less diverse compared to rural built environment microbiomes in the United States (Dannemiller, Gent, Leaderer, & Peccia, 2016). Overall, it is likely that the shift from rural to urban environments will result in decreased human exposures to microbial biomass and diversity. Determining the impact of this decrease in microbial exposures on emotional health and well-being is an important objective for future studies.

4.3 Influences of Building Design on the Microbiome of the Built Environment

Architecture and building design have also been noted to heavily influence the complex bacterial and fungal communities within the built environment. Ventilation, a key feature in building design, is of major consideration in shaping the microbial community indoors, particularly with respect to the source and flow rate of ventilation air, and ambient factors like temperature and humidity. Buildings and residences with passive ventilation, or outdoor air infiltration, have complex microbial compositions that more closely resemble those from the outdoor environment than the less diverse microbiomes found within indoor environments that are mechanically ventilated with filtrated air (Kembel et al., 2014; Meadow et al., 2014). It follows that in large buildings the architectural design informs a "biogeography" indoors, where rooms and spaces furthest from access to the outdoor environment have microbial communities that are less diverse, have less biomass, and have less resemblance to rooms with outdoor access (Adams et al., 2014; Kembel et al., 2014; Poza et al., 2012). In addition to dynamic engineered components within the built environment, such as ventilated air and potable running water, the building materials selected for these systems can also influence the diversity and extent of microbial growth found within the built environment. A number of building materials can influence microbial communities, and important considerations include parameters like material composition (Rintala, Nevalainen, & Suutari, 2002; Torvinen et al., 2006) and moisture content (Dedesko & Siegel, 2015; Rintala et al., 2002). Incorporating what has so far been highlighted on the dynamic microbe-host-built environment interactions, a movement on "bioinformed design" is emerging in both developing and developed communities, with a focus on mindful building design and materials, as well as occupant behaviors that promote healthy indoor environments (Green, 2014) (Fig. 2).

4.4 Influences of Indoor Water Sources on the Microbiome of the Built Environment

Indoor water from distribution systems and premise plumbing is an important influence on the built environment microbiome. Pipes that feed water to houses and buildings, despite disinfection efforts, are inevitably lined with biofilms that contribute microbes to the water as it runs over the microbial colonies; an 8 ounce glass of drinking water is estimated to contain an average of 24 million bacterial cells while, during an 8 min shower, up to 6 billion microbes can come in contact with our skin (Lautenschlager, Boon, Wang, Egli, & Hammes, 2010; Pinto, Xi, & Raskin, 2012). In addition to ingestion and skin contact, drinking water can contribute to the microbiome of humans and the built environment in a number of ways, including

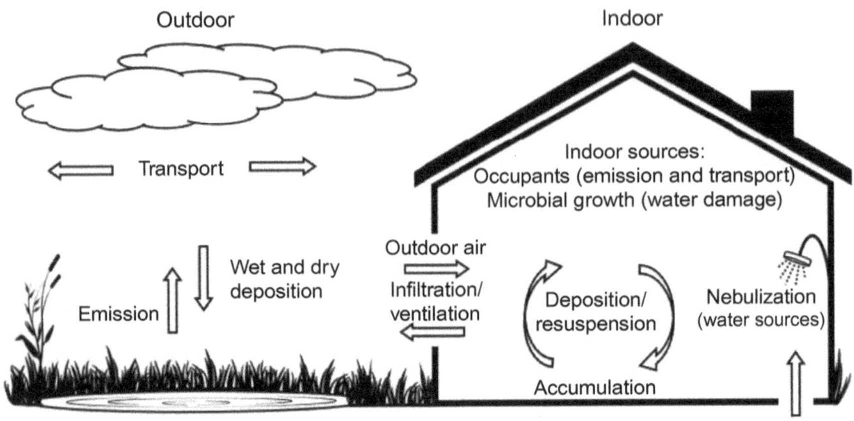

Fig. 2 Sources, sinks, and physical processes of biological aerosols (bioaerosols) contributing to, and interacting with, the indoor environment. Bioaerosols encountered indoors (whole or fragmented) include pollen, bacteria, and fungi that can enter from the outdoors. Outdoor bioaerosols can be generated from proximal sources like water, vegetation, and soil, as well as those bioaerosols that transport vast distances and undergo numerous interactions with polluted environments and atmospheric processes. Outdoor bioaerosols can enter the indoor environment through ventilation systems, poorly sealed homes, and open windows. Microbes in outdoor bioaerosols also can be tracked in by occupants. Bioaerosols can be directly produced, and accumulate indoors, through emissions from occupants (shedding and coughing), microbial growth from water damage, and nebulization from water sources like shower heads and sinks.

(1) inhalation by human occupants of aerosolized water droplets from sinks and showers (Nazaroff, 2016) and (2) contributing to biofilms that form on shower curtains and other surfaces present indoors (Kelley, Theisen, Angenent, St Amend, & Pace, 2004). The diversity and complexity of microbial communities found within the drinking water distribution systems, from water treatment facilities to the faucet, depend on a number of factors including drinking water treatment methods and water quality parameters (Pinto et al., 2012). This is not a reason for alarm as the microbial constituents of these biofilms are largely benign (and may even be beneficial) with exception of a few opportunistic microbes, such as *Legionella* spp., that are mainly of concern to immunocompromised individuals. Much of the variation in microbial communities found in water distribution systems is based on the water chemistry (Ji, Parks, Edwards, & Pruden, 2015; Wang, Masters, Edwards, Falkinham, & Pruden, 2014), posttreatment disinfection methods (Baron, Vikram, Duda, Stout, & Bibby, 2014), and distance traveled in distribution systems (Henne, Kahlisch, Brettar, & Hofle, 2012). Also to be considered are built environment endpoints like:

condition and operation of water heaters (Lau & Ashbolt, 2009; Wang, Masters, Falkinham, Edwards, & Pruden, 2015), building plumbing (Wang et al., 2014), the age of the water pipes (Henne et al., 2012), the disinfection techniques and additives to the water system (Baron et al., 2014), the distance of the building from treatment facilities (Henne et al., 2012), the local building's filtration methods (Ji et al., 2015), and the amount of water stagnation within the pipes (Lautenschlager et al., 2010).

The extent to which biofilms present in drinking water systems and the built environment may be harmful or beneficial is still under debate and is largely dependent on the location and conditions in which these biofilms form and potentially harbor and liberate opportunistic microbes. For example, *Legionella* spp. have been noted to be present in biofilms in pipes and shower heads (Lau & Ashbolt, 2009); similarly, other opportunistic pathogens like *Sphingomonas* spp. and *Methylobacterium* spp. have been documented growing in shower curtain biofilms (Kelley et al., 2004). Biofilms may serve as reservoirs for opportunistic microorganisms to enter flowing water. Some studies have identified the presence of *Legionella* spp. in aerosol liberated during showering (Deloge-Abarkan, Ha, Robine, Zmirou-Navier, & Mathieu, 2007) and *Mycobacterium* spp. in aerosol liberated from hot tubs (Angenent, Kelley, St Amend, Pace, & Hernandez, 2005). Further, changes in water chemistry and other parameters like temperature can serve to exacerbate an otherwise harmless situation involving opportunistic pathogens. It is important to characterize the bacterial communities within drinking water systems because of the presence of potentially pathogenic bacteria as well as evidence that microbial communities can lead to increased levels of heavy metals found in drinking water (White, Tancos, & Lytle, 2011).

It should be emphasized that a diverse microbial community of bacteria in drinking water and water aerosols does not always imply negative health outcomes. The microbial diversity present in these systems may actually have health benefits. It is through the very nature of recognizing the microbial diversity present in the built environment that we can take a "probiotic approach" in engineering our indoor environment to promote and facilitate the growth of beneficial microorganisms (Green, 2014; Wang, Edwards, Falkinham, & Pruden, 2013). Indeed, indoor water sources serve as a reservoir of mycobacteria inside the home, and environmental mycobacteria have been shown to confer protection from asthma (Rook et al., 2004; Zuany-Amorim, Sawicka, et al., 2002), have antidepressant-like behavioral effects (Lowry et al., 2007), reduce anxiety and improve cognitive function (Matthews & Jenks, 2013), and promote stress resilience in mice

(Reber et al., 2016), in part through induction of regulatory T cells (Treg) (Reber et al., 2016; Zuany-Amorim, Manlius, et al., 2002). Potable water is undoubtedly a major factor when examining the microbiome of the built environment. With the variety of factors influencing the water-borne microbial communities and the frequency and diversity of sources of contact that humans have with water, there are numerous future research opportunities with potential health applications in the built environment.

4.5 Implications of Urbanization on Human Health

Studies are beginning to reveal differences between the microbiomes of agricultural and rural dwellers on the one hand from urban dwellers on the other, with implications for health. Many studies have shown a gradient in atopic sensitization with urban dwellers on the high end and active traditional farmers on the low end, while rural dwellers and modern farmers fall in-between (Zekveld et al., 2006). Furthermore, studies by Holbreich and colleagues have shown reduced asthma and atopic sensitization in children of Amish communities actively practicing traditional farming techniques and Swiss communities practicing modern farming techniques as compared to nonfarming Swiss communities. Evidence suggests that a farming lifestyle is protective against asthma and has a marginal effect on atopic sensitization with rates being (asthma (%), atopic sensitization (%)): Amish (5.2, 7.2), Swiss-farm (6.8, 25.2), and Swiss-nonfarm (11.2, 44.2) (Holbreich et al., 2012). Interestingly, inoculation studies in mice suggest that dust from Amish barns and houses was more protective against asthma than dust from modern farming houses (Gozdz et al., 2015). Anecdotal observations had preceded these studies and suggested that growing up in and around farms can have a protective effect against asthma and allergies, so much so that the phrase "the farm effect" has emerged as a term to describe this phenomenon (Lauener et al., 2002; Valkonen et al., 2015; von Mutius, 2016; Wlasiuk & Vercelli, 2012).

Could this observed gradient in asthma and atopy be due in part to differences in exposure to environmental microbiomes? As previously mentioned in this review, the outdoors, particularly green (fields, farms, parks, forests, general open space) and blue (oceans, ponds, lakes, rivers, other bodies of water) spaces and agricultural land, are associated with higher microbial biodiversity and biomass. This increase in biodiversity is effectively transferred to the agricultural built environment with increased biodiversity and relative abundance of livestock-associated microbes in the house and mattress dust (Ege et al., 2011; Valkonen et al., 2015).

Some practices common to Western life are being questioned by microbiologists because of observed unintended consequences on health and quality of life. Medical professionals are beginning to acknowledge this and are promoting a reduction in the all-out warfare against microbes and replacing this approach with one of managed coexistence (Al-Ghalith & Knights, 2015). Antibiotics, when used in excess, can devastate the intestinal microbiota (Becattini, Taur, & Pamer, 2016), promote antibiotic resistance (Modi, Collins, & Relman, 2014), and have lasting effects in terms of asthma (Sun, Svendsen, Karmaus, Kuehr, & Forster, 2015) and brain development and behavior in children (Diaz Heijtz, 2016). Within the built environment, humans have devised many ways to disinfect surfaces using chemicals, UV light, and antimicrobial surface materials, with varying degrees of effectiveness. However, even in hyperclean environments, such as a neonatal intensive care unit, there are some types of bacteria like *Enterobacteriaceae*, a common cause of nosocomial infection, that are not affected by disinfection (Bokulich, Mills, & Underwood, 2013). In addition, opportunistic, potentially pathogenic microbes, were observed to recolonize surfaces shortly after disinfection (Bokulich et al., 2013). More research is needed in terms of sanitization within the built environment and antibiotic use, especially around the prenatal and postnatal time periods, to better inform the scientific community, public and health officials, and general public on best practices.

While more research is needed in order to understand the impacts of current sanitization practices on microbial communities in the built environment, research on certain aspects of the built environment that enhance microbial diversity has revealed the potential for beneficial effects of the microbiome of the built environment on human health. Earlier, we briefly mentioned that opening windows and naturally ventilating rooms makes indoor air more similar to outdoor air (Kembel et al., 2014; Meadow et al., 2014). The outdoors in general, including the air, has more biodiversity and biomass than indoor air. Exposure to a larger diversity and overall more microbes, especially early in life, has been shown to be protective against asthma and atopic sensitization and markers of chronic inflammation such as C-reactive protein (McDade, Rutherford, Adair, & Kuzawa, 2010; Wlasiuk & Vercelli, 2012). Another element of life in the built environment that can greatly increase the number of outdoor microbes, biodiversity, and biomass is pets (dogs more than others) (Aichbhaumik et al., 2008; Dunn et al., 2013; Fujimura et al., 2010; Ownby et al., 2002). Finally, living in close proximity to green and blue spaces is known to increase the biodiversity and biomass in and around a built environment (Fuller, Irvine, Devine-Wright, Warren, & Gaston, 2007; Ruokolainen et al., 2015). Nonmicrobiome studies have shown that green and blue spaces have an array of beneficial mental and physical health effects (Astell-Burt, Feng, & Kolt, 2014; Fuller et al., 2007; Hanski et al., 2012; Logan, 2015; Mitchell & Popham, 2008; Rook, 2013; Ruokolainen et al., 2015; Shwartz, Turbe, Simon, & Julliard, 2014). Keeping in mind that there are many confounding variables associated with green and blue spaces including increases in physical activity, increases in socioeconomic status, better dietary habits, and overall esthetics (Hanski et al., 2012; Lautenschlager et al., 2010; Ulrich, 1984; von Hertzen et al., 2015). On the contrary, gray space (urban concrete environments) have been shown to decrease biodiversity and have negative effects on mental and physical health (Logan, 2015).

Built environment research has shown that the outdoor environment and occupants are unidirectional vectors for microbial deposition into the built environment. Yet we know that this relationship must be bidirectional, and therefore the void of research of the built environment as a microbial vector for occupants and the juxtaposed outdoor environment needs to be filled. In addition, we have been so concerned with abolishing microbes from our environment that we have put little thought into the potential for symbiotic relationships between microbes originating in the built environment and human occupants. One controversial, but important, starting point for built environment microbiome research is water distribution pipes, which we know already contain biofilms with potentially harmful yet also beneficial microbes. There may be benefits in examining the possibility of managing microbial community structure in water distribution systems through smart design, such that we can facilitate the growth of beneficial microbial communities and mitigate the formation of pathogenic or harmful microbial communities. The addition of additives to drinking water has been accepted chemically for the past 70 years in Westernized nation via fluoride treatment (Ripa, 1993).

5. THE MICROBIOME OF THE BUILT ENVIRONMENT, IMMUNOREGULATION, AND HUMAN BEHAVIOR

5.1 Mechanisms Through Which Environmental Microbes Influence Host Physiology and Behavior

To the extent that the microbiome of the built environment may affect immunoregulation in human occupants, evidence suggests that immunoregulation in turn affects human behavior. Although these links are tentative, there are several lines of evidence that support this hypothesis. Studies in mice have shown that immunization with environmental bacteria, specifically a heat-killed preparation of the soil-derived bacterium, *Mycobacterium vaccae*, increases immunoregulation (Reber et al., 2016; Zuany-Amorim, Manlius, et al., 2002), attenuates inflammation (Reber et al., 2016; Rook et al., 2004; Zuany-Amorim, Manlius, et al., 2002), decreases anxiety (Matthews & Jenks, 2013; Reber et al., 2016), and improves cognitive function (Matthews & Jenks, 2013) in mice deficient in exposures to environmental bacteria. Humans living in Westernized urban environments also have microbiomes that are deficient in environmental Actinobacteria. Actinobacteria are phylogenetically closely related to Firmicutes (Hug et al., 2016) and, together, Firmicutes and Actinobacteria, by far, account for the majority of known probiotic species (Hoisington, Brenner, Kinney, Postolache, & Lowry, 2015; Lowry et al., 2016). Whereas Firmicutes are abundant in the gut microbiome, Actinobacteria are more abundant on the skin and upper airways (Flores et al., 2014), and the Actinobacteria genus *Mycobacteria*, specifically, are abundant in the mouth and upper airways (Macovei et al., 2015). Of potential interest, many Actinobacteria, especially *Streptomyces* and *Mycobacteria*, have larger genomes and biosynthetic clusters relative to Firmicutes, such as *Lactobacilli*, suggesting that they are genetically more complex and are able to produce a greater diversity of microbial antigens and metabolites (Fig. 3).

We have previously reviewed molecular mechanisms, including bacterial antigens and metabolites, that influence immunoregulation and thus levels of inappropriate inflammation that have been associated with stress-related psychiatric disorders, such as posttraumatic stress disorder and major depressive disorder (Lowry et al., 2016). We have also reviewed in detail specific probiotics, mainly belonging to the phyla Actinobacteria and Firmicutes that have been shown to confer mental health benefits (Allen et al., 2012; Hoisington, Brenner, et al., 2015; Lowry et al., 2016). Immunoregulation, indicated by a balanced expansion of effector T-cell populations and Treg, are known to be driven by microbial signals. Here, we highlight diverse cellular mechanisms through which these bacteria can influence host immunoregulation.

6. INFLAMMATION ANERGIC MACROPHAGES

Human intestinal macrophages display profound inflammation anergy (an inability to respond to an inflammatory stimulus with an inflammatory

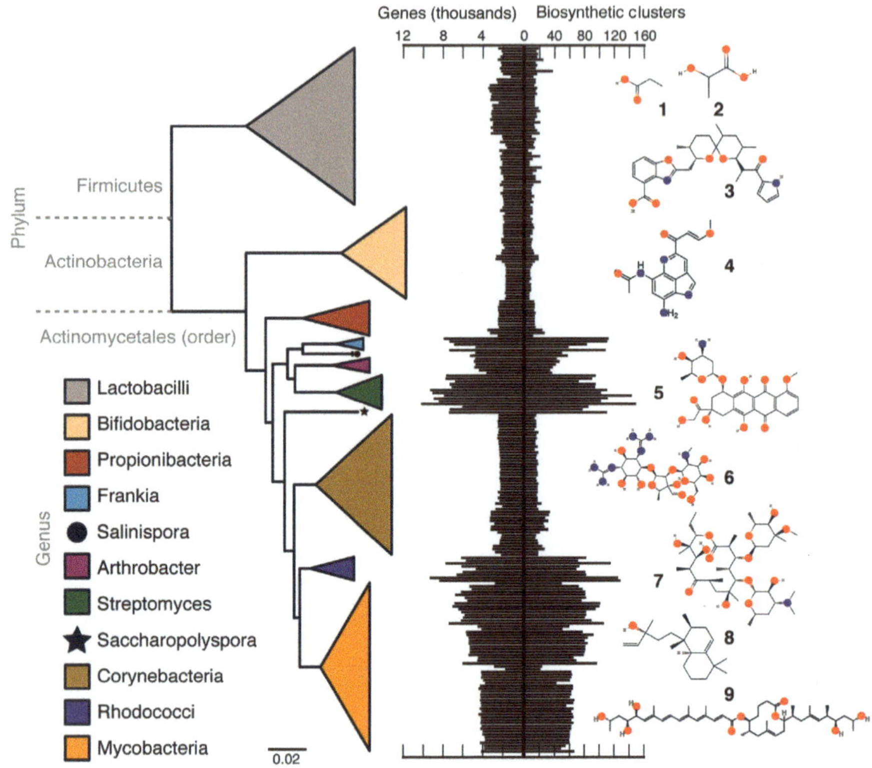

Fig. 3 Actinobacteria as a source of therapeutic small molecules. A phylogenetic tree was constructed for members of the phylum Actinobacteria for which completed genome sequences were available in the IMG database as of May 2014. *Lactobacilli* and members of the phylum Firmicutes were used as an outgroup to root the tree. Each clade is collapsed in diagrammatic representation and colored with respect to the genus of each strain. Biosynthetic clusters in each genome sequence were predicted using ClusterFinder within the IMG framework. The number of predicted biosynthetic clusters is plotted along with the genome size for each strain represented in the phylogenetic tree, demonstrating that *Streptomyces*, *Frankia*, *Salinispora*, *Arthrobacter*, *Saccharopolyspora*, *Rhodococci*, and *Mycobacteria* are among the highest producers of bioactive compounds within the phylum Actinobacteria. Examples of known bioactive secondary metabolites produced by Actinobacteria are shown on the far right with the *red* and *blue* circles representing oxygen and nitrogen, respectively. 1—Propionic acid, a fermentation product of *Propionibacteria*; 2—lactic acid, a fermentation product of *Lactobacilli*; 3—demethyl C-11 cezomycin, an ionophore produced by *Frankia* species; 4—lymphostin, an immunosuppressant produced by *Salinispora* species; 5—doxorubicin, a chemotherapeutic produced by *Streptomyces* species; 6—streptomycin, an antibacterial agent produced by *Streptomyces griseus*; 7—erythromycin, an antibacterial agent produced by *Saccharopolyspora erythraea*; 8—isotuberculosinol, a pathogenic agent produced by *Mycobacterium tuberculosis*; 9—mycolactone, a toxin produced in a variety of pathogenic *Mycobacteria*.

response), despite the fact that they retain phagocytic and bacteriocidal activity (Smith, Ochsenbauer-Jambor, & Smythies, 2005; Smythies et al., 2005). Specifically, intestinal macrophages fail to produce proinflammatory cytokines, including IL-1, IL-6, IL-10, IL-12, regulated on activation, and tumor necrosis factor, in response to diverse inflammatory stimuli (Smythies et al., 2005). This effect, in part, is due to downregulation of the adapter proteins myeloid differentiation primary response gene 88 (MyD88) and Toll interleukin receptor 1 domain-containing adapter-inducing interferon β, which together mediate all Toll-like receptor (TLR) MyD88-dependent and -independent nuclear factor kappa-light-chain-enhancer of activated B cells signaling (Smythies et al., 2010). Downregulation of these inflammatory mechanisms is due in part to TGF-β derived from intestinal extracellular matrix (Smythies et al., 2010). Consequently, "old friends" that induce TGF-β, such as *M. vaccae* (Zuany-Amorim, Sawicka, et al., 2002), have the potential to suppress inflammatory responses in macrophages. Several lines of evidence suggest commensal bacteria modulate intestinal inflammation via effects on pattern-recognition receptor signaling mechanisms. MyD88-deficient mice have increased vulnerability to dextran sulfate sodium (DSS)-induced colitis, a model of IBD, suggesting that commensal bacteria are recognized by TLRs under normal conditions to initiate host-protective responses (Rakoff-Nahoum, Paglino, Eslami-Varzaneh, Edberg, & Medzhitov, 2004). Consistent with this hypothesis, depletion of gut microbes using antibiotics increases the mouse's vulnerability to DSS-induced colitis, and this effect can be corrected by feeding either lipopolysaccharide (a TLR4 ligand characteristic of Gram-negative bacteria) or lipoteichoic acid (a TLR2 ligand that is a major constituent of the cell wall of Gram-positive bacteria) via drinking water (Rakoff-Nahoum et al., 2004). A number of intracellular negative regulators of TLR signaling have been described, including peroxisome proliferator-activated receptor gamma (PPARγ) (Shibolet & Podolsky, 2007).

7. REGULATORY MACROPHAGES

Microorganisms with immunoregulatory properties can directly influence macrophage activity. Helminth infections induce population expansion of alternatively activated macrophages, driven by type two helper T cells (Th2) rather than type one helper T cells (Th1) as other types of infections normally do (Mosser & Edwards, 2008). Such macrophages secrete IL-10 and TGF-β rather than IL-12, are able to inhibit lymphocyte proliferation in a contact-dependent manner (Elliott & Weinstock, 2012; Loke, MacDonald, Robb, Maizels, & Allen, 2000), and may be responsible for preventing inflammation in mucosal surfaces such as the lung. However, some helminths drive types of regulatory macrophages that are distinct from alternatively activated macrophages (Elliott & Weinstock, 2012). For example, several species of filarial nematodes secrete cystatin, a cysteine protease inhibitor that induces macrophages to produce IL-10 and IL-12 p40 through activation of intracellular signaling pathways. These two interleukins can prevent allergic sensitization and airway hyperresponsiveness (Schnoeller et al., 2008). Similarly, a colon-infiltrating macrophage population induced by *Schistosoma* infection prevents colitis in mice in a T cell-independent manner (Smith et al., 2007).

PPARγ is a ligand-activated nuclear receptor with potent anti-inflammatory properties that modulates inflammatory responses (Chinetti, Fruchart, & Staels, 2003). PPARγ primes human monocytes to differentiate into alternative M2 macrophages with antiinflammatory properties and correlates with expression of IL-10 in these cells (Bouhlel et al., 2007; Odegaard et al., 2007). These M2 macrophages in turn exert paracrine antiinflammatory effects on M1 macrophages (Bouhlel et al., 2007). Ligands of PPARγ include an endogenous prostaglandin J2 metabolite, 15-deoxy-Δ12;14-prostaglandin J2, a derivative of arachidonic acid (Kliewer et al., 1995), and naturally occurring polyunsaturated fatty acids and their metabolites, including conjugated linoleic acids (Hwang, 2000). Similar effects have been described in peripheral mononuclear cells (Kim et al., 2011). Feeding mice conjugated linoleic acid protects against experimentally induced IBD and colorectal cancer through activation of PPARγ (Bassaganya-Riera et al., 2004; Evans et al., 2010). Of potential importance, conjugated linoleic acids are bacterial metabolites and produced by several genera, including *Lactobacillus*. *Lactobacillus reuteri* and *Lactobacillus plantarum* mediate the conversion of dietary linoleic acid into immunomodulatory conjugated linoleic acids (Lee et al., 2003; Ogawa et al., 2005). Most of the conjugated linoleic acid produced is located in the extracellular space (~98%) (Lee et al., 2003; Roman-Nunez, Cuesta-Alonso, & Gilliland, 2007), suggesting that bacterially derived conjugated linoleic acids may be metabolic signaling molecules that modulate the host immune response. Probiotic administration increased colonic conjugated linoleic acid concentrations in concert with decreased macrophage accumulation in the mesenteric lymph nodes of mice with colitis; deletion of PPARγ in myeloid cells prevented the protective effects of probiotics (Bassaganya-Riera et al., 2012).

7.1 Regulatory Dendritic Cells

A particularly important immunoregulatory function of microbial signals is the generation of regulatory dendritic cells (DCreg) that tend to drive regulatory rather than inflammatory responses. This is important because DCreg can process gut contents, autoantigens and allergens, and thereby downregulate responses to the target antigens of the major groups of chronic inflammatory disease. It is likely that normal health status requires the presence of a certain background proportion of DCreg. This background of DCreg could be regarded as functioning as a "Treg adjuvant." Diverse "old friends," including commensal bacteria, can induce DCreg. A mixture of several probiotic organisms (including four *Lactobacilli* species, three *Bifidobacteria* species, and one *Streptococcal* species) were found to ameliorate recurrent Th1-mediated murine colitis by inducing IL-10 and TGF-β^+ Treg (Di Giacinto, Marinaro, Sanchez, Strober, & Boirivant, 2005). In vitro this preparation caused human dendritic cells (DC) to release more IL-10 and inhibited their ability to induce Th1 cells (Hart et al., 2004). Other bacterial probiotic strains (Braat et al., 2004; Smits et al., 2005) and a ubiquitous environmental saprophyte often present in healthy soils and water (Le Bert, Chain, Rook, & Noursadeghi, 2011), can also modulate human DC function in vitro so as to induce T cell responses with a more regulatory bias, and potentially become or function more like DCreg. In the gut, some of these DCreg express the integrin alpha chain CD103 and have unique immunoregulatory properties (Jaensson et al., 2008). The chief ligand for CD103 is E-cadherin, an adhesion molecule found on epithelial cells (Hadley, Bartlett, Via, Rostapshova, & Moainie, 1997). CD103 is involved in de novo conversion of $Foxp3^-CD4^+$ cells to $Foxp3^+$ Treg cells (Sun et al., 2007). Conversion of DC to this tolerogenic phenotype is driven locally by TGF-β and retinoic acid (RA). $CD103^+$ DCs express *aldh1a2*, the gene encoding Raldh2. This enzyme is involved in conversion of dietary retinal to RA, which enhances development of $FoxP3^+$ T cells rather than Th17 cells (Sun et al., 2007). Some probiotic *Lactobacillus* strains, such as *L. plantarum* WCFS1 induce migration of these $CD103^+$ DCreg to the spleen, and bias the immune response toward Treg (Smelt et al., 2012). As is the case with the commensal and environmental microorganisms described earlier, "old infections" or pathobionts such as *Helicobacter pylori* also can induce DCreg so as to induce T cell responses with a more regulatory bias, resulting in systemic immunoregulatory responses (Arnold, Dehzad, et al., 2011; Arnold et al., 2012).

7.2 Regulatory B Cells

Regulatory B cells (Breg), specifically IL-10-secreting $CD1d^{hi}CD5^{+}$ B cells can suppress experimental autoimmune encephalomyelitis (Matsushita, Yanaba, Bouaziz, Fujimoto, & Tedder, 2008) and have recently been designated B10 cells (Yanaba, Bouaziz, Haas, et al., 2008). They act in part by increasing the number of pulmonary $CD4^{+}CD25^{+}Foxp3^{+}$ Treg in the lungs (Amu et al., 2010). IL-10^{+} Breg are also found in humans (Yanaba, Bouaziz, Matsushita, et al., 2008).

Helminths are one example of "old infections" that induce Breg. *Schistosoma mansoni* (*S. mansoni*) infections prevent anaphylaxis in a mouse model, and this suppression of the effector phase of the allergic response is mediated by IL-10-secreting B cells (Mangan et al., 2004). Drug depletion of human B cells can occasionally exacerbate Th1-mediated conditions, suggesting that the removal of a B cell-mediated regulatory mechanism (Yanaba, Bouaziz, Matsushita, et al., 2008). IL-10 production by B cells is increased in multiple sclerosis patients with intestinal helminth infections, but not in patients infected by *Trypanosoma cruzi* (Correale, Farez, & Razzitte, 2008). The helminths also increase circulating Treg, as discussed later.

7.3 Regulatory T Cells

Ultimately, many of these immunoregulatory mechanisms result in induction of Treg, whether secondary to changes in macrophages, DC, B cells, or combined influences of these mediators. However, some "old friends" release molecules that specifically expand Treg populations. The gut commensal *Bacteroides fragilis* releases a polysaccharide antigen that acts directly on Treg to drive expansion of Treg via TLR2 (Round et al., 2011). The helminth, *Heligmosomoides polygyrus*, directly drives Treg expansion via ligation of the TGF-β receptor (Grainger et al., 2010). Treatment with oral *L. reuteri* for 9 days increased the percentage and total number of $CD4^{+}CD25^{+}Foxp3^{+}$ T cells in the spleens of experimental animals (Karimi, Inman, Bienenstock, & Forsythe, 2009). Colonization of mice by commensal *Clostridium* strains increased TGF-β levels and numbers of $Foxp3^{+}$ Treg in the colon (Atarashi et al., 2011).

8. URBAN UPBRINGING AND CITY LIVING AFFECT NEURAL SOCIAL STRESS PROCESSING

Meta-analyses suggest that individuals living in cities have increased risk for anxiety and affective disorders (21% and 39% increased risk, respectively) (Peen, Schoevers, Beekman, & Dekker, 2010). These findings suggest elevated risk of mental illness associated with urban upbringing and urban living. Unique aspects of social stress processing associated with urban living have been proposed to contribute to the greater risk of mental illness in this environment (Krabbendam & van Os, 2005; Peen et al., 2010; Selten & Cantor-Graae, 2005; van Os, Pedersen, & Mortensen, 2004). Indeed, human imaging studies, using functional magnetic resonance imaging (fMRI) during social evaluative threat have demonstrated that urban upbringing and city living impact social evaluative stress processing in humans. Being brought up in an urban setting was inversely correlated with gray matter volume (Haddad et al., 2015), specifically in males, and was associated with increased functional responses to stressors, in the perigenual anterior cingulate cortex (Lederbogen et al., 2011), an important structure for control of amygdala function, negative emotions (Pezawas et al., 2005), and stress (Diorio, Viau, & Meaney, 1993). Urban upbringing was also associated with decreased connectivity between the perigenual anterior cingulate cortex and amygdala (Lederbogen et al., 2011). Meanwhile, urban living, by itself, was associated with increased stress-associated amygdala responses, relative to rural living (Lederbogen et al., 2011). Although these findings were considered most likely to be due to differences in social stress in rural and urban environments, possibly related to a greater instability in hierarchical position in social networks in urban environments (Lederbogen et al., 2011), an alternative hypothesis is that differential microbial exposures underlies these effects, or differential microbial exposures interact with differences in social stress environments to alter social stress neural processing.

Meanwhile, those born and raised in cities have approximately double the risk of developing schizophrenia (Krabbendam & van Os, 2005; March et al., 2008; Pedersen & Mortensen, 2001; van Os et al., 2004; Vassos, Pedersen, Murray, Collier, & Lewis, 2012). Urban upbringing has been negatively correlated with gray matter volume in the dorsolateral prefrontal cortex (Haddad et al., 2015), a region with anatomical and functional alterations in schizophrenic patients (Fusar-Poli, Radua, McGuire, & Borgwardt, 2012), observations that led Haddad and colleagues (2015) to conclude that differences in social conditions along urban–rural gradients might account for the effects of urban upbringing on both neural structure and schizophrenia risk. As with anxiety and affective disorders, an alternative hypothesis is that differential microbial exposures underlies these effects, or interact with differences in social stress environments to alter brain structure and schizophrenia risk. Consistent with this alternative hypothesis, a seminal study by Tillisch et al. (2013) demonstrated that oral administration of a fermented milk product with probiotic to healthy women for 4 weeks resulted in reduced task-related responses, using a standardized emotional faces attention task for fMRI, in neural systems including affective, viscerosensory, and somatosensory cortices, as well as changes in resting state activity in specific circuits (Tillisch et al., 2013). The network responsive to the emotional attention task included the midbrain periaqueductal gray, insular and somatosensory cortices, and the prefrontal cortex. Resting state alterations following probiotic administration included a periaqueductal gray-seeded resting state network that included sensory regions including thalamus and insula, limbic regions including the cingulate gyrus, amygdala, hippocampus, parahippocampal gyrus, the basal ganglia, and cortical modulatory regions, including medial and dorsolateral prefrontal cortex. This network with resting state alterations following probiotic administration strongly resembles the sensory spinothalamocortical network originating in lamina I of the spinal cord that is implicated in homeostatic control of emotion. This system consists of sensory afferents arising from lamina I of the spinal cord to the periaqueductal gray, thalamus, primary somatosensory cortex, viscerosensory (insular) cortex, and anterior cingulate cortex (Craig, 2003). Furthermore, immunization with heat-killed preparations of bacteria has been shown to induce long-term changes in gene expression in serotonergic neurons, located in the ventral periaqueductal gray in mice (Reber et al., 2016). Finally, Cryan and colleagues have recently demonstrated that germ-free mice have upregulation of genes related to myelination and myelin plasticity, as well as neural activity-induced pathways, in the prefrontal cortex. Importantly, these alterations could be reversed by colonization of germ-free mice with conventional microbiota following weaning (Hoban et al., 2016). Altogether, these studies demonstrate that microbial inputs, either during development, or during adulthood, are sufficient to modulate neural systems implicated in anxiety and affective disorders, and neurodevelopmental disorders, including schizophrenia. Evidence points toward potential involvement of spinothalamocortical pathways involved in homeostatic control of emotion.

9. CONCLUSIONS

It is important to understand the mechanisms through which microbes associated with the microbiome of the built environment, including environmental microorganisms, influence immunoregulation. The built

Burr, M. L., Butland, B. K., King, S., & Vaughan-Williams, E. (1989). Changes in asthma prevalence: Two surveys 15 years apart. *Archives of Disease in Childhood, 64*, 1452–1456.
Chinetti, G., Fruchart, J. C., & Staels, B. (2003). Peroxisome proliferator-activated receptors: New targets for the pharmacological modulation of macrophage gene expression and function. *Current Opinion in Lipidology, 14*, 459–468.
Clemente, J. C., Pehrsson, E. C., Blaser, M. J., Sandhu, K., Gao, Z., Wang, B., et al. (2015). The microbiome of uncontacted Amerindians. *Science Advances, 1*, 3.
Contreras, M., Costello, E. K., Hidalgo, G., Magris, M., Knight, R., & Dominguez-Bello, M. G. (2010). The bacterial microbiota in the oral mucosa of rural Amerindians. *Microbiology, 156*, 3282–3287.
Correale, J., Farez, M., & Razzitte, G. (2008). Helminth infections associated with multiple sclerosis induce regulatory B cells. *Annals of Neurology, 64*, 187–199.
Craig, A. D. (2003). A new view of pain as a homeostatic emotion. *Trends in Neurosciences, 26*, 303–307.
Dannemiller, K. C., Gent, J. F., Leaderer, B. P., & Peccia, J. (2016). Influence of housing characteristics on bacterial and fungal communities in homes of asthmatic children. *Indoor Air, 26*, 179–192.
De Filippo, C., Cavalieri, D., Di, P. M., Ramazzotti, M., Poullet, J. B., Massart, S., et al. (2010). Impact of diet in shaping gut microbiota revealed by a comparative study in children from Europe and rural Africa. *Proceedings of the National Academy of Sciences of the United States of America, 107*, 14691–14696.
Dedesko, S., & Siegel, J. A. (2015). Moisture parameters and fungal communities associated with gypsum drywall in buildings. *Microbiome, 3*, 71.
DeLeon-Rodriguez, N., Lathem, T. L., Rodriguez, R., Barazesh, J. M., Anderson, B. E., Beyersdorf, A. J., et al. (2013). Microbiome of the upper troposphere: Species composition and prevalence, effects of tropical storms, and atmospheric implications. *Proceedings of the National Academy of Sciences of the United States of America, 110*, 2575–2580.
Deloge-Abarkan, M., Ha, T. L., Robine, E., Zmirou-Navier, D., & Mathieu, L. (2007). Detection of airborne Legionella while showering using liquid impingement and fluorescent in situ hybridization (FISH). *Journal of Environmental Monitoring, 9*, 91–97.
Diaz Heijtz, R. D. (2016). Fetal, neonatal, and infant microbiome: Perturbations and subsequent effects on brain development and behavior. *Seminars in Fetal and Neonatal Medicine*. Epub ahead of print. http://dx.doi.org/10.1016/j.siny.2016.04.012.
Di Giacinto, C., Marinaro, M., Sanchez, M., Strober, W., & Boirivant, M. (2005). Probiotics ameliorate recurrent Th1-mediated murine colitis by inducing IL-10 and IL-10-dependent TGF-beta-bearing regulatory cells. *Journal of Immunology, 174*, 3237–3246.
Dietert, R. R., & Dietert, J. M. (2008). Potential for early-life immune insult including developmental immunotoxicity in autism and autism spectrum disorders: Focus on critical windows of immune vulnerability. *Journal of Toxicology and Environmental Health Part B, Critical Reviews, 11*, 660–680.
Diorio, D., Viau, V., & Meaney, M. J. (1993). The role of the medial prefrontal cortex (cingulate gyrus) in the regulation of hypothalamic-pituitary-adrenal responses to stress. *The Journal of Neuroscience, 13*, 3839–3847.
Downs, S. H., Marks, G. B., Mitakakis, T. Z., Leuppi, J. D., Car, N. G., & Peat, J. K. (2001). Having lived on a farm and protection against allergic diseases in Australia. *Clinical and Experimental Allergy, 31*, 570–575.
Dunn, R. R., Fierer, N., Henley, J. B., Leff, J. W., & Menninger, H. L. (2013). Home life: Factors structuring the bacterial diversity found within and between homes. *PLoS One, 8*, e64133.
Ege, M. J., Mayer, M., Normand, A. C., Genuneit, J., Cookson, W. O., Braun-Fahrlander, C., et al. (2011). Exposure to environmental microorganisms and childhood asthma. *The New England Journal of Medicine, 364*, 701–709.

Elliott, D. E., & Weinstock, J. V. (2012). Helminth-host immunological interactions: Prevention and control of immune-mediated diseases. *Annals of the New York Academy of Sciences*, *1247*, 83–96.

Emerson, J. B., Keady, P. B., Brewer, T. E., Clements, N., Morgan, E. E., Awerbuch, J., et al. (2015). Impacts of flood damage on airborne bacteria and fungi in homes after the 2013 Colorado Front Range flood. *Environmental Science & Technology*, *49*, 2675–2684.

Evans, G. W., & McCoy, J. M. (1998). When buildings don't work: The role of architecture in human health. *Journal of Environmental Psychology*, *18*, 85–94.

Evans, N. P., Misyak, S. A., Schmelz, E. M., Guri, A. J., Hontecillas, R., & Bassaganya-Riera, J. (2010). Conjugated linoleic acid ameliorates inflammation-induced colorectal cancer in mice through activation of PPARgamma. *The Journal of Nutrition*, *140*, 515–521.

Flores, G. E., Bates, S. T., Knights, D., Lauber, C. L., Stombaugh, J., Knight, R., et al. (2011). Microbial biogeography of public restroom surfaces. *PLoS One*, *6*, e28132.

Flores, G. E., Caporaso, J. G., Henley, J. B., Rideout, J. R., Domogala, D., Chase, J., et al. (2014). Temporal variability is a personalized feature of the human microbiome. *Genome Biology*, *15*, 531.

Frankel, M., Beko, G., Timm, M., Gustavsen, S., Hansen, E. W., & Madsen, A. M. (2012). Seasonal variations of indoor microbial exposures and their relation to temperature, relative humidity, and air exchange rate. *Applied and Environmental Microbiology*, *78*, 8289–8297.

Fujimura, K. E., Johnson, C. C., Ownby, D. R., Cox, M. J., Brodie, E. L., Havstad, S. L., et al. (2010). Man's best friend? The effect of pet ownership on house dust microbial communities. *The Journal of Allergy and Clinical Immunology*, *126*, 410–412.

Fuller, R. A., Irvine, K. N., Devine-Wright, P., Warren, P. H., & Gaston, K. J. (2007). Psychological benefits of greenspace increase with biodiversity. *Biology Letters*, *3*, 390–394.

Fusar-Poli, P., Radua, J., McGuire, P., & Borgwardt, S. (2012). Neuroanatomical maps of psychosis onset: Voxel-wise meta-analysis of antipsychotic-naive VBM studies. *Schizophrenia Bulletin*, *38*, 1297–1307.

Gorvitovskaia, A., Holmes, S. P., & Huse, S. M. (2016). Interpreting Prevotella and Bacteroides as biomarkers of diet and lifestyle. *Microbiome*, *4*, 15.

Gozdz, J., Holbreich, M., Metwali, N., Thorne, P., Sperling, A., Martinez, F. D., et al. (2015). *Opposite effects of farming on asthma: Mice exposed to Amish and Hutterite environmental products recapitulate asthma protection and risk.* B.21. TRANSLATIONAL ASTHMA RESEARCH. New York, NY: American Thoracic Society. A2489.

Graham-Rowe, D. (2011). Lifestyle: When allergies go west. *Nature*, *479*, S2–S4.

Grainger, J. R., Smith, K. A., Hewitson, J. P., McSorley, H. J., Harcus, Y., Filbey, K. J., et al. (2010). Helminth secretions induce de novo T cell Foxp3 expression and regulatory function through the TGF-beta pathway. *The Journal of Experimental Medicine*, *207*, 2331–2341.

Grantham, N. S., Reich, B. J., Pacifici, K., Laber, E. B., Menninger, H. L., Henley, J. B., et al. (2015). Fungi identify the geographic origin of dust samples. *PLoS One*, *10*, e0122605.

Green, J. L. (2014). Can bioinformed design promote healthy indoor ecosystems? *Indoor Air*, *24*, 113–115.

Haddad, L., Schafer, A., Streit, F., Lederbogen, F., Grimm, O., Wust, S., et al. (2015). Brain structure correlates of urban upbringing, an environmental risk factor for schizophrenia. *Schizophrenia Bulletin*, *41*, 115–122.

Hadley, G. A., Bartlett, S. T., Via, C. S., Rostapshova, E. A., & Moainie, S. (1997). The epithelial cell-specific integrin, CD103 (alpha E integrin), defines a novel subset of alloreactive CD8+ CTL. *Journal of Immunology*, *159*, 3748–3756.

Hanski, I., von Hertzen, L., Fyhrquist, N., Koskinen, K., Torppa, K., Laatikainen, T., et al. (2012). Environmental biodiversity, human microbiota, and allergy are interrelated. *Proceedings of the National Academy of Sciences of the United States of America, 109*, 8334–8339.

Hart, A. L., Lammers, K., Brigidi, P., Vitali, B., Rizzello, F., Gionchetti, P., et al. (2004). Modulation of human dendritic cell phenotype and function by probiotic bacteria. *Gut, 53*, 1602–1609.

Henne, K., Kahlisch, L., Brettar, I., & Hofle, M. G. (2012). Analysis of structure and composition of bacterial core communities in mature drinking water biofilms and bulk water of a citywide network in Germany. *Applied and Environmental Microbiology, 78*, 3530–3538.

Hoang, C. P., Kinney, K. A., Corsi, R. L., & Szaniszlo, P. J. (2010). Resistance of green building materials to fungal growth. *International Biodeterioration & Biodegradation, 64*, 104–113.

Hoban, A. E., Stilling, R. M., Ryan, F. J., Shanahan, F., Dinan, T. G., Claesson, M. J., et al. (2016). Regulation of prefrontal cortex myelination by the microbiota. *Translational Psychiatry, 6*, e774.

Hoisington, A. J., Brenner, L. A., Kinney, K. A., Postolache, T. T., & Lowry, C. A. (2015). The microbiome of the built environment and mental health. *Microbiome, 3*, 1–12.

Hoisington, A., Maestre, J. P., Kinney, K. A., & Siegel, J. A. (2015). Characterizing the bacterial communities in retail stores in the United States. *Indoor Air*. Epub ahead of print. http://dx.doi.org/10.1111/ina.12273.

Holbreich, M., Genuneit, J., Weber, J., Braun-Fahrlander, C., Waser, M., & von Mutius, E. (2012). Amish children living in northern Indiana have a very low prevalence of allergic sensitization. *The Journal of Allergy and Clinical Immunology, 129*, 1671–1673.

Hospodsky, D., Qian, J., Nazaroff, W. W., Yamamoto, N., Bibby, K., Rismani-Yazdi, H., et al. (2012). Human occupancy as a source of indoor airborne bacteria. *PLoS One, 7*, e34867.

Hospodsky, D., Yamamoto, N., Nazaroff, W. W., Miller, D., Gorthala, S., & Peccia, J. (2015). Characterizing airborne fungal and bacterial concentrations and emission rates in six occupied children's classrooms. *Indoor Air, 25*, 641–652.

Hug, L. A., Baker, B. J., Anantharaman, K., Brown, C. T., Probst, A. J., Castelle, C. J., et al. (2016). A new view of the tree of life. *Nature Microbiology, 1*, 16048.

Hwang, D. (2000). Fatty acids and immune responses—A new perspective in searching for clues to mechanism. *Annual Review of Nutrition, 20*, 431–456.

Jaensson, E., Uronen-Hansson, H., Pabst, O., Eksteen, B., Tian, J., Coombes, J. L., et al. (2008). Small intestinal CD103+ dendritic cells display unique functional properties that are conserved between mice and humans. *The Journal of Experimental Medicine, 205*, 2139–2149.

Ji, P., Parks, J., Edwards, M. A., & Pruden, A. (2015). Impact of water chemistry, pipe material and stagnation on the building plumbing microbiome. *PLoS One, 10*, e0141087.

Karimi, K., Inman, M. D., Bienenstock, J., & Forsythe, P. (2009). Lactobacillus reuteri-induced regulatory T cells protect against an allergic airway response in mice. *American Journal of Respiratory and Critical Care Medicine, 179*, 186–193.

Kelley, S. T., & Gilbert, J. A. (2013). Studying the microbiology of the indoor environment. *Genome Biology, 14*, 202.

Kelley, S. T., Theisen, U., Angenent, L. T., St Amend, A., & Pace, N. R. (2004). Molecular analysis of shower curtain biofilm microbes. *Applied and Environmental Microbiology, 70*, 4187–4192.

Kembel, S. W., Jones, E., Kline, J., Northcutt, D., Stenson, J., Womack, A. M., et al. (2012). Architectural design influences the diversity and structure of the built environment microbiome. *The ISME Journal, 6*, 1469–1479.

Kembel, S. W., Meadow, J. F., O'Connor, T. K., Mhuireach, G., Northcutt, D., Kline, J., et al. (2014). Architectural design drives the biogeography of indoor bacterial communities. *PLoS One*, *9*, e87093.

Kemp, P. F., & Aller, J. Y. (2004). Bacterial diversity in aquatic and other environments: What 16S rDNA libraries can tell us. *FEMS Microbiology Ecology*, *47*, 161–177.

Kim, D. I., Kim, K. H., Kang, J. H., Jung, E. M., Kim, S. S., Jeung, E. B., et al. (2011). Trans-10, cis-12-conjugated linoleic acid modulates NF-kappaB activation and TNF-alpha production in porcine peripheral blood mononuclear cells via a PPARgamma-dependent pathway. *The British Journal of Nutrition*, *105*, 1329–1336.

Kliewer, S. A., Lenhard, J. M., Willson, T. M., Patel, I., Morris, D. C., & Lehmann, J. M. (1995). A prostaglandin J2 metabolite binds peroxisome proliferator-activated receptor gamma and promotes adipocyte differentiation. *Cell*, *83*, 813–819.

Korkalainen, M., Taubel, M., Naarala, J., Kirjavainen, P., Koistinen, A., Hyvarinen, A., et al. (2016). Synergistic proinflammatory interactions of microbial toxins and structural components characteristic to moisture-damaged buildings. *Indoor Air*. Epub ahead of print. http://dx.doi.org/10.1111/ina.12282.

Krabbendam, L., & van Os, J. (2005). Schizophrenia and urbanicity: A major environmental influence—Conditional on genetic risk. *Schizophrenia Bulletin*, *31*, 795–799.

Lau, H. Y., & Ashbolt, N. J. (2009). The role of biofilms and protozoa in Legionella pathogenesis: Implications for drinking water. *Journal of Applied Microbiology*, *107*, 368–378.

Lauener, R. P., Birchler, T., Adamski, J., Braun-Fahrlander, C., Bufe, A., Herz, U., et al. (2002). Expression of CD14 and Toll-like receptor 2 in farmers' and non-farmers' children. *Lancet*, *360*, 465–466.

Lautenschlager, K., Boon, N., Wang, Y., Egli, T., & Hammes, F. (2010). Overnight stagnation of drinking water in household taps induces microbial growth and changes in community composition. *Water Research*, *44*, 4868–4877.

Lax, S., Nagler, C. R., & Gilbert, J. A. (2015). Our interface with the built environment: Immunity and the indoor microbiota. *Trends in Immunology*, *36*, 121–123.

Lax, S., Smith, D. P., Hampton-Marcell, J., Owens, S. M., Handley, K. M., Scott, N. M., et al. (2014). Longitudinal analysis of microbial interaction between humans and the indoor environment. *Science*, *345*, 1048–1052.

Le Bert, N., Chain, B. M., Rook, G., & Noursadeghi, M. (2011). DC priming by *M. vaccae* inhibits Th2 responses in contrast to specific TLR2 priming and is associated with selective activation of the CREB pathway. *PLoS One*, *6*, e18346.

Lederbogen, F., Kirsch, P., Haddad, L., Streit, F., Tost, H., Schuch, P., et al. (2011). City living and urban upbringing affect neural social stress processing in humans. *Nature*, *474*, 498–501.

Lee, S. O., Kim, C. S., Cho, S. K., Choi, H. J., Ji, G. E., & Oh, D. K. (2003). Bioconversion of linoleic acid into conjugated linoleic acid during fermentation and by washed cells of Lactobacillus reuteri. *Biotechnology Letters*, *25*, 935–938.

Leung, M. H., & Lee, P. K. (2016). The roles of the outdoors and occupants in contributing to a potential pan-microbiome of the built environment: A review. *Microbiome*, *4*, 21.

Ley, R. E., Backhed, F., Turnbaugh, P., Lozupone, C. A., Knight, R. D., & Gordon, J. I. (2005). Obesity alters gut microbial ecology. *Proceedings of the National Academy of Sciences of the United States of America*, *102*, 11070–11075.

Little, A. E., Robinson, C. J., Peterson, S. B., Raffa, K. F., & Handelsman, J. (2008). Rules of engagement: Interspecies interactions that regulate microbial communities. *Annual Review of Microbiology*, *62*, 375–401.

Logan, A. C. (2015). Dysbiotic drift: Mental health, environmental grey space, and microbiota. *Journal of Physiological Anthropology*, *34*, 23.

Loke, P., MacDonald, A. S., Robb, A., Maizels, R. M., & Allen, J. E. (2000). Alternatively activated macrophages induced by nematode infection inhibit proliferation via cell-to-cell contact. *European Journal of Immunology*, *30*, 2669–2678.

Lowry, C. A., Hollis, J. H., de Vries, A., Pan, B., Brunet, L. R., Hunt, J. R., et al. (2007). Identification of an immune-responsive mesolimbocortical serotonergic system: Potential role in regulation of emotional behavior. *Neuroscience*, *146*, 756–772.

Lowry, C. A., Smith, D. G., Siebler, P. H., Schmidt, D., Stamper, C. E., Hassell, J. E., Jr., et al. (2016). The microbiota, immunoregulation and mental health: Implications for public health. *Current Environmental Health Reports*, *3*, 270–286.

Luongo, J. C., Barberan, A., Hacker-Cary, R., Morgan, E. E., Miller, S. L., & Fierer, N. (2016). Microbial analyses of airborne dust collected from dormitory rooms predict the sex of occupants. *Indoor Air*. Epub ahead of print. http://dx.doi.org/10.1111/ina12302.

Lymperopoulou, D. S., Adams, R. I., & Lindow, S. E. (2016). Contribution of vegetation to the microbial composition of nearby outdoor air. *Applied and Environmental Microbiology*, *82*, 3822–3833.

Macovei, L., McCafferty, J., Chen, T., Teles, F., Hasturk, H., Paster, B. J., et al. (2015). The hidden 'mycobacteriome' of the human healthy oral cavity and upper respiratory tract. *Journal of Oral Microbiology*, *7*, 26094.

Mangan, N. E., Fallon, R. E., Smith, P., van Rooijen, N., McKenzie, A. N., & Fallon, P. G. (2004). Helminth infection protects mice from anaphylaxis via IL-10-producing B cells. *Journal of Immunology*, *173*, 6346–6356.

March, D., Hatch, S. L., Morgan, C., Kirkbride, J. B., Bresnahan, M., Fearon, P., et al. (2008). Psychosis and place. *Epidemiologic Reviews*, *30*, 84–100.

Martinez, I., Stegen, J. C., Maldonado-Gomez, M. X., Eren, A. M., Siba, P. M., Greenhill, A. R., et al. (2015). The gut microbiota of rural papua new guineans: Composition, diversity patterns, and ecological processes. *Cell Reports*, *11*, 527–538.

Matsushita, T., Yanaba, K., Bouaziz, J. D., Fujimoto, M., & Tedder, T. F. (2008). Regulatory B cells inhibit EAE initiation in mice while other B cells promote disease progression. *The Journal of Clinical Investigation*, *118*, 3420–3430.

Matthews, D. M., & Jenks, S. M. (2013). Ingestion of *Mycobacterium vaccae* decreases anxiety-related behavior and improves learning in mice. *Behavioural Processes*, *96*, 27–35.

McDade, T. W., Rutherford, J., Adair, L., & Kuzawa, C. W. (2010). Early origins of inflammation: Microbial exposures in infancy predict lower levels of C-reactive protein in adulthood. *Proceedings of the Biological Sciences*, *277*, 1129–1137.

Meadow, J. F., Altrichter, A. E., Kembel, S. W., Kline, J., Mhuireach, G., Moriyama, M., et al. (2014). Indoor airborne bacterial communities are influenced by ventilation, occupancy, and outdoor air source. *Indoor Air*, *24*, 41–48.

Miletto, M., & Lindow, S. E. (2015). Relative and contextual contribution of different sources to the composition and abundance of indoor air bacteria in residences. *Microbiome*, *3*, 61.

Miller, A. H., & Raison, C. L. (2016). The role of inflammation in depression: From evolutionary imperative to modern treatment target. *Nature Reviews. Immunology*, *16*, 22–34.

Miller, J. D., Rand, T. G., & Jarvis, B. B. (2003). Stachybotrys chartarum: Cause of human disease or media darling? *Medical Mycology*, *41*, 271–291.

Mitchell, R., & Popham, F. (2008). Effect of exposure to natural environment on health inequalities: An observational population study. *Lancet*, *372*, 1655–1660.

Modi, S. R., Collins, J. J., & Relman, D. A. (2014). Antibiotics and the gut microbiota. *The Journal of Clinical Investigation*, *124*, 4212–4218.

Mosser, D. M., & Edwards, J. P. (2008). Exploring the full spectrum of macrophage activation. *Nature Reviews. Immunology*, *8*, 958–969.

Nazaroff, W. W. (2016). Indoor bioaerosol dynamics. *Indoor Air*, *26*, 61–78.

Obregon-Tito, A. J., Tito, R. Y., Metcalf, J., Sankaranarayanan, K., Clemente, J. C., Ursell, L. K., et al. (2015). Subsistence strategies in traditional societies distinguish gut microbiomes. *Nature Communications*, *6*, 6505.

Odegaard, J. I., Ricardo-Gonzalez, R. R., Goforth, M. H., Morel, C. R., Subramanian, V., Mukundan, L., et al. (2007). Macrophage-specific PPARgamma controls alternative activation and improves insulin resistance. *Nature*, *447*, 1116–1120.

Ogawa, J., Kishino, S., Ando, A., Sugimoto, S., Mihara, K., & Shimizu, S. (2005). Production of conjugated fatty acids by lactic acid bacteria. *Journal of Bioscience and Bioengineering*, *100*, 355–364.

Ownby, D. R., Johnson, C. C., & Peterson, E. L. (2002). Exposure to dogs and cats in the first year of life and risk of allergic sensitization at 6 to 7 years of age. *JAMA*, *288*, 963–972.

Pedersen, C. B., & Mortensen, P. B. (2001). Evidence of a dose-response relationship between urbanicity during upbringing and schizophrenia risk. *Archives of General Psychiatry*, *58*, 1039–1046.

Peen, J., Schoevers, R. A., Beekman, A. T., & Dekker, J. (2010). The current status of urban-rural differences in psychiatric disorders. *Acta Psychiatrica Scandinavica*, *121*, 84–93.

Pezawas, L., Meyer-Lindenberg, A., Drabant, E. M., Verchinski, B. A., Munoz, K. E., Kolachana, B. S., et al. (2005). 5-HTTLPR polymorphism impacts human cingulate-amygdala interactions: A genetic susceptibility mechanism for depression. *Nature Neuroscience*, *8*, 828–834.

Pinto, A. J., Xi, C., & Raskin, L. (2012). Bacterial community structure in the drinking water microbiome is governed by filtration processes. *Environmental Science & Technology*, *46*, 8851–8859.

Poza, M., Gayoso, C., Gomez, M. J., Rumbo-Feal, S., Tomas, M., Aranda, J., et al. (2012). Exploring bacterial diversity in hospital environments by GS-FLX Titanium pyrosequencing. *PLoS One*, *7*, e44105.

Qian, J., Hospodsky, D., Yamamoto, N., Nazaroff, W. W., & Peccia, J. (2012). Size-resolved emission rates of airborne bacteria and fungi in an occupied classroom. *Indoor Air*, *22*, 339–351.

Rakoff-Nahoum, S., Paglino, J., Eslami-Varzaneh, F., Edberg, S., & Medzhitov, R. (2004). Recognition of commensal microflora by toll-like receptors is required for intestinal homeostasis. *Cell*, *118*, 229–241.

Reber, S. O., Siebler, P. H., Donner, N. C., Morton, J. T., Smith, D. G., Kopelman, J. M., et al. (2016). Immunization with a heat-killed preparation of the environmental bacterium *Mycobacterium vaccae* promotes stress resilience in mice. *Proceedings of the National Academy of Sciences of the United States of America*, *113*, E3130–E3139.

Rintala, H., Nevalainen, A., & Suutari, M. (2002). Diversity of streptomycetes in water-damaged building materials based on 16S rDNA sequences. *Letters in Applied Microbiology*, *34*, 439–443.

Ripa, L. W. (1993). A half-century of community water fluoridation in the United States: Review and commentary. *Journal of Public Health Dentistry*, *53*, 17–44.

Roman-Nunez, M., Cuesta-Alonso, E. P., & Gilliland, S. E. (2007). Influence of sodium glycocholate on production of conjugated linoleic acid by cells of Lactobacillus reuteri ATCC 55739. *Journal of Food Science*, *72*, M140–M143.

Rook, G. A. (2010). 99th Dahlem conference on infection, inflammation and chronic inflammatory disorders: Darwinian medicine and the 'hygiene' or 'old friends' hypothesis. *Clinical and Experimental Immunology*, *160*, 70–79.

Rook, G. A. (2013). Regulation of the immune system by biodiversity from the natural environment: An ecosystem service essential to health. *Proceedings of the National Academy of Sciences of the United States of America*, *110*, 18360–18367.

Rook, G. A., Adams, V., Hunt, J., Palmer, R., Martinelli, R., & Brunet, L. R. (2004). Mycobacteria and other environmental organisms as immunomodulators for immunoregulatory disorders. *Springer Seminars in Immunopathology*, *25*, 237–255.

Rook, G. A., Raison, C. L., & Lowry, C. A. (2014). Microbial 'old friends', immunoregulation and socioeconomic status. *Clinical and Experimental Immunology*, *177*, 1–12.

Ross, A. A., & Neufeld, J. D. (2015). Microbial biogeography of a university campus. *Microbiome*, *3*, 66.

Round, J. L., Lee, S. M., Li, J., Tran, G., Jabri, B., Chatila, T. A., et al. (2011). The Toll-like receptor 2 pathway establishes colonization by a commensal of the human microbiota. *Science*, *332*, 974–977.

Ruokolainen, L., von Hertzen, L., Fyhrquist, N., Laatikainen, T., Lehtomaki, J., Auvinen, P., et al. (2015). Green areas around homes reduce atopic sensitization in children. *Allergy*, *70*, 195–202.

Rusin, P., Orosz-Coughlin, P., & Gerba, C. (1998). Reduction of faecal coliform, coliform and heterotrophic plate count bacteria in the household kitchen and bathroom by disinfection with hypochlorite cleaners. *Journal of Applied Microbiology*, *85*, 819–828.

Schnoeller, C., Rausch, S., Pillai, S., Avagyan, A., Wittig, B. M., Loddenkemper, C., et al. (2008). A helminth immunomodulator reduces allergic and inflammatory responses by induction of IL-10-producing macrophages. *Journal of Immunology*, *180*, 4265–4272.

Seifried, J. S., Wichels, A., & Gerdts, G. (2015). Spatial distribution of marine airborne bacterial communities. *MicrobiologyOpen*, *4*, 475–490.

Sela, D. A., & Mills, D. A. (2010). Nursing our microbiota: Molecular linkages between bifidobacteria and milk oligosaccharides. *Trends in Microbiology*, *18*, 298–307.

Selten, J. P., & Cantor-Graae, E. (2005). Social defeat: Risk factor for schizophrenia? *The British Journal of Psychiatry*, *187*, 101–102.

Shibolet, O., & Podolsky, D. K. (2007). TLRs in the Gut. IV. Negative regulation of Toll-like receptors and intestinal homeostasis: Addition by subtraction. *American Journal of Physiology. Gastrointestinal and Liver Physiology*, *292*, G1469–G1473.

Shwartz, A., Turbe, A., Simon, L., & Julliard, R. (2014). Enhancing urban biodiversity and its influence on city-dwellers: An experiment. *Biological Conservation*, *171*, 82–90.

Smelt, M. J., de Haan, B. J., Bron, P. A., van Swam, I., Meijerink, M., Wells, J. M., et al. (2012). *L. plantarum*, *L. salivarius*, and *L. lactis* attenuate Th2 responses and increase Treg frequencies in healthy mice in a strain dependent manner. *PLoS One*, 7, e47244.

Smith, P., Mangan, N. E., Walsh, C. M., Fallon, R. E., McKenzie, A. N., van Rooijen, N., et al. (2007). Infection with a helminth parasite prevents experimental colitis via a macrophage-mediated mechanism. *Journal of Immunology*, *178*, 4557–4566.

Smith, P. D., Ochsenbauer-Jambor, C., & Smythies, L. E. (2005). Intestinal macrophages: Unique effector cells of the innate immune system. *Immunological Reviews*, *206*, 149–159.

Smith, M. I., Yatsunenko, T., Manary, M. J., Trehan, I., Mkakosya, R., Cheng, J., et al. (2013). Gut microbiomes of Malawian twin pairs discordant for kwashiorkor. *Science*, *339*, 548–554.

Smits, H. H., Engering, A., van der Kleij, D., de Jong, E. C., Schipper, K., van Capel, T. M., et al. (2005). Selective probiotic bacteria induce IL-10-producing regulatory T cells in vitro by modulating dendritic cell function through dendritic cell-specific intercellular adhesion molecule 3-grabbing nonintegrin. *The Journal of Allergy and Clinical Immunology*, *115*, 1260–1267.

Smythies, L. E., Sellers, M., Clements, R. H., Mosteller-Barnum, M., Meng, G., Benjamin, W. H., et al. (2005). Human intestinal macrophages display profound inflammatory anergy despite avid phagocytic and bacteriocidal activity. *The Journal of Clinical Investigation*, *115*, 66–75.

Smythies, L. E., Shen, R., Bimczok, D., Novak, L., Clements, R. H., Eckhoff, D. E., et al. (2010). Inflammation anergy in human intestinal macrophages is due to Smad-induced

IkappaBalpha expression and NF-kappaB inactivation. *The Journal of Biological Chemistry, 285*, 19593–19604.

Sonnenburg, E. D., & Sonnenburg, J. L. (2014). Starving our microbial self: The deleterious consequences of a diet deficient in microbiota-accessible carbohydrates. *Cell Metabolism, 20*, 779–786.

Strachan, D. P. (1989). Hay fever, hygiene, and household size. *BMJ, 299*, 1259–1260.

Sun, C. M., Hall, J. A., Blank, R. B., Bouladoux, N., Oukka, M., Mora, J. R., et al. (2007). Small intestine lamina propria dendritic cells promote de novo generation of Foxp3 T reg cells via retinoic acid. *The Journal of Experimental Medicine, 204*, 1775–1785.

Sun, W., Svendsen, E. R., Karmaus, W. J., Kuehr, J., & Forster, J. (2015). Early-life antibiotic use is associated with wheezing among children with high atopic risk: A prospective European study. *The Journal of Asthma, 52*, 647–652.

Suzuki, T. A., & Worobey, M. (2014). Geographical variation of human gut microbial composition. *Biology Letters, 10*, 20131037.

Szponar, B., & Larsson, L. (2000). Determination of microbial colonisation in water-damaged buildings using chemical marker analysis by gas chromatography-mass spectrometry. *Indoor Air, 10*, 13–18.

Tillisch, K., Labus, J., Kilpatrick, L., Jiang, Z., Stains, J., Ebrat, B., et al. (2013). Consumption of fermented milk product with probiotic modulates brain activity. *Gastroenterology, 144*, 1394–1401.

Torvinen, E., Meklin, T., Torkko, P., Suomalainen, S., Reiman, M., Katila, M. L., et al. (2006). Mycobacteria and fungi in moisture-damaged building materials. *Applied and Environmental Microbiology, 72*, 6822–6824.

Ulrich, R. S. (1984). View through a window may influence recovery from surgery. *Science, 224*, 420–421.

United Nations. (2014). *World Urbanization Prospects: The 2014 revision*. New York, NY: United Nations.

Valkonen, M., Wouters, I. M., Taubel, M., Rintala, H., Lenters, V., Vasara, R., et al. (2015). Bacterial exposures and associations with atopy and asthma in children. *PLoS One, 10*, e0131594.

van Os, J., Pedersen, C. B., & Mortensen, P. B. (2004). Confirmation of synergy between urbanicity and familial liability in the causation of psychosis. *The American Journal of Psychiatry, 161*, 2312–2314.

Vassos, E., Pedersen, C. B., Murray, R. M., Collier, D. A., & Lewis, C. M. (2012). Meta-analysis of the association of urbanicity with schizophrenia. *Schizophrenia Bulletin, 38*, 1118–1123.

von Hertzen, L., Beutler, B., Bienenstock, J., Blaser, M., Cani, P. D., Eriksson, J., et al. (2015). Helsinki alert of biodiversity and health. *Annals of Medicine, 47*, 218–225.

von Mutius, E. (2016). The microbial environment and its influence on asthma prevention in early life. *The Journal of Allergy and Clinical Immunology, 137*, 680–689.

von Mutius, E., Martinez, F. D., Fritzsch, C., Nicolai, T., Roell, G., & Thiemann, H. H. (1994). Prevalence of asthma and atopy in two areas of West and East Germany. *American Journal of Respiratory and Critical Care Medicine, 149*, 358–364.

Wang, H., Edwards, M. A., Falkinham, J. O., III, & Pruden, A. (2013). Probiotic approach to pathogen control in premise plumbing systems? A review. *Environmental Science & Technology, 47*, 10117–10128.

Wang, H., Masters, S., Edwards, M. A., Falkinham, J. O., III, & Pruden, A. (2014). Effect of disinfectant, water age, and pipe materials on bacterial and eukaryotic community structure in drinking water biofilm. *Environmental Science & Technology, 48*, 1426–1435.

Wang, H., Masters, S., Falkinham, J. O., III, Edwards, M. A., & Pruden, A. (2015). Distribution system water quality affects responses of opportunistic pathogen gene markers in household water heaters. *Environmental Science & Technology, 49*, 8416–8424.

White, C., Tancos, M., & Lytle, D. A. (2011). Microbial community profile of a lead service line removed from a drinking water distribution system. *Applied and Environmental Microbiology*, *77*, 5557–5561.

Wlasiuk, G., & Vercelli, D. (2012). The farm effect, or: When, what and how a farming environment protects from asthma and allergic disease. *Current Opinion in Allergy and Clinical Immunology*, *12*, 461–466.

Wu, G. D., Chen, J., Hoffmann, C., Bittinger, K., Chen, Y. Y., Keilbaugh, S. A., et al. (2011). Linking long-term dietary patterns with gut microbial enterotypes. *Science*, *334*, 105–108.

Yamamoto, N., Hospodsky, D., Dannemiller, K. C., Nazaroff, W. W., & Peccia, J. (2015). Indoor emissions as a primary source of airborne allergenic fungal particles in classrooms. *Environmental Science & Technology*, *49*, 5098–5106.

Yanaba, K., Bouaziz, J. D., Haas, K. M., Poe, J. C., Fujimoto, M., & Tedder, T. F. (2008a). A regulatory B cell subset with a unique CD1dhiCD5+ phenotype controls T cell-dependent inflammatory responses. *Immunity*, *28*, 639–650.

Yanaba, K., Bouaziz, J. D., Matsushita, T., Magro, C. M., St Clair, E. W., & Tedder, T. F. (2008b). B-lymphocyte contributions to human autoimmune disease. *Immunological Reviews*, *223*, 284–299.

Yatsunenko, T., Rey, F. E., Manary, M. J., Trehan, I., Dominguez-Bello, M. G., Contreras, M., et al. (2012). Human gut microbiome viewed across age and geography. *Nature*, *486*, 222–227.

Yunginger, J. W., Reed, C. E., O'Connell, E. J., Melton, L. J., III, O'Fallon, W. M., & Silverstein, M. D. (1992). A community-based study of the epidemiology of asthma. Incidence rates, 1964-1983. *The American Review of Respiratory Disease*, *146*, 888–894.

Zekveld, C., Bibakis, I., Bibaki-Liakou, V., Pedioti, A., Dimitroulis, I., Harris, J., et al. (2006). The effects of farming and birth order on asthma and allergies. *The European Respiratory Journal*, *28*, 82–88.

Zuany-Amorim, C., Manlius, C., Trifilieff, A., Brunet, L. R., Rook, G., Bowen, G., et al. (2002). Long-term protective and antigen-specific effect of heat-killed *Mycobacterium vaccae* in a murine model of allergic pulmonary inflammation. *Journal of Immunology*, *169*, 1492–1499.

Zuany-Amorim, C., Sawicka, E., Manlius, C., Le Moine, A., Brunet, L. R., Kemeny, D. M., et al. (2002). Suppression of airway eosinophilia by killed *Mycobacterium vaccae*-induced allergen-specific regulatory T-cells. *Nature Medicine*, *8*, 625–629.

CHAPTER FIFTEEN

The Importance of Diet and Gut Health to the Treatment and Prevention of Mental Disorders

S.L. Dawson[*,†,1], **S.R. Dash**[*,‡,1], **F.N. Jacka**[*,§,¶,||,2]

[*]Food and Mood Centre, IMPACT SRC, Deakin University, School of Medicine, Geelong, VIC, Australia
[†]Early Life Epigenetics Group, Murdoch Childrens Research Institute (MCRI), Royal Children's Hospital, Parkville, VIC, Australia
[‡]Collaborative Research Centre for Mental Health, Carlton, VIC, Australia
[§]Centre for Adolescent Health, Murdoch Children's Research Institute (MCRI), Royal Children's Hospital, Parkville, VIC, Australia
[¶]Royal Melbourne Hospital, University of Melbourne, Melbourne, VIC, Australia
[||]Black Dog Institute, Prince of Wales Hospital, Randwick, NSW, Australia
[2]Corresponding author: e-mail address: felicejacka@gmail.com

Contents

Abstract

The departure from traditional lifestyles and the rising disease burden of mental disorders are increasing global health concerns. Changes in diet around the world mean that populations are now increasingly reliant on highly processed, poor quality foods, which

[1] Joint first author.

International Review of Neurobiology, Volume 131
ISSN 0074-7742
http://dx.doi.org/10.1016/bs.irn.2016.08.009

have been linked to increased risk for mental disorder. Conversely, a nutrient-rich diet is understood to be protective of mental health, and researchers are now aiming to understand the biological underpinnings of this relationship. The gut microbiota has been proposed as a key mediator of this link, given its association with both diet and mental health. Importantly, several critical "windows of opportunity" for prevention and intervention have been identified, particularly early life and adolescence; these are periods of rapid development and transition that provide a foundation for future health. Strategies that promote overall diet quality, high in fiber and nutrients, have been linked to increased microbial diversity and gut health. Improving diet quality and subsequent gut health may have benefits for individuals' mental health, as well as the mental health of future generations. Here we discuss specific, targeted dietary and gut focused strategies for the prevention and treatment of mental disorder.

1. INTRODUCTION

Mental disorders, specifically anxiety and depression, account for a significant proportion of global disease burden (Whiteford et al., 2013). Lifestyle has changed dramatically through globalization and urbanization, and the shift away from more traditional lifestyles has been linked to an increase in mental disorders (Logan & Jacka, 2014). Diet has recently been shown to be one of the leading risk factors for early mortality (Global Burden of Disease Study Collaborators et al., 2015) and is a shared risk factor for both physical and mental disorders (Jacka, Sacks, Berk, & Allender, 2014). The relationship between diet and mental disorders is evident across countries, cultures, and age groups and there is emerging evidence highlighting some of the relevant pathways for this relationship (Berk et al., 2013; Jacka, Cherbuin, Anstey, Sachdev, & Butterworth, 2015). One pathway of particular interest relates to the gut and its resident microbiota, as the microbiome is strongly affected by diet and also appears to influence multiple factors related to the risk for mental disorders (Dash, Clarke, Berk, & Jacka, 2015). Given the modifiable nature of the diet and the microbiome, this new knowledge base affords the potential for new prevention and treatment strategies for mental health and has significant implications for public health and clinical treatment.

2. DIET AND MENTAL HEALTH ACROSS THE LIFESPAN

While diet quality has long been understood as important to physical health, recognition of its relevance to mental health has been slower to develop (Sarris et al., 2015). Early nutritional psychiatry research focused largely on individual nutrients, such as omega-3s or folate, and their relationship with mental health. However, given the synergistic nature of nutrients in food, the field has since moved toward a "whole of diet" approach that better captures these complexities (Hu, 2002). Dietary patterns in both affluent and developing countries now consist primarily of high-energy, nutrient poor foods, and have shifted away from more traditional diets of the past, comprised of higher intakes of plant foods and quality proteins (Logan & Jacka, 2014). Over the last decade, evidence has accumulated to support the relationship between diet quality and mental health and these associations have been consistently observed in adults, adolescents, and children across a multitude of different countries and cultures (Jacka, Mykletun, & Berk, 2012).

2.1 Early Life, Childhood and Adolescence

A robust evidence base suggests that diet is important to mental health across the lifespan, although there are particular periods of rapid development or transition that present critically important windows of opportunity. Diet quality before and during pregnancy is important to the mental health of the mother (Baskin, Hill, Jacka, O'Neil, & Skouteris, 2015), but also has implications for the future mental health (Jacka, Ystrom, et al., 2013; Pina-Camacho, Jensen, Gaysina, & Barker, 2015; Steenweg-de Graaff et al., 2014) and cognition (Barker, Kirkham, Ng, & Jensen, 2013) of her offspring. This suggests the potential to target the nutritional status of a woman during pregnancy in order to positively influence future mental health outcomes in children (O'Neil et al., 2014).

As in other domains of health, childhood experiences are highly impactful on future health (Heim, Newport, Mletzko, Miller, & Nemeroff, 2008). Unfortunately, many studies have shown that young people are eating well below dietary recommendations; the most recent Nutrition Survey in Australia found that less than 1% of children met the minimum recommendations for the five major food groups (Australian Bureau of Statistics (ABS), 2016), and similar trends are seen globally (Kontogianni et al., 2008; Larson, Neumark-Sztainer, Hannan, & Story, 2007; Northstone & Emmett, 2005). Young people are increasingly reliant on nutrient poor, high sugar foods such as sweets, soft drink, snacks, and baked goods (Adair & Popkin, 2005). These dietary patterns have linked to obesity and noncommunicable disease, but the extant evidence suggests that what children and adolescents are eating is critically important to both the brain and mental health as well. Consuming a diet of high sugar and "snack-like foods" has been linked to behavioral and emotional problems in children (Jacka, Ystrom, et al., 2013; Kohlboeck et al., 2012; Overby & Hoigaard, 2012) which have been linked to mental disorder in adulthood (Becker, El-Faddagh, Schmidt, & Laucht, 2007; Copeland, Shanahan, Costello, & Angold, 2009).

By adolescence, young people have greater independence in food choices and begin to establish dietary habits that will carry through to adulthood (Mikkila, Rasanen, Raitakari, Pietinen, & Viikari, 2004). This period is also critical, as it is typically during this time that mental disorders emerge for the first time (Kessler et al., 2005). Researchers around the world have consistently demonstrated that the diet quality of young people is important to mental health during the transition to adulthood (Jacka et al., 2011, 2010; Jacka, Rothon, Taylor, Berk, & Stansfeld, 2013; Kohlboeck et al., 2012; Oellingrath, Svendsen, & Hestetun, 2014; Overby & Hoigaard, 2012; Weng et al., 2012). Although requiring updating, existing systematic reviews confirm a relationship between diet quality and mental health in children and adolescence (O'Neil et al., 2014). This evidence suggests the importance of establishing healthy habits in early life and encouraging replacement of snack-like foods with a diverse and nutrient-rich diet that supports brain development and lays the foundation for a healthy adulthood (Cutler, Flood, Hannan, & Neumark-Sztainer, 2009).

2.2 Diet and Mental Health in Adults and the Elderly

Dietary patterns in adulthood can be influenced by many social, demographic, and individual factors, and tend to be fairly firmly established (Mikkila, Rasanen, Raitakari, Pietinen, & Viikari, 2005). This is an important period for lowering disease risk or managing health conditions through lifestyle behaviors (Beaglehole et al., 2011; Stampfer, Hu, Manson, Rimm, & Willett, 2000) and these health behaviors remain central to brain and mental health during aging. A recent systematic review and metaanalysis concluded that a healthy diet is significantly associated with reduced odds for depression (Lai et al., 2014). Similarly, a metaanalysis of 22 studies investigating the protective effects of adherence to a Mediterranean-style diet on brain diseases demonstrated that higher adherence was associated with a reduced risk for depression as well as cognitive decline (Psaltopoulou et al., 2013).

Elderly individuals often experience changes in living situation, weakening social networks and are frequently managing chronic physically conditions, all of which contribute to the risk of depression (Huang, Wang, Li, Xie, & Liu, 2011; Vink, Aartsen, & Schoevers, 2008). Age-related changes to income, mobility, and nutrient absorption are important to nutrition status in old age (Dean, Raats, Grunert, & Lumbers, 2009; Tiihonen, Ouwehand, & Rautonen, 2010), and several studies have demonstrated the importance of overall diet, particularly consumption of fruits and vegetables, to mental health in later life (Payne, Steck, George, & Steffens, 2012; Tsai, Chang, & Chi, 2012), while unhealthy dietary patterns are also related to an increased risk of depression in this age group (Jacka, Cherbuin, Anstey, & Butterworth, 2014). The components of a good quality diet may support natural changes to gut microbiota and cognitive function associated with mental health (Claesson et al., 2012; Feart et al., 2009), whereas higher intakes of unhealthy foods and lower intakes of healthy foods have both been linked to reduced hippocampal volume—a brain region important to mental health as well as cognitive function (Jacka, Cherbuin, Anstey, Sachdev, et al., 2015).

While the observational evidence for a relationship between diet quality and depression, in particular, is highly consistent and fulfills Bradford Hill criteria for causality (Dash, O'Neil, & Jacka, 2016; Jacka et al., 2012), the nature of the association remains complex and is certainly bidirectional (Jacka, Cherbuin, Anstey, & Butterworth, 2015). However, recent intervention studies offer support for a causal relationship and support dietary approaches to prevention and treatment. In particular, two recent randomized controlled trials indicate the efficacy of dietary improvement as a strategy for the prevention of depression (Sanchez-Villegas & Martinez-Gonzalez, 2013; Stahl, Albert, Dew, Lockovich, & Reynolds, 2014), and a recent world-first trial has demonstrated the efficacy of diet as a treatment strategy for depression (O'Neil et al., 2013); these have important implications for the way we prevent, manage, and treat common mental disorders. Given the strength and consistency of the diet–mental health association in observational (and animal) studies, and the new intervention data supporting causality, researchers are now investigating the plausible biological pathways that may mediate this relationship.

A primary focus of interest is now the gut and its resident microbiota, given the strong relationship of diet to gut health and the emerging evidence that gut microbiota play a role in influencing mood and behavior and the following section will focus on gut health in mental disorder and the imperative to take a dietary approach to improving gut-related factors.

3. THE IMPORTANCE OF DIET FOR GUT HEALTH ACROSS THE LIFESPAN

Gut disturbances are common to mental and neurodevelopmental disorders (Dash et al., 2015). Gastrointestinal dysfunction and complaints are observed in children with autism spectrum disorder (ASD) (de Magistris et al., 2010) and microbial dysbiosis is reported in depression and anxiety (Foster & McVey Neufeld, 2013), ASD (MacFabe, 2015; Rosenfeld, 2015), behavioral disorders (Lee & de La Serre, 2015; Shin, Whon, & Bae, 2015), and bowel disease (Casén et al., 2015). Microbial dysbiosis refers to disturbance in microbiota or abnormal host–microbiota interactions (Lee & de La Serre, 2015; Petersen & Round, 2014) and may be characterized by loss of beneficial microbiota (Petersen & Round, 2014), low microbial diversity (Lee & de La Serre, 2015; Petersen & Round, 2014), and an increase in pathogens, or proliferation of proinflammatory bacteria (such as Proteobacteria) (Petersen & Round, 2014; Shin et al., 2015). Dysbiosis increases intestinal permeability, which may serve to increase circulating levels of lipopolysaccharide (LPS) (Cani & Delzenne, 2011). LPS is a metabolic endotoxin that triggers a neuroinflammatory response and promotes low-grade inflammation (Qin et al., 2007). Inflammation is a proposed underlying pathway for mental disorders (Frommberger et al., 1997; Kivimäki et al., 2013; Pasco et al., 2010; Raison, Capuron, & Miller, 2006).

Across all ages, diet strongly determines microbiota composition (Claesson et al., 2012; David et al., 2014; Laursen et al., 2016; Wu et al., 2011). Importantly, diet is modifiable and microbiota composition responds rapidly to dietary change (David et al., 2014). Dietary improvement toward a healthy and diverse high fiber diet is thus of critical importance for promoting gut health (O'Keefe et al., 2015).

3.1 The Influence of Diet on Early Life Microbiota

The greatest changes to the gut microbiota composition occur during early establishment in infancy (from around birth until around the age of three) (Yatsunenko et al., 2012), and in old age as dietary control and general health conditions change (Claesson et al., 2012). In animals, poor quality maternal gestational diets (such as high fat or low dietary fiber) reduces microbial diversity in the offspring (Ma et al., 2014; Myles et al., 2013; Sonnenburg et al., 2016) and increases colonic inflammation (Ma et al., 2014). Animal data extensively support the relevance of the gut and early microbial colonization for the development of: the immune system (Arrieta et al., 2015; Olszak et al., 2012; Thorburn et al., 2015); metabolism (Cox et al., 2014); normal development of the HPA stress response (Sudo et al., 2004); and brain barrier structure (Braniste et al., 2014). The same data do not yet exist for humans, however modifying the maternal gut microbiota during pregnancy through probiotics modulates the expression of Toll-like receptor genes in the placenta and in the fetal gut, indicating a link with immune development (Rautava, Collado, Salminen, & Isolauri, 2012). Many women do not meet dietary guidelines nor improve their diets during pregnancy (de Jersey, Nicholson, Callaway, & Daniels, 2013; Hure, Young, Smith, & Collins, 2009; Kaiser & Allen, 2002; Malek, Umberger, Makrides, & Zhou, 2015) and it is unclear how this may impact on infant microbial acquisition.

In animals, poor quality diets have an intergenerational effect on gut microbiota (Sonnenburg et al., 2016). Low-fiber diets result in an intergenerational degradation of gut microbiota, such that dietary correction is not possible after four generations of exposure to a western style diet (Sonnenburg et al., 2016). How this relates to humans is unclear, however those consuming industrialized diets have lower microbial diversity and they are missing specific bacteria with genes for degrading plant polysaccharides and xylans compared to those consuming ancestral style diets (De Filippo et al., 2010; Schnorr et al., 2014). Although this does not attest to an intergenerational influence, it supports the contention that microbial diversity may be at risk as dietary quality diminishes.

Delivery mode and breastfeeding status influence the early microbiota (Azad et al., 2013). Breast milk contains human milk oligosaccharides (HMOs), which have prebiotic potential. These HMOs are synergistically metabolized by Lactobacillus and Bifidobacterium and this selectively promotes their growth (Koropatkin, Cameron, & Martens, 2012). Importantly, breast milk transfers maternal mucosal memory to the infant via maternal sIGA, and this may improve tolerance of commensals and resistance to pathogens as the infant gut microbiota establishes (Maynard, Elson, Hatton, & Weaver, 2012). Dietary variety introduced at weaning introduces additional glycans for microbial fermentation (Koropatkin et al., 2012). Indeed, gut microbiota composition at 9 and 18 months is strongly influenced by breastfeeding duration, the composition of the complementary diet and the transition to family foods (Laursen et al., 2016). These factors were more important than vertical transmission from the mother, birth mode, gestational age, or maternal obesity in determining microbial composition and diversity (Laursen et al., 2016). Foods high in protein and fiber increased alpha diversity and altered microbial composition, while fat intake negatively correlated with diversity.

3.2 The Western Diet Has a Detrimental Influence on Adult Gut Health

The importance of diet for gut health is demonstrated by O'Keefe and colleagues' cross-over dietary study (O'Keefe et al., 2015). This study highlights the detrimental effect of the western diet on colonic health. Twenty African Americans and 20 rural South Africans exchanged diets for 2 weeks: both groups consumed a western diet consisting of 14 g/day of fiber and a plant-based rural African diet consisting of 28 g/day of fiber, plus 38 g/day resistant starch (O'Keefe et al., 2015). The western diet reduced microbial diversity, reduced butyrate production, and significantly increased gut mucosal inflammation (O'Keefe et al., 2015). Conversely the high fiber, rural African diet promoted colonic health, characterized by a reduction in inflammation and also increased saccharolytic fermentation and production of SCFAs (particularly butyrate due to the resistant starch load) (O'Keefe et al., 2015). The beneficial effect on colonic health attributed to the plant-based diet is likely due to butyrate, given its role in resolving inflammation, reducing oxidative stress, maintaining the colonic epithelium, and intestinal mucous barrier (Hamer et al., 2008; Maslowski et al., 2009).

Low intake of fiber and high intakes of fat and protein are characteristics of the western diet that are particularly damaging to gut health. In mice, high fat feeding (72% fat, 28% protein, and <1% carbohydrate) altered the gut microbiota, increased intestinal permeability, and promoted metabolic endotoxemia (Cani et al., 2008). In humans, high animal fats and protein intakes influence gut microbiota composition toward bilophilic species, such as the sulfite-reducing species *Bilophila wadsworthia*, which is implicated in inflammatory bowel disease (David et al., 2014; Flint, Duncan, Scott, & Louis, 2014). High protein, low carbohydrate diets are also implicated in increasing N-nitrosamine production, a carcinogen related to colon-cancer risk (Russell et al., 2011).

3.3 Beneficial Influence of Plant-Based Diets on Adult Gut Health

The carbohydrate content of plant-based diets is important for promoting gut health due to the presence of nondigestible dietary fibers. Carbohydrates classify as a "prebiotic" when they are "a nondigestible compound that, through its metabolization by microorganisms in the gut, modulates composition, and/or activity of the gut microbiota, thus conferring a beneficial physiological effect on the host" (Bindels, Delzenne, Cani, & Walter, 2015). Examples of prebiotics are inulin and fructooligosaccharides (found in foods such as asparagus, leek, onion, artichoke, and wheat; Eswaran, Muir, & Chey, 2013), transgalactooligosaccharides (a manufactured product), and human breast milk (Bindels et al., 2015). These are fermented by gut microbiota to result in SCFA production (Bindels et al., 2015). Foods containing prebiotic fiber (such as inulin, fructooligosaccharides) may also improve intestinal barrier function and decrease intestinal permeability and metabolic endotoxemia by increasing endogenous production of glucagon-like peptide-2 (GLP-2) (Cani & Delzenne, 2011; Russo et al., 2012). Moreover, plant polyphenols (i.e. present in tea, berries, and other fruit) have prebiotic potential and modulate gut ecology by inhibiting the growth and colonic adherence of specific pathogenic bacteria, and stimulating the growth of species associated with gut mucous layer thickness (such as Lactobacillus spp., Bifidobacterium spp., and *Akkermansia muciniphila*) (Anhê et al., 2015; Parkar, Stevenson, & Skinner, 2008).

In industrialized societies, the Mediterranean diet exemplifies a healthy dietary pattern that is beneficial to gut health (De Filippis et al., 2015). High-level adherence to the Mediterranean diet, consisting of fresh plant-based foods, moderate in fish, olive oil, and eggs and low in saturated fat and red meat, is associated with a healthier microbiota and higher levels of short chain fatty acids when compared to a western diet (De Filippis et al., 2015). The Mediterranean diet has a low inflammatory potential (Schwingshackl & Hoffmann, 2014), and promotes an abundance of saccharolytic microbiota and, consequently, high fecal concentrations of butyrate, propionate, and acetate (De Filippis et al., 2015; Gutierrez-Diaz, Fernandez-Navarro, Sanchez, Margolles, & Gonzalez, 2016). These mechanisms may partially explain the relationship between adherence to a quality Mediterranean diet and reduced risk of depression and cognitive impairment (Psaltopoulou et al., 2013).

4. OPPORTUNITIES FOR PREVENTION AND TREATMENT OF MENTAL HEALTH DISORDERS

4.1 Dietary Strategies for the Prevention of Mental Disorders: A Public Health Perspective

The rise in mental disorder burden has been met with campaigns to reduce stigma and increase awareness, however few specific, action-oriented prevention strategies have been developed or implemented (Jacka, Reavley, et al., 2013). Psychiatric epidemiology has focused primarily on understanding the etiology of common mental disorders and has been slow to develop primary prevention strategies to be implemented at a population level (Galea, 2013; Jacka, Reavley, et al., 2013), despite the identification of several key prevention opportunities (Shonkoff, Boyce, & McEwen, 2009). Given the shared risk factors and high comorbidity of physical and mental conditions, there have been calls for a shared framework for the prevention of common mental and noncommunicable disorders (O'Neil et al., 2015). It is essential that mental health prevention is integrated within existing programs for disease prevention and control, for example obesity or cardiovascular disease, in order to provide a cost effective strategy that is likely to benefit both mental and physical disorders. Taking lessons from previous public health campaigns, a top-down approach that implements change at a government and policy level is important to success (Brownson, Chriqui, & Stamatakis, 2009). While this is important, policy change can be slow to develop, and we must also develop clear and specific recommendations that can be implemented at a community, organizational, and individual level. Such recommendations are beginning to emerge (Dash et al., 2016; Opie et al., 2015).

From a developmental origins of health and disease perspective, it is important to implement prevention strategies in the earliest stages of human life (Barker, 2015; Gillman, 2005; Hanson & Gluckman, 2011). The early life period is characterized by high developmental plasticity and is influenced by in utero environmental factors (Hanson & Gluckman, 2011). This period is therefore an important target for health interventions aiming to influence predisposition to noncommunicable diseases (Hanson & Gluckman, 2011). Nutrition is a key in utero factor influencing brain development and subsequent disease risk (Bale et al., 2010; Schlotz & Phillips, 2009), and thus addressing prenatal diet quality is important. Dietary prevention or treatment strategies may improve maternal nutrition, but these improvements do not always transfer to children (Vähämiko et al., 2013). To address the developmental origins of health and disease, policy strategy and prevention initiatives must focus on improving health literacy and fostering nutritional empowerment early in life (Barker, 2015; Hanson & Gluckman, 2011). The focus for nutritional interventions may best be shifted away from women of childbearing age to girls and young women because early attainment of healthy diet may promote life-long sustainable dietary improvement (Barker, 2015). Children and young women may then be empowered to make positive dietary choices, and these behaviors are anticipated to influence their families and social groups. Over time this may gain momentum such that politicians and commercial organizations are prompted to respond and improve the food environment (Barker, 2015; Hanson & Gluckman, 2011).

4.2 Targeting Early Life Gut Microbiota

Targeting prevention strategies toward the maternal gut microbiota is relevant for influencing in utero microbial exposure and early life microbial colonization (Rautava et al., 2012). Prenatal probiotic supplementation has been shown to influence maternal and infant gut microbiota in some trials (Grönlund, Grześkowiak, Isolauri, & Salminen, 2011; Lahtinen et al., 2009) but not all (Wright & Starkweather, 2015). Given that diet rapidly influences the composition of the microbiota (David et al., 2014), we speculate that targeting maternal gut health through a dietary intervention delivered during pregnancy may modulate maternal and infant microbiota. Increasing intake of fibrous plant-based foods (grains, vegetables, beans and legumes, fruit, and nuts) and fermented foods, and reducing intakes of refined foods are central to gut health, therefore interventions may center around the Mediterranean diet, or other forms of healthy diet (Opie et al., 2015). Dietary improvement is protective of maternal mental health (Opie et al., 2015) and as such may be superior to prebiotic or probiotic supplementation alone.

In infancy, the establishment of a healthy gut microbiota may be relevant for neurodevelopmental outcomes (Borre et al., 2014). In animals, microbial modulation of mice during weaning improves symptoms of neurodevelopmental disease (Hsiao et al., 2013), and emerging evidence suggests that targeting infant microbiota through probiotics reduces the risk of attention deficit hyperactivity disorder (ADHD) and Asperger syndrome (Pärtty et al., 2015) although further research is required. Certainly birth mode and feeding strategies shape early microbial composition, however transition to the family diet appears to be an important determinant of microbial diversity in infancy (Laursen et al., 2016). Therefore, care should be taken during complimentary feeding to transition to a high quality diet that includes good sources of dietary fiber and protein (Laursen et al., 2016).

4.3 Dietary Targeting of Mothers for Reducing the Risk of Mental Disorders in Children

Diet quality is generally poor during pregnancy because the dietary guidelines are rarely met (de Jersey et al., 2013; Hure et al., 2009; Kaiser & Allen, 2002; Malek et al., 2015). Poor quality diets are also associated with increased maternal depressive symptoms (Baskin et al., 2015), and they are independently related to poorer emotional, cognitive (Barker et al., 2013), and behavioral outcomes in young children (Jacka, Ystrom, et al., 2013; Pina-Camacho et al., 2015; Steenweg-de Graaff et al., 2014). Interventions aiming to treat prenatal depression should include a dietary focus (Barker et al., 2013). Opie and colleagues' dietary recommendations for preventing depression could be incorporated into dietary interventions with a focus on traditional dietary patterns, increased intake of plant-based foods and omega-3 polyunsaturated fatty acids, along with reducing intake of refined foods (Opie et al., 2015). These recommendations are expected to promote a healthy microbiome, because are congruous with dietary patterns that have been consistently associated with a healthier gut microbiota (De Filippo et al., 2010; Martínez et al., 2015; Obregon-Tito et al., 2015; O'Keefe et al., 2015; Schnorr et al., 2014; Yatsunenko et al., 2012). Thus, targeting gut health in pregnant women by improving diet may have important benefits to the gut health of children, given the transmission of microbiota to infants before and during birth.

Poor diets also drive the pathogenesis of inflammatory metabolic conditions such as obesity, type 2 and gestational diabetes. These conditions are risk factors for neurodevelopmental disorders (Krakowiak et al., 2012; Suren et al., 2014; Van Lieshout, Taylor, & Boyle, 2011; Van Lieshout & Voruganti, 2008), therefore mothers at risk of gestational diabetes mellitus (GDM) and obesity should be targeted for treatment. During pregnancy, dietary counseling with probiotic supplementation of *Lactobacillus rhamnosus* GG and *Bifidobacterium lactis* has been seen to improve maternal glucose regulation and reduce the incidence of GDM (Laitinen, Poussa, & Isolauri, 2008). When GDM was present, dietary counseling reduced the risk of large birth size (Luoto, Laitinen, Nermes, & Isolauri, 2010), this may improve the chances of vaginal delivery which is beneficial for healthy microbial acquisition (Dominguez-Bello et al., 2010).

Similarly, managing obesity by targeting weight loss and gut health during pregnancy may be relevant for reducing childhood risk of cognitive problems, ADHD symptoms, adolescent eating disorders, and adulthood psychotic disorders (Van Lieshout et al., 2011). Dietary interventions delivered during pregnancy are effective for reducing gestational weight gain and reducing cesarean section incidence (Tanentsapf, Heitmann, & Adegboye, 2011). During pregnancy, dietary strategies may be adopted to treat the low-grade exdotoxemia and metabolic inflammation that is characteristic of pregnant obese women (Basu et al., 2011; Dewulf et al., 2013). To limit infant exposure to maternal immune activation, the inclusion of prebiotic dietary inulin-type fructans in prospective dietary interventions may modulate microbiota and assist with the management of low-grade exdotoxemia metabolic inflammation (Dewulf et al., 2013). Managing obesity during pregnancy may improve infant gut health because compared to lean controls, the microbiota of infants born to obese mothers differs and has greater proinflammatory activity (Collado, Isolauri, Laitinen, & Salminen, 2010). However, later transition to family foods appears to exert a greater influence over infant microbiota than maternal obesity (Laursen et al., 2016).

4.4 Diet as a Treatment for Mental Health

Medication and psychological intervention remain the primary forms of treatment for mental disorders, although evidence suggests the efficacy of these strategies has been overestimated (Cuijpers, Smit, Bohlmeijer, Hollon, & Andersson, 2010; Gelenberg, 2010). "Lifestyle Medicine" targets modifiable risk factors for mental disorders as a low risk, cost effective, and integrative strategy for the treatment of mental illness (Sarris, O'Neil, Coulson, Schweitzer, & Berk, 2014). This treatment approach requires the development of appropriate diagnostic tools that incorporate assessment of health behaviors, for example physical activity and diet quality, in order to identify populations particularly at risk and encourage early intervention. Dietary and lifestyle (i.e. exercise and smoking) recommendations as well as dietetic services should be a standard component of care for at risk patients and as well as those with current MDD and this understanding has been incorporated into recent updates to Clinical Practice Guidelines in Australia (Malhi et al., 2015).

Results of a recent randomized controlled trial support dietary strategies for the treatment of depression, demonstrating highly significant results in a relatively short period of intervention (O'Neil et al., 2013). It will be important to replicate these key findings, however, given the strength and consistency of the evidence for the diet–depression association, it is now critical to translate evidence to action (Dash et al., 2016). Furthermore, continuing explication of the key biological mechanisms of action that underpin the diet–depression association, such as the gut microbiota, is important for developing targeted treatment and prevention strategies (Schmidt, Shelton, & Duman, 2011).

5. CONCLUSION

This chapter highlights the importance of diet to mental and gut health across the lifespan. The emerging and established evidence now strongly supports taking a dietary approach to the prevention and management of highly prevalent mental disorders. Future work in this new field of nutritional psychiatry research should focus on replication, the scaling up of interventions, and further identification of the pathways that mediate the impact of dietary improvement on depressive illness. The evidence clearly supports the gut and its resident microbiome as a key target for research and intervention. Public health approaches and messages should now focus on the importance of diet for mental as well as physical health, while clinicians should promote the benefits of dietary improvement and facilitate access to dietary support for their patients. Finally and critically, given the likely causal role of diet in depressive illness and the global burden of this disabling disorder, policy makers should urgently make the necessary changes to improve access to quality food and reduce access to the unhealthy and processed food products that are made ubiquitous and heavily promoted by industry.

REFERENCES

Adair, L. S., & Popkin, B. M. (2005). Are child eating patterns being transformed globally? *Obesity Research*, *13*, 1281–1299.

Anhê, F. F., Varin, T. V., Le Barz, M., Desjardins, Y., Levy, E., Roy, D., et al. (2015). Gut microbiota dysbiosis in obesity-linked metabolic diseases and prebiotic potential of polyphenol-rich extracts. *Current Obesity Reports*, *4*(4), 389–400.

Arrieta, M.-C., Stiemsma, L. T., Dimitriu, P. A., Thorson, L., Russell, S., Yurist-Doutsch, S., et al. (2015). Early infancy microbial and metabolic alterations affect risk of childhood asthma. *Science Translational Medicine*, 7, 307ra152.

Australian Bureau of Statistics (ABS). (2016). *Australian Health Survey: Consumption of food groups from the Australian Dietary Guidelines*. Canberra: Australian Bureau of Statistics.

Azad, M. B., Konya, T., Maughan, H., Guttman, D. S., Field, C. J., Chari, R. S., et al. (2013). Gut microbiota of healthy Canadian infants: Profiles by mode of delivery and infant diet at 4 months. *CMAJ*, *185*, 385–394.

Bale, T. L., Baram, T. Z., Brown, A. S., Goldstein, J. M., Insel, T. R., McCarthy, M. M., et al. (2010). Early life programming and neurodevelopmental disorders. *Biological Psychiatry, 68*, 314–319.

Barker, M. (2015). Developmental origins, behaviour change and the new public health. *Journal of Developmental Origins of Health and Disease, 6*, 428–433.

Barker, E. D., Kirkham, N., Ng, J., & Jensen, S. K. (2013). Prenatal maternal depression symptoms and nutrition, and child cognitive function. *The British Journal of Psychiatry: The Journal of Mental Science, 203*, 417–421.

Baskin, R., Hill, B., Jacka, F. N., O'Neil, A., & Skouteris, H. (2015). The association between diet quality and mental health during the perinatal period. A systematic review. *Appetite, 91*, 41–47.

Basu, S., Haghiac, M., Surace, P., Challier, J. C., Guerre-Millo, M., Singh, K., et al. (2011). Pregravid obesity associates with increased maternal endotoxemia and metabolic inflammation. *Obesity (Silver Spring), 19*, 476–482.

Beaglehole, R., Bonita, R., Horton, R., Adams, C., Alleyne, G., Asaria, P., et al. (2011). Priority actions for the non-communicable disease crisis. *Lancet, 377*, 1438–1447.

Becker, K., El-Faddagh, M., Schmidt, M. H., & Laucht, M. (2007). Is the serotonin transporter polymorphism (5-HTTLPR) associated with harm avoidance and internalising problems in childhood and adolescence? *Journal of Neural Transmission, 114*, 395–402.

Berk, M., Williams, L. J., Jacka, F., O'Neil, A., Pasco, J. A., Moylan, S., et al. (2013). So depression is an inflammatory disease, but where does the inflammation come from? *BMC Medicine, 11*, 200.

Bindels, L. B., Delzenne, N. M., Cani, P. D., & Walter, J. (2015). Towards a more comprehensive concept for prebiotics. *Nature Reviews. Gastroenterology & Hepatology, 12*(5), 303–310.

Borre, Y. E., O'Keeffe, G. W., Clarke, G., Stanton, C., Dinan, T. G., & Cryan, J. F. (2014). Microbiota and neurodevelopmental windows: Implications for brain disorders. *Trends in Molecular Medicine, 20*, 509–518.

Braniste, V., Al-Asmakh, M., Kowal, C., Anuar, F., Abbaspour, A., Toth, M., et al. (2014). The gut microbiota influences blood-brain barrier permeability in mice. *Science Translational Medicine, 6*, 263ra158.

Brownson, R. C., Chriqui, J. F., & Stamatakis, K. A. (2009). Understanding evidence-based public health policy. *American Journal of Public Health, 99*, 1576–1583.

Cani, P. D., Bibiloni, R., Knauf, C., Waget, A., Neyrinck, A. M., Delzenne, N. M., et al. (2008). Changes in gut microbiota control metabolic endotoxemia-induced inflammation in high-fat diet-induced obesity and diabetes in mice. *Diabetes, 57*, 1470–1481.

Cani, P. D., & Delzenne, N. M. (2011). The gut microbiome as therapeutic target. *Pharmacology and Therapeutics, 130*, 202–212.

Casén, C., Vebø, H. C., Sekelja, M., Hegge, F. T., Karlsson, M. K., Ciemniejewska, E., et al. (2015). Deviations in human gut microbiota: A novel diagnostic test for determining dysbiosis in patients with IBS or IBD. *Alimentary Pharmacology and Therapeutics, 42*(1), 71–83.

Claesson, M. J., Jeffery, I. B., Conde, S., Power, S. E., O'Connor, E. M., Cusack, S., et al. (2012). Gut microbiota composition correlates with diet and health in the elderly. *Nature, 488*, 178–184.

Collado, M. C., Isolauri, E., Laitinen, K., & Salminen, S. (2010). Effect of mother's weight on infant's microbiota acquisition, composition, and activity during early infancy: A prospective follow-up study initiated in early pregnancy. *American Journal of Clinical Nutrition, 92*, 1023–1030.

Copeland, W. E., Shanahan, L., Costello, E. J., & Angold, A. (2009). Childhood and adolescent psychiatric disorders as predictors of young adult disorders. *Archives of General Psychiatry, 66*, 764–772.

Cox, L. M., Yamanishi, S., Sohn, J., Alekseyenko, A. V., Leung, J. M., Cho, I., et al. (2014). Altering the intestinal microbiota during a critical developmental window has lasting metabolic consequences. *Cell, 158*, 705–721.

Cuijpers, P., Smit, F., Bohlmeijer, E., Hollon, S. D., & Andersson, G. (2010). Efficacy of cognitive-behavioural therapy and other psychological treatments for adult depression: Meta-analytic study of publication bias. *The British Journal of Psychiatry: The Journal of Mental Science, 196*, 173–178.

Cutler, G. J., Flood, A., Hannan, P., & Neumark-Sztainer, D. (2009). Major patterns of dietary intake in adolescents and their stability over time. *The Journal of Nutrition, 139*, 323–328.

Dash, S., Clarke, G., Berk, M., & Jacka, F. N. (2015). The gut microbiome and diet in psychiatry: Focus on depression. *Current Opinion in Psychiatry, 28*, 1–6.

Dash, S. R., O'Neil, A., & Jacka, F. N. (2016). Diet and common mental disorders: The imperative to translate evidence into action. *Frontiers in Public Health, 4*, 81.

David, L. A., Maurice, C. F., Carmody, R. N., Gootenberg, D. B., Button, J. E., Wolfe, B. E., et al. (2014). Diet rapidly and reproducibly alters the human gut microbiome. *Nature, 505*, 559–563.

De Filippis, F., Pellegrini, N., Vannini, L., Jeffery, I. B., Storia, A. L., Laghi, L., et al. (2015). High-level adherence to a Mediterranean diet beneficially impacts the gut microbiota and associated metabolome. *Gut*, 1–10, gutjnl-2015-309957.

De Filippo, C., Cavalieri, D., Di Paola, M., Ramazzotti, M., Poullet, J. B., Massart, S., et al. (2010). Impact of diet in shaping gut microbiota revealed by a comparative study in children from Europe and rural Africa. *Proceedings of the National Academy of Sciences, 107*, 14691–14696.

de Jersey, S. J., Nicholson, J. M., Callaway, L. K., & Daniels, L. A. (2013). An observational study of nutrition and physical activity behaviours, knowledge, and advice in pregnancy. *BMC Pregnancy and Childbirth, 13*, 115.

de Magistris, L., Familiari, V., Pascotto, A., Sapone, A., Frolli, A., Iardino, P., et al. (2010). Alterations of the intestinal barrier in patients with autism spectrum disorders and in their first-degree relatives. *Journal of Pediatric Gastroenterology and Nutrition, 51*, 418–424.

Dean, M., Raats, M. M., Grunert, K. G., & Lumbers, M. (2009). Factors influencing eating a varied diet in old age. *Public Health Nutrition, 12*, 2421–2427.

Dewulf, E. M., Cani, P. D., Claus, S. P., Fuentes, S., Puylaert, P. G. B., Neyrinck, A. M., et al. (2013). Insight into the prebiotic concept: Lessons from an exploratory, double blind intervention study with inulin-type fructans in obese women. *Gut, 62*, 1112–1121.

Dominguez-Bello, M. G., Costello, E. K., Contreras, M., Magris, M., Hidalgo, G., Fierer, N., et al. (2010). Delivery mode shapes the acquisition and structure of the initial microbiota across multiple body habitats in newborns. *Proceedings of the National Academy of Sciences, 107*, 11971–11975.

Eswaran, S., Muir, J., & Chey, W. D. (2013). Fiber and functional gastrointestinal disorders. *American Journal of Gastroenterology, 108*, 718–727.

Feart, C., Samieri, C., Rondeau, V., Amieva, H., Portet, F., Dartigues, J. F., et al. (2009). Adherence to a Mediterranean diet, cognitive decline, and risk of dementia. *JAMA, 302*, 638–648.

Flint, H. J., Duncan, S. H., Scott, K. P., & Louis, P. (2014). Links between diet, gut microbiota composition and gut metabolism. *Proceedings of the Nutrition Society, 74*, 13–22.

Foster, J. A., & McVey Neufeld, K.-A. (2013). Gut–brain axis: How the microbiome influences anxiety and depression. *Trends in Neurosciences, 36*, 305–312.

Frommberger, U. H., Bauer, J., Haselbauer, P., Fräulin, A., Riemann, D., & Berger, M. (1997). Interleukin-6-(IL-6) plasma levels in depression and schizophrenia: Comparison between the acute state and after remission. *European Archives of Psychiatry and Clinical Neuroscience, 247*, 228–233.

Galea, S. (2013). An argument for a consequentialist epidemiology. *American Journal of Epidemiology, 178*, 1185–1191.

Gelenberg, A. J. (2010). A review of the current guidelines for depression treatment. *The Journal of Clinical Psychiatry, 71*, e15.

Gillman, M. W. (2005). Developmental origins of health and disease. *The New England Journal of Medicine, 353*, 1848–1850.

Global Burden of Disease Study Collaborators, Forouzanfar, M. H., Alexander, L., Anderson, H. R., Bachman, V. F., Biryukov, S., et al. (2015). Global, regional, and national comparative risk assessment of 79 behavioural, environmental and occupational, and metabolic risks or clusters of risks in 188 countries, 1990–2013: A systematic analysis for the Global Burden of Disease Study 2013. *Lancet, 386*(10010), 2287–23235, 5.

Grönlund, M.-M., Grześkowiak, Ł., Isolauri, E., & Salminen, S. (2011). Influence of mother's intestinal microbiota on gut colonization in the infant. *Gut Microbes, 2*, 227–233.

Gutierrez-Diaz, I., Fernandez-Navarro, T., Sanchez, B., Margolles, A., & Gonzalez, S. (2016). Mediterranean diet and faecal microbiota: A transversal study. *Food & Function*, 7(5), 2347–2356.

Hamer, H. M., Jonkers, D., Venema, K., Vanhoutvin, S., Troost, F. J., & Brummer, R. J. (2008). Review article: The role of butyrate on colonic function. *Alimentary Pharmacology and Therapeutics, 27*, 104–119.

Hanson, M., & Gluckman, P. (2011). Developmental origins of noncommunicable disease: Population and public health implications. *American Journal of Clinical Nutrition, 94*(6 Suppl.), 1754S–1758S.

Heim, C., Newport, D. J., Mletzko, T., Miller, A. H., & Nemeroff, C. B. (2008). The link between childhood trauma and depression: Insights from HPA axis studies in humans. *Psychoneuroendocrinology, 33*, 693–710.

Hsiao, E., McBride, S., Hsien, S., Sharon, G., Hyde, E., McCue, T., et al. (2013). Microbiota modulate behavioral and physiological abnormalities associated with neurodevelopmental disorders. *Cell, 155*, 1451–1463.

Hu, F. B. (2002). Dietary pattern analysis: A new direction in nutritional epidemiology. *Current Opinion in Lipidology, 13*, 3–9.

Huang, C. Q., Wang, Z. R., Li, Y. H., Xie, Y. Z., & Liu, Q. X. (2011). Cognitive function and risk for depression in old age: A meta-analysis of published literature. *International Psychogeriatrics/IPA, 23*, 516–525.

Hure, A., Young, A., Smith, R., & Collins, C. (2009). Diet and pregnancy status in Australian women. *Public Health Nutrition, 12*, 853–861.

Jacka, F. N., Cherbuin, N., Anstey, K. J., & Butterworth, P. (2014). Dietary patterns and depressive symptoms over time: Examining the relationships with socioeconomic position, health behaviours and cardiovascular risk. *PLoS One, 9*, e87657.

Jacka, F. N., Cherbuin, N., Anstey, K. J., & Butterworth, P. (2015). Does reverse causality explain the relationship between diet and depression? *Journal of Affective Disorders, 175*, 248–250.

Jacka, F. N., Cherbuin, N., Anstey, K. J., Sachdev, P., & Butterworth, P. (2015). Western diet is associated with a smaller hippocampus: A longitudinal investigation. *BMC Medicine, 13*, 215.

Jacka, F. N., Kremer, P. J., Berk, M., de Silva-Sanigorski, A. M., Moodie, M., Leslie, E. R., et al. (2011). A prospective study of diet quality and mental health in adolescents. *PLoS One, 6*, e24805.

Jacka, F. N., Kremer, P. J., Leslie, E. R., Berk, M., Patton, G. C., Toumbourou, J. W., et al. (2010). Associations between diet quality and depressed mood in adolescents: Results from the Australian Healthy Neighbourhoods Study. *The Australian and New Zealand Journal of Psychiatry, 44*, 435–442.

Jacka, F. N., Mykletun, A., & Berk, M. (2012). Moving towards a population health approach to the primary prevention of common mental disorders. *BMC Medicine, 10*, 149.

Jacka, F. N., Reavley, N. J., Jorm, A. F., Toumbourou, J. W., Lewis, A. J., & Berk, M. (2013). Prevention of common mental disorders: What can we learn from those who have gone before and where do we go next? *The Australian and New Zealand Journal of Psychiatry, 47*, 920–929.

Jacka, F. N., Rothon, C., Taylor, S., Berk, M., & Stansfeld, S. A. (2013). Diet quality and mental health problems in adolescents from East London: A prospective study. *Social Psychiatry and Psychiatric Epidemiology, 48*, 1297–1306.

Jacka, F. N., Sacks, G., Berk, M., & Allender, S. (2014). Food policies for mental and physical health. *BMC Psychiatry, 14*, 132.

Jacka, F. N., Ystrom, E., Brantsaeter, A. L., Karevold, E., Roth, C., Haugen, M., et al. (2013). Maternal and early postnatal nutrition and mental health of offspring by age 5 years: A prospective cohort study. *Journal of the American Academy of Child and Adolescent Psychiatry, 52*, 1038–1047.

Kaiser, L. L., & Allen, L. (2002). Position of the American Dietetic Association: Nutrition and lifestyle for a healthy pregnancy outcome. *Journal of the American Dietetic Association, 102*, 1479–1490.

Kessler, R. C., Berglund, P., Demler, O., Jin, R., Merikangas, K. R., & Walters, E. E. (2005). Lifetime prevalence and age-of-onset distributions of DSM-IV disorders in the National Comorbidity Survey Replication. *Archives of General Psychiatry, 62*, 593–602.

Kivimäki, M., Shipley, M. J., Batty, G. D., Hamer, M., Akbaraly, T. N., Kumari, M., et al. (2013). Long-term inflammation increases risk of common mental disorder: A cohort study. *Molecular Psychiatry, 19*, 149–150.

Kohlboeck, G., Sausenthaler, S., Standl, M., Koletzko, S., Bauer, C. P., von Berg, A., et al. (2012). Food intake, diet quality and behavioral problems in children: Results from the GINI-plus/LISA-plus studies. *Annals of Nutrition & Metabolism, 60*, 247–256.

Kontogianni, M. D., Vidra, N., Farmaki, A. E., Koinaki, S., Belogianni, K., Sofrona, S., et al. (2008). Adherence rates to the Mediterranean diet are low in a representative sample of Greek children and adolescents. *The Journal of Nutrition, 138*, 1951–1956.

Koropatkin, N. M., Cameron, E. A., & Martens, E. C. (2012). How glycan metabolism shapes the human gut microbiota. *Nature Reviews. Microbiology, 10*, 323–335.

Krakowiak, P., Walker, C. K., Bremer, A. A., Baker, A. S., Ozonoff, S., Hansen, R. L., et al. (2012). Maternal metabolic conditions and risk for autism and other neurodevelopmental disorders. *Pediatrics, 129*, e1121–e1128.

Lahtinen, S. J., Boyle, R. J., Kivivuori, S., Oppedisano, F., Smith, K. R., Robins-Browne, R. M., et al. (2009). Prenatal probiotic administration can influence Bifidobacterium microbiota development in infants at high risk of allergy. *Journal of Allergy and Clinical Immunology, 123*, 499–501.

Lai, J. S., Hiles, S., Bisquera, A., Hure, A. J., McEvoy, M., & Attia, J. (2014). A systematic review and meta-analysis of dietary patterns and depression in community-dwelling adults. *The American Journal of Clinical Nutrition, 99*, 181–197.

Laitinen, K., Poussa, T., & Isolauri, E. (2008). Probiotics and dietary counselling contribute to glucose regulation during and after pregnancy: A randomised controlled trial. *British Journal of Nutrition, 101*, 1679.

Larson, N. I., Neumark-Sztainer, D., Hannan, P. J., & Story, M. (2007). Trends in adolescent fruit and vegetable consumption, 1999–2004: Project EAT. *American Journal of Preventive Medicine, 32*, 147–150.

Laursen, M., Andersen, L., Michaelsen, K., Mølgaard, C., Trolle, E., Bahl, M., et al. (2016). Infant gut microbiota development is driven by transition to family foods independent of maternal obesity. *Msphere, 1*, e00069–00015.

Lee, S. H., & de La Serre, C. B. (2015). Gut microbiome–brain communications regulate host physiology and behavior. *Journal of Nutritional Health & Food Science, 3*, 1–12.

Logan, A. C., & Jacka, F. N. (2014). Nutritional psychiatry research: An emerging discipline and its intersection with global urbanization, environmental challenges and the evolutionary mismatch. *Journal of Physiological Anthropology, 33*, 22.

Luoto, R., Laitinen, K., Nermes, M., & Isolauri, E. (2010). Impact of maternal probiotic-supplemented dietary counselling on pregnancy outcome and prenatal and postnatal growth: A double-blind, placebo-controlled study. *British Journal of Nutrition, 103*, 1792–1799.

Ma, J., Prince, A. L., Bader, D., Hu, M., Ganu, R., Baquero, K., et al. (2014). High-fat maternal diet during pregnancy persistently alters the offspring microbiome in a primate model. *Nature Communications, 5*, 3889.

MacFabe, D. F. (2015). Enteric short-chain fatty acids: Microbial messengers of metabolism, mitochondria, and mind: Implications in autism spectrum disorders. *Microbial Ecology in Health and Disease, 26*, 28177.

Malek, L., Umberger, W., Makrides, M., & Zhou, S. J. (2015). Adherence to the Australian dietary guidelines during pregnancy: Evidence from a national study. *Public Health Nutrition, 19*, 1155–1163.

Malhi, G. S., Bassett, D., Boyce, P., Bryant, R., Fitzgerald, P. B., Fritz, K., et al. (2015). Royal Australian and New Zealand College of Psychiatrists clinical practice guidelines for mood disorders. *The Australian and New Zealand Journal of Psychiatry, 49*, 1087–1206.

Martínez, I., Stegen James, C., Maldonado-Gómez Maria, X., Eren, A. M., Siba Peter, M., Greenhill Andrew, R., et al. (2015). The gut microbiota of rural papua new guineans: Composition, diversity patterns, and ecological processes. *Cell Reports, 11*(4), 527–538.

Maslowski, K. M., Vieira, A. T., Ng, A., Kranich, J., Sierro, F., Yu, D., et al. (2009). Regulation of inflammatory responses by gut microbiota and chemoattractant receptor GPR43. *Nature, 461*, 1282–1286.

Maynard, C. L., Elson, C. O., Hatton, R. D., & Weaver, C. T. (2012). Reciprocal interactions of the intestinal microbiota and immune system. *Nature, 489*, 231–241.

Mikkila, V., Rasanen, L., Raitakari, O. T., Pietinen, P., & Viikari, J. (2004). Longitudinal changes in diet from childhood into adulthood with respect to risk of cardiovascular diseases: The Cardiovascular Risk in Young Finns Study. *European Journal of Clinical Nutrition, 58*, 1038–1045.

Mikkila, V., Rasanen, L., Raitakari, O. T., Pietinen, P., & Viikari, J. (2005). Consistent dietary patterns identified from childhood to adulthood: The cardiovascular risk in young finns study. *The British Journal of Nutrition, 93*, 923–931.

Myles, I. A., Fontecilla, N. M., Janelsins, B. M., Vithayathil, P. J., Segre, J. A., & Datta, S. K. (2013). Parental dietary fat intake alters offspring microbiome and immunity. *Journal of Immunology, 191*, 3200–3209.

Northstone, K., & Emmett, P. (2005). Multivariate analysis of diet in children at four and seven years of age and associations with socio-demographic characteristics. *European Journal of Clinical Nutrition, 59*, 751–760.

O'Neil, A., Itsiopoulos, C., Skouteris, H., Opie, R. S., McPhie, S., Hill, B., et al. (2014). Preventing mental health problems in offspring by targeting dietary intake of pregnant women. *BMC Medicine, 12*, 208.

Obregon-Tito, A. J., Tito, R. Y., Metcalf, J., Sankaranarayanan, K., Clemente, J. C., Ursell, L. K., et al. (2015). Subsistence strategies in traditional societies distinguish gut microbiomes. *Nature Communications, 6*, 6505.

Oellingrath, I. M., Svendsen, M. V., & Hestetun, I. (2014). Eating patterns and mental health problems in early adolescence—A cross-sectional study of 12–13-year-old Norwegian schoolchildren. *Public Health Nutrition, 17*, 2554–2562.

O'Keefe, S. J. D., Li, J. V., Lahti, L., Ou, J., Carbonero, F., Mohammed, K., et al. (2015). Fat, fibre and cancer risk in African Americans and rural Africans. *Nature Communications, 6*, 6342.

Olszak, T., An, D., Zeissig, S., Vera, M. P., Richter, J., Franke, A., et al. (2012). Microbial exposure during early life has persistent effects on natural killer T cell function. *Science, 336*, 489–493.

O'Neil, A., Berk, M., Itsiopoulos, C., Castle, D., Opie, R., Pizzinga, J., et al. (2013). A randomised controlled trial of a dietary intervention for adults with major depression (the "SMILES" trial: Study protocol). *BMC Psychiatry, 13*(1), 114, submitted.

O'Neil, A., Jacka, F. N., Quirk, S. E., Cocker, F., Taylor, C. B., Oldenburg, B., et al. (2015). A shared framework for the common mental disorders and non-communicable disease: Key considerations for disease prevention and control. *BMC Psychiatry, 15*, 15.

O'Neil, A., Quirk, S. E., Housden, S., Brennan, S. L., Williams, L. J., Pasco, J. A., et al. (2014). Relationship between diet and mental health in children and adolescents: A systematic review. *American Journal of Public Health, 104*, e31–e42.

Opie, R. S., Itsiopoulos, C., Parletta, N., Sanchez-Villegas, A., Akbaraly, T. N., Ruusunen, A., et al. (2015). Dietary recommendations for the prevention of depression. *Nutritional Neuroscience*, 1–11.

Overby, N., & Hoigaard, R. (2012). Diet and behavioral problems at school in Norwegian adolescents. *Food & Nutrition Research, 56*, 17231.

Parkar, S. G., Stevenson, D. E., & Skinner, M. A. (2008). The potential influence of fruit polyphenols on colonic microflora and human gut health. *International Journal of Food Microbiology, 124*, 295–298.

Pärtty, A., Kalliomäki, M., Wacklin, P., Salminen, S., & Isolauri, E. (2015). A possible link between early probiotic intervention and the risk of neuropsychiatric disorders later in childhood: A randomized trial. *Pediatric Research, 77*, 823–828.

Pasco, J. A., Nicholson, G. C., Williams, L. J., Jacka, F. N., Henry, M. J., Kotowicz, M. A., et al. (2010). Association of high-sensitivity C-reactive protein with de novo major depression. *British Journal of Psychiatry, 197*, 372–377.

Payne, M. E., Steck, S. E., George, R. R., & Steffens, D. C. (2012). Fruit, vegetable, and antioxidant intakes are lower in older adults with depression. *Journal of the Academy of Nutrition and Dietetics, 112*, 2022–2027.

Petersen, C., & Round, J. L. (2014). Defining dysbiosis and its influence on host immunity and disease. *Cellular Microbiology, 16*, 1024–1033.

Pina-Camacho, L., Jensen, S. K., Gaysina, D., & Barker, E. D. (2015). Maternal depression symptoms, unhealthy diet and child emotional–behavioural dysregulation. *Psychological Medicine, 45*, 1851–1860.

Psaltopoulou, T., Sergentanis, T. N., Panagiotakos, D. B., Sergentanis, I. N., Kosti, R., & Scarmeas, N. (2013). Mediterranean diet, stroke, cognitive impairment, and depression: A meta-analysis. *Annals of Neurology, 74*, 580–591.

Qin, L., Wu, X., Block, M. L., Liu, Y., Breese, G. R., Hong, J. S., et al. (2007). Systemic LPS causes chronic neuroinflammation and progressive neurodegeneration. *Glia, 55*, 453–462.

Raison, C. L., Capuron, L., & Miller, A. H. (2006). Cytokines sing the blues: Inflammation and the pathogenesis of depression. *Trends in Immunology, 27*, 24–31.

Rautava, S., Collado, M. C., Salminen, S., & Isolauri, E. (2012). Probiotics modulate host–microbe interaction in the placenta and fetal gut: A randomized, double-blind, placebo-controlled trial. *Neonatology, 102*, 178–184.

Rosenfeld, C. S. (2015). Microbiome disturbances and autism spectrum disorders. *Drug Metabolism and Disposition, 43*, 1557–1571.

Russell, W. R., Gratz, S. W., Duncan, S. H., Holtrop, G., Ince, J., Scobbie, L., et al. (2011). High-protein, reduced-carbohydrate weight-loss diets promote metabolite profiles likely to be detrimental to colonic health. *American Journal of Clinical Nutrition, 93*, 1062–1072.

Russo, F., Linsalata, M., Clemente, C., Chiloiro, M., Orlando, A., Marconi, E., et al. (2012). Inulin-enriched pasta improves intestinal permeability and modifies the circulating levels of zonulin and glucagon-like peptide 2 in healthy young volunteers. *Nutrition Research, 32*, 940–946.

Sanchez-Villegas, A., & Martinez-Gonzalez, M. A. (2013). Diet, a new target to prevent depression? *BMC Medicine, 11*(3), 1–11.

Sarris, J., Logan, A. C., Akbaraly, T. S., Amminger, G. P., Balanzá-Martínez, V., Freeman, M. P., et al. (2015). Nutritional medicine as mainstream in psychiatry. *The Lancet Psychiatry, 2*, 271–274.

Sarris, J., O'Neil, A., Coulson, C. E., Schweitzer, I., & Berk, M. (2014). Lifestyle medicine for depression. *BMC Psychiatry, 14*, 107.

Schlotz, W., & Phillips, D. I. (2009). Fetal origins of mental health: Evidence and mechanisms. *Brain, Behavior, and Immunity, 23*, 905–916.

Schmidt, H. D., Shelton, R. C., & Duman, R. S. (2011). Functional biomarkers of depression: Diagnosis, treatment, and pathophysiology. *Neuropsychopharmacology: Official Publication of the American College of Neuropsychopharmacology, 36*, 2375–2394.

Schnorr, S. L., Candela, M., Rampelli, S., Centanni, M., Consolandi, C., Basaglia, G., et al. (2014). Gut microbiome of the Hadza hunter–gatherers. *Nature Communications, 5*, 3654.

Schwingshackl, L., & Hoffmann, G. (2014). Mediterranean dietary pattern, inflammation and endothelial function: A systematic review and meta-analysis of intervention trials. *Nutrition, Metabolism, and Cardiovascular Diseases, 24*, 929–939.

Shin, N.-R., Whon, T. W., & Bae, J.-W. (2015). Proteobacteria: Microbial signature of dysbiosis in gut microbiota. *Trends in Biotechnology, 33*(9), 496–503.

Shonkoff, J. P., Boyce, W. T., & McEwen, B. S. (2009). Neuroscience, molecular biology, and the childhood roots of health disparities: Building a new framework for health promotion and disease prevention. *JAMA, 301*, 2252–2259.

Sonnenburg, E. D., Smits, S. A., Tikhonov, M., Higginbottom, S. K., Wingreen, N. S., & Sonnenburg, J. L. (2016). Diet-induced extinctions in the gut microbiota compound over generations. *Nature, 529*, 212–215.

Stahl, S. T., Albert, S. M., Dew, M. A., Lockovich, M. H., & Reynolds, C. F., 3rd. (2014). Coaching in healthy dietary practices in at-risk older adults: A case of indicated depression prevention. *The American Journal of Psychiatry, 171*, 499–505.

Stampfer, M. J., Hu, F. B., Manson, J. E., Rimm, E. B., & Willett, W. C. (2000). Primary prevention of coronary heart disease in women through diet and lifestyle. *The New England Journal of Medicine, 343*, 16–22.

Steenweg-de Graaff, J., Tiemeier, H., Steegers-Theunissen, R. P. M., Hofman, A., Jaddoe, V. W. V., Verhulst, F. C., et al. (2014). Maternal dietary patterns during pregnancy and child internalising and externalising problems. The generation R study. *Clinical Nutrition (Edinburgh, Scotland), 33*, 115–121.

Sudo, N., Chida, Y., Aiba, Y., Sonoda, J., Oyama, N., Yu, X. N., et al. (2004). Postnatal microbial colonization programs the hypothalamic–pituitary–adrenal system for stress response in mice. *Journal of Physiology, 558*, 263–275.

Suren, P., Gunnes, N., Roth, C., Bresnahan, M., Hornig, M., Hirtz, D., et al. (2014). Parental obesity and risk of autism spectrum disorder. *Pediatrics, 133*, e1128–e1138.

Tanentsapf, I., Heitmann, B. L., & Adegboye, A. R. A. (2011). Systematic review of clinical trials on dietary interventions to prevent excessive weight gain during pregnancy among normal weight, overweight and obese women. *BMC Pregnancy and Childbirth, 11*, 81.

Thorburn, A. N., McKenzie, C. I., Shen, S., Stanley, D., Macia, L., Mason, L. J., et al. (2015). Evidence that asthma is a developmental origin disease influenced by maternal diet and bacterial metabolites. *Nature Communications*, *6*, 7320.

Tiihonen, K., Ouwehand, A. C., & Rautonen, N. (2010). Human intestinal microbiota and healthy ageing. *Ageing Research Reviews*, *9*, 107–116.

Tsai, A. C., Chang, T. L., & Chi, S. H. (2012). Frequent consumption of vegetables predicts lower risk of depression in older Taiwanese—Results of a prospective population-based study. *Public Health Nutrition*, *15*, 1087–1092.

Vähämiko, S., Isolauri, E., Poussa, T., & Laitinen, K. (2013). The impact of dietary counselling during pregnancy on vitamin intake and status of women and their children. *International Journal of Food Sciences and Nutrition*, *64*, 551–560.

Van Lieshout, R. J., Taylor, V. H., & Boyle, M. H. (2011). Pre-pregnancy and pregnancy obesity and neurodevelopmental outcomes in offspring: A systematic review. *Obesity Reviews*, *12*, e548–e559.

Van Lieshout, R. J., & Voruganti, L. P. (2008). Diabetes mellitus during pregnancy and increased risk of schizophrenia in offspring: A review of the evidence and putative mechanisms. *Journal of Psychiatry and Neuroscience*, *33*, 395.

Vink, D., Aartsen, M. J., & Schoevers, R. A. (2008). Risk factors for anxiety and depression in the elderly: A review. *Journal of Affective Disorders*, *106*, 29–44.

Weng, T. T., Hao, J. H., Qian, Q. W., Cao, H., Fu, J. L., Sun, Y., et al. (2012). Is there any relationship between dietary patterns and depression and anxiety in Chinese adolescents? *Public Health Nutrition*, *15*, 673–682.

Whiteford, H. A., Degenhardt, L., Rehm, J., Baxter, A. J., Ferrari, A. J., Erskine, H. E., et al. (2013). Global burden of disease attributable to mental and substance use disorders: Findings from the Global Burden of Disease Study 2010. *Lancet*, *382*, 1575–1586.

Wright, M. L., & Starkweather, A. R. (2015). Antenatal microbiome: Potential contributor to fetal programming and establishment of the microbiome in offspring. *Nursing Research*, *64*(4), 306–319.

Wu, G. D., Chen, J., Hoffmann, C., Bittinger, K., Chen, Y.-Y., Keilbaugh, S. A., et al. (2011). Linking long-term dietary patterns with gut microbial enterotypes. *Science*, *334*, 105–108.

Yatsunenko, T., Rey, F. E., Manary, M. J., Trehan, I., Dominguez-Bello, M. G., Contreras, M., et al. (2012). Human gut microbiome viewed across age and geography. *Nature*, *486*, 222–227.

INDEX

Note: Page numbers followed by "*f*" indicate figures, and "*t*" indicate tables.

J

K

L

N

Q

Volume 42

Volume 43

Volume 44

Volume 45

Volume 46

Volume 47

Volume 48

Volume 49

Volume 50

Volume 51

Volume 52

Volume 53

Volume 54

Volume 55

Volume 56

Volume 57

Volume 58

Volume 59

Volume 60

Volume 61

Volume 62

Volume 63

Volume 64

Volume 65

Volume 66

Volume 67

Volume 68

Volume 69

Volume 70

Volume 71

Volume 72

Volume 73

Volume 74

Volume 75

Volume 76

Volume 77

Volume 78

Volume 79

Volume 80

Volume 81

Volume 82

Volume 83

Volume 84

Volume 85

Volume 86

Volume 87

Volume 88

Volume 89

Volume 90

Volume 91

Volume 92

Volume 93

Volume 94

Volume 95

Volume 96

Volume 97

Volume 98

Volume 99

Volume 100

Volume 101

Volume 102

Volume 103

Volume 104

Volume 105

Volume 106

Volume 107

Volume 108

Volume 109

Volume 110

Volume 111

Volume 112

Volume 113

Volume 114

Volume 115

Volume 119

Volume 120

Volume 121

Volume 122

Volume 123

Volume 124

Volume 125

Volume 126

Volume 127

Volume 128

Volume 129

Volume 130

HUMAN RIGHTS, EQUALITY AND DEMOCRATIC RENEWAL IN NORTHERN IRELAND

Human Rights, Equality and Democratic Renewal in Northern Ireland

Edited by
COLIN J HARVEY
Professor of Constitutional and Human Rights Law
University of Leeds

HART PUBLISHING
OXFORD – PORTLAND
2001

Hart Publishing
Oxford and Portland, Oregon

Published in North America (US and Canada) by
Hart Publishing c/o
International Specialized Book Services
5804 NE Hassalo Street
Portland, Oregon
97213-3644
USA

Distributed in the Netherlands, Belgium and Luxembourg by
Intersentia, Churchillaan 108
B2900 Schoten
Antwerpen
Belgium

Hart Publishing is a specialist legal publisher based in Oxford, England. To order further copies of this book or to request a list of other publications please write to:

Hart Publishing, Salter's Boatyard,
Folly Bridge, Abingdon Road, Oxford OX1 4LB
Telephone: +44 (0)1865 245533 or Fax: +44 (0)1865 794882
e-mail: mail@hartpub.co.uk
WEBSITE: http//www.hartpub.co.uk

British Library Cataloguing in Publication Data
Data Available
ISBN 1–84113–119–9 (hardback)

Typeset by Hope Services (Abingdon) Ltd.
Printed and bound in Great Britain on acid-free paper by
Biddles Ltd, www.biddles.co.uk

This book is dedicated to all those killed in the conflict and to those who worked so hard to bring it to an end.

> *"The tragedies of the past have left a deep and profoundly regrettable legacy of suffering. We must never forget those who have died or been injured, and their families. But we can best honour them through a fresh start, in which we firmly dedicate ourselves to the achievement of reconciliation, tolerance, and mutual trust, and to the protection and vindication of the human rights of all" (Good Friday Agreement 1998).*

Contents

Preface

This book springs from my time in Belfast from 1997–2000 and from various interactions with the people that I met and worked with there in all spheres of public life. The work tries to bring together in one place a number of critical essays on a variety of aspects of politics and law in Northern Ireland. At the heart of this collection lies the Good Friday Agreement 1998 and all that it has brought in its wake. I had the great privilege to be around for the period of the negotiation and adoption of the Agreement. It will I believe come to be seen as a monumental event in Irish history, and despite the many problems it remains I think a marvellous achievement.

Growing up in Derry in the 1970s and 1980s brought the realities of the conflict into full view. The consequences of the "long war" were a dominant feature of day-to-day life. This never, however, extinguished the legacy of non-violent political struggle. The names associated with the civil rights movement were a recurring feature of my childhood years. Politics was debated with an intensity that seems to have gone out of fashion in our consensual times. It was impossible not to be politicised in one way or another (detachment is of course a political position) and the ideals of the civil rights movement inspired me and many others. The ideals held out the promise of emancipation as a realisable human achievement which we all had responsibility for working towards in dialogue with others. In this sense there is little particularly new about dialogical models of politics. The challenge, then as now, is to win the argument for solutions based upon dialogue and not force. Despite its flaws, the Good Friday Agreement 1998 may well come to be seen as the partial embodiment of that 1960s radicalism and the ideals it served. So now that we are in new times, and the only struggle worth talking about is the political one, it is perhaps worth reminding ourselves of the narratives and ideals that were buried in thirty years of conflict. These ideals have returned to become the common sense of the current political process. This collection of essays is advanced as a critical contribution to the ongoing dialogue about the transformation of Northern Ireland. All the contributors have, by thought and deed, shown their commitment to progressive change. The essays speak for themselves in this regard.

There are too many people to thank individually, you know who you are. As ever Gerry Cross was a great source of friendship and support. Kieran McEvoy listened to my ramblings and during my time in Belfast I found his enthusiasm for scholarly ideas and his practical commitment to political struggle inspiring. I would like to thank all the contributors for delivering "product" and for putting up with an at times impatient editor. Editing a book, like everything else these days, is a process. I have learned this lesson well. Thanks to all at Hart Publishing and to Richard Hart for his enthusiasm for this project.

In particular, I would like to thank Lisa King for her insights on equality in practice and for reminding me of the formidable obstacles to real change. She does not simply "talk the talk". She lived this book too and many of the arguments were radically improved as a result of my ongoing conversations with her.

COLIN J HARVEY
September 2000

List of Contributors

Gordon Anthony is a Lecturer in the School of Law at the Queen's University Belfast.

Andrew Evans is a Reader in the Institute of European Studies at the Queen's University of Belfast.

Colin J Harvey is Professor of Constitutional and Human Rights Law at the University of Leeds.

John Jackson is Professor of Public Law at the Queen's University of Belfast.

Stephen Livingstone is Professor of Human Rights Law at the Queen's University of Belfast.

Christopher McCrudden is Professor of Human Rights Law at the University of Oxford.

Kieran McEvoy is a Reader in the School of Law at the Queen's University of Belfast.

Linda Moore is an Investigations Worker with the Northern Ireland Human Rights Commission.

John Morison is Professor of Jurisprudence at the Queen's University of Belfast.

Mary O'Rawe is a Lecturer in the School of Public Policy, Economics and Law at the University of Ulster.

1

Northern Ireland in Transition: An Introduction

COLIN J HARVEY

THE GOOD FRIDAY Agreement of 1998 (or Belfast Agreement) promises the people of Northern Ireland a fresh start underpinned by guarantees on human rights, equality and participation.[1] After three decades of violent conflict the Agreement holds out the potential for radical transformation and a transition to a new democratic polity. It contains commitments on: new democratic structures (North, South, British and Irish); criminal justice; policing; human rights (North and South); equality; prisoner releases; decommissioning, and victims. The Agreement is a relatively comprehensive document which addresses complex issues of public policy and ongoing interactions between Britain and Ireland. Given the history, what is remarkable is its explicit basis in the progressive political values of human rights, equality and democratic governance. The reality of life in Northern Ireland may be a long way from the ideal, but the normative basis for a new beginning is clearly established. The issue now, and for many years ahead, is to ensure that the promise of the Agreement is effectively delivered. As many of the contributions in this book suggest, this will be a significant challenge and the project has implications beyond Northern Ireland. It will, for example, require a re-think of orthodox understandings of constitutional law and politics in the United Kingdom. If developments in Northern Ireland are to be properly understood, a new paradigm of constitutional law is required which has both explanatory and critical

[1] *Agreement Reached in the Multi-Party Negotiations* (Cm 3883, 1998). The full text of the Agreement is printed in the Annex. The Agreement was translated into legislation in the form of the Northern Ireland Act 1998. For a useful link to the parliamentary stages of the legislation and all the legal preparations for devolution see http://www.nio.gov.uk/implemact.htm.

potential. The challenge is to recognise diversity while promoting cooperation and partnership in all the relations that currently exist within and between Britain and Ireland. As I argue, process-based models of law and democracy offer potentially the most attractive basis for analysing ongoing developments. What these conversational models encourage is more sophisticated thinking on new forms of decentred democratic relations, based upon models of rational discourse.[2] Fragmentation then becomes an opportunity to erode the patriarchal nature of traditional UK constitutional law, while maintaining a rich notion of law and democracy. Although the general trends in British constitutionalism are noted, the main focus of this collection is on Northern Ireland. The developments in the last decade merit such detailed and focused treatment.

This collection of essays, written by scholars who have all made a contribution in their own ways to the process of change, presents a critical perspective on how the transition is progressing and identifies key themes in the current law and politics of Northern Ireland. There are lessons, both good and bad, for Britain, Ireland and beyond in the recent history of Northern Ireland. On human rights and equality there are innovative new mechanisms that will, for example, be of interest to those in Ireland and Britain seeking to achieve progressive change. For public lawyers, the new constitutional order created in the wake of the Agreement cannot simply be fitted neatly into traditional notions of UK constitutional law. For example, I would argue that the Agreement has a constitutive force which should be given full recognition in legal and political life. To deny the legitimacy of the Agreement in law is to neglect its historic significance for Ireland, North and South. Such neglect is standard in British constitutionalism, but can no longer be justified. This should test the traditional understanding of the unitary nature of the United Kingdom. It represents a constitutional moment which must be accorded due weight by constitutional lawyers. There are distinct aspects of this new order which should encourage some rethinking of traditional positions. It is, of course, in the nature of tradition to perpetuate itself and simply absorb novelty. And this may well happen in this instance. British constitutionalism has a habit of absorbing change and moving on largely unaltered in its basics. It will be up to those seeking a new beginning, however, to ensure that this does not happen.

[2] See Jürgen Habermas "A Short Reply" (1999) 12 *Ratio Juris* 445, at 449–50.

Each of the chapters addresses the above themes in different ways. Colin J Harvey, in Chapter 2, examines the nature of constitutionalism in Northern Ireland and defends his position that the Agreement has a constitutive force in law and politics. He argues that events should be viewed in context. In particular, he highlights broader developments in constitutional law. He suggests that analysis of Northern Ireland should acknowledge the role of political violence and coercion in both its construction and for much of its history. As is noted, what has followed the Agreement demonstrates that democratic structures can be created through law which pay due regard to local contexts. Continuing with the theme of constitutionalism, Gordon Anthony and Andrew Evans explore the European Union (EU) dimension. Their argument is that the traditional understanding of UK constitutional law may cause problems for the new institutional arrangements in Northern Ireland. The chapter provides a detailed overview of Northern Ireland's role in EU policy formation with a call for a move "beyond orthodoxy" in UK constitutional law thinking.

While there are different understandings of the term, it is clear that the concept of equality underpins the Agreement. Christopher McCrudden offers, in Chapter 4, both an outline of how equality gained such an important place, as well as detailed treatment of developments since the Agreement. Considerable thought has been given in Northern Ireland to making paper commitments mean something in practice in this area. McCrudden has played a leading role in suggesting mechanisms to make equality meaningful in practice, as well as in law. As he shows, the political pressure led to creative thinking and eventually to the innovative provisions of the Northern Ireland Act 1998. The practical impact of the new equality schemes will be watched closely. As he states, at present we are mainly talking about promises. There is some way to go before these become a reality.

Equality stands alongside the human rights aspects of the Agreement. In Chapter 5 Colin J Harvey explores the role of the Northern Ireland Human Rights Commission. While it is too early to make a definitive assessment it is argued that it has a fundamental role to play in enriching political democracy. Harvey thus departs from rights-based thinking which leads only to the worship of judicial enforcement of rights. He is interested in all the mechanisms that can be effectively used to ensure that rights are protected within a flourishing deliberative democracy. Although much work has been done to rehabilitate the judiciary in recent years, he remains sceptical about their ability to take the lead in promoting progressive law and politics. This is why national institutions for the protection of human rights are so important.

Stephen Livingstone examines the role of the judiciary and the legal profession during the conflict. He notes the lack of involvement of lawyers in public debates in this period. A variety of reasons are advanced for this, including the linkage made between professionalism and neutrality. He rejects the argument that the lack of involvement of lawyers was of no consequence and suggests that when they did speak out they were listened to by the judges. On the judiciary he is critical of the failure of the Northern Ireland courts, in the period prior to direct rule, to develop a constitutional jurisprudence which would challenge abuses of power. He concludes with an examination of three areas: anti-terrorism law and policy; judicial review; and anti-discrimination law and policy. Livingstone recognises that law is Janus-faced, it has both repressive and reforming dimensions. To understand the judicial role you have to acknowledge both. What is interesting is the way that the judges and legal profession are more likely to respond to criticism from "one of their own". This "discursive community" has its own systemic logic and will respond to argumentation framed by institutional actors within it. While it is never completely closed, it is a discursive community which seeks to perpetuate its own systemic logic. This is important, as the chapter by McEvoy also reveals, in understanding how change is achieved within this community in practice.

Livingstone ends with a call, rightly I think, to the judges to see the Agreement as part of a new constitutional order which requires the development of a distinctive constitutional jurisprudence. Whatever the arguments about the judicial role in a political democracy, the judges will have a part to play. It is essential that in discharging their responsibilities they pay due regard to the constitutive nature of the Agreement and the progressive values which underpin it.

John Jackson looks at the criminal justice system in Chapter 7. As he argues, this is an area that can truly be said to be undergoing a period of transition. As a member of the Criminal Justice Review he refrains from detailed critique of its recommendations, however, the chapter includes useful insights into the problems and prospects. In particular, Jackson notes the advantage of a review process that was able to take a global view of criminal justice and criticises reviews in England and Wales which have tended to take a piecemeal approach.

Policing is highly contested in Northern Ireland. Linda Moore and Mary O'Rawe examine in detail the proposals for a new beginning to policing contained in the Patten Report. They argue that there are many truths surrounding the history and work of the RUC. Problems have, however, arisen when one truth is given particular priority, for example, when the organisation is described as a model police force, or the most professional force in the world. As they note, such inaccurate, and highly partial accounts, do little to advance the debate. Their chapter contains detailed treatment of the substantial problems facing police reform. This is based on their own comparative research on situations of conflict elsewhere. While not uncritical of the Patten Report, their principal argument is for swift and effective implementation.

After noting the successes in relation to state abuses, Kieran McEvoy looks at the influence which human rights and humanitarian interventions have had on the actions of the paramilitaries. McEvoy presents a fascinating insight (based on his extensive empirical research) into paramilitary activity in Northern Ireland. This chapter should make useful reading for those who prefer moral condemnation to rational argument. He is interested in what works in practice and how interventions have impacted on paramilitaries. The focus is the pragmatic one of achieving meaningful change for people in local communities and thus makes a refreshing change from some of the literature in this area. In drawing clear distinctions between republican and loyalist paramilitaries, he notes the sensitivity of republicans to interventions by respected human rights groups. For a movement concerned about international publicity criticism by organisations such as Amnesty International was not welcome. McEvoy shows that what tended to work was engagement with paramilitaries on the basis of sound contextual understanding in, to put it crudely, a language they could understand. Again this takes us back to the idea of a discursive community discussed above. Interventions must be based on a solid knowledge of the intricacies of the community. In the example he gives he argues that this sort of engagement made a real difference to the lives of affected individuals and groups. It was therefore more effective on the ground than some of the more high profile "interventions". If, however, the internal dynamics of the group is the determining factor it can prove extremely difficult to measure impact precisely. This raises the issue of whether in practice it is the internal dynamics which shifts in accordance with internal strategic imperatives rather than external influences. No discursive community is communicatively "closed", however, republicans had their own systemic logic which did prove remarkably resilient. An optimistic reading is that communication changed the terms of the conversation within republicanism (particularly the Hume-Adams initiative). Dialogue has its own dynamics and cannot be reduced to purely internal or external strategic manipulation. As McEvoy notes, the period examined was one when republican paramilitaries were seeking, through political means, disengagement from their "policing" role and the "armed struggle". The interesting issue here is how the political language that made this disengagement possible was shaped. I would suggest that the language of political and legal compromise originated in the exhaustive work of the social democratic tradition in Ireland. As he notes, a rather vague "rights and equality" agenda has become a defining feature of modern Irish republicanism. Of course, prior to this late conversion to exclusively democratic means the extensive and destructive use of political violence was the defining characteristic. But he is correct to observe how the language of rights and equality has now been internalised in the republican movement.

The final contribution, by John Morison, examines the role of the voluntary sector in the democratic renewal of Northern Ireland. Morison draws upon the work of Michel Foucault to argue for the importance of the governmentality approach. He wants to downgrade the significance of the state and focus attention instead on the informal networks of the public sphere. He is explicitly critical of the nature of the Northern Ireland Assembly and looks to the community and voluntary sectors as sources of constitutional renewal. He notes the scale and wider role of this sector in Northern Ireland. Although sceptical about the new democratic structures, he is extremely critical of direct rule. He draws on previous collaborative work with his reference to "communicative constitutionalism". An interesting issue here, not taken up, is how this commitment to a dialogical model combines with the governmentality approach. "Communicative constitutionalism" has strong echoes of the recent resurgence of interest in deliberative democracy. However, Foucault, with his stress on highly localised forms of resistance, would have been sceptical about the grand claims made by some deliberative democrats. There is the additional issue of the institutionalisation of civil society. Arato, for example, argues that civil society must be securely institutionalised if it is to become a key site within a participatory politics.[3] The Civic Forum is, in the

[3] Andrew Arato *Civil Society, Constitution, and Legitimacy* (Oxford, Rowman & Littlefield Publishers Inc, 2000). See also J Cohen and A Arato *Civil Society and Political Theory* (Cambridge Mass, MIT Press, 1992); Iris Marion Young *Inclusion and Democracy* (Oxford, Oxford University Press, 2000) p. 194.

Northern Ireland context, an example of this institutionalisation. The point is that formal institutions matter and are an important part of shaping a deliberative democracy. Morison is correct however to place the emphasis on the informal public sphere as essential to the project of democratising democracy. The challenge is to embed a process of reflective interaction between the formal and informal public spheres.

The transition promised by the Agreement will take time. After thirty years of violent conflict, and a considerably longer period of instability, this is to be expected. It is, however, understandable that this can cause impatience. The main problem in the years ahead will be ensuring that the progressive elements of the Agreement are realised in practice. The forces of conservatism have a firm grip on public life in Northern Ireland. These forces can be found in unlikely places. This fact holds out the challenge of continuing struggle on the part of those committed to the new beginning. The law and politics of Northern Ireland will remain the site of continuing contestation. In the disagreements that will arise, and in ongoing political and legal conversations, it is essential that the constitutive force of the Agreement is accorded due recognition. British constitutionalism must accommodate the distinctive nature of the process of conflict resolution in Ireland, North and South. The transition is not simply another example of devolution and constitutional lawyers must come to acknowledge this fact. In Northern Ireland the growing literature on transitional justice may in fact be a more appropriate source of insight and a better starting point.

2

The New Beginning: Reconstructing Constitutional Law and Democracy in Northern Ireland

COLIN J HARVEY

AN ANALYSIS OF constitutionalism in Northern Ireland must pay due regard to the role of political violence and the responses to it. This goes beyond reflection upon the pure wilfulness of the original constitutional moment,[1] although in this case it is of fundamental significance.[2] Northern Ireland was born from political violence. Coercion

[1] Constitutionalism can be viewed as the ongoing attempt to bury the coercive and wilful legacy of its origins. It implies limits, even though the memory of the constitutive moment remains to haunt the present. This "constructivist" exercise in reconstruction frequently underplays (deliberately?) the fictitious nature of the constitutional traditions upon which it rests. The distrust shown by the framers of constitutions of "the people" perhaps relates to this basic fact.

[2] In relation to the use of force in the formation of Northern Ireland note the following comment, Harry Calvert *Constitutional Law in Northern Ireland: A Study in Regional Government* (London and Belfast, Steven and Sons Ltd and Northern Ireland Legal Quarterly Inc, 1968) p. 33: "Finally, it will be appreciated that should the issue [Westminster-inspired secession of Northern Ireland from the United Kingdom] ever arise, it is likely that considerations of legality will play but a small part in the determination of it. Unionist sentiment has on previous occasions consistently demonstrated that in its view there are higher considerations than legality". He was referring specifically to the threat of force from the Ulster Volunteer Force in the 1912–1914 period. An organisation which had A V Dicey on its side. See Kevin Boyle, Tom Hadden and Paddy Hillyard *Law and State: The Case of Northern Ireland* (London, Martin Robertson, 1975) p. 6: "The Northern Protestants had armed themselves to resist 'home rule' as early as 1912. When in 1920 and 1921 the predominantly Roman Catholic population of the southern counties of Ireland took up arms against the British and eventually gained their independence as the Irish Free State. . ., the Northern Protestants under the leadership of the Unionist Party opted for a measure of local autonomy within the United Kingdom as provided in the Government of Ireland Act 1920".

has been an ever present aspect of its troubled existence. It is a constitutional project which has consistently failed. The ghosts of its violent birth have continued to haunt the political process and have made compromise and consensus difficult to achieve. The constitutional settlement under examination in this chapter is a sophisticated attempt to lay the ghosts of coercion to rest. While the quest for legitimacy underpins the political process, one argument in this chapter is that it is the legitimacy of contested constitutionalism which has been secured. The status of Northern Ireland remains contested by the very nature of its normative base. The consent principle, so often presented as the guarantee of unionist hegemony, is in fact the deliberative basis for productive instability from a nationalist/republican perspective. It may be correct to term this a "constitutional settlement" but by the nature of the system it may only be for now. The basis of this settlement is, however, full acceptance by the participants that the only legitimate struggles in future will be political ones.

The events of the last decade in Northern Ireland have resulted in significant legal and political change. This has taken place in the context of general constitutional change in the United Kingdom. The unitary conception of the constitution may remain even though some of its traditional elements are clearly being reconsidered and/or redescribed.[3] While it is possible to view developments in Northern Ireland as the result of purely internal dynamics, it makes more sense to acknowledge the broader constitutional and other trends which impact upon it. There have been many attempts to secure a negotiated constitutional settlement in Northern Ireland. This has largely been a history of failure. Until the most recent political process a settlement which could command widespread support in Ireland was unattainable. The Good Friday Agreement 1998, and the new structures that have been created, are of significance beyond Northern Ireland. One of the main arguments of this chapter is that the Agreement, and what has followed, demonstrates the strengths and weaknesses of tailoring democratic structures to local contexts. While the thinking which underpins the new order in Northern Ireland is not novel, it does illustrate that complex structures can be created to facilitate deliberative law and democracy.

[3] See Neil Walker "Beyond the Unitary Conception of the United Kingdom Constitution?" [2000] *Public Law* 384.

THE POLITICS OF LEGAL ANALYSIS

Constitutional law is constituted by the political and this is the major component in its ongoing development. In addition to the politics of new beginnings, the ongoing life of constitutional law is also shaped by politics. The constitutional lawyer must take a political position on legal analysis.[4] One of the aims in this chapter is to examine the new constitutional structures created and argue that a critical understanding of legality matters in Northern Ireland. In other work I have defended a critical tradition in legal scholarship which takes legality seriously.[5] Legal analysis is too often the subject of caricature. For example, legalism is on occasion used as an accusation in studies of Northern Ireland.[6] This is perfectly understandable and at times highly accurate. However, it pays scant regard to the advances made by the critical tradition in modern law and to the contribution of legal analysis to our understanding of social and political life. There is still potential in the concept of legality. But this must be a concept of legality without illusions. My argument is that a critical understanding of legality is essential for understanding this and other constitutional law contexts.[7] The critical legal tradition drawn upon here is firmly grounded in a conception of democratic law with a firm commitment to deliberative democracy. The primary right in this conception is the right to democratic governance and its corollary, the right to participate.[8] My

[4] See J A G Griffith "The Brave New World of Sir John Laws" (2000) 63 *MLR* 159, at 176, for a strong argument on the politics of law.

[5] Colin Harvey "The Politics of Legality" (1999) 50 *NILQ* 528; Colin Harvey "Governing After the Rights Revolution" in Colin Harvey, John Morison and Jo Shaw (eds) *Voices, Spaces, and Processes in Constitutionalism* (Oxford, Blackwell, 2000) p. 61; Colin Harvey "Complex Conversations: Legality, Politics and Constitutionalism in Northern Ireland" (2000) 3 *Contemporary Issues in Irish Law and Politics* 70.

[6] Brendan O'Leary "The Nature of the Agreement" (1999) 22 *Fordham International Law Journal* 1628.

[7] See Colin Harvey "Legality, Legitimacy, and Democratic Renewal: The New Assembly in Context" (1999) 22 *Fordham International Law Journal* 1389.

[8] This right is increasingly recognised at the international level see Thomas M Franck "Democracy and the Democratic Entitlement" in Gregory H Fox and Brad R Roth (eds) *Democratic Governance and International Law* (Cambridge, Cambridge University Press, 2000) p. 25: "Attention must . . . be paid to democracy as a right protected by international law and institutions. Democracy does not provide a guarantee against civil war. It merely provides the only known process by which a genuine social discourse can proceed among persons legitimately representing the spectrum of opinions and interests in a community or *polis*. Without it, there can be decisions. There can even be negotiation and discourse. But there can never be a genuine social convergence".

argument is that critical legal analysis can be an important tool in the service of democratic experimentalism.[9] Those who are understandably critical of legalism mistake partial and distorted forms of argumentation as definitive models. Attaching importance to the value of legality does not necessarily imply rigid adherence to mechanistic formalism and an overly technocratic approach. Critics of legalism are not afraid of extensive engagement with law when significant legislative change is proposed. In this sense there is a performative contradiction at work. Even if everything is up for grabs the contest will take place within the realm of legal argumentation. Engagement in legal argumentation must indicate some belief that the substantive content of the law matters even if this remains open to further dialogue. Against this there are reasons why one should have no illusions about legality. Law is rooted in political and social contexts which determine its content and effectiveness. Its effectiveness in promoting good governance depends on how legal discourse is deployed by actors in the public sphere. The coercive role of law is as important as the enabling. Its disciplining functions can be written off as a form of systemic oppression, but there is more to it than that. Political communities can opt for repressive law when, for example, they wish to eradicate racial discrimination or punish corporations for environmental damage. In other words, legal regulation can help create the conditions for the existence of a functioning democratic polity. The goal of critical legal scholarship should remain emancipation and thus the interest rests in the role that law can play in achieving this objective.[10] As argued here, law has emancipatory potential which can be realised through political engagement. I suggest that in practice the democratic state should remain an important focus for individuals and groups, both in its

[9] See Roberto Mangabeira Unger *What Should Legal Analysis Become?* (London, Verso, 1996) pp. 180–1: "An underlying theme in this book is that our problem is less that we have too many programs than that we have just one program: the only political program with authority in the modern world, the program of democratic experimentalism from the eighteenth century to the present day, the program that liberals share with socialists. Its central commitment is to lift the grid of social division and hierarchy weighing upon our practical, passionate, and cognitive dealings with one another".

[10] One should not of course underestimate the problems facing anyone wishing to develop a critical theory with practical intent. The experience of the Frankfurt School is instructive. The focus on domination and instrumental reason led Theodor Adorno and Max Horkheimer to a totalising critique of Western thought. Viewed in historical context this rejectionist, and cautious, scholarly approach was understandable, but one wonders about the consequences for social movements seeking practical change today. See Martin Jay *The Dialectical Imagination: A History of the Frankfurt School and the Institute of Social Research 1923–1950* (Berkeley, University of California Press, 1973).

coercive and enabling guises.[11] This is an argument in support of the continuing importance of the formal public sphere.

The law and politics of institutional design in Northern Ireland are evidence that such experimentalism is possible. The complex arrangements are a sophisticated attempt to give due recognition to the bi-national nature of the conflict, while also encouraging a movement away from absolutist notions of sovereignty. This signals *inter alia* the influence of postnational thought on the developing process.[12] The settlement is founded on legal guarantees of human rights, equality and participation. Although the meaning of these terms differs, depending upon the specific context, there is a strong deliberative component in the constitutional model that has been adopted. One of the arguments advanced is that it is the deliberative component which will be of most interest in the years ahead. At the core of the Agreement is a concept of self-government with equality and participation at its heart. It is profoundly social democratic in both substance and design.

CONSTITUTIONALISM IN CONTEXT

Contestation, constitutionalism and coercion

An examination of constitutionalism must be rooted in a theory of democracy, for this is the only way to move coherently through the multiplicity of facts about political and legal life. I have defended the importance of deliberative democracy elsewhere, and this is the theoretical premise upon which much of the argument in this chapter is

[11] See Edward Broadbent "Social Democracy: Past and Future" (1999) Fall *Dissent* 45; Anthony Giddens *The Third Way: The Renewal of Social Democracy* (Cambridge, Polity Press, 1998) p. 48: "Markets cannot replace government in any of these areas, but neither can social movements or other kinds of non-governmental organization . . . no matter how significant they have become". See also Anthony Giddens *The Third Way and its Critics* (Cambridge, Polity Press, 2000) pp. 164–5: "The state continues to have a fundamental role to play in economic life as in other areas. It cannot replace either markets or civil society, but needs to intervene in each . . . Government must maintain a regulatory role in many contexts, but as far as possible it should become a facilitator, providing resources for citizens to assume responsibility for the consequences of what they do".

[12] See Joseph Ruane and Jennifer Todd "Irish Nationalism and the Conflict in Northern Ireland" in David Miller (ed) *Rethinking Northern Ireland* (London, Longman, 1998) p. 55, at 68.

based.[13] This critical perspective is not content with noting the establishment of the institutions of liberal constitutionalism. Rather the concern is with the promotion of the principles which underpin the concept of deliberative democracy, such as the right to participate. Central to this perspective is the view that the public sphere is not confined to the institutions of constitutional law and that the informal public sphere has a vital role in a democracy.

The problem which all theorists of constitutionalism face is the "wilfulness" of the process of constitution-building and its origins in coercion.[14] Constitutional moments can be far from democratic and the ghosts of coercion can return to haunt the present. As noted, Northern Ireland was born from violence and it has proven difficult to bury the memory of this. Viewed in context the partition of Ireland seems more an exercise in avoidance than polity formation. One could, for example, argue that Northern Ireland never "constituted" itself legitimately as a political entity. This takes us to the heart of what is meant by constitutionalism. There is much talk of constitutional settlements in Northern Ireland without reference to what precisely is meant by "constitution". Preuss talks of a constitution as an "operational framework" or as a state of being constituted.[15] He states: "What it can and in fortunate instances does do is create such institutional conditions as are suited to exert a beneficial pressure on society to rationalize and improve itself".[16]

A constitution thus can create the conditions for societal development and reflective political practice.[17] It is noteworthy that Preuss views this constitutive process as based on normative principles. He

[13] Cf. John S Dryzek *Deliberative Democracy and Beyond: Liberals, Critics, Contestations* (Oxford, Oxford University Press, 2000), he makes a distinction between deliberative and discursive democracy. His concern is that deliberative democracy should not lose its critical edge (p. 3): "I refer to this more critical strand of deliberative democracy as discursive democracy, which (contrary to much current usage, which tends to use 'deliberative' and 'discursive' interchangeably) I distinguish from a model confined to politics in the vicinity of liberal constitutionalism".

[14] See Ruti Teitel *Transitional Justice* (New York, Oxford University Press, 2000) p. 191, she notes that the central dilemma is "how to reconcile the concept of constitutionalism with revolution". She proposes an alternative "constructivist" paradigm in which transitional constitutionalism not only constitutes the political order but is constitutive of political change.

[15] Ulrich K Preuss *Constitutional Revolution: The Link Between Constitutionalism and Progress* (New Jersey, Humanities Press, 1995) p. 109.

[16] *Ibid.*

[17] See Thomas Paine *The Rights of Man* (Harmondsworth, Penguin Books Ltd, 1969) p. 207: "A constitution is not the act of a government, but of a people constituting a government; and government without a constitution is power without a right".

further argues: "A society is constituted when it must constantly confront itself in suitable institutional forms and in normatively directed processes of adjustment, resistance and self-correction".[18] This captures both the fact of the constitutive moment and the ongoing processes of dialogue which it must facilitate and encourage.[19] Jürgen Habermas argues: "A constitution can be thought of as a historical project that each generation of citizens continues to pursue".[20]

The dynamic and open-ended nature of constitutionalism explains the ongoing struggles for recognition and inclusion which take place in democratic societies. But to what extent is this ongoing process of dialogue based on consensual constitutionalism and how do we respond when is it actively used by anti-modern elements in society?[21] Modern trends in political theory, associated with the work of Habermas and Rawls, tend towards the consensual. However much pluralism is welcomed and celebrated, it is the achievement of consensus which underpins their work.[22] Habermas is particularly interested in the consensual nature of communicative action. This work is opposed by radical democratic theorists who prefer "agonistic" approaches to the political. The thinking can be traced to the work of Nietzsche and Arendt, but can also be found in Marx. Modern representatives of this school of thought include William Connolly and Chantal Mouffe.[23] Villa summarises their concerns as follows:

> "These theorists worry that modern democracies are hardly democratic at all; that the bureaucratic edifice of the state has usurped the space of the political, rendering citizens the passive recipients of policy decisions; and

[18] *Supra* n. 15.

[19] See Hannah Arendt *On Revolution* (London, Penguin Books, 1973) p. 145: "The word 'constitution' obviously is equivocal in that it means the act of constituting as well as the law or rules of government that are 'constituted', be these embodied in written documents or, as in the case of the British Constitution, implied in institutions, customs, and precedents".

[20] Jürgen Habermas *The Inclusion of the Other: Studies in Political Theory* (Ciaran Cronin and Pablo De Greiff (eds) Cambridge Mass, The MIT Press, 1998) p. 203.

[21] See Ulrich Beck *The Reinvention of Politics: Rethinking Modernity and Global Social Order* (Cambridge, Polity Press, 1997) p. 42, as he notes "not just social movements outside the political institutions, but also social counter-movements inside the political system know how to use the spaces of reflexive democracy".

[22] See John Rawls "The Idea of an Overlapping Consensus" (1987) 7 *Oxford Journal of Legal Studies* 1.

[23] See William E Connolly *The Terms of Political Discourse* (3rd edn, Oxford, Blackwell, 1993); Chantal Mouffe "Carl Schmitt and the Paradox of Liberal Democracy" in David Dyzenhaus (ed) *Law as Politics: Carl Schmitt's Critique of Liberalism* (Durham and London, Duke University Press, 1998) p.159; Chantal Mouffe *The Return of the Political* (London, Verso, 1993).

> that liberal theory has contributed to this state of affairs by promoting a conception of politics which is essentially juridical/administrative, one which seeks ways of diminishing, if not eradicating, the contest and debate that is the lifeblood of a robust democratic politics".[24]

The attempt to eradicate contestation can have unanticipated consequences. In particular, fundamentalist positions can thrive in such an environment. The "agonistic" turn in political theory reminds us that conflict has an important place in democratic life and in the social development of constitutionalism. This theoretical position may appear out of place in Northern Ireland. Surely, one might suggest, continuing contestation is the last thing that is needed in Northern Ireland? This concern is misplaced, and here I think we have to be selective. Agonistic politics seriously underestimates the role of consensus and a politics of accommodation. However, it highlights the silences in the public sphere and assists attempts to fracture unreflective/false consensus. Arendt presents the most attractive view in this regard by stressing contestation based upon judgement and a sense of public-spiritedness.[25] The "agonistic turn" helps remind those interested in deliberative democracy of the importance of action and thus political struggle. The problems start when we place too much faith in purely expressive politics and reify the idea of conflict. Consensus and compromise must assume a central place in democratic life. At its most basic, certain levels of autonomy must be secured through legal and political practice in order to allow for inclusive dialogue to occur in the first place. Purely expressive politics neglects the importance of rational debate in the public sphere and the conditions which make this happen. There is also a tendency to see legalism as part of the problem. Careful examination would reveal that debates over legality are an important part of the continuing contest in the public sphere and that law can play a part in facilitating deliberative democracy. Law has a role in the retrieval of public space and in the development of the public sphere. In particular, law operationalises those procedural mechanisms which facilitate the democratic realisation of societal norms.[26]

[24] Dana R Villa *Politics, Philosophy and Terror: Essays on the Thought of Hannah Arendt* (Princeton, Princeton University Press, 1999) p. 108.

[25] *Ibid.* p. 124.

[26] See Seyla Benhabib "Models of Public Space: Hannah Arendt, the Liberal Tradition, and Jürgen Habermas" in Craig Calhoun (ed) *Habermas and the Public Sphere* (Cambridge Mass, The MIT Press, 1992) p. 73, at 87: "Public space is not understood *agonistically* as a space for competition for acclaim and immortality among a political elite; it is viewed democratically as the creation of procedures whereby those

Continuing dialogue does not mean non-stop conflict. I think Habermas has demonstrated the consensual underpinnings of communicative action and the fact that the search for this consensus is embedded within our own discursive practices. When someone, for example, criticises an agreement because it is based on a false consensus she must have some idea what a fair consensus would look like.

Dryzek has expressed some concern that deliberative democracy is losing its critical edge.[27] This has led him to defend what he terms discursive democracy. He draws a sharp distinction between discursive and deliberative democracy. One of the problems he correctly identifies is that critical theory has been compromised by an ever closer association with liberal theory. On renewing its critical bite he states: "One way to do this is to ground deliberative democracy in a strong critical theory of communicative action, and to re-emphasize oppositional civil society and public spheres as sources of democratic critique and renewal".[28] I would agree with this argument, and that is why the "agonistic" turn provides some useful thoughts on how this might be achieved. I am, however, not convinced that there is a meaningful distinction to be made between deliberative and discursive democracy.

Learning from history

We are in an age of reflexive law and politics. Learning processes are therefore central to a modern understanding of constitutionalism.[29] Devolution is not new in the Northern Ireland context. The failure of the original devolutionary model was one of the reasons for the disturbances in Northern Ireland which eventually led to the imposition of direct rule in 1972.[30] Although the conflict of the last three decades has focused attention on Northern Ireland, there have been problems since

affected by general social norms and collective political decisions can have a say in their formulation, stipulation and adoption . . . [T]here may be as many publics as there are controversial general debates about the validity of norms".

[27] Dryzek *supra* n. 13.

[28] Dryzek *supra* n. 13 at p. 4.

[29] See Andrew Arato *Civil Society, Constitution, and Legitimacy* (Oxford, Rowman & Littlefield Publishers Inc, 2000) p. 254: "Many of the projects of constitution making of the last twenty years have been characterized by increasing reflexivity. All major efforts indicate learning from the repeated failure of revolutionary democratic constitution making and the need to satisfy the demands of democracy and legality in a new way".

[30] For analysis of direct rule phase one see Brigid Hadfield *The Constitution of Northern Ireland* (Belfast, SLS Legal Publications, 1989) pp. 100–1.

the formation of the "state". As noted, the process of state formation buried the seeds of conflict (not necessarily violent) by neglecting a substantial legitimacy deficit.[31] The minority nationalist/republican community were effectively sealed within a state where their minority status was assured. In the governing constitutional context it was foreseeable that this community would become constructed as the "enemy within".[32] The new Government also viewed itself as under threat from the administration in the South of Ireland.[33] While there were continuing problems, it was not until the late 1960s that the practical implications of the compromises of the 1920s became readily apparent. This portrayal of events injects a level of historical inevitability that is perhaps misplaced. It is hard to deny that the circumstances which surrounded the creation of Northern Ireland were hardly encouraging. The distinction between constitutionalism and violence was blurred in this context. Given the history of coercion, the legitimacy of the arrangements concluded in the 1920s has been consistently questioned. With the passage of time it is perhaps easier to try to avoid this issue, and following the referenda on the Agreement, is possibly now less relevant. However, the fact remains that the original constitutive moment lacked the principled elements of a newly constituted political community.[34]

The Government of Ireland Act 1920 proposed the establishment of Parliaments in Dublin and Belfast with powers of local self-government.[35] The Act effectively legislated for partition along the six county border.[36] The option of a unified all-Ireland Parliament was never taken. The Act established a bicameral legislature with a fifty-two-

[31] Colin Harvey "Democracy in Transition: Mainstreaming Human Rights and Equality in Northern Ireland" (1999) 4 *Journal of Civil Liberties* 307, at 308.

[32] See Fionnuala Ní Aoláin *The Politics of Force: Conflict Management and State Violence in Northern Ireland* (2000, Belfast, Blackstaff); Colm Campbell *Emergency Law in Ireland 1918–1925* (Oxford, Clarendon Press, 1994).

[33] Kieran McEvoy and Ciaran White "Security Vetting in Northern Ireland: Loyalty, Redress and Citizenship" (1998) 61 *MLR* 341, at 343.

[34] See Shane O'Neill "Pluralist Justice and its Limits: The Case of Northern Ireland" (1994) 42 *Political Studies* 363, in a detailed critique of Walzer's communitarianism he argues that justice in Northern Ireland will arise only when participants develop a critically-reflexive stance towards their identities.

[35] J J Lee *Ireland 1912–1985: Politics and Society* (Cambridge, Cambridge University Press, 1989) p. 43. See Claire Palley "The Evolution, Disintegration and Possible Reconstruction of the Northern Ireland Constitution" (1972) *Anglo-American Law Review* 368, at 383–406.

[36] Government of Ireland Act 1920, s. 1(2): the counties of Antrim, Armagh, Derry, Down, Fermanagh and Tyrone. Northern Ireland Act 1998, s. 2, repeals the Government of Ireland Act 1920.

member House of Commons and a twenty-six-member Senate.[37] The Act provided for excepted, reserved and transferred matters. The method of dividing legislative powers has remained and can be found in the Northern Ireland Act 1998. The UK Parliament remained supreme, but due to the quick emergence of a constitutional convention governing the relationship between Westminster and Stormont, was "freed from legislative and executive responsibility over [Northern Ireland] and hence from the need to find time at Westminster to deal with and discuss Northern Ireland matters".[38] Representation at Westminster was reduced to thirteen seats.[39] This devolution scheme remained in place until events in 1968 began to call into question the "state" in Northern Ireland. The problems with the model are well-known. McCrudden argues that the settlement was not designed with local conditions in mind and reflected British experience and practice.[40] The British model of constitutionalism was thus imposed inappropriately in this local context. As McCrudden notes, the British constitutional experience failed Northern Ireland.[41] The failures of this constitutional order led to a serious breakdown of the system in the late 1960s. It is speculation, but one wonders how a differently constituted legal and political order might have fared.

The courts and the legal profession are dealt with in another contribution to this book.[42] The surprising fact is that in the period until the mid-1960s the courts were not used to challenge government action in controversial cases.[43] Palley argues:

> "The probable causes for this failure are a combination of the unavailability of legal aid until 1965, some unadventurousness by the local legal profession, an unawareness of the opportunities court machinery creates for manipulation of the political process, and a general tendency on the part of the political opposition to dismiss the courts as manifestations of the unionist establishment and as therefore unlikely to deliver judgments adverse to the interests of that establishment [footnotes omitted]".[44]

[37] Hadfield *supra* n. 30 at p. 49.

[38] Brigid Hadfield "Committees of the House of Commons and Northern Ireland Affairs" (1981) 32 *NILQ* 199, at 201.

[39] Hadfield *supra* n. 30 at p. 59.

[40] Christopher McCrudden "Northern Ireland and the British Constitution" in Jeffrey Jowell and Dawn Oliver (eds) *The Changing Constitution* (3rd edn, Oxford, Oxford University Press, 1994) p. 323.

[41] McCrudden *supra* n. 40 pp. 331–2.

[42] See Chapter 6.

[43] Palley *supra* n. 35 at p. 390.

[44] *Ibid.*

While some of the reasons for scepticism remained, those engaged in various forms of political struggle only became attuned to the potential of legal process much later.

The civil rights movement of the 1960s highlighted the discriminatory nature of the system and gave voice to demands for human rights protection and equality.[45] It was the reaction to these calls for reform which was ultimately to bring an end to the traditional order in Northern Ireland. Some attempts were made to recognise the grievances of the nationalist/republican community, for example, in the conclusions of the report of the Cameron Commission 1969.[46] The devolutionary scheme of the 1920s ended in chaos with the suspension of government under the Northern Ireland (Temporary Provisions) Act 1972.[47] The period from 1972 onwards was marked by various attempts to reinvigorate constitutionalism in Northern Ireland within the context of direct rule from Westminster.[48] In response to increased political violence there was a steady "securitisation" of society, with resulting state abuse of human rights.[49] These human rights violations have been a continuing source of controversy.[50] The security response was an inappropriate one given the nature of the conflict and did little if anything to assist the attempts to establish a stable political order. In

[45] *Disturbances in Northern Ireland: Report of the Commission appointed by the Governor of Northern Ireland* (Cmnd 532, 1969), para. 12 after referring to the Campaign for Social Justice and the Northern Ireland Civil Rights Association it states: "These organisations concern themselves with immediate social reforms, such as appointments to jobs and housing discrimination by unionists, support for universal adult franchise in local government elections and fairer electoral boundaries in local government. These are not concerned, as organisations, with altering the constitutional structure of Northern Ireland, and in this sense represent a quite new development amongst Catholic activists".

[46] *Ibid.* para. 127: "The weight and extent of the evidence which was presented to us concerned with social and economic grievances or abuses of political power was such that we are compelled to conclude that they had substantial foundation in fact and were in a very real sense an immediate and operative cause of the demonstrations and consequent disorders after 5th October 1968".

[47] R F Foster *Modern Ireland 1600–1972* (London, Penguin Books, 1988) p. 587.

[48] For an early example see *Northern Ireland Constitutional Proposals* (Cmnd 5259, 1973). From the beginning the British Government refused to impose a "rigidly defined and detailed scheme" (para. 13). The report stated: "There cannot be a 'government settlement', only a 'community settlement', and its full achievement will be a matter of years". Note that Chapter 4 of the Report is headed "A Charter of Human Rights".

[49] Note Boyle, Hadden and Hillyard *supra* n. 2 at p. 181: "But if there is a lesson to be learned from the five years of conflict from 1969 to 1974 it is that the choice of a security response to a crisis of intercommunal relations is likely to impede the restoration of stability".

[50] See, e.g., Amnesty International *Political Killings in Northern Ireland* (London, Amnesty International, 1994).

this context the emphasis was on the promotion of consensual constitutionalism between the parties, in a context arguably ill-suited to such a proceduralist orientation.[51] This was not aided by the attempts of successive British Governments to portray their role as essentially neutral. Serious constitutional thought on Northern Ireland did, however, shift away from majoritarian democracy towards consociationalism.

The effort to secure a settlement had varying levels of success. The Northern Ireland Assembly Act 1973 made provision for elections to a new unicameral Assembly of seventy-eight members elected on the basis of proportional representation.[52] The Northern Ireland Constitution Act 1973 contained detailed provisions on the powers of the Assembly. The Act also included anti-discrimination provisions confined to direct discrimination only.[53] The Sunningdale Agreement, concluded in December 1973 made provision for an Irish dimension with the proposed Council of Ireland.[54] On 1 January 1974 legislative responsibility was devolved to the Northern Ireland Assembly. The power-sharing Assembly was brought down by the Ulster Workers' Council strike in 1974, thus heralding the beginning of an extended period of direct rule.[55] During this period efforts were made to make the system more accountable and responsive. It was however generally regarded as inadequate and left a large democratic deficit in Northern Ireland.[56]

The attempts to reintroduce power-sharing were largely unsuccessful. The Northern Ireland Act 1982 was aimed at securing "rolling devolution" and in this regard failed.[57] Pending devolution the aim of the legislation was to subject direct rule to local scrutiny and thus inject a Northern Irish element into the process.[58] The Assembly which did

[51] Cf. McCrudden *supra* n. 40.

[52] Hadfield *supra* n. 30 at p. 104.

[53] See Northern Ireland Constitution Act 1973, ss. 17(1) and 19. The Act also established the Standing Advisory Commission on Human Rights to advise "the Secretary of State on the adequacy and effectiveness of the law for the time being in force in preventing discrimination on the ground of religious belief or political opinion and in providing redress for persons aggrieved by discrimination on either ground" (s. 20(1)(a)).

[54] Hadfield *supra* n. 30 at pp. 112–13.

[55] See Northern Ireland Act 1974.

[56] See Brigid Hadfield "The Northern Ireland Constitution" in Brigid Hadfield (ed) *Northern Ireland Politics and the Constitution* (Buckingham, Open University Press, 1992) pp. 1–12, at 6, she describes the consequences of direct rule as "at best disquieting and at worst deplorable".

[57] See C O'Leary, S Elliott and R Wilford *The Northern Ireland Assembly 1982–1986* (London, Hurst, 1988); Conor Gearty "The Northern Ireland Act 1982" [1982] *Public Law* 518.

[58] Gearty *ibid.* at p. 523: "If worked properly by the politicians, this new scheme would most probably improve direct rule".

function was boycotted by the SDLP and Sinn Féin, thus ensuring that it did not have cross-community support. It was dissolved in June 1986. Hadfield argues that it fulfilled a worthwhile function and did make direct rule more accountable.[59] There were, however, serious problems with this form of devolution.[60]

The 1980s were also to see increased focus on the Irish dimension. This was given formal expression in the Anglo-Irish Agreement 1985.[61] The Agreement established an Intergovernmental Conference to deal with political, security and related matters, legal matters including the administration of justice, and the promotion of cross-border cooperation. The Agreement attracted fierce opposition from the unionist community. It did however give formal recognition to the bi-national nature of the conflict and thus the importance of an Irish dimension to any constitutional settlement in Northern Ireland.

The UK and international dimensions

Northern Ireland does not exist in a vacuum, its problems are not unique, and it is not immune to wider national and international developments. The changing trends in British constitutionalism have also played a significant part. The election of a Labour Government in May 1997 was undoubtedly a factor in the success of the process. The new Government had a substantial majority in the House of Commons and thus was not reliant on the votes of unionist MPs. The Government could be confident of securing the passage of its legislative measures through Parliament. This was combined with a willingness to be pragmatic in the search for a settlement, evident in the work of the new Secretary of State for Northern Ireland, Majorie (Mo) Mowlam. As a democratic socialist political party it was comfortable with the progressive language of human rights and equality. The enactment of the Human Rights Act 1998 is an example of the willingness to legislate

[59] Hadfield *supra* n. 30 at p. 167.

[60] See P D H Smith "The Northern Ireland Assembly 1982–1986: The Failure of an Experiment" (1987) *Parliamentary Affairs* 482, at 499: "The Assembly did not succeed in its appointed task and it is difficult to avoid the conclusion that, even in 1982, the chances of promoting reconciliation via the debating chamber were slight".

[61] Tom Hadden and Kevin Boyle *The Anglo-Irish Agreement: Commentary, Text and Official View* (Dublin and London, Edwin Highel Ltd and Sweet and Maxwell Ltd, 1989); Brigid Hadfield "Anglo-Irish Agreement 1985—Blue Print or Green Print?" (1986) 37 *NILQ* 1.

positively for human rights protection. The decentralisation of power was also part of the constitutional reform agenda and thus devolution to Scotland, Wales and Northern Ireland was actively pursued. In Northern Ireland there was nothing new about the attempt to secure devolution, but now it could also be presented as part of a process of constitutional change in the United Kingdom generally. Constitutional reform in the United Kingdom did rule some options out. Reintegration, a position argued for by some unionists, no longer made sense in the new era of decentralisation.

The international dimension must be accorded due weight, although it can prove difficult to assess in precise terms.[62] The role of the USA was significant, a fact demonstrated by increasing resort by political actors to this audience. It is yet to be seen how Northern Ireland will fare in relation to some of the less desirable aspects of globalisation and American hegemony.[63]

Postnational thinking in the European Union was also important in shaping the discourses which made compromise possible. The human rights movement in Northern Ireland in particular saw the significance of using international networks to lend weight to arguments for change. Although part of general trends in transnational communication, it required political action to bring this about. No matter how insular Northern Ireland may appear at times, it exists within an expanding network of transnational institutions and processes. One challenge for the future will be to assess precisely the impact of these networks.

A CONSTITUTIONAL SETTLEMENT

Talking about peace

Many of the ideas which were to form the basis for the settlement in 1998 were around for some time. What had altered by this point was the ideology of some of the dominant political actors. As Todd notes,

62 See Michael Cox "Northern Ireland: The War that came in from the Cold" (1998) 9 *Irish Studies in International Affairs* 73; Joseph Ruane and Jennifer Todd *The Dynamics of Conflict in Northern Ireland* (Cambridge, Cambridge University Press, 1996) pp. 266–9.

63 While openness to American influence has been a key aspect of the political process, resistance will also be important in the future, see Fredric Jameson "Globalization and Political Strategy" (2000) *New Left Review* (July/August) 49.

this is particularly true of nationalist/republican ideology.[64] Devolution has been the preference of some of the main political parties for a considerable period.[65] In the 1990s substantially more progress was made in securing a settlement which moved in this direction.

The dominant principle of the talks process of the 1990s was "inclusivity".[66] The idea here was, at some point and under specified conditions, to bring all political parties together to reach a negotiated settlement. This would be a deliberative process with clear structuring rules. It is evident that channels of public and private communication with all groups, including paramilitaries, had existed for some time. There were also important discussions between the political parties. John Hume, the leader of the SDLP, initiated a dialogue with Gerry Adams, the President of Sinn Féin, in January 1988. Todd is correct to note the fundamental role of John Hume in crafting the political language that made compromise possible.[67] The purpose of this, and subsequent talks, was to persuade the republican movement of the illegitimacy of continuing armed struggle. In April 1993 they issued a joint statement on their interpretation of the right to national self-determination and in September 1993 adopted a report which contained proposals for further discussion. The Hume-Adams agreement was a significant development. A response came from the British and Irish Governments in the form of the Downing Street Declaration 1993. The Declaration committed the Governments to working towards a new political agreement founded on the principle of consent, to encompass arrangements in Northern Ireland, the whole of Ireland, and between Britain and Ireland.[68] The three-stranded approach, agreed in the early 1990s, was thus given explicit recognition in the document. On self-determination the British Government agreed that it was for the people of Ireland, by agreement between the two parts, to exercise their right of self-determination on the basis of consent, freely and concurrently given to bring about a united Ireland if they wished.[69] The Irish Government accepted that the right of self-determination could only be

[64] See Jennifer Todd "Nationalism, Republicanism and the Good Friday Agreement" in Joseph Ruane and Jennifer Todd (eds) *After the Good Friday Agreement: Analysing Political Change in Northern Ireland* (Dublin, University College Dublin Press, 1999) p. 49.

[65] Hadfield *supra* n. 30 at p. 229.

[66] Paul Bew, Henry Patterson and Paul Teague *Between War and Peace: The Political Future of Northern Ireland* (London, Lawrence and Wishart, 1997) p. 203.

[67] *Supra* n. 64 at p. 53.

[68] Downing Street Declaration 1993, para. 2.

[69] *Ibid.* para. 4.

achieved subject to the consent of a majority of people in Northern Ireland.[70] In the result of an overall settlement the Irish Government agreed to bring forward proposals for a change in the Constitution of Ireland that reflected the principle of consent.[71] The two Governments also committed themselves to give institutional recognition to the links between the people of Britain and Ireland as part of the "totality of relationships".[72] In August 1994 the PIRA declared a cease-fire and the Combined Loyalist Military Command followed in October. This was another major contributing factor to the eventual settlement.

The next major step in the process was the publication of the Framework Document in February 1995. This contained the following documents: "A Framework for Accountable Government in Northern Ireland"; and "A New Framework for Agreement". The documents contained detailed proposals on a new structure of governance. In particular, these provided for: a single unicameral Assembly of ninety members elected by proportional representation; a separate panel of three elected in Northern Ireland to supplement the work of the Assembly; a committee system constituted in line with party strength to oversee departmental activity; a system of detailed checks and balances; mechanisms to ensure compliance with the United Kingdom's EU and international obligations; establishment of a North/South body with consultative, harmonising and executive powers; and a standing Intergovernmental Conference to acknowledge the importance of East/West relationships. On the protection of human rights, the document recognised the consensus that existed on the significance of rights protection in Ireland as a whole. It encouraged representatives from the North and South of Ireland to adopt a Charter or Covenant which would endorse and reflect protections for everyone living on the island of Ireland. This was an important development and many of the ideas were to find their way into the final Agreement.

In order to overcome problems with the process a "twin-track" initiative was launched in November 1995. The aim was to try to make progress with all-party talks in parallel with decommissioning. In order to address the issue of decommissioning of paramilitary arms the Government established an International Body on Decommissioning. It issued its first report in January 1996. The report included six principles of democracy and non-violence which were drafted to address

70 *Ibid*. para. 5.
71 *Ibid*. para. 7.
72 *Ibid*. para. 9.

the problems resulting from the reasonable concerns of both sides. The issue of decommissioning continued to impact on the process. In particular, it caused problems in relation to devolution. However, the lack of agreement on decommissioning impeded progress throughout the process.

In June 1996 multi-party talks commenced in accordance with rules formulated by the two Governments and after elections. The aim was to begin substantive negotiations around the "three-stranded approach" with the ambition of securing a settlement.[73]

Securing the Good Friday Agreement

On entering the talks all the political parties committed themselves to the Mitchell Principles of democracy and non-violence. These were set out in the report of the International Body on Decommissioning. The rules of procedure for the talks were agreed in July 1996 and amended in September 1997.[74] Substantive negotiations were held up again by the decommissioning issue, and it was not until September 1997 that substantive negotiations began on the three strands. In January 1998 the two Governments advanced a series of "Propositions on Heads of Agreement" intended to give added focus to the ongoing talks. In the same month they circulated proposals on Strands Two and Three, after meeting with the participants in London. The talks intensified, with significant levels of involvement from the British and Irish Governments, until the Agreement was concluded on 10 April 1998.

The three-stranded approach proved to be a useful device. In fact it structures significant sections of the Good Friday Agreement 1998.[75] The Agreement was concluded on 10 April 1998. In the referenda which followed on 22 May 1998 it secured a 71.12 per cent "yes" vote in Northern Ireland and a 94.4 per cent "yes" vote in the Republic of

[73] There were eight key political parties at the talks: the Alliance Party; the Northern Ireland Women's Coalition; the Progressive Unionist Party; Sinn Féin; the Social Democratic and Labour Party; the Ulster Democratic Party; the Democratic Unionist Party; and the Ulster Unionist Party. Sinn Féin was admitted in September 1997 because of a renewed PIRA cease-fire. The DUP left in July 1997 and did not return to the talks. There were a number of other brief exclusions during the process.

[74] The three independent Chairpersons at the talks were: Senator George Mitchell; General John de Chastelain; and Harri Holkeri.

[75] (Cm 3883, 1998). See Brigid Hadfield "The Belfast Agreement, Sovereignty and the State of the Union" [1998] *Public Law* 599.

Ireland.[76] It was translated into law in the form of the Northern Ireland Act 1998 where it is clearly stated that the purpose is to implement the Agreement. Both its content, and the popular endorsement, give it a level of legitimacy not secured for any previous settlement. While the process of drafting the text was largely confined to the political parties, and the British and Irish Governments, people in the North and South of Ireland were given the option of accepting or rejecting it. The legitimacy of the arrangement stems not simply from the popular endorsement it received but also its substantive provisions. What is important is not only that a majority of people on the island of Ireland supported the deal, but that the content reflects a recognition of local conditions and is based on normative principles of good governance.

Much of the content of the Agreement borrows from old ideas and themes. What mattered here was the changing relationship between participants and the movement made by key political actors. The Agreement is structured as follows: Declaration of Support; Constitutional Issues; Strand One; Strand Two; Strand Three; Rights, Safeguards and Equality of Opportunity; Decommissioning; Security; Policing and Justice; Prisoners; Validation, Implementation and Review. There is an emphasis on new beginnings and a fresh start. For example, the Declaration of Support states:

> "The tragedies of the past have left a deep and profoundly regrettable legacy of suffering. We must never forget those who have died or been injured, and their families. But we can best honour them through a fresh start, in which we firmly dedicate ourselves to the achievement of reconciliation, tolerance, and mutual trust, and to the protection and vindication of the human rights of all".

Elsewhere in the Declaration the participants pledge to work in good faith to ensure the success of the new arrangements. On Constitutional Issues the participants endorsed the commitment of the British and Irish Governments to a new British-Irish Agreement which would replace the Anglo-Irish Agreement. The Agreement sets out principles which should be recognised. Key to this is what is termed the consent principle. This means that the constitutional status of Northern Ireland is dependent on the wishes of the majority of people there. Any change to the status of Northern Ireland is thus tied to the consent of a majority

[76] See Michael O'Neill "'Appointment with History': The Referenda on the Stormont Peace Agreement, May 1998" (1999) 22 *West European Politics* 160; Sydney Elliott, "The Referendum and Assembly Elections in Northern Ireland" (1999) 14 *Irish Political Studies* 138.

of its people. This has traditionally attracted criticism from those who question the legitimacy of the "state in Northern Ireland". The border was created in order to ensure unionist hegemony and this fact has been used to question the ethical basis of the consent principle. The position, while based on historical facts, reflects a questionable view of the rigidity of community identification. A modern "republican" analysis, borrowing from deliberative democratic thinking, might lend rather more weight to this deliberative component of the constitution of Northern Ireland. By basing the constitutional position of Northern Ireland on the principle of consent the United Kingdom is clearly opening the door to a future alteration in this status. The aim of bringing about a united Ireland is recognised as legitimate, where this right of self-determination is based on the consent of the people of Ireland freely and concurrently given. Again, the right is tied to the consent of the majority in Northern Ireland. Where the right of self-determination is carried out in accordance with the Agreement there is a binding obligation on both Governments to introduce legislation to give effect to this wish. Whatever choice is made the Agreement commits the Government exercising jurisdiction to "rigorous impartiality" and "full respect for, and equality of, civil, political, social and cultural rights". The choice of individuals in Northern Ireland to be identified as either Irish or British is also to be respected. The Annexes to this section include draft clauses for incorporation in British legislation and draft legislation for the Irish Government with respect to amendment of the Irish Constitution. In relation to the Irish Government the Agreement contains new Articles to be included in the Constitution.

The other strands of the Agreement are discussed in more detail below. This section on Constitutional Issues provides a framework for the rest of the Agreement by establishing the mechanisms which define the constitutional status of Northern Ireland. By giving express recognition to both British and Irish citizenship it respects the bi-national nature of the problem. The Agreement does not go as far as, for example, a joint sovereignty solution might go in recognising this. In constitutional law terms Northern Ireland remains, for now, a part of the United Kingdom.[77] And as I have noted this is a system undergoing change. There is a certain level of instability built into this relationship

[77] Northern Ireland Act 1998, s.1(1): "It is hereby declared that Northern Ireland in its entirety remains part of the United Kingdom and shall not cease to be so without the consent of a majority of the people of Northern Ireland voting in a poll held for the purposes of this section in accordance with Schedule 1".

precisely because this is subject to the consent principle.[78] This principle can be portrayed as a form of solid guarantee of the constitutional status of Northern Ireland. But this is a misreading. The current political preferences need not be perpetuated into the future and one can foresee a time when the constitutional status of Northern Ireland will be re-opened. In other words, the constitutional status of Northern Ireland remains open to contestation, but this is a conflict confined in the future to something approaching the force of the better argument. For example, at some future date significant sections of the unionist community may decide that their interests are served best within an all-Ireland constitutional system. My point here is that the contest over the status of Northern Ireland has not ended. It is the way that the contest will be conducted that has fundamentally altered. While one option may be to hope that the status issue dissolves into the mists of post-nationalism, there is little evidence to suggest that it will fade away.

Argument over the legitimacy of the "state" in Northern Ireland aside for a moment, what is significant in the legal recognition of the consent principle is the strong deliberative component. A proportion of the population of the United Kingdom is given the opportunity to opt out if it wishes. For those interested in the unification of Ireland the struggle becomes an essentially political contest to secure this through agreement. In addition, to state baldly that Northern Ireland is, in constitutional law terms, just like any other part of the United Kingdom is another misreading of the Agreement. Irish people in Northern Ireland are accorded rights which attach to their identity and ethos which go beyond that extended elsewhere in the United Kingdom. There are substantial bi-national elements to the Agreement which move us beyond traditional thinking about constitutional law and practice. This is linked to the constitutive nature of the Agreement. It is not evident that this is fully understood, for example, in the treatment of the Irish population in Northern Ireland. The devolutionary scheme in Northern Ireland is recognition of past failings and is based on explicit normative principles and constraints which govern devolution. Most significantly it is distinct, as it is part of a constitutional settlement which followed a long period of violent conflict. To try to fit this neatly into the

[78] *Ibid*. s. 1(2): "But if the wish expressed by a majority in such a poll is that Northern Ireland shall cease to be part of the United Kingdom and form part of a united Ireland, the Secretary of State shall lay before Parliament such proposals to give effect to that wish as may be agreed between Her Majesty's Government in the United Kingdom and the Government of Ireland". See also *ibid*. Sch 1.

traditional mould is questionable. In terms of deliberative democratic thought, the Agreement is a new constitutive act with a normative status which confronts traditional UK constitutional law directly.

REACHING AGREEMENT

The theatre of the absurd: decommissioning and devolution

The idea of devolution is not new in Northern Ireland. The "state" which emerged in the 1920s was a devolved administration. It is the nature of devolution, and in particular the structuring rules, which have altered over time. Strand One of the Agreement contains detailed provisions on Democratic Structures in Northern Ireland which are examined in this section. On a number of occasions it appeared as if devolution might not take place. The principal problem was the linkage made between the decommissioning of paramilitary arms and devolution of power. The Agreement provided for transitional arrangements to ensure that several organisational matters were effectively dealt with.[79] The Assembly could thus operate in shadow form but the necessary devolution order could only be laid before Parliament if the Secretary of State believed that "sufficient progress [had] been made in implementing the Belfast Agreement".[80] It was with the implementation of the Agreement that problems arose.

The establishment of the International Body has already been mentioned. The Agreement contains a section on decommissioning which states:

> "All participants reaffirm their commitment to the total disarmament of all paramilitary organisations. They also confirm their intention to continue to work constructively and in good faith with the Independent Commission, and to use any influence they may have, to achieve the decommissioning of all paramilitary arms within two years from endorsement in referendums North and South of the agreement and in the context of the implementation of the overall settlement".

The interpretation of this section has been exhaustively debated. It appears that the use of "vague language" was an attempt to keep Sinn

[79] Strand One para. 35.
[80] Northern Ireland Act 1998, s. 3.

Féin in the talks.[81] The matter was complicated by a letter from Tony Blair to David Trimble confirming the unionist interpretation of this paragraph.[82] This letter is thought to have been influential in the UUP's acceptance of the Agreement. The two years which followed the adoption of the Agreement saw several attempts to devolve power in an acceptable and stable way. The Omagh bomb attack on 15 August 1998 by republicans, which killed twenty-nine people, was a grim reminder that the physical force tradition had not gone away.

On 25 June 1998 Assembly elections were held by proportional representation from the existing eighteen Westminster constituencies and 108 members returned.[83] On 1 July 1998 the Assembly met for the first time in shadow form. David Trimble (UUP) and Séamus Mallon (SDLP) were jointly elected as First and Deputy First Minister designate (FM-DFM). The shadow Assembly made progress on a number of important areas during this phase. Significant aspects of the Agreement were translated into law in the form of the Northern Ireland Act 1998. The problem with devolution was finding an appropriate time to make the Devolution Order. A first devolution deadline was set for 10 March 1999 but was not met. The talks continued between the UUP and Sinn Féin without success. The two Governments issued a Declaration (the Hillsborough Declaration) on 1 April 1999 which was presented to the political parties as the best way forward towards implementation. The Declaration stated that: "[T]here is agreement among all parties that decommissioning is not a precondition but is an obligation deriving from their commitment in the Agreement".

The document proposed that on a date to be established nominations would be made of Ministers to take up office. On a date proposed by the Independent Commission, but not later than one month after nomination, a collective act of reconciliation would take place. This would see

[81] Paul Bew and Gordon Gillespie *Northern Ireland: A Chronology of the Troubles 1968–1999* (Dublin, Gill and Macmillan, 1999) p. 358.

[82] *Ibid.*: "I understand [that] your problem with Paragraph 25 of Strand I is that it requires decisions on those who should be excluded or removed from office in the Northern Ireland Executive to be taken on a cross-community basis. This letter is to let you know that if, during the course of the first six months of the Shadow Assembly or the Assembly itself, these provisions have been shown to be ineffective, we will support changes to those provisions to enable them to be made properly effective in preventing such people from holding office. Furthermore, I confirm that in our view the effect of the decommissioning section of the Agreement, with decommissioning schemes coming into effect in June, is that the process of decommissioning should begin straight away".

[83] The elected representatives are known as Members of the Legislative Assembly (MLA). See Northern Ireland (Elections) Act 1998.

paramilitary arms put beyond use on a voluntary basis. The Commission would report on progress within one month of nominations and if progress had not been made then the nominations would not be confirmed by the Assembly. The Declaration was unsuccessful and the talks continued. The result was the imposition of a further deadline of 30 June 1999 as a date for agreement on the terms which would lead to the devolution of power. The aim of the British Government was to have devolution in Northern Ireland, Scotland and Wales at roughly the same time.[84] Again this failed and the result was another document. The "Way Forward Document" was circulated to the political parties after five days of talks at the end of June and the beginning of July. The document contained proposals to bring about devolution by running the D'Hondt mechanism for executive formation on 15 July 1999. The ambition was for the Independent Commission to report in September and December 1999 and in May 2000 on progress. The document contained reference to a "fail-safe" mechanism and much effort was made to ensure that this passed through its parliamentary stages expeditiously. This initiative again ended in failure, as the UUP did not engage in the process of executive formation when the Assembly met on 15 July 1999. In line with last minute alterations to the Assembly's standing orders the executive was not validly constituted.

A "tightly focused" review process[85] began on 6 September 1999 and ended on 19 November 1999, again chaired by Senator George Mitchell. After this review the Secretary of State announced that he would convene a meeting of the Assembly on 29 November 1999. The Assembly met and nominated ten Ministers, as well as Chairpersons, Deputy Chairpersons and members of the ten. Departmental Committees. Powers were devolved to the Northern Ireland Assembly on 2 December 1999. Devolution was not to last and on 11 February 2000 it was suspended by the Secretary of State.[86] The reasoning behind

[84] Power was devolved to the Scottish Parliament and the Welsh Assembly on 1 July 1999.

[85] The mandate was as follows: "The review will take as its starting point the three principles agreed by all pro-Agreement parties on 25 June: an inclusive Executive; decommissioning of all paramilitary arms by May 2000; decommissioning to be carried out in a manner determined by the International Commission on Decommissioning; and will determine how to overcome the difficulties which exist in the practical implementation of those principles. This will be its only focus".

[86] Northern Ireland Act 2000. The Act provides for the making of a "Restoration Order" by the Secretary of State. It provides for the automatic re-appointment of Ministers and Chairpersons and Deputy Chairpersons of Committees to their previous positions after the making of the Restoration Order.

the suspension given by the British Government was the protection of the Agreement. In particular, its aim was the political survival of the First Minister, David Trimble in the light of the lack of progress at the time on decommissioning. The decision to suspend devolution was criticised by the SDLP and Sinn Féin.[87] Various attempts were again made following this to reach an agreement on the issue.

On 5 May 2000 the British and Irish Governments issued a statement which provided that "subject to a positive response to this statement the British government will bring forward the necessary order to enable the Assembly and Executive to be restored by May 22, 2000". The statement went on to call on paramilitary organisations to "put their arms completely and verifiably beyond use" and made positive reference to the continuing role of the Independent International Commission on Decommissioning. On 6 May 2000 the two Governments sent a letter to all the political parties in Northern Ireland setting out proposals necessary to secure full implementation of the Good Friday Agreement. This included a commitment by the Irish Government to establish the Irish Human Rights Commission in July 2000 as well a detailed commitment on security, policing, justice and prisoners. On the same day the PIRA issued a statement which committed the organisation to initiating a process "that will completely and verifiably put IRA arms beyond use". The organisation agreed to resume contact with the Commission on Decommissioning and allow a number of arms dumps to be inspected by agreed third parties. These third parties would report back to the Commission and regularly re-inspect the dumps. In the afternoon of 6 May 2000, in a Joint Statement by the Taoiseach and the Prime Minister, it was announced that Martii Ahtisaari[88] and Cyril Ramaphosa[89] would be the "agreed third parties". This succeeded in securing the restoration of devolved power.

Does the Agreement have a nature?

Northern Ireland has generated a massive literature and many interpretations and solutions have been proposed.[90] One issue worth

[87] *The Irish Times*, 14 February 2000.

[88] Chairperson of the International Crisis Group and the former President of Finland.

[89] A board member of the International Crisis Group and former Secretary-General of the African National Congress.

[90] See generally John Whyte *Interpreting Northern Ireland* (Oxford, Clarendon Press, 1990).

examining is which of the proposed solutions does the Agreement most closely resemble. In other words, does the Agreement have a nature? It is therefore time to step back from the detail of the process and the Agreement. What is the nature of the Agreement? What sort of conception of democracy does it represent? These are questions which demand some response. It is well established that any empirical work in law and politics has a theoretical basis. Criticising theory makes no sense, as any selection of relevant legal or political facts will be based on some theoretical premises. Routine rejection of theory by legal scholars or practitioners is an exercise in deception and concealment. Even pragmatism comes in a variety of guises. This is moving away from the main issue, but a defence of the importance of theory to legal analysis is needed in the Northern Ireland context.

At the most general level the Agreement is an attempt to transform a violent conflict into an exclusively political conflict. The aim is not to eradicate conflict entirely. In human history conflict is the norm and not the exception.[91] People disagree about almost every aspect of social, political and cultural life. Northern Ireland is thus no exception to this general rule. The Agreement is an example of a mechanism aimed at giving practical recognition to a well-established violent conflict in Northern Ireland. The end result is a settlement with substantial similarities to previous proposals. The main difference rests not in the substance of the settlement but in the fact that those engaged in political violence ended their campaigns and became, on the whole, participants in the political process. The solution is a long way from Irish unity and a substantial distance from reintegration. In constitutional law Northern Ireland remains part of the United Kingdom. On one reading the complex mechanisms created are an attempt to blur absolutist notions of sovereignty and thus make this basic fact more palatable to the nationalist/republican community. This is a rather misleading reading of the sudden taste for postnationalism in official circles in Northern Ireland. While it is correct to recognise that it is the republican movement which has travelled the furthest down the road of political compromise, as I argue here Northern Ireland remains contested terrain. The focus should be on the deliberative aspects of the settlement. My argument is that contrary to what is often said, the

[91] See Claire Palley "Constitutional Devices in Multi-Racial and Multi-Religious Societies" (1968) 19 *NILQ* 377, at 380: "Throughout recorded history conflict has been the natural condition of mankind, the only difference between various epochs being that dominating groups have justified their aggression by reference to differing criteria".

status of Northern Ireland remains contested by the nature of the normative principles upon which it is based.

What we have is democratic governance tailored to the particular needs of a divided society. Given Northern Ireland's past it was likely that the conflict between the two main political communities would dominate any settlement. This has proved to be the case. Although the Agreement is not always consistent on this, there is a strong commitment to express recognition of the identity and ethos of "both communities". The criticism of this approach is that it legitimises and thus perpetuates existing divisions. Rather than transcending difference the model legitimises it. It is said to be deficient in other, connected, areas also. For example, the politics of gender, and thus differences within "communal blocks", can be seriously downplayed.

O'Leary has provided a decisive answer to the question posed. He argues that it is a consociational agreement.[92]

> "In other words, it is a political arrangement that meets all four of the criteria laid down by that doyen of political science, the Dutchman Arend Lijphart: cross-community executive power-sharing, proportionality rules applied throughout the relevant governmental and public sectors, community self-government or autonomy and equality in cultural life and veto rights for minorities [footnotes omitted]".[93]

He recognises that it is not only a consociational agreement and that it has important external dimensions.[94] Much of his argument is convincing, but he is prone to use the word legalism rather too often as an accusation. What he presumably means to criticise is orthodox legal positivism, which does not pay due regard to law's role as a tool in the service of democratic experimentalism. It is beyond the bounds of this chapter to defend fully a theory of legality which would meet O'Leary's criticisms. Suffice to say that there are values embedded in legalism which radicals and conservatives alike would not wish to lose.

The concept of democracy which the Agreement resembles is that of consociational democracy, as O'Leary argues. But this is far from the only element. For example, in the longer term the deliberative

[92] See Brendan O'Leary "The Nature of the Agreement" (1999) 22 *Fordham International Law Journal* 1628. Cf. Rick Wilford "Designing the Northern Ireland Assembly" (2000) 53 *Parliamentary Affairs* 577; Rick Wilford "The Good Friday Agreement Revisited" (2000) 3 *Contemporary Issues in Irish Law and Politics* 1.

[93] *Ibid*. p. 1630.

[94] *Ibid*. p. 1631.

component may be the one that has the lasting impact.[95] In particular, the stress on the right to participate is of fundamental significance. I would argue that the Agreement is informed by a conception of deliberative democracy and that the focus on participation is evidence of this.

THE NORTHERN IRELAND SETTLEMENT

The approach taken here is first to outline the provisions of the Agreement in relation to the new government structures. While the Agreement's legal status is a matter for debate, the legal framework that has emerged is intended to give legal life to its provisions. It has a foundational status. But like all processes of constitution-building it has a dual aspect. There is first the foundational moment, and secondly, the continuing process of interpretation and application. Given the concept of constitutionalism I have defended here, one of the intriguing issues is the extent to which the constitutive moment itself should be "up for grabs" and subject to renegotiation and thus possible outright rejection.

I argue in this chapter that the Agreement has a constitutive status. While it has been translated into law in the Northern Ireland Act 1998, I deliberately refer principally in this section to the Agreement. The aim is to reinforce the argument that the Agreement underpins all consideration of the new constitutional structures in Northern Ireland.

Executive authority

The Agreement is clear on the fact that the prime source of authority on all devolved responsibilities rests with the Northern Ireland Assembly.[96] The Agreement provides that "executive authority [is] to be discharged on behalf of the Assembly by the First Minister and Deputy First Minister and up to ten Ministers with Departmental Responsibilities".[97] The Office of the First Minister and Deputy First

[95] On the position prior to the Agreement see John Morison and Stephen Livingstone *Reshaping Public Power: Northern Ireland and the British Constitutional Crisis* (London, Sweet and Maxwell, 1995).

[96] Strand One para. 4.

[97] Strand One para. 14; Northern Ireland Act 1998, Part III. Contrast the Agreement with the Northern Ireland Act 1998, s. 23(1): "The executive power in Northern Ireland

Minister (OFM-DFM), a Department of the Northern Ireland executive, is thus central to the new governmental structures. Two junior Ministers have been appointed to assist the FM-DFM.[98] The First Minister and Deputy First Minister and the ten Ministers form the Executive Committee.[99] The former are elected jointly by the Assembly on a cross-community basis[100] and the latter positions are allocated to the political parties on the basis of the d'Hondt system.[101] The Executive Committee is a key feature of the new governmental structures. It provides a forum for discussion of important issues and is involved in prioritising executive and legislative proposals and in recommending common positions.[102] The Executive Committee is responsible for agreeing a Programme for Government which is subject to the approval of the Assembly and following scrutiny in the Assembly Committees.[103] It had its first meeting following the restoration of devolution on 1 June 2000. An Agenda for Government was soon adopted and in October 2000 the first draft Programme for Government was published.

Ministers and departments

There are effectively eleven Departments.[104] The number of Departments, the Implementation Bodies and the Matters for Co-operation were agreed on 18 December 1998. It is for the FM-DFM to determine the number of Ministerial Offices and the functions to be

shall continue to be vested in Her Majesty". Detailed provision for the practical operation of the Assembly can be found in the Standing Orders.

98 See Northern Ireland Act 1998, s. 19: they are Denis Haughey (SDLP) and Dermot Nesbitt (UUP).

99 Strand One para. 17; Northern Ireland Act 1998, s. 20.

100 Strand One para. 15 and para. 5(d); Northern Ireland Act 1998, s. 16. The voting mechanism is parallel consent: "a majority of those members present and voting, including a majority of the unionist and nationalist designations present and voting".

101 Strand One para. 16. This is a system that allocates places proportionally according to the strength of the political party.

102 Strand One para. 19.

103 Strand One para. 20.

104 Northern Ireland Act 1998, ss. 17–18; Departments (Northern Ireland) Order 1999, SI 1999/283. The Departments are: Office of the First Minister and Deputy First Minister; Agriculture and Rural Development; Culture, Arts and Leisure; Education; Enterprise, Trade and Investment; Environment; Finance and Personnel; Health, Social Services and Public Safety; Higher and Further Education, Training and Development; Regional Development; Social Development.

exercised by the office holders.[105] The Ministers are allocated according to the d'Hondt formula. The Agreement states that all Ministers should liaise regularly with their Committee.[106] The power of Ministers is limited in similar ways to the general restrictions on legislative competence.[107] A Minister does not have the power to do any act which "aids or incites another person to discriminate against a person or class of person on that ground".[108] As a condition of appointment all Ministers are required to affirm a Pledge of Office.[109] This Pledge of Office includes the obligation:

> "(c) to serve all the people of Northern Ireland equally, and to act in accordance with the general obligations on government to promote equality and prevent discrimination".[110]

There is, in addition to this, a Ministerial Code of Conduct.[111] This includes an obligation to "operate in a way conducive to promoting good community relations and equality of treatment".[112] Ministers have full executive authority in their respective areas but this is authority within the Programme for Government agreed by the Executive Committee and endorsed by the Assembly as a whole.[113] There is a mechanism for the removal of a Minister from Office:

> "if (s)he loses the confidence of the Assembly, voting on a cross-community basis, for failure to meet his or her responsibilities including, inter alia, those set out in the Pledge of Office".[114]

Exclusion is the most extreme form of sanction. Nevertheless it is an option and there is provision for the exclusion of Ministers or junior Ministers for twelve months.[115] A Minister could be excluded for a failure to observe terms of her or his Pledge of Office, thus giving the

[105] Northern Ireland Act 1998, s. 17(1).
[106] Strand One para. 22.
[107] Northern Ireland Act 1998, s. 24.
[108] *Ibid*. s. 24 (1)(d).
[109] Strand One para. 23 and Annex A.
[110] Annex A. See Northern Ireland Act 1998, s. 18(8): "A Northern Ireland Minister shall not take up office until he has affirmed the terms of the pledge of office".
[111] Annex A.
[112] *Ibid*.
[113] Strand One para. 24.
[114] Strand One para. 25.
[115] Northern Ireland Act 1998, s. 30(1).

Pledge added significance.[116] The Assembly has the power to extend a period of exclusion or to bring it to an end.[117]

Committees

For each of the new Departments there is an Assembly Committee.[118] The Chair and Deputy Chair of the Committees are allocated proportionately, again using the d'Hondt mechanism.[119] The aim is to ensure that membership of the Committees is allocated according to party strengths.[120] The Committees have four broad roles: scrutiny; policy development; consultation; and in relation to the initiation of legislation.[121] Their specific powers include: consideration and advice on Departmental budgets and plans; approval of secondary legislation; the power to call persons and papers; and to initiate enquiries and make reports.[122] In addition to the Departmental Committees there are other Standing Committees such as the Committee of the Centre, the Audit Committee and the Committee on Standards and Privileges.[123] The Committee of the Centre was established to examine and report on the work of the OFM-DFM in several areas including: Economic Policy Unit; Equality Unit; Civic Forum; Victims; Women's Issues; and Freedom of Information.[124]

Legislative powers

Even following devolution the Westminster Parliament retains the legal authority to legislate for Northern Ireland. In traditional UK constitutional law terms this remains the position. However, constitutional

[116] It is not only Ministers who can be excluded. Provision is made for the exclusion of a political party because: "it is not committed to non-violence and exclusively peaceful and democratic means . . . or . . . because it is not committed to such of its members as are or might become Ministers or junior Ministers observing the other terms of the pledge of office".

[117] Northern Ireland Act 1998, s. 30(3) and (4).

[118] Strand One para. 8; Northern Ireland Act 1998, s. 29.

[119] Strand One para. 8.

[120] *Ibid.*

[121] Strand One para. 9.

[122] *Ibid.*

[123] Also Committee on Procedures, Business Committee and Public Accounts Committee.

[124] The first meeting of the Committee of the Centre took place on 26 January 2000.

conventions can emerge quite quickly to govern the relationship between Westminster and the Northern Ireland Assembly. For example, as Palley notes in relation to the Government of Ireland Act 1920:

> "[S]oon after powers had been conferred on the Parliament of Northern Ireland, constitutional conventions developed as to the exercise of Westminster's concurrent powers. It became convention that Parliament at Westminster would not legislate in respect of transferred powers . . . without the consent of the Government of Northern Ireland".[125]

It is likely that similar constitutional conventions will develop in relation to the current devolutionary scheme also.

The Assembly has the authority to pass primary legislation for Northern Ireland in the devolved areas.[126] The laws of the Assembly are to be known as Acts[127] and to become law they must be passed by the Assembly and receive Royal Assent.[128] This legislative power is subject to a number of controls, including human rights protections.[129] Issues are divided into: excepted matters;[130] reserved matters;[131] and transferred matters.[132] Excepted matters are non-devolved issues, while reserved matters are those which may at some future time be devolved. Transferred matters are those which are within the jurisdiction of the Assembly. The Secretary of State has the power, with the consent of the Assembly, to make a reserved matter a transferred one and *vice versa*.[133]

The Agreement clearly envisages disputes over legislative competence being decided by the courts.[134] But detail is lacking. The Northern Ireland Act 1998 is clearer on this issue. For example, where a court or tribunal decides that a provision of an Act is not within the legislative competence of the Assembly it may make an order which removes or limits the retrospective effect of the decision or suspends its

[125] See Palley *supra* n. 35 at p. 385.

[126] Strand One para. 26; Northern Ireland Act 1998, Part II.

[127] Northern Ireland Act 1998, s. 5(1).

[128] *Ibid.* s. 5(2).

[129] Strand One para. 26(a).

[130] Northern Ireland Act 1998, Sch 2. This includes: tax; nationality, immigration and asylum; elections; and the appointment and removal of judges.

[131] *Ibid.* Sch 3. This includes: criminal law; consumer safety in relation to goods; civil defence; and human genetics.

[132] *Ibid.* s. 4(1).

[133] *Ibid.* s. 4(2) and (3).

[134] Strand One para. 28: "Disputes over legislative competence will be decided by the Courts".

effect until the defect is corrected.[135] The court or tribunal must take into account the extent to which third parties would be adversely effected by the decision.[136] If it is considering making such an order it must give notice to the Attorney General.[137] Other rules apply where the matter is a "devolution issue".[138] Where there is a choice as to permissible interpretations on the issue of competence, there is an interpretative obligation to read an Act of the Assembly "in a way which makes it within that competence or, as the case may be, does not make it invalid by reason of that section".[139]

The final court of appeal on a devolution issue is the Judicial Committee of the Privy Council.[140] It is clear from all the devolution schemes that this body may have an important role to play in resolving issues that arise. The term devolution issue is defined in Schedule 10. It means:

> "(a) a question whether any provision of an Act of the Assembly is within the legislative competence of the Assembly;
> (b) a question whether a purported or proposed exercise of a function by a Minister or Northern Ireland department is, or would be, invalid by reason of section 24;
> (c) a question whether a Minister or Northern Ireland department has failed to comply with any of the Convention rights, any obligation under Community law or any order under section 27 so far as relation to such an obligation; or
> (d) any question arising under this Act about excepted or reserved matters".[141]

There is detailed provision for dealing with proceedings in Northern Ireland, England, Wales and Scotland in relation to a devolution issue.[142] The position within Northern Ireland is of relevance here. The courts in Northern Ireland are empowered to refer any devolution issue which does arise to the Court of Appeal in Northern Ireland.[143] The Court of Appeal in Northern Ireland is entitled to refer a devolution issue to the Judicial Committee,[144] but not in cases where a reference

135 Northern Ireland Act 1998, s. 81(1) and (2).
136 *Ibid*. s. 81(3).
137 *Ibid*. s. 81(4)(a).
138 *Ibid*. s. 81(4)(b).
139 *Ibid*. s. 83(2).
140 *Ibid*. s. 82 and Sch 10.
141 *Ibid*. Sch 10 para. 1.
142 *Ibid*. Sch 10 Parts II, III and IV.
143 *Ibid*. Sch 10 para. 7.
144 *Ibid*. Sch 10 para. 9.

has been made to it from a court or tribunal in Northern Ireland. In such cases an appeal may lie to the Judicial Committee, but only with the leave of this court or the Court of Appeal in Northern Ireland.[145] There is provision for direct references to the Judicial Committee. For example, the FM-DFM acting jointly may require a court or tribunal to refer any devolution issue which is not the subject of proceedings to the Judicial Committee.[146]

Given the stress throughout the Agreement on self-government and participation it is surprising to find such reliance on the courts to resolve questions over devolution issues and legislative competence. It may be a convenient way to reach a decision on a controversial matter, but it sits uneasily with the participatory nature of the Agreement.[147] This objection applies irrespective of the composition of the judiciary and even after a process of judicial reform. The objection targets the appropriateness of giving this institution such an enhanced role in political life. The politics of human rights has done much to infuse the judiciary with a new sense of legitimacy in recent years. Despite the wave of euphoria that has greeted the Human Rights Act 1998 there is still good reason to question uncritical faith in judicial activism. The real issue remains whether we are promoting the core value of self-government and thus the human right to participate in the decisions that shape our social existence. Should we really be giving the last word on this to the Judicial Committee?[148] The argument can be made for a new constitutional court, but careful attention would need to be paid

[145] Northern Ireland Act 1998, Sch 10 paras 7, 8 and 10.

[146] *Ibid.* Sch 10 para. 34.

[147] See Colin Harvey "Questions about the Culture of the Legal System in the North" *Irish News*, 10 January 2000. Although referring generally to the development of the British constitution, the following is of interest in this context also, K D Ewing "The Politics of the British Constitution" [2000] *Public Law* 405, at 437: "For much of the century the courts have been one of the major forces of conservatism, above all concerned in a one dimensional way with national safety, public order and political stability. It is only in the latter part of the century that the courts have shown sympathy for and commitment to the values underlying the liberal constitution".

[148] See A W Bradley and K D Ewing *Constitutional and Administrative Law* (12th edn, London and New York, Longman, 1997) p. 371 in discussing its role prior to the devolutionary schemes they state: "The Committee's role is now of special significance in relation to those constitutions which include protection for fundamental human rights, but the degree of protection which the Committee has given in this role has been very uneven. In particular, the Committee has fluctuated between adopting a strict and legalistic approach to fundamental rights provisions and a broader, more purposive approach that recognises the constitution as a living instrument. Not surprisingly, the resulting case-law on issues such as freedom of expression and the death penalty is not wholly satisfactory".

to its composition, accountability and precise role and how it would facilitate the new participatory structures that have been created.

Human rights, equality and the politics of voting

Although the Assembly is the prime source of authority on all devolved matters this power is not unlimited and there are important safeguards. These safeguards must be viewed in the context of the continuing role of the Westminster Parliament. The reasoning behind the safeguards is clearly spelt out in the Agreement.

> "There will be safeguards to ensure that all sections of the community can participate and work together successfully in the operation of these institutions and that all sections of the community are protected".[149]

The safeguards are not there solely to protect individual human rights, they are also there to ensure participation. Human rights protection is one aspect of this, but not the only one. The basic safeguards listed in the Agreement are: (1) the allocation of Committee Chairs; (2) Ministers and Committee membership in proportion to party strengths; (3) the human rights components including the European Convention on Human Rights, any Bill of Rights for Northern Ireland which supplements it and the creation of the Human Rights Commission; (4) the proofing of key decisions to ensure compliance with the European Convention and any Bill of Rights; (5) cross-community voting, rules involving parallel consent and weighted majorities; (6) and the creation of the Equality Commission.[150] These safeguards are not exhaustive.

In order to build-in safeguards the enabling Act ties the legislative competence of the Assembly to *inter alia* compatibility with "Convention rights"[151] and Community law.[152] The Assembly has no competence to do anything which "discriminates against any person or class of persons on the ground of religious belief or political opinion".[153] The Minister in charge of a Bill must make a statement of legislative competence.[154] There is also a role here for the Presiding

[149] Strand One para. 5.

[150] Strand One para. 5(a)–(e).

[151] Northern Ireland Act 1998, s. 6(2)(c). This has the same meaning as in the Human Rights Act 1998, see Northern Ireland Act 1998, s. 98(1).

[152] *Ibid.* s. 6(2)(d).

[153] *Ibid.* s. 6(2)(e).

[154] *Ibid.* s. 9.

Officer to ensure that any Bill introduced is within the legislative competence of the Assembly.[155] The Attorney General for Northern Ireland may refer the issue of legislative competence to the Judicial Committee for a decision.[156] In addition, there are what are termed "entrenched enactments".[157] These include the Human Rights Act 1998 and various provisions of the Northern Ireland Act 1998.[158] Detailed provision is made in the Act for what the standing orders must contain,[159] for example, that the Presiding Officer send a copy of each Bill to the Northern Ireland Human Rights Commission.[160]

Within the OFM-DFM there is a Community Relations, Human Rights and Victims Division, a Human Rights Unit and an Equality Unit. It has already been noted that the Committee of the Centre will have an important role in scrutinising this work. The main function of the Human Rights Unit is not legal advice but information provision to Government on the Human Rights Act 1998 and in the development of a culture of rights and responsibilities. In relation to the Human Rights Act 1998 the Directorate (as it then was) took its lead from the Home Office Human Rights Task Force. In future the new Unit will monitor departmental performance, raise awareness of human rights issues and liaise with other human rights organisations, in particular, the Northern Ireland Human Rights Commission. It has the additional role of providing advice and support for the Northern Ireland Assembly in relation to the United Kingdom's responsibilities arising under other international instruments. There is a separate Human Rights Unit within the Northern Ireland Office (NIO).

What is notable is that human rights and equality considerations are structuring the way that work is being done within public administration in Northern Ireland. This is a considerable change from the past. These norms are now structuring the way that institutions are being organised and how they do their business. This is a beginning only, and the results of all this restructuring must await detailed examination.

155 Northern Ireland Act 1998, s. 10.

156 *Ibid.* s. 11.

157 *Ibid.* s. 7. These "shall not be modified by an Act of the Assembly or subordinate legislation made, confirmed or approved by a Minister or Northern Ireland Department".

158 *Ibid.* s. 7(1)(b) and (c).

159 *Ibid.* s. 41(1) and (2), providing that the proceedings of the Assembly shall be regulated by standing orders and that they cannot be made, amended or repealed without cross-community support. See Northern Ireland Assembly Standing Orders (as at 3 February 2000).

160 Northern Ireland Act 1998, s. 13(4)(a).

Relations with other bodies

The Assembly exists within an increasingly complex set of constitutional arrangements. These are based not only on primary and secondary legislation but also what has been called "concordatry".[161] It is important to be clear about the practical operation of the new relations and their legal and non-legal underpinnings.

The response to external relationships is one of the duties of the FM-DFM.[162] This response must include cross-community involvement.[163] Provision is made in the Agreement to ensure effective coordination between the Assembly and the Government of the United Kingdom on the effective input of Ministers to national policy-making.[164] The Secretary of State for Northern Ireland retains an important role.[165] He or she: is responsible for NIO matters which are not devolved; will lay before the Westminster Parliament any Assembly legislation on reserved matters; represent Northern Ireland interests in the UK Cabinet; and has the right to attend the Assembly at its invitation.[166] She or he has the power to direct that action be taken in relation to the act of a Minister or Department, if it would be "incompatible with international obligations, with the interests of defence or national security or with the protection of public safety or public order".[167] A power also exists to direct that action be taken if required for the purposes of giving effect to these matters.[168]

The power of the Westminster Parliament to enact legislation for Northern Ireland remains, as the Agreement states, "unaffected".[169] For example, it will continue to legislate for non-devolved issues[170] and to scrutinise the responsibilities of the Secretary of State.[171] In strict constitutional law terms the Westminster Parliament remains supreme,

[161] Richard Rawlings "Concordats of the Constitution" (2000) 116 *LQR* 257, at 258: "[It] constitutes one of the main pillars of the new devolutionary architecture of the United Kingdom".

[162] Strand One para. 18.

[163] Strand One para. 30.

[164] Strand One para. 31.

[165] Strand One para. 32.

[166] Strand One para. 32 (a)–(d).

[167] Northern Ireland Act 1998, s. 26(1).

[168] *Ibid.* s. 26(2).

[169] Strand One para. 33; Northern Ireland Act 1998, s. 5(6).

[170] Strand One para. 33(a).

[171] Strand One para. 33(d).

even after devolution, and has the authority to legislate for the devolved administrations. Constitutional conventions may emerge quickly to govern this relationship, particularly on the issue of legislating for transferred matters. The principles which underpin the relationship, in the devolved context, can be found in the Memorandum of Understanding and the concordats.[172] The emphasis is on cooperation, communication and partnership. While the legal position leans heavily towards Westminster, the documents reflect a less patriarchal approach to devolution.[173] These documents were agreed between the UK Government, the Scottish Ministers and the Cabinet of the National Assembly of Wales on 1 October 1999. Provision is made for a new Joint Ministerial Committee consisting of: Ministers of the UK Government; Scottish Ministers; members of the Cabinet of the National Assembly of Wales; and Ministers of the Northern Ireland Executive. The Executive Committee of the Northern Ireland Assembly has considered and agreed to them with one addition. The documents are not legally binding in nature and the overall purpose is to ensure that cooperation takes place on, for example, formulating a single UK policy approach to the EU.

The informal public sphere has played an important part in political and social life in Northern Ireland. The community and voluntary sectors continue to make a substantial contribution to developing a culture of civility.[174] This role is recognised in the Agreement with the proposal for the establishment of a Civic Forum.[175] This is a consultative mechanism on social, economic and cultural issues.[176] The OFM-DFM is responsible for the provision of administrative support to this body and drew up the guidelines for the selection of representatives.[177] It includes representatives of the business, trade union and voluntary sectors.[178]

[172] (Cm 4444, 1999). This contains: Memorandum of Understanding; Supplementary Agreement on the establishment of a Joint Ministerial Committee; Concordat on Co-ordination of EU policy issues; Concordat on Financial Assistance and Industry; Concordat on Industrial Relations; and a Concordat on Statistics.

[173] Rawlings *supra* n. 161 at p. 262.

[174] See Chapter 10.

[175] Strand One para. 34.

[176] *Ibid.*; Northern Ireland Act 1998, s. 56. It has sixty members.

[177] *Ibid.*

[178] *Ibid.*

ALL-ISLAND STRUCTURES: NORTH-SOUTH RELATIONS AND THE AGREEMENT

Analysis of all-Ireland structures here is on the basis of my previous argument about the contested nature of the constitutional status of Northern Ireland. The consent principle, often assumed to bring certainty with it, in practice carries the seeds of healthy democratic instability. The inclusion of an "Irish dimension" was essential to securing the support of the nationalist/republican community for the Agreement. The unionist community has consistently resisted this. The Agreement, however, contains a clear "Irish dimension". Strand Two of the Agreement deals with North-South relations in the form of a new North/South Ministerial Council. The negotiations on Strand Two proved problematic and caused severe problems in the final stages of the process.[179] George Mitchell has expressed concern about the working methods of the British and Irish Governments in relation to this section of the Agreement.[180]

The Agreement provides for the establishment of a North/South Ministerial Council to bring together those with executive responsibility to "develop consultation, co-operation and action within the island of Ireland . . . on matters of mutual interest within the competence of the Administrations, North and South".[181] The Annex to Strand Two lists twelve areas for North/South cooperation and others may be considered by the Council.[182] Participation in the Ministerial Council is one of the essential responsibilities attaching to Ministerial office.[183] It is for the FM-DFM, acting jointly, to make nominations to attend the Council.[184] They are also responsible for ensuring cross-community attendance at Council meetings: "It is understood that the North/South Ministerial Council and the Northern Ireland Assembly are mutually inter-dependent, and that one cannot successfully function without the other."[185]

[179] See George Mitchell *Making Peace* (London, William Heineman, 1999) pp. 161–76.

[180] Mitchell *ibid.*

[181] Strand Two para. 2; *Agreement between the Government of the United Kingdom of Great Britain and Northern Ireland Establishing a North/South Ministerial Council* (Cm 4294, 1999).

[182] On 18 December 1998 the following areas were identified for cooperation: transport; agriculture; education; health; environment; and tourism.

[183] Agreement *supra* n. 181; Northern Ireland Act 1998, s. 52(2).

[184] *Ibid.* s. 52.

[185] Strand Two para. 13.

It is to meet in plenary session twice a year[186] and in sectoral formats on a more regular basis.[187] The Council is, for example, to use its best endeavours to reach agreement on common policies "in areas where there is a mutual cross-border and all-island benefit".[188] Each side involved in the Council is accountable to the Assembly and the Oireachtas respectively.[189] The Agreement provides for the establishment of implementation bodies.[190] There are six such bodies[191] and their task is to implement, on a cross-border and all-island basis, policies agreed in the Council.[192] The Council is supported by a joint Secretariat which is staffed by civil servants from the Northern Ireland Civil Service and the Irish Civil Service.[193] Consideration is to be given to a joint parliamentary forum for discussion of matters of mutual interest and concern,[194] as well as a joint consultative forum made up of the social partners and individuals with expertise in social, cultural, economic and other issues.[195]

The inaugural plenary meeting of the Council took place in Armagh on 13 December 1999. At this meeting a Memorandum of Procedure was adopted relating to the practical operation of the Council. The Council agreed that it should meet in sectoral formats in relation to the Implementation Bodies and the Matters for Co-operation.

THE BRITISH-IRISH DIMENSION

While the North/South dimension was of particular concern to the nationalist/republican community, the East/West dimension might be

[186] Strand Two para. 3(i).

[187] Strand Two para. 3(ii).

[188] Strand Two para. 3(iii).

[189] Strand Two para. 6.

[190] Northern Ireland Act 1998, s. 55.

[191] In the areas of inland waterways, food safety, trade and business development, special European Union programmes, language, and aquaculture and marine matters. The bodies are: Waterways Ireland; The Food Safety Promotion Board; The Trade and Business Development Body; The Special EU Programmes Body; The North-South Language Body; The Foyle, Carlingford and Irish Lights Commission. See *Agreement between the Government of the United Kingdom of Great Britain and Northern Ireland Establishing the Implementation Bodies* (Cm 4293, 1999).

[192] Strand Two para. 11; Northern/South Co-operation (Implementation Bodies) Order 1999, SI 1999/859.

[193] Strand Two para. 16.

[194] Strand Two para. 18.

[195] Strand Two para. 19.

thought to have been the focus of the unionist community. This is a simplification. For many in the unionist community the only relation that matters is the continuance of Northern Ireland as part of the United Kingdom and thus the supremacy of the Westminster Parliament.

The legal basis for this dimension rests on bilateral agreements between Ireland and the United Kingdom. As with the other elements this originates in practice in the Agreement. Strand Three of the Agreement addresses the British-Irish dimension. It provides for the establishment of a British-Irish Council "to promote the harmonious and mutually beneficial development of the totality of relationships among the peoples of these islands".[196] It is made up of representatives of the British and Irish Governments, the devolved institutions in Northern Ireland, Scotland and Wales with representatives of the Isle of Man and the Channel Islands.[197] The Council will meet at summit level twice a year and in sectoral formats on a more regular basis.[198] The aim is to encourage discussion and to reach agreement on matters of mutual interest.[199] The Council may adopt common policies or common actions and it is open to two or more members of the Council to develop bilateral or multilateral arrangements.[200] These arrangements do not require the approval of the Council and may operate independently of it.[201] The Council held its inaugural summit on 17 December 1999. At this meeting the Council agreed a Memorandum on Working Procedures, adopted a list of issues for early discussion and decided which administrations would take the lead on sectoral areas.

There is to be a new British-Irish Intergovernmental Conference which will effectively subsume the institutions created by the Anglo-Irish Agreement 1985.[202] The aim is to bring the British and Irish Governments together on matters of mutual interest and to promote cooperation.[203] The Agreement provides for regular meetings of the

[196] Strand Three "British-Irish Council" para. 1; *Agreement between the Government of the United Kingdom of Great Britain and Northern Ireland Establishing a British-Irish Council* (Cm 4296, 1999).

[197] Strand Three "British-Irish Council" para. 2.

[198] Strand Three "British-Irish Council" para. 3.

[199] Strand Three "British-Irish Council" para. 5.

[200] Strand Three "British-Irish Council" para. 10.

[201] Strand Three "British-Irish Council" para. 10.

[202] Strand Three "British-Irish Intergovernmental Conference" para. 1; *Agreement between the Government of the United Kingdom of Great Britain and Northern Ireland Establishing a British-Irish Intergovernmental Conference* (Cm 4295, 1999).

[203] Strand Three "British-Irish Intergovernmental Conference" para. 2.

Conference in relation to non-devolved matters and the Irish Government is permitted to put forward views.[204] The Conference will facilitate cooperation on security matters and deal in particular with rights, justice, prisons and policing in Northern Ireland.[205] There is recognition that these issues may be devolved to Northern Ireland and will thus not be of further concern to the Conference.[206] The Conference will keep the operation of the British-Irish Agreement under review and publish a report three years after the Agreement enters into force.[207] This report is to include the views of the Northern Ireland Assembly.[208]

CONCLUSION

Violence and coercion have scarred law and politics in Northern Ireland since its inception. To ignore this is to miss the fundamental significance of the Good Friday Agreement. The people of Ireland were presented with a chance to give expression to their democratic will. The message they sent was clear. The support for the Agreement represents a firm commitment to a dialogical form of open-ended democracy. Northern Ireland is contested terrain, but the days of insularity and exceptionalism are gone. It must now add its own voice to the ongoing conversation about legal and political life in Ireland, the United Kingdom and beyond.

The argument advanced here is that the Agreement, and what has followed, reflects a commitment to the primary value of self-government. It is the right to participate, and the practical eradication of the factors impeding this, which underpins the Agreement. The conception of democracy is therefore a strongly deliberative one. Those who crafted the Agreement have shown how democratic structures can be created which ensure inclusion. The Agreement demonstrates that democratic processes can be moulded to fit local contexts and promote self-government. There are a number of "weak links". The first is the continuing influence that traditional constitutional law thinking in the United Kingdom exerts over the new arrangements. I argue in this

[204] Strand Three "British-Irish Intergovernmental Conference" para. 5.
[205] Strand Three "British-Irish Intergovernmental Conference" para. 6.
[206] *Ibid.*
[207] Strand Three "British-Irish Intergovernmental Conference" para. 9.
[208] *Ibid.*

chapter that Northern Ireland cannot be neatly fitted into orthodox versions of UK constitutional law. In particular, the Agreement has a constitutive force which requires express recognition. The second is the broad and ill-defined role given to the courts. It is not to the courts, even if substantial reform takes place, that those genuinely interested in deliberative democracy should be looking. The other main difficulty is that much is off the agenda for the Assembly. While this respects the need for specific protection and constitutional realities it can appear as the familiar condescension of Westminster-style governance.

The Agreement and what has followed opens the way for the reinvigoration of the public sphere in Northern Ireland. While the informal public sphere was already operational, the creation of local structures creates a local focus. The challenge will now be to replenish the resources of argumentation in the public sphere and combat the impoverishment of political discourse. As I have argued, the constitution is both a constitutive act and a basis for continuing dialogue. Contestation must be at the heart of the new polity in Northern Ireland. We must begin to accept that in a democracy worth the name everything must remain "up for grabs".[209] The constitutional vision that the courts, among others, must promote is one that keeps a critical dialogue going. A healthy public sphere (formal and informal) is at the heart of this new constitutional order.

[209] See Jeremy Waldron *Law and Disagreement* (Oxford, Clarendon Press, 1999) pp. 302–6.

3

Northern Ireland, Devolution and the European Union

GORDON ANTHONY and ANDREW EVANS

INTRODUCTION

THE BELFAST AGREEMENT of 1998 is, as other contributions to this collection have made clear, widely understood to have put in place in Northern Ireland the beginnings of an all-inclusive, democratic form of government.[1] This understanding stems from the fact that the Belfast Agreement and related legislation address various contentious issues that have destabilised political relationships within Northern Ireland and between the United Kingdom and Ireland more generally. There are, for example, provisions in the Agreement dealing with the formation of a Northern Ireland Assembly headed by a multi-party executive; the creation of a North/South Ministerial Council; the creation of cross-border implementation bodies; the creation of a British-Irish Council; equality; policing; prisoner releases, and the decommissioning of terrorist weapons.[2] Although the process of implementing these provisions has proved problematic, there nevertheless exists an expectation that their collective implementation will further the equal protection and representation of competing political interests. Thus, while notions of democracy and legitimacy previously were contested in the context of Northern Ireland, the Belfast Agreement is now argued to have progressed legal and political discourse beyond its

[1] See Chapter 2. For a cautionary analysis, however, see D Kennedy "Dash for Agreement: Temporary Accommodation or Lasting Settlement?" (1999) 22 *Fordham International Law Journal* 1440.

[2] See further J Morison and G Anthony "An Agreed Peace? The Institutions of Democratic Governance in the Belfast Agreement 1998 and the Northern Ireland Act 1998" (1999) 11 *ERPL* 1313.

more familiar confines.[3] In particular, the creation of the North/South Ministerial Council and the British-Irish Council seems to have the potential to foster new political relationships which will depart from core constitutional understandings of the balance of power within the United Kingdom.[4]

Little attention, however, has been paid to the significance of the European Union for the creation of new institutional relationships within and without Northern Ireland. Although it has long been recognised that European integration has exerted some influence on institutional relationships within the United Kingdom[5], the devolution of power to Northern Ireland now raises many more specific questions about the role which Northern Ireland institutions should, and can, play in EU policy-making.[6] Consideration of these questions is essential to understanding how far the institutions established pursuant to the Belfast Agreement will, in practice, enjoy policy-making power and influence. Many of the areas of competence devolved to the Northern Ireland Assembly fall broadly within the competence of the EU institutions,[7] and there likewise exists overlap between the responsibilities of the cross-border implementation bodies and various EU policy areas.[8] The central question about the Northern Ireland institutions, therefore, is whether they will play an active role in EU policy formation or whether they will serve merely to implement decisions taken by other institutional actors. If the institutions perform an active role, it might be expected that Northern Ireland will start to enjoy effective and rep-

[3] See, e.g., C Harvey "Legality, Legitimacy, and Democratic Renewal: The New Assembly in Context" (1999) 22 *Fordham International Law Journal* 1389.

[4] See V Bogdanor "The British-Irish Council and Devolution" (1999) 34 *Government and Opposition* 287.

[5] See, e.g., A Scott, J Peterson and D Millar "Subsidiarity: a Europe of the Regions v. The British Constitution" (1994) 32 *JCMS* 47.

[6] This issue has already received considerable attention in the Scottish context. See, e.g., T St John Bates "Devolution and the European Union" in T St John Bates (ed.) *Devolution to Scotland: The Legal Aspects* (Edinburgh, T&T Clark, 1997) p. 63; A Cygan "Scotland's Parliament and European Affairs: Some Lessons from Germany" (1999) 24 *ELRev* 483; and G Clark "Scottish Devolution and the European Union" [1999] *PL* 504.

[7] See, e.g., Arts. 32–38 (ex Arts. 37–46) EC relative to the Agriculture Ministry; Art. 137 (ex Art. 118) EC relative to the Social Development and Higher Education, Training and Employment Ministries; and Art. 174 (ex 130r) EC relative to the Environment Ministry.

[8] The North/South Implementation Bodies are: Waterways Ireland; the Food Safety Promotion Board; the Trade and Business Development Body; the Special European Union Programmes Body; the Foyle, Carlingford and Irish Lights Commission; and the North/South Language Body.

resentative government. Equally, however, if the institutions perform only a limited role relative to EU policy formation, the significance of Northern Ireland's new institutions might be expected to be largely symbolic. Indeed, under this latter scenario, the understanding that the implementation of the Belfast Agreement will lead to representative government in Northern Ireland may, in real terms, come to be seen as misguided and unduly optimistic.[9]

In the following pages this chapter explores more fully the legal framework for Northern Ireland participation in EU policy-making. It does so by reference to the imagery of multi-level governance. Multi-level governance incorporates an understanding that it is possible for different tiers of government to interact in a non-hierarchical manner which maximises opportunities for regional and sub-national participation in the policy-making process.[10] This chapter suggests that, although there now exists the potential for non-hierarchical interaction which engages the Northern Ireland Assembly, the realisation of a form of multilevel governance presently may be constrained by core constitutional understandings in the United Kingdom *and* in the European Union.[11] Specifically, it is argued that the centralising tendency of the UK constitutional tradition, coupled with EU law's understanding of Member State sovereignty as the basis for the Union, may act as a brake on the Northern Ireland institutions' participation in EU policy formation. Interaction of the kind envisaged in multi-level

[9] It might be noted that the Northern Ireland Assembly does not have a European Ministry. Rather, the responsibility for European Affairs is divided within the Office of the First and Deputy First Ministers (a full listing of the Northern Ireland Ministries is available at http://www.ni-assembly.gov.uk/ministers.htm).

[10] "Political arenas are interconnected rather than nested. While national arenas remain important for the formation of state executive preferences, the multi-level model rejects the view that sub-national actors are nested exclusively within them. Instead, sub-national actors operate in both national and supranational arenas, creating transnational associations in the process. States do not monopolise links between domestic and European actors, but are one among a variety of actors contesting decisions that are made at a variety of levels": G Marks, L Hooghe and K Blank "European Integration from the 1980s: State-Centric v. Multiple-Level Governance" (1996) 34 *JCMS* 341, 346. See further, A Benz and B Eberlein "The Europeanization of Regional Policies: Patterns of Multi-level Governance" (1999) *JEPP* 329.

[11] It should be noted that this chapter is written very much from the perspective of UK law. It does not purport to forward any authoritative points about multi-level governance from the perspective of other legal orders. For contributions from the perspective of other national orders see, e.g., C Jeffrey "Towards a 'Third Level' in Europe? The German Länder in the European Union" (1996) 44 *Political Studies* 253 and E Bomberg and J Peterson "European Union Decision Making: The Role of Sub-national Authorities" (1998) 46 *Political Studies* 219.

governance literature requires that sub-national actors and institutions are able to influence developments at the EU level through the occasional by-passing of pre-existing State structures and preferences. This chapter argues, however, that opportunities for the Northern Ireland institutions to by-pass State structures may be limited. The core constitutional understanding which underpins devolution, for example, is that the Northern Ireland institutions are subordinate bodies which are without sovereign powers.[12] Similarly, while the European Union seeks to maximise regional involvement in the policy-making process, EU law does not require Member States to recast their "internal" understandings of the sovereign basis of the State by way of accommodating regional preferences.[13] Rather, the question of regional representation at the EU level is solely a matter for the internal law of the Member States.[14] Consequently, while there is the imperative of maximising Northern Ireland's participation at the EU level, it may be that orthodox constitutional understandings will frustrate the potential for interaction which inheres in the "new" institutional dynamic of the Belfast Agreement.

The chapter begins, by exploring more closely some of the "internal" tensions that may arise as a result of the coexistence of the Belfast Agreement's new institutional relationships and orthodox understandings of the institutional balance of the United Kingdom. Here, the chapter considers on the one hand how far there exists within the

[12] e.g., Northern Ireland Act 1998, s. 6, provides that primary Acts of the Northern Ireland Assembly may be quashed on the grounds that they are *ultra vires* the Assembly's competence. "(1) A provision of an Act is not law if it is outside the legislative competence of the Assembly. (2) A provision is outside that competence if any of the following paragraphs apply—(a) it would form part of the law of a country or territory other than Northern Ireland, or confer or remove functions exercisable otherwise than in or as regards Northern Ireland; (b) it deals with an excepted matter and is not ancillary to other provisions . . . dealing with reserved or transferred matters; it is incompatible with any of the Convention rights; it is incompatible with Community law; it discriminates against any person or class of person on the ground of religious belief or political opinion". See further *ibid. s.* 24. With regard to the position in Scotland and Wales, see the Scotland Act 1998, ss 29, 57, and the Government of Wales Act 1998, s. 107.

[13] A Evans "Regionalist Challenges to the EU Decision-Making System" (2000) 6 *EPL*, 377.

[14] A feature which would support "State-centric" views of the EU process. The seminal statement of State-centric (liberal intergovernmental) approaches to European integration can be found in A Moravcsik "Preferences and Power in the European Community: A Liberal Intergovernmentalist Approach" (1993) 31 *JCMS* 473 and *The Choice for Europe: Social Purpose and State Power from Messina to Maastricht* (Cornell University Press, 1998). For criticism of the State-centric (or liberal intergovernmental) approach see Marks *et al*, *supra* n. 10.

general framework for devolution to Northern Ireland, Scotland and Wales a clear understanding that the collective interests of the United Kingdom should be afforded primacy over any divergent regional interests.[15] On the other hand, the chapter also considers how far there exist within the Belfast Agreement opportunities for the Northern Ireland Assembly to develop the kind of institutional role that may contradict strict observance of the political conventions associated with devolution. These opportunities, which are most apparent in relation to the British-Irish Council and the North/South Ministerial Council, may allow the Northern Ireland Assembly to perform a more active EU policy-making role than that envisaged by conventional UK constitutional understandings. The point which will be made in this context, therefore, is, that while UK constitutional orthodoxy limits the autonomy of the Northern Ireland institutions, some aspects of the Belfast Agreement do not. Thus, the chapter argues that it will be through emphasising the significance of the British-Irish Council and North/South Ministerial Council that the Northern Ireland institutions may come to enjoy a greater degree of influence at the EU level.

The remainder of the chapter then considers the EU law approach to the role of regional institutions in EU policy-making. Here, the chapter focuses on the principle of "subsidiarity", opportunities for "regional" institutions to participate in EU Council decision-making, and on the importance to the EU process of the EU's Committee of the Regions. Although these features of the EU decision-making process are seen as important relative to developing a system of non-hierarchical institutional interaction, the chapter argues that EU law will, in its present form, remain unable to guarantee proper regional representation at the EU level. This argument is developed primarily by reference to the fact that EU law adopts a "neutral" position relative to the internal constitutional arrangements of the Member States. This neutrality, which is apparent both on the face of the EC Treaty and in the jurisprudence of the European Court of Justice, is argued not only to prevent the EU identifying regions for its own purposes but also from engaging properly in a process of dialogue with regional actors. The chapter concludes, therefore, by considering whether EU law and UK law could develop in a manner that would accommodate more fully the concerns and interests of sub-national institutions.

[15] See, e.g., the Common Annex to the *Concordat on Co-ordination of European Union Policy Issues* (Cm. 4444, 1999).

THE BELFAST AGREEMENT, UK CONSTITUTIONALISM AND EU POLICY-MAKING

The limiting influence of orthodoxy

The centralising tendency of the UK constitutional tradition is, at source, attributable to domestic public law's historical emphasis on the doctrine of the sovereignty of the Westminster Parliament.[16] This doctrine, at its most basic, holds that the Westminster Parliament is free to make or unmake any law whatsoever.[17] Although the core elements of the doctrine have been subjected to much criticism in recent years,[18] the logic of the doctrine clearly has informed the programme of constitutional reform that has included the devolution of power to Northern Ireland.[19] The legality of primary Acts of the Northern Ireland Assembly, for example, can be challenged on the grounds that they transgress the boundaries of power devolved by the sovereign Westminster Parliament.[20] Likewise, and of more immediate importance in the given context, section 26(2) of the Northern Ireland Act 1998 provides that the UK Parliament may legislate when this is deemed necessary to ensure that the United Kingdom's international obligations are met in respect of Northern Ireland.[21] In other words,

[16] An understanding most famously associated with the writings of A V Dicey. See *An Introduction to the Study of the Law of the Constitution* (10th edn, London, Macmillan, 1959).

[17] For a judicial statement to this effect see, e.g., *Madzimbamuto v. Lardner-Burke* [1969] 1 AC 645, 723 (per Lord Reid). For more recent academic commentary to the same effect see, e.g., C Forsyth "Of Fig Leaves and Fairy Tales: The *Ultra Vires* Doctrine, the Sovereignty of Parliament and Judicial Review" (1996) 55 *CLJ* 122.

[18] See, e.g., in relation to the process of European integration, M Hunt *Using Human Rights Law in English Courts* (Oxford, Hart Publishing, 1997) chs 1–3. And for doubts about its explanatory force relative to the process of devolution, at least insofar as relates to Scotland, see V Bogdanor "Devolution: The Constitutional Aspects" in J Beatson, C Forsyth and I Hare (eds) *Constitutional Reform in the United Kingdom: Practice and Principles* (Oxford, Hart Publishing, 1998) p. 9.

[19] Other initiatives contained in New Labour's programme of reform have included the incorporation in domestic law of (most of) the ECHR; the re-introduction of an elected authority for London; the introduction of freedom of information legislation; and House of Lords reform. See further R Brazier "New Labour, New Constitution" (1998) 49 *NILQ* 1. For a critical account see J Morison, 'The Case Against Constitutional Reform?' (1998) 25 *JLS* 510.

[20] See *supra* n.12.

[21] And see further para. 33(b) of Strand One of the Belfast Agreement. "The Westminster Parliament (whose power to make legislation for Northern Ireland would remain unaffected) will . . . legislate as necessary to ensure that the United Kingdom's international obligations are met in respect of Northern Ireland".

the Northern Ireland Act provides that the UK Parliament may, in accordance with the demands of EU policy-making, override the Northern Ireland Assembly, thereby occasionally negating the devolution of powers envisaged by the Belfast Agreement. The absence of fuller autonomy for the Northern Ireland Assembly, therefore, can be seen to inhere in the text of the very Act that purports to establish and empower local institutions.[22]

The limiting influence of orthodoxy is also apparent at the political level. Here, there exists a Concordat which seeks to ensure that the Westminster Government and the Northern Ireland Assembly (the Concordat also deals with the Westminster Government's relations with the Scottish and Welsh institutions) adopt a unified UK approach to the formulation and implementation of EU policies.[23] Although it is unclear whether the Concordat will enjoy legal force,[24] it is evident that it will enjoy strong political force.[25] The Concordat does, for example, emphasise the imperative of inter-institutional confidentiality during the formulation of EU policy preferences, and it also emphasises the need to give full effect to the outcome of any negotiations finalised on behalf of the UK Government at the EU level. This emphasis on confidentiality has been criticised by Richard Rawlings who considers that it may limit the bargaining power of sub-national institutions when there exist competing interests.[26] More significantly, Rawlings also considers that adherence to a single policy line might lead to central

[22] The argument that it is too easy for the UK Government to prorogue Northern Ireland's local institutions is addressed at greater length in Chapter 10.

[23] The terms of these Concordats were presaged in part by para. 31 of Strand One of the Belfast Agreement: "terms will be agreed between appropriate [NI] Assembly representatives and the Government of the UK to ensure effective coordination and input by Ministers to national policy-making, including on EU issues". It should be noted that, at the time of writing, the terms of the Concordat were still being debated by the Northern Ireland Assembly. Nevertheless, it is assumed that the Concordat will be similar in content to those agreed between the Westminster Government and the Scottish Parliament and Welsh Assembly. See the Common Annex to the *Concordat on Co-ordination of European Union Policy Issues* (Cm. 4444, 1999).

[24] See further Richard Rawlings "Concordats of the Constitution" (2000) 116 *LQR* 257.

[25] It should be noted that the Concordat(s) on European Union Policy Issues form only one part of a much larger Memorandum of Understanding which seeks to structure relations between the Westminster Government and the devolved authorities (Cm. 4444, 1999)). There are other Concordats dealing e.g. with financial assistance to industry; international relations; and government statistics, and the Memorandum also makes provision for a Joint Ministerial Council which will meet to consider issues of mutual concern.

[26] *Supra* n. 24 pp. 272–4.

domination in the representation of "UK" interests. Consequently, while the official function of the Concordat is to ensure that regional Ministers are engaged in the formulation of the United Kingdom's collective policy preferences, Rawlings considers that regional input may, in effect, be minimal: "the devolved administrations have only indirect access to the UK cabinet system of government. The EU concordat stands for a relationship of give and take: it is the devolved administrations that stand to be taken for granted".[27]

At one level, of course, the Concordat's emphasis on a uniform sense of purpose and duty is eminently reasonable. The United Kingdom is, from the perspective of the EU, one state with one set of responsibilities, and a disjointed approach to the implementation of EU policies may not only frustrate the Union's objectives, but also create legal difficulties for the UK Government.[28] Cast in terms of the need for a form of multi-level governance, however, the Concordat clearly has the potential to militate against effective representation of Northern Ireland interests at the EU level. Literature on the "coupling" of decision-making arenas,[29] for example, emphasises the need for one decision-making arena to be sensitive to the needs and concerns of other decision-making arenas.[30] Although it might be argued that the Concordat is, on its face, couched in the language of coupling, the related constitutional framework for devolution is not. The result in practice, therefore, may be for UK participation in EU policy making to remain highly centralised, with the corresponding role of the Northern Ireland Assembly being reduced to one of implementing EU enactments adopted by other actors not immediately representative of Northern Irish concerns.

[27] *Ibid*. p. 273.

[28] e.g., if the Northern Ireland Assembly failed to introduce legislation to give effect to a Community Directive in a policy area coming within its competence, this might lead the European Commission to initiate Art. 226 EC enforcement proceedings against the UK Government (although the UK Government might be able to pre-empt such action by invoking the Northern Ireland Act 1998, s. 26(2). It should also be noted that the UK Government has already sought to absolve itself of financial responsibility for breaches of EU law by the devolved institutions: see para. 20 of the Memorandum of Understanding (Cm. 4444, 1999)).

[29] A Benz and B Eberlein *supra* n. 10.

[30] According to the literature, each arena enjoys "semi-autonomy", and decision-making in one arena sets the context for decision-making in other arenas. The emphasis in the relationship between the arenas, therefore, is on information exchange and negotiation. See further *ibid*. p. 11. Cf. I J Sand "The Changing Preconditions of Law and Politics: Multilevel Governance and Mutually Interdependent, Reflexive and Competing Institutions in the EU and the EEA" *Arena Working Papers* 97/29.

The Belfast Agreement and the straining of orthodoxy

The institutional understandings attributable to orthodoxy may appear to be complicated, however, by the creation of the British-Irish Council (BIC) and the North/South Ministerial Council (N/SMC). These bodies, which were established primarily for purposes of accommodating the competing political aspirations of the unionist and nationalist communities in Northern Ireland, clearly have the potential to engender new institutional relationships within and without the United Kingdom.[31] The BIC, which is not a formal decision-making body, will provide a discussion forum that brings together representatives of the UK Government, the Irish Government, the devolved institutions, the Isle of Man and the Channel Islands. Similarly, the N/SMC conjoins members of the Northern Ireland Assembly and the Irish Government for purposes of identifying and developing policy preferences in areas of mutual concern.[32] Indeed, under the terms of the Belfast Agreement, the N/SMC is obliged to use "best endeavours to reach agreement on the adoption of common policies, in areas where there is mutual cross-border and all-island benefit, and which are within the competence of both Administrations, North and South, making determined efforts to overcome any disagreements".[33]

The suggestion that the workings of these institutions could strain orthodox understandings of how UK policy on the EU should be developed is, at this stage, largely speculative. The BIC, for example, has not, at the time of writing, been convened, and the workings of the N/SMC are at an early stage. Nevertheless, it is arguable that there is within these institutions at least the opportunity for the Northern Ireland Assembly to further its EU preferences in a manner that might

[31] On the function of these bodies see further Chapter 2. And see also Morison and Anthony *supra* n. 2 at pp. 1325–30.

[32] The Annex to Strand Two of the Agreement suggested that areas of mutual concern might include: agriculture (animal and plant health); education (teacher qualifications and exchanges); transport (strategic transport planning); environment (environmental protection, pollution, water quality, and waste management); waterways (inland waterways); social security/social welfare (entitlements of cross-border workers and fraud control); tourism (promotion, marketing, research, and product development); relevant EU programmes (SPPR, INTERREG, Leader II and their successors); inland fisheries; aquaculture and marine matters; health (accident and emergency services and other related cross-border issues); and urban and rural development.

[33] Strand Two, N/SMC, para. 5 (ii). Agreed policies are to be implemented by the cross-border implementation bodies envisaged under Strand Two of the Agreement.

contradict the essential demands of the Concordat. In relation to the BIC, for example, the Belfast Agreement expressly permits "two or more members to develop bilateral or multilateral arrangements between them . . . These arrangements will not require the prior approval of the BIC as a whole and will operate independently of it".[34] Given that the BIC may discuss "approaches to EU issues",[35] it is conceivable that the Northern Ireland Assembly and other members will begin to develop a more intense relationship around EU issues which are of particular interest to them. Under these circumstances, the difficulty facing orthodoxy would be whether it should seek to override "informal" arrangements that have been agreed within the framework of an institution established jointly by two sovereign governments. In particular, it may be difficult to reconcile confidentiality requirements in the Concordat with the debate anticipated in the BIC.

The potential which the N/SMC offers for the straining of orthodoxy is even more striking. According to paragraph 17 of Strand Two of the Belfast Agreement, the N/SMC is to "consider the EU dimension of relevant matters, including the implementation of EU policies and programmes and proposals under consideration in the EU framework. Arrangements [are] to be made to ensure that the views of the Council are taken into account and represented appropriately at relevant EU meetings".[36] Although the Agreement does not specify what force the views of the N/SMC will have, or indeed who will represent those views, it seems that the Northern Ireland Assembly will enjoy opportunities to seek to influence the Irish Government in its dealings with the EU. This is highly significant insofar as it allows the Northern Ireland Assembly to further its own EU policy preferences within the framework of its relations with another sovereign State. The nature of those preferences may, of course, occasionally be different from those "agreed" on behalf of the United Kingdom within the framework of the Concordat. But that does not mean that the pursuit of those preferences necessarily should be seen as politically disingenuous. On the contrary, the pursuit of those preferences might better be said to be indicative of the Northern Ireland Assembly's willingness to capitalise on its unique constitutional and political position. The Belfast

[34] Belfast Agreement, Strand Three, BIC, para. 10.

[35] *Ibid*. para. 5.

[36] Paragraph 3(iii) of Strand Two further states that the Council is to meet "in an appropriate format to consider institutional or cross-sectoral matters (including in relation to the EU) and to resolve disagreement". The first sectoral meeting on Special EU Programmes was held in Dublin on 16 June 2000.

Agreement has put in place institutional relationships that reflect the unique political dynamics of Northern Ireland. More significantly, these institutional relationships seemingly offer the Northern Ireland Assembly the opportunity to step outside restrictive state structures by way of influencing developments at the EU level. It remains to be seen, therefore, whether the Northern Ireland Assembly will emphasise its position as an institution of the United Kingdom or as an institution that enjoys various legally founded relationships which go far beyond those associated with UK constitutional orthodoxy.

REGIONALISM, EU CONSTITUTIONALISM AND EU POLICY-MAKING

The understanding that "orthodoxy" can limit regional representation is, somewhat surprisingly, also informative in relation to EU law. Although the Maastricht Treaty introduced reforms that reflected the demands of regionalism, it is arguable that EU law remains ill-equipped fully to accommodate the concerns and preferences of regional institutions.[37] The Maastricht reforms included introduction of the "subsidiarity" principle, amendment of rules regarding the composition of the Council of the Union,[38] and establishment of the Committee of the Regions. Closer inspection of the wider constitutional context of these reforms, however, reveals that the opportunities they afford sub-national actors may be limited. The reforms were introduced into a constitutional order that emphasises the primacy of relations between the EU and the Member States.[39] This point of emphasis, which stems from an (arguable) understanding of the EU as a "delegate" of the Member States,[40] has meant that EU law does not

[37] M P Chiti "Regionalismo comunitario e regionalismo interno: due modelli da ricomporre" (1992) *Rivista Italiano di Diritto Pubblico Comunitario* 33, 35.

[38] This is the name given to what used to be called the "Council of Ministers". See Declaration 93/591 (OJ 1993 L281/18) concerning the name to be given to the Council following the entry into force of the Maastricht Treaty.

[39] e.g., the ECJ, in its seminal judgment in *Flamino Costa v. ENEL*, stated that the EU institutions benefited from the "transfer of powers from the [Member] States": Case 6/64 [1964] ECR 585, 593. See also Case 26/62 *NV Algemene Transport-en Expeditie Onderneming van Gend en Loos* v. *Nederlandse Administratie der Belastingen* [1963] ECR 1, 12.

[40] G Majone "Europe's 'Democratic Deficit': the Question of Standards" (1998) *ELJ* 5. The primary importance of the Member States can also be seen in the fact that they enjoy the ultimate power of decision relative to the process of EC Treaty amendment: see Art. 48 TEU. At a more general level, however, it should be noted that the question of whether the EU ultimately is driven by its own dynamics or by intergovernmental

concern itself with the internal constitutional arrangements of Member States, but rather works with Member States however defined from within. Consequently, while it might be expected that the Maastricht reforms would benefit regional and sub-national interests, the "neutrality" of EU law relative to domestic constitutional matters means that this may not always be so. In particular, the fact that EU law does not interfere with the internal constitutional arrangements of the United Kingdom seemingly means that the limiting influence of UK constitutional orthodoxy can enjoy legitimacy under EU law.

Subsidiarity

The subsidiarity principle is embodied in both Article 1 of the Maastricht Treaty and Article 5 of the EC Treaty.[41] The principle, which has had different meanings in different contexts,[42] essentially seeks to prevent an over-centralisation of power at the level of the EU institutions. According to Article 1 TEU, decisions should be "taken as closely as possible to the citizen". Article 5 of the EC Treaty states: "in areas which do not fall within its exclusive competence, the Community shall take action, in accordance with the principle of subsidiarity, only if and in so far as the objectives of the proposed action cannot be sufficiently achieved by the Member States and can therefore, by reason of the scale or effects of the proposed action, be better achieved by the Community".[43]

In terms of the need for effective sub-national institutional participation in the EU decision-making process, these provisions might be argued to pull in different directions. On the one hand, the wording of

considerations has long occupied a central position in academic debate. See, most famously, the contrasting accounts of liberal intergovernmentalists (e.g., Moravcsik *supra* n.14) and neo-functionalists, e.g., E Haas *The Uniting of Europe: Political, Economic and Social Forces, 1950–57* (London, Stevens, 1958); L Lindberg *The Political Dynamics of European Economic Integration* (Stanford University Press, 1963); and J Tranholm-Mikkelsen "Neo-functionalism: Obstinate or Obsolete? A Reappraisal in the Light of the New Dynamism of the EC" (1991) 20 *Millennium* 1.

[41] See also the Protocol on the application of the principles of subsidiarity and proportionality.

[42] See N Emiliou "Subsidiarity: An Effective Barrier Against the 'Enterprises of Ambition'?" (1992) 17 *ELRev* 383.

[43] *Cf* the distinction between substantive and procedural subsidiarity in A Scott *et al supra* n. 5. See also K Neunreither "Subsidiarity as a Guiding Principle for European Community Activities" (1993) *Government and Opposition* 206.

Article 1 TEU seemingly lends itself to an understanding that the existence of "autonomous regional bodies endowed with sufficient powers and resources"[44] is a condition for achievement of EU objectives.[45] On the other hand, however, the wording of Article 5 EC might be said to strengthen the existing nexus between EU and Member State constitutional law, thereby preserving Member State freedom to centralise or decentralise as preferred.[46] Article 5 EC adopts a more narrow approach to subsidiarity by referring only to relations between the European Community and the Member States. This formulation falls short of previous formulations that urged that the subsidiarity principle be applied generally in relations between EU, national and regional institutions.[47] These latter formulations, which are often associated with the demands of the German *Länder*, apparently have been rejected by the EU on the grounds that "it is not for the Community to interfere in the distribution of powers between the central, regional or local authorities in the Member States".[48] By emphasising the internal sovereignty of the Member States, therefore, EU law seems determined to "respect the national identities of its Member States",[49] even though such an approach could, in relation to centralised constitutions, frustrate regional and sub-national interests.

Council of the Union

EU law's inability to guarantee effective sub-national participation in the EU decision-making process can also be seen in relation to the workings of the Council of the Union. Article 203 EC provides that the Council shall consist of a representative of each Member State at ministerial level, authorised to commit the government of that Member

[44] European Parliament Resolution of 18 November 1993 (OJ 1993 C329/279) on the Participation and Representation of the Regions in the Process of European Integration: the Committee of the Regions, para. E of the Preamble.

[45] See Arts. 158–162 EC.

[46] V Bogdanor *Devolution in the United Kingdom* (Oxford, Oxford University Press, 1999) pp. 277–8.

[47] See, e.g., Resolution of 18 November 1993 (OJ 1993 C329/279) on the Participation and Representation of the Regions in the Process of European Integration: the Committee of the Regions, para. 4.

[48] Commission Reply by Mr Delors to WQ E–3099/93 (OJ 1994 C289/27) by Víctor Arbeloa Muru. *Cf* the approach of the CFI in Case T–465/93 *Consorzio Gruppo di Azione Locale Murgia Messapica* v. *EC Commission* [1994] ECR II–361.

[49] See Art. 6(3) TEU. See further U Everling "Reflections on the Structure of the European Union" (1992) *CMLRev*. 1053, 1071.

State. Although, in theory, this provision permits a regional Minister to represent a Member State in the Council, such representation is, in practice, problematic. From the perspective of the UK Government, for example, regional representation may be problematic, given that a Northern Ireland Minister may not be responsible to an institution representative of his Member State as a whole.[50] But the problem of representation is exacerbated by EU law's failure to oblige the UK Government to address from within the issue of representation. Once again, the neutrality of EU law can be seen in the fact that, while Article 203 EC conceives of a Member State being formally represented by one member on the Council, it does not make any suggestions as to how that member is to be chosen.[51] The result of this neutrality is that Member States may maintain highly centralised arrangements for the internal distribution of power. Such arrangements may deny regional Ministers the power of formal representation in the Council, or they may limit more generally the role of regional institutions within a national delegation in the Council. Either way, there exists under Article 203 EC the potential for regional interests to be denied full representation.

The practical consequences of EU law's reliance on internal constitutional arrangements may be illustrated by reference to the "localisation" principle in Sweden and the "speciality" principle in France. According to the former principle, regional institutions may only deal with matters which pertain to their region and which do not fall within the competence of the state, another regional institution or any other

[50] It is uncertain whether the problem of accountability could be solved through holding the United Kingdom's central institutions responsible for the conduct of a Northern Ireland Minister as a Council member. It has already been envisaged, e.g., that a Scottish Minister may speak for the United Kingdom in the Council with the "UK lead Minister" retaining overall responsibility for the negotiations: see *Scotland's Parliament* (Cm 3658, 1997) p. 17. But such a solution doubtless would highlight rather than resolve the further problem that, as a representative of the United Kingdom, a Northern Ireland Minister might be prevented from expressing Northern Ireland interests divergent from those pursued by the central institutions of the United Kingdom: see V Bogdanor *Devolution in the United Kingdom* (Oxford, Oxford University Press, 1999) p. 280.

[51] The European Parliament has called on Member States which have regions with exclusive legislative competence under their national constitutional law to facilitate the participation of regional representatives in Council consideration in matters falling within their competence: see Resolution of 18 November 1993, *supra* n. 47, para. I of the Preamble. It should also be noted that, in practice, regional ministers from Austria, Belgium, Germany, and Spain have participated in the Council. See M Westlake *The Council of the European Union* (London, Cartermill, 1995) p. 57. However, they participate as "common representatives putting forward collective sub-national positions". See V Bogdanor *supra* n. 46 at p. 280.

institution.[52] According to the latter principle, regional institutions may only deal with matters which pertain specifically to the region rather than being of national interest.[53] The limitations represented by these principles may be reinforced by the tendency of some Member States to treat "European matters" as foreign policy matters[54] and to adopt "realist"[55] conclusions as to the need for national positions in EU decision-making to be unified.[56] In Sweden, for example, municipalities may not interfere with state "foreign policy"[57] or undertake commitments that are binding on the state (unless they have been specifically authorised to do so).[58] Under such circumstances, therefore, members of regional institutions may be excluded by national law from participating in national delegations in the Council of the Union.[59] Consequently, while the traditional distinction between domestic and international may be eroded as a result of EU membership, the neutrality of EU law entails that it need not be fundamentally undermined.[60]

[52] Art. 2(1) of the *Kommunallag* (SFS 1991:900).

[53] R Letteron "Les Aides des collectivités territoriales aux services publics" (1993) *AJDA* 437.

[54] M Keating "Europeanism and Regionalism" in B Jones and M Keating (eds) *The European Union and the Regions* (Oxford, Oxford University Press, 1995) pp. 1 and 12.

[55] See, e.g., J M Grieco *Cooperation among Nations* (Ithaca, NY, Cornell University Press, 1990).

[56] See, e.g., the Common Annex to the *Concordat on Co-ordination of European Union Policy Issues* (Cm. 4444, 1999) as considered above. See also H C Jones *The National Assembly for Wales and the European Union* (Welsh Office, Cardiff, 1998) p. 14. According to F Snyder "Ideologies of Competition in European Community Law" (1989) 52 *MLR* 149, 172, different, conflicting and often contradictory interests may be expressed as "unified" national interests in the Council of the Union.

[57] *Kommunal medverkan i internationella frågor*, Ds C 1986:10. *Cf* the significance for German regions of the importance attached to "imperative reasons of foreign and integration policy" in *Bayerische Staatsregierung* v. *Bundesregierung* [1990] 1 CMLR 649, 651.

[58] Prop. 1977/78:44 *om godkännande av överenskommelse mellan Danmark, Finland, Norge och Sverige om kommunalt samarbete över nordiska riksgränser.*

[59] Indeed, it has been argued that the neutrality of EU law has also served to prejudice the role of established regional institutions where a Member State is negotiating at the EU level, i.e. the process of EU negotiations may lead to a process of "recentralisation": see, e.g., J Biancarelli "La Communauté européenne et les collectivités locales: une double dialectique complexe" (1991) *Revue Française d'Administration Publique* 515, 526. See also M V Agostini "The Role of the Italian Regions in Formulating Community Policy" (1990) *The International Spectator* 87. On the position in the United Kingdom see Bogdanor *supra* n. 46 at p. 278.

[60] P G Cerny "Globalization and other Stories: the Search for a New Paradigm for International Relations" (1996) *International Journal* 617, 624.

The Committee of the Regions

The Committee of the Regions, perhaps more so than any other of the reforms effected by the Maastricht Treaty, may be thought to offer the Northern Ireland Assembly the opportunity to participate directly in EU policy-making. The Committee was established by Article 263 of the EC Treaty and, under Article 265 of the same Treaty, it is to be consulted by the Council or Commission in relation to decisions taken in the fields of education,[61] culture,[62] public health,[63] guidelines for trans-European networks,[64] and cohesion[65] (the Committee, which may also be consulted by the European Parliament, is further to be consulted in all other cases where the Council or Commission considers consultation appropriate). At the same time, the Committee may take the initiative and issue an opinion in cases where it considers such action appropriate. It is thus described as "an institution through which regional and local bodies can officially be involved in drawing up and implementing Community policies"[66] and "included in the legislative process".[67] Indeed, according to the Committee itself, its establishment "enables regional and local bodies to participate, via the Committee of the Regions, in the decision-making process of the European Union".[68]

However, centralised representation of regional institutions in such a body may not necessarily constitute an adequate basis for regional participation in EU policy-making. At a practical level, the Committee of the Regions has argued that it should be included in various other committees that are consulted by the Commission. Regional representatives are often excluded from day-to-day EU policy-making,[69] hence the value of such consultation is understood to be that it would ensure

[61] Article 149 EC.

[62] Article 151 EC.

[63] Article 152 EC.

[64] Article 156 EC.

[65] Articles 159, 161 and 162 EC.

[66] Preamble to Declaration 94/209 (OJ 1994 L103/28) winding up the Consultative Council of Regional and Local Authorities.

[67] European Parliament Resolution of 18 November 1993, *supra* n. 47, recital G in the Preamble.

[68] Opinion of 17 May 1994 (OJ 1994 C217/10) on the Draft Notice from the Commission Laying Down Guidelines for Operational Programmes which Member States are Invited to Establish in the Framework of a Community Initiative Concerning Urban Areas, para. 4 of the Preamble.

[69] e.g., with regard to the use of EU funds. See J Scott *Development Dilemmas in the European Community* (Buckingham, Open University Press, 1995) p. 34.

that the views of individual regions affected by EU policies would be more readily "heard".[70] But beyond such practical considerations, there exists the much more fundamental problem that regional institutions may not even be guaranteed representation in the Committee of the Regions. In particular, the requirement proposed by the Commission that membership of the Committee should be limited to those holding elective office at regional or local level[71] was excluded from the final version of the EU Treaty.[72] The reason for the exclusion of the requirement was, according to the Commission, that "the Member States have very different institutional structures which are exclusively within their purview".[73] The result of this respect for state sovereignty, however, is that the "autonomy" of members of the Committee in relation to the central institutions of their Member State may be uncertain.[74] Once again, therefore, the neutrality of EU law seems to have the potential to frustrate Northern Ireland's fuller participation in the EU decision-making process.

Possible future reforms

The fact that the institutional reforms effected by Maastricht have failed fully to guarantee regional and sub-national participation in the policy-making process raises the question of how EU law may be reformed in future. The dilemma facing the EU is essentially one of finding the most appropriate point of institutional balance between the need for representative government and the need for effective government. Regional participation in each and every decision taken in areas of regional concern would likely "overload" the institutional process to such an extent that effective decision-making would become

[70] Opinion of 17 May 1994 (OJ 1994 C217/26) on the Proposal for a Decision Laying Down a Series of Guidelines on Trans-European Energy Networks, para. A.5.1. See, more particularly, regarding EU decisions in the field of fisheries policy, the Opinion of 16 July 1998 on the future of peripheral areas in the EU (CdR 23/98) para. 8.6.

[71] Article 198a(2) of the draft TEU prepared by the Commission (Bull. EC, Supp. 2/91, 178). See, similarly, the position of the European Parliament in its Resolution of 18 November 1993, *supra* n. 47, para. 9.

[72] The members of the earlier Consultative Council of Regional and Local Authorities were required to be so qualified. See Art. 3(1) of Declaration 88/487 (OJ 1988 L247/23) setting up the Council.

[73] Commission Reply by Mr Millan to WQ 1405/92 (OJ 1993 C16/15) by Mr Sotiris Kostopoulos.

[74] M Vellano "Coesione economica e sociale e ripartizione di competenze: le nuove iniziative comunitarie" (1995) *RDE* 193, 195.

impossible. But equally, the fact that contrasting regional opinions may not fully be heard suggests that, in the absence of further reform, the EU may face an increased legitimacy crisis.[75]

One option is EC Treaty amendment. In particular, consideration may be given to amendment of Article 230 EC. Article 230 EC provides for review of the legality of measures adopted by EU institutions. According to this provision, Member States enjoy "privileged" standing before the European Court of Justice. Regional institutions, however, are "non-privileged" applicants which may challenge only those EU decisions that are specifically addressed to them or are of "direct and individual concern" to them. The European Courts' approach[76] to the standing of non-privileged applicants has, on balance, tended to be restrictive,[77] and the result has been for regional access to the review process to be limited.[78] Indeed, the problems encountered by regional institutions may be exaggerated because, in the absence of a clear mandate from the EU legislature for the European Courts to do otherwise, the Courts tend to "individualise" the effects of EU enactments by reference to market interests.[79]

Article 230 EC may need to be amended, therefore, to entitle regional institutions to challenge before the European Courts the legality of EU enactments which disregard regional interests recognised by EU law.[80] The understanding that review proceedings can provide an alternative "democratic" forum has long been recognised in other contexts,[81] and

[75] G de Búrca "The Quest for Legitimacy in the European Union" (1996) 59 *MLR* 349.

[76] Review actions brought by non-privileged applicants have, since 1993, been heard by the Court of First Instance: see Council Decision 93/350/EEC, OJ 1993 L144/21. The ECJ continues to hear actions initiated by privileged applicants.

[77] An approach most famously associated with Case 25/62 *Plaumann & Co* v. *Commission* [1963] ECR 95. For criticism of the ECJ's approach see, e.g., C Harlow "Towards a Theory of Access for the European Court of Justice" (1992) 12 *YEL* 213.

[78] Case 222/83 *Municipality of Differdange* v. *EC Commission* [1984] ECR 2889. It should be noted that regional institutions may have the opportunity to intervene in proceedings before the European Courts (see, e.g., Case T–194/95 *Intv I Area Cova SA* [1996] ECR II–591). However, given that an intervener has to accept the case as he finds it at the time of the intervention (Art. 93(4) of the Rules of Procedure of the ECJ (OJ 1991 L176/7)), a regional institution may be precluded from raising arguments of specific relevance to the region.

[79] Case C–321/95P *Stichting Greenpeace Council (Greenpeace International)* v. *EC Commission* [1998] ECR I–1651.

[80] e.g. in relation to cohesion or environmental protection. See, e.g., Case C–2/90 *Commission* v. *Belgium* [1992] ECR I–4431, I–4480. See in the same connection, Case C–155/91 *EC Commission* v. *EC Council* [1993] ECR I–939.

[81] See, e.g., D Feldman "Public Interest Litigation and Constitutional Theory in Comparative Perspective" (1992) 55 *MLR* 44.

a system of open access for regional institutions might, without overloading the broader institutional framework, provide for a form of residual participation.[82] More importantly, the knowledge that such challenges would be possible might be expected to lead central institutions of Member States to take account of regional interests when participating in EU decision-making. Under those circumstances, the result would be for regional institutions to enjoy something approximating the kind of autonomy assumed by the literature on multilevel governance and the coupling of decision-making arenas.

The need for "amendment" is also apparent beneath the level of the EC Treaty. EU legislation governing the Structural Funds, for example, provides for decision-making under the legislation to be based on a partnership.[83] This partnership is defined in the legislation as "close consultation . . . between the Commission and the Member State, together with the authorities and bodies designated by the Member State within the framework of its national rules and current practices, namely: the regional and local authorities and other competent public authorities, the economic and social partners, and any other competent bodies within this framework". It must, according to the legislation, "be conducted in full compliance with the respective institutional, legal and financial powers of each of the partners".[84] The legislation thus seeks to facilitate participation of regional institutions in EU decision-making, but it also seeks to avoid challenging national law provisions which may preclude their effective participation.[85]

The desirability of reassessing the EU's understanding of partnership has been recognised by the European Commission. According to the Commission, the Council wrongly watered down its 1993 proposals for greater participation by regional actors.[86] In particular, the

[82] Although, of course, there remains the question of whether the European Courts would wish to become involved in essentially political disputes. On the ECJ's role as an institutional actor see, e.g., G de Búrca "The Principle of Subsidiarity and the Court of Justice as an Institutional Actor" (1998) 36 *JCMS* 217.

[83] Article 8 of Regulation 1260/1999 (OJ 1999 L161/1).

[84] *Ibid*. Art. 8(1).

[85] See, regarding "unequal" partnership, the COR Opinion of 19 November 1997 (OJ 1998 C64/5) on the Views of the Regions and Local Authorities on Arrangements for European Structural Policy After 1999, para. 3.6.1. Note also that this latter feature of EU legislation is tacitly recognised by the ECJ which defines partnership as "a kind of dialogue . . . between the Commission and the Member State concerned": Case C–303/90 *France* v. *EC Commission: Code of Conduct* [1991] ECR I–5315, I–5338 (per AG Tesauro). Such a definition does not guarantee, or imply, a role for regional institutions.

[86] Explanatory Memorandum to the Re-examined Proposals of 15 July 1993 (COM(93)379) p. 2.

Commission criticised the Council for stipulating that regional participation must take place "within the framework of each Member State's national rules and current practices" and "in full compliance with the respective institutional, legal and financial powers of each of the partners". For the Commission, partnership should ensure that decisions regarding the use of EU funds are made collectively by all those involved, notably the Commission, the Member States, and the regional institutions concerned.[87] However, the placing partnership within the framework of EU-Member State relations, means that the "mobilisation" of regional institutions, social partners, and non-profit-making organisations has often been inadequate and always too formal.[88] Consequently, while the Structural Funds may be of great importance to Northern Ireland, the Northern Ireland institutions may have only limited opportunities to participate in the system of "partnership" associated with the Funds.

In this context, therefore, EU legislation might be amended to protect such decision making against unduly restrictive national rules and practices. At the very least, the application of national legislation restricting regional participation in such decision-making should have to be justified by reference to much more than simply the requirements of state sovereignty. Possible justifications might still be found in the obligation of the EU to respect the national identities of Member States.[89] To the extent that no such justification was established, however, the Commission and regional institutions should be free to adopt decisions on cohesion unhindered by national legislation restricting the role of regional institutions in such decision-making. To the same extent, the relationship between such decision-making and Council decision-making might become horizontal rather than being determined by the hierarchical supremacy of Council decision-making.

[87] *Community Structural Funds 1994–9* (COM(93)67) p. 21. In the view of the Commission, the partnership should also incorporate the economic and social partners, such as chambers of commerce, employers, and trades unions, all of whom have a valuable role to play in the success of a region's economy: see Bruce Millan's address to the European League for Economic Cooperation's Conference on "Developing Europe's Regions after 1992: New Ideas for the New Europe" Cleveland, 10 July 1992 (IP92/563).

[88] *Community Structural Assistance and Employment* (COM(96)109) p. 26. Even Interreg programmes may be left to be administered by central institutions. See, e.g., E Meehan "Britain's Irish Question: Britain's European Question? British-Irish Relations in the Context of European Union and the Belfast Agreement" (2000) *Review of International Studies* 83, 92.

[89] Art. 6(3) TEU.

CONCLUSION

This chapter has sought to provide an overview of the legal framework governing the Northern Ireland Assembly's role in EU policy formation. It has suggested that orthodox constitutional and institutional considerations in the United Kingdom and the European Union might serve to limit the Assembly's ability to participate in the policy-making process. Specifically, the chapter has suggested that orthodoxy may deny the Northern Ireland Assembly the degree of autonomy that is prescribed by literature on multilevel governance in the European Union. Multilevel governance literature is premised upon an understanding that different levels of government may interact in a non-hierarchical manner which maximises opportunities for regional and sub-national participation in the policy-making process. Although interaction of this kind was identified as necessary if the Northern Ireland Assembly is to perform a truly representative function, it was suggested that the centralising tendency of UK constitutional law and the neutrality of EU law militate against such interaction. The chapter's core argument, therefore, was that there is a need for reforms that will move the existing legal framework "beyond orthodoxy".

It may be many years before a new orthodoxy emerges in UK law and EU law. Indeed, the emergence of a new orthodoxy in EU law would seem to depend in part upon the emergence of new constitutional understandings within the United Kingdom. The extent to which the Member States control the process of EC Treaty amendment, for example, would suggest that, in the absence of UK constitutional law adopting a more decentralised perspective, it is arguable whether the UK Government would endorse the suggested amendment of Article 230 EC. UK policy is clearly premised upon the need to maintain a centralised policy line, and greater access to Article 230 EC actions would only weaken the centralised approach by allowing "the Northern Ireland institutions" to raise issues of specific regional concern. For that reason, central government enthusiasm for amendment of Article 230 EC may be limited.

The development of an enhanced institutional role for the Northern Ireland Assembly, therefore, clearly depends upon internal developments in UK law. In this regard, the chapter has suggested that the potential to move beyond orthodoxy might already be said to inhere in the British-Irish and North/South Ministerial Councils. The chapter's argument in this regard may, of course, prove to be ill-founded.

However, if the argument is not ill-founded, the creation of the Councils, when set beside related institutional developments in the United Kingdom, may, through time, lead to a demand for a recasting of orthodoxy. The Belfast Agreement that gave rise to these bodies is only one part of a much wider process of change. The devolution of power to Scotland and Wales, for example, has created other institutional dynamics that will interact with, and maybe overtake, the specific dynamics of the Northern Ireland settlement. Indeed, it has already been argued by other commentators that devolution has initiated a process which can only lead to fundamental change in the institutional balance of the United Kingdom.[90] Consequently, while the objective of instruments such as the *Concordat on Co-ordination of European Union Policy Issues*[91] is the maintenance of a unified UK policy line, it might be expected that such centralised approaches will become increasingly difficult to justify and sustain in future years.

[90] e.g., R Brazier "The Constitution of the United Kingdom" (1999) 58 *CLJ* 96.
[91] (Cm 4444, 1999).

4

Equality

CHRISTOPHER MCCRUDDEN

INTRODUCTION

IN NORTHERN IRELAND talk of equality and human rights has often been ignored or marginalised. It has been perceived by too many in positions of power as divisive, as ignoring "the real problems", even as subversive. During 1998 something remarkable happened. Discussions about equality and human rights moved from the margins into the mainstream. The Good Friday Agreement, drawing on the best international and European practice, identified equality and human rights as a central element in a new constitutional settlement. The purpose of this chapter is to discuss this sea change and what happened subsequently. Following the Agreement, there was a real danger that equality would be pushed back to the margins. However, as we will see, a coalition of the disadvantaged and politicians ensured that this did not happen.

The chapter focuses on four aspects: first, the emergence and development of the equality agenda in Northern Ireland from the initial anti-discrimination legislation to the development of "mainstreaming" as a political and legal approach to equality; secondly, the convergence of the political and equality agendas in the negotiations leading up to the Agreement; thirdly, the strengthening of the equality agenda in the Agreement and its subsequent incorporation into the Northern Ireland Act 1998, the legal basis for the new constitutional settlement in Northern Ireland; fourthly, the major developments in the area of equality from the enactment of this legislation until the second anniversary of the Agreement. But first, what does mainstreaming equality mean?

In essence "mainstreaming" is the principle that equality be seen as an integral part of all public policy-making and implementation, rather than something separated off in a policy or institutional ghetto. The concept has emerged from several sources, of which the most important are debates about how best to advance women's equality. One early, 1980s, source was the attempt to integrate gender issues into policy-making in the area of development assistance, such as lending by the World Bank or decision-making in the United Nations Development Programme.[1] Since then, the concept has been adopted in ever expanding areas. Mainstreaming was adopted as a policy goal at the Fourth United Nations world conference on women in 1995.[2] More recently, the European Commission has become involved in developing such approaches in Europe.[3] The Council of Europe convened a group of specialists on mainstreaming in February 1996; their report in March 1998 presented a conceptual framework, a methodology for conducting mainstreaming, and a discussion of "good practice" in the area.[4] At the national level also there are examples of mainstreaming policies, some already in existence, some in embryo, in the Netherlands, Sweden, Denmark, Flanders, Portugal, Finland, Ireland, Canada, Australia, and New Zealand.[5]

We shall see below that the concept is not free of ambiguities and problems, some of which are brought into greater clarity by the experience of constitutionalising mainstreaming in Northern Ireland. The Northern Ireland model is unusual, if not unique, in two respects. First, the mainstreaming undertaken goes beyond gender. In particular, it focuses attention on equality between the two religio-political communities in Northern Ireland as well as several other groups. Secondly, it is underpinned by a firm legal foundation. How did this come about? To understand fully the development of mainstreaming in Northern Ireland, we need to begin the story much earlier, with the civil rights movement of the late 1960s.

[1] Shahra Razavi and Carol Miller *Gender Mainstreaming: A Study of Efforts by the UNDP, the World Bank and the ILO to Institutionalize Gender Issues* (Geneva, United Nations Research Institute for Social Development, 1995).

[2] United Nations Report of the Fourth World Conference on Women (UN DOC A/Conf.177/20, 1995).

[3] European Commission *Equal Opportunities for Women and Men in the European Union 1996* (1997) pp. 15–20.

[4] Council of Europe, Rapporteur Group on Equality between Women and Men, *Gender Mainstreaming* (GR-EG (98), 1 March 26 (1998)).

[5] Christopher McCrudden "Mainstreaming Equality in the Governance of Northern Ireland" (1999) 22 *Fordham International Law Journal* 1696–1775.

DEVELOPING AN ANTI-DISCRIMINATION AGENDA IN NORTHERN IRELAND

The Northern Ireland civil rights campaign of the 1960s focused on the need to eradicate discrimination between Catholics and Protestants.[6] This movement led to some action by the then Northern Ireland Government, but anti-discrimination legislation as such began after the Northern Ireland Government was suspended in 1972 and "direct rule" was introduced. The Northern Ireland Constitution Act 1973 made it unlawful for a public authority carrying out functions relating to Northern Ireland to discriminate, or to aid or incite another person to discriminate, against a person or class of person on the ground of religious belief or political opinion. Two features of the Act's approach are important. First, it protected from discrimination only in the religio-political context. Secondly, it protected only from direct discrimination, that is (to put it somewhat over-simplistically) discrimination which arises from an intentional act. There was, as a consequence, little litigation under these provisions.

The second major development in anti-discrimination legislation was in the area of employment. A government committee (the Van Straubenzee Committee) considered the question of discrimination in the private sector of employment in 1973 and produced a penetrating report.[7] The Fair Employment Act 1976 partially implemented this report, and addressed also employment in the public sector. A Fair Employment Agency was established to enforce the legislation in 1977. However, the legislation had little effect on employers' practices. Research carried out by the Policy Studies Institute in 1987 showed that the vast majority of employers believed that the Act had made little, if any, impact on their behaviour.[8] The research also confirmed the extent of the economic inequality between the two communities in Northern Ireland. According to the PSI study, for example, Catholic male unemployment, then at 35 per cent, was two and a half times that

[6] *Disturbances in Northern Ireland: Report of the Commission appointed by the Governor of Northern Ireland* (Cmd 532, 1969).

[7] *Working Party on Discrimination in the Private Sector of Employment: Report and Recommendations* (the Van Straubenzee Report) (Belfast, Ministry of Health and Social Services, HMSO, 1973).

[8] David Smith and Gerald Chambers *Inequality in Northern Ireland* (Oxford, Clarendon Press, 1991).

of Protestant male unemployment, and continued at this level despite there being over 100,000 job changes a year.

From the mid-1980s, inequality of opportunity between Catholics and Protestants again became a key political issue, largely due to pressure from outside Northern Ireland. A campaign was begun in the USA to bring pressure to bear on American corporations, state legislatures and municipal governments with investments in Northern Ireland to adopt a tougher set of anti-discrimination principles (called the "MacBride Principles") and sought to encourage employers to engage in affirmative action.[9] The MacBride campaign met with opposition from the British Government, but proved popular with American state and city legislators. A number of states enacted legislation requiring American companies in which they invested to ensure fair employment practices in their Northern Ireland subsidiaries. This American campaign began to fill, however partially and inadequately, the vacuum caused by the failure of institutions in Northern Ireland to address the issue adequately.

Partly in response, in 1986 the Northern Ireland Department of Economic Development proposed new legislation that offered hope of a more robust approach, but the proposals, emphasising voluntary compliance, fell short of what was likely to be effective.[10] In October 1987 the Standing Advisory Commission on Human Rights (SACHR) published a major report providing a comprehensive and authoritative analysis of the problem and a detailed set of proposals for legislation and other government initiatives.[11] Crucially the report shifted the terms of the debate from the eradication of prejudiced discrimination to the reduction of unjustified structural inequality in the employment market, whether caused by discrimination or not. In December 1988 the Government responded by publishing new legislation. After significant amendments this was passed in July 1989.[12] The new Fair Employment Act 1989 marked a departure from existing approaches, emphasising compulsory rather than voluntary compliance, giving

[9] Christopher McCrudden "Human Rights Codes for Transnational Corporations: What Can the Sullivan and MacBride Principles Tell Us?" (1999) 19 *Oxford Journal of Legal Studies* 167–201.

[10] Department of Economic Development *Equality of Opportunity in Northern Ireland: Future Strategy Options* (Belfast, HMSO, 1986).

[11] SACHR *Religious and Political Discrimination and Equality of Opportunity in Northern Ireland: Report on Fair Employment* (Cmd 327, 1987).

[12] Christopher McCrudden "The Evolution of the Fair Employment (Northern Ireland) Act 1989 in Parliament" in R J Cormack and R D Osborne (eds) *Discrimination and Public Policy in Northern Ireland* (Oxford, Oxford University Press, 1991).

broader powers to the enforcement agency (the Fair Employment Commission), and requiring limited affirmative action and compulsory monitoring. The Act adopted many, but by no means all of the SACHR Report's Recommendations.

In some respects, the requirements of the legislation were far-reaching and certainly ahead of their time in comparison with the rest of the United Kingdom. The main features of the legislation, until substantially amended in 1998 (see below) were that individual complaints of religious and political discrimination in employment were made to the Fair Employment Tribunal, a specialised wing of the industrial (now employment) tribunal system in Northern Ireland. Unlawful discrimination was defined to include both direct and indirect discrimination. The Fair Employment Commission (FEC) might assist complainants in taking such complaints. Remedies included damages (except for unintentional indirect discrimination) and recommendations for action by the employer for reducing the adverse effect on the complainant of any unlawful discrimination.

There were several exceptions to the general prohibition of employment discrimination, including exceptions for various types of affirmative action: training to remedy under-representation; encouraging applications from an under-represented community; and agreed redundancy schemes to preserve progress made under affirmative action measures. Another important exception provided that discrimination which was necessary for safeguarding national security or for protecting public safety or public order was not unlawful. The Act provided that the Secretary of State's certificate was conclusive evidence that an act was done for these purposes.

In addition to providing a means of resolving complaints of unlawful discrimination, the legislation established a detailed regulatory structure to ensure that employers took action without the need for a complaint to trigger action. Many private sector employers (those specified in delegated legislation) were required to register with the Fair Employment Commission. Public sector employers were deemed to be automatically registered with the Commission. Registered employers were required to monitor the religious composition of their full-time workforce. Larger employers (those with more than 250 employees) and all public sector employers were required to monitor the religious composition of applications for employment. A monitoring return had to be completed yearly. Employers had to classify existing (and where relevant, prospective) employees by sex, religion and occupation.

Religion might be determined either by reference to the school(s) attended or by directly asking the employee or applicant, or by using other specified methods. This monitoring return must be submitted to the Fair Employment Commission. Although not provided for by the Act, the Fair Employment Commission decided soon after the Act came into force to publish regularly an overview of the results of monitoring.

Employers were also under a duty periodically (every three years) to review their employment practices (excluding redundancy) for the purpose of determining whether members of each community were enjoying, and were likely to continue to enjoy, fair participation in employment in the concern. Where fair employment was not evident, employers were required to engage in affirmative action. Affirmative action was also enforceable by the Commission, as was the setting of goals and timetables against which to measure progress. Government contracts and grants might be withdrawn in cases of persistent and recalcitrant behaviour.

The FEC had the power to review patterns and practices in employment and where necessary to issue directions which were enforceable on employers. Appeals against such directions were heard by the Fair Employment Tribunal. The Commission was also able to accept binding agreements from employers which were enforceable if not complied with. The Commission was able to revise the Code of Practice, the first version of which was produced by the Department of Economic Development.

This basic structure remained substantially unchanged between 1990 and 1998. However, several changes of detail were introduced. In particular, in 1991, the 1989 Act was amended to introduce a revised approach to the confidentiality of monitoring information. The revised approach generally preserved the confidentiality of monitoring information, but permitted the disclosure of this otherwise confidential information to several statutory bodies (including the FEC), and the Fair Employment Tribunal. The legislation was also amended in 1994 to remove the limit on compensation which might be awarded by the Fair Employment Tribunal.

Aside from these provisions regarding discrimination and employment equality between the two religio-political communities, from 1976 Northern Ireland adopted equivalent measures to those dealing with sex discrimination and equal pay as those adopted in the rest of the United Kingdom. Provisions dealing with discrimination on the basis of race were much longer delayed. Although the main legislation addressing racial discrimination in the rest of the United Kingdom had been passed in 1976, it was not until 1997 that the equivalent legislation was enacted for Northern Ireland.

"POLICY APPRAISAL AND FAIR TREATMENT" (PAFT)

Anti-discrimination law (even of the breadth of the fair employment legislation) was, however, gradually perceived as insufficient to achieve the substantial change that the 1987 SACHR Report had defined as necessary. In its Second Report in 1990, SACHR argued that the Government should establish machinery that would monitor the impacts of legislation, policy and administration on equality of opportunity and on relations between the two sections of the community.

Another development at this time involved the reform of "community relations" policy making within the Northern Ireland Office. In September 1987 Tom King, then Secretary of State for Northern Ireland, announced the establishment of a Central Community Relations Unit within the Central Secretariat of the Northern Ireland Office. The purpose of this reorganisation, according to the announcement, was to ensure that in "every decision we take, whether it is in the fields of housing, education, planning or employment, or any other fields of government, . . . we have taken into account any community relations aspects there may be".[13] The new unit would coordinate all Northern Ireland policy-making. In discussions with the Northern Ireland Office before the new initiative was announced, SACHR was informed that it was intended that a senior officer in each department would be made responsible for examining policies and proposals in relation to their community impact. If, in the view of that officer, any such policy or proposal might have a disparate community impact, the matter could be taken to the Permanent Secretary and ultimately to the Secretary of State for decision.

More generally, British administrative policy in the rest of the United Kingdom was becoming more favourably disposed to attempts to engage systematically in "policy appraisal", and to "mainstream" other policies in government.[14] Since the 1980s regulatory impact

[13] "Secretary of State takes Direct Responsibility for Community Relations Matters", Northern Ireland Information Service, 8 September 1987.

[14] McCrudden *supra* n. 5.

assessments had often been required, as had occasional cost/benefit analyses of proposed projects. In addition, "proofing" government policy proposals to ensure compliance with certain obligations was becoming more common.

All these elements contributed to the announcement by the Government in 1990 that a non-statutory policy of "equality proofing" would be introduced in Northern Ireland. A circular was issued giving advice to all Northern Ireland departments about the need to consider discrimination in relation to religious affiliation, political opinion and gender.[15] This was coordinated with an initiative launched in the United Kingdom by the ministerial group on women's issues that encouraged all government departments to develop basic guidance on equality proofing throughout the United Kingdom.

There were several years of controversy over the content of the guidelines in Northern Ireland, including the failure to cover areas such as race, disability and age. Revised and more inclusive guidelines, renamed the Policy Appraisal and Fair Treatment guidelines (PAFT), came into effect in January 1994. The groups coming within the scope of its guidelines went beyond the two religious communities, and included people of different gender, age, ethnic origin, marital and family status and sexual orientation, and the disabled. PAFT was an attempt to establish a procedure within government decision-making by which the principles of equity and equality could be made effective. "Equality and equity", it said, "are central issues which must condition and influence policy making in all spheres and at all levels of Government activity".[16] We can see here the crucial shift from an anti-discrimination to a mainstreaming approach. But little detailed guidance was given to departments or other public bodies as to how to accomplish this task, although a commitment was subsequently given that the Annual Report on PAFT implementation by the Central Community Relations Unit (CCRU) would be published, providing a degree of transparency to the process.

There were, moreover, a number of unresolved ambiguities. First, it was unclear how far the Government was willing to go beyond action of an anti-discrimination kind. Secondly, it was unclear whether the initiative was much more than "window dressing" in response to political pressure, particularly from the USA. Thirdly, many aspects of the

[15] Central Secretariat Circular, *Equal Opportunity Proofing: Guidelines* (January 1990).

[16] Central Secretariat Circular, *Policy Appraisal and Fair Treatment* (May 1993).

guidelines—the stress on the UK context, their inclusiveness, the use of international human rights language and concepts—seem to have been designed to make more acceptable to civil servants and public opinion an initiative whose primary rationale was the need to tackle Catholic disadvantage.

The inclusive, broad and radical-sounding nature of the initiative raised expectations that proved difficult to satisfy in practice. Unlike the equivalent guidelines in the rest of the United Kingdom, the PAFT guidelines were available from government on request and were widely circulated by NGOs among the relevant groups. Perhaps naively, they took the PAFT guidelines at face value, expected fairness, and behaved accordingly. When the promise was not delivered, unsurprisingly, they mobilised. The guidelines were soon embroiled in public controversy. Unison, the public sector trade union, took judicial review proceedings against one of the public bodies that intended to privatise its services on the grounds that to do so discriminated against women. The judicial review was ultimately unsuccessful, but two things emerged. First, the PAFT guidelines had not formally been issued to the public body concerned, which was a considerable embarrassment for the government. Secondly, the court held that, had the guidelines been issued properly, the public body would have been legally required to take them into account. This appeared to give the guidelines a legal status, something that had hitherto not been clear.

It became clear later, however, that while departments had to take the guidelines into account, once they did so, it would be difficult to contest their decision legally, whatever the result of that consideration—in short that PAFT was legally enforceable procedurally, but not substantively. However, the effect of all this was to raise the political status of the guidelines in the eyes of both the public bodies and departments to which they applied and the campaigning groups. In a sustained attempt to encourage groups to use the guidelines, the Committee on the Administration of Justice (CAJ), a Northern Ireland human rights NGO, organised briefing sessions on the guidelines for a range of interested voluntary and community organisations. The NGOs responded with enthusiasm. A loose coalition was born that was dedicated to putting PAFT into effect.

Meanwhile, another factor played an important role in making PAFT a major focus of political interest. During the passage of the Fair Employment Act 1989, the Government committed itself to conducting a formal review of the operation of the legislation and other government policy in this area within five years of its commencement. Originally, this task was given to the CCRU within the Northern Ireland Office, the government department responsible for Northern Ireland, but responsibility was later transferred to SACHR. SACHR commissioned research into several areas of government policy as part of its enquiry. Particularly important among this research was a short but highly critical piece on the operation of PAFT, which showed that PAFT appeared to be largely ignored by substantial sections of the policy-making apparatus of government.[17] Increasingly, the focus of political attention shifted from concern about the operation of the Fair Employment Act narrowly conceived, to the ineffectiveness of the policy that had been seen as a necessary complement to the legislation—PAFT.

EMERGENCE OF ALTERNATIVE MODELS OF MAINSTREAMING

The potential for a mainstreaming approach to impact significantly on inequality, combined with evidence of the lack of such impact in practice, contributed to pressure for reform. Unison, the union involved in the initial judicial review, commissioned the author to prepare a study on reform of PAFT. A discussion paper, *Mainstreaming Fairness*, was produced which set out various options and raised questions for further consideration.[18] A possible model for a statutorily-based PAFT was tentatively suggested to stimulate debate. The proposals included provision for a statutory duty to be imposed on the Secretary of State and on public bodies to ensure that material inequalities between certain groups should be progressively reduced.

In November 1996 the Committee on the Administration of Justice (CAJ), the human rights NGO which had taken up the PAFT issue, circulated the paper extensively among opinion formers, trade unions, voluntary groups, lawyers, politicians, and civil servants in Northern

[17] Robert Osborne, Anthony Gallagher, and Robert Cormack, with Sally Shortall "The Implementation of the Policy Appraisal and Fair Treatment Guidelines in Northern Ireland', in Eithne McLaughlin and Padraic Quirk, (eds), *Policy Aspects of Employment Equality in Northern Ireland,* Employment Equality in Northern Ireland Series, vol. 2 (Belfast, Standing Advisory Commission on Human Rights, 1996) pp. 127–52.

[18] Christopher McCrudden, Committee on the Administration of Justice, *Mainstreaming Fairness? A Discussion Paper on "Policy Appraisal and Fair Treatment"* (Belfast, Committee on the Administration of Justice, 1996).

Ireland, and requested comments. Significantly, one of the earliest responses to the paper was in the form of an extensive discussion by one of the researchers within SACHR.[19] This developed the "Mainstreaming Fairness" proposal further, and suggesting the establishment of an Equality Commission to oversee public sector application of equality proofing mechanisms.

In early 1997 the Government responded to both papers with a detailed critique.[20] Prominent civil servants had earlier expressed concern at the growing pressure to give legislative force to PAFT.[21] Now, a sustained attack was mounted against the idea. It was argued that the proposals would "effectively constitutionalize [*sic*]" the equality aspiration in a manner which "would seek to dictate the socio-economic policies of future governments, irrespective of electoral mandates or budgetary constraints",[22] and that the emphasis on consultation and external participation could undermine representative democracy. Other concerns were also raised, including the bureaucratic burden and extra costs in implementing PAFT.

Despite these objections, the Mainstreaming Fairness proposal was substantially taken up by SACHR and became one of its central recommendations for reform. SACHR reported in June 1997, criticising the existing implementation of PAFT and making detailed recommendations for a revised scheme.[23] The Report recommended, as a minimum, that a number of measures should be incorporated into the PAFT system. These included, "effective political control over, and responsibility for, the policy on both direct and indirect effects on equality generally and community differentials in particular", "adequate monitoring of both the direct and indirect impacts of policy on community differentials, and other equality measures", "full consideration of alternative policies which might give effect to government

[19] Nigel Hutson *Policy Appraisal and Fair Treatment in Northern Ireland: A Contribution to the Debate on Mainstreaming Equality* (Belfast, Standing Advisory Commission on Human Rights, 1996).

[20] *Commentary on "Mainstreaming Fairness?—A Discussion Paper on Policy Appraisal and Fair Treatment" by Dr Christopher McCrudden and on "Policy Appraisal and Fair Treatment in Northern Ireland: A Contribution to the Debate on Mainstreaming Equality"* (Belfast, Committee on the Administration of Justice, 1997).

[21] David Watkins "Comments", in Equal Opportunities Commission for Northern Ireland *Working Towards Equality in the Twenty-First Century: Report of a Conference held on 23 October 1996* (Belfast, Equal Opportunities Commission for Northern Ireland, 1997).

[22] *Ibid.*

[23] SACHR *Employment Equality: Building for the Future* (Cmd 3684, 1997).

objectives but reduce or avoid unwelcome effects on equality generally and community differentials in particular", "greater transparency in the manner in which government policy is assessed" and "greater accountability in the manner in which the civil service and public bodies fulfil their remit to promote equality". More far-reaching still, the SACHR Report recommended that the policy on PAFT be given legislative form, with enforcement based on an internal NIO unit, such as a strengthened CCRU.

THE GOOD FRIDAY AGREEMENT AND EQUALITY

The Good Friday Agreement had a crucial impact on the development of these issues. Up to then the debate had only indirect relevance for, or input into, the search for a constitutional settlement to the conflict. By 1997, however, a new politics was emerging in Northern Ireland that meant that previous approaches to resolving the constitutional problem were to be supplemented with a new concentration on equality. The debate on equality issues, including PAFT, previously separate from the political negotiations, now became entangled with them. In particular, both the revision of the Mainstreaming Fairness proposal, and the British Government's response to the SACHR Report, have to be seen in the context of the peace negotiations which culminated in the Good Friday Agreement.

Several developments affected the ultimate outcome. First, in May 1997, a new Labour Government was elected, committed to breathing new life into the constitutional talks, unencumbered by a unionist veto, and backed by a substantial majority in the House of Commons. Soon after, the IRA resumed its cease-fire. Suddenly it seemed as if a peace settlement might actually emerge. The two issues—equality and the search for a settlement—now became intertwined. For the two Governments, equality issues were perceived as an important part of "confidence building" in the Catholic/nationalist community and, from then on, policy proposals on equality were affected significantly by the talks process.

Secondly, while earlier attempts at establishing peace in Northern Ireland had addressed questions of discrimination and human rights, this time the talks involved parties which had not participated previously and which viewed equality and human rights issues as particularly salient, including Sinn Féin and the various fringe loyalist parties. For these parties, failure to address successfully equality and human rights issues important to their communities would make it much more difficult for them to "sell" any agreement. Once human rights were identified as an area that was important, particularly to Sinn Féin, it then became important for those who wanted to keep Sinn Féin "on board" to include it in their own proposals. The SDLP, Sinn Féin and the PUP all embraced a reform of PAFT as part of its strategy on equality. So too did the Women's Coalition, which played an important role in keeping the issues to the fore in the negotiations. For the Ulster Unionist Party, equality was either an issue that it considered it could not oppose, or did not consider sufficiently important to make a priority.

Thirdly, there developed outside the formal talks process what the journalist, Mary Holland, called a "parallel peace process", involving an informal coalition of such bodies as the Committee on the Administration of Justice, Unison, the Women's Support Network, and many other community and NGO groups in Northern Ireland.[24] This loose network had contacts within the talks process, in particular through the Women's Coalition, and succeeded in getting at least part of the human rights and equality agenda into the negotiations both before and during the final frenzied days.

Fourthly, the new Labour Government was more comfortable with a strong "rights" approach than the earlier Conservative Government had been. The new Secretary of State, Mo Mowlam, whilst in opposition, had supported draft legislation on equality, and was much closer to the "parallel peace process" than any of her predecessors had been. Before the election she publicly announced that she intended "to make it a statutory duty for government bodies to take equality of opportunity into account through more rigorous enforcement of the Policy Appraisal and Fair Treatment guidelines".[25]

Negotiations on the equality agenda in the talks took place largely in the months of December 1997 to April 1998. In January 1998 the British and Irish Governments published a joint statement—the so-called "Heads of Agreement" paper—setting out their best guess on the bare bones of a settlement. The statement included a paragraph on human rights and equality that envisaged provisions to safeguard the rights of both communities in Northern Ireland "to achieve full respect for the

[24] Mary Holland "Latest Plan to Tackle Inequality Crucial to North Peace" *The Irish Times*, 12 March 1998.

[25] Majorie Mowlam "Towards Genuine Consent in Ulster" *The Independent*, 25 February 1997.

principles of equity of treatment and freedom from discrimination, and the cultural identity and ethos of both communities". The use of the term "equity", rather than the stronger term "equality", met with a hostile response from both nationalist commentators and human rights advocates. This appears to have concentrated Irish Government minds further on the equality issue.[26]

In February 1998 the Committee on the Administration of Justice published a revised proposal by the present author, *Benchmarks for Change*.[27] It proposed replacing PAFT with a statutory obligation to promote equality of opportunity and establishing a strong mechanism within the Northern Ireland civil service to monitor and enforce this obligation. It also envisaged a high degree of involvement by those outside government in the assessment and development of equality issues, including those affected by policy proposals and the statutory equality agencies. The proposal received extensive support across the range of groups most affected. Several of the political parties picked up aspects of it, as did the Irish Government.

The British Government's proposals on equality were set out in the White Paper, *Partnership for Equality*, published in March 1998. From the perspective of what subsequently occurred, two proposals were particularly important. First, the Government proposed a new statutory framework to supersede the PAFT's administrative guidelines. There would be a statutory obligation on Northern Ireland "public sector bodies" (including district councils and UK governemtn departments operating in Northern Ireland, as well as the Northern Ireland departments) to ensure that "consistent with their other responsibilities", their various functions "are carried out with due regard to the need to promote equality of opportunity in those areas covered by the current PAFT guidelines". Each public body would be required to adopt a statutory scheme setting out "how it proposed to take regard of its new statutory obligations in its day-to-day work".

[26] Niall O'Dowd, "Delicate Balance has been Upset by Paper's Pro-unionist Slant" *The Irish Times*, 17 January 1998; Christopher McCrudden "Replacing 'Equality' with 'Equity' shows the Difference a Word can Make" *The Irish Times*, 22 January 1998; Bertie Ahern Speech at Trinity College Dublin, 3 February 1998; Bertie Ahern Statement of An Taoiseach, Mr Bertie Ahern TD on Northern Ireland, Dáil Éireann, 4 February 1998.

[27] Christopher McCrudden, Committee on the Administration of Justice *Benchmarks for Change: Mainstreaming Fairness in the Governance of Northern Ireland* (Belfast, Committee on the Administration of Justice, 1998).

Secondly, the Government proposed, subject to public consultation, to create a new unified statutory authority bringing together the existing Northern Ireland equality agencies: the Fair Employment Commission, the Equal Opportunities Commission, the Commission for Racial Equality and the Northern Ireland Disability Council. The intention was to provide an institutional mechanism to monitor and enforce the new statutory duty. The new body, external to the civil service, would provide the various public bodies with the assistance necessary to enable them to implement the duty effectively. The proposed amalgamation of the existing bodies was to prove the most controversial element of the White Paper, giving rise to considerable unease amongst the equality bodies themselves (with the exception of the Fair Employment Commission) and many of the NGO groups.

One of the key issues urged on the parties to the negotiations was the centrality of the human rights and equality issues to the success of the peace process. Moreover a consensus emerged that discussion should go beyond a classical, narrow definition of rights centred on political and civil rights, to include social, cultural and economic rights, and that equality issues should be mainstreamed. In the months of negotiation, it also emerged that there was significant support across the political spectrum for the CAJ and SACHR approach to the replacement for PAFT. It was feared that the British Government would seize the opportunity of the negotiations to bolster its, apparently more limited, White Paper proposals with the authority of a peace agreement. Arguments were put to several of the parties to the negotiations to try to prevent this, and to have stronger proposals inserted in the final text.

All this activity was reflected in the "Mitchell Document" that was presented by the Chairmen of the talks at the beginning of April 1998 as a draft paper for discussion. The bones, and much of the flesh of the ultimate Agreement, were in the Mitchell Document, including the sections on equality and human rights. However, some significant changes regarding equality were made to these aspects of the document in the run-up to final agreement on 10 April. Also, the Agreement departed from the White Paper proposals in some important respects.

Two equality agendas were addressed in the Agreement, one national, the other social. The national equality agenda—equal respect for the two different allegiances, Irish and British—was reflected in the institutional provisions designed to ensure fair representation in the Assembly and Executive, and in the establishment of North-South institutions. Beyond these arrangements, however, the parties affirmed a list of important rights; the right of free political thought: the right to freedom and expression of religion; the right to pursue democratically national and political aspirations; and the right to seek constitutional change by peaceful and legitimate means. A new Bill of Rights, supplementing the European Convention on Human Rights, was envisaged to reflect the principles of "mutual respect for the identity and ethos of both communities and parity of esteem". In addition, there were new obligations on government to encourage the use of the Irish language.

The Agreement was equally forthright and inclusive on social equality and it is on this area that I will concentrate in the remainder of the chapter. The parties affirmed "the right to equal opportunity in all social and economic activity, regardless of class, creed, disability, gender or ethnicity; . . . and the right of women to full and equal political participation". (The references to disability, ethnicity and participation by women were inserted by the negotiators during the final days.)

Provisions governing Ministers in the new Executive Authority, in particular the Pledge of Office, required Ministers "to serve all the people of Northern Ireland equally, and to act in accordance with the general obligations on government to promote equality and prevent discrimination". (This was also inserted in the last few days of the negotiations.) Under the Code of Conduct for Ministers, Ministers were required to "operate in a way conducive to promoting good community relations and equality of treatment". Moreover, an individual "may be removed from office following a decision of the Assembly taken on a cross-community basis, . . . for failure to meet his or her responsibilities including, inter alia, those set out in the Pledge of Office" which included the duty of equality and impartiality.

Pending the devolution of powers to a new Northern Ireland Assembly, the British Government committed itself to pursuing policies for sustained economic growth and stability in Northern Ireland and for promoting social inclusion, including in particular community development and the advancement of women in public life. Subject to public consultation, the British Government would also develop a new regional development strategy for Northern Ireland, for consideration in due course by the Assembly. This would aim to tackle the problems of a divided society and social cohesion in urban, rural and border areas. The Government also planned to introduce "a new more focused Targeting Social Need initiative" and a range of measures aimed at combating unemployment and progressively eliminating the differential in unemployment rates between the two communities by targeting objective need.

A new Northern Ireland Human Rights Commission would be established, with membership "reflecting the community balance". This would be established by Westminster legislation, independent of government, with an extended and enhanced role beyond that exercised by the Standing Advisory Commission on Human Rights. Its duties would include keeping under review the adequacy and effectiveness of laws and practices, making recommendations to government, providing information and promoting awareness of human rights, considering draft legislation referred to it by the new Assembly and "in appropriate cases" bringing court proceedings or providing assistance to individuals doing so. It would consult and advise on "the scope for defining, in Westminster legislation, rights supplementary to those in the European Convention on Human Rights, to reflect the particular circumstances of Northern Ireland, drawing as appropriate on international instruments and experience". It would also give consideration to offering "a clear formulation of the rights not to be discriminated against and to equality of opportunity in both the public and private sectors". In this respect, the Agreement kept open the possibility that a Northern Ireland Bill of Rights would include the concept of "indirect discrimination" in any new anti-discrimination duty applying to the actions of public bodies in Northern Ireland, an idea which was rejected in the earlier White Paper.

The Agreement noted that "[s]ubject to the outcome of public consultation underway, the British Government intends as a particular priority, to create a statutory obligation on public authorities in Northern Ireland to carry out all their functions with due regard to the need to promote equality of opportunity in relation to religion and political opinion; gender; race; disability; age; marital status; dependants; and sexual orientation". Under the Agreement, "[p]ublic bodies [would] be required to draw up statutory schemes showing how they would implement this obligation". As part of the equality duty, they would be required to include "arrangements for policy appraisal, including an assessment of impact on relevant categories". The references to impact assessment and information were added at a late stage of the negotiations. In these respects the Agreement went further than the White Paper.

The Agreement additionally proposed "arrangements to provide that key decisions and legislation are proofed to ensure that they do not infringe the ECHR and any Bill of Rights for Northern Ireland". The Assembly "may appoint a special Committee to examine and report on whether a measure or proposal for legislation is in conformity with equality requirements, including the ECHR/Bill of Rights". This had not been included in the White Paper. The Assembly "shall then consider the report of the Committee and can determine the matter in accordance with the cross-community consent procedure". It would be "open to the new Northern Assembly to consider bringing together its responsibilities for these matters into a dedicated Department of Equality". These elements, too, were added at a late stage.

The Agreement noted that the British Government proposed to create a new statutory Equality Commission to replace the Fair Employment Commission, the Equal Opportunities Commission (NI), the Commission for Racial Equality (NI) and the Disability Council. Such a unified Commission would "advise on, validate and monitor the statutory equality obligation and will investigate complaints of default". However this proposal, which did not have the support of the negotiating parties or the Irish Government, was "subject to the outcome of public consultation currently underway", a condition which had not been included in the original Mitchell document.

The British Government also undertook, subject to public consultation, to "make rapid progress" with the measures on employment equality included in the White Paper, "and covering the extension and strengthening of anti-discrimination legislation, [and] a review of the national security aspects of the present fair employment legislation . . .".

The new British-Irish Intergovernmental Conference was also given significant human rights responsibilities. Strand Three provided that "[i]n recognition of the Irish Government's special interest in Northern Ireland and of the extent to which issues of mutual concern arise in relation to Northern Ireland, there will be regular and frequent meetings of the Conference concerned with non-devolved Northern Ireland matters, on which the Irish government may put forward views and proposals. These meetings, to be co-chaired by the Minister for Foreign Affairs and the Secretary of State for Northern Ireland, would also deal with all-island and cross-border co-operation on non-devolved issues. The Conference . . . will address . . . the areas of rights . . . in Northern Ireland (unless and until responsibility is devolved to a Northern Ireland administration)".

LEGISLATING THE EQUALITY ASPECTS OF THE AGREEMENT

The Agreement thus contained strong provisions on equality. The question was whether these would be adequately reflected in the implementing legislation. For campaigners, the issue became how to translate what appeared to be a breakthrough at the political level into legislative text. This was to prove a difficult task, particularly in ensuring the effective incorporation of the public sector equality duty that would replace PAFT. The period between the conclusion of the Agreement in April, and the publication of the Bill implementing that Agreement in July, saw 123 submissions on the Government's White Paper of March, *Partnership for Equality*. All but two of these were received after the Agreement was concluded and many took account of the Agreement in their submission. Most endorsed the principle that equality of opportunity should be placed on a statutory basis, though a substantial proportion questioned whether the government proposals would achieve their stated objective. Many also opposed the creation of a unified Equality Commission. These two issues, on which this section concentrates, increasingly dominated public discussion of the equality agenda in this period.

In May, an alternative proposal to that adopted in the White Paper on mainstreaming was submitted to the Secretary of State suggesting that, in the light of the Agreement, several features of the White Paper's proposals should be revised.[28] It argued that the enforcement and monitoring of the equality of opportunity duty on the public sector should be carried out by establishing an effective internal monitoring and enforcement mechanism within the Northern Ireland Civil Service/Executive. This should be complemented by mechanisms for increased public participation and a role for the existing equality commissions. The powers that the White Paper recommended for the Secretary of State to intervene, where there was a breach of the equality duty by a public body, should be strengthened and clarified. Further research was necessary to assess the effect of the amalgamation of existing equality bodies into one equality body (or into a new Human Rights Commission) before a decision on amalgamation should take place.

[28] Christopher McCrudden, Committee on the Administration of Justice *Equality: A Proposal in the Light of Multi-party Talks Agreement* (Belfast, Committee on the Administration of Justice, 1998).

The Secretary of State announced her decisions on the equality aspects of the Bill on 10 July. Her announcement made it clear that the campaign to modify the Government's proposals on amalgamating the equality commissions into a new Equality Commission had been unsuccessful. The Equality Commission envisaged by the White Paper would be set up. At the same time she was concerned to reassure critics that the White Paper proposals "seemed to pass responsibility for the promotion and oversight of the equality of opportunity obligation to an external body" and stressed that "internal arrangements for co-ordinating, promoting and monitoring the activities of Government Departments and public bodies must also be rigorous and effective".[29] To meet the criticism that issues of religious equality would dominate the working of the new body, she indicated that the legislation would require the Equality Commission to devote appropriate resources to gender, race and disability issues. It would also allow the Commission to establish consultative councils on these issues.

The Secretary of State also responded to public consultations on the White Paper on the matter of the form and content of the equality duty on public authorities. She indicated that several criticisms of the proposals contained in the White Paper "were based on misapprehensions". It was not the intention "to leave substantial areas of discretion to those in the public sector". "To remove any ambiguity, the requirements on the public sector to carry out appraisals of policies, including equality impact assessments, to consult with representatives of interests which might be affected, and to publish information on appraisals, will all be clarified in the Bill".

The Government introduced the Northern Ireland Bill into the House of Commons on 15 July. It received its second reading on 20 July. Clause-by-clause consideration of the contents of the Bill took place (first in Committee, then at Report stage) between then and 31 July when the Bill was given its Third Reading. During the summer, further intensive consultations took place between the Government and interested groups (including the Northern Ireland political parties). The Bill was then given its second reading in the House of Lords on 5 October (again the debate focused on the principles underlying the Bill). This was followed by detailed clause-by-clause consideration of the Bill at the Committee and Report stages on 26 October and 10–11 November respectively. The Lords Third Reading debate took place on

[29] "Secretary of State announces Equality White Paper Decisions", Northern Ireland Information Service, 10 July 1998.

17 November. The next day the Commons considered the Lords amendments, and agreed them. The Bill received the Royal Assent on 19 November 1998 and became law.

The politics of the Bill's passage is important for an understanding of what transpired during the parliamentary phase. The Government commanded a sizeable majority in the House of Commons and there was never any doubt that it could push the legislation through in any form that it wished. Nor was there any serious prospect that the Conservative majority in the House of Lords would choose the legislation implementing the Agreement as a basis for attacking the Government. Indeed, neither in the Lords nor in the Commons did the Conservative opposition appear to have played much of a role in the negotiations surrounding the Bill's passage. The Government Ministers involved were Paul Murphy, MP, in the Commons, and Lord Williams and Lord Dubs in the Lords. Behind Mr Murphy stood the Central Community Relations Unit of the Northern Ireland Office (based in Belfast) and the Northern Ireland Office civil servants in London. The other principal actors on the parliamentary stage, as regards the equality aspects of the Bill, were the Northern Ireland MPs in the Commons, and a few Labour backbenchers. In particular Kevin McNamara, MP, in the Commons and Lord Archer of Sandhill in the Lords both played key roles. Finally, Lord Lester of Herne Hill was the Liberal Democrat front-bench spokesperson on the equality aspects of the Bill in the Lords.

Outside the Houses of Parliament, the main actors were the Northern Ireland political parties, the Irish Government, the statutory equality agencies, and the loose coalition (including such bodies as the Northern Ireland Council on Ethnic Minorities, Unison and Disability Action) that took its cue largely from the briefings of the CAJ. Behind the scenes (only when the Bill was in its final stages was there any attempt to "go public" on these issues), the CAJ and members of the coalition briefed influential figures in the United States Administration and Congress, British parliamentarians, the Irish Government and other NGOs, whilst also being consulted directly by Paul Murphy. Particular attention was paid by the CAJ and the statutory agencies to constructing a sufficient consensus across the Northern Ireland parties to make it difficult for others to argue that the Government should not "take sides" between the parties on these issues. This resulted in several amendments being jointly supported by Ulster Unionists, the SDLP, and Liberal Democrats.

In the Commons Second Reading debate, considerable attention was given to the issue of amalgamation of the existing equality commissions, rather than the issues surrounding the equality duty on public authorities. Both the Secretary of State and Mr Murphy made a strong commitment to engage in further consultations during the summer months after the Bill had left the Commons and to amend the equality aspects of the Bill before consideration in the Lords. During the summer, Paul Murphy consulted extensively on the equality and human rights aspects of the Bill with the political parties represented in the Assembly, the chairs of the existing equality commissions, the CAJ, members of the coalition, and others. As a result of these meetings, it became clear that the Government was prepared to introduce extensive amendments on the equality aspects of the Bill.

On 14 October, a few days before the Committee stage was to begin, Mr Murphy announced the Government's response to the summer consultations on the equality issues in the Bill, and the type of amendments that it would support in the Lords. First, during the summer, confusion had arisen as to the allocation of responsibility for equality issues once powers were devolved to the new Assembly and Executive. The Murphy announcement proposed that the provisions of the Bill on equality (basically the Equality Commission and the equality duty) would be reserved matters. That meant that the Secretary of State would continue to have responsibility for them, although the Assembly would be able to legislate on these issues with the permission of the Secretary of State. However, the existing bodies of law on fair employment, gender equality, race relations and disability discrimination in Northern Ireland would become transferred matters, on which the Assembly would have legislative responsibility. The Bill would be amended, in addition, to ensure that the Assembly would be kept more closely informed on the enforcement of the new statutory equality duty.

Secondly, there was further clarification of the equality duty. The obligation would apply to UK government departments operating in Northern Ireland (including the Northern Ireland Office) as well as to Northern Ireland departments and public bodies. Public bodies would be required to produce equality schemes, without being requested by the Equality Commission to do so, as had been the proposal until then. Greater detail would be included on what would be required to assess the impact of various policies on equality of opportunity. Assessments would include, for instance, consideration of alternatives that would better promote equality of opportunity. Public bodies would also be required to review their equality schemes on a five-yearly basis.

Some of the amendments introduced by the Government at Committee stage reflected this announcement in a straightforward manner. However, others did not. First, although the amended legislation now imposed a duty on most public authorities to produce schemes, it included the awkward provision that the Equality Commission would have to ask new public authorities to produce a scheme. Secondly, the amendments meant that the Equality Commission would be able to specify that only some functions of a public authority would be affected by the requirement to produce a scheme. Thirdly, there were no amendments requiring several other important aspects of impact analysis, including specification of the aims and purposes of the policy under assessment, specification of alternatives, specification of consultation procedures prior to decision-making and specification of the reasons for the policy eventually adopted by the authority. Nor did the amendments include requirement of a five-yearly review by public authorities of the measures taken to comply with the equality duty. Taken together, the Government's amendments seemed to reflect neither the summer consultations nor the Murphy announcement.

Several other issues had arisen during the summer that needed clarification. First, the consultations indicated that an amendment would be forthcoming which permitted an "affirmative action" exception to the "equality of opportunity" duty on public authorities. There was a concern that, without this, the equality of opportunity duty could be used to argue against measures that aimed at the reduction of disadvantage. There was some evidence that the PAFT guidelines had been mistakenly interpreted by some government departments to undermine just such provisions.[30] There was also a precedent for such an amendment. The Fair Employment Act 1976, which included an equality of opportunity provision, was amended in 1989 to include protection for affirmative action measures. Yet no such amendment appeared. Secondly, the schedule that included the details of the enforcement procedures on the equality duty was potentially ambiguous as to whether the impact of all policies would have to be assessed, or just those policies specifically concerned with equality of opportunity.

[30] Department of Health and Social Services *Policy Development and Review Unit, Review of Charging Policy for Non-Residential Personal Social Services* (Belfast, HMSO, 1997) annex 8.

The Lords Committee stage debate saw a detailed consideration of all the outstanding equality issues. The main challenge to the Government was led by Lord Archer and Lord Lester, both briefed extensively by the Committee on the Administration of Justice, the coalition, the statutory agencies and others. The outcome was that between the Committee stage and Report stage in the House of Lords, the Government came up with many amendments or statements of clarification which met the concerns of those arguing for a more explicit approach, particularly on the equality duty on public authorities. Moreover, where government felt an amendment was unnecessary, interpretative statements by Ministers often indicated why that was so.

THE NORTHERN IRELAND ACT 1998 AND EQUALITY

What, then, was the result of all these amendments and commitments? What does the Northern Ireland Act 1998, as finally passed, require? In this section the provisions of the Act concerning equality are considered. (Other chapters in this book consider the remaining human rights aspects of the Act.) The Act establishes a new Equality Commission for Northern Ireland, to consist of not less than fourteen nor more than twenty Commissioners appointed by the Secretary of State. The Secretary of State is to appoint one Commissioner as Chief Commissioner, and at least one Commissioner as Deputy Chief Commissioner. In making appointments, the Secretary of State is required, as far as practicable, to secure that the Commissioners, as a group, are representative of the community in Northern Ireland.

The Commission takes over the functions of the Fair Employment Commission for Northern Ireland, the Equal Opportunities Commission for Northern Ireland, the Commission for Racial Equality for Northern Ireland, and the Northern Ireland Disability Council, which are abolished. In exercising its functions the Equality Commission is required to aim to secure an appropriate division of resources between the functions previously exercisable by each of these bodies, and to have regard to advice offered by a "consultative council". This will be a group of people selected by the Commission to advise in relation to the functions in question.

The provisions of the Bill on equality—basically the Equality Commission and the equality duty—are to be reserved matters for which the Secretary of State will have responsibility. The Assembly will be able to legislate on these matters with the permission of the Secretary of State. The existing bodies of law on fair employment, gender equality, race relations and disability discrimination in Northern Ireland will become transferred matters, on which the Assembly will have legislative responsibility. The Bill was amended to reflect this demarcation of responsibility. In addition, further amendments were introduced to ensure that the Assembly will be kept closely informed on the enforcement of the new statutory equality duty.

Important safeguards are included in the Act. All Ministers must affirm the terms of the Pledge of Office set out in the Agreement. This includes a commitment to uphold the Code of Conduct also set out in the Agreement. Potentially, any legislation in the equality area can be made subject to the condition of cross-community support. Also, the Assembly cannot legislate in a way that is incompatible with rights under the European Convention on Human Rights or EU law (which is particularly relevant to gender discrimination issues). Nor may the Assembly legislate in a way that discriminates directly on grounds of religious belief or political opinion. In addition, if the Assembly legislates in a way incompatible with the United Kingdom's international obligations, the Secretary of State may decide not to submit such a Bill for Royal Assent.

Section 75 of the Act provides that each "public authority" is required, in carrying out its functions relating to Northern Ireland, to have "due regard" to the need to promote equality of opportunity between certain different individuals and groups. Many Northern Ireland bodies (Northern Ireland departments, local authorities, and quangos) are automatically included. Other bodies operating in Northern Ireland will have to be "designated" by the Secretary of State, including such bodies as the police, educational institutions, and UK government departments. The relevant categories included are persons of different religious belief, political opinion, racial group, age, marital status or sexual orientation; between men and women generally; persons with a disability and persons without; and persons with dependants and persons without. Without prejudice to these obligations, a public authority in Northern Ireland is also, in carrying out its functions, to have "regard" to the desirability of promoting good relations between persons of different religious belief, political opinion or racial group. Schedule 9 to the Act makes detailed provisions for the enforcement of these duties.

All public authorities included within the definition of public authority are required to submit an equality scheme to the Equality Commission. Only where a public authority has been notified in writing by the Commission that it does not need to, is it exempted from producing such a scheme. In Parliament, concern was expressed at the apparently open-ended power of exemption granted to the Equality Commission. In response, the Government made it clear that it was only in very limited circumstances that the Government envisaged such exemptions being granted by the Commission either to a body entirely, or with regard to particular functions of a body.[31]

An equality scheme must show how the public authority proposes to fulfil the duties imposed by section 75 in relation to the relevant functions, and specify a timetable for measures proposed in the scheme. The Act sets out in some detail (without being exhaustive) what an equality scheme must contain in order to be in compliance with the legislation. The list includes the authority's arrangements for assessing compliance with the duties under section 75, for consulting on matters to which a duty under that section is likely to be relevant (including details of the persons to be consulted), for assessing and consulting on the likely impact of policies adopted or proposed to be adopted by the authority on the promotion of equality of opportunity, for monitoring any adverse impact of policies adopted by the authority on the promotion of equality of opportunity, for publishing the results of such assessments and such monitoring, for training staff, and for ensuring, and assessing, public access to information and to services provided by the authority. In addition, an equality scheme must conform to any guidelines as to form or content which are issued by the Equality Commission. These guidelines are subject to the approval of the Secretary of State.

The legislation also details what is required in an authority's publishing of its assessments. It must state the aims of the policy to which the assessment relates, details of any consideration given by the authority to measures which might mitigate any adverse impact of that policy on the promotion of equality of opportunity, and alternative policies which might better achieve the promotion of equality of opportunity. Also, in making any decision with respect to a policy adopted or proposed to be adopted by it, the authority is required to take into account any such assessment and consultation carried out in relation to the

[31] Paul Murphy, House of Commons, Official Report, 18 November 1998, cols 1068–9.

policy. The Government also made it clear that it expected consultation "to embrace those directly affected by a policy as well as non-governmental organisations and relevant statutory bodies".[32]

What happens after a scheme is submitted for approval to the Equality Commission depends on what type of public body is involved. A distinction is made between Northern Ireland departments and public bodies, and UK-wide public bodies. A "public authority" is defined to include any department, corporation or body listed in Schedule 2 to the Parliamentary Commissioner Act 1967 and designated for the purposes of that section by order by the Secretary of the State. The inclusion of these latter bodies, being mainly UK-wide government departments, has resulted in special arrangements being devised relating to the procedures with which they must comply regarding equality schemes.

We describe first what happens in the former case. On receipt of the scheme the Commission must either approve it, or refer it to the Secretary of State. Where the Commission refers the scheme to the Secretary of State, the Commission is required to notify the Northern Ireland Assembly in writing that it has done so and send the Assembly a copy of the scheme. Where a scheme is referred to the Secretary of State he or she has three options: to approve it, to request the public authority to make a revised scheme, or to make a scheme for the public authority. Where the Secretary of State requests a revised scheme, or makes a scheme himself or herself, he or she shall notify the Assembly in writing. Where the Secretary of State has made a scheme for the public authority, he or she is required also to send the Assembly a copy of the scheme.

Certain of these provisions do not apply in the case of UK-wide departments. On receipt of a scheme submitted by a UK government department the Commission shall approve it, or itself request a department to make a revised scheme. Where such a request is made, the government department must, if it does not submit a revised scheme to the Commission in the time provided, send to the Commission a written statement of the reasons for not doing so. The provisions relating to the notification of the Assembly do not apply. Nor do the provisions empowering the Secretary of State to make schemes for the public body directly. These provisions are intended to "avoid a situation where the

[32] Alfred Dubs, House of Lords, Official Report, 11 November 1998, col. 810.

Secretary of State much reach a decision or issue a direction in a case involving her Department or that of a Cabinet colleague".[33]

If the Commission receives a complaint, made in accordance with certain formalities of failure by a public authority to comply with an equality scheme approved by the Commission or made by the Secretary of State, it is required to investigate the complaint, or give the complainant reasons for not investigating. The formalities with which complaints must comply are that the complaint must be made in writing by a person who claims to have been directly affected by the failure. A complaint must also be sent to the Commission during the period of twelve months starting with the day on which the complainant first knew of the matters alleged. Before making a complaint the complainant must bring the complaint to the notice of the public authority, and give the public authority a reasonable opportunity to respond.

In addition to investigating on the basis of a complaint, however, it appears that the Equality Commission itself has power to carry out an investigation into the compliance by the public authority with an equality scheme without having received a valid complaint. Although there is room for doubt, the power to carry out such an investigation appears to be derived from the Equality Commission's general duty to keep under review the effectiveness of the duties imposed by section 75 of the Act. Paragraph 11 of the Schedule, in addition, provided explicitly for the same conditions to be applied to investigations which arise from complaints as investigations which are "carried out by the Commission where it believes that a public authority may have failed to comply with a scheme".

What happens to the results of these investigations depends on the type of public authority involved. Again, a distinction is drawn between Northern Ireland and UK-wide public bodies. In the case of the former, the Commission is required to send a report of both types of investigation to the public authority concerned, the Secretary of State, the Assembly, and the complainant. If a report recommends action by the public authority concerned and the Commission considers that the action is not taken within a reasonable time, the Commission may refer the matter to the Secretary of State. Where a matter is referred to the Secretary of State, the Secretary of State may give directions to the public authority in respect of any matter referred to him. Where the Commission refers a matter to the Secretary of Stare

[33] Paul Murphy, House of Commons, Official Report, 18 November 1998, cols 1068–9.

it shall also notify the Assembly in writing that it has done so. Where the Secretary of State gives directions to a public authority, he or she shall notify the Assembly in writing that he or she has done so.

Somewhat different provisions apply in the case of UK-wide bodies. Certain of these provisions do not apply, particularly those empowering the Secretary of State to give directions to the public authority in respect of its failure to present a scheme. Instead, the Commission may lay before Parliament and the Assembly a report of any investigation regarding compliance with an equality scheme by such a department.

DELIVERING ON THE EQUALITY PROVISIONS OF THE AGREEMENT

The equality provisions of the Agreement represent a major development in the efforts to create a more equal society in Northern Ireland. They constitute a shift from an anti-discrimination to a mainstreaming approach. Before looking at the problems that may arise in implementing this, it is important to stress their innovative nature both in Northern Ireland and international terms.

Governments in North America, Western Europe and the Commonwealth have sought to address the disadvantaged position of ethnic groups, women and other by developing anti-discrimination law in specific areas such as employment or housing, particularly in the private sector. In all countries of Western Europe, and much of the Commonwealth, such legislation is now in place. The Northern Ireland experience suggests, however, that while such legislation is necessary, it is insufficient by itself. The legislation is essentially negative, aiming to prevent discrimination, rather that positively to promote equality. There is, moreover, growing concern in many countries about the extent to which anti-discrimination norms are practically effective.[34] Attempts have been made to develop mechanisms to ensure greater compliance, for instance by creating specialised bodies tasked with enforcement, but these often have little effect on key government decisions. There have also been attempts to develop policies that bring the weight of government to bear more directly, for example, by making government contracts and grants to the private sector conditional on implementing equality policies. However, their influence touches only a limited sphere of activity.

[34] Alfred W Blumrosen *Modern Law: The Law Transmission System and Equal Opportunity* (Madison, Wisconsin, University of Wisconsin Press, 1993).

In this chapter we have been examining an attempt to go several steps further to require government and public bodies to weave policies of equality and non-discrimination into the fabric of decision-making across all spheres of government—in short, to "mainstream" fairness issues in public policy. This attempt is particularly important if the problem is defined, as it increasingly is, as involving not only the problem of "discrimination", but the larger issue of unacceptable inequalities affecting women and particular minority groups, whether caused by discrimination or not.

This is not to underestimate the importance of securing the effectiveness of traditional anti-discrimination legislation. One of the first commitments in the Agreement to be honoured following the Northern Ireland Act 1998 was the promised reform of the fair employment legislation. In December 1998, a new Fair Employment and Treatment (Northern Ireland) Order was published and laid before Parliament, implementing many of the proposals announced in the White Paper. As an Order made under the Northern Ireland Act it was not subject to the degree of parliamentary scrutiny to which ordinary legislation would be subject. It was debated briefly on 7 December 1998 in the House of Lords,[35] and considered by the House of Commons Standing Committee on Delegated Legislation.[36] In several respects, the new legislation more closely resembled that which SACHR had advocated in its 1987 Report that had the 1989 legislation.

In brief, the new Order[36a] repealed the 1976 and 1989 Acts whilst substantially re-enacting major parts of the previous scheme, although with those significant changes set out in the White Paper, *Partnership for Equality*. The major changes enacted included provisions which broaden the scope of the legislation to prohibit unlawful discrimination on the grounds of religious beliefs or political opinion in goods, facilities and services, including the sale of land (subject to the limitation that land sales not publicly advertised are excluded from coverage). The scope of monitoring returns and triennial reviews was broadened in several respects: by increasing the number of employers which must register with the FEC; by including part-time workers within the scope of a new definition of employee; by requiring all reg-

[35] House of Lords, Official Debates, vol. 595, 7 December 1998, cols 755–74.

[36] House of Commons, Fourth Standing Committee on Delegated Legislation, Draft Fair Employment and Treatment (Northern Ireland) Order 1998.

[36a] Fair Employment and Treatment (Northern Ireland) Order 1998, SI 1998 No. 3162 (NI 21).

istered employees to include details of those applying for work; by requiring public authorities to include details of those leaving employment in monitoring returns; by permitting regulations to prescribe only one principal method for assessing community affiliation; by requiring triennial reviews to include assessing community affiliation; and by requiring triennial reviews to include redundancy issues. (New monitoring regulations were published in March 1999 to come into operation on 1 January 2001.)[37] The scope of the affirmative action exceptions was broadened slightly: by permitting employers to train non-employees of one religion where that religion is under represented in the workforce; by permitting employers to recruit from those not in employment; and by modifying the affirmative action provisions relating to redundancy by omitting the need to have an agreed procedure. New powers and duties were given to the Equality Commission, including a duty to keep the workings of the Order under review, and new powers were given to deal with persistent discrimination. The scope of the employment provisions was broadened to include partnerships and barristers. The individual complaints procedure was modified somewhat: by enabling tribunals to make recommendations for reducing the adverse effect of any unlawful discrimination on someone other than the complainant; by providing for tribunals to have the power to award damages for unintentional indirect discrimination; and by empowering the Labour Relations Agency to draw up arbitration schemes. Finally, following the judgement by the European Court of Human Rights in the *Tinnelly* and *McElduff* cases, both the Northern Ireland Act 1998[38] and the new Order[39] provided the right of appeal against a national security certificate, signed by the Secretary of State, to a new tribunal.

These new provisions have yet, of course, to prove their effectiveness. In its first assessment of the legislation, the House of Commons Northern Ireland Affairs Committee, reporting in August 1999,[40] considered that the previous legislation had made an important contribution to improving the degree of fairness in employment in Northern Ireland, though it agreed that there was still much to be done. It was an indication of the new consensus that appeared to be emerging around

[37] Fair Employment (Monitoring) Regulations (Northern Ireland) 1999.

[38] Northern Ireland Act 1998, ss. 90–92.

[39] 1989 Order, arts 79–80.

[40] House of Commons, Northern Ireland Affairs Committee, *The Operation of the Fair Employment (Northern Ireland) Act 1989: Ten Years On: Report and Proceedings of the Committee* (Session 1998–99, HC 95-I).

equality issues in Northern Ireland that a Committee comprising Northern Ireland MPs as diverse as Eddie McGrady (SDLP), Jeffery Donaldson (UUP), Roy Beggs (UUP), and Peter Robinson (DUP), could assent to a report which stressed the need for strong enforcement of equality legislation.

We might have hoped, then, that a major section of this chapter would have analysed the achievements of the new Executive and Assembly in the area of equality. However, as other chapters relate, the provisions of the Northern Ireland Act relevant to the formation and operation of the new Executive and Assembly (with the exception of the section 75 public sector equality duty, which we consider below) had hardly become operational at the end of 1999, before being suspended a matter of weeks later. Even so, we can see even in this short time, that equality issues became an issue of central importance to the new administration.

The operation of the new Targeting Social Need (new TSN) programme had been announced in the White Paper published in March 1998, and identified by the Agreement as one of the British Government's key socio-economic commitments. New TSN was formally launched in July 1998. Every Northern Ireland department and the Northern Ireland Office was to develop an Action Plan describing how it intended to implement New TSN over a three-year period. Departments were expected to consult on their Action Plans before adopting them. The first New TSN Annual Report was published in November 1999 and included the consultative policy areas covered by the Action Plans passed to the new Executive.

In deciding which new Ministries to create, responsibility for equality issues (including new TSN) was allocated, not to a new Department of Equality (as the Agreement had floated as a possibility), but rather to the Office of the First and Deputy-First Minister (OFM-DFM, headed by David Trimble (UUP) and Seamus Mallon (SDLP)). This resulted in the transfer of the equality functions of the Department of Economic Development (in relation to such issues as anti-discrimination legislation) to OFM-DFM, as well as the staff and functions of the Central Community Relations Unit (CCRU) within the Northern Ireland Office (including the coordination of the new TSN).

Within the OFM-DFM, a new Equality Unit was created, headed by two junior Ministers, one an Ulster Unionist, the other from the SDLP, as part of a wider Directorate on Equality, Human Rights and Community Relations. The unit was intended by the SDLP to be central mechanism for the ensuring the success of the section 75 public sector duty, and the major player in ensuring that equality issues gained a higher public political profile. Arrangements were in place prior to suspension, for example, to ensure that the Unit would have staff heading it at a more senior level that hitherto (Grade 3 rather than Grade 5), in order to ensure that its views should carry weight with the other departments. It was also envisaged that the Equality Unit staff would service the main inter-departmental committee discussing equality-related issues, the Social Steering Group, composed of senior departmental civil servants from each department. The Northern Ireland Office published its first Circular on the section 75 duty and the importance of departments ensuring that the duty was taken seriously.[41]

Progress had also been made, within the Executive, on taking forward a review of the composition of the Northern Ireland civil service. Long a source of concern, particularly in the areas of religio-political and gender under-representation, the Department of Finance and Personnel, headed by Mark Durkan (SDLP) was given the task of establishing a review that would look at progress to date and at how legislation might be changed to enable greater representativeness to be achieved, particularly at senior levels. With suspension, this issue was not taken forward.

The coming into operation of the Equality Commission and the public sector equality duties was independent of whether devolution comes into effect. It was always likely to prove a complex task amalgamating the existing equality bodies into one, and so it proved. A Working Group was established by the Government, chaired by an independent outside expert, consisting of representatives from each of the equality bodies to be amalgamated into the Commission. It was asked to consider how best to proceed with the mechanics of amalgamation, including the appropriate executive structure, staffing issues, accommodation, and resources. It also commissioned a draft set of guidelines on the section 75 equality duties (from the author) that would be ready for the new Commission to consider when it came into office. The Group reported in March 1999.[42] In the rest of the chapter, we shall consider the operation of the equality duty.

Three issues regarding the equality duties have predominated to date. The first was the content of the guidelines to be issued by the

[41] Office of the First and Deputy First Minister, Circular 1/00 *Northern Ireland Act 1998 Section 75 Statutory Equality Obligation* (January 2000).

[42] *Report of Equality Commission Working Group* (March 1999).

Equality Commission. The second involved the issue of the designation by the Secretary of State of those public authorities that were not automatically covered by the Northern Ireland Act provisions. The third issue was whether the Equality Commission would grant extensive exemptions to public bodies, delaying the coming into effect of the duty to submit an equality scheme to the Commission. At the time of writing, other issues are surfacing, such as the adequacy of the draft equality schemes being put out for consultation, the burdens being faced by NGOs that are being consulted on the schemes, and the criteria that the Equality Commission should use in deciding whether or not to approve the schemes, but we shall concentrate on the first three issues.

A common theme links the three issues to be discussed: the question of timing. For those bodies automatically subject to the duties, the equality duty came into effect from 1 January 2000. The legislation provided that a draft equality scheme must be submitted by such bodies within six months of this date, by 30 June 2000. We have seen that the Equality Working Group had prepared a set of draft guidelines that the incoming Commission could consider. This was included in the published report of the Working Group. Following the first formal meeting of the new Commission, this draft was issued for consultations. Following this round of consultations, the draft was revised and sent to the Secretary of State in late December 1999 and were simultaneously sent for legal advice. Following receipt of legal advice, an amended version was sent for formal approval in mid-January 2000. The letter giving formal approval was received on 23 March, and the Guidelines were formally launched on 31 March.[43] The one major change to emerge as a result of the consultation process, apart from the restructuring of the content, was that public authorities were permitted to set out their *arrangements* for determining which specific policies would be subject to impact assessment, rather that have to identify the specific policies in the scheme itself.

The length of time taken to secure approval by the Secretary of State, however, meant that some public authorities subject to the requirement to produce an equality scheme began to worry that they would be unable to prepare a draft scheme, consult effectively, redraft the scheme and submit it to the Equality Commission in time to meet the statutory deadline. In particular, they argued to the Equality Commission that the

[43] Equality Commission of Northern Ireland, *Guide to Statutory Duties: A Guide to the Implication of the Statutory Duties on Public Authorities Arising from Section 75 of Northern Ireland Act 1998* (March 2000).

Commission could and should exercise its statutory power to exempt these bodies from the duty to submit the draft schemes by 30 June. The Commission considered, however, on the basis of legal advice and taking into account the statement by the responsible Minister in Parliament at the time of enactment that such exemptions would be rare, and refused to grant any exemptions on this basis.

Timing was also relevant for the third issue: the designation of those public bodies by the Secretary of State not included automatically under section 75. Initially, it would appear, the Northern Ireland Office has simply "invited" departments and public bodies to submit themselves to the section 75 duty. Perhaps not surprisingly, few did and when the Equality Commission was notified of those few that had, it requested the Secretary of State to consider whether, in the public interest, other bodies should be included. Political endorsement for a much more comprehensive designation had come from the Office of First and Deputy First Minister (when still operating). There was increasing concern among the NGOs both at the delay in designation itself, but also at the range of functions that could escape the statutory duty if the relevant UK body were not be designated. By the time of writing (June 2000), no designation order had yet been made, but it would seem likely that an initial designation of some few bodies would be made sooner rather than later, with a potentially longer list being designated later.

A PRELIMINARY ASSESSMENT

Perhaps at this point we should remind ourselves why mainstreaming is important. How does mainstreaming differ from traditional anti-discrimination approaches? The most important difference is that it concentrates on achieving equality rather than simply eliminating discrimination. Mainstreaming involves government proactively taking equality in to account. It is intended to be anticipatory rather than retrospective, to be extensively participatory rather than limited to small groups of the knowledgeable, and to be integrated into activities of those involved in policy-making. The motivation for mainstreaming lies in the realisation that unless special attention is paid to equality in policy-making, it becomes too easily sidelined and submerged in the day-to-day concerns of policy-makers who do not view equality as central to their concerns. Mainstreaming, by definition, attempts to address this problem by requiring all government departments to engage directly with equality issues.

Mainstreaming will have other, more indirect, consequences. One of these is to encourage greater transparency in decision-making since it necessitates defining the likely impact of policies at an earlier stage of policy-making, more systematically and to a greater extent than is usually contemplated. It will also encourage greater participation in policy-making. Unlike more traditional mechanisms of consultation, mainstreaming as now to be practised in Northern Ireland requires impact assessments of a degree of specificity that establishes a clear agenda for discussion between policy-makers and those most affected. In combination, impact assessment and participation will develop links between government and "civil society", encouraging greater participation in decision-making by marginal groups and lessening the democratic deficit.

There are dangers and limitations to mainstreaming. In particular, it may result in the over-fragmentation of equality policy, especially if it becomes an alternative to traditional anti-discrimination and other equality mechanisms. If all public bodies have responsibility for equality, there is a danger that none will regard it as an important part of their function. There is a need, therefore for some centralised responsibility within government to ensure that mainstreaming is consistently applied, according to common standards. Nor should one overlook the fact that building such a requirement into civil service decision-making will require considerable cultural change. Apart from practical issues, there are problems of departmental exclusiveness and collective responsibility. Mainstreaming may well cut across the working practices, and potentially the ethos, of civil service bureaucracy. The dismal experienced in Northern Ireland of non-statutory PAFT approach to mainstreaming before the reforms introduced by the Northern Ireland Act 1998 is eloquent testimony to this.

The implication is that a strong political commitment to mainstreaming is absolutely crucial and must drive a new approach being taken by departments and other public bodies. But it means more than that. It means also that the legal status of mainstreaming needs to be considered. It is noticeable that many of the jurisdictions that have introduced mainstreaming have done so without according it any clear legal status. Mostly, mainstreaming has been introduced administratively by circular, without any formal legal underpinnings. At best, the status of mainstreaming in many countries is that of "soft law". The

Northern Ireland experience points to the inadequacy of a "soft law" approach. Whether a "hard law" approach will be any more successful in Northern Ireland remains to be seen. It is to that issue that we now turn.

What are the prospects of the legal mainstreaming approach adopted in Northern Ireland being effective? The provisions of the Act are promises, not reality. They are a necessary part of the process of achieving substantive equality, fairness and justice. But neither the provisions of the Agreement not the Act itself delivers such change directly. This delivery will require political will at all levels. The provisions of the Act, in other words, represent the potential for change. The provisions will reframe the debate. But we must ensure that change actually occurs, particularly in the area of greatest disadvantage.

The provisions will need to be put into effective operation. In this context there is a real difficulty. Ultimately, those who will have to operate this system day-to-day, particularly in the absence of devolved government, are the civil service and other public servants. The response of parts—and I stress parts—of the public service to these initiatives has been problematic in the past. Often it has been ungenerous and lacking in imagination. Sometimes, it seems that it has been actively opposed to necessary change. If the Agreement is to mark a new beginning for Northern Ireland, as is the wish of the vast majority of the population, all institutions have the obligation to change and adapt. The public service cannot be an exception to this, however difficult it must be for some to give up the almost unrestrained power they were able to exercise for a generation. For its own sake, as well as that of Northern Ireland as a whole, the civil service must not be seen as obstructive to this aspect of the Agreement. The Equality Commission can no doubt play a role in assisting the public service to adapt, but ultimately the responsibility will lie with the public service itself, and of course the members of the Executive and the Assembly, now re-established.

Much depends on the quality of the new Equality Commission and on its effectiveness in managing the transition from four separate bodies into one. Its first test was its skill in drawing up effective guidelines as to the criteria to be followed to comply with their statutory duty. Now, it will be important to ensure that the guidelines are adhered to by the public authorities in their day-to-day practice. There will be a substantial opportunity for the groups most affected to insert themselves into the policy-making process.

How far the promise of the Agreement's equality provisions is delivered will depend therefore, on the commitment, determination and skill of all the political parties, on a strong well-financed, and independent Equality Commission, effective NGOs, and crucially, on the political will to place equality at the heart of decision-making. Using the new tools will be a challenge for politicians to ensure that human rights and equality remain central to political life, for civil service and public authorities to incorporate a culture of human rights into administration, and for civil society to use these tools imaginatively and persistently. A lasting peace depends upon them all.

5

Building a Human Rights Culture in a Political Democracy: The Role of the Northern Ireland Human Rights Commission

COLIN J HARVEY

HUMAN RIGHTS SHOULD be effectively protected at the national level and all other mechanisms must be essentially aimed at achieving this basic objective. The argument in this chapter is that national level human rights protection has a priority status, even with the steady proliferation of international and regional standards and institutions. This is the starting point for the arguments presented here. The Northern Ireland Human Rights Commission is one of the important consequences of the adoption of the Agreement. The creation of this new statutory body provided formal recognition of the importance of human rights protection. The aim is to supplement the existing mechanisms of protection with a public body dedicated to raising awareness of human rights issues. The stated aim is the construction of a human rights culture in Northern Ireland, whatever this might be taken to mean. This chapter examines the role of the Human Rights Commission in the new democratic structures of Northern Ireland. First, I raise some critical questions about conceptions of rights and democracy. Secondly, I explore the context, including the international and national dimensions, and finally, I analyse the functions and powers of the Commission. The aim in this chapter, however, is not simply to offer a guide to the powers and functions of the Commission. The primary purpose is to encourage critical legal thinking about the role of a public body like this in a functioning democratic polity. My argument is that national institutions for protecting human rights are important precisely because they do not suffer from the problems surrounding the judicial protection of human rights. It is much too early to make a comprehensive assessment, so this chapter concentrates on providing an overview and on raising critical questions about the enforcement of human rights.

ENRICHING POLITICAL DEMOCRACY: HUMAN RIGHTS, LAW AND MISTRUST

Before embarking on an outline of the work of the Commission it is worth raising some questions about the protection of human rights. There is a tendency in the literature of human rights law to glorify judicial protection. The courtroom, so it is argued, is the forum of principle where rational argumentation can take place. In such a forum rights are, it is supposed, more likely to be protected by judges who are not subject to the demands of populism and majoritarianism. This is used very often in arguments about minority rights. It is said that judicial protection is important precisely because rights matter regardless of the majority view. This justification of judicial intervention is convenient for judges also, because here they find a suitably noble justification for their role in a modern democracy. There is, however, something deeply disturbing about the reasoning which underpins such arguments. It reveals a mistrust of democracy and of the citizens who function within that democratic structure. The obvious response is that the experience of democracy in Northern Ireland has not been a good one. And this is undoubtedly correct. However, the reasoning contains the seeds of a profound pessimism about humanity. It is effectively ruling out the possibility of reshaping democracy, and the capacity of people to alter their behaviour to reflect human rights considerations. The surprising thing is that the logic conflicts with the reasons offered for thinking in terms of rights in the first place.[1] Rights-thinking is said to matter because of a belief in the autonomy of individuals. Thinking in rights-terms is thought to be important because we value the autonomy of individuals. Often when the move is made to consider enforcement this respect for autonomy turns into a profound distrust. The autonomous individual, who provided the justification for the theory

[1] See Jeremy Waldron *Law and Disagreement* (Oxford, Clarendon Press, 1999) chs 10–13; Jürgen Habermas *The Inclusion of the Other: Studies in Political Theory* (Ciaran Cronin and Pablo De Greiff (eds), Cambridge Mass, The MIT Press, 1998) p. 203.

of rights, now becomes the main danger. Can the autonomous individual be trusted to do the "right thing", we might say. The point I would make is that it is perfectly possible to believe in human rights protection but be sceptical of the merits of exclusively judicial enforcement and thus to believe in the primacy of political democracy.[2] We should be sceptical or "cynical" about placing our faith in the judiciary to promote progressive politics.[3] As a body of men (and it usually is men) they have not shown any great attachment to progressive politics.[4] This is why innovations, such as Human Rights Commissions, can be so important. In Northern Ireland the Human Rights Commission idea formed part of the final settlement, thus giving added legitimacy to its role. It is, however, still a means to an end, and not an end in itself.

The stakes are raised in this debate when we then talk about constitutional rights, or those rights which are granted special protection within a legal system. This has become an issue in Northern Ireland with the proposed adoption of a Bill of Rights. The assumption tends to be that fundamental rights deserve special protection and thus should be removed from the normal channels of legislative amendment. In effect, the argument is that such rights should be given a higher status and thus shielded from normal processes of deliberation. This may seem, at first glance, a wise option given the problems of Northern Ireland. Where there have been human rights abuses, and severe disagreements about the meaning of rights, it is perhaps understandable that closure is sought in the courtroom. When political life appears indeterminate and chaotic, law can seem to bring the certainty of decision. This decisionism is often lauded as one of law's chief merits. We are told that at some point there must be a reasoned decision and who else but the modern judiciary would we turn to? This idea of detachment is superficially appealing but few can now doubt the

[2] See Paul Mahoney "Marvellous Richness of Diversity or Invidious Cultural Relativism" (1998) 19 *HRLJ* 1, at 3: "The Convention is grounded on a certain political philosophy, namely that political democracy is the best system of government for ensuring respect of fundamental freedoms and human rights. Any theory of interpretation or review by the Court must be compatible with that basic underpinning of political theory", illustrating that it is possible in a human rights instrument to respect the fundamental importance of political democracy.

[3] For a definition of the cynic see Friedrich Nietzsche *Beyond Good and Evil* (Marion Faber (trans.), Oxford, Oxford University Press, 1998) p. 27: "I mean by the so-called cynic, those people who simply acknowledge what is animal-like, common, the 'rule' about themselves and still have enough spirituality and excitability to need to speak about themselves and their kind *in front of witnesses*".

[4] See Colin Harvey "Agreement's Rights Promise Must be Fulfilled" *The Irish Times*, 24 February 2000.

situatedness of judges and lawyers in the political struggles that rage in modern democracies. Nietzsche talked of philosophers as "advocates who refuse the name" whose idea of truth was an "abstract version of their heart's desires".[5] The judge, like the philosopher, is such a situated institutional actor whose judgments are again tied to his "heart's desires". The difference springs from the institutional context of judging and not the act itself.

There is another way to present this. In a political community like Northern Ireland there are established struggles for recognition and democratic inclusion. Underpinning these struggles is the right to participate and one wonders how much this is compromised by removing issues from public deliberation. If reasonable people disagree about the meaning of human rights then reasonable judges will also. The next step is to ask why we should then be beholden to the meaning of rights adopted in this forum. In making this assessment surely it does make a difference that we are talking about an unelected group of elite decision-makers. It may not be a welcome argument at present, but it remains a persuasive one. As is evident from what I have said here, the starting point for this chapter is caution about the judicial role in the human rights sphere. If we place the judiciary on a pedestal we will be disappointed. My suggestion is that more attention should be focused on all the other legal and political channels which promote change. The Human Rights Commission is one good example. The intention here is not, of course, to exclude the courts. One could not do this without offending basic tenets of political democracy. They have a role to play. The ambition here is simply to provoke a critical dialogue. There is a danger that some human rights talk encourages an unreflective approach to politics and law and thus impoverishes political dialogue in the public sphere.[6] And that is what is not needed in Northern Ireland. The purpose of human rights in Northern Ireland is to ensure both participation and protection.

[5] Friedrich Nietzsche *supra* n. 3 at p. 8. Cf. Richard Rorty *Philosophy and Social Hope* (London, Penguin Books, 2000) p. 98: "Dewey the romantic would have been delighted that the courts sometimes tell the politicians and the voters to start noticing that there are people who have been told to wait for ever until a consensus emerges—a consensus within a political community from which these people are effectively excluded". Later in this work (at p. 119) Rorty argues that Habermas, for example, is effectively reviving Dewey's social democratic approach.

[6] Cf. Mary Ann Glendon *Rights Talk: The Impoverishment of Political Discourse* (New York, Free Press, 1991); Allan C Hutchinson *Waiting for Coraf: A Critique of Law and Rights* (Toronto, University of Toronto Press, 1995).

THE CONTEXT

There is both an international and local context to consider. First, I will consider the international context. Legal and political discourse in the human rights field is increasingly transnational in nature. What I mean by this is that discussions between politicians, lawyers, judges and NGOs take place not only within the state system but within transnational networks of legal and political communication. Changes in national policy, and judicial decisions, can be quickly transmitted by these networks. This can, in some instances, amount to internationalism or transnationalism in the most superficial sense. Case law from other states can be used as justification for decisions which have already been made. The mere fact that this occurs does not tell us very much. That such networks exist and transmit information does not have any necessary results. For example, the practical impact of this is far from certain. One impact that can be measured is the rise in number of national institutions dedicated to the protection of human rights.[7] There is now considerable support for the creation of these institutions and the United Nations has taken an active role since 1991.[8] Support for such institutions is evidence of the weaknesses of the existing international mechanisms.[9] It indicates that national protection remains of primary significance. Livingstone states:

[7] See Stephen Livingstone "The Northern Ireland Human Rights Commission" (1999) 22 *Fordham International Law Journal* 1465, at 1466–70; Anne Gallagher "Making Human Rights Treaty Obligations a Reality: Working with New Actors and Partners" in Philip Alston and James Crawford (eds) *The Future of UN Human Rights Treaty Monitoring* (Cambridge, Cambridge University Press, 2000) p. 201, at 202: "National human rights institutions may be defined for present purposes as independent entities which have been established by government under the constitution or by a law and entrusted with special responsibilities in terms of promotion and protection of human rights". See also Mac Darrow and Philip Alston "Bills of Rights in Comparative Perspective" in Philip Alston (ed), *Promoting Human Rights Through Bills of Rights* (Oxford, Oxford University Press, 1999) p. 465, at 523: "In recent years, a number of new constitutions have made explicit provision for the establishment of a national human rights commission. Examples include: Ethiopia, South Africa, Malawi, Uganda and the Philippines. Mandating the creation of such an institution sends a strong symbolic message, gives bureaucratic clout to it, and accords a degree of protection from parliamentary or executive manipulation. While many such institutions will probably fail at least in the short term, the record so far has been more encouraging than might have been anticipated".

[8] Sarah Spencer and Ian Bynoe *A Human Rights Commission: The Options for Britain and Northern Ireland* (London, IPPR, 1998) p. 45. The UN High Commissioner for Human Rights, Mary Robinson, has a Special Adviser on National Human Rights Institutions.

[9] *Ibid.*

"Although such institutions go under a variety of names, there is a certain degree of uniformity in their structure, functions and powers. Nearly all are established and funded by government, though composed of members of civil society rather than politicians or bureaucrats."[10]

The Paris Principles, adopted by the UN General Assembly in 1993, provide international guidance relating to the mandate and independence of these bodies. The national institutions have been endorsed on many occasions since the formal adoption of these Principles.[11] Livingstone argues that the institutions can play an important role in strengthening the protection of human rights.[12] He does, however, note a recent study which suggests that little evaluative work has been done.[13] These bodies have potential, but their impact will need to be carefully assessed in relation to their effectiveness at the national level. Their importance will rest on the ability to bridge gaps between human rights developments and national political communities. They will also contribute to the concrete application of norms to local contexts. The development is a significant one for the human rights movement.

The other context of relevance is the local. The Human Rights Commission is not the first such body and it does not function within a political or legal vacuum. Its predecessor was the Standing Advisory Commission on Human Rights (SACHR).[14] The body was hampered from its inception by a limited mandate and the absence of sufficiently strong powers.[15] While it undoubtedly did some useful work one commentator has concluded: "Overall . . . SACHR has had a somewhat limited influence on policy-making and implementation as regards the protection of human rights in Northern Ireland".[16]

SACHR was clearly a weak body which was, for a variety of reasons, unable to deliver effective results in the overall protection of human rights in Northern Ireland. Given this it was unsurprising that thoughts turned to the creation of a new institution. The initiative was taken on

[10] Livingstone *supra* n. 7 at p. 1467.

[11] Spencer and Bynoe *supra* n. 8 at p. 47. They note that at the World Conference on Human Rights in Vienna the national institutions were granted separate speaking rights from their governments.

[12] Livingstone *supra* n. 7 at p. 1469.

[13] *Supra* n. 8

[14] For some examples of SACHR's work, of relevance in the context of this chapter, see SACHR *Bill of Rights: A Discussion Paper* (1976); SACHR *The Protection of Human Rights by Law in Northern Ireland* (Cmnd 7009, 1977).

[15] Northern Ireland Constitution Act 1973, s. 20(1).

[16] Livingstone *supra* n. 7 at p. 1474.

this by the NGO sector. As the peace process developed, so the opportunities of feeding this thinking into the process increased. The election of a British Government more attuned to human rights talk also had an important impact. The context was one conducive to arguments for a Human Rights Commission. In a paper commissioned by SACHR, Brice Dickson (now the Chief Commissioner) advanced four strong reasons for creating a Commission.[17] First, new laws to protect human rights were about to be enacted.[18] With the changes in law it was sensible, he argued, to create a body to oversee their implementation.[19] Secondly, as a society emerging from conflict, Northern Ireland could use a Commission.[20] Dickson refers to the problems faced by societies in transition and the help that might be offered by a national Commission.[21] The South African experience is drawn upon as a useful example. The third argument is interesting. Here Dickson argues that a Commission should be expected to take a position on the issue of balancing conflicting rights.[22] He states:

> "By confronting such potential and actual conflicts a Human Rights Commission could help to demonstrate to society that everyone in Northern Ireland has duties and responsibilities as well as rights. This should in turn assist the development of more harmonious relations between different elements of the society. At the very least it should help to defuse potentially explosive confrontations".[23]

This is ambitious. It requires the Commission to take the lead in demonstrating how conflicting rights might be balanced. One can, however, see problems in this argument. Is it not more likely that the conflicts which are evident in the political community will also appear within the Commission? This is particularly so given that it is intended to be representative of the community. If reasonable people disagree about the meaning of rights, then reasonable human rights commissioners will also disagree. Often in discussion of rights the issue revolves around balance, and it is here that substantive disagreement comes to the fore. My argument is that this balancing is inherently political in that it takes place within the structures of a position

[17] Brice Dickson *Creating an Effective Human Rights Commission for Northern Ireland* (May 1998).

[18] *Ibid.* pp. 6–7.

[19] *Ibid.* p. 7.

[20] *Ibid.*

[21] *Ibid.*

[22] *Ibid.*

[23] *Ibid.* p. 8.

(acknowledged or unacknowledged) within political theory. A transparent approach to balancing rights is thus more difficult and demanding than is often assumed. At the very least it requires a clear statement of the normative framework within which the balance has been struck. For it is this which will in practice determine the outcome of the balancing exercise. The final argument advanced is that "this is the way the world is going".[24] In other words, Northern Ireland should join a growing international trend. Dickson also acknowledges the problems and the paper lists potential difficulties. Again he mentions four: first, the fact that there is no equivalent body in Britain. As he notes, this is not really a strong argument against as there is no reason why, for example, England could not have one. Secondly, there is the argument that there are too many other quangos in Northern Ireland already. Here again the argument can be met by ensuring that the Commission is appointed in a transparent manner and that specific criticisms of quangos are addressed. Another issue he identifies is the problem of disputes between related bodies. As with the other difficulties this is one that is possible to address. The final problem he mentions is that a Commission may raise "people's expectations unduly".[25] As he notes, a clear list of specific aims and objectives would reduce the problem. The rest of the paper, which deals with the powers and duties that a Human Rights Commission should have, is not examined in any detail here. It is worth noting, however, that on the duty to investigate human rights issues Dickson states the following:

> "It could include the powers to summon and enforce the attendance of witnesses, to examine witnesses under oath, to compel disclosure of documents and even perhaps (as in South Africa) the power to enter and search premises".

The functions and powers of the Commission are examined below, but it is notable that the power of investigation does not include the matters listed here.

FUNCTIONS AND POWERS

The Agreement refers to the establishment of a new Northern Ireland Human Rights Commission "with membership from Northern Ireland

[24] *Ibid.*
[25] *Ibid.* p. 10.

reflecting the community balance".[26] The Agreement states that it should have "an extended and enhanced role beyond that currently exercised by the Standing Advisory Commission on Human Rights".[27] The Human Rights Commission thus forms a part of the overall settlement. It is to be an independent safeguard on the protection of rights in Northern Ireland.

The functions and powers of the Human Rights Commission are contained in the Northern Ireland Act 1998, which reflects the terms of the Agreement. The Commission is composed of a Chief Commissioner (Professor Brice Dickson) and nine part-time commissioners.[28] The Commission has a general obligation to keep the adequacy and effectiveness of Northern Ireland law and practice relating to the protection of human rights under review.[29] This function relates not only to the content of law and practice but its effectiveness and thus its practical impact. The Commission has initiated reviews of the law and practice in Northern Ireland affecting: older persons; people who are lesbian, gay, bisexual or transgendered; and people who are mentally ill. It will provide advice to the Secretary of State and the Executive Committee of the Assembly on measures which should be taken to protect human rights.[30] This is following a general or specific request for advice or on other occasions when it thinks it appropriate.[31] The Commission has offered advice to the Secretary of State, and to MPs in London, on proposed new laws dealing with asylum and immigration, freedom of information, terrorism, and equality for people with disabilities. This advisory role also involves work on specific pieces of Northern Ireland legislation. The Commission has the job of advising the Assembly if a Bill is compatible with human rights.[32] The Commission has agreed a Protocol with the OFM-DFM with a view to obtaining early access to draft laws and policies proposed by the

[26] "Rights, Safeguards and Equality of Opportunity" para. 5.

[27] *Ibid.*

[28] Professor Christine Bell, Margaret Ann-Dinsmore, Tom Donnelly, Rev Harold Good, Francis McGuinness, Professor Tom Hadden, Angela Hegarty, Patricia Kelly, Inez McCormack. The Commission has a staff of twelve people in addition to this and the Chief Executive is Paddy Sloan.

[29] Northern Ireland Act 1998, s. 69(1).

[30] *Ibid.* s. 69(3).

[31] *Ibid.* s. 69(3)(a) and (b).

[32] *Ibid.* s. 69(4). For the purposes of this section of the Act "human rights" includes Convention rights. In other words, when assessing effectiveness and giving advice the Commission is not confined to the European Convention on Human Rights. This presents its own problems given the proliferation of human rights standards in the last fifty years.

Assembly and Executive so that it can comment on the human rights implications. On UK-wide activities the Commission has participated in the work of the UK task force for the implementation of the Human Rights Act 1998.

The Commission has a role in giving assistance to individuals[33] and has the power to bring proceedings.[34] It has wide powers with respect to assistance to individuals. The Commission may grant assistance in relation to proceedings raising human rights issues when: the case involves an issue of principle; when it would be unreasonable not to do so because of, for example, the complexity of the case; or when special circumstances arise.[35] If it decides to grant assistance it can arrange for the provision of legal advice and/or make arrangements for legal representation.[36] The Commission has a broad discretion in these areas which will permit a more flexible approach. The Commission has developed and applied its own criteria for assisting individuals. It has received over one hundred applications for assistance and has granted assistance in some of these. It has also developed criteria for applying to court for permission to intervene in proceedings as a third party. The purpose is to assist the court with the rules and principles of human rights law which are relevant to the proceedings in question. The Commission has been granted permission to intervene in several instances, including a case before the European Court of Human Rights.[37]

The power to investigate human rights violations is important for national institutions.[38] The Commission may "conduct such investigations as it considers necessary and expedient".[39] It does not, however, have the power to compel the disclosure of documents. The power is thus rendered less effective than it might be, although it is still a valuable aid to its work. The Commission has developed criteria for conducting investigations into matters of human rights concern in Northern Ireland. Thus far it has investigated the juvenile justice system and the effects of the policing of parades. Its research has also been

[33] Northern Ireland Act 1998, s. 69(5)(a).
[34] *Ibid.* s. 69(5)(b).
[35] *Ibid.* s. 70(2).
[36] *Ibid.* s. 70(3).
[37] *McKerr and others* v. *UK*, App. No. 28883/95, 4 April 2000.
[38] See Paris Principles 1993 "Methods of Operation": "Within the framework of its operation, the national institution shall: . . . (b) Hear any person and obtain any information and any documents necessary for assessing situations falling within its competence".
[39] Northern Ireland Act 1998, s. 69(8).

useful in identifying areas where an investigation might be necessary. For example, the Commission has conducted research into the use of plastic baton rounds in Northern Ireland with a view to deciding whether to conduct a formal investigation.

The Commission has been involved in a large number of other activities. It has submitted comments to the UN Committee on the Elimination of Discrimination Against Women and to the Council of Europe's Committee on Social Rights and organised a series of "Training for Trainers" seminars in additional to education seminars on human rights issues. Members of the Commission have attended a large number of national and international events.

The Commission has the potential to be much more effective in the protection of human rights than SACHR. A problem that will arise relates to resources. The Commission has a broad remit, yet it has limited resources.[40] Its ambitions may be curbed by lack of resources. It is therefore fortunate that the Commission must report to the Secretary of State within two years on recommendations for improving *inter alia* its effectiveness.[41] There will be other matters to address, but one suspects that the issue of resources will figure prominently in this report.

The Commission published a draft Strategic Plan for consultation in September 1999.[42] The Plan set out *inter alia* the Commission's core values, its resources and grounds for choosing possible options and its working methods. The core values listed are: independence; fairness; openness; accessibility; accountability; participation, and equality.[43] The possible options included: a Bill of Rights for Northern Ireland; the promotion of a human rights culture; and reference to specific areas such as victims' rights and criminal justice and policing.[44] The section on working methods set out its proposed position on: approaches from individuals and groups; forming strategic alliances; international work, and the exercise of its statutory functions.[45] The Commission, for example, emphasises the importance of international work and thus it has made representations to international committees which monitor the United Kingdom's adherence to international human

[40] It has an annual budget of £750, 000.

[41] Northern Ireland Act 1998, s. 69(2)(a).

[42] Northern Ireland Human Rights Commission *Draft Strategic Plan 1999–2002* (issued for consultation on 30 September 1999).

[43] *Ibid*. pp. 9–10.

[44] *Ibid*. pp. 15–30.

[45] *Ibid*. pp. 31–49.

rights treaties. The Commission adopted a final version of the Plan, to cover the period 2000–2002, in May 2000.

The functions and powers of the Commission are not as extensive as one would have hoped. The resources are clearly inadequate and the power of investigation is a limited one. Nevertheless the Commission has the potential to take a lead on issues of human rights protection in Northern Ireland.

HUMAN RIGHTS PROTECTION ON THE ISLAND OF IRELAND

The Agreement's provisions are not confined to one jurisdiction. Its adoption has implications for the protection of human rights on the island of Ireland as a whole. This will happen both formally and informally. The formal commitments are described below. Informally there are already networks in the human rights sphere which function on an all-Ireland basis. Less tangible is the spill-over impact of the Agreement. It is difficult to imagine that the human rights transition in the North will not have an impact on the South. There will be opportunities for NGOs in the South to make active use of the principles and practices being developed in the North.

The Agreement commits the Irish Government to taking "steps to further strengthen the protection of human rights in its jurisdiction".[46] It is to bring forward measures, drawing upon constitutional reform proposals,[47] to strengthen and underpin the constitutional protection of human rights.[48] These proposals are to draw upon the European Convention on Human Rights and other international human rights instruments.[49] These steps are to lead to "at least an equivalent level of protection of human rights as will pertain in Northern Ireland".[50] In addition to this, it committed itself to: the establishment of a Human Rights Commission with a mandate and remit equivalent to that established in Northern Ireland; proceeding, as quickly as possible, with the

[46] "Rights, Safeguards and Equality of Opportunity" para. 9.

[47] See Constitutional Review Group *Report of the Constitutional Review Group* (Dublin, Stationery Office, 1996). See also the work of the All-Party Oireachtas Committee on the Constitution. It has thus far published four progress reports including: *The President* (3rd progress report); and *The Courts and The Judiciary* (4th progress report).

[48] *Supra* n. 46.

[49] "Rights, Safeguards and Equality of Opportunity" para. 9.

[50] *Ibid.*

ratification of the Framework Convention on National Minorities; implementing enhanced employment equality legislation; introducing equal status legislation; and "continue to take further active steps to demonstrate its respect for the different traditions in the island of Ireland".[51] The Agreement makes provision for cooperation between the two Commissions.[52] It envisages the establishment of a joint committee of representatives of the two Commissions "as a forum for consideration of human rights issues in the island of Ireland".[53] In addition:

> "The joint committee will consider, among other matters, the possibility of establishing a charter, open to signature by all democratic political parties, reflecting and endorsing agreed measures for the protection of the fundamental rights of everyone living in the island of Ireland".[54]

The reference to the rights of everyone living in the island of Ireland is of some significance given the heated debates in Ireland on immigration and asylum in recent years.[55] There will be difficult issues to address, including how this instrument would be distinct from, and relate to, the proliferating body of human rights norms. At the time of writing the Irish Government is in the process of appointing the Commissioners to its Commission. It had also announced a time-scale for incorporating the European Convention on Human Rights. The Agreement will have an impact on political and legal life in the North and South of Ireland.

A BILL OF RIGHTS FOR NORTHERN IRELAND

The Human Rights Commission has placed significant weight on the importance of drafting a Bill of Rights for Northern Ireland. The adoption of a Bill of Rights would be a good signal of the fresh start brought about by the Agreement. There has been discussion of a Bill of Rights for Northern Ireland for some time.[56] It was recommended as an

[51] *Ibid.* para. 9.

[52] *Ibid.* para. 10.

[53] *Ibid.*

[54] *Ibid.*

[55] See Colin Harvey "Right to Seek and Enjoy Asylum is a Human Right" *The Irish Times*, 25 March 2000.

[56] See CAJ *Making a Bill of Rights Stick: Options for Implementation in Northern Ireland* (September 1997); CAJ *Making Rights Count* (October 1990). The CAJ has also published its own Bill of Rights for Northern Ireland.

appropriate response by a number of bodies in the past. While there appears to have been widespread consensus among the political parties on the need for such an instrument, it never materialised in practice. There are a variety of reasons for this. The most convincing explanation is that the British Government was opposed to it. The Conservative Government was not as open to the discourse of human rights and equality as the Labour Government which came to power in May 1997. In addition, there was a belief that a Bill of Rights might have to await a constitutional settlement for Northern Ireland.

One of the more important functions of the Commission is its work on a Bill of Rights for Northern Ireland. The Bill of Rights exercise follows the conclusion of a constitutional settlement for Northern Ireland. It is from this settlement, and thus the Agreement, that the exercise derives its legitimacy. A vote for the Agreement can be read as consent for the adoption of a Bill of Rights. It is the content and form that this should take which is to be decided. The Agreement does not spell out precisely which rights should be included. An understanding of the nature of the Agreement is thus essential to consideration of the Bill of Rights exercise. The Agreement is a complex framework which attempts to accord due recognition to the bi-national nature of the conflict in Northern Ireland. It is a rejection of one form of majoritarian democracy and has been described as a consociational settlement. Although recognising the importance of the identity and ethos of the two main communities, it does not sanction action or behaviour which would undermine the principle of democratic inclusion upon which the whole edifice is based. In this sense the consociational and human rights elements perhaps do not conflict as directly as is sometimes thought.

The emphasis in the Agreement is on new beginnings and a fresh start. In moving away from the past this section suggests a fresh start with the "vindication of the human rights of all" as a major part of this. To what extent should the Bill of Rights be seen as part of this new beginning and fresh start? It might be suggested that an unduly narrow approach would not reflect this particular commitment. Other aspects of the Agreement are not so clear on this. The fact that the new constitutional structures reflect the divided nature of Northern Ireland is understandable, given the history of discrimination and exclusion. These sections of the Agreement attempt to ensure that all sections of the community can effectively participate in the governance of Northern Ireland. Human rights guarantees are one way to guarantee that impediments to full participation in the polity are removed and that one group is not permitted to dominate others. The structures created place great weight on securing trust between the two main political communities.

Reference to the Bill of Rights is included in the section on "Rights, Safeguards and Equality of Opportunity". On human rights, the parties to the Agreement affirm the importance of the following rights: the right of free political thought; the right to freedom and expression of religion; the right to pursue, democratically, national and political aspirations; the right to seek constitutional change by peaceful and legitimate means; the right to choose freely one's place of residence; the right to equal opportunity in all social and economic activity, regardless of class, creed, disability, gender or ethnicity; the right to freedom from sectarian harassment; and the right of women to full and equal political participation. The Bill of Rights for Northern Ireland is mentioned under the heading "United Kingdom Legislation". The relevant section merits full citation:

> "The new Northern Ireland Human Rights Commission . . . will be invited to consult and to advise on the scope for defining, in Westminster legislation, rights supplementary to those in the European Convention on Human Rights, to reflect the particular circumstances of Northern Ireland, drawing as appropriate on international instruments and experience. These additional rights to reflect the principles of mutual respect for the identity and ethos of both communities and parity of esteem, and—taken together with the ECHR—to constitute a Bill of Rights for Northern Ireland. Among the issues for consideration by the Commission will be:
>
> the formulation of a general obligation on government and public bodies fully to respect, on the basis of equality of treatment, the identity and ethos of both communities in Northern Ireland; and
>
> a clear formulation of the rights not to be discriminated against and to equality of opportunity in both the public and private sectors".

This section of the Agreement structures the debate. What is notable is the mixture of consociational and other elements in this passage. Rather than follow the language of a new beginning or a fresh start, thus permitting the Human Rights Commission to take a dynamic approach, the participants have come up with a formulation which enables and limits. On the enabling side it is evident that the passage provides for an instrument which goes beyond the European Convention. The Convention is recognised to be weak in several areas, on a right to equality for example. It is also primarily confined to civil and political rights. The section also enables the Commission to take account of international instruments and experience.

Arguments have been made for and against an expansive approach to this exercise. Those wishing to see the best possible Bill of Rights are hampered by the limiting language of the Agreement. However, the language of the Agreement is not as limiting as might be thought. This section must be read as part of the overall stress on a new beginning and a fresh start and the importance of the vindication of the human rights of all. It would scarcely accord with the new beginning if a narrow approach was adopted here. There is an additional problem. The voluminous literature on Northern Ireland has not provided us with a consensus position on the "particular circumstances of Northern Ireland". It is precisely this "context" which is contested. To focus on this phrase, and give it undue weight, may result in an unproductive and divisive debate. Given this, the issue should be which rights a modern Bill of Rights must contain, given current developments and international experience. This raises its own problems, in particular, the rights to be included and the mechanism of enforcement. For example, it has been suggested that a new constitutional court might be necessary given the particular circumstances of Northern Ireland.[57] This raises old and familiar questions about the role of the judges and human rights enforcement in a political democracy.

The Commission launched the Bill of Rights exercise on 1 March 2000 and it has placed great weight on the process. The belief is that an open and transparent process, which includes as many people as possible, will secure the legitimacy of any instrument that is drafted. In its work the Commission can draw upon the knowledge and experience of existing social movements in Northern Ireland as well as comparative experience of other countries. It has made clear that it wants to achieve the best possible Bill of Rights within the context of the Agreement. It may be impeded in this by those who want to see this exercise rigidly defined and restrictively implemented. The Agreement is not without its ambiguities and there is scope for a narrow and restrictive approach to be taken. As I have suggested, this would, however, conflict with the spirit of the Agreement which is firm in its commitment to a new beginning and the protection and the vindication of the human rights of all. The Human Rights Commission has sought to make the process an

[57] See CAJ (1997) *supra* n. 56 at pp. 14–16.

inclusive one. In order to draw upon existing expertise it established working groups in the areas of: social and economic rights; children and young people; criminal justice; cultural rights and community identity; language; victims' rights; equality; education; and implementation.[58] The working groups were tasked with making recommendations to the Commission on the rights that might be included in its draft Bill of Rights. The Commission has also produced a training manual as part of its Bill of Rights exercise.

The process of drafting a Bill of Rights for Northern Ireland will be instructive. After years when Northern Ireland was known mainly for the bad aspects of human rights abuse it may now attempt to lead the way in the United Kingdom and Ireland on human rights protection. There are difficult issues of principle and practice to be addressed. As I argue, the Commission has an important role in enriching political democracy. The issue now is to what extent a Bill of Rights can make a real contribution in this regard also.

CONCLUSION

It is too early to reach any conclusions about the work of the Human Rights Commission. This chapter raises some issues about the role of the Commission in promoting progressive political values in Northern Ireland. I have argued that the creation of the Commission moves us away from a court-centred approach to rights protection. I have deliberately been provocative in my comments on the role of the judiciary, and in the thoughts presented on rights-talk. It seems to me that a critical perspective is needed now more than ever. This is largely because everyone is at least trying to "talk the talk" in Northern Ireland. In this it is easy to lose sight of the reasons why we engage in human rights discourse in the first place. My suggestion is that not all uses of rights discourse enrich political democracy and that in some cases it can impoverish the language of political dialogue. On some occasions it has a hegemonic status and thus a detrimental impact on other important political values. This is not an argument against human rights (quite the reverse). The political struggle to create a just political democracy cannot be confined to rights discourse alone. The Human Rights

[58] The Commission produced a series of pamphlets on rights that are not adequately protected by the European Convention. See also Northern Ireland Human Rights Commission *First Annual Report 1999–2000*, HC 715.

Commission thus has a role to play in renewing and enriching rights discourse in Northern Ireland and taking decisive action when the language of rights is patently being abused.

6

And Justice for All? The Judiciary and the Legal Profession in Transition

STEPHEN LIVINGSTONE

INTRODUCTION

THE LAST THIRTY years has witnessed regular use of the law by all parties to the Northern Ireland conflict. The state has made the most extensive use of law, whether it is through emergency powers to repress political violence or fair employment legislation to bring about changes in the distribution of economic resources. However those critical of the state have also had recourse to the use of law, whether through civil actions to challenge the actions of the police and army or judicial reviews to question the actions of the prison service. Political parties have resorted to the courts to continue their disputes by other means and victims on all sides have sought to invoke the law to gain redress. This use of law has been symbolic as well as instrumental. The state has often pointed to the fact that it deals with its opponents through the use of legal process to establish the legitimacy of its actions and distinguish democratic politics from terrorism. However critics have also sought to invoke the law, usually the "higher law" of public international law or international human rights law, to problematise that claim and argue that what has prevailed in Northern Ireland is more a rule of force than a rule of law.

The law therefore has been a contested site for most of the past thirty years. This has posed particular challenges for Northern Ireland's lawyers and judges in terms of the work they do and their own perception of their role. This chapter seeks to explore how they have responded to these challenges as a precursor of a discussion of what the role of judges and lawyers may be in the future. The Good Friday Agreement and the Northern Ireland Act resulting from it have created a range of new legal institutions and legal provisions. Further reforms resulting from the Patten Commission and the Criminal Justice Review are likely to make even more alterations in the legal landscape. In other words, Northern Ireland's new constitutional structure is heavily underwritten by law and will pose new challenges for lawyers as to how it is operated. Some guidance to their response in the future may come from what they have done in the recent past. The chapter begins with an exploration of the role of the legal profession, before moving on to an assessment of the judiciary from whom they are drawn.

LAWYERS OF THE CONFLICT

Many non-lawyers, both inside Northern Ireland and without, assume that the conflict has been a major boon for lawyers. As the number of criminal trials rose with the growth of political violence from the early 1970s onwards it would seem axiomatic that all lawyers must have gained financially, especially as the defendants in such trials were all legally aided. However, although the number of lawyers in Northern Ireland has risen over the past thirty years[1] it appears the closest many have got to conflict related work is the processing of criminal injury claims. Though there is a dearth of research on the issue, what evidence does exist suggests that only a relatively small number of lawyers, less than 5 per cent of the total legal profession in Northern Ireland, has become engaged with work related to the conflict. One barrister, speaking to Morison and Leith in 1990, refers to a "Crumlin Road Mafia" and suggests that a core of about twelve counsel were regularly involved in such cases.[2] Jackson and Doran, in their study of the operation of the Diplock courts, also observed that a "small core" of counsel appeared regularly for prosecution and defence in these trials.[3] Many of the same lawyers appear to be involved in judicial review and

[1] The number of barristers in Northern Ireland rose from sixty to 330 and solicitors from 500 to 1,000 over the period 1965–93. See B Dickson *The Legal System of Northern Ireland* (3rd edn, Belfast, SLS Publications, 1993) p. 98.

[2] See J Morison and P Leith *The Barrister's World* (Milton Keynes, Open University Press, 1992) p. 40. At the time all Diplock court trials took place at Belfast's Crumlin Road courthouse.

[3] See J Jackson and S Doran *Judge Without a Jury* (Oxford, Oxford University Press, 1995) p. 83.

civil actions relating to the conduct of the police or the prison service. Although a broader range of solicitors than barristers have been engaged in such work most of it has been the preserve of six to eight firms. The UN Special Rapporteur on the Independence of Judges and Lawyers, in the report of his 1997 mission to Northern Ireland, suggested only about twenty to thirty solicitors were actively involved in representing the accused in politically sensitive trials.[4]

As regards the approach taken by lawyers involved in the defence of those accused of terrorist crimes or the judicial review of official policy, most have adopted a fairly traditionalist stance. There has been little in the way of activist, let alone radical lawyering.[5] Even those lawyers involved in conflict-related work have generally presented themselves as offering as good as possible a service to their clients, without endorsing those clients causes or methods. They have generally been content to operate the law as given rather than see it as part of their calling to alter it. One of the few studies existing of barristers involved in the defence of those accused of terrorist charges in Northern Ireland found them firmly wedded to traditional concepts of the cab rank principle and the need to represent all clients equally, whatever their political views or actions.[6] Even those lawyers frequently involved in cases challenging the content or application of emergency laws (including cases before international human rights tribunals) have generally avoided making public statements on them outside the courtroom,[7] indeed perhaps the most important public contribution by a private lawyer to discussion of the justice system was produced by an English QC.[8] Although Northern Ireland has witnessed a significant growth of human rights NGOs, which have played an important role in publicising human rights concerns in Northern Ireland and placing them on the international agenda,[9] practising

[4] *Report of the Special Rapporteur on the Independence of Judges and Lawyers, Param Cumaraswamy, on a Mission to Great Britain and Northern Ireland* (E/CN.4/1998/39/Add.4).

[5] For a discussion of the differences between traditional, activist and radical lawyering See, e.g., S Scheingold *The Politics of Rights: Lawyers, Public Policy and Political Change* (New Haven, Yale University Press, 1974) pp. 171–99, P Marguiles "Progressive Lawyering and Lost Traditions" (1995) 73 *Texas LR* 1139.

[6] B Jorgensen "Defending the Terrorists: Queens Counsel before the Courts of Northern Ireland" (1982) 9 *J of Law and Society* 115

[7] For one prominent exception see the pamphlet by Belfast solicitor Paddy McGrory *Law and the Constitution: Present Discontents* (Derry, Field Day, 1986).

[8] Lord Gifford QC's report on the Supergrass system, which will be discussed further below: T Gifford *Supergrasses: The Use of Accomplice Evidence in Northern Ireland* (London, Cobden Trust, 1984).

barristers and solicitors have not generally been to the fore in such organisations, just as they were not generally in leadership positions in the civil rights movement of the late 1960s. Lawyers professional organisations have, as will be discussed in more detail later, been somewhat muted when it comes to expressing concern about aspects of the justice system; and although for many years the chair of the Standing Advisory Commission on Human Rights (precursor to the present Northern Ireland Human Rights Commission) was a practising lawyer, this organisation too adopted a fairly low-key approach to advancing its calls for change. Nor have lawyers themselves formed many *ad hoc* organisations to scrutinise the legal system. Those that were created, such as the Northern Ireland Socialist Lawyers, had a fairly short life. Even where it comes to lawyers who work on human rights cases there is little sense of them developing a strategy, seeking the best case to advance a particular cause, as opposed to responding to circumstances as they present themselves.

Overall therefore there is little evidence of even those lawyers involved in human rights work in Northern Ireland pursuing an activist approach, of seeing it as part of the lawyer's calling to engage actively in public debate designed to uphold the values of the rule of law. There is almost no evidence of lawyers adopting a more radical approach, involving a rejection of the separation between lawyer and client as well as that between law and politics. Lawyers in Northern Ireland have not been prepared to turn trials into political theatre. Indeed, they encouraged republican defendants to move away from the early 1970s strategy of refusing to recognise the courts, as some lawyers in the USA did in the 1960s.[10] While there have been vague and unsubstantiated allegations that some lawyers have offered assistance beyond the scope of their professional obligations to clients involved in terrorism, a matter to be explored in more detail below, it is clear that none of Northern Ireland's lawyers have emulated some of those who defended the Red Army Faction in Germany by explicitly endorsing their client's aims and means.

Perhaps because of this approach there has been little formal official interference with the work of lawyers. Apart from a proposal from the

[9] See P Mageean and M O'Brien "From the Margins to the Mainstream: Human Rights and the Good Friday Agreement" (1999) 22 *Fordham International Law Journal* 1499.

[10] Notably, William Kunstler's involvement in the Chicago Eight Trial in 1969. For a discussion see J Auerbach *Unequal Justice: Lawyers and Social Change in Modern America* (New York, Oxford University Press, 1976) pp. 288–90.

Independent Commissioner for the Holding Centres in 1994 that a list of government-approved lawyers be established to provide legal representation for those accused of terrorist offences, a proposal dropped in the face of widespread hostility from the legal profession, there have been no moves to restrict a suspect's right to counsel of their choosing. Legal aid on a fairly generous scale has been available for all those charged of politically related offences and has also funded many of the judicial review challenges to decisions of the police or prison service. Lawyers acting for those accused of terrorist offences have not faced any disciplinary action from their professional bodies unlike, for example, American lawyers representing Communists in the 1950s.[11] Informally, however, the situation has been rather different.

Although solicitor Patrick Finucane, murdered by the Ulster Freedom Fighters at his home in 1989, was not the first lawyer to be killed in the conflict[12] his death raised particular concerns as to the extent to which the state endorsed the right of all lawyers to practice freely and independently. Several weeks before Finucane's murder, then Minister of State Douglas Hurd had observed, in a debate on the Prevention of Terrorism Act, that certain lawyers were "unduly sympathetic to the cause of the IRA".[13] After his murder it emerged that clients of Finucane, who represented many republican defendants, and several other lawyers claimed that the police had threatened their lawyers in the course of interrogations. A year after Finucane's death a former military intelligence informant claimed that he had told his handlers that loyalist paramilitaries were targeting the lawyer on several occasions, once very shortly before his death, but that nothing was done in respect of this case.[14] Further information has subsequently come to light that a police informant, who has recently been charged in respect of his involvement in Finucane's killing, claims that he had passed on information on the assassination plan to his Special Branch handlers, again without any action being taken to warn or protect the solicitor.

Finucane's murder, the allegations of security force collusion in it and claims of threats made to the clients of other lawyers occurred at a time when the police were denying immediate access for lawyers to

[11] See Auerbach, *supra* n. 10 at pp. 237–40.

[12] For example, barrister and law lecturer Edgar Graham, a prominent supporter of the use of supergrasses, was murdered by the IRA at Queens University in 1983.

[13] *Hansard*, 17 January 1989, Standing Committee B, col. 508.

[14] For a summary of these events see Lawyers Committee for Human Rights *Human Rights and Legal Defense in Northern Ireland* (1993) pp. 42–61.

clients arrested under anti-terrorist legislation in over 50 per cent of cases.[15] They also routinely refused to allow solicitors to be present during interrogations. Although anti-terrorist legislation permitted delays of access for up to forty-eight hours this was limited to circumstances where the police had reasonable grounds for suspecting that it might lead to interference with evidence or the alerting of a suspect, a narrow exception which one would not expect to occur in more than half of the cases in which people were arrested on suspicion of involvement in terrorism. The combination of these factors led to growing international interest in the circumstances of the relatively small number of lawyers in Northern Ireland involved in the defence of those suspected of terrorist offences. First raised by the Claire Palley, the UK nominee to the UN Sub Commission on Human Rights, in 1992 the issue was taken up by the Lawyers Committee on Human Rights in a 1993 report and subsequently led to a mission to Northern Ireland by the UN Special Rapporteur on the Independence of Judges and Lawyers in 1997. All recommended an independent inquiry into the circumstances of Patrick Finucane's death and the end of restrictions on the access of lawyers to their clients. They also noted that despite allegations in some journalistic accounts of lawyers abusing professional privilege to assist terrorists, no lawyer had been charged with a criminal offence in respect of such actions, nor had the police ever notified the Law Society of any allegations with a view to the pursuing of disciplinary action. However the calls for an independent inquiry, either into Finucane's murder or into the circumstances of lawyers generally, has yet to meet with a positive response from government. Nor has there been any such inquiry into the murder of Rosemary Nelson in 1999, a lawyer who had represented a number of clients in politically sensitive cases in the Portadown area, and who had testified to the US House of Representatives on police harassment of herself and her clients.

LAWYERS AND THE CONFLICT

As noted earlier, most of Northern Ireland's lawyers have, whether through accident or choice, eschewed work related to the conflict. However the issue of threats to lawyers suggested the rather more dis-

[15] The UN Special Rapporteur, *supra* n. 4 at para. 40, observed that access was deferred in 58 per cent of all PTA detentions in 1987–91.

turbing scenario that they were hostile or at least indifferent to the fate of those who did pursue legal practice in this area. Although the Law Society issued a statement condemning Finucane's murder it did not follow this up with calls for an inquiry or measures to protect the independence of solicitors. The Bar Council, apparently taking the view that such threats only concerned solicitors, said nothing on the issue. Nor have these organisations been particularly vocal in respect of changes in the law which, in the opinion of many domestic and international human rights observers, have brought the law into disrepute by undermining the extent to which it complies with rule of law principles and international human rights standards. The Law Society and Bar Council have failed even to make submissions to some of the inquiries looking at anti-terrorist law in Northern Ireland and even where they have taken a strong position, for example on the restriction of the right of silence in 1988, their advocacy has not assumed a particularly vocal form. The Lawyers Committee on Human Rights, in their 1993 report, commented on the "startling lack of concern accorded the small minority of defense solicitors by their colleagues in civil practice"[16] and described the Law Society's response to the Finucane murder as "tepid". Similarly the UN Special Rapporteur argued that both the Law Society and Bar Council should have been more vocal in their defence of lawyers' rights and argued that by not doing so they were failing to live up to their obligations under Principle 25 of the UN Basic Principles on the Role of Lawyers.[17]

These criticisms appear to have had some impact. In May 1999 a special AGM of the Law Society overturned a Law Society Council decision not to press for an independent judicial inquiry into the Finucane case and also passed a motion calling for a similar inquiry into the case of Rosemary Nelson.[18] However the vote on the Nelson case in particular was only reached after an acrimonious debate amid allegations of the discussion descending into a sectarian headcount.[19] That it has taken so long for Northern Ireland's lawyers collectively to reach a decision to call for fuller investigation into the deaths of their colleagues (only shortly before the Law Society meeting did the Bar Council endorse the need for an inquiry) raises important questions as

[16] See *supra* n. 14 at p. 41.

[17] See *supra* n. 15 at para. 36.

[18] *The Irish Times*, 12 May 1999.

[19] See E Moloney "Northern Solicitors Demand Inquiries" *Sunday Tribune*, 16 May 1999

to lawyers' attitudes to those involved in conflict-related work. A number of reasons can be advanced for their relative quiet on such issues.

One is the structure of the legal profession which, as we have seen, is made up largely of sole practitioners and small firms.[20] Such a structure encourages a sense of independence among lawyers and does not lend itself easily to collective work or the forming of collective opinions. Fear is undoubtedly a second major influence and fear comes in several forms. One is physical fear for the safety of the lawyer and their family. The murders of Graham, Finucane and Nelson, plus the threats relayed to other lawyers, have no doubt discouraged others from taking a more public stance on issues of the justice system. However there is also the fear, not uncommon in other professions in a divided society such as Northern Ireland, that being seen to take a public stand on certain issues is somehow unprofessional and may undermine one's credibility and professional reputation. This may be one reason why, despite the growing number of Catholics entering the legal profession, exactly the sort of articulate people from a nationalist background one might have expected to be critical of repressive legal measures, there has been little criticism of the justice system from lawyers as a whole.[21] As Finnouala O'Connor comments on the growing number of Catholic lawyers with nationalist sympathies in Northern Ireland, although some may feel guilt at saying nothing publicly:

> "In the main the most affluent and those on the next few rungs down have neither opted for posts that would expose them to IRA attack, nor offered any criticism of the system in which they prosper. Emergency legislation has repeatedly tinkered with conditions of arrest, powers to detain and the conduct of trials. The new wave of Catholics in the legal profession, like most of their predecessors, have with very few exceptions been content to work the system in silence".[22]

Raising explicitly "political" concerns may be seen as unsettling a professional, meritocratic, consensus that helps ensure reputation and

[20] For further discussion of the structure of the profession in Northern Ireland see M Fox and J Morison "Lawyers in a Divided Society: Legal Culture and Legal Services in Northern Ireland" (1992) 19 *J of Law and Society* 124.

[21] Since the legal profession is effectively exempt from Fair Employment Act 1989 monitoring, estimates of the religious make-up of the profession is guesswork at best. However as Fox and Morison *supra* n. 20 at p.139, observe one indicator is that the graduating class from Queens University's Law School, the main source of entrants to the profession, has been over 50 per cent Catholic for the last twenty years.

[22] F O'Connor *In Search of a State: Catholics in Northern Ireland* (Belfast, Blackstaff, 1993) pp. 27–28.

financial security for all lawyers. More positively some may argue that eschewing such issues has ensured that the legal profession is not dragged into Northern Ireland's sectarian maelstrom and that the law therefore remains open to all, no matter how unpopular their cause. While the idea that lawyers "held the line" against a descent into barbarism, especially in 1970s, may not be without merit it has the appearance of an argument thought up after the event rather than one which has been argued through against alternatives.

A third factor is that Northern Ireland's lawyers, trained mostly in Britain and Ireland, share the positivistic legal culture of most lawyers in these islands. This draws heavily on British traditions which, as Hunt has recently observed, espouse an essentially positivist approach to law and, via a stress on the sovereignty of Parliament, a rejection of the idea that law reflects particular values.[23] In this paradigm the lawyer's role is to remain essentially *neutral* and to shelve any personal political views when he or she enters the professional arena. Several of those arguing against the Law Society endorsing inquiries into the Finucane and Nelson cases claimed that this was departing from the Society's traditional *neutrality*. The argument is even more clearly made in a letter from the Lord Chief Justice, Sir Robert Carswell, to the Lord Chancellor on the issue of whether the declaration new QCs are required to make should be altered to ensure it was more open to all sections of the community. Carswell stated that: "I have little doubt myself that this is all part of an ongoing politically based campaign to have the office of Queens Counsel replaced by a rank entitled Senior Counsel or something like that".[24] The clear implication is that seeking change is "political" and that "political" activity is not something lawyers should be involved in.

This paradigm of neutrality, plus fear and a sense of independence, would appear to have constrained Northern Ireland's lawyers from contributing fully to the very extensive public debate which has raged on the institutions of justice. Given that such a debate has occurred, with extensive participation from political parties and NGOs, the question may be asked as to whether the lawyers' greater presence would have made a difference. I would suggest that it would in that, in a society without effectively functioning political institutions, the

[23] See M Hunt "The Human Rights Act and Legal Culture: The Judiciary and the Legal Profession" (1999) 26 *J of Law and Society* 86.

[24] An extract from the letter is quoted by Kerr J in *In re Treacy and McDonald's Application*, 2 May 2000, QBD (unreported).

only frame of reference for most judges was the views of their peers. When lawyers did express a more public view, for example in respect of the Supergrass trials, it appears to have been a significant influence on judicial decisions that ultimately gave effect to defendant's rights. Perhaps similar action in respect of areas such as confessions, the use of lethal force by the state or the abrogation of the right of silence might also have had a significant impact. More interesting still is the issue of whether this quietism is likely to continue. I would suggest it would not. The debate within the Law Society on the Finucane and Nelson inquiries plus the decision, after some internal debate, of the Bar Council to support a challenge to the QCs oath, would seem to presage a change. The Good Friday Agreement's commitment to parity of esteem seems likely to lead to further challenges to the exclusively British character of most of Northern Ireland's legal institutions. As the spectre of political violence recedes it is likely that more people within Northern Ireland, including lawyers, will become more comfortable with asserting their political allegiances and with demanding that these are respected. With the end of terrorism, the focus of legal action is likely to shift from the reactive arena of the criminal trial to the proactive civil sphere of judicial review and human rights actions, where lawyers have to craft arguments which require greater empathy with their client's objectives. In any case, with the advent of the Human Rights Act 1998, plus a potential Bill of Rights for Northern Ireland, lawyers' training and legal culture is likely to alter in a way which emphasises the idea of law having a particular normative content and of lawyers as having an obligation to uphold this. All in all, the legal profession in Northern Ireland is likely to face pressure to move away from one which eschews any public commitment to particular values or politics towards one which recognises and accommodates diverse views. The extent to which it achieves this is of importance not just for lawyers but also for the judiciary from whom they are drawn.

THE ROLE OF THE JUDICIARY

Background

The civil rights movement that developed in Northern Ireland in the 1960s took its name, some of its strategy and many of its slogans from the civil rights movement in the USA.[25] One thing it did not take was a belief in seeking to pursue change through the courts. While this may again reflect the under-representation of lawyers in Northern Ireland's public life it may also in large part be due to the fact that in the years between 1920 and 1968 the judiciary had given little indication of being sympathetic to change. Although the Government of Ireland Act 1920 had placed the courts in the (for a British court) constitutionally unusual position of being able to review and invalidate the legislation of the devolved legislature, there was little experience of this having been successfully invoked,[26] even less evidence of it being employed to challenge the inequalities of Northern Ireland society.

There were a number of reasons for this. The Government of Ireland Act 1920 was hardly a Bill of Rights. Although section 5 prohibited the Northern Ireland Parliament from making a law to "directly or indirectly establish or endow any religion. . . . or give a preference, privilege or advantage, or impose any liability or disadvantage, on account of religious belief" this was narrowly drawn and fell short of being a provision which prohibited discrimination on grounds of religion. It was therefore powerless to prohibit the indirectly discriminatory legislation, such as that abolishing PR for local elections or requiring long periods of residence to qualify for benefits, which entrenched sectarianism in the Northern Ireland state.[27] Moreover much of the entrenching of inequality, such as the reserving of jobs or council houses for Protestants, occurred through government administrative action or private sector recruitment rather than legislation. As campaigners throughout the common law world have found, the common law itself is not the best instrument for combating such discrimination. Some of it was also due to broader nationalist reluctance to engage with the institutions of the state. In a 1947 debate the Attorney General conceded that the Education Act 1930 (which he was in the process of repealing) was probably in contravention of section 5.[28] However, in its eighteen years of existence it had never been challenged before the courts.

To some extent therefore the lack of judicial engagement with minority concerns can be traced to factors beyond the control of the judiciary. However there were also aspects of the courts themselves

[25] See, generally, B Purdie *Politics in the Streets* (Belfast, Blackstaff, 1990).

[26] See H Calvert *Constitutional Law in Northern Ireland* (London, Stevens, 1968) p. 289.

[27] For a summary of these measures see P Buckland *A History of Northern Ireland* (Dublin, Gill and Macmillan, 1981) pp. 61–2.

[28] See Calvert *supra* n. 26 at p. 289.

which made them an unappealing vehicle of civil rights concern. For one thing the composition of the higher judiciary was, for most of the period of devolved government, almost entirely Protestant and unionist. Several had served as unionist MPs or as Attorney-General before joining the bench.[29] For another, on the few occasions that people had resorted to the courts to challenge actions of the executive in relation to matters of constitutional controversy, they had shown themselves to be essentially deferential to those holding elected office. In an early case on the Civil Authorities (Special Powers) Act 1922 the High Court had refused to question whether any evidence existed that a detainee had been involved in conduct prejudicial to peace or the maintenance of order.[30] The court indicated that its role terminated once it had satisfied itself that the regulation under which the detention occurred was validly made by the Minister of Home Affairs. A similar attitude prevailed in *Forde* v. *McEldowney* where the Court of Appeal, despite a spirited dissent from MacDermott LCJ, took the view that it could not rule *ultra vires* the same Minister's decision to prohibit a "Republican Club or any like club howsoever described".[31] To Boyle, Hadden and Hillyard this decision served to significantly undermine the position of those advocating peaceful change in the Northern Ireland state.[32]

The judges involved in these decisions may well have taken the view that they were acting consistently with British law at the time which, before the development of judicial review from the 1970s onwards, encouraged substantial deference to elected officials. The *Forde* v. *McEldowney* decision was indeed upheld by the House of Lords.[33] However, as McCrudden and others have pointed out, understandings of the judicial role predicated on a deference to elected officials as the expressions of a democratic will were hardly appropriate when those officials were part of a permanent government.[34] The failure of

[29] K Boyle, T Hadden and P Hillyard *Law and State: The Case of Northern Ireland* (Oxford, Martin Robertson, 1975) p. 12 observed that of twenty High Court judges appointed between 1920–70, fifteen were openly associated with the unionist Party.

[30] *R (O'Hanlon)* v. *Governor of Belfast Prison* (1922) 56 ILTR 170.

[31] [1970] NI 11.

[32] See Boyle, Hadden and Hillyard *supra* n. 29 at pp. 14–15. Harry Calvert "The Republican Clubs Case" (1970) 21 *NILQ* 191, 193 commented that such was the judicial capitulation in this case that "even moderate men may start looking to the streets".

[33] [1969] 2 All ER 1053.

[34] See, generally, C McCrudden "Northern Ireland and the British Constitution" in J Jowell and D Oliver (eds) *The Changing Constitution* (3rd edn, Oxford, Oxford University Press, 1994) p. 323; J Morison and S Livingstone *Reshaping Public Power: Northern Ireland and the British Constitutional Crisis* (London, Sweet and Maxwell, 1995) ch. 3.

Northern Ireland's courts to develop a constitutional jurisprudence which challenged abuses of power by an entrenched unionist Government rendered them at best irrelevant, at worst hostile, to the developing politics on the streets in the late 1960s. However those protesting often appealed, at least initially, to ideas of law and to the equal application of British legal standards in particular. As a British Government took over direct rule of Northern Ireland in the 1970s and appealed to such standards, including ideas of democracy and the rule of law, rather than communal loyalty, as underpinning the legitimacy of their authority, so the role of the courts became increasingly significant.

THE JUDICIARY IN THE NORTHERN IRELAND CONFLICT 1969–99

Compared to the situation of lawyers, there is almost a glut of writing on the Northern Ireland judiciary, some it even from judges themselves.[35] Most has concentrated on the judicial response to anti-terrorist legislation and the extent to which the Northern Ireland courts have upheld individual human rights. While this is a valuable perspective and one which must be included in any assessment of judicial performance in Northern Ireland, as I have argued elsewhere a full picture of the role of the law in the Northern Ireland conflict must embrace both its repressive and reforming dimensions.[36] Since the late 1960s successive governments in Northern Ireland have sought to use law not just to prevent and punish political violence but also to produce political and social change which might bring about greater social consensus. A comprehensive analysis of the role of the judiciary requires assessment of their engagement with these forms of law too. Therefore in this section I look at the judiciary in respect of three areas—anti-terrorist law and policy, judicial review and anti-discrimination law—

[35] See, e.g., B Dickson "Northern Ireland's Troubles and the Judges" in B Hadfield (ed.) *Politics and the Constitution* (Buckingham, Open University Press, 1992) p. 131; B Dickson "The European Convention in Northern Irish Courts" [1996] *EHRLR* 496; S Lee and C Hill "Without Fear or Favour? Judges and Human Rights in Northern Ireland" in *Standing Advisory Commission on Human Rights: Eighteenth Report* (Cm 739, HMSO, 1993) p. 81; S Livingstone "The House of Lords and the Northern Ireland Conflict" (1994) 57 *MLR* 333; Lord Lowry "Civil Proceedings in a Beleaguered Society" [1987] *Denning LJ* 109.

[36] See S Livingstone "Using Law to Change a Society: The Case of Northern Ireland" in S Livingstone and J Morison (eds) *Law, Society and Change* (Dartmouth, Aldershot, 1990) p. 51.

before examining debates on the composition of the judiciary and the challenges facing it in the wake of the Good Friday Agreement.

Anti-terrorist law and policy

As noted earlier, special legislation to deal with political violence is hardly a new phenomenon either in Northern Ireland, or in Ireland generally.[37] However the past thirty years have witnessed the most extensive emergency legislation in Northern Ireland and, perhaps because of the rise of a number of lawyers specialising in this type of work, the most extensive efforts to challenge its scope and application. Therefore a significant body of jurisprudence has arisen which largely tracks changes in anti-terrorist policy. In the early 1970s internment and its fall out in terms of the treatment of suspects preoccupied the courts. Throughout most of the 1970s and into the 1980s there was extensive examination of the circumstances in which confessions could be admitted before trial courts. The issue of Supergrasses predominated in the mid-1980s, giving way in the 1990s to concerns relating to the curtailment of the right of silence and access to defence lawyers. Throughout these thirty years there have been recurrent issues of the scope of emergency powers and the legality of the use of force, including lethal force, by the security forces.

In responding to these issues judicial attitudes appear to be shaped by a number of influences. One is a clear abhorrence of terrorism and a belief in the right of society to prevent and punish it. This may be seen as hardly surprising given that Northern Ireland's judges have always been high on the target list of republican paramilitary organisations and five have been murdered since 1973.[38] Several others have survived assassination attempts and all have lived lives subject to extensive security precautions. Second is an approach to statutory interpretation and the development of the common law that is largely shaped by British traditions of positivism. A third is a strong commitment to judicial independence and a view that judges must not simply become rubber stamps for the executive. In examining their response to a number of

[37] See C Townshend *Political Violence in Ireland* (Oxford, Oxford University Press, 1983).

[38] Lord Justice Gibson in 1987, County Court judges William Doyle in 1983 and Rory Conaghan in 1974, Magistrates William Saunders in 1973 and Martin McBirney in 1974.

issues in detail we shall see that these do not always pull in the same direction.

Consider, for example, the issue of confessions. In the immediate wake of internment the police developed a number of specialised interrogation centres which, coupled with detention over a number of days, they hoped would produce confessions from terrorist suspects. In *R* v. *Flynn and Leonard* Lowry LCJ concluded that such confessions could not be regarded as "voluntary" since they were produced by a "set up officially organised and operated to obtain information . . . from persons who would otherwise have been less than willing to give it".[39] Within a year the Westminster Parliament, acting on the recommendations of the Diplock Report that the approach of Lowry was "hampering the course of justice",[40] replaced the voluntariness test with the Northern Ireland (Emergency Provisions) Act 1973 requirement that confessions be admitted unless there was *prima facie* evidence of torture, inhuman or degrading treatment.[41] In the face of a clear statutory injunction to admit confessions that they would previously not have done the Northern Ireland courts retrenched. In *R* v. *McCormick* McGonigal LJ, in a judgment that would subsequently be the focus of substantial criticism, indicated that the new standard "leaves it open to an interviewer to use a moderate degree of physical maltreatment for the purpose of inducing a person to make a statement".[42] Although it is worthwhile noting that McGonigal's comments came directly from the European Commission's opinion in the *Greek* case,[43] two years later Lowry LCJ in *R* v. *O'Halloran* appeared to take the view that such retrenchment went too far.[44] He stated that the court would be reluctant to find that any physical violence would be relevant to an interrogation and hence that it would amount to *prima facie* evidence of inhuman and degrading treatment. The courts also indicated that the new 1973 Act test had not abolished the judicial discretion to exclude statements where, in all the circumstances, it was fair to do so.[45] However this discretion was rarely to be exercised and the courts

[39] [1972] 5 NIJB 1.

[40] *Report of the Commission to Consider Legal Procedures to Deal with Terrorist Activities in Northern Ireland*, Cmnd 5185 (1982) (The Diplock Report) para. 87.

[41] Originally the Northern Ireland (Emergency Provisions) Act 1973, s. 6, now modified and contained in the Northern Ireland (Emergency Provisions) Act 1991, s. 11.

[42] [1977] NI 105, 111.

[43] (1969) 12 *Yearbook of the European Convention on Human Rights*1, 501.

[44] (1979) 2 NIJB 1.

[45] See, e.g., *R* v. *Milne* [1978] NI 110.

also showed a reluctance to find that treatment had amounted to inhuman or degrading treatment in the absence of physical, as opposed to verbal, mistreatment. Many of the cases in the early 1980s focused not on allegations of physical violence but on lengthy periods of interrogation coupled with allegations of verbal abuse or threats.[46] Only in the rarest cases were these confessions excluded and the new consensus appears to be reflected in the revised test included in the Northern Ireland (Emergency Provisions) Act 1987, which expressly preserved the discretion and added violence or threat of violence as a new ground for exclusion.

The judicial approach to the admissibility of confessions has come in for significant criticism.[47] Critics note that in the two years between the decision in *McCormick* and that in *O'Halloran* complaints against the police relating to interrogations doubled, leading to an Amnesty International mission and eventually to the Bennett inquiry which recommended significant changes in interrogation practices.[48] They also note that in its first report on the holding centres the European Committee for the Prevention of Torture recommended substantial immediate improvements or closure if this could not be achieved.[49] However the confessions cases reveal a clear tension between the three factors identified earlier as influencing judicial decisions in this area. Preserving a sense of judicial independence leaned in the direction of exercising close supervision over interrogation practices, lest trials become rubber stamps of what had happened in the interrogation rooms.[50] On the other hand it was clear that obtaining confessions proved effective in jailing those involved in terrorist activity[51] and that Parliament had clearly, on a number of occasions, endorsed legislation explicitly designed to ensure confessions would be admitted more

[46] See, e.g., *R* v. *Dillon and Gorman* [1984] NI 292; *R* v. *Cowan* [1987] 1 NIJB 15; *R* v. *Howells* [1987] 5 NIJB 10.

[47] See, e.g., D Walsh *The Use and Abuse of Emergency Legislation in Northern Ireland* (London, Cobden Trust, 1983) p. 53.

[48] *Report of the Committee into Police Interrogation Practices in Northern Ireland* (Cmnd 7497, HMSO, London, 1979).

[49] *Report to the Government of the United Kingdom on the Visit to Northern Ireland by the European Committee for the Prevention of Torture, Inhuman or Degrading Treatment or Punishment from 20 to 29 July 1993* (CPT/Inf (94) 17) at para. 45.

[50] See D Greer "The Admissibility of Confessions under the Northern Ireland (Emergency Provisions) Act" (1980) 31 *NILQ* 205.

[51] E Mallie and P Bishop *The Provisional IRA* (London, Heinemann,1987) p. 255, note of the interrogation techniques used in the mid-1970s: "it was a development that produced immediate and potentially disastrous results for the Provisionals".

easily. At least it can be said that the Northern Ireland cases, in contrast to those of Israel and South Africa where even physical violence was not always deemed sufficient to rule a confession inadmissible, reveal a judiciary struggling with these tensions.

Similar issues have arisen in relation to the curtailment of the right to silence. Like the replacement of the voluntariness rule for confessions, the Criminal Evidence (Northern Ireland) Order 1988 (SI 1987/NI 20) struck at a well established common law principle. Defence lawyers sought to ensure that any such derogation was as minimal as possible and received some early support when courts ruled that inferences could only be drawn where a *prima facie* case was already made out against the defendant.[52] However the Northern Ireland courts had already allowed a somewhat broader interpretation of the Order before first the House of Lords and then the European Court of Human Rights in the *Murray* case suggested that courts should use a "common sense" approach to the drawing of inferences.[53] The *Murray* decision also had implications for the denial of access of solicitors to their clients in police detention. This had not generally been raised by lawyers until the late 1980s, despite the fact that the police regularly exercised a discretion given to them by the emergency provisions legislation to delay detainee's access to a solicitor for forty-eight hours[54] and never allowed solicitors to be present during interviews. This changed once the additional risk of a client's silence in those initial forty-eight hours being used against them was added.

The first wave of cases focused on the denial of immediate access. These were at least successful in forcing the police to offer some more specific reasons for denial of access in each case.[55] When the police subsequently responded by arguing that they feared solicitors would be asked to pass on messages which (unknown to the lawyers themselves) contained coded instructions for other terrorists, some solicitors

[52] See J Jackson "Curtailing the Right to Silence: Lessons from Northern Ireland" [1991] *Crim LR* 404.

[53] *Murray* v. *DPP* (1993) 97 Cr App R 151; *Murray* v. *United Kingdom* (1996) 22 EHRR 29.

[54] In his study, Walsh *supra* n. 47 at p. 66 found that none of the detainees he interviewed were granted access within the first forty-eight hours. The discretion to deny access can only be exercised on a number of specified grounds, which relate to things like interfering with evidence or alerting other suspects. This evidence suggested it was being exercised in a blanket fashion.

[55] See, e.g., *Re McNearney's Application*, September 1991 (unreported). However, in *R* v. *Harper* (1990) 4 NIJB 75 the Court of Appeal refused to exclude a confession after the defendant had been denied access to a solicitor.

offered undertakings that they would not communicate to anyone what passed between them and their client. However the courts concluded that not even that was enough in some circumstances and accepted police arguments that a solicitor giving such an undertaking might nevertheless be kidnapped and tortured by the IRA to reveal information.[56] This was despite an affidavit from one of the solicitors in these cases that:

> "The firm in which I am a partner has acted for over twenty years for many people alleged to have committed serious terrorist crimes. I personally have been in a position to give advice in these matters for a period of seven years and during that time have advised many hundreds of people . . . It has not been my personal experience nor has it been the experience of any member of this firm that any pressure has been brought to bear by members of terrorist organisations to divulge confidential information".[57]

Although the litigation was largely unsuccessful, it did appear to have a significant impact on police practice and denials of immediate access dropped dramatically from 1992 onwards.[58] Attention then shifted to the issue of whether solicitors had a right to be present during interviews, especially in the wake of the European Court of Human Rights decision in *Murray* that denial of access, coupled with inferences drawn from silence, could amount to a breach of Article 6 of the Convention. Within Northern Ireland the issue was most extensively considered by a three judge court of Queens Bench in *Re Russell's Application*.[59] In a decision replete with concerns as to the nature of the terrorist threat and Parliament's right to specify special measures to deal with it Hutton LCJ rejected the idea that the common law conferred a right to participate but added that even if it did:

> "The court has to decide these applications in a situation where there has been a very grave terrorist threat in this jurisdiction and where Parliament has passed the relevant section to give effect to the recommendations of the Diplock report".[60]

[56] See *Re McKenna and McKenna*, February 1992 (unreported); *Re Kenneway*, 1992 (unreported).

[57] Affidavit of Barra McGrory in *Re McKenna and McKenna*, *supra* n. 56.

[58] The UN Special Rapporteur *supra* n. 4 at para. 40, notes that the rate of deferral fell from 58 per cent in 1987–91, to 26 per cent in 1992, 14 per cent in 1993 and by 1996 had dropped to 3 per cent.

[59] [1996] NI 311.

[60] *Ibid.* at p. 346.

Hutton also indicated a view that the PACE code provisions (which generally allowed solicitors to be present but contained an exception for those arrested under anti-terrorist provisions) were consistent with Article 6. He noted that even in *Murray* the Court had not granted a right to participate and had indicated that restrictions could be placed on lawyer's contact with their client for good cause. The following year his view was upheld by the House of Lords in the *Begley* case.[61]

Many of the themes raised in the confessions cases appear again in this saga. However the statutory authority was weaker (Parliament had not expressly provided for solicitors to be excluded from sitting in on interviews) and arguably the need for judicial independence stronger, as these restrictions impugned the integrity of officers of the court. The judiciary could have scrutinised police justifications for denial of access more thoroughly[62] and a view that Parliament must have intended that solicitors be denied the right to sit in interviews only for cause would have been consistent with the Strasbourg Court's approach to Article 6. However perhaps the lack of public complaint from lawyers collectively, as opposed to those specifically involved in defence of those accused of scheduled offences, that these provisions prevented lawyers from doing their jobs effectively left the courts inclined to see the restrictions as justified.

The theme of a need to ensure judicial independence perhaps comes across most strongly in the Supergrass episode, which convulsed the courts in the mid-1980s. Between November 1981 and November 1983 a total of around 600 suspects were arrested on the evidence of nineteen republican and eight loyalist "supergrasses".[63] Initially the courts displayed a willingness both to believe the evidence of the supergrass, Kelly J describing Christopher Black as "one of the best witnesses I have ever heard", and to hold that it was safe to convict on their uncorroborated evidence alone.[64] However, beginning with the successful appeal of Charles McCormick (ironically a police officer implicated by

[61] R v. *Chief Constable RUC, ex parte Begley* [1997] NI 275, [1997] 4 All ER 833.

[62] Or balanced these more fully against the interests of suspects. *In re Floyds Application* [1997] NI 414, Carswell LCJ saw no particular reason why immediate access should be granted in respect of a seventeen-year-old suspect who had recently given birth. It is unclear after this when police officers would be found to have exercised their discretion to deny access unreasonably.

[63] S Greer *Supergrasses: A Study in Anti-Terrorist Law Enforcement in Northern Ireland* (Oxford, Oxford University Press, 1995) p. 57.

[64] Thirty-one of fifty-six defendants convicted in the first three trials in 1983 were convicted on the uncorroborated evidence of the supergrass, see Greer *ibid.* at p. 252.

one of his own informers) in January 1984,[65] the use of supergrasses went into decline and by the end of 1986 over 75 per cent of those originally charged on supergrass evidence had been acquitted. A substantial number of these were acquitted on appeals where the Northern Ireland Court of Appeal concluded that their judicial brethren had either underestimated the credibility problems of such witnesses or misapplied the corroboration requirements. These appeal judgments involved a particularly exacting scrutiny of the fact-finding processes of trial judges and, as Greer comments, "legal criteria even more strict than those of the regular common law were eventually observed".[66]

While it is difficult to know what exactly provoked this change in judicial approach it appears not insignificant that, unusually in the history of anti-terrorist law in Northern Ireland, the judicial performance was subject to significant criticism from within the legal profession. In January 1984 Lord Gifford QC published a highly critical report, which crucially argued that supergrass evidence would not have been accepted by juries, and other public criticisms were made by members of the Northern Ireland Bar.[67] While the use of supergrasses was clearly highly efficient at putting those suspected of involvement in terrorism behind bars it also brought judicial independence into substantial doubt. Perhaps crucially, unlike other aspects of anti-terrorist strategy that have been reviewed by the courts, it had no explicit parliamentary approval.

Parliament has also had little to say about the use of force by the police and army in Northern Ireland, yet here judicial scrutiny has been rather less exacting. Over 350 people have been killed by the security forces during the Northern Ireland conflict, mostly by the army.[68] A significant proportion of these killings have occurred in circumstances which cast suspicion on claims that the force used was reasonable but to date only four security force members have been convicted of murder for killings committed while on duty. Although this low rate of convictions is primarily due to the limited number of prosecutions brought, especially in respect of killings resulting from planned interception operations, the judicial role is also worth examining. Acquittal rates for security force personnel in these cases run at over 80 per cent

[65] *R* v. *McCormick* [1984] NI 50.

[66] See Greer *supra* n. 63 at p. 273.

[67] T Gifford *Supergrasses: The Use of Accomplice Evidence in Northern Ireland* (London, Cobden Trust 1984). See also E Grant "The Use of Supergrass Evidence in Northern Ireland 1982–5" (1985) 135 *NLJ* 1125.

[68] A total of 294 of 357 deaths can be attributed to the army.

and arguably the way in which the judiciary has interpreted the key legal provision at issue, section 3(1) of the Criminal Law Act 1967, has discouraged the bringing of prosecutions in the first place.[69] In the 1975 case of *R* v. *Naughton*, Lord Lowry LCJ took the view that "the security forces are operating in conditions with which the ordinary law was not designed to cope and in regard to which there are no legal precedents".[70] The Attorney-General for Northern Ireland subsequently sought greater clarity from the House of Lords as to what standards should apply in the special circumstances of Northern Ireland but their lordships simply responded that it was essentially a matter for the jury (despite the fact that juries had been suspended on the recommendation of one of its members).[71] Some of their dicta however indicated that where security force personnel reasonably suspected they were engaging terrorists courts should be loath to second-guess their actions.

Bereft of such guidance therefore Northern Ireland's judges have continued to adjudicate on a number of criminal and civil cases relating to killings by the state. In some they have displayed a commendable willingness to examine the facts thoroughly and to scrutinise the evidence of security force personnel as closely as anyone else. However it is probably fair to observe that this has been in cases where those killed clearly had no connection with terrorism and where even some security force members were prepared to testify against their colleagues.[72] Moreover such cases are arguably offset by those where the courts have displayed considerable sympathy for the police officers or soldiers involved.[73] The most notorious was the comments of Gibson LJ who, acquitting several police officers of murder of alleged republican paramilitaries, commended them for bringing the suspects to justice, "in their case the final courts of justice".[74]

In addition to the criminal trials of security force personnel, issues of lethal force have also engaged judicial attention via inquests. In those

[69] This provides that "a person may use such force as is reasonable in the circumstances in the prevention of crime or in affecting or assisting in the lawful arrest of offenders or suspected offenders or of persons unlawfully at large".

[70] [1975] NI 203, 208.

[71] *Attorney-General for Northern Ireland's Reference* (No 1 of 1975) [1977] AC 105.

[72] As, e.g., in the cases of Clegg or Fisher and Wright. However even these cases, where convictions ensued, were followed by extensive campaigns to seek early release and a review of the conviction. Clegg's murder conviction was subsequently quashed.

[73] See *R* v. *Robinson* [1984] 4 NIJB 1; *R* v. *Montgomery*, 1984 (unreported); *R* v. *Elkington*, 1992 (unreported).

[74] See *R* v. *Montgomery supra* n. 73. These killings and the cover-up which emerged at the trial subsequently became the focus of the Stalker inquiry.

cases where a criminal prosecution is not brought, and even some of those where it is, the inquest becomes often the only form of public examination of how the death occurred. Inquests in Northern Ireland have been extensively criticised in respect of a number of structural flaws, notably the lack of legal aid for victim's relatives and the attenuated form of verdict available.[75] However those flaws have arguably been exacerbated by a number of judicial decisions. In *McKerr* the Northern Ireland Court of Appeal had at least indicated that a coroner could compel security force members involved in the killing to testify at the inquest, but this was subsequently reversed by the House of Lords.[76] More recent Northern Ireland decisions have upheld the issuing of public interest immunity certificates and the screening of soldiers giving evidence which only increase the difficulties for the inquest to perform its function.[77]

These cases on use of lethal force by the security forces, together with some on excessive use of non-lethal force,[78] are perhaps the most disturbing in the area of anti-terrorist law and policy. By suggesting that the police or army enjoy some implicit level of immunity for their actions when dealing with suspected terrorists they undermine the judges' own ideology of the law being equally applicable to all. They also undercut efforts to suggest that the law, far from being simply an emanation of a unionist state, is open for appropriation by all communities in Northern Ireland. Throughout the history of Northern Ireland nationalist support for illegal armies has always in part been underwritten by fears that the law will not check the use of force by unionists against them. While Northern Ireland's judges have scarcely shown themselves unwilling to convict loyalist paramilitaries, this reluctance to ensure the effective accountability of the security forces fails to assuage such fears.

Overall the judicial engagement with anti-terrorist law and policy suggest that images of a judiciary wholly subservient to the executive, still less to a unionist agenda, are misleading. Instead we have seen a judiciary which clearly shares the state's abhorrence of terrorist

[75] See, e.g., Amnesty International *Northern Ireland: Killings by Security Forces and Supergrass Trials* (London, 1988).

[76] *McKerr* v. *Armagh Coroner* [1990] 1 All ER 865.

[77] See *Re Ministry of Defence's Application* [1994] NI 279; *Re Jordan's Application* [1995] NI 308.

[78] Though there the courts have given some important verdicts criticising ill treatment, see, e.g., *Re Gillen* [1988] 1 NIJB 47 (habeas corpus granted in respect of mistreatment of suspect); *Adams* v. *Chief Constable*, 1998 (unreported) (£30,000 awarded for assault on suspect).

violence but which also struggles to reconcile ideas of judicial independence with fidelity to traditional British constitutional notions of the sovereignty of Parliament. Neither the House of Lords, which has displayed an almost total deference to the views of the executive in respect of Northern Ireland,[79] nor Parliament, which has regularly supplemented emergency law to overturn any unfavourable judicial decisions,[80] have made its task easy. Given that the anti-terrorist law regime has primarily been administered against the nationalist community in Northern Ireland it would be hard for any judiciary, no matter how rigorously independent, to preserve the image of equal justice. Moreover failure to firmly enforce what Parliament has commanded in this area would risk losing the confidence of other sections of the community. What is perhaps most conspicuously missing from this jurisprudence is what Kirchheimer has called "the most awesome as well as the most creative part of the judicial experience, the entertaining of a small but persistent grain of doubt in the purposes of his own society".[81] There is no sense of wonder as to whether anything has ever been wrong in the state of Northern Ireland, no sense of any alternative perspective to the evil of terrorism and the right of the state to combat it, even in such a divided polity. A greater interest in international human rights norms might have provided at least some sort of partial alternative discourse to the official language of Westminster and Whitehall but although such norms have been referred to more frequently recently the courts have largely done so only to dismiss them.[82]

Judicial review

In Northern Ireland, as in the United Kingdom generally, judicial review has mushroomed since the mid-1980s.[83] Some of the reviews

[79] See Livingstone *supra* n. 35; C Gearty "The Cost of Human Rights: English Judges and the Northern Ireland Troubles" [1994] *Current Legal Problems* 19.

[80] Beginning with the immediate passing of retrospective legislation to annul a decision of the Northern Irish courts that special powers allowing the army to make arrests were *ultra vires* the Government of Ireland Act 1920, see *R (Hume)* v. *Londonderry Justices* [1972] NI 91 and B Hadfield "A Constitutional Vignette—From SR & O 1970/214 to SI 1989/509" (1990) 41 *NILQ* 54.

[81] O Kirchheimer *Political Justice: The Use of Legal Procedure for Political Ends* (Princeton, Princeton UP, 1961) p. 233

[82] See B Dickson "The European Convention in Northern Irish Courts" [1996] *EHRLR* 496.

[83] See B Hadfield and E Weaver "Judicial Review in Perspective" (1995) 46 *NILQ* 113.

taken relate to aspects of anti-terrorist law and policy, such as access to solicitors or inquests, which has already been explored in this chapter. However there are also many more which engage different issues of controversy. Whereas criminal defence is a necessity for clients, the invocation of judicial review represents a choice to engage with the legal system and one that may owe as much to the ingenuity of lawyers as to the demands of their clients. Although the majority of judicial reviews in relation to matters of controversy have been launched from a nationalist perspective there are an increasing number from unionists[84] and even different parts of government itself.[85] This section focuses on three areas where the courts' intervention into matters of significant public debate has been sought.

The first relates to the prisons.[86] Although the domestic courts were largely by-passed during Northern Ireland's major prison controversy, the protests and hunger strikes of 1976–81,[87] things changed from the early 1980s onwards. In part this may have been due to lawyers' perception of a greater availability of judicial review in the prison context following the *St Germain* decision,[88] in part to prisoners seeing it as a more economical way of challenging the authority of prison staff. Whatever the reason prisoners, primarily republican prisoners, became the most enthusiastic exponents of judicial review in Northern Ireland from the mid-1980s onwards.[89] These challenges originally focused on the disciplinary powers of prison governors and Boards of Visitors but moved wider to deal with matters such as visits, release schemes and equal treatment between men and women in prisons. In relation to discipline powers prisoners had some significant victories. The Northern Ireland Court of Appeal took the lead in the United Kingdom in declaring that governor's adjudications were subject to judicial review[90] and

[84] See e.g. *Re Austin* [1998] NI 329 (loyalist prisoner challenging temporary release conditions); *Re Dallas* [1996] NI 276 (unionist mayor challenging censure actions by local council); *Re Williamson*, 2000 (unreported) (victim of IRA bombing challenging early release of prisoner convicted of the offence).

[85] See *Re Secretary of State for the Home Department* [1999] NIJB 68 (Home Secretary challenging Sentence Review Commissioners' early release decision).

[86] For a general summary of Northern Ireland prison cases see B Dickson "Judicial Review and Prisoners in Northern Ireland" [1998] *Public Law* 57.

[87] Although the European judicial institutions were engaged, see *McFeeley* v. *United Kingdom* (1980) 3 EHRR 161.

[88] *R* v. *Board of Visitors of Hull Prison, ex parte St Germain* [1979] QB 425; this decision rejected the argument that prisons were beyond the judicial review jurisdiction.

[89] Hadfield and Weaver *supra* n. 83 at p.119, note that prisoners cases reached 53 per cent of all judicial review cases by 1991.

[90] *McKiernan* v. *Governor HM Prison Maze* (1985) 6 NIJB 6.

subsequently imposed more stringent procedural requirements on the conduct of some of these hearings.[91] Most of the cases on other issues have been lost[92] but the court's willingness to grant leave for review of such matters may have done something to reassure those in prison that avenues for redress of grievances existed.

A second area is cases concerning local councils. With the rise of Sinn Féin as a political force from the early 1980s its members began to be elected in numbers to local councils. Unionist councillors in a number of councils then sought either to exclude such councillors altogether or place significant barriers in the way of their effective participation in council business. Most of these actions led to judicial reviews and most were ruled *ultra vires* by judges who, while not hiding their distaste for Sinn Féin's support for the IRA,[93] nevertheless came to the conclusion that unionists had exceeded their powers.[94] Such cases, coming as they did only a year after the height of the Supergrass saga, may well have done something to assuage nationalist fears of unionist judicial bias and certainly cleared the way for greater political participation by Sinn Féin.

A third and most recent set of cases concern what might be seen as issues relating to "parity of esteem between the two communities". One manifestation of this has been cases relating to the parades issue. In *Re Farrell's Application*, Nicholson LJ concluded that the Parades Commission had given a sufficiently clear statement of its reasons, which were based on relevant criteria, to uphold its decision to permit a parade down Garvaghy Road.[95] More recently in *Re White's Application* (a case which will be discussed in greater detail below) Carswell LCJ rejected a challenge to the composition of the Commission.[96] Another manifestation of this issue is the case on the

[91] See, e.g., *Re O'Hare* [1989] 1 NIJB 1 (delay in charging); *Re Murphy's Application* [1988] 8 NIJB 94 (content of charges).

[92] See, e.g., *McCartney* v. *Governor HM Prison Maze* (1987) 11 NIJB 94 (visits of Sinn Féin councillor prohibited).

[93] Hutton J in *In re Curran and McCann's Application* [1985] 7 NIJB 22 expressly took judicial notice of the fact that when Sinn Féin representatives took part in local councils this "is just one plank of their policy, the other plank being the unambivalent support of murder and other acts of terrorist violence committed to overthrow democratic government in Northern Ireland".

[94] See *In re Curran and McCann supra* n. 93; *In re French and others Application* [1985] 7 NIJB 48; *In re Neeson's Application* [1986] 13 NIJB 24. Indeed it has sometimes seemed that judges' revulsion for Sinn Féin has been exceeded only by their exasperation at the conduct of unionist councillors.

[95] [1999] NIJB 143.

[96] 18 May 2000 (unreported).

QCs' declaration, which has been discussed earlier.[97] Although Kerr J rejected all of the applicants' claims of discrimination in that case his recognition that the issue of the declaration was a matter of controversy on which the Lord Chancellor would wish to consult marks a welcome acknowledgement of deep-seated conflicts of identity and allegiance within Northern Ireland.

Overall the development of the judicial review jurisdiction in Northern Ireland is a matter worthy of note. While some applicants may have been left disappointed by decisions against them, the willingness of courts to consider such issues has provided a significant official avenue for their airing of grievances in a society hitherto bereft of political institutions. Moreover the values of judicial review, such as transparency and the need for consultation, are of particular relevance in a polity where one community complains of an historic exclusion from decision-making. Some of those involved in prisons cases have subsequently played a part in other judicial reviews after their release. They are unlikely to have done so had they felt the courts were closed to them. In this way judicial review may have served as something of a counter-balance to the unequal appearance of anti-terrorist law.

Fair employment

The same may be said of the development of fair employment law, especially since the introduction of the Fair Employment Act 1989. Clearly the judges have not been in the vanguard of dealing with religious discrimination. We have already seen how they failed to offer much in the way of a remedy to those campaigning on the issue in the 1960s. Rather this has been a legislative effort, with significant political pressure from Ireland and the USA, as well as within the United Kingdom; combining to produce perhaps Europe's most advanced anti-discrimination law.[98] One of the themes of this development has been to avoid judicial enforcement of anti-discrimination law, first through the use of the Fair Employment Agency in a quasi-adjudicative capacity, later through the creation of Fair Employment Tribunals to deal with individual claims of religious discrimination.

[97] *In re Treacey and MacDonald's Application*, 2 May 2000 (unreported).

[98] See S Rose and D Magill "The Development of Fair Employment Legislation in Northern Ireland" in Magill and Rose (eds) *Fair Employment Law in Northern Ireland: Debates and Issues* (SACHR, 1995) p. 1.

However this legislation still allows the Tribunal to state a case for the Court of Appeal and it is here that the courts must engage with questions as to what amounts to discrimination and the legality of measures to combat it. Although commentators have observed some differences between court and tribunal interpretations these have not been stark.[99] The Court of Appeal has delivered a number of enlightened and progressive judgments on issues such as the burden of proof in discrimination issues,[100] sectarian harassment[101] and affirmative action.[102] At the very least it can be said that there is no evidence of the courts significantly undermining progressive equality legislation, as has been observed in relation to courts in Great Britain or the USA in dealing with race and sex equality law. In respecting the clearly expressed wishes of Parliament the performance of the Northern Ireland courts in these cases may have assuaged the fears of those who saw them as implacably committed to a unionist agenda. As with the judicial review cases these decisions may have contributed something to republicanism's turn to constitutionalism by demonstrating the possibility of a justice equally open to all.

COMPOSITION OF THE JUDICIARY AND CHALLENGES FOR THE FUTURE

Unlike the police or the civil service, the Northern Ireland judiciary has faced few calls for structural change over the past thirty years. Efforts by the Irish Government to explore the notion of three judge trial courts at the time of the Anglo-Irish Agreement in 1985 were swiftly rebuffed, apparently after opposition from the judiciary in Northern Ireland. In the Good Friday Agreement, while most of Northern Ireland's other public institutions were cited for change, the judiciary appeared as the one that got away. Whether this is testament either to political parties' satisfaction with their work, their own ability to resist change or a lack of political interest in what happened in the courts is difficult to say. However the Agreement did contain a sting in the tail, the provision for a Criminal Justice Review whose terms of reference included "the arrangements for making appointments to the judiciary

[99] See C Bell "The Case-Law of the Fair Employment Tribunal" in Magill and Rose *ibid.* at pp. 71, 93.

[100] *Belfast Port Employers* v. *Fair Employment Commission* [1994] NIJB 36.

[101] *Smyth* v. *Croft Inns* [1995] NI 292.

[102] *Hall* v. *Shorts Missile Systems* [1996] NI 214.

and magistracy, and safeguards for protecting their independence". In the end the issue of judicial appointments proved to be one of the most contentious of the Review, with newspaper reports indicating that its publication had been delayed due to concern within both the Northern Ireland judiciary and the Lord Chancellor's office as to its recommendations for an independent element in the appointments mechanism.

The Review keeps strictly to its terms of reference and eschews any discussion of the role or performance of judges in Northern Ireland over the past thirty years in favour of a concentration on appointments and training. While stressing that "merit . . . must in our view continue to be the key criterion in determining appointments"[103] it goes on to add that there is a need for action to ensure a judiciary which is "reflective of Northern Ireland society".[104] To achieve these objectives it advocates the establishment of a Judicial Appointments Commission, whose members would include both judges and lay members, relaxation of eligibility criteria and efforts at outreach to communities who are currently under-represented among the judiciary. However it retains the idea that the judiciary should be drawn exclusively from legal practitioners and rejects arguments for applying Fair Employment Act 1989 monitoring procedures to judicial appointments or applications. Moreover its key recommendation of a Commission is dependent on control of justice matters being devolved to the Northern Irish Assembly and Executive.

Whether the Review's recommendations will ever come to fruition is therefore subject to a number of considerations, most notably the progress of devolved government and the speed at which it is deemed safe to devolve further functions to it. However, the very fact of the Review and its recommendations has served to shed light on an institution which seemed almost beyond public scrutiny. Some have expressed dismay that implementation of its recommendations would lead to a highly politicised judiciary, with lawyers jockeying to see if they can get their cases before a "nationalist" or "unionist" judge. However the Review indicates that its aim is not to produce a "representative" judiciary, in the sense of one which actively pursues the interests of particular groups, but a "reflective" one. While acknowledging that judges must continue to apply the law independently of the will of the executive or political parties it acknowledges that the confi-

[103] *Review of the Criminal Justice System in Northern Ireland* (HMSO, 2000) para. 6.84.

[104] *Ibid*. para. 6.85.

dence of the whole of the public in the administration of justice may be undermined if the judges are all Protestant men. Currently there are greater concerns in respect of gender (only one of Northern Ireland's twenty-five most senior judges is female) than community background (although no public information exists there appears to be a balance at High Court level, less so in the County Court). A more reflective judiciary would ensure that a greater range of experiences and insights are brought to bear on judicial decisions, a matter commentators throughout the United Kingdom have seen as of increasing importance as judges grapple with more policy laden choices in interpreting the Human Rights Act 1998.[105] It would also bring the judiciary more into line with other public institutions in Northern Ireland.

However, as the earlier discussion of the traditionalist character of Northern Ireland's legal profession (from which judges will continue to be drawn) suggested, it would not be impossible to produce a bench which is reflective in terms of community background and gender but overwhelmingly conservative. For some this may seem entirely appropriate. With the restoration of devolved government, and on a power-sharing basis, it might seem desirable that the judiciary maintains a traditionalist stance of deference to elected officials and does not encourage what Brendan O'Leary has referred to as "the temptations of legalism".[106] However this, it seems to me, would risk a return to the constitution as a framework for pragmatic interest bargaining, an approach which has already been described as inappropriate for Northern Ireland and which undermines the Good Friday Agreement's innovative blend of communal and individual rights. The approval of the Agreement in a referendum suggests that it is more than a deal between political elites but rather the development of a set of principles that will govern relations between all citizens of Northern Ireland. The task of policing the application of these principles remains one for the courts.

This insight returns us to the question of the paradigm of judicial interpretation referred to earlier in this chapter. This reflected traditional British constitutional practice whereby the absolute sovereignty of Parliament was recognised and deference was owed to the actions of elected officials. This in turn was based on the idea that the outcome of parliamentary elections reflected expressions of the democratic will in

[105] See K Malleson *The New Judiciary: The Effects of Expansion and Activism* (Aldershot, Ashgate, 1999) pp. 31–2.

[106] B O'Leary "The Nature of the Agreement" (1999) 22 *Fordham ILJ* 1628, 1659.

response to competing political programmes. Only the electorate could subsequently decide that the government had erred, the courts lacked the legitimacy to offer an alternative vision. It was argued that, however appropriate such a paradigm was for the United Kingdom as a whole, it failed to reflect either the lack of real electoral challenge to unionist government between 1921–72 or the existence of any electoral challenge to direct rule in 1972–2000. With the Agreement, the passing of the Northern Ireland Act 1998 and the return of devolved government the courts now face a choice about which paradigm they pursue. They can treat the new constitutional arrangements purely as the outcome of Westminster's passing of the Northern Ireland Act 1998, in which case legislation of the new Assembly is to be approached much in the same way as that of the Stormont Parliament was—Westminster legislation is entitled to a higher level of deference and judicial review of administrative action will be shaped largely by developments in the English courts. Alternatively, and more imaginatively, they can recognise that the Northern Ireland Act was, as its preamble states, "for the purpose of implementing the agreement reached at multi-party talks on Northern Ireland set out in Command Paper 3383" and treat that Agreement too as part of the constitutional paradigm, in other words that the Agreement is the founding of a new constitutional order within Northern Ireland and that the exercise of public power must reflect it. This would not be to authorise a highly interventionist judiciary but rather to encourage the development of a more supple constitutional jurisprudence.

The content of this constitutional jurisprudence is obviously a matter for more extensive reflection and future development but some initial outlines can be sketched here. One would be that, given the cross-community support requirements and pre-legislative human rights scrutiny provisions, legislation of the Assembly should be entitled to a high level of deference by the courts. The voting arrangements should diminish concerns that "discrete and insular minorities" will be ignored by the legislature.[107] However the Agreement and the Act also envisage the upholding of individual rights. Where Assembly legislation strikes at the very essence of such rights (for example were it to abolish non-religious schools, contrary to Article 2 of Protocol 1 of the Convention) then a court must clearly intervene. A second is that the

[107] The phrase is taken from Justice Stone's footnote 4 in his judgment in *United States* v. *Carolene Products* (1938) 304 US 144. See also J Ely *Democracy and Distrust* (Cambridge, Harvard University Press, 1980).

Agreement and the Act clearly envisage leaving the development of policy on a significant number of controversial issues to various commissions rather than the legislature. These include policing, parades, human rights and equality. However a clear theme running through the establishment of all these bodies is that they must be "representative of the community". Since representativeness is key to their legitimacy the courts have a clear duty to ensure it is maintained. In this respect the recent decision of Carswell LCJ in *Re White*, that a Parades Commission lacking any female members remained representative, is not an encouraging precedent. Thirdly, when it comes to administrative action, the need to ensure equality of opportunity, reflected in section 75 of the Northern Ireland Act 1998, is a key principle. The courts should not see the devising of Equality Schemes, required under section 75(4) of the Act, as exhausting this requirement. It is clearly reflective of the spirit of the Agreement and therefore should impact on its implementation. Finally, the courts face an interesting challenge as to how to deal with actions of the UK Government in respect of Northern Ireland. Westminster legislation is of course exempt from the competence requirements of the Northern Irish Assembly and most UK government departments acting in Northern Ireland seem likely not to be designated for the purposes of section 75. Arguably, therefore, they should be governed by "old British" and not "new Northern Ireland" constitutional paradigms. However given that the UK Government is also a signatory to the Agreement it can be argued that its content has implication for the actions of the British Government too, in so far as they relate purely to Northern Ireland.

Adoption of such a new constitutional paradigm might also have implications for how Northern Ireland courts view the Human Rights Act 1998. On one level this can be seen as purely Westminster legislation, responding to UK-wide concerns. However the explicit reference to it in the Agreement offers the courts an opportunity to see it as also responding specifically to concerns as to inadequate human rights protection in Northern Ireland, though this will obviously be an easier step to take in respect of any specific Northern Ireland Bill of Rights. Viewing it in this way might provide the courts with a way into what has previously been argued is missing from their decisions, a sense of concern about what has gone wrong in the past. Certainly, in the absence of any comprehensive truth commission for Northern Ireland, they may well find that people invoke Human Rights Act provisions as a way of reopening debates about what has happened in the past. One can think especially of incidents involving lethal force where the Act could presage changes in inquests, civil liability and discovery. Dealing with the past while devising paradigms to govern the future—such, as Teitel has observed, is the lot of any court system in a time of transition.[108]

[108] See R Teitel "Transitional Jurisprudence: The Role of Law in Political Transformation" (1997) 106 *Yale LJ* 2009, 2014 on the need for legal institutions in times of transition to be both forward and backward looking.

7

Shaping the Future of Criminal Justice

JOHN JACKSON

INTRODUCTION

THE CRIMINAL JUSTICE system in Northern Ireland can truly be said to be currently experiencing a transition from a period when it had to deal with intense political violence towards a period of relative calm which is bringing with it its own challenges. While the troubles have left their mark on the Irish criminal process throughout the nineteenth and twentieth centuries, the past thirty years have put a particular burden on the criminal justice system in Northern Ireland.[1] This is largely because during the 1970s and 1980s there was an intensification of political violence against the institutions of state within Northern Ireland resulting in a need for considerable security precautions to be taken to protect those who work within the system, including the police, prosecutors and the judiciary. But it has also been because the UK Government's response to the violence, particularly from the mid-1970s, has been to adopt a policy of "criminalisation" within the legal process rather than resorting, as in the past, to extra-legal, emergency measures outside the process.[2]

When the UK Government imposed direct rule on Northern Ireland in 1972, it immediately established a Commission under the chairmanship of a British judge, Lord Diplock, to consider what legal procedures

[1] See J Jackson and S Doran "The Judicial Role in Criminal Cases in Ireland" in N Dawson, D Greer and P Ingram (eds) *One Hundred and Fifty Years of Irish Law* (Belfast, SLS Legal Publications, 1996) p. 69.

[2] For a brief account of the "criminalisation" strategy, see T Hadden, K Boyle and C Campbell, "Emergency Law in Northern Ireland: The Context" in A Jennings (ed), *Justice under Fire: The Abuse of Civil Liberties in Northern Ireland* (London, Pluto Press, 1988) pp. 8–10.

were necessary to deal with terrorist activities within the criminal justice system.[3] As a result of the recommendations of this Commission radical departures were made to ordinary criminal procedure such as the introduction of "Diplock" courts which provided for cases connected with the troubles to be tried by judge rather than jury. Those suspected of political violence have for the most part ever since been processed through the criminal justice system rather than through any extra-judicial system.[4]

This emergency legislation has remained largely intact in Northern Ireland throughout the last thirty years. That said, there has been a considerable reduction in the numbers of suspects being processed though the Diplock courts since the declaration of the paramilitary cease-fires in 1994 and this has inevitably led to calls for the legislation to be repealed.[5] Despite the cease-fires, however, paramilitary groups have continued to maintain a strong presence in certain areas of Belfast and elsewhere in Northern Ireland and this has enabled them to "police" and dispense their own form of paramilitary justice in these areas.[6] Paramilitaries and certain political parties have also continued to challenge the legitimacy of the criminal justice system, pointing to the clear over-representation of the Protestant or unionist population in the police and alleging that this community is also over-represented in other parts of the system including the judiciary.

It is against this background that the parties to the Good Friday Agreement in 1998 saw a clear need for a review to take place of the entire criminal justice system. Such a review was, of course, only one of the many issues dealt with in the Agreement and many other issues dealt with under the Agreement such as the mechanisms for devolution, prisoner releases and decommissioning have captured much larger headlines. But in terms of establishing confidence in the institutions of government there are few areas more important than the justice system. One of the central themes of the Agreement has been the commitment which it makes to partnership, equality and mutual self-

[3] See Lord Diplock, *Report of the Commission to Consider Legal Procedures to Deal with Terrorist Activities in Northern Ireland* (Cmnd 5185, 1972).

[4] The use of internment, the detention of suspects without trial, was eventually phased out in 1975, although the mechanism for its revival remained in the statute book until 1998.

[5] See *Statistics on the Operation of the Northern Ireland (Emergency Provisions) Act 1996*, prepared by the Statistics and Research Branch of the Criminal Justice Policy Division of the Northern Ireland Office.

[6] Paramilitary "punishment beatings" have been given considerable media attention, see *Belfast Telegraph*, 18 February 1999.

respect. In this context the establishment of a Human Rights Commission and a new Equality Commission under the Northern Ireland Act 1998 has been hailed as an important step in creating a Northern Ireland where prominence is given to human rights and equality. Clearly if these principles are to have any meaning they must be seen to penetrate the justice system. The specific concerns about policing were addressed by the establishment of an independent commission into policing under Chris Patten, but in parallel with this Commission the parties to the Agreement called for a criminal justice review to be carried out by the British Government "through a mechanism with an independent element".[7]

This chapter will focus on the unique features of the Criminal Justice Review, the approach adopted by the Review and some of its most significant recommendations.[8] As the author was a member of the Review, the chapter will not make an exhaustive critique of its recommendations. Some of these are examined elsewhere in this collection. Rather it will attempt to show how the context in which the Review was established guided its approach and led to its conclusions. It will be argued that its terms of reference and approach have led to a package of important recommendations which have the capacity to make fundamental changes to the criminal justice system and the legal system, but that even if the reforms are implemented they will not inevitably lead to greater confidence in the system without other issues, which the Review was unable to consider, being addressed as well.

THE SCOPE OF THE CRIMINAL JUSTICE REVIEW

The Criminal Justice Review was the most comprehensive review ever to be carried out into the criminal justice system in Northern Ireland. There have, of course, been a number of reviews on specific issues, chiefly of late on the operation of emergency powers considered necessary for dealing with the political and paramilitary violence. The most famous of these was the review mentioned above carried out by Lord Diplock in 1972 into the powers that would be necessary to enable terrorist or paramilitary offenders to be dealt with by the ordinary criminal justice system. A number of other periodic reviews have been

[7] Belfast Agreement p. 23.

[8] See Criminal Justice Review Group, *Review of the Criminal Justice System in Northern Ireland* (2000).

carried out into emergency legislation since then but they have all more or less endorsed the need for the package of measures which Lord Diplock recommended.[9] What these reviews never did, however, was to look at the operation of the criminal justice system as a whole. There have been some reviews into specific aspects of the criminal system: for example, there was a report in 1968—the Hunt Report—which reviewed the role of the police,[10] there was also a review in 1972 of the prosecution system and there was a report in the early 1970s into the organisation of the magistrates' courts and county courts.[11] But there had been no overall review of the system as a whole.

Although this Review was more comprehensive than any of its predecessors, the review group was given very clear guidance on how to approach its task. Often criminal justice reviews are set up in the wake of particular crises or at least shortcomings that have been revealed about the criminal justice system. The two Royal Commissions on criminal procedure and justice in England and Wales, for example, were established, one in 1979 and the other in 1991, as a result of concern about particular miscarriages of system in the English system.[12] Such reviews often involve looking in a rather technical way at the processes and procedures that led to the shortcomings. As explained above, the Northern Ireland review was set up in a different context, notably the need to ensure fairness and equality and human rights throughout the entire system. The Agreement as a whole affirms the need to establish institutions on the basis of partnership, mutual respect and human rights and this philosophy was brought very firmly to bear on the criminal justice review. The terms of reference of the Review explicitly required the Review to take into account the aims of the criminal justice system as laid down by the participants to the Agreement. These were to deliver a fair and impartial system of justice to the community, to be responsive to the community's concerns, encouraging community involvement

[9] See, e.g., *Report of a Committee to consider, in the Context of Civil Liberties and Human Rights, Measures to Deal with Terrorism in Northern Ireland* (Cmnd 5847, 1975); *Review of the Operation of the Northern Ireland (Emergency Provisions) Act 1978* (Cmnd 7497, 1984); *Review of the Northern Ireland (Emergency Provisions) Acts 1978 and 1987* (Cmnd 115, 1990).

[10] Hunt Committee, *Report of the Advisory Committee on Police in Northern Ireland* (Cmnd 535, 1969).

[11] *Report of the Working Party on Public Prosecutions* (Cmnd 554, 1971); *Report of the Committee on County Courts and Magistrates' Courts* (Cmnd 5824, 1974).

[12] See *Royal Commission on Criminal Procedure: Report* (Cmnd 8092, 1981); *Royal Commission on Criminal Justice: Report* (Cm 226, 1993).

where appropriate, to have the confidence of all parts of the community and to deliver justice efficiently and effectively.

At one level this gave the reviewers an opportunity to take what the Agreement described as a "wide-ranging" look at the whole system.[13] The Review was being asked to look ahead, rather than look behind at any past inequities; to set out a vision for the future, rather than dwell on past failures. The terms of reference required the reviewers to address the structure, management and resourcing of publicly-funded elements of the criminal justice system and six specific issues were then highlighted.[14] Four of these issues were limited to criminal justice, namely the arrangements for the organisation and supervision of the prosecution process, measures to improve the responsiveness and accountability of, and lay participation in the criminal justice system, the scope for structured cooperation between the criminal justice agencies on both parts of the island and the structure and organisation of criminal justice functions that might be devolved to an Assembly. But the two other issues were not limited to criminal aspects of justice, namely the arrangements for making appointments to the judiciary and magistracy and mechanisms for dealing with law reform. Clearly any proposals concerning the appointment of judges could not be limited purely to criminal cases, and similarly mechanisms for addressing law reform, even if limited to a consideration of criminal law reform, could have an impact for the way in which law reform as a whole should be addressed.

So much for the wide-ranging scope of the terms of reference. It must also be noted, however, that there were important limiting features in the terms of reference as well. Mention has already been made of the fact that the Agreement set up another commission to look at policing and the criminal justice reviewers were specifically required to exclude policing from their review. In addition the Review could not consider the significant array of emergency powers that were given to the security forces to counter paramilitary violence and which remain largely in the statute book despite the paramilitary cease-fires. This has been considered separately by the British Government. A Home Office paper into legislation against terrorism for the whole of the United Kingdom reported in December 1998 and a specific Diplock review for Northern Ireland was set up in December 1999 and reported in May 2000.[15]

[13] Belfast Agreement p. 22.

[14] See *Review of the Criminal Justice System supra* n. 8, Annex B; Belfast Agreement p. 24.

[15] See Home Office, *Legislation Against Terrorism; A Consultation Paper* (Cm 4178, 1998).

At one level the exclusion of policing and emergency powers seemed sensible. It is generally acknowledged that policing has been one of the most controversial and divisive issues in Northern Ireland society and few could therefore object to the parties to the Agreement seeing a need for this issue to be handled separately, justifying an independent commission in its own right. An excessive examination of the emergency powers which have largely dominated the recent history of the criminal justice system in Northern Ireland could also have had a debilitating effect on the Criminal Justice Review which was designed to look forward to a future when such powers would no longer be necessary.

The exclusion of these issues, however, undoubtedly prevented the review taking a holistic view of criminal justice as a whole. Policing is such an important part of criminal justice that any criminal justice review which excluded it would inevitably be less than complete. The Policing Commission and the Criminal Justice Review did not work in complete isolation from each other and each was aware in broad terms of what the other was considering. But as the Policing Commission was independent and the Criminal Justice Review was government-led, it was not considered appropriate for the two reviews to share each other's thinking.[16] The fact that the Policing Commission was given until summer 1999 to report while the Criminal Justice Review did not have to report until the autumn of that year did give the Criminal Justice Review an opportunity to take the Policing Commission's recommendations into account and to develop these recommendations.[17] One example of this was the way in which the Review developed the Policing Commission's proposal to establish district policing partnership boards to represent the views of local people and to monitor the performance of the police at a local level. The review group took the view that there was much to be said for combining the functions of community safety and policing. Policing was an important aspect of community safety but not the only aspect and it recommended instead the establishment of local community safety and policing partnership boards.

It would clearly have been more difficult for the review group to disagree with the Policing Commission in any fundamental way, especially when Mo Mowlam, the Secretary of State, accepted the Commission's Report in principle on its publication in September 1999

[16] See *supra* n. 8 at para. 1.23.

[17] The Commission reported on 9 September 2000. See Independent Commission on Policing, *A New Beginning: Policing in Northern Ireland* (1999).

and when her successor, Peter Mandelson, announced in January 2000 after a four month consultation period that the Government accepted many of its recommendations.[18] A government-led criminal justice review, albeit one with an independent element, would have found it very difficult to dissent from these recommendations after this announcement.

The relationship between the Criminal Justice Review and the Policing Commission was therefore a rather curious one. Although policing was outside the remit of the Review, in practice there is a clear overlap between policing and other aspects of the criminal justice system. Each review was in theory free to come up with very different recommendations but as the Policing Commission reported first the Review was inevitably somewhat bound by the approach of the Commission. Both reviews were, of course, bound by the principles of the Agreement and reporting second did not in practice appear to inhibit the recommendations of the Criminal Justice Review. But the completely independent nature of the Policing Commission and the fact that it reported first gave the impression that the Review was of secondary importance to the Commission and served to emphasise the limit of its remit.

The exclusion of policing and emergency legislation did more than merely restrict the scope of the Criminal Justice Review. We have seen that the review was charged with prescribing arrangements which would enhance confidence in the criminal justice system. Yet the exclusion of policing and emergency legislation meant that the Review was unable to address the very issues which have done more than anything to prevent all sections of Northern Ireland society having confidence in the criminal justice system. The Government's own community attitudes survey has consistently shown that Catholic confidence in the police lags considerably behind confidence in other areas of the criminal justice system (for example judges) and that Catholics have less confidence in the disposal of terrorist cases than ordinary cases.[19] In its own consultation exercise the Review found that for many the experience and perceptions of the criminal justice system were influenced by views on policing and emergency legislation. Asking the Review to address the issue of confidence in the context of future criminal justice arrangements without looking at policing and emergency legislation was therefore rather like asking an architect to design a house without

[18] HC Debs cols 845–8 (19 January 2000).

[19] *Community Attitudes Survey Sixth Report*, NISRA Occasional Paper No 10 (Central Survey Unit, 1999).

walls and a roof. The review group recognised this limitation when it rather more circumspectly stated that it was conscious of the linkages between the three areas of policing, criminal justice and emergency legislation, and that "its efforts to develop proposals for a fair rights-based and effective criminal justice system which inspired the confidence of the community as a whole could not be divorced from the outcome of those separate reviews".[20] To look to the recommendations of the Review in the expectation that they will by themselves boost the confidence of all parts of the community in the criminal justice system is therefore somewhat misplaced, as the Review itself accepted.

FORWARD-LOOKING, NOT BACKWARD-LOOKING

Mention has already been made of the forward-looking remit of the Review. This gave the review group an important opportunity to draw a line under the past and recommend solutions for a new era of peace. But it is also important to recognise that its very forward-looking remit disabled it from acting as effectively as it might as a bridge between those considered part of the system and those whose confidence had to be inspired in it. It has been said that questions of history are integral to processes of conflict resolution and political transition, with political reconciliation requiring some form of "historical audit" whereby participants have an opportunity to commit their experiences to a history of the conflict which allows grievances to be aired and experiences to be acknowledged.[21] Although established as part of a "peace" agreement, however, the Review set itself firmly against any attempt to go back into history. It heard a number of calls for it to investigate past events, but like the policing commission it was not set up as a committee of inquiry with legal powers to call for papers and question witnesses. This meant that it was unable to make judgements about the performance of particular organisations and groups such as the Department of the Director of Public Prosecutions and the judiciary, although it did (somewhat more than the Policing Commission) pay tribute to all those who had paid with their lives the ultimate price for

[20] *Supra* n. 8 at para. 1.22.

[21] A Mulcahy, "Policing History: The Official Discourse and Organizational Memory of the Royal Ulster Constabulary" (2000) 40 *Brit J Criminol* 68, 85.

upholding the rule of law and serving the course of justice during the recent troubles.

This did not mean that the Review was not able to take account of the views of those who made representations to it. One of the significant features of the Review was the open nature of the consultation that took place during it, in contrast to many other criminal justice reviews that have taken place elsewhere, certainly within the United Kingdom. It was required to engage in "wide" consultation and during the course of its work it received ninety written submissions, it held over seventy meetings with interested groups and organisations and in order to hear the views of those who operate on the ground it held a series of nine seminars across Northern Ireland. This threw up issues which were not mentioned in the consultation exercise. For example, the Review reported that early in the consultation process, it became clear that a number of groups were anxious for the Review to examine ways of dealing with juvenile crime and the arrangements for managing and delivering juvenile justice.

The Review was also able to take account of the views it received about past events. As it said, it was important to understand those points of view in order to make recommendations that would inspire confidence in the future. What, however, it could not do was to try to reach an accommodation of the very different views that it heard with a view to reconciling these differences. The Review was required to take as its guidance the principles of human rights, mutual respect and partnership laid down in the Agreement and could not act as a "go-between" between various interests. So, for example, when it came to the vexed question of the ethos of the courts, it is not surprising that it heard "mixed views" about the flying of flags, about the proclamation of "God save the Queen" as the judges enter some courts and about symbols and emblems.[22] The Review recommended that symbols such as the royal coat of arms should continue to be displayed in the exterior of courthouses in recognition of the constitutional fact that as the courts remain within the United Kingdom they should continue to be called the "royal courts of justice". At the same time it considered that there was no need for these signs to be displayed in the interior of courtrooms where justice was actually dispensed and where all must feel comfortable.

At first blush this smacks of a convenient compromise, trying to give something to both parts of the divided community. In fact, however,

[22] *Supra* n. 8 at para. 8.30.

the recommendations are better seen as informed by the principles of the Agreement which recognises the constitutional position of the courts in the United Kingdom but which also recognises the need for all parts of the community to feel part of the institutions of the state and to feel comfortable about playing their full part in them. A similar approach was taken towards the question of what kind of oath judges should take on appointment to office. The Review satisfied itself on the basis of legal advice that there was no constitutional impediment to a politically neutral oath in modern language with no reference made to the Crown but which simply required members of the judiciary to swear that they will "well and faithfully serve office" and "do right to all manner of people without fear or favour, affection or ill will according to the law of the land".[23] The Review concluded that this was the best option to take in order to make nationalist members of the community fully comfortable about sitting as judges.

GOVERNMENT-LED

A further limiting feature of the Review was the fact that it was a government review, albeit one with an "independent element", a curious combination of government and independent members. Technically, the five "independent assessors" consisting of members of the legal profession, academia and the voluntary sector appeared able to out-vote the four senior civil servants on the Review who represented the three government departments most involved in the criminal justice issues to be considered, the Northern Ireland Office, the Attorney-General's Office which has responsibility for prosecutions and the Lord Chancellor's Department which has responsibility for courts and the appointment of judges. The way in which the independent assessors interacted with the government members of the group was summed up as follows:[24]

> "We were an unusual group in that we were a mix of civil servants representing the Government and independent members who played a full part in all aspects of the review. As a result, we were not wholly a creature of government, nor were we entirely independent, as was the Independent Policing Commission for Northern Ireland. But we were given freedom to address the task before us in the way we chose".

[23] *Ibid.* para. 6.128.
[24] *Ibid.* para. 1.8.

The last sentence is important as the group was given the freedom to develop its own dynamic without government Ministers breathing down its neck. Nevertheless, the fact that civil servants representing the Government played a prominent role raises questions about the extent to which the Review could recommend matters contrary to existing British government policy. Of course, at a general level there could be no conflict between the principles of the Agreement and government policy as the government was a signatory to the Agreement. Similarly, the commitment the Government has given to human rights by signing various human rights instruments and now by the incorporation of the European Convention on Human Rights into UK law goes hand-in-hand with the commitment in the Agreement towards human rights. But at a level of greater detail it would have been hard for the Review to overturn specific aspects of government policy provided these were sufficiently human rights proofed.

One example suffices to make this point. The Review reported that a number of groups had raised doubts about the legislation which enables inferences to be drawn from a suspect's failure to answer police questions and a failure to testify. The Review's response to this issue was to look rather minimally at the Strasbourg rulings on the right to silence. It referred to the case of *Murray* v. *United Kingdom*[25] which held that in the circumstances of that case there was no violation of Article 6 of the Convention by reason of the application of the law permitting inferences to be drawn from silence. The review group noted that there were other applications currently before the European Court raising the same issue but instead of dealing with the issue as a matter of principle, it merely stated that it would be necessary for the Government to monitor these cases and if necessary take remedial action, although it did recommend that there should be research into the impact of the rules governing police questioning including a review of the cautions which are issued to suspects warning them of the consequences of failing to answer police questions.

MODERNISING THE CRIMINAL JUSTICE SYSTEM

So far we have dwelt upon the limitations of the Review but within the limitations set, the review group proposed a number of changes which

[25] [1996] 22 EHRR 29.

have gone beyond current government policy in the rest of the United Kingdom. Apart from its commitment to human rights, one of the broad themes of government policy has been the need to modernise both government and the courts and instil greater openness, transparency, accountability and accessibility into the process. As the Lord Chancellor put it in his preface to the government White Paper on *Modernising Justice*,[26] the Government was elected on a radical agenda to modernise the country, exposing all its institutions and services to scrutiny. Those that are out of date, he said, or inefficient or unaccountable to the people would not survive unchanged. In another context the Home Secretary has viewed the Human Rights Act 1998 as at the heart of the Government's citizenship agenda whereby the state and civil society should act in partnership, each to facilitate but also to control the other, what Giddens has called the politics of the "third way".[27]

But there are serious questions about how far the Government has gone in its policy to effect this active citizen culture within the justice system. Despite the Lord Chancellor's rhetoric about access to justice, the Government's legal aid reforms seem designed to cut costs rather than deliver greater access to justice.[28] Changes in prosecution practice are coming about as a result of the Macpherson and Glidewell reviews but there is as yet no right for victims to be given reasons when prosecutions are not brought.[29] Belatedly, the Government has established a review of the criminal courts system but the review is focused more on managerial concerns than on concerns about citizens in the courts.[30] Indeed, the Government's policy has gone in the direction of *less* citizen involvement in its proposal to abolish the right of defendants to elect for trial by jury when they are charged with either-way offences. There has also been little to encourage greater transparency and openness. The Freedom of Information Act excludes victims learning details about criminal investigations. Concern about the closed nature of

[26] *Modernising Justice* (1998).

[27] A Giddens, *The Third Way: The Renewal of Social Democracy* (Cambridge, Polity Press, 1998).

[28] See M Zander, "The Government's Plans on Legal Aid and Conditional Fees" (1998) 61 *MLR* 538.

[29] *The Stephen Lawrence Inquiry, Report of an Inquiry by Sir William Macpherson of Cluny* (Cm 4262, 1999); *The Review of the Crown Prosecution Service* (Cmnd 3960, 1999).

[30] The Criminal Courts Review by Lord Justice Auld was announced by the Lord Chancellor on 14 December 1999. See M Zander, "What on Earth is Lord Justice Auld Supposed To Do?" [2000] *Crim LR* 419.

judicial appointments has been met by accepting the idea of a commission to review judicial appointments rather than by a full-blown judicial appointments commission.[31] Some attempts have been made to monitor the impact of race and gender in the criminal justice system but there remains much to be done in this area.[32] Finally, the need to reduce juvenile crime has led to talk of encouraging greater civic responsibility in juveniles by means of restorative justice principles but the Government's reforms seem destined to widen the gap between intent and achievement.[33] Principles of restorative justice involve making restoration to the victim, reintegrating the offender into the community and bringing home the consequences of offending to the offender. But it is questionable whether these objectives will be met through the Government's two flagship measures in this area, the Crime and Disorder Act 1998 which focuses on the prevention of crime through a series of orders that may be enforced through the civil courts and the Youth Justice and Criminal Evidence Act 1999 which has followed these initiatives up with a new sentence of referral to a youth offending panel.

By contrast the Northern Ireland Criminal Justice Review's reforms seem more radical and here the advantages of a review able to take a global view of substantial areas of criminal justice outshine the piecemeal approach to reform adopted in England and Wales. Apart from engaging in an active consultation process, the review group was able to take a strong comparative approach towards its work. In order to acquaint itself with structures and arrangements in other countries, the group commissioned a number of research reports and visited a range of jurisdictions, including England and Wales, Scotland and the Republic of Ireland but also countries in Western Europe and North America, South Africa and New Zealand. One of the general lessons brought home was the different ways in which systems tackle crime and organise their criminal justice systems. But the Review also found that the present Government did not have a monopoly on the themes of accountability, transparency, openness and citizen involvement and

[31] The idea of an independent scrutiny of the appointments process was recommended by the Peach Report. See *An Independent Scrutiny of the Appointment Processes of Judges and Queen's Counsel in England and Wales: A Report to the Lord Chancellor by Sir Leonard Peach* (1999).

[32] See the various publications under s 95 of the Criminal Justice Act 1991 cited *supra* n. 8 at para. 3.41.

[33] See C Ball, "A Significant Move towards Restorative Justice, or a Recipe for Unintended Consequences?" [2000] *Crim LR* 211.

responsibility. A number of jurisdictions were also rallying to these themes with the result that prosecutors were more prepared to give reasons to victims for not prosecuting cases, agencies were publishing annual reports, there were more open appointment procedures as regards the judiciary and there was a trend towards greater supervision or control by prosecuting authorities over police investigations. The group reported that a number of countries were also developing restorative justice schemes whereby young people were being dealt with outside the formal justice system by, for example, family conferencing schemes.

Driven by the values of the Good Friday Agreement and drawing on best practice elsewhere, the group came up with a wide-ranging package of reforms dealing with fundamental issues such as prosecution, judicial appointments and youth justice and how to ensure fairness and accountability throughout the system. A new independent Criminal Justice Inspectorate was recommended, responsible for the inspection of all aspects of the criminal justice system other than the courts. The review also recommended that a strategy for equity monitoring be developed throughout the criminal justice system so that the effect of decisions on people can be monitored according to categories such as community background, gender, race, ethnic origin, sexual orientation and disability. The outcome of this monitoring should also be published on a regular basis to demonstrate that problems have been identified and action is being taken.

The principles of independence, accountability, transparency and fairness informed many of the specific proposals. For example, a newly named prosecution service was proposed to take responsibility for prosecuting all criminal cases through the courts with extended powers to direct that specific matters are investigated by the newly-created Police Ombudsman where the prosecutor is not happy with the response of the police. On devolution the head of this service should no longer act under the direction of a politically-appointed Attorney-General. There is to be a presumption in favour of giving reasons to victims and other interested parties for not prosecuting cases and there are a number of recommendations for a published annual report, a code of practice explaining how decisions are taken, a published code of ethics, inspection arrangements and complaints mechanisms.

The group considered that under new devolved arrangements an independent Judicial Appointments Commission could better safeguard the need for appointments to be insulated from political interference than a system whereby appointments were in the hands of a Minister. A commission would also provide greater transparency than the present arrangements. While appointments should be made strictly on merit, there was a need to make the judiciary reflective of all sections of the community in terms of community background and gender and the review proposed widening the present eligibility requirements to draw upon as wide a pool of legally qualified persons as possible.

To make the courts less remote and give all sections of the community a feeling that they have a stake in the system, the group proposed a number of practical measures such as consulting victims about decisions affecting their cases, as well as ending the practice of lawyers and judges wearing wigs. In this context the proposal already mentioned to abolish coats of arms within courthouses was also seen as a necessary measure to make all people as comfortable as possible in the courtroom environment.

Finally, youth justice was given specific priority as indeed it has in England and Wales. But instead of recommending the coercive model proposed in the Prevention of Crime and Disorder Act 1998 and the Youth Justice and Criminal Evidence Act 1999 which seeks to prevent crime by means of a variety of sanctioning orders, the review group considered that the philosophy of restorative justice should be integrated into the juvenile justice system using a conference model (called a "youth conference") which would be available for all juveniles subject to the full range of human rights safeguards. The group considered that there was also a role for community restorative justice schemes to deal with low-level crime but it considered that such schemes should receive referrals from statutory agencies and be accredited according to human rights standards.

CONCLUSION

Within the spirit of inclusivity and fairness which imbues the Good Friday Agreement, the review group came up with a vision which will bring the legal system truly into the twenty-first century, safeguarding traditional values such as independence but recognising also the need for greater accountability, transparency and civic participation in the system. When the Review reported in March 2000, the British Government announced a six-month consultation period and in October 2000 it reported that a full timetable for reform is to be completed by April 2001. This augurs well for those believe that the Northern Ireland criminal justice system needs to be reformed along modern human rights grounds. Some anxiety has been expressed about the fact that some of the reforms are contingent upon the devolution of criminal justice functions to the Northern Ireland Executive. But with the exception of the establishment of the judicial appointments commission, few of the reforms are in fact contingent in this way. Of course, the passage of legislation is only a first step towards realising the vision set by the Review. There is a need for all the agencies and professionals involved to aspire themselves towards the civic culture promoted by the Review and this may take time.

What is less clear is the extent to which the Review will help towards resolving the conflict in Northern Ireland. Here we return to the limitations on the terms of reference and the task given to the Review. The parties clearly believed that there was a need for a new vision for criminal justice in a post-conflict situation. If the vision of a rights-based, accountable criminal justice system is realised this can only help heal some of the wounds that have been opened up by an excessively closed system in the past. The recommendation that in future the Director of Public Prosecutions should give reasons to victims or the relatives of victims of criminal or alleged criminal activity should help to inspire confidence in the newly named prosecution system. But these and the other reforms are prospective, not retrospective, and the question is whether they will be enough to heal the rift that has emerged between those deemed part of the criminal justice system and those at the receiving end of it during the times of the troubles.

There has been some discussion about the desirability of a truth and reconciliation commission in Northern Ireland. There is, of course, currently an important inquiry being conducted by Lord Saville into the events on "Bloody Sunday" and the Victims Commissioner has issued a report on the victims of the troubles.[34] But there have as yet been no announcements to investigate other disputed deaths or shootings such as the killing of the lawyers Pat Finucane and Rosemary Nelson.[35] At the end of this chapter it is not proposed to examine how far it is necessary to investigate past events of this kind. The point is

[34] The Saville Inquiry was announced on 29 January 1998. Sir Kenneth Bloomfield, the Victims Commissioner, published his report, *We Will Remember Them*, in May 1998.

[35] Pat Finucane was murdered in 1989 by the Ulster Freedom Fighters and Rosemary Nelson was murdered in 1999 by the Red Hand Defenders and there have been allegations of security force collusion in both cases.

that it may not be enough to reshape the institutions of criminal justice. In some manner there is also a need to reshape the attitudes of those inside and outside the system. One way in which this may be attempted is by more open dialogue. The Criminal Justice Review tried to encourage dialogue by holding seminars to which all those with an interest in criminal justice were invited. These never received the publicity given to the open meetings held by the Policing Commission but they did spark some lively exchanges between those with very different experiences of the criminal justice system. In the new era opened up by the reforms proposed by the Criminal Justice Review it will be important to create as many opportunities for dialogue as possible so that all can begin to feel part of the changes involved.

8

A New Beginning for Policing in Northern Ireland?

LINDA MOORE and MARY O'RAWE

INTRODUCTION

SINCE NORTHERN IRELAND was first created under the Government of Ireland Act 1920, policing has been a key site where issues around the legitimacy of the state have been played out. The place accorded to policing in the Good Friday Agreement of 1998 is indicative of a sense in some quarters that if policing can somehow be "got right" many of the other pieces of the jigsaw will slot into place.

The complex web of identities, loyalties and allegiances tied up in how and by whom Northern Ireland is policed is symptomatic of wider difficulties to be overcome in terms of moving Northern Ireland towards a stable and enduring peace. In signing up to the Agreement, which was subsequently endorsed in a referendum by 71 per cent of the population of Northern Ireland, the political parties recognised that policing in Northern Ireland is a highly emotive subject, invoking great hurt for many people, including police officers and their families.[1] They agreed it to be essential that policing structures and arrangements are such that the police service is "professional, effective and efficient, fair and impartial, free from partisan political control; accountable, both under the law for its actions and to the community it polices; and operates within a coherent and co-operative criminal justice system, which conforms with human rights norms".[2] As such the historical and continuing debate around changes to policing institutions in the North is not, and cannot be, a simple matter of modernisation and professionalisation within a managerialist framework. This said, the extent

[1] Belfast Agreement p. 22.
[2] Belfast Agreement p. 22.

of the challenge set by the Agreement is evidenced in the deep-seated fears and emotions which surface at the mere mention of policing arrangements, past, present or future. Despite the supposedly consensual wording of the Agreement, the policing debate has continued to prove divisive. Alternate displays of everything from intense pride, defensiveness and ownership of the Royal Ulster Constabulary (RUC) by largely unionist sections of the population, to complete antipathy and alienation from the same police force by other parts of the community (increasingly working class loyalist as well as the more traditional republican and nationalist constituencies) continue to characterise public exchanges. In some senses, things have been brought to a head by the current discourse around the changes to policing proposed by the Patten Commission,[3] set up under the Good Friday Agreement of 1998 to look at future arrangements for policing in Northern Ireland. The Disband the RUC/Save our RUC dialectic continues to play itself out and must be sensitively unpacked at both a substantive and symbolic level in order to evaluate whether we really are on the threshold of a new and better beginning for policing in Northern Ireland. By the same token policing change cannot be allowed to be held hostage to political fortune.

It is not that the methods and policies of policing Northern Ireland have remained static over the years, nor a case of the Royal Ulster Constabulary being all good or all bad. On one level, therefore, the dialectic and polarisation of views is based on a false premise. However, attempts to grapple with vital issues of accountability, representativeness and respect for human rights have long been stymied by the narrowness of the official security-oriented political agenda which has never completely taken on board the extent of the polarisation of views or how deeply people have been affected and alienated from the police and/or each other by the differential ways in which policing has been experienced by different sections of the community in Northern Ireland. There are various truths around what the RUC has done and how it has contributed to the containment or exacerbation of violent conflict in Northern Ireland. However, officialdom has, time and again, fallen into the trap of prioritising or validating one discourse and one truth over any other, i.e. that the RUC is the best and most professional police force in the world and has acted as nothing other than a neutral arbiter in a sectarian feud. This approach has many attendant

[3] Report of the Independent Commission on Policing for Northern Ireland, *A New Beginning: Policing in Northern Ireland* (September 1999) (the Patten Report).

implications which are not of benefit to moving us towards more accountable and acceptable policing arrangements.

Analysis of some of the core features of the policing experience and the discourse which fuels and is fuelled by it, therefore has merit not just in terms of what it can say about policing, but also in terms of understanding some of the reasons behind the "one step forward, ten steps back" *danse macabre* which has characterised and dogged the current peace process.

Against this backdrop, this chapter will examine whether the recent report of the Patten Commission, and the current Government's method of implementing the Patten recommendations contain enough to cut through the alienation to create policing arrangements which are representative, accountable and respectful of human rights. The chapter will further consider the potential of the human rights paradigm to build and maintain consensus around policing in Northern Ireland, and whether it has been fully utilised in the current process. Finally, the chapter will argue the prize to be gained by getting things right this time far outweighs the price to be paid if the *status quo* is merely re-invented under a new and better-packaged guise.

THE CURRENT CONTEXT

It would be the contention of this chapter that the Patten Report presents a new and innovative programme for change. It certainly attempts to put human rights, equality and accountability at the heart of policing arrangements. However, in many respects it stops short of articulating how these goals can be reached, and in its failure to follow through coherently the logic of the human rights paradigm on some key issues, it contains inherent contradictions which risk undermining much that is positive about the Report. In part, this has enabled government to reclaim its traditional space in the policing debate. At the time of writing this chapter, the signs are that far from building on the Patten Report's recommendations, they have been severely diluted by the Government in drafting legislation purportedly designed to implement the Report.[4] Of Patten's 175 recommendations, less than one-third have been included within the legislation. The rest have been

[4] Police (Northern Ireland) Act 2000. The Act received Royal Assent in November 2000. However the timetable for it coming into operation is unclear and is essentially in the hands of the Secretary of State for N. Ireland.

drawn into an "implementation plan" devised by a civil service "Patten Action Team", reflecting the views of, and allocating responsibility within, a narrow group essentially consisting of the RUC, the Northern Ireland Office and the current Police Authority.

In part, the dilution of the Patten recommendations has been a result of concessions to the Ulster unionists and linked to fraught political circumstances during the period of consultation on the Patten Report. Unionist opposition to certain of Patten's proposals—most notably the proposed change in the name and badge of the Royal Ulster Constabulary to the Northern Ireland Police Service[5]—won the promise of a reconsideration of these recommendations.[6] Not surprisingly, nationalists were aggrieved at this dilution of the Patten recommendations, particularly given Patten's insistence that there should be no "cherry-picking" of the Report. And once again we were into the hurly-burly of using human rights as bargaining chips and construing the Patten recommendations in terms of what they individually appeared to be conceding to one community or taking away from another. Several hundred amendments were made to the draft legislation as it went through the Westminster Parliament. However the final legislation has still fallen short of convincing nationalist and republican political parties that a "new beginning" is really being delivered. Because they have not so far been persuaded to give up support to the government's plans by nominating party members to sit on the Policing Board created by the Police (NI) Act 2000, veiled threats have been emanating from the Secretary of State's direction that if the SDLP and Sinn Féin are not prepared to endorse the "new dispensation" all plans for implementing Patten's proposals will be shelved for the forseeable future.

Tying the implementation of the Patten Report so closely into the political climate in Northern Ireland creates an inherent uncertainty about the extent to which Patten will be delivered. The on-off nature of devolved government in Northern Ireland raises problems for the implementation of some aspects of the Report. At the time of writing the devolved Assembly and Executive have reconvened. However, the situa-

[5] Patten Report para. 17.6.

[6] The Police (Northern Ireland) Act gives the Secretary of State for Northern Ireland power to make regulations prescribing the design of a flag and emblem for the force, and to make provision for a new name after consultation with the Policing Board. Unionist and government amendments to clause 1 in Committee stage as to whether and to what extent the Police Service of Northern Ireland should specifically be deemed to incorporate the RUC in its "title deeds" led to further political controversy without addressing underlying issues of concern.

tion remains uncertain and there is no guarantee that this government will have a long life.[7] Furthermore, the authority vested in the Secretary of State to speed up or, more likely, slow down the pace of change, on the security advice of the Chief Constable is a potential threat to the process in that the police institution or, the whim of an individual Secretary of State, can come to be seen to dictate the pace of change.

These factors are important in themselves. However, it can be argued that, for the most part, the watering down of the Patten Report in the legislation has less to do with local party politics or legitimate security concerns and is, instead, infinitely more connected on one hand to a broader British state agenda, and on the other to opposition to change within the policing and security department of the Northern Ireland Office.

Policing policies and practice are under challenge throughout the United Kingdom. The admission of institutionalised racism in the Metropolitan Police following the findings of the Macpherson Report[8] is just one indication of the poor relations between sections of the public and the police. Throughout the United Kingdom, the police leadership and their political cohorts struggle to remain in control of the process of change.

In this respect, Northern Ireland is no exception. The whole process of change has to date been overseen by those responsible for policing in the past—the Secretary of State, the Northern Ireland Office and the Chief Constable. This triumvirate has ensured that their own powers in relation to policing remain virtually unaffected. Patten had recommended that an Oversight Commissioner be appointed "as soon as possible" with responsibility for "supervising the implementation of our recommendations".[9] The failure to appoint this person until *after* the consultation process on the Patten Report, when legislation had been drafted and debated in Parliament, and an implementation plan had already been drawn up by civil servants, is one of the most disgraceful aspects of the post-Patten process.[10] Already it would seem that the old order is reasserting itself. To understand why this is so problematic and why the process of change is as vital as the final product in

[7] In the event of a further suspension or collapse of the Executive, the British Government has included reserve powers in the legislation to establish a Policing Board with members selected by the Secretary of State.

[8] *The Stephen Lawrence Inquiry: Report of an Inquiry by Sir William Macpherson of Cluny* (Cm 4262, 1999) (the Macpherson Report).

[9] Patten Report para. 19.4.

[10] Tom Constantine, former head of the United States Drug Enforcement Agency was selected as Oversight Commissioner in June 2000.

terms of making policing more effective in Northern Ireland, we must look further at the history of policing in Northern Ireland.

POLICING IN NORTHERN IRELAND: THE BACKGROUND

As previously stated, policing and the rule of law have been heavily contested issues since the creation of the Northern Ireland state. From its establishment in 1922, the Royal Ulster Constabulary was seen by large sections of the Catholic/nationalist community as closely identified with the unionist/Protestant cause.[11]

During five decades of Stormont rule, the RUC was directly accountable to the Minister of Home Affairs in the unionist government who ensured policing conformed to the dominant unionist agenda.[12] Changes in policing and the justice system were core demands of the civil rights movement in the 1960s. The policing of the civil rights protests and televised pictures of demonstrators being beaten by police created international consternation. The Cameron Commission and the Scarman Tribunal, investigating aspects of the "disturbances" were critical of aspects of policing and particularly of the sectarian conduct of the B Specials.[13] In 1969 the Hunt Committee produced recommendations aimed at normalising and professionalising policing in Northern Ireland.[14]

However, even after the British Government took direct responsibility for running Northern Ireland in 1972, and after the implementation of the Hunt Report, the RUC was still perceived as a partisan body by

[11] The RUC has always been overwhelmingly Protestant in membership. Initially a quota system was created, with the intention that one-third of officers would be Catholic; one-third Protestant ex-Royal Irish Constabulary (The Royal Irish Constabulary was the police force for the whole island of Ireland before partition) and one-third drawn from the mainly Protestant Ulster Special Constabulary, a body that had been established in 1920 as an alternative to the paramilitary Ulster Volunteer Force (UVF). In reality, many recruits to the new force came from the UVF. The quota for Catholics was never filled and indeed the proportion of Catholics in the RUC actually declined from a peak of 21.1 per cent in 1923 to the present figure of just under 8 per cent. Historical figures from J D Brewer and K Magee, *Inside the RUC* (Oxford, Clarendon Press, 1991), current figure of approximately 8 per cent from Patten Report para. 14.1.

[12] For example, within three months of being formed, a ban on police officers joining the Orange Order was lifted and in 1923 a specifically police lodge was formed. Members of the RIC had not been permitted to be members of political or religious organisations including the Orange Order. See Brewer and Magee *supra* n. 11 at p. 2 for details.

[13] A part-time armed branch of the RUC.

[14] Recommendations included the abolition of the B Specials, the disarming of the RUC and the creation of a Police Authority as a source of accountability. Although the police were disarmed they were soon after rearmed and the Police Authority never became the type of effective oversight body envisaged by Hunt.

many Catholics. This response stemmed not only from the force's mainly Protestant composition[15] but also from the increasingly central role that the police played in the Government's "counter-terrorist" strategy. This role brought the police into regular conflict with large sections of the Catholic population, but particularly with Catholic working class communities.

Ironically, far from eliminating political violence, the attitude and behaviour of the security forces has often served to inflame tensions. The sense of grievance in Catholic working class communities was exacerbated through repeated negative experience of policing; day-to-day harassment; use of plastic bullets; accusations of police collusion with loyalist paramilitaries; and alleged "shoot to kill" policies.[16]

The RUC has never operated as a "normal" police force. Since partition, it has relied on draconian special powers.[17] Police stations are highly fortified and the police routinely armed. The role of the army (both British and local regiments) has given particular cause for concern. Throughout the conflict the RUC patrolled nationalist areas accompanied by large patrols of soldiers. The police themselves were under constant danger of attack. Over the past thirty years, 302 officers have been killed and around 8,500 wounded and disabled.[18] This immense human tragedy should never be discounted or minimalised, but neither should it blind us to the wrongdoing by members of the same force which has inflicted similar suffering on other families.

While the challenge to the RUC has come mainly from nationalists, alienation between some working class Protestant communities and the police has increasingly been a problem for the force. In part this has been because of loyalist hostility to the RUC's role since the signing of the Anglo-Irish Agreement in 1985. More recently the force's role in upholding Parades Commission decisions to re-route certain Orange marches has been particularly resented by unionists. However, aspects

[15] Despite initial quotas (being) laid down to ensure one-third Catholic membership of the RUC, this never reached more than 21 per cent and has long remained under 10 per cent.

[16] For further details on these allegations of human rights abuse in policing see for example: M O'Rawe and L Moore *Human Rights on Duty—Principles for Better Policing: International Lessons for Northern Ireland* (Belfast, Committee on the Administration of Justice, 1997); R McVeigh, *It's Part of Life Here: the Security Forces and Harassment in Northern Ireland* (Belfast, Committee on the Administration of Justice, 1994); Human Rights Watch/Helsinki, *To Serve without Favor: Policing, Human Rights and Accountability in Northern Ireland* (HRW, 1997).

[17] "Emergency" legislation has been on the statute books continually since Northern Ireland first came into being.

[18] Figures from Patten Report para. 8.18.

of "normal" inner city policing—for example regularly stopping and searching young people—have also led to conflict between Protestant communities and police.[19]

As long as the various paramilitary campaigns continued, any talk of policing reform was rejected by the state and by most unionists. The new situation created by the republican and loyalist cease-fires, however, placed the issue of policing high on the political agenda.

The signing of the Good Friday Agreement in April 1998 and support for this in the public referendum provided further momentum for change, culminating in the Patten Commission proposals.

The Patten Commission and its work

The Agreement tasked the Policing Commission with inquiring into policing in Northern Ireland and making proposals for future policing structures and arrangements. These proposals were to focus on composition, recruitment, training, culture, ethos and symbols as well as accountability to the law and to the community. They were also to include recommendations on issues such as re-training and job placement for existing police officers during the period of transition. The Commission was also tasked with finding means of encouraging widespread community support for policing arrangements.[20]

The Agreement provided a mandate for change. Despite protestations from some quarters about unwelcome interference into the "best force in the world", no change was not an option. The crisis of legitimacy for policing in Northern Ireland had been an important aspect of the conflict, and the place given to policing in the Agreement was recognition of the need to address the problems. On the other hand, given the requirement for the Commission to find means of encouraging a broad base of support for their proposals it is unlikely that disbandment of the RUC was ever a real consideration either.

The Commission was established in June 1998 to report in the summer of 1999. The Commission was headed by Chris Patten, previously ousted by the British electorate as a Conservative MP and more recently former Governor of Hong Kong. The seven other Commissioners included a senior female police officer based in Boston; a Canadian criminologist; a local management expert; a former Inspector of Constabulary and an

[19] Research documenting this phenomenon includes McVeigh *supra* n. 16; A Hamilton, L Moore and T Trimble *Policing a Divided Society: Issues and Perceptions in Northern Ireland* (Coleraine, Centre for the Study of Conflict, 1995).

[20] Belfast Agreement p. 23.

academic from John Jay College of Criminal Justice, New York. Two local Commissioners, Dr Maurice Hayes (former Northern Ireland Ombudsman) and Peter Smith QC (barrister) were seen as representing the nationalist and unionist perspectives (although these constituencies would not necessarily have seen them in that light).

The Commission used a wide variety of means for gathering information. It held sixty days of plenary meetings with interested individuals and groups.[21] Advertisements in the press invited the public to write to the Commission with their opinions on policing. Political parties, church groups and others with an interest in policing were also approached for their views. Every police station in Northern Ireland was visited and non-commissioned officers consulted. By the end of the process the Commission had received approximately 2,500 individual written submissions.[22] A series of public meetings was held in every District Council area of Northern Ireland in autumn 1998. In all, more than 10,000 people attended these meetings, with over 1,000 people speaking.[23] The Commission carried out research through focus groups and conducted a public attitude survey. Commissioners carried out field visits to other jurisdictions including the Republic of Ireland, Canada, South Africa, Spain and the USA. A cultural audit of the RUC was also carried out.[24]

The Patten Report notes that attitudes in public meetings tended to fall into one of three categories: "meetings held in strongly Protestant/unionist areas where participants expressed strong, often unqualified, support for the RUC; meetings in strongly Catholic/nationalist areas where participants expressed strong criticism of the RUC and demanded a new police service; and mixed meetings at which a range of views were expressed".[25]

In some cases meetings became rather like truth commission hearings as victims of state violence or families of RUC officers who had been killed or injured told their harrowing stories. The Commission had clearly hoped that public meetings would provide a useful forum for testing possible recommendations. It was clear from early on that this would not be possible.[26]

[21] Patten Report para. 2.3. [22] *Ibid.* para. 2.3.
[23] *Ibid.* para. 2.3. [24] *Ibid.* para. 2.3. [25] *Ibid.* para. 3.16.
[26] Journalist Bea Campbell describes one of the first meetings in West Belfast: "*A huge public meeting, comprised of hundreds of people turning out to have their conversation about policing and peace, with representatives of the Commission who had come into their community to listen to them. What, of course, they gave to the Policing Commission was a litany of grief. Very politely, about two thirds of the way through that meeting . . . the chair of the Policing Commission said politely, 'Thank you for telling us*

Despite the very divided responses to the Commission, Patten reports that there was broad consensus of views on some issues. They noted similarities in views expressed by "people of both traditions in lower income, high unemployment areas":

> "Commissioners found the concerns expressed in, for example, the Shankill Road area of West Belfast very similar to those expressed on the Falls Road; and the same was true in North Belfast and elsewhere around Northern Ireland. People were concerned about youth crime and antisocial behaviour, about the threat to their children from a rapidly growing drug problem and about paramilitary thuggery. They wanted a police service with a permanent local presence to deal with these threats, and officers they recognized and could identify with".[27]

Focus group research demonstrated that Protestants from lower income groups could be as strongly alienated from the police as were their Catholic counterparts.[28]

Overall Patten found:

> "broad acceptance across the communities, albeit with differences of emphasis, that the composition of the police service should be more representative of the population, in particular that there should be more Catholic and women officers".[29]

Given the consensus on these important issues, it is a pity that the Patten Report did not frame its proposals in a way that was seen to build on and reflect back this consensus. For example, the Report is weak on the non-controversial area of increased gender representation within the force. Highlighting the changes and opportunities necessary in this area could have given a very different slant to the subsequent debate and provided a more holistic framework in which proposals to bolster numbers of Catholics and nationalists were seen only as part of a much bigger and much more comprehensive equality agenda. This kind of approach could have been all the more significant given the predictability of the media and many political leaders wanting to focus on a very limited set of recommendations, that is those most likely to provoke divisive reactions. Media attention has focused on proposals to

about your experiences, and if more of you want to tell us about your experiences, fine—but we would also be very interested in positive suggestions about future arrangements'. Now, I have no doubt that the Policing Commission learned instantly that it had made a mistake in saying that": Committee on the Administration of Justice, *The Agreement and a New Beginning to Policing in Northern Ireland* (Belfast, CAJ, June 1999) p. 71.

[27] Patten Report para. 3.19.

[28] *Ibid.* para. 3.20.

[29] *Ibid.* para. 3.23.

change the name of the RUC to the Northern Ireland Police Service and to change the police badge. The proposal to give local councils the power to raise additional money for policing was also controversial as it was alleged this would lead to councils hiring paramilitary groups to conduct policing activities, despite Patten's reassurances that this was not a possibility.

The Patten Report's proposals

In all there are 175 recommendations in the Patten Report. The Commission made clear that these must be treated as a package and "cherry-picking" was not an option:

> "The 'significant change' in policing should not be a cluster of unconnected adjustments in policy that can be bolted or soldered onto the organisation that already exists. The changes we propose are extensive and they fit together like pieces in a jigsaw puzzle".

It was recognised that the recommendations could not all be introduced at once and Patten proposed that while some should be introduced immediately others must be gradual.[30] Furthermore, "there are still some areas where change will, for the time being, be constrained by the threats to police and community security".[31] As noted above, the Secretary of State emphasised that the pace and extent of change would be dependent in many cases on the security situation. While the safety of officers and security concerns are a key determinant, it is also important that change is not blocked or delayed unnecessarily, particularly as human rights organisations have consistently argued that the militarised nature of policing in Northern Ireland has served to fuel rather than ameliorate the conflict.

During consultations human rights organisations had urged the Commission to base its recommendations on human rights principles as these provide a bottom line standard for the treatment of all citizens in a society. Despite the tendency, not least by unionists, to link human rights issues with nationalism, it is important that human rights are not seen as the preserve of one community. The Patten Report tried to take on that challenge and to that end stated clearly that: "we have not tried to balance what may be politically acceptable to this group against

[30] *Ibid*. para. 1.8. [31] *Ibid*. para. 8.3.

what is reckoned to be acceptable to that".[32] However, what has happened since would seem to indicate that the Government's response to the Patten Report is attempting to do precisely that.

Keeping policing very much in the realm of a political football may seem like a good strategy in the short-term. However, until we move beyond the traditional, tried and tested and shown-to-fail responses, we cannot deal with the conflicting legacies of the past nor lay proper foundations for the future.

The Patten Commission set its own questions for testing its recommendations:[33]

- Does this proposal promote effective and efficient policing?
- Will it deliver fair and impartial policing, free from partisan control?
- Does it provide for accountability, both to the law and to the community?
- Does it protect and vindicate the human rights and human dignity of all?

As the Commission notes, these tests are to some degree a matter of judgement, and not a precise science. However, these are good benchmarks and we use them here in assessing both the Patten Report and proposed legislative and administrative arrangements for implementation of the Patten recommendations. The key principles providing a focus for this chapter are, thus, the need for respect for human rights, equality and accountability.

Human rights

The language in the Patten Report is couched quite deliberately in terms of human rights and the Report's first chapter is devoted specifically to the subject. Given the emphasis on human rights in the Report, it is particularly disappointing that the Police (NI) Act 2000 has failed to give a statutory basis to most of the recommendations explicitly relating to human rights.

According to the Report the fundamental purpose of policing should be "the protection and vindication of the human rights of all".[34] Commission consultation showed widespread agreement that people

[32] Patten Report para. 1.10.
[33] *Ibid.* para. 1.10.
[34] *Ibid.* para. 4.1.

want police to protect their human rights from infringement by others, and to respect their human rights in the exercise of that duty.

The Patten Report proposed a raft of recommendations in pursuance of this goal:[35]

- a programme of action to focus policing on a human-rights based approach;
- a new oath/declaration expressing commitment to upholding human rights;
- a code of ethics incorporating the European Convention on Human Rights;
- a human rights dimension integrated into every module of police training, for recruits and experienced officers;[36]
- respect for human rights to be an important aspect of police officers' appraisal;
- appointment of a lawyer with specific expertise in human rights to the police legal services;[37]
- performance of police service as a whole in respect of human rights to be monitored by the new Policing Board.[38]

Despite this, in the sixty-five pages of legislation which comprise the Police (Northern Ireland) Act 2000, the words "human rights" appear only three times, twice in mention of the Human Rights Act 1998 and once in relation to the wording of the proposed new declaration of office. Aspects such as police training and the comprehensive programme of action are not referred to in the Act but are relegated for inclusion in the Implementation Plan. Worryingly, in the implementation plan, responsibility for these important aspects of the Patten Report is vested with the Northern Ireland Office, the Chief Constable and the Policing Board. The extent of the Policing Board's involvement in developing the programmes is dubious given the time-scale suggested. The programme of action is to be published by April 2001, the very month that the Policing Board is expected to become fully operational.[39]

[35] Recommendations all from Patten Report, ch. 4.

[36] The Commission notes that human rights training in the RUC lags behind other police organisations. Even in the new curriculum of 700 training sessions there were only two dedicated to human rights.

[37] This lawyer would be consulted about proposed police operations that raise human rights considerations.

[38] Patten Report para. 4.12.

[39] See NIO Implementation Plan, June 2000.

Alongside, the chapter in the Patten Report devoted to human rights the main body of the Report also contains recommendations with the potential to increase protection in this area e.g.:

- immediate closure of the holding centres at Castlereagh, Gough Barracks and Strand Road;[40]
- the appointment of a commissioner and Tribunal to oversee all intrusive surveillance policing.[41]

This strong emphasis on human rights is positive. It is disappointing, however, that other recommendations and omissions in the Report contradict this human rights ethos. And it is perhaps this lack of follow through which has allowed for such dilution of the human rights aspects of the Report in the legislation. Particularly disturbing aspects are the Report's coverage of emergency law and plastic baton rounds, its lack of consideration of human rights issues in relation to the military and its failure to tackle the issue of past human rights abuse within the force.

Emergency legislation

Policing arrangements cannot be considered in isolation from the legislation which gives police their powers. The UK Government has come under considerable international pressure in relation to emergency legislation in policing Northern Ireland. The European Court of Human Rights has on several occasions found the Government to have breached its obligations under the European Convention on Human Rights.[42] The United Nations Committee on the Rights of the Child has expressed concern about the application of emergency legislation to children[43] and the UN Committee Against Torture has called for the closure of the detention centres.[44]

In their study of the future of policing in Northern Ireland, McGarry and O'Leary conclude that there is a good case, given the cease-fires, for emergency laws to be repealed, "An emergency legal regime, involving draconian police powers, inevitably produces excesses by members

[40] Patten Report para. 8.15. [41] *Ibid*. paras 6.44 and 6.45.

[42] See *Ireland* v. *United Kingdom*, Series A, No 25 (1979–80) 2 EHRR 25; *Brogan* v. *United Kingdom* No 1 Series A No 145-B (1989) 11 EHRR 117 (and (1991) 13 EHRR 439); *Murray (John)* v. *United Kingdom* (1996) 222 EHRR 29, (1996) 17 HRLJ 39.

[43] Committee on the Rights of the Child, *Consideration of Reports submitted by States Parties under Article 44 of the Convention* (CRC/C/15/Add. 34 (January 1997)).

[44] *Conclusions and Recommendations of the United Nations Committee against Torture* (UN Doc.A/53/44 (1998)).

of the security forces. It may sow dragons' teeth rather than respect for the legal system".[45]

As Robbie McVeigh notes in his study of security force harassment "training may help to address some illegal and inappropriate security force behaviour but the existence of emergency legislation means a whole range of police behaviour that is often perceived to be harassing occurs *inside* rather than *outside* the law".[46]

Despite serious concerns, emergency legislation merits just two paragraphs in the Patten Report. Rather than recommending that the Government repeal emergency legislation, Patten suggests that when the threat of terrorism has diminished to a point where no additional powers are necessary to combat it, the law in Northern Ireland should be the same as the rest of the United Kingdom with a single piece of anti-terrorist legislation.[47] This process is already underway with the introduction of the Terrorism Act 1999. As well as the fact that the Government proposes to keep a whole raft of special powers specifically for Northern Ireland for what could be another five years, this piece of legislation will in effect mean normalising a whole series of draconian, former emergency powers throughout the United Kingdom on a permanent basis. The scope of this chapter does not extend to the thorough examination warranted by a piece of legislation of this type. Suffice to say, it will have serious implications for policing in Britain as well as Northern Ireland, which do not necessarily bode well for the enhanced protection of human rights. If the lessons and experience of Northern Ireland are anything to go by, such legislation will have many negative consequences with the potential to jeopardise police-community relations to the point where they lead to less effective policing.

Positively, Patten recommended that records be kept of all stops and search and other actions taken under emergency powers.[48] Unfortunately, the Report did not explicitly state that these records should

[45] J McGarry and B O'Leary *Policing Northern Ireland: Proposals for a new Start* (Belfast, Blackstaff, 1999) p. 78. Emergency law allows police to stop and search without reasonable suspicion, initially hold detainees for forty-eight hours and then, with political authorisation for up to a total of seven days, and to deny access to a solicitor for lengthy periods. Combined with removal of right to silence, right to jury trial, the weight which can be placed on confession evidence alone, such powers lead to human rights abuse.

[46] McVeigh *supra* n. 16 at p. 192.

[47] Patten Report para. 8.14. Patten's failure to engage seriously with this issue is "justified" by the fact that a government review of emergency legislation was conducted separately. This is a significant flaw compounded by a separate attempt to look at the whole criminal justice system as somehow separate from policing.

[48] *Ibid*. para. 8.14.

include the religion of the subject. The Macpherson Report on the handling of the Stephen Lawrence murder inquiry by the Metropolitan Police has endorsed the need to record stop and search incidents in relation to the self-defined ethnic identity of the subject and to monitor and publish analysis of these records.[49] Translated to our own situation it would seem important to collect and analyse patterns of stop and search in terms of the religion of the person being stopped. This issue has not been addressed in specific follow-up to Patten, though the section 75 duty imposed by the Northern Ireland Act 1998 would seem to require nothing less.

Plastic bullets

The use of plastic bullets in Northern Ireland is another matter which has attracted international attention. The European Parliament condemned the use of plastic baton rounds as long ago as 1982. More recently, the United Nations Committee against Torture recommended the abolition of plastic bullets. The Patten Report states "all of us began our work wanting to be able to recommend that [plastic baton rounds] be dispensed with straight away".[50] Disappointingly, however the Commission did not make this recommendation. Patten recommended restrictions on their use, readily available guidance for officers and urgent research into alternative forms of crowd control. The Report further recommended that the Policing Board and Ombudsperson monitor the performance of the police in public order situations and follow up any concerns.[51]

While no one can doubt that police officers frequently find themselves in dangerous situations during disturbances, an important flaw in the Patten Report is that the Report seems to view the question of plastic bullets in security terms alone. Both the human rights dimension and the particular history of Northern Ireland in respect of these weapons are ignored. The Report appears to assume that plastic bullets are only fired in riot situations. Yet most of those killed have not been rioters.[52]

The United Campaign against Plastic Bullets estimates that since the British Army first used a baton round in August 1970, well over 100,000

[49] Macpherson Report, recommendations 61 and 62.
[50] Patten Report para. 9.15.
[51] *Ibid.* paras. 9.12–9.20.
[52] For details of the circumstances of all deaths, see United Campaign Against Plastic Bullets, *A Report on the Misuse of the Baton Round in the North of Ireland*, Submission to the Mitchell Commission on Arms Decommissioning (January 1999).

have been fired.[53] Seventeen people—eight of them children—have been killed by rubber or plastic bullets since the start of the conflict.[54] The scale of the problem is much larger than official figures suggest. Her Majesty's Inspectorate of Constabulary recorded some twenty plastic bullet injuries between 1 January and 25 August 1996. However, in a recently published medical article, doctors reported treating 155 patients for plastic bullet injuries received during the week of 8–14 July 1996. In one week in 1996, on the RUC's own figures, over 6,000 baton rounds were fired during public disturbances in relatively small areas of Northern Ireland. A breakdown of these statistics would further indicate that plastic bullets have been resorted to disproportionately against nationalists.

The Army

Another important gap in the Patten Report is the absence of any serious discussion about the role of the army including the Royal Irish Regiment in Northern Ireland. The army has been at the centre of many concerns about human rights. Robbie McVeigh noted that "while the RUC structures maintain a semblance of liaison and democratic control—this is completely missing in the case of the RIR and other regiments of the British Army".[55] Yet, apart from stating that the army should be used as little as possible,[56] the Patten Report largely ignores this important question. As long as the army is or potentially will be deployed in support of the police, the same standards in terms of training, accountability and openness should apply to all of the security forces.

Dealing with the past

Perhaps one of the most worrying aspects of the Patten Report's treatment of human rights is the Commission's failure to provide mechanisms to identify and deal with officers responsible for committing human rights abuses in the past in terms of ensuring such abuses do not

[53] *Ibid.* pvii.
[54] *Ibid.*
[55] McVeigh *supra* n. 16.
[56] Patten Report paras 8.11 and 8.12.

recur in the "new" dispensation. Serious allegations have been made over many years by reputable international human rights organisations, the United Nations and other international bodies about the abuse of human rights by RUC officers. Charges of involvement in harassment in local communities, ill-treatment of people held in detention, shoot-to-kill incidents and sectarianism have been made as have charges of inaction in the face of wrongdoing by colleagues. A recent example which indicates that this issue is more than academic is the case of Bernard Griffen. This young man was arrested, beaten and subjected to sectarian abuse by RUC officers who then planted a coffee jar bomb at his home and attempted to frame him for a number of criminal offences, including assault on police officers. The incident only came to light because another officer who had initially gone along with the treatment, decided that things had gone a bit too far. He subsequently reported the matter, but not before Mr Griffen had served a number of months in custody on remand in respect of the hoax charges against him. It was only possible to prosecute and convict the officers involved because their colleague and a soldier who had also been involved in the incident turned Queen's evidence. The whistleblowers in this case were also convicted and fined for their part, which in itself raises a number of questions about protection for whistleblowers, another issue that the Patten Report did not address.

The Griffen case is unusual in that abuse in this case was uncovered and punished. Otherwise, officers have rarely been held accountable for these types of offences. Successive police complaints mechanisms have not substantiated one complaint of assault made by any person arrested under "emergency" law. Even where lethal force has been used in controversial circumstances, there has been a dearth of prosecutions and no police officer has ever been convicted in this regard. To permit this culture to permeate the new policing service is dangerous. In our comparative study of policing in transition, the case of El Salvador was instructive on this matter.[57] There the old police force had been disbanded and a new service created based on a quota system of 20 per cent ex-security force members, 20 per cent former guerrillas and 60 per cent civilians. Given that the police was being created anew, it was agreed to move one of the criminal investigation units en masse into the new police on the grounds that their expertise would be needed and it would take time to train new recruits. This decision proved disastrous.

[57] O'Rawe and Moore *supra* n. 16.

A United Nations evaluation later found evidence of serious corruption within the new CID units including involvement in death-squad type murders. It became necessary to reorganise the whole department, screen all members of the CID, purge some, send them all back for retraining and disperse most to other units within the police.[58]

In Northern Ireland the RUC Special Branch has been the subject of many allegations of human rights abuse. The Patten Report recommended that Special Branch and Crime Branch be brought together under an amalgamated command and that the support units of Special Branch be amalgamated into the wider police service. It further recommended that officers should not spend such long periods in security work as has been common in the past and after five years or so an officer should be posted elsewhere.[59] These recommendations, while positive in themselves, would not appear sufficient to break down the culture of Special Branch that has built up over years and to create a new beginning based on human rights for this "force within a force".

Although the right of individuals to change, reform and be rehabilitated must be acknowledged, this needs to happen in conjunction with some type of vetting and purging procedures aimed at weeding out recalcitrant officers/would be officers who are likely to continue to be involved in serious human rights abuse.

The Patten Report failed to recognise the significance of failure to deal creatively and coherently with this issue. The Report stated simply: "we do not, in this report, make judgements about the extent to which the RUC may or may not have been culpable in the past of inattention to human rights or abuse of human rights".[60] The Report went on to state that "bad apples" should be "dealt with" but did not propose any scheme for tackling the issue.[61] This reluctance to tackle human rights abuse within the force threatens the goal of a new beginning. The officers who are alleged to have made death threats to the defence solicitor Rosemary Nelson prior to her murder by loyalists will still be able to serve, as will those accused of having watched from the safety of their landrover while Robert Hamill was kicked to death by a loyalist mob.[62] It will be difficult for people who have been victims of

[58] O'Rawe and Moore *supra* n. 16 pp. 227–34.

[59] Patten Report paras 12.12–12.16.

[60] *Ibid.* para. 4.4.

[61] *Ibid.* para. 5.19. In light of the evidence, it would also seem that the "bad apple" thesis is unconvincing. This is not simply a matter of individual wrongdoing and structural issues need to be addressed. Patten appears to have missed this boat entirely.

[62] For summary of case see McGarry and O'Leary *supra* n. 45 at p. 39.

security force harassment or violence to believe that this really is a fresh start when those who abused them remain unaccountable.

In any case, the Government has been even less forthcoming in respect of dealing with deviant culture and sub-culture within the force. It has even determined that, while accepting the recommendations in principle, the Patten Report's recommendations on Special Branch must wait as long as a "significant threat from terrorism continues".[63]

The issues above would seem to indicate that the human rights challenge was not sufficiently met in the Patten Report, despite efforts by the Commission to firmly bed human rights principles at the heart of policing in Northern Ireland. Even more so, the Government's interpretation of the Report threatens the change process in its failure to look at human rights in more than minimalist terms. No mention is made in either the legislation or implementation plan of a wide range of international human rights norms and standards with particular relevance to policing. Instead, the emphasis of the Bill continues to be on "efficiency and effectiveness", with no indication that these terms are to be construed in a way which recognises that no police service can be either efficient or effective unless it conforms to human rights standards in theory and practice.

Equality

McGarry and O'Leary have argued that

> "The police should, as far as possible, be representative of all the minorities in Northern Ireland: unionists (of the 'yes' and 'no' persuasions), nationalists and others; and of Protestants, Catholics, agnostics, atheists and those of other religious faiths; and of the majority as well as the minority in Northern Ireland, that is, women and men. Above all, the police must be nationally representative. This criterion is the most important benchmark for change".[64]

Throughout its history the RUC has been a predominantly Protestant and male force. Figures from the Patten Report show that in 1998 only about 8 per cent of its officers were Catholic, while Catholics make up more than 40 per cent of the population in Northern Ireland. Only 12.6

[63] Northern Ireland Office, *Report of the Independent Commission on Policing for Northern Ireland, Implementation Plan* (June 2000).

[64] McGarry and O'Leary *supra* n. 45 at p. 45.

per cent of officers were female (a third of whom are in the Part Time Reserve).[65] Undoubtedly, intimidation by republican paramilitaries has been a factor affecting Catholic recruitment.[66] However, McGarry and O'Leary argue that "emphasising 'intimidation' as the primary explanation of Catholic and nationalist dispositions towards the RUC is simply not historically convincing. Catholics have not joined the RUC in large numbers at any time, even when there was relatively little or no republican violence".[67] They further make the point (accepted by the Commission) that it is important to ensure that nationalists join the police as well as Catholics. The two categories are not always interchangeable.

The Commission was of the view that "real community policing is impossible if the composition of the police service bears little relationship to the composition of the community as a whole".[68] One of the most difficult challenges faced by the Commission was to produce recommendations aimed at increasing Catholic and nationalist recruitment to the police.

The logistics which faced the Commission illustrate the difficulty. The force currently has around 11,500 officers including full-time, permanent and reserve officers. While significant downsizing is necessary to create a viable peacetime policing service, there is a simultaneous need to increase Catholic recruitment to the police in order to create a more representative service.

The Patten Report made a number of recommendations in pursuit of these twin goals. The Commission set a target of downsizing to a police service of 7,500 full-time officers over the next ten years.[69] This is less than the figure of 7,905 projected by the RUC in its Fundamental Review[70] but substantially more than that the estimate by economist Paul Teague that "a peaceful Northern Ireland would only require a

[65] Patten Report para. 13.9.

[66] A 1996/97survey carried out by the Police Authority for Northern Ireland found that when asked about any factors that might deter Catholics from joining the RUC, some 71 per cent of respondents mentioned fear of intimidation or attack. See PANI, *Policing . . . a New Beginning: a Submission by the Police Authority for Northern Ireland to the Independent Commission on Policing* (1998) p. 43.

[67] McGarry and O'Leary *supra* n. 45 at p. 16.

[68] Patten Report para. 14.2.

[69] *Ibid*. para. 14.9.

[70] Cited in Police Authority for Northern Ireland *Shaping Our Future Together: a Response from the Police Authority for Northern Ireland to the Report of the Independent Commission on Policing* (Belfast, 1999, PANI) p. 30.

police force of 3500–4000".[71] Patten's recommendations were based on packages for voluntary early retirement.

Recommendations aimed at increasing recruitment of under-represented groups included more effective liaison between schools and universities and the police service; the support of community leaders, including political party leaders, church leaders and others in removing disincentives to members of their communities joining the police;[72] lay involvement in recruitment panels; and emphasis on advertising aimed at attracting under-represented groups.[73]

Most controversially the Patten Report proposed the introduction of quotas for Catholic and Protestant new recruits. All candidates would be required to reach a specified standard of merit and would then enter a pool from which the required number of candidates would be drawn. An equal number of Protestants and Catholics would then be drawn from the pool of qualified candidates. The ratio of recruits would be kept to 50:50 for ten years. An opinion survey conducted by the PANI found that a strong majority of respondents agreed with the Report's recommendation on quotas.[74] The Commission concluded that such a process would require amendment to existing domestic legislation but would not be incompatible with existing European legislation.[75] Recommendations in terms of increasing female recruitment and retention were neither hard-hitting nor far-reaching. Priority for opportunities for part-time work, job-sharing, career breaks and child care vouchers are proposed. However, though this is obviously an issue at the heart of moving policing from the domain of a male, and often macho, preserve, overall this is an area where the Report, and consequently the implementation measures proposed, are weak.[76]

The Patten Report's recommendations on recruitment were accepted by the Government, although, at the time of writing, wrangling continues about how the issue is dealt with in the Police (Northern Ireland) Act. While the legislation allows for the introduction of quotas for Catholic/Protestant recruitment, the time-scale

[71] Paul Bew, Henry Patterson and Paul Teague *Northern Ireland Between War and Peace: The Political Future of Northern Ireland* (London, Lawrence and Wishart, 1997) p. xx.

[72] The repeal of the Gaelic Athletic Association's rule 21, which prohibits members of the police in Northern Ireland from being members of the Association, was recommended in pursuit of this.

[73] Patten Report paras 15.2–15.8.

[74] PANI *supra* n. 70 at p. 35.

[75] Patten Report para. 15.11.

[76] See Patten Report ch. 14 "Composition of the Police Service".

suggested for continuing this programme is not as generous as recommended in the Report.[77]

It is questionable whether Patten's recommendations will provide the impetus needed to produce the increase in representation of Catholics and other under-represented groups needed to create confidence.[78] In some ways it is a chicken and egg situation. Many community leaders would not be comfortable promoting policing as a career within their community if it appears that the legislation and other implementation measures will not deliver a sufficiently human rights and equality proofed organisation. However, without movement towards a critical mass of women or nationalists or whoever, it will be difficult for the organisation to show that it is capable of change in these respects.

Within this context, it must also be remembered that increasing the presence of under-represented groups in the police is about much more than just a numbers game. People from different backgrounds must feel safe and welcome in the organisational environment or they will quickly leave. Previous research carried out by consultants for the RUC had found strong evidence of sectarian and sexual harassment of Catholic and female officers. The Patten Report proposed changes aimed at creating a more neutral working environment, for example that the Union flag should no longer be flown from police buildings and that on those occasions when it is appropriate to fly a flag, the flag should be that of the police service which should be free from associations with the British or Irish states.[79]

It was also proposed that the name of the force be changed to the Northern Ireland Police Service. This was amended by Peter Mandelson to renaming the force as the Police Service of Northern Ireland. The change to the name, and particularly the loss of the prefix "Royal" has been bitterly contested by unionists and some policing and police relatives' organisations. Also contested was Patten's proposal

[77] However, in a speech by Secretary of State for Northern Ireland, Peter Mandelson MP, Second Reading of Patten Report, House of Commons, 6 June 2000, the Government indicated it was prepared to amend the legislation in this respect in the face of protest.

[78] McGarry and O'Leary for example had proposed that in order to downsize to a peacetime force of 8,000, the Government should aim to cut the number of currently serving officers to 5,000 over a medium period (about four years) while recruiting 3,000 new officers. During this transition period if Catholic recruitment was at 93 per cent the result would be a force that was 40 per cent Catholic: McGarry and O'Leary *supra* n. 45 at pp. 53–4.

[79] Patten Report para. 17.6.

for a new badge which would be "entirely free from any association with either the British or Irish states".[80] Mandelson accepted that the badge should be altered but suggested that consideration be given to incorporating the George Cross in the design. As noted above, this is one area where concessions were granted to unionists and police organisations, such as the Police Federation.

In response to concerns about joint membership of the RUC and loyal orders, the Patten Report proposed a register of police interests, rather than a ban on officers joining secret organisations.[81]

The most fundamental failure of the Patten Report in respect of equality is, perhaps, its unwillingness to address the question of whether there is institutionalised sectarianism within the RUC. The Commission insisted it was not a Commission of inquiry and this is true. However, without confronting the issue of sectarianism and uncovering the extent of it, a rottenness may remain at the core of the organisation. Just as Macpherson unravelled the various threads of racism running through the processes by which the Metropolitan Police dealt with Stephen Lawrence's family, so we need a similar investigation of the RUC. The obvious parallel with the case of Robert Hamill proves the problem if proof were needed.

Despite the Patten Report's failure seriously to address the issue, it would not appear that the Government is minded to establish a taskforce independent of police to investigate and tackle the issue of sectarianism within the police—though it would appear this is exactly what is required.

In terms of an overall equality agenda, it is surprising that the Patten Report appears to have ignored the equality implications of the Northern Ireland Act 1998 for public authorities. Under section 75 of the Act a new duty has been placed on public authorities requiring them to have due regard to the need to promote equality of opportunity. This duty relates not just to people of different religion/political belief but also racial group, age, marital status, sexual orientation, gender and disability. Public authorities are required to produce equality schemes which indicate how they intend dealing with and monitoring these issues in relation to their service.

[80] Patten Report para. 17.6.

[81] *Ibid.* para. 15.16. There are issues too in terms of how this recommendation has been translated into the legislation—who has access to the information, for what purposes and how long such information should be kept after someone has left the service.

The implications for the police in complying with this scheme are profound, and extend far beyond the need to monitor the religious/political background of people being stopped and searched, to examine the implications of policy and practice in terms of equality. The Police have been designated as a public authority for the purposes of the Northern Ireland Act 1998, so it is probably here rather than in respect of Patten's specific proposals that the equality debate will be sited in the foreseeable future.

Legal and democratic accountability

It is key to the transformation of policing that an appropriate framework of legal accountability is set in place. The rule of law has suffered untold damage in Northern Ireland, and it is vital that trust in the law be built across the community. As Crawshaw has argued:

> "In the event of a serious breakdown of public order, perhaps even involving armed insurgency or acts of terrorism, police and members of other security agencies are faced with great personal danger and formidable challenges to their professional expertise. For these, and other, reasons they almost invariably feel justified in breaching legal and ethical standards which would constrain them under other, less daunting, circumstances. When they do so they risk undermining the democratic and legal principles on which the legitimacy of the state they are defending and their own legitimacy are based".[82]

The Patten Report recognised this to a certain extent but missed the boat on such issues as repeal of emergency legislation; the promotion of enhanced legislative protection for whistleblowers, or the creation of a duty on police officers to report misconduct by colleagues. Nor does the Report seem able to countenance the part that the security lobby has played in shaping the law around police powers to date. Boyle and Hadden have argued that:

> "in many respects the law has been tailored to comply with security requirements rather than the other way round . . . The essentials of the structures put in place in 1973 and 1974 have not been changed. The most significant alterations have not been to limit or restrict police or army

[82] Ralph Crawshaw, former UK police officer and advisor to Council of Europe. For further expansion of this argument, see Crawshaw, Devlin and Williamson *Human Rights and Policing: Standards for Good Behaviour and a Strategy for Change* (The Hague, Kluwer Law International, 1998).

> powers but to increase them whenever that has been demanded by the security authorities. In the review of the legislation prior to its re-enactment in 1989, for example, Lord Colville recommended the introduction of a number of significant limitations . . . and a few extensions . . . None of the limitations or safeguards were accepted and all the extensions were approved".[83]

Before looking at the specific laws relating to police accountability, it should be recognised that some broader legislative developments will have an impact on the policing process. The incorporation of the European Convention into UK law in the form of the Human Rights Act 1998 will provide citizens with recourse to the courts when there is an alleged breach of human rights under the Convention. However, given the conservativeness of European jurisprudence to date, particularly in the amount of leeway given to states claiming that a state of emergency exists, this will in no way be a panacea for human rights protection. Section 75 of the Northern Ireland Act 1998 remains to be tested for its potential to safeguard equality and non-discrimination at a range of levels. Meanwhile, as previously stated, the Government continues to persist with a somewhat Jekyll and Hyde approach to the notion of human rights protection. What is given with one hand appears to be taken away with the other in the continued insistence of government on the need for extensive, and in many ways excessive, anti-terrorist powers. In the political arena, one can clearly see the competing needs of security and civil liberties jockeying for position. Unfortunately the translation of this conflict to the legal arena has rarely seen an acceptable balance being achieved. More often, narrow security needs are prioritised at the expense of a broader, more holistic vision of community safety which does not necessarily equate effective policing with erosion of rights.

As this narrow security-led approach has secured permanent anti-terrorist legislation on the statute books, much has already been lost. El Salvador further proves instructive in this regard. Where the legal framework under which police operate is not itself changed to give human rights protection real meaning, the best police structures and policy manuals and codes of conduct will not prevent a return to the same practices previously used to abuse human rights.

The new legislation envisaged to give effect to the Patten Report will obviously be contaminated to some extent by the breadth of police

[83] Kevin Boyle and Tom Hadden *Northern Ireland: The Choice* (London, Penguin Books, 1994) p. 97.

powers provided for in other laws. However, as far as possible it should aim to meet the tests proposed by Patten in terms of its ability to secure human rights, equality, impartiality, transparency and accountability.

The legislative framework is not the be all and end all and will achieve nothing without political will and commitment to the process of change. A test of how much the will to change really does exist is how quickly and how far the police organisation has moved to implement the measures already within its gift. Some evidence of movement is obvious in, for example, the removal of permanent police security on the road to the International Airport, the reduction of vehicle checkpoints and joint army patrols and the decision to close Castlereagh Holding Centre. However, the fact remains that much more could and should be done on a range of issues.

In respect of the proposed framework for improved legal accountability, there is still a long way to go. Maurice Hayes had recommended a new model for dealing with complaints against the police in his report *A Police Ombudsman for Northern Ireland?*.[84] The government claimed to accept his proposals and translated their interpretation of them into legislation prior to the Patten Commission reporting.

The Patten report made it clear by its need to re-endorse Hayes' proposals that the Police (Northern Ireland) Act 1998, hastily and ill-advisedly put through Parliament while Patten was still deliberating, had been inadequate in putting the new Ombudsoffice on an appropriate statutory footing. It would have been expected then that Part VII of the 1998 Act either be extensively amended or repealed and fully redrafted to meet the concerns of both Patten and Hayes.

This has not happened. The new Act does not enhance the Ombudsperson's powers in any respect. Rather the draft legislation sought to circumscribe her powers still further. Despite amendments to the final legislation, the legal provision for this office is still problematic.

Any examination of the 1998 Act and Police (Northern Ireland) Act 2000 shows a failure to measure up to the deeply felt need in the Patten Report that this important office must be staffed and resourced accordingly. As well as acting on her own initiative, compiling data on complaint trends and patterns and having a dynamic and cooperative

[84] (1987, Belfast, HMSO). For further details on this area of the debate see Moore and O'Rawe "Accountability and Police Complaints in Northern Ireland: Leaving the Past Behind?" in A Goldsmith and C Lewis (eds) *Civilian Oversight of Policing* (Oxford, Hart Publishing, 2000).

relationship with other agencies involved in community safety, a key recommendation of Patten was that the Ombudsperson should have the power to investigate and comment on police policies and practice. This has not been given. Similarly, investigatory powers in respect of matters deemed by the Secretary of State to be in the past are essentially rendered null and void.

Despite the Patten Report's failure to deal with serious human rights abusers within the police service the issue will not go away, and in many senses the Ombudsperson appears to be the only possible mechanism whereby past abuse might be addressed. If she is curtailed in this respect, the legacy of a number of high profile cases (e.g. the Stalker and Stevens inquiries), together with a failure to establish patterns of previous abuses which have gone unpunished, will undoubtedly return to haunt the "new" beginning.

Two other important areas where legislative guidance is still lacking are those of the extent of the operational independence/responsibility of the Chief Constable and the whole notion of what is meant by "national security". If these issues are left to be interpreted by politicians and police officers, participative democracy and open government will undoubtedly continue to suffer. It is right that the Chief Constable of the Police Service of Northern Ireland should be free from partisan political control. Northern Ireland still bears testimony to the legacy of such control in the past. It is also correct that managerial, day-to-day, operational decisions be taken by the same Chief Constable. However, often in the past the distinction has been blurred as to where operations end and policy begins. The Patten Report deals with the issue on one level by requiring that even where a Chief Constable does make an operational decision, this does not mean that there should not be *ex post facto* accountability. However, this does not deal with how a dispute might be mediated where the Chief Constable contends that an issue is operational and the Policing Board disagrees. Similar concerns arise over national security. Again, such a concept is so nebulous and the protections it offers so open to abuse that its limits require to be defined in legislation. From an accountability point of view it would appear that the Government feel the Chief Constable should be responsible to the Secretary of State for matters pertaining to national security. However, if the Policing Board is unable to hold the Chief Constable to account on issues so fundamental to community safety, an important level of oversight is removed.

There are many other issues as regards potential legislation that space does not allow us to deal with here. One final issue that should be highlighted is the Government's decision that only new officers should be required to take a new oath/declaration wherein human rights is explicitly mentioned. Mandelson has stated in Parliament that to require serving officers to take this oath would involve "legal difficulties". Unpacking of these difficulties would actually appear to put them more in the category of political ones. In any event, it seems a shame that while the tenor of the Patten Report's report is founded on notions of human rights, the symbolism involved in all members of the policing organisation making an explicit commitment to human rights will be lost. Rather than a new beginning, this has unfortunate echoes of a new beginning for some but not all. Given that the Home Office is considering similar wording for all other UK forces, it would seem sensible to enact one piece of UK-wide legislation that would enable all police officers to express publicly their commitment to human rights.

The Patten Report describes democratic accountability as the way in which "elected representatives of the community tell the police what sort of service they want".[85] The weakness of accountability structures over policing have been a key area of concern from the establishment of the RUC to the present.[86] The Patten Report recognises the inadequacy of the current tripartite system.[87] The Report notes that the Police Authority for Northern Ireland had deficient powers but also criticises the tendency of the Authority to defend police in relation to alleged wrongdoing before such allegations have been properly investigated.[88]

Policing Board

Patten proposed that the PANI be replaced by a new Policing Board, with the primary statutory function of "[holding] the Chief Constable and the police service publicly to account".[89] This Board would have a majority elected membership with nineteen members, ten of whom are Assembly members, drawn from parties comprising the new executive,

[85] Patten Report para. 5.4.

[86] For a more detailed discussion of the weakness of current accountability structures see O'Rawe and Moore *supra* n. 16.

[87] Patten Report para. 5.5.

[88] *Ibid*. para. 5.12.

[89] *Ibid*. para. 6.3.

selected by the d'Hondt system. The remaining nine members would be appointed by the Secretary of State in consultation with First and Deputy First Minister and should be representative of business, unions, legal profession, community/voluntary sector.[90] While the emphasis on increased democratic involvement is welcome, the proposed basis for membership means that the smaller parties will be squeezed out. It is also a weakness that the independent representatives will not be elected but selected.

Some of the powers recently removed from PANI under the Police (Northern Ireland) Act 1998 are to be returned to the Policing Board. There is a welcome emphasis on transparency with the proposal that the Policing Board meet in public once a month to receive a report from the Chief Constable.[91]

The Policing Board will have responsibility for overall monitoring of police performance including performance on human rights and public order policing. They will be involved in developing medium to long-term policing plans.[92]

One area of the Police (Northern Ireland) Bill which has caused consternation is the Government's interpretation of the powers given to the Policing Board in relation to inquiries and reports. Patten had recommended that the Board should have reasonably strong powers in relation to retrospective accountability and should be able to require the Chief Constable to report on any issue pertaining to the performance of the police service. The Board would be able to follow up his/her reports by initiating an inquiry with which all officers would be required to cooperate.[93] The Patten Report stressed that "the grounds on which the Chief Constable might question this requirement should be strictly limited to issues such as those involving national security, sensitive personnel matters and cases before the courts".[94]

However, the draft legislation gives the Chief Constable much wider grounds for refusing to provide a Report. Grounds include when the issue relates to a matter being investigated by a statutory authority, or if it would be likely to "prejudice the prevention or detection of crime, the apprehension or prosecution of offenders or the administration of justice".[95] Ultimately the Secretary of State is given power to determine

[90] *Ibid.* paras 6.11–6.14.
[91] *Ibid.* para. 6.36.
[92] *Ibid.* para. 6.23.
[93] *Ibid.* paras 6.22 and 6.23.
[94] *Ibid.* para. 6.22.
[95] Police (Northern Ireland) Bill, clause 55(c), (d).

whether a report should be produced or whether an inquiry should go ahead and whether the body or individual recommended by the Policing Board should be allowed to carry it out. This carries dangerous messages in terms of how "trusted" the Board will be to carry out its functions. It would appear that the legislation envisages the Secretary of State as very much the dominant partner in the supposedly tripartite arrangements for accountability. Little can actually be done by the Policing Board without guidelines, approval or goodwill from the Secretary of State. This does not bode well for attracting high calibre people to the Board, or allowing the Board itself to fulfil a real and independent oversight role.[96]

Amongst the voices raised in horror at the Government's implementation of the Patten Report in relation to the Policing Board, even the normally compliant Police Authority for Northern Ireland has condemned the Bill as "set to render [the] Policing Board ineffective".[97]

Accountability at local council level

The Patten Report proposed the establishment of District Policing Partnership Boards (DPPBs) based in local councils and established as a committee of the Council. These would have a majority elected membership with the remaining members being independent people selected by Council. Each DPPB was to be broadly representative in terms of religion, gender, age, and cultural background, which will be no mean feat given the make-up of most local councils in Northern Ireland.

These bodies were proposed as advisory, explanatory and consultative. Government amended this proposal to state that at council level these accountability bodies be called District Policing Partnerships. The Police Bill leaves the bodies as mainly advisory and with no meaningful power. Patten had proposed that DPPBs be able to raise additional funding for policing which could include private policing. However, this raised the spectre in the media of paramilitaries being employed to carry out policing functions and the Government has

96 For example, the draft legislation not only stipulates that in order to instigate an investigation, a weighted majority—twelve out of nineteen—of Policing Board members must support the recommendation, but even where this is achieved, the Secretary of State still has power to stop the inquiry taking place.

97 Police Authority for Northern Ireland, Press Release, 19 May 2000.

deferred the decision on whether District Policing Partnerships in the future will be able to raise additional funds.[98]

A particular issue of concern in relation to human rights is the exclusion under the Police (Northern Ireland) Act of people convicted of scheduled offences becoming independent members of District Policing Partnerships.[99] In the case of the Policing Board the Secretary of State has the power to remove either a political representative or independent member from the Board on the grounds of that member having been convicted of a criminal offence.[100] If the political process is to move forward, it is crucial that ex-prisoners are reintegrated into society and into the process itself. The continued exclusion of people with political convictions from political life will be a bar to the process of change. This is quite apart from human rights concerns about the safety of some convictions under emergency legislation and the juryless Diplock courts.

Local accountability

Perhaps the biggest disappointment in terms of democratic accountability is in relation to accountability at community level. The Patten Report represented a historic opportunity to do something creative in relation to accountability and policing at a local level but that opportunity has been missed. District Policing Partnerships will replace council-based Community Police Liaison Committees (CPLCs). However, Patten does not recommend the replacement of other CPLCs despite the problems with these in terms of representation, effectiveness and transparency.

"Local communities and police should be encouraged to develop consultative forums on lines that suit them and their neighourhoods".[101] It should be the aim of every police beat manager to have such a forum in his or her patrol area. There is a danger that this process will be police led, organised on police terms.

Community safety is much broader than formal policing. The Patten Report made some attempt to address this issue in terms of talking about policing rather than the police. However, this aspect of the

98 NIO Implementation Plan recommendation 32.
99 Police (Northern Ireland) Act, Sch 3(8).
100 *Ibid*. Sch 1.
101 Patten Report para 6.34.

Report was under-developed and has been well and truly ignored by government. To add to the disappointment in this level of accountability, the Police (Northern Ireland) Act fails to address the issue of local consultative fora, apart from placing a duty on District Policing Partnerships to facilitate police-community consultation at a local level.[102]

Process of change

The above commentary indicates the authors' concerns that the new beginning to policing will not be so new after all. Of course there is much that may constitute improvement in policing arrangements, but similarly there is much that has been lost in failing to open up the process of change to a much wider constituency.

To date interpretation of the Patten Report and how it should be implemented has been entirely in the hands of the same civil service and police organisation (albeit through newly constituted Patten Action Teams) that have failed so miserably in the past to address human rights concerns in relation to policing. In traditional style, much of what was innovative in the Patten Report has been clawed back and the parameters of the debate resituated on familiar turf. This has led to a situation where, rather than seeking to build on and improve the Report, the focus for stakeholders who do not concur with government policy in this area, has switched to trying to reclaim the basis of the Patten recommendations themselves.

The only glimmer of hope that the process might open up to the point that some substantial rethinking take place, was that provided by the role of Oversight Commissioner. Patten's vision of the role was that the Commissioner would "provide more than a stocktaking function". Rather, the "review process would provide an important impetus to the process of transformation".[103] The Government's interpretation of the role appears to be minimalist. The Implementation Plan envisages the Oversight Commissioner conducting progress meetings with Ministers and NIO officials, the Chief Constable, the Police Authority, and in due course the Policing Board, at least three times a year.[104] Although it was heartening that, in the first weeks of his appointment,

[102] Police (Northern Ireland) Bill, Part II, clause 15.

[103] Patten Report para 19.5.

[104] NIO Implementation Plan p. 93.

Mr Constantine sought meetings with groups outside of government, the Oversight Commissioner has already stated himself unwilling to second guess elected and appointed officials. He does not see his role being to monitor the implementation of the Patten Report or intervene proactively in the process. Instead it would appear that he intends to hold the appropriate agencies to account in terms of how they have interpreted that the Report should be implemented, and in respect of the actions and time-scales they set for themselves.

Here, comparisons with the role of the Metropolitan Police in implementing the recommendations of the Macpherson Report are instructive. In their critique of the Macpherson Report, McLaughlin and Murji note that the response of the Metropolitan Police was dominated by a carefully orchestrated reformist discourse.[105] While this means that the police accept a degree of change and cooperate with it, McLaughlin and Murji note that "in a classic police move, the reformers are attempting to claim the report as 'theirs' and pull the debate back inside the organisation. Consultation and change will happen, if the reformers have their way, on terms laid down by the force".[106]

This should ring alarm bells here. The trend for the change process to be led by the current leadership of RUC, PANI, and the NIO must be strongly contested. There is a need for external input beyond these bodies as well as strong independent input from the Overright Commissioner from the earliest opportunity.

The Patten Report was not the end of the exercise but only the beginning. Despite the weaknesses identified in this chapter, many of the changes proposed in the Report are urgently needed. The process of implementing these will be in many ways as important as the substance of the recommendations. That process must be based on transparency, accountability and the participation of the policed as well as the police. Continued failure to recognise this can only result in a not so very new beginning for policing in Northern Ireland.[107]

[105] Eugene McLaughlin and Karim Murji "After the Stephen Lawrence Report" (1999) 19 *Critical Social Policy* 371.

[106] *Ibid*. 379.

[107] At the time of going to press, John Reid has recently taken over from Peter Mandelson as Secretary of State for Northern Ireland. It remains to be seen whether he can undo any of the damage already done by Mandelson's dogged pursuit of a narrow and short-sighted political agenda as regards policing in Northern Ireland.

9

Human Rights, Humanitarian Interventions and Paramilitary Activities in Northern Ireland

KIERAN McEVOY*

INTRODUCTION

FOR THE ALMOST three decades of the Northern Ireland conflict human rights activists lobbied and campaigned on the basis that state human rights abuses were central to the origins and continuance of the conflict and that the protection of rights were key to its resolution.[1] Such arguments appear to have been taken on board by the British and Irish Governments and the pro-Agreement parties who shaped the substantial rights and equality provisions of the Good Friday Agreement.[2] The "mainstreaming" of such issues in future conduct of *state* agencies is, at least in part, the result of a highly successful range of interventions by key human rights actors over a large number of years and the significance of those interventions is beginning

* The author would like to express his appreciation for those who commented on earlier versions of this chapter including Maggie Beirne, Martin O'Brien, David Petrasek, Fionnuala Ní Aoláin, and Tom Hadden. I would also like to thank all of those who attended the Council on Human Rights Policy meeting in Geneva in August 1999 on Holding Armed Groups Accountable at which I presented the first version of this chapter—their comments and contributions have improved it considerably. I would also like to express my appreciation to all of those who agreed to be interviewed for this piece of work and finally to Colin Harvey for his encouragement and patience which went well above and beyond the call of duty. All views expressed and remaining errors are the sole responsibility of the author.

[1] For a summary of these arguments see e.g. Committee on the Administration of Justice *et al Declaration on Human Rights, The Northern Ireland Conflict and Peace Process* (Belfast, CAJ, 1994).

[2] See C Harvey and S Livingstone "Human Rights and the Northern Ireland Peace Process" [1999] *EHRLR* 162–77.

to attract serious scholarly analysis.[3] In a similar fashion, this chapter seeks to examine the shape and influence, if any, which similar groups and individuals may have had upon the conduct of the non-state actors to the conflict, the paramilitaries.

This chapter thus examines the interventions of various human rights and humanitarian activists who have sought to mitigate or eradicate paramilitary abuses in Northern Ireland. All have been motivated by an opposition to paramilitary violence although their methodologies have differed considerably. For reasons which will become apparent below I have chosen to focus upon a range of different *styles* of intervention, which include but are not limited to the frameworks provided by international human rights and humanitarian law. In particular, the notion of *humanitarian* interventions is broader than the comparatively narrow principles of humanitarian law, if not its philosophical origins.[4]

I have developed four heuristic typologies of intervention with paramilitaries in Northern Ireland. All are based on actual projects. They are designed to capture distinct modes of attempting to influence paramilitary behaviour. These are characterised as (a) interventions by international human rights non-governmental organisations based upon humanitarian law principles; (b) interventions by political lobby groups using a human rights framework; (c) direct engagement by humanitarian groupings seeking to change paramilitary behaviour: (d) state focused human rights groups engaged in the creation of "a human rights culture". Before considering each typology, it is necessary first to offer some brief background on the nature and extent of paramilitary abuses and the context in which they occurred.

[3] P Mageean and M O'Brien "From the Margins to the Mainstream: Human Rights and the Good Friday Agreement" (1999) 22 *Fordham International Law Journal* 1499–538. Some commentators have referred to the actions of human rights groups and other elements of civil society as "the parallel peace process". For a discussion of this process concerning the equality elements of the Agreement and post-Agreement era see C McCrudden "Mainstreaming Equality in the Governance of Northern Ireland" (1999) 22 *Fordham International Law Journal* 1696–776.

[4] "We shall utterly fail to understand the true character of the law of war unless we are to realize that its purpose is almost entirely humanitarian in the literal sense of the word, namely to prevent or mitigate suffering and, in some cases, to rescue life from the savagery of battle and passion. This, and not the regulation and direction of hostilities, is its essential purpose": H Lauterpacht "The Problem of the Revision of the Law of War" (1950) *British Yearbook of International Law* 360, 364. The phrase "humanitarian intervention" in this chapter is therefore defined broadly to refer to attempts to prevent or mitigate suffering at the hands of the paramilitaries in Northern Ireland.

BACKGROUND TO PARAMILITARY ABUSES IN NORTHERN IRELAND

Republican violence

Since 1969 over 3,600 people have been killed and over 40,000 people injured in Northern Ireland. In addition, the conflict has cost the British Exchequer several billion pounds.[5] Since 1973 police figures suggest that approximately 2,300 people have been the victims of paramilitary punishment shootings (usually in the knees, thighs, elbows, ankles or a combination) and since 1983, approximately 1,700 people have been the victim of paramilitary punishment beatings, often involving attacks with baseball bats, hurling sticks studded with nails, iron bars and other heavy implements.[6]

This chapter focuses primarily on the IRA as the largest and most active republican grouping.[7] Interrupted by three major cease-fires, (the most recent called in July 1997 and still in place), since 1969 the IRA have carried out bombings and shootings in Northern Ireland, Britain and Europe.[8] They have attacked the security forces (including the British Army and Royal Ulster Constabulary), political and judicial figures, loyalist paramilitaries and civilians. The IRA has also bombed commercial targets in order to apply economic and political pressure for a British withdrawal.

The IRA has killed or injured civilians by deliberate targeting, for example of those adjudged guilty (by the IRA) of "informing" or those considered guilty of anti-social activity such as drug-dealing. They have also been responsible for numerous "mistakes" wherein civilians have been erroneously or negligently killed in botched attacks on economic, security force or loyalist paramilitaries. In the late 1980s

[5] See M T Fay, M Morrissey, and M Smyth *Northern Ireland's Troubles: The Human Costs* (London, Pluto Press, 1998).

[6] RUC Website 2000. http://www.ruc.police.uk/. The RUC only began systematically to keep figures on punishment beatings by paramilitaries in 1983. Their figures suggest a marked increase in beatings following the IRA and loyalist cease-fires in 1994 as paramilitaries apparently, initially at least, became less ready to use firearms for such attacks.

[7] Other violent republican groupings have included the Official IRA, the Irish National Liberation Army (INLA), Irish People's Liberation Organisation (IPLO) and a number of groupings which have come to prominence in the wake of the IRA cease-fires, the "Continuity" IRA and the "Real" IRA.

[8] J Bowyer-Bell *The Secret Army: The IRA* (Dublin, Poolbeg, 1979); P Bishop and E Mallie *The Provisional IRA* (London, Corgi, 1989); P Taylor *Provos: The IRA and Sinn Fein. The Book of the BBC TV Series* (London, Bloomsbury, 1997).

deliberate attacks on civilians increased after the IRA decided to extend its range of "legitimate targets" to include those involved in service provision to the security forces.[9]

The IRA are a highly centralised and relatively disciplined paramilitary grouping. After reorganising in the mid-1970s into a cellular structure, they have proved to be a ruthless and persistent paramilitary organisation, consistently recognised in military and intelligence circles as the most effective terrorist organisation in the world.[10] They are well armed and apparently highly motivated.[11] While there has been some leakage of IRA personnel and weapons to dissident republican groupings since the cease-fires (including the "Real IRA" who killed twenty-nine civilians in a bomb attack in Omagh in 1998), they appear at the time of writing to remain a cohesive and organised group.

As noted above, violent attacks have been carried out not solely as a result of the IRA's "military" campaign, but also because of what are described as "policing" activities. These punishment attacks have often been the focus of most attention by human rights and humanitarian actors. The IRA has a distinct section within its structure known as the "civil administration" which is tasked with the policing of anti-social activities.[12] The system requires a considerable logistical commitment in order to hear complaints, investigate, make recommendations and carry out the attacks. As the system has become routinised over time, relationships have developed with professional agencies (for example social workers or youth workers) and individuals who have tried to intervene regarding those under threat (discussed below).

Loyalist violence

Loyalist paramilitaries have committed acts of violence, in their case in support of the maintenance of the Union and the perceived failure by the state to "deal with" republican terrorism.[13] The two main loyalist

[9] J Bowyer Bell *The Irish Troubles: A Generation of Violence* (Dublin Gill & Macmillan, 1993).

[10] P Wilkinson *Terrorism and the Liberal State* (2nd edn, London, Macmillan, 1986).

[11] In 1980 and 1981, ten republican prisoners starved themselves to death over a five-month period in protest over the Government's attempts to force them to accept the status of criminal rather than political prisoners.

[12] M O' Doherty *The Trouble with Guns: Republican Strategy and the Provisional IRA* (Belfast, Blackstaff, 1998).

[13] S Bruce *The Red Hand: Protestant Paramilitaries in Northern Ireland* (Oxford, Oxford University Press, 1992).

paramilitary groupings are the Ulster Volunteer Force and the Ulster Defence Association (also sometimes referred to as the Ulster Freedom Fighters), with other smaller groups forming during the peace process including the Loyalist Volunteer Force.

These groups have traditionally targeted Catholic civilians, economic or civilian targets in the Irish Republic or republican activists.[14] They consider themselves driven to the use of political violence because of the IRA's campaign against the British state and their community. [15] Loyalists argue that with fewer clear targets (other than occasional identified republicans) they have been forced to attack the Catholic community as a whole. Other than through information garnered through collusion with the security forces,[16] much of their violence has been described as sectarian and indiscriminate.[17] Loyalists have been considerably less successful than republicans in garnering an international constituency in support of their activities.

Loyalists too have engaged in punishment violence. While such punishments are also directed against anti-social activities by alleged criminals, the phenomenon is arguably more complex than on the republican side. Loyalists share the republicans' sense of responsibility regarding anti-social criminal behaviour in their communities. In addition, however, loyalist punishments also appear concerned with the internal disciplining of their own members, territorial disputes between rival paramilitary factions and drug-related disputes, sometimes between differing factions.[18]

Loyalist punishments are less formalised and systematised.[19] There are no specialist loyalist squads dealing with punishments, activists are also responsible for "military" attacks on nationalists. Warnings tend not to be given, tariff scales are unclear, there is no central bureaucracy

[14] J Cusack and H McDonald *UVF* (Poolbeg, Dublin, 1997).

[15] J McAuley *The Politics of Identity : A Loyalist Community in Belfast* (Aldershot, Avebury, 1994).

[16] e.g., Human Rights Watch *To Serve Without Favour: Policing, Human Rights and Accountability in Northern Ireland* (New York, Human Rights Watch, 1997) especially ch. six.

[17] S Bruce "Re-appraising Loyalist Violence" in A O'Day (ed) *Terrorism's Laboratory: The Case of Northern Ireland* (Aldershot, Dartmouth, 1995) pp. 115–37.

[18] T Winston "Alternatives to Punishment Shootings and Beatings in a Loyalist Community in Belfast" (1997) 8 *Critical Criminology* 122–8.

[19] C Bell "Alternative Justice in Ireland" in N Dawson, D Greer and P Ingram (eds) *One Hundred and Fifty Years of Irish Law* (Belfast, SLS Legal Publications, 1996) pp. 145–67.

with much local autonomy, and punishments tend to be carried out swiftly.[20]

Loyalist command structures are considerably looser than their republican counterparts, and they are widely viewed by academics, security and prison staff as less organised and less disciplined. While a number of capable political and community leaders have emerged from the ranks of former loyalist paramilitaries, most academic commentators suggest that in general they attract a lesser calibre of recruits than the IRA in particular.[21] Such structural and personnel factors mean that interventions with loyalist paramilitaries are all the more problematic.

TYPOLOGIES OF INTERVENTION

Interventions by international human rights NGOs based upon humanitarian law

Given that the origins of the modern human rights movement are often traced to the horror of the Second World War, it is understandable that the focus of most human rights non-governmental organisations (NGOs) has traditionally been directed at state abuses. In recent years however, a heated debate has emerged as to whether the monitoring role of such groups should be expanded to include abuses carried out by non-state entities such as paramilitary groupings.[22] Major organisations such as Amnesty International and Human Rights Watch have broadened their remit to include abuses perpetrated by such groupings.

The arguments which led to that expansion are complex and are not the primary focus of this chapter which is more concerned with the nature and impact of such monitoring.[23] In the Northern Ireland con-

[20] Interview with former UVF prisoner and manager of prisoners' reintegration project which is also involved in establishing a restorative justice scheme in a loyalist area designed to avoid punishment attacks, 29 June 1999.

[21] Steve Bruce, in the primary study on loyalist paramilitarism, suggests this is because those members of the Protestant/loyalist community who wish to fight to maintain the Union may do so legitimately by joining the British Army or the local police, leaving only less able recruits for the loyalist paramilitaries: Bruce *supra* n. 13 at pp. 272–3.

[22] See, e.g., Special Issue of (1996) 6 *International Commission of Jurists Review*; D Petrasek *Heroes and Demons: Human Rights Approaches to Armed Groups* (Geneva, International Council on Human Rights Policy, 2000).

[23] For current purposes, the arguments concerning the expansion of the role of human rights groups to include monitoring of groups by paramilitaries may be summarised as

text, in the 1990s both Amnesty International and Human Rights Watch began to criticise the activities of paramilitaries in Northern Ireland primarily under the rubric of international humanitarian law, in effect defining it as "a conflict not of international character" under the 1949 Geneva Conventions.[24] Both groups focused in particular upon the protections contained in Common Article Three of the Conventions.[25]

falling under a number of headings. These include a sense of genuine moral outrage at the heinous abuses of civilians by some non-state entities (see R Kogod Goldman "International Humanitarian Law: Americas Watch's Experience in Monitoring Internal Armed Conflicts" (1993) 9 *American University Journal of International Law and Policy* 49–94); the view that some human rights groups were "stung by the taunts" that by ignoring the actions of armed groups they were not acting impartially (see N Rodley "Can Armed Oppositional Groups Violate Human Rights ?" Paper presented at Conference on Human Rights in the Twenty First Century: A Global Challenge, Alberta, Canada, 1990); more recently the suggestion that traditional notions of "the state" and its functions, the human rights/humanitarian law divide and "armed conflicts" were dated concepts (see, e.g., A Clapham *Human Rights in the Private Sphere* (Oxford, Clarendon Press, 1993); P Baer "Amnesty International and its Self Imposed Limited Mandate" (1993) 12 *Netherlands Quarterly on Human Rights* 5); and the related view that social and economic rights were not best enhanced by an exclusive focus on governments in an era of powerful multinationals wielding considerable economic influence (see C Jochnick "Confronting the Impunity of Non-State Actors: New Fields for the Promotion of Human Rights" (1999) 21 *Human Rights Quarterly* 56–80).

[24] It is important to acknowledge that there is a considerable debate in the academic literature as to whether the Northern Ireland context fell within the notion of an "internal armed conflict". See F NT Aoláin *The Politics of Force: Conflict Management and State Violence in Northern Ireland* (Belfast, Blackstaff, 2000) esp. ch. 5. A strong view has been argued by some academic commentators that it did not (see G Hogan and C Walker *Political Violence and the Law in Ireland* (Manchester, Manchester University Press, 1989); K Boyle and C Campbell *Human Rights in Situations of Armed Conflict and Political Violence* (Santiago, Sinergos Consultores Ltda, 1992). While some of the arguments concerning the non-applicability of humanitarian law are compelling (such as the failure by paramilitaries to exercise effective control over territory), the fact remains that both Human Rights Watch and Amnesty International have applied such principles to the Northern Ireland context regardless.

[25] Common Article 3 states :

"In the case of armed conflict not of an international character occurring in the territory of one of the high contracting parties, each party to the conflict shall be bound to apply, as a minimum, the following provisions :

(I) Persons taking no active part in the hostilities, including members of the armed forces who have laid down their arms and those placed 'hords de combat' by sickness, wounds, detention, or any other cause, shall in all circumstances be treated humanely, without any adverse distinction founded on race, colour, religion or faith, sex, birth or wealth, or any similar criteria. To this end the following acts are and shall remain prohibited at any time and in any place whatsoever with respect to the above mentioned persons:

(a) violence to life and person, in particular murder of all kinds, mutilation, cruel treatment and torture;

(b) taking of hostages;

(c) outrages upon personal dignity, in particular humiliating and degrading treatment;

(d) the passing of sentences and the carrying out of executions without previous judgement

Human Rights Watch/Helsinki

In 1991 Helsinki Watch[26] conducted its first major review of human rights abuses in Northern Ireland. As well as outlining a range of state human rights abuses, its report focused upon loyalist and republican paramilitary abuses. The details of that Report and some of the arguments advanced are worth reproducing in brief.

The Report outlined republican attacks on both military (British Army and RUC) targets and civilians between 1969 and 1990, including paramilitary members who had been killed for actions such as passing information to the security forces.[27] They also detailed "human bomb" attacks, wherein civilians described as "collaborators" were forced by the IRA to drive vehicles loaded with explosives to British Army checkpoints where the bombs exploded. They compared those killed by the IRA (52.7 per cent security forces, 9 per cent IRA members, and 23.6 per cent were Protestant civilians, of whom 1.1 per cent were loyalist paramilitaries) with loyalist killings (89.6 per cent civilians, of whom 80.06 per cent were Catholic). Finally the Report made reference to the scale of punishment shootings carried out by republican and loyalist paramilitaries.

The Report referred to international humanitarian law standards and in particular Common Article 3 of the Geneva Conventions regarding internal armed conflicts, while acknowledging that "the application of Common Article 3 to the conflict in Northern Ireland is debatable".[28] That report also referred to the International Red Cross Rules of 1990 regarding "International Humanitarian Law Governing the Conduct of Hostilities in Non-International Armed Conflicts", the Declaration of Minimum Humanitarian Standards (the "*Turko*" Standards 1990) and Protocol 1 of the Geneva Conventions (not then ratified by Britain but regarded by Helsinki Watch as a principle of international law of general application).

Helsinki Watch concluded by condemning the use of violence by paramilitary groups against civilians, the taking of hostages and the

pronounced by a regularly constituted court, afforded all the judicial guarantees which are recognised as indispensable as civilised peoples.

(II) The wounded and the sick shall be collected and cared for".

[26] After their early reports on Northern Ireland Helsinki Watch changed their name to Human Rights Watch Helsinki.

[27] Helsinki Watch *Human Rights in Northern Ireland* (New York, Human Rights Watch, 1991).

[28] Helsinki Watch *ibid*. p. 111.

conducting of "punishment shootings". In addition, the killings of security forces and opposing paramilitary groups were condemned on the grounds that they constituted a "resort to perfidy".[29] That criticism was repeated in their 1992 *World Report* and in 1993 and 1994. In 1993 and 1994, Helsinki Watch called for all paramilitaries in Northern Ireland to refrain from the use of violence for political ends and called for an end to punishment shootings, assaults and banishments.[30] In effect, in these various reports Helsinki Watch concluded by making no ultimate distinction between attacks on civilians and combatants (because of the "perfidy" argument), and interpreted the notions of customary norms of international humanitarian law to preclude all politically motivated violence by paramilitaries in Northern Ireland.

Helsinki Watch has also written a number of reports which have focused more specifically upon the question of paramilitary punishment attacks and banishments. In 1992, the relevant section of a report on Children in Northern Ireland argued that the activities of the paramilitaries breached customary standards of international humanitarian law.[31] Similarly, in the Human Rights Watch/Helsinki 1997 report on policing in Northern Ireland, a chapter on paramilitary "policing" concluded by arguing that "paramilitary punishment assaults and shootings thus violate the right to life, freedom from humiliating and degrading treatment, the right to due process and the guarantee of a fair trial as codified in Common Article 3".[32] Neither of these reports made reference to the broader contention advanced elsewhere by Human Rights Watch that paramilitaries should desist from the use of political violence per se in Northern Ireland.

[29] Helsinki Watch *ibid.* pp. 114–16. Protocol 1 refers to international armed conflicts including those "in which peoples are fighting against colonial domination and alien occupation". Article 37 notes that "It is prohibited to kill, injure or capture an adversary by resort to perfidy". Perfidy is defined in Article 37(c) as the feigning of civilian, non-combatant status. Helsinki Watch referred to the fact that paramilitaries in Northern Ireland do not bear arms openly and frequently kill by assassinations and claimed that therefore such activities may be properly characterised as being carried out "by resort to perfidy".

[30] See Human Rights Watch *World Report 1992: An Annual Review of Developments and the Bush Administration Policy on Human Rights World-wide* (1992); Helsinki Watch (1993) 5 *Northern Ireland: Human Rights Abuses by All Sides* (May, No. 6); Human Rights Watch Helsinki (1994) 6 *Northern Ireland: Continued Abuses on All Sides* (March, No. 4).

[31] Helsinki Watch *supra* n. 27 at p. 54.

[32] Human Rights Watch *supra* n. 16.

Amnesty International

Following its change of policy in 1991, Amnesty International began to monitor the activities of both republican and loyalist groupings in Northern Ireland. In July 1992 Amnesty International condemned the killing by the IRA of three alleged IRA "informers", drawing a parallel with the similar killings by the UFF and UVF in the same year. While Amnesty continued to be very active on allegations of state human rights abuses,[33] it was not until the publications of its major report on political killings in Northern Ireland in 1994 that the issue of paramilitary monitoring was given substantial consideration.

In that Report, Amnesty International "calls upon armed political groups to observe minimum humane standards". The Report cites norms of international humanitarian law which apply certain minimum limitations on all parties to internal conflict.[34] Unlike the equivalent Human Rights/Helsinki Watch reports, Amnesty do not cite specific instruments to support their assertion regarding the applicability of such standards to Northern Ireland. Rather they suggest that "above all international humanitarian law forbids governments and their opponents alike to torture any person, to deliberately kill civilians, to harm those who are wounded, captured or seeking surrender, or to take hostages. These acts can never be justified. It also forbids the passing of sentences and carrying out of sentences without due process of law. Amnesty International, of course, opposes executions without qualifications, under any circumstances". There is no discussion of the notion of "perfidy" as customary international humanitarian law in the Amnesty Report.

Amnesty lists a range of killings and injuries for which republican groups are responsible. Confusingly however, clear distinctions are not made by Amnesty between those military targets regarded as acceptable and civilian targets regarded as unacceptable under humanitarian law. For example, accounts are given of IRA bomb and sniper attacks on British soldiers manning checkpoints, police officers on patrol, attacks on off duty soldiers and police as well as the killing of civilians, including a nun caught up in a bomb attack, but no clear distinctions are suggested. The only clear assertion is where an IRA bomb attack on

[33] e.g. see Amnesty International *Fair Trial Concerns in Northern Ireland: The Right of Silence*. (London, Amnesty International, 1992).

[34] Amnesty International *Political Killings in Northern Ireland* (London, Amnesty International, 1994) p. 34.

a military hospital in which two soldiers are killed and a number of civilians are injured is quite indisputably described as "a blatant violation of humanitarian law".[35] Attacks in which civilians are either mistakenly or deliberately killed or injured are described *as* "in breach of minimum humane standards".[36] Amnesty also noted punishment beatings, shootings and banishments carried out by republican paramilitaries and condemned one such attack where a young man was shot in the legs three times for refusing to allow the IRA to use his car.

With regard to loyalist killings and injuries, Amnesty noted that ". . . the main victims of Loyalist killings are ordinary members of the Catholic Community"[37] and noted that loyalists too were responsible for punishment attacks. Amnesty concluded this section of their report by urging the leadership of armed political groups to take steps to ensure that their members "don't torture, don't kill prisoners, don't kill civilians, don't take hostages".[38]

In examining this important Report, it is difficult to conclude other than the failure by Amnesty to make distinctions between different types of attacks on military or security force personnel (other than a definitive statement regarding the unacceptability of the attack on a military hospital) is informed by an understandable political sensitivity regarding the notion of a "legitimate target" in Northern Ireland. Amnesty simply list a range of killings and injuries, acknowledge that loyalists kill predominantly Catholic civilians and conclude by urging that basic principles of humanitarian law are upheld. That vagueness was repeated the following year in the Amnesty Summary of Human Rights Concerns when after detailing the numbers of deaths and injuries, as well as a number of recent punishment attacks, the relevant section concluded by stating "Amnesty International opposes human rights abuses carried out by armed political groups, namely the torture or killing of prisoners, other deliberate and arbitrary killings and hostage taking".[39]

The political difficulties posed by observing more closely the requirements of humanitarian law have been highlighted by a number of other

[35] Amnesty International *ibid* p. 36.

[36] *Ibid.* p. 37.

[37] *Ibid.* p. 42.

[38] *Ibid.* p. 48.

[39] Amnesty International *Summary of Human Rights Concerns* (London, Amnesty International, 1995) p. 15. For a discussion on whether armed groups can be guilty of human rights abuses see T Meron "When Do Acts of Terrorism Violate Human Rights?" (1990) *Israel Yearbook of Human Rights* 271–9.

Amnesty International interventions. For example, following their resumption of violence after the 1994 cease-fire, an IRA car bomb at Thiepval Barracks Lisburn (the Headquarters of the British Army) killed one soldier and injured almost thirty civilians.[40] Amnesty released a statement condemning the injuries to the civilians but were immediately branded partial and biased by unionist Security Spokesperson, Ken Magginis, who noticed their failure to explicitly condemn the killing of the British solider. Such criticism mirrored those made by the former Secretary of State Sir Patrick Mayhew when he responded to Amnesty's 1994 Report by saying:

> "Why do you call on the paramilitaries only to desist from killing civilians? Are human rights denied by you, as well as by the terrorists, to police officers? . . . Will you now additionally call on the paramilitaries to stop killing police officers and the military who support them?" [41]

Before considering the impact of such interventions in Northern Ireland upon the paramilitaries, there are a number of issues of broader relevance which arise regarding this style of intervention.

The notion of a "legitimate" target

One key problematic regarding the extension of a monitoring role to paramilitaries in the Northern Ireland context is the distinction between military and civilian targets which is at the very core of humanitarian law principles.

Human Rights Watch has attempted to steer around this difficulty by (arguably) stretching the concept of "perfidy" as customary humanitarian law, to rule out all paramilitary violence whether it is directed against the security forces or civilians. In effect, Human Rights Watch collapsed humanitarian law into pacifism in their general reports on human rights in Northern Ireland. Such a view masks the difficult political reality that republican violence is more likely (although by no means universally) to be deemed permissible by humanitarian law than that carried out by loyalists. Amnesty on the other hand has oscillated somewhat between vague formulations which obscure the distinctions between civilian and military targets (e.g. condemning deliberate and

[40] "Blasts Rock Army HQ: Two Fight for Lives After Double Attack" *Irish News*, 8 October 1996.

[41] "Watchdog's Company Requested But Mixed Welcome is Expected" *The Irish Times*, 13 February 1999.

arbitrary killings) and more specific condemnations which have in turn attracted negative political reactions.

In a conflict which involves two sets of paramilitary protagonists one of whom generally targets military and security force personnel and the other civilians, the application of humanitarian law is inherently problematic. The very concept of a "legitimate" target is anathema to a sizeable section of the community. On the other hand to expand customary humanitarian law to such an extent that it levels all acts of political violence onto a single plain is both illogical and self-defeating. It leaves protagonists with a choice of either stopping political violence entirely or simply ignoring any international strictures since all violent activities are deemed to breach humanitarian standards.

International NGOs as "cease-fire monitors"

Even the condemnations of paramilitary punishment beatings and shootings have not been without their political difficulties in the Northern Ireland context. The increased political significance of punishments in the wake of the republican and loyalist cease-fires of 1994[42] brought its own pressures to Amnesty International and Human Rights Watch.

In February 1999, Unionist and Conservative politicians were seeking to link an end to punishment attacks to the decommissioning of paramilitary weapons and a "freezing" of the early release of prisoners. Amnesty International and Human Rights Watch were called upon for "help" by unionist leader David Trimble. Mr Trimble requested that they should "give us the help needed to ensure the conditions of non-violence they [Sinn Féin] signed up to in the Agreement".[43] The invitation (part of a carefully orchestrated exercise) was preceded by a similar editorial and commentary from a British Sunday newspaper, and was extensively covered in the British print and broadcast media.[44] While Human Rights Watch were able to reply that "we do not normally send a delegation to investigate one single aspect of a complex

[42] The first IRA cease-fire ended in February 1996 and was renewed in July 1997.

[43] David Trimble, "Letter from First Minister (Designate) to Secretary General of Amnesty International and Executive Director of Human Rights Watch" (Stormont Belfast, First Minister (Designate) Press Office, 1 February 1999).

[44] Mr Trimble had in fact failed to reply to a letter from Amnesty International six months previously suggesting a mission to cover the entire remit of Amnesty's human rights concerns.

human rights situation".[45] However, amnesty found that the public perception of their long planned mission to Northern Ireland had been totally skewed by the Trimble invite.

While all human rights NGOs carry out their work in a political environment, it was a deeply invidious position for major human rights organisations to be placed in the role of "cease-fire monitor". Amnesty and Human Rights Watch understandably resisted such a role. Nonetheless, their intervention was described to the author by one local human rights activist as "politically naive, playing into the hands of the most reactionary political forces in Britain and Ireland".[46] Critical human rights reports conducted by international NGOs are often utilised for propagandist purposes and NGOs cannot be held responsible for the political uses to which various groupings will put their work. Nonetheless, the Northern Ireland experience is a salutary reminder of the potential for the monitoring of non-state actors in particular to be used in such a fashion.

International NGOs as norm creators

The third key issue to consider with regard to the Amnesty International and Human Rights Watch remit, is that international weight and credibility afforded to such organisations is in itself norm-creating. As one human rights activist told the author:

> "Groups like Amnesty, Human Rights Watch and so on to a large extent define what human rights are to the public. For example, despite continuously stressing the narrowness of the organisation's remit, when Amnesty used to work almost solely on prisoners of conscience issues, people did not understand that economic, social and cultural rights were also human rights. Now Amnesty works on paramilitary abuses, therefore for a lot of people that is what human rights abuses are. These organisations define what human rights are wittingly or unwittingly, and local groups have to deal with that".[47]

The key aspect in this context is that international NGOs should recognise that there are implications at local levels for human rights NGOs from the policy decisions taken by the larger groupings.[48]

[45] "Rights Group Declines Trimble's Invitation" *The Irish Times*, 10 February 1999.

[46] Interview, 23 June 1999.

[47] Interview, Maggie Beirne, Research and Policy Officer CAJ, 28 June 1999.

[48] The author is unaware of any consultation with local human rights groups in Northern Ireland before the change to Amnesty's mandate in 1991 and has been unable to discern to what extent such consultation took place with groups in the other countries.

Clearly human rights organisations must have the freedom to develop organisational policies as they see fit. However, in a context of increasingly global NGO networks, that freedom must be exercised in a spirit which recognises the responsibilities which accompany hard won international respect and prestige to ensure sensitivity to local consequences by an ongoing process of consultation and discussion with local groups.

The impact of monitoring by international NGOs on paramilitaries in Northern Ireland

Based upon the interviews conducted with paramilitaries and former paramilitaries for this and other research,[49] it is my contention that the use of humanitarian law by NGOs to condemn "military" attacks carried out by republicans and loyalists on security forces or civilians had no discernible impact on the targeting policy of either set of protagonists. In both instances, such policy was dictated by the ideology, practices and operational capacities of the respective organisations. However, with regard to the monitoring of paramilitary abuses such as punishment beatings and shootings, it is arguable that the work of Amnesty and Human Rights Watch contributed to the increased public awareness and political importance of the phenomenon in the 1990s. In my view, it is possible to discern different impacts between the loyalist and republican paramilitaries.

As discussed above, with a looser organisational structure, differing range of reasons for such activities and less developed international political constituency, it is likely that humanitarian law monitoring had a less significant effect on loyalist paramilitaries. In addition, Loyalists may have been less sympathetic to interventions from such groupings as Amnesty International to begin with given their active and regular criticisms of state abuses.[50]

[49] e.g., K McEvoy *Resistance, Management and Power in Prisons: Paramilitary Imprisonment in Northern Ireland* 1969–2000 (Oxford, Oxford University Press, forthcoming 2001); K McEvoy and H Mika "Punishment, Politics & Praxis: Restorative Justice and Non-Violent Alternatives to Paramilitary Punishments in Northern Ireland" *Policing and Society* (forthcoming 2001); J Auld, B Gormally, K McEvoy and M Ritchie *Designing a System of Restorative Community Justice in Northern Ireland: A Discussion Document* (Belfast, 1997).

[50] "Amnesty historically never really had much credibility in relation to the loyalist community. They were seen as an arm of the republican movement with regard to human rights abuses, being used in a propaganda war. There was a major credibility gap

Republicans on the other hand may have been more sensitive to criticisms from such groupings. They have often drawn upon critical international human rights reports highlighting state abuses to underline their position about the illegitimacy of the British state in Ireland. Indeed as one republican spokesperson acknowledged, the IRA leadership appeared acutely aware of the change in Amnesty's position. Using the phrase "the monitoring of NGEs" (a technical phrase "non-governmental entity" usually associated with Amnesty International) he explained:

> "When Amnesty changed their line I think we had been expecting it. They had a certain amount of credibility because they had highlighted abuses in the past at Castlereagh [interrogation centre] and places and they were not seen as simply anti-republican . . .Their criticisms probably had an impact at the leadership levels because the leadership did not want to be dealing with pressure from large numbers of beatings. It probably added to a tightening up of procedures within the movement [IRA] further down the line".[51]

The lessons for broader applicability would appear to be that monitoring of non-state entities is likely to be most effective where credibility has been established, where there is a more politicised and more disciplined organisation and where the political expression of that movement has an eye to the nurturing of an international constituency.

Interventions by political lobby groups using a human rights framework

Throughout the 1990s in particular several groups emerged in Northern Ireland which, describing themselves as human rights groups, campaigned against abuses by paramilitary groupings and by the IRA in particular. These included groups which highlighted republican attacks on economic targets such as the cross-border railway (Peacetrain), attacks on Protestant security force members and their families in border areas (FAIR, Families Acting for Innocent Relatives) and a group called the "Long March" which invited "all sections of the Protestant and unionist population to march from Derry to Garvaghy

between Amnesty International and the loyalist paramilitaries": interview, former UVF prisoner, 29 June 1999.

[51] Interview, republican spokesperson, 1 July 1999.

Road in Portadown",[52] in effect combining a focus on republican terrorism with contested Orange marches through nationalist areas.

The best known such lobbying group was FAIT (Families Against Intimidation and Terror) which was operational for nine years before dissolving in 1999. It is FAIT which can best serve as a heuristic device to examine some of the broader themes concerning interventions by such groupings on paramilitary abuses. FAIT was established in 1990 by a former member of the Official IRA[53] and the mother of a petty criminal who had been the victim of an IRA punishment attack. Although initially focused upon punishment attacks by the IRA, it expanded its mandate to include criticism of loyalist violence as well. FAIT described themselves as: "an anti-sectarian, non political group which is committed to the preservation of human rights with their main focus being geared towards an end to all forms of terrorist beatings shootings and intimidation".[54] FAIT's principal strategy has been to use the media to embarrass the political wings of republicanism and loyalism by highlighting the abuses of the victims of punishment attacks. Their tactics included almost daily press releases, placarding political meetings, shadowing figures such as Sinn Féin President Gerry Adams on visits to the USA, bringing victims to major speaking engagements and generally creating maximum embarrassment for such figures by contributing to the "wall to wall coverage"[55] of the punishment beatings issue during the peace process era.

Although FAIT describes itself as a human rights grouping, there are clearly difficulties in this regard. First, they make no usage of international human rights or humanitarian law standards in their literature or public pronouncements.[56] FAIT were, for example, strongly opposed

[52] "Plans for Long March Unveiled" *Belfast Telegraph*, 12 June 1999.

[53] The Official IRA were a more Marxist-oriented republican grouping who were involved in a split with the Provisional IRA in 1970. The Provisional IRA quickly became the larger of the two groupings and over time inherited the mantle of "the IRA". While the officials declared a cease-fire in 1972, they continued to engage in armed actions including killings, kneecappings and robberies at least until the mid-1990s. The schism also lead to a vicious political struggle between the political wings of the two movements, the Workers Party (OIRA) and Sinn Féin (PIRA).

[54] FAIT Website (2000) Website of Families Against Intimidation and Terror http://www.fait.org.

[55] M Tomlinson "Walking Backwards into the Sunset: British Policy and the Insecurity of Northern Ireland" in D Miller (ed.) *Rethinking Northern Ireland: Culture, Ideology and Colonialism* (London, Longman, 1998) pp. 94–122.

[56] In an interview with author, it was clear that FAIT's Director who has been with the organisation from the outset was unaware of the legal detail of either set of standards, the differences between them or their respective usages or limitations. Interview, Sam Cushnahan, Director of FAIT, 1 July 1999.

to any notion of distinguishing between military and civilian targets.[57] Secondly, they did not work on state abuses of rights, the primary focus of international human rights and indeed humanitarian standards. Thirdly, their established international networks were not generally with the international human rights community but rather with other victims and anti-terrorist lobbying groups in Spain (Association for the Victims of Terrorism), Algeria (Algerian Collective) and France (SOS Attentats). Fourthly, they were almost entirely funded by government, a relationship which most non-governmental human rights organisations avoid to ensure maximum credibility and impartiality.

Instead, it would be more accurate to locate their work in terms of a political lobbying group working in a human rights framework. The origins of FAIT (and other similar groups),[58] their work programmes, their operational methods and their objectives, clearly have a political objective—which is too put an end to all paramilitary activity. Human rights groups work within the political realm to promote human rights objectives and this requires that they not challenge the state, nor indeed government, per se but instead challenge any behaviour that is in violation of human rights standards. The category of groups that FAIT belongs to must, of logical necessity, challenge the very existence of paramilitaries and they are therefore operating to a preeminently political agenda, albeit addressing human rights or humanitarian concerns en route to that goal.

The impact of political lobby groups using a human rights framework

While FAIT has been highly successful in attracting media publicity to the issue of paramilitary punishment attacks, it too has been the object

[57] Interview, Sam Cushnahan *ibid*.

[58] Other actors have made similar interventions in Northern Ireland while describing themselves as human rights groups. For example, one prominent academic has run a number of campaigns directed primarily against violent punishment attacks carried out by republicans. His tactics have included criticising the policy of other human rights groups while eulogising the efforts of FAIT, standing in the general election against Gerry Adams after forming a group called "Human Rights 1997" focused on the sole issue of punishment attacks and calling for the exclusion of Sinn Féin and loyalist parties from negotiations under the rubric of another organisation called the Northern Ireland Human Rights Association. See L Kennedy "Nightmares within Nightmares: Paramilitary Repression within Working Class Communities" in L Kennedy (ed) *Crime and Punishment in West Belfast* (The Summer School, West Belfast, 1995) pp. 67–80 and "New Candidate Accuses Sinn Fein over Human Rights Abuses" *Irish News*, 26 July 1997.

of considerable negative publicity. As a state-funded lobbying group, it has been consistently criticised by nationalist and republican commentators in particular as being a propagandist organisation for the Northern Ireland Office.[59]

FAIT's high media profile has also ensured that the group's organisational and personnel difficulties have been extensively documented by the local media. As the organisation's last Director colourfully suggested: "if someone breaks wind in the FAIT office, you can guarantee some journalist is going to write about it".[60] Despite the groups ostensible non-political stance, several leading staff members have been dismissed because of explicitly political activities.[61] The group has also been accused of breaching the confidentiality of the victims of terrorist abuses.[62] When asked why the group had such a chequered

[59] As one Sinn Féin Councillor suggested after FAIT allegations of IRA involvement regarding a shooting incident in West Belfast in which a young man was shot in both legs: "It is clear from FAIT's comments that they are more interested in scoring cheap political points against republicans than in dealing with the issues of the attacks on the nationalist people by criminal elements. FAIT have made this claim without a single shred of evidence to substantiate it. Those genuinely involved in the real work on the ground within the community to alleviate problems arising out of a policing vacuum will question the sole aim and only purpose of this group, those behind it and those who fund it", cited in "FAIT Denies Scoring Points off Sinn Fein" *Irish News*, 16 February 1996.

[60] Interview with Sam Cushnahan, Director of FAIT, 1 July 1999. The group's co-founder Nancy Gracey was forced to resign from the executive after she allegedly used FAIT resources for a three week paid holiday in the USA, after which she promptly established a rival group initially called "United Against Intimidation and Violence" and later renamed "OUTCRY". See "Nancy Gracey Sets Up New Group" *Belfast Telegraph*, 14 May 1996; "Outcry to Tell First Lady of Attacks" *Irish News* 29 August 1997. Further financial and managerial irregularities also led to several investigations by their statutory funder the Northern Ireland Central Community Relations Unit and the RUC: "Union Asks for NIO to Investigate FAIT" *Belfast Telegraph*, 21 October 1998.

[61] Former Spokesperson Glyn Roberts, an Alliance Party activist, resigned after accusations that he was using the group to build a political platform for himself ("FAIT is Rocked by Former Director's Attack" *Belfast Telegraph*, 28 September 1998). His replacement, a controversial figure called Vincent McKenna who had himself previously stood as independent candidate, was in turn dismissed after he had explicitly aligned the group with the unionist anti-Agreement camp, campaigned against prisoner releases and named a number of suspects he claimed were responsible for the Omagh bombing ("Yes Voters Cheated: FAIT Man Could Not Back Deal" *Newsletter*, 1 January 1999). McKenna then also established his own lobby group "the Northern Ireland Human Rights Bureau", which, despite a highly critical BBC documentary focused on McKenna's credibility, still make sporadic interventions on paramilitary abuses in the media until his conviction for sexually abusing his daughter in late 2000.

[62] "FAIT Faces Breakup After US Scandal: Future is Bleak Say Members" *Irish News*, 29 July 1997. FAIT's Director Sam Cushnahan acknowledged that there was a tension between service provision to those under threat to paramilitaries and encouraging such individuals to criticise paramilitaries publicly at FAIT organised press conferences. However he argued that while ultimately it was for the victim themselves to

organisational and personnel history, Director Sam Cushenanan candidly linked its failings to the fixation with accessing the media:

> "There were two downsides to our success in gaining such access to the media. Firstly was that it encouraged an obsessive interest in us so that all of our mistakes were aired very publicly. Secondly that the guaranteed access attracted a certain kind of people to our staff, people on an ego trip or building their own political career, and they did a lot of damage to our credibility . . . We probably get more respect now internationally than we do at home".[63]

In sum, its interventions were often explicitly politically partial (e.g. in aligning themselves with the anti-Agreement lobby), almost entirely media focused and engendered a common scathing attitude regarding the group's lack of credibility in both republican[64] and loyalist communities. On the loyalist side, one former activist also highlighted their lack of credibility and suggested that FAIT may have exacerbated the situation in some instances with regard to the loyalist paramilitaries.[65] Despite such difficulties, they succeeded in their objectives of raising the political profile of punishment attacks and proving a political irritant to both Sinn Féin and the loyalist parties.

A number of lessons of more general applicability arise from the activities of groups such as FAIT in the Northern Ireland context. First, the lack of substantive engagement in the human rights or humanitarian law framework arguably removes a necessary anchor for any group which is operating in such a highly politically contentious arena. Without such an external reference point, any "human rights" group will be susceptible to the charge of being politically motivated, of

decide, he felt that often such publicity could provide "an extra suit of armour" around such victims: Interview, Sam Cushnahan, Director of FAIT, 1 July 1999.

[63] Interview, Sam Cushnahan, Director of FAIT, 1 July 1999.

[64] "They [FAIT] were only interested in anti-republican propaganda. Our ones [the IRA] would have thought they were a bit of a joke. They did not provide any realistic alternative to punishments other than saying support the RUC". Interview, republican spokesperson, 1 July 1999.

[65] Q. "How would the UVF have viewed FAIT ?" A. "They would laugh at them . . . They are sitting in a city centre office, generally middle class people that haven't experienced life in working class neighbourhoods, trying to moralise on issues that affect them without realising that it is the community people themselves who are approaching the paramilitaries . . . They did everything in a very public way. They used the media a lot for condemnations, but you are never going to get honest dialogue with the paramilitaries when you take that kind of approach. In actual fact some people would say that it had the opposite effect where the paramilitaries say stuff them, well shown them": interview, former UVF prisoner, 29 June 1999.

building alliances with the groups who are in opposition to the paramilitary groups, and this may diminish their own credibility as an impartial player in the political arena. Secondly, while the media will often happily reproduce criticisms of the activities of non-state groups, over-exposure to the media may have negative consequences on the quality of staff and the organisational image. Without an explicitly impartial framework such as international humanitarian law, journalists too will begin to question an organisation's motivation and political allegiances. Even with such a framework, any organisation which continuously find the media amenable to presenting their claims-making efforts can expect to pay a price in close media scrutiny of their own affairs. Thirdly, a *modus operandi* which means that all work is carried out through the media imposes obvious limitations on an organisation's effectiveness. The vast majority of individuals who came to FAIT for assistance were referred to another project BASE 2 (discussed below). No attempts at pragmatic dialogue with paramilitaries or their political wings were attempted. A refusal to engage in such a fashion may be indicative of a lack of organisational confidence.[66] FAIT's only engagement with Sinn Féin and the loyalist parties had been in a blatant attempt politically to embarrass those groupings, where the media were brought along with them to the meetings and the parties were invited to sign an anti-violence charter condemning punishment attacks.[67] While such interventions can have an impact upon the political context within which non-state entities operate, they provide little in terms of influencing the policies or practices of such groupings, nor indeed provide such groupings with viable alternatives to their abusive practices.

Direct engagement by humanitarian groupings seeking to change paramilitary behaviour

In the past thirty years a variety of community, voluntary and statutory services have developed policies and working practices around the contours of the political conflict and the material reality of paramilitaries

[66] When asked why FAIT had not tried such pragmatic engagement, FAIT's former Director suggested that the ability of paramilitary groups and their political representatives to "co-opt or usurp the focus of community or lobby groups" had been highlighted in other settings and FAIT did not wish to take such a risk. Interview, Sam Cushenhan.

[67] "Fait Urges Sinn Fein to Endorse New Charter" *Belfast Telegraph*, 23 March 1998.

in many working class areas. Indeed it has often been necessary for groups in such areas to have at least the acquiescence of the paramilitaries in order properly to carry out their functions. For example, those who work with offenders, in crime prevention projects, who provide diversionary services to young people and who work in a range of other community development initiatives inevitably come into contact with both paramilitaries, their supporters and those under threat from the paramilitaries. In such circumstances, a number of programmes have emerged to provide services to those under paramilitary threat and to seek to mitigate the nature and consequences of paramilitary behaviour. These have included cross-community programmes (aimed at republicans and loyalists simultaneously) as well as initiatives focused on republicans and loyalists respectively.

BASE 2: an intervention with both republicans and loyalists

In 1990 a programme called Base 2 was established by the Northern Ireland Association for the Care and Resettlement of Offenders (NIACRO), a largely state-funded voluntary agency. Base 2 was designed to intervene with those under threat from paramilitaries and to provide such individuals with a discreet and confidential service.[68] Regarding themselves as a "humanitarian crisis intervention response",[69] they have largely eschewed publicity or media focus upon their work. Base 2 handled an average of 200 clients a year up until 1994 and experienced a dramatic increase in numbers after the cessations as paramilitaries relied more upon banishments rather than physical punishments. Taking referrals from a variety of sources, Base 2's role has been to clarify with loyalist and republican groupings the nature of the threat, provide advice, transport and accommodation out of the community to those under threat and to serve as a reintegration service if a period of paramilitary banishment is time limited. While staff clarified the nature of a threat with paramilitaries, no negotiation with regard to tariffs (for example shooting in one leg as opposed to two) was permitted and all dealings were premised on the organisation's implacable opposition to punishment violence.

BASE 2's work has clearly had an impact in preventing well over 2,000 individuals from being beaten, shot or killed over the past nine

[68] P Conway "A Response to Paramilitary Policing in Northern Ireland" (1997) 8 *Critical Criminology* 109–22.

[69] Base 2 *Annual Report of the Base 2 Project* (Belfast, NIACRO, 1997) p. 1.

years.[70] The paramilitaries accept that they have increasingly come to view it as "a resource" which prevents them from otherwise carrying out punishment attacks.[71] For the range of statutory, voluntary and community organisations who work with those who are under threat, many suppressed their professional and ideological misgivings that such an initiative (funded largely by statutory monies) was colluding with a violent and unlawful system of paramilitary violence. The fact that the alternative was to leave many such individuals at the mercy of the paramilitaries has, until recently, proved a more persuasive argument.

Community restorative justice : an intervention with republicans

In 1996 three criminal justice practitioners and the current author[72] were approached by a number of republicans to provide a training programme on issues relating to "informal justice". That programme was conducted over a seven-week period and included international and historical examples of alternative and restorative justice, human rights, crime prevention, mediation and non-violence. It was premised on a continued emphasis by the trainers of their opposition to violent punishments. This training programme was followed by a residential course wherein the trainers and trainees drafted the outline of a model of non-violent and lawful community-based alternatives to punishment beatings and shootings. That model was written up into a draft report, circulated to a number of relevant bodies, and a revised version published in December 1997.[73]

That document was fully endorsed by Sinn Féin. Following publication of the report, funding for four pilot projects in republican areas and a coordinator position for three years was achieved in September 1998 under the management of NIACRO. The IRA has expressed its support for these community-based restorative justice projects as

[70] Base 2 *Annual Report of the Base 2 Project* (Belfast, NIACRO, 1999).

[71] A republican spokesperson interviewed for this research suggested that their willingness to engage with the project in 1990 happily coincided with a decision in the late 1980s that republicans should be more actively engaged with individuals and agencies outside the movement, reasoning (probably correctly) that such engagement probably carried significantly more professional risks for such bodies than for the republican movement: Interview, republican spokesperson, 1 July 1999.

[72] All of those involved were either then or had previously been employed by the Northern Ireland Association for the Care and Resettlement of Offenders.

[73] Auld *et al supra* n. 49.

mechanisms for their "responsible disengagement" from punishment attacks.[74] Sinn Féin stressed community restorative justice as one of their five key demands in their submission to the Patten Commission on Policing[75] and restorative justice is now invariably cited as the response to questions on punishment beatings and shootings. The pilot projects are now operational. Over 200 people have gone through the introductory six to seven week training course, and there are demands for further training as well as for the establishment of further pilot projects in ten to twelve areas across Northern Ireland.[76]

The "alternatives programme": an intervention with loyalists

In 1996 a former UVF prisoner was commissioned by NIACRO to carry out research on punishment attacks by loyalist paramilitaries in the Greater Shankill area of West Belfast.[77] The research was facilitated by EPIC, the Ex-Prisoners Interpretative Centre, a self-help reintegration project which works with UVF and Red Hand Commando (RHC) prisoners.[78] The researcher's position as an ex-prisoner and the support of the reintegration project facilitated crucial access to the UVF as well as their political wing, the Progressive Unionist Party.[79] The research revealed that there was a willingness by the UVF leadership and the Progressive Unionist Party to explore viable alternatives to

[74] "The IRA want these programmes to work because for almost 15 years now republicans have been saying that punishment shootings are not a solution to the problems of petty crime and anti-social behaviour . . . This is a community problem and while IRA action may arrest it temporarily, ultimately the community must take a lead in tackling anti-social behaviour . . . We want people to support the Restorative Justice approach by bringing their problems to the dedicated and highly trained workers operating the programmes rather than to the IRA": IRA spokesperson interviewed in "IRA Pledges Support For Community Justice Plan" *Andersontown News,* 29 March 1999.

[75] Sinn Féin *Sinn Fein Submission to the Independent Commission on Policing* (Belfast, Sinn Fein, 1998).

[76] For further critical reflection on these developments see K McEvoy and H Mika *supra* n. 49.

[77] T Winston "Alternatives to Punishment Shootings and Beatings in a loyalist Community in Belfast" (1997) 8 *Critical Criminology* 122.

[78] The Red Hand Commandos are a small loyalist paramilitary grouping who have always been closely aligned to the larger Ulster Volunteer Force.

[79] "An outsider couldn't have done it. Absolutely not. For a start they wouldn't have been talking to anyone, no one [from the UVF] would speak to them. It had to come from the ranks of former combatants of the organisation, it had to come from people who had credibility within the organisation, people who wouldn't have been seen as suspicious, whose motives were not in question": Interview, former UVF prisoner, 29 June 1999.

punishment attacks, albeit with certain exceptions imposed by the former. These included interfactional disputes or drug-related activities, internal disciplinary matters, sexual offences and particularly violent attacks.[80]

With those limitations, a project based on the principles of restorative justice has been established known as Greater Shankill Alternatives directed by the former prisoner who conducted the original research. With some differences, the project programme is not dissimilar to that in republican areas wherein complaints are investigated, mediation sessions established between victims, offenders and families, restitution or reparative work agreed, networking with available statutory and community provision (who are represented on the management committee) and a system of mentoring and support for offenders going through the system. The project liaises directly with the relevant paramilitary groups. This project is also now considering expansion to a number of other different venues across Northern Ireland.[81]

The impact of humanitarian groupings directly engaging with paramilitary groupings

Perhaps unsurprisingly, the impact of humanitarian groups that intervened in this fashion are the easiest of the intervention styles to assess. In each instance, material social structures have been developed which have directly and incontrovertibly prevented or mitigated human suffering at the hands of the paramilitaries. A number of features of broader interest is shared by the approach of these three programmes and are key to their obvious success in preventing humanitarian abuses by paramilitary groupings.

Pragmatic engagement with paramilitaries

One key feature which the BASE 2, Community Restorative Justice and Shankill Alternatives share is a willingness to engage pragmatically with paramilitary groupings. All such engagements have been done privately outside the glare of the media spotlight. While all the projects

[80] Interview, former UVF prisoner, 29 June 1999.
[81] Interview, former UVF prisoner, 29 June 1999.

have shared an opposition to punishment violence, and this has been made clear to the paramilitary groupings, dialogue has still taken place which accepts as a reality the place of such paramilitaries in local communities and the more complex dialectic between the paramilitaries and their communities. As noted earlier, such a relationship is not viewed simply as one of repression, control and fear (although such factors are clearly relevant in certain circumstances) but also acknowledges notions of "responsibility" amongst paramilitaries for protection of their communities and a parallel culture of dependency in the community that anti-social crime should be "sorted out" by the paramilitary organisations. They are also premised upon an acknowledgement to a greater or less extent, that formal state policing has been problematic in both loyalist and republican communities.[82]

Such engagement requires, particularly for professional organisations, a willingness to engage in politically sensitive and clear organisational "risk-taking" activities. NIACRO, which has directly managed two of the projects and has been an important player in the third, is a large NGO with most of it income coming from state sources. For any such group to engage directly with paramilitaries requires considerable organisational self-confidence (compared for example to FAIT), a clear rationale based on agreed principles for the engagement, and a willingness to counter any inevitable negative media or governmental comment in pursuit of those objectives.

The localisation of the paramilitary constituency

Another key theme which emerges in the work of these projects is a more clear understanding of the nature of paramilitary constituencies. The most familiar methodology for human rights and humanitarian activists is to seek to embarrass governments about their human rights violations in the eyes of significant "stakeholders". Such stakeholders may include international fora, important trading nations, local and international media or oppositional groupings within the state in question.

While such a focus is important with regard to the political wings of paramilitary groups such as Sinn Féin who have developed a significant national and international constituency, it is important to remember that for paramilitary groupings those whose views are considered sig-

[82] See R McVeigh *Harassment by the Security Forces: Its Part of Life Here* (Belfast, CAJ, 1994).

nificant are often considerably more localised. Their primary interaction is with other paramilitaries and with members of their own community, many of whom may be supportive of their activities. Traditional techniques and strategies of embarrassment, encouragement or affirmation designed to moderate or alter paramilitary behaviour may have to be adapted to suit this much narrower and more localised canvas.

For example, one of the key strategies employed by the Community Restorative Justice project has been to utilise existing community fora, organisations, residents groupings and other community structures to explain the concepts of restorative justice and the content of the programmes. Literally hundreds of meetings have been held in community centres, parish halls and other local settings. A key concern for republicans, which has informed the work of the CRJ activities, has been to underline that they are not "abandoning" their responsibility to protect the community from anti-social or criminal activity[83] but rather empowering local communities to tackle these issues themselves.

To counter such concerns, numerous articles and features have appeared in local community newspapers, republican newsheets, housing association newsletters and other outlets explaining the republican movement's support for Restorative Justice and its usefulness in the community. In one of the project areas, every household of several thousand families were distributed with a copy of the *Community Charter*, an explanation of the project and an invitation to become involved. Such a strategy is a patient and time-consuming exercise which takes on board the concerns of the paramilitaries that they are seen to "disengage responsibly" and address those concerns by ensuring that their constituency is both encouraged along the same pathway and feels some sense of ownership over that process.

Similarly with regard to the notion of "embarrassment" amongst loyalist paramilitaries, criticisms focused solely on the brutality of punishments are less likely to be impactive in communities which support

[83] Republican dissidents opposed to the peace process have viewed the IRA's commitment to ending punishment violence as a vulnerable political flank of the mainstream movement. "Community Restorative Justice is British double speak for collaboration with Crown Forces . . . NIACRO is dedicated to recruiting ex-prisoners into a new police force which will serve as an auxiliary wing of the RUC. . .It is clear that the establishment of a new British police force in the guise of community justice is the initiative of a British colonial agency operating from Stormont": cited in "Blue Book for New British Police", *Saoirse*, September 1998. Saoirse is the magazine of republican Sinn Féin, the political wing of the Continuity IRA, a republican organisation not currently observing a ceasefire.

such actions. However, the involvement of their immediate paramilitary rival groupings, the UVF and RHC, in a project on the Shankill designed to intervene on paramilitary punishments, has clearly impacted upon the decision of the UDA also to begin the process of establishing a similar programme. All the paramilitary organisations are involved in a relationship with the same community in a relatively small area and clearly one organisation does not wish to be viewed as out of step with the activities of others, in particular when such a programme is gaining increased credibility at a local and national level.

In sum, by identifying in a cold-eyed and pragmatic fashion the needs and fears of the paramilitary groups in relation to their own very localised constituencies, more nuanced strategies can be developed to build upon humanitarian interventions.

The role of former prisoners in establishing and maintaining credibility

A third point which has been crucial in the development of the Community Restorative Justice and Greater Shankill Alternatives programme has been the involvement and support of former prisoners in such initiatives. After a thirty year conflict, there are literally thousands of formers prisoners in republican and loyalist communities in Northern Ireland.[84] Many such individuals are highly respected in their own communities, they have returned from prison with enhanced educational and organisational abilities and play a crucial role in the process of peace-building at both a community and political level.[85] Crucially, as outlined above, such former combatants also have credibility within the paramilitary organisations and have proved an invaluable resource in both gaining access to such groupings and lobbying within paramilitary circles for less violent forms of actions.

In sum therefore, a willingness to engage pragmatically with paramilitaries outside the glare of the media; a sensitivity to the complexity of the relationship between paramilitaries and their communities; and the utilisation of the skills and credibility of former combatants; these have been the key ingredients in the success of these interventions to date.

[84] Republicans estimate that there are over 15,000 ex-republican prisoners alone: *Cosite na n-iarchimí Annual Report 1999* (Belfast, Cosite na n-iarchimí, 1999).

[85] K McEvoy "The Agreement, Prisoner Release and the Political Character of the Conflict" (1999) 22 *Fordham International Law Journal* 145–81.

Creating a human rights culture

The fourth style of practical work which has arguably impacted on the behaviour of paramilitary groupings in Northern Ireland has been that carried out by human rights organisations which have, perhaps paradoxically, been primarily focused on the activities of the state. While opposed to the use of political violence, such groups have not prioritised paramilitary abuses and focused instead on the primacy of the state's responsibility to ensure the effective protection of human rights. Whether this can be properly described as an intervention is discussed below. While there are many groups who could be considered, the best example of this style of work is Northern Ireland's primary human rights NGO, the Committee on the Administration of Justice (CAJ).

CAJ was established in 1981 and has worked on a wide variety of rights issues including prisoners' rights, emergency laws, children's rights, gender equality, racism, disability, a campaign for a Bill of Rights, policing and the operation of the criminal justice system.[86] The debate regarding CAJ's role with regard to the behaviour of paramilitaries was initially located almost exclusively within the humanitarian law framework.

In 1991, in the context of Amnesty International's extension of their remit to non-state entities, CAJ underwent considerable internal debate as to whether it should follow the Amnesty example.[87] The

[86] CAJ has well developed relations with other domestic and international human rights groups such as Amnesty International, the Lawyers Committee for Human Rights and Human Rights Watch. CAJ has a cross-community membership of over 300, takes no position on the constitutional status of Northern Ireland and is unequivocally opposed to the use of political violence. The organisation has been awarded a number of international human rights prizes, including the Reebok Human Rights Award and the 1998 Council of Europe Human Rights Prize. The author has been involved on the management committee of CAJ since 1992 and served in a range of capacities including as Chairperson between 1997–9.

[87] The debate was reflected in the various positions argued in the Committee's newsletter *Just News*. Summarising for the sake of brevity, the case for a change in the remit was that CAJ should follow Amnesty's example; that the embarrassment of publicity might pin the blame on those guilty of hostage-taking and assassinations; that fewer people would be "put off" by the apparent one-sidedness of human rights groups; and that a sole focus upon abuses by the state was only part of the picture and therefore imbalanced. See, e.g., (1992) 7 *Just News* "Humanitarian Law: The Case for a Change of Remit" (February No 2). The case for retaining the CAJ position of non-monitoring was that humanitarian law could not be applied to the Northern Ireland conflict; that to do so would offer some legitimacy to some paramilitary protagonists and actions and that the notion of a "legitimate target" ran contrary to the organisation's essentially pacifist

issue was finally decided at an extraordinary general meeting of the CAJ membership in late 1991. After considering the respective arguments, the membership voted overwhelmingly to reiterate its opposition to political violence but to refrain from action on non-state abuses. CAJ's position has remained unaltered to the present day and its spokespersons remain convinced that that decision is the correct one for a human rights NGO in the Northern Ireland context.[88]

The impact of NGOs focused upon creating a human rights culture

Despite its position of non-monitoring paramilitary violence (other than condemning such violence when asked by the media), CAJ strongly dispute that its position represents an abdication of responsibility or that it had no impact upon the behaviour of paramilitary groupings in Northern Ireland.[89] Their experience also raises a number of thematic points of broader relevance to understanding how such groups may nonetheless have an impact.

Political transition and internalisation of rights discourses

In its attempts at creating "a human rights culture in Northern Ireland", CAJ argue that it has been very successful in developing networks and relationships with a wide variety of political parties, including the political representatives of republicanism and loyalism. It suggests that the adoption of human rights discourses by such group-

position of *complete* opposition to political violence; that the application of humanitarian law would lead to accusations of selective condemnation in accepting attacks on the security forces; and finally that an extension of the mandate would dilute CAJ's focus on the state and its culpability in the conflict in persistently eroding the rule of law. See, e.g., "Humanitarian Law; Not as Simple as it Seems" *Just News* (December 1991).

[88] "Applying humanitarian standards would mean that there were legitimate targets and there were illegitimate targets. That would have draw CAJ automatically into a situation where we would be criticising the killing of civilians but not those of soldiers or police officers. Because CAJ is opposed to all violence, we don't have to make those kind of distinctions. Furthermore, making those kind of distinctions would mean criticising loyalists much more that republicans . . . From the public perspective we would have been slating the loyalists and not the republicans, I can't really see how we could have continued to function. We would have ended up criticising the state and loyalist paramilitaries, and saying to republicans that it is OK to kill soldiers and police—it would have been crazy": interview, Maggie Beirne, Research and Policy Officer CAJ, 28 June 1999.

[89] Interview, Maggie Beirne *ibid*.

ings has not only altered the political landscape regarding human rights generally in Northern Ireland, but *may* also have had some impact on the behaviour of the paramilitary groups themselves:

> "Regarding the Good Friday Agreement, if it wasn't that Sinn Fein and the Loyalist parties had made such a big deal about human rights, then its difficult to believe that you would have got such strong human rights protection in the Agreement . . . it is difficult to say the extent to which it [the human rights debate concerning the agreement] may have influenced the behaviour of republicans and Loyalists on the ground".[90]

I have argued elsewhere that the assertion of the "rights and equality agenda" has become a defining characteristic of modern Irish republicanism.[91] That transition has seen Sinn Féin (and to some extent the loyalist parties) draw extensively upon materials produced by CAJ and similar groups in putting forward their respective positions on rights and equality.[92] The adoption of such political parties of the human rights agenda has arguably done much to both highlight the disparity between their own political rhetoric regarding such discourses and the practices of their military wings, as well as leaving such groupings more amenable to seeking alternative and more humane ways of conducting their affairs. While critics might argue that this is an overly slow process of internalising a respect for rights by osmosis, a view that republicans would strongly dispute,[93] it does represent a significant contribution to changing the paramilitary "environment" at least.

Other than their willingness to engage in the restorative justice initiatives in the area of punishment attacks, a further illustrative example of this point has been the question of the republican movement's position with regard to the "disappeared".[94] In urging the IRA to

[90] Interview, Maggie Beirne *ibid.*

[91] For a discussion on changes amongst paramilitaries with regard to legality see K McEvoy "Law, Struggle and Political Transformation in Northern Ireland" (2000) *Journal of Law and Society* 27, 4 542–571.

[92] See Mageean and O'Brien *supra* n. 3.

[93] Republicans argue that the desire for justice has always been part of their campaign. They point to the involvement of republicans in the 1960s civil rights campaign and argue that it was the unionist Government's violent reaction to peaceful protest for civil and political liberties which led to republicans being "forced" to engage in armed struggle: interview, republican spokesperson, 1 July 1999.

[94] This issue refers to a number of individuals, some of them members of the republican movement, who were abducted by the IRA in the 1970s and early 1980s, interrogated and murdered and their bodies buried and not returned to their families. Following a high profile campaign for the return of the bodies, legislation was introduced North and South of the border guaranteeing that the bodies would not be tested for forensic

respond to the plight of the families, albeit under severe political pressure, Gerry Adams and other Sinn Féin spokespersons began to refer to the matter as "a human rights issue".[95] For republican leaders (renowned for the care of their language) to suggest that the matter was "a human rights issue" was indeed significant. Normally abuses by republicans have been either justified, described as wrong, contextualised or otherwise explained by Sinn Féin leaders. To the author's knowledge, this is the first time that a senior republican figure has suggested that a republican act constituted a human rights abuse. This is not to suggest any careful distinction between human rights and humanitarian law per se, but rather that such language reflects some degree of genuine internalisation of human rights discourses beyond the rhetorical claims making process which Sinn Féin have so successfully pressed as part of their political struggle.

Acknowledging the limitations of the human rights paradigm

The experiences of an organisation like CAJ are also instructive in highlighting the need by human rights actors to acknowledge the limitations of the human rights paradigm. In a conflict situation, there is sometimes a tendency to conflate human rights with conflict resolution. While human rights abuses clearly may add to a political conflict, and creating a just and fair society is an intrinsic part of peace-building, it would be wrong to overstate the peace-making potential of the human rights framework.

For example, in the Northern Ireland context, many commentators felt that the early release of paramilitary prisoners was a crucial component to the conflict resolution process.[96] However CAJ did not take a position on the issue because it is not contained within international human rights standards. Similarly with regard to the issue of disputed marches, while it may be conceptually useful to remind both marchers

evidence and establishing a commission. The IRA announced that a specialist unit tasked with pulling together information on the matters had garnered knowledge on the whereabouts of nine bodies. At the time of writing, after a massive excavation operation at a number of sites in the Irish Republic, only three bodies have been uncovered so far. See "Nine Victims The IRA Admit To Taking And Killing" *Belfast Telegraph*, 28 May 1999; "Disappeared Legislation Within Weeks" *Belfast Telegraph*, 16 April 1999.

[95] "Alliance Slams Adams 'Human Rights' Remarks" *Irish News*, 31 May 1999.

[96] For an overview see K McEvoy "Prisoner Release and Conflict Resolution: International Lessons for Northern Ireland" (1998) 8 *International Criminal Justice Review* 33–61.

and residents groups that one element of their dispute is a clash of rights, the human rights framework can make no pretence to have all the answers in such a dispute.

Human rights actors tend by their very nature to be dynamic and energetic in seeking to resolve the problems of their and other societies. Such clear strengths need to be tempered and focused, however, by an acknowledgement that both human rights and humanitarian law are limiting frameworks within which to achieve those objectives. Many actors will understandably and laudably wish to focus their time and energy on peace-building or conflict resolution efforts beyond the human rights or humanitarian law framework. However, rather than stretching and expanding that framework to achieve a task for which it is at best only partially suited, consideration might be given to addressing such matters under the explicit rubric of peace or conflict resolution groupings and enjoying the considerably increased flexibility that such a move would bring.

CONCLUSION

Based upon the Northern Ireland experience, I would contend that humanitarian groupings which have engaged directly with paramilitary groupings have played perhaps the most significant role in affecting the behaviour of the latter. Clearly the most impactive of these interventions have occurred in a political context where both republicans and loyalists were actively seeking alternatives to political violence. That said, they have clearly saved lives and protected individuals from broken or mangled limbs. In terms of the other typologies of interventions discussed, their influence has been less directly tangible and perhaps more significant with regard to altering the political "environment" in which paramilitaries operated rather than their actual behaviour. Such a conclusion is perhaps to be expected.

The standard techniques of human rights groups involve campaigning work with the media, with civil society, with other political actors and with the international community. Except where it is necessary for the safety of the individuals involved (in extremely repressive situations), most human rights NGOs work in a very transparent and open way. Indeed human rights groups often secure their safety and their integrity by being deliberately open in their exchanges across the political spectrum. In contrast, for example, to mediation and other groups, their work is one of using public avenues of influence and pressure, rather than confidential and intensely private lobbying. These latter techniques are normally foresworn for fear that they could be mis-used to challenge the *bona fides* and integrity of the criticisms made.

One could imagine, on the basis of the typologies discussed earlier, for example comparing the work of FAIT, to Base 2, that the techniques likely to be used by groups working on humanitarian principles may well be of a very different nature. The basis of intervention is one of humanitarianism; if it is not to be exploited for political gain by any of the parties to the conflict, it seems that interventions would be more effective if treated with absolute confidence. Looking back to all the interventions (often by individuals rather than groups) in the developments of the cease-fires, on the matter of the disappearances, or on the matter of punishment beatings, most success seems to be recorded when the avenues of influence are "back channels" or otherwise not subjected to intense public scrutiny.

The parallels (though obviously uneven) between the International Committee of the Red Cross and major international human rights organisations (Amnesty International, Human Rights Watch etc.) spring to mind. It is not accidental that the techniques used to promote humanitarian standards have been of a different order to those used to promote human rights. Perhaps we need to examine whether specific national parallels of the ICRC approach would be worth promoting to complement the work that local human rights groups are doing. Certainly the current emphasis, which appears to be on human rights groups extending their remit to work on humanitarian law, appears somewhat illogical, given that the skills, principles and tactics of engagement are of a different order. It might therefore be useful to consider what an effective "humanitarian" NGO could usefully do; what operating principles would be important; and how would/could work with others such as human rights NGOs. In the final analysis, humanitarian interventions may well require a distinctive philosophy and series of techniques and practices in encouraging non-state entities such as paramilitary groups to end or reduce their abuses.

10

Democracy, Governance and Governmentality: The Role of the Voluntary Sector in the Democratic Renewal of Northern Ireland

JOHN MORISON*

INTRODUCTION

This chapter seeks to explore the possibilities for constitutional renewal in Northern Ireland below the level of large-scale institutional settlement. It will draw upon the governmentality approach associated with the later writings of Michel Foucault,[1] to reinforce in the circumstances of Northern Ireland the general argument that:

> "if you try to analyse power not on the basis of freedom, strategies, and governmentality, but on the basis of the political institution, you can only conceive of the subject as a subject of law. One then has a subject who has or does not have rights, who has had those rights either granted or removed by

* Professor of Jurisprudence, School of Law, Queen's University Belfast. The author would like to acknowledge helpful advice and assistance from a range of figures in government and in the voluntary sector including particularly Seamus McAleavey and Jacqui Irwin of NICVA as well as from Don Harley and Anne O'Keeffe from the Voluntary Activity Unit. I am grateful also to Ray Geary with whom I have had useful discussions and to Tim Cunningham and to the editor, Colin Harvey. A version of the paper appears in 30 *Oxford Journal of Legal Studies* (2001).

[1] See, e.g., M Foucault "Governmentality" in G Burchell, C Gordon and P Miller (eds) *The Foucault Effect: Studies in Governmentality* Hemel Hempstead, Harvester Wheatsheaf, 1991) pp. 87–104 and L Martin, H Gutman and P Hutton (eds) *Technologies of the Self: A Seminar with Michel Foucault* (London, Tavistock, 1988). As Rose expresses

> the institution of political society; and all this brings us back to a legal concept of the subject. On the other hand, I believe that the concept of governmentality makes it possible to bring out the freedom of the subject and its relationship to others—which constitute the very stuff [matière] of ethics".[2]

The insights of this governmentality approach put less emphasis on ideas of high constitutionalism, where settlements are brokered and structures are imposed, and stress instead the importance of the active subject as the entity through which and by means of which power is actually exercised beyond traditional state boundaries. All of the work associated with the governmentality approach engages with the idea that, as Hunt and Wickham observe, there were dramatic changes in the techniques (and objectives) of government developed in the Western world from the eighteenth century onwards. As they see it, within Foucault's approach, "modernity . . . is marked by the emergence of 'government' and 'governmentality' ". Foucault is deploying "the term 'government' in a very different sense from the conventional idea of state executives and legislatures" and in a way that is consistent with his downgrading of the importance of the state.[3] Instead of state

it, "to analyse political power through the analytics of governmentality is not to start from the apparently obvious historical or sociological question: what happened and why? It is to start by asking what authorities of various sorts wanted to happen, in relation to problems defined how, in pursuit of what objectives, through what strategies and techniques": *Powers of Freedom: Reframing Political Thought* (Cambridge, Cambridge University Press, 1999) p. 20. The governmentality approach has had an impact across a whole range of social science research. See, for example, N Rose and P Miller, "Power Beyond the State: Problematics of Government" (1992) 43 *British Journal of Sociology*, 173; A Barry, T Osbourne, and N Rose (eds) *Foucault and Political Reason: Liberalism, Neo-Liberalism and Rationalites of Government* (Chicago, University of Chicago Press, 1996); A Davidson (ed.) *Foucault and his Interlocutors* (Chicago, Chicago University Press, 1997); C O'Farrell (ed.) *Foucault: The Legacy* (Brisbane, Queensland University of Technology, 1997); G Pavlich *Justice Fragmented: Mediating Community Disputes under Post-modern Conditions* (London, Routledge, 1996); D Garland "'Governmentality' and the Problem of Crime: Foucault, Criminology and Sociology" (1997) 1 *Theoretical Criminology* 173; M Dean, *Governmentality: Power and Rules in Modern Society* (London, Sage, 1999).

[2] M Foucault "The Ethic of the Concern for Self as a Practice of Freedom" in P Rabinow (ed.), *Ethics: Subjectivity and Truth* (Harmondsworth, Penguin, 2000) p. 300.

[3] A Hunt and G Wickham *Foucault and Law: Towards a Sociology of Law and Governance* (London, Pluto Press, 1994) p. 52. Cf. V Tadros, "Between Governance and Discipline: The Law and Michel Foucault" (1998) 18 *Oxford Journal of Legal Studies* 75–103. Recently Rose has characterised the contribution of the governmentality approach as having "reframed the role to be accorded to the 'the state' in analyses of control and regulation. Centres of political deliberation and calculation have to act through the actions of a whole range of other authorities , and through complex technologies, if they are to be able to intervene upon the conduction of persons, activities, spaces and

action (or rather in addition to it), there is the important quality of the freedom of the subject. Governmental action by itself cannot attain its own ends; it requires the willing cooperation of the individual subject participating in their own governance. As Foucault expresses it, government is "not a matter of imposing laws on men, but rather of disposing of the laws themselves as tactics". This involves understanding how power is actually distributed and organised in a whole variety of networks and strategies beyond the formal structures of the state as traditionally conceived. Within this approach constitutionalism is about ethics – the way we live together – and not simply about legal structures. As Foucault puts it, "the concept of governmentality . . . cover[s] the whole range of practices that constitute, define, organize, and instrumentalize the strategies that individuals in their freedom can use in dealing with each other". Everyone who seeks to govern, "to control, determine, and limit the freedom of others", are themselves "free individuals" who have "at their disposal certain instruments they can use to govern others".[4] Government involves not just the forms of rule by which authorities govern populations but the "technologies of the self" through which people shape their own subjectivity. A proper understanding of power must acknowledge an idea of freedom, of individuals "making themselves up" as active subjects or as citizens capable of bearing a regulated freedom within complex chains of constraints, calculations of interests, patterns and habits, and obligations and fears. Without this, government is simply the imposition of the sovereign will on variously compliant or recalcitrant subjects.

The structures of the Belfast Agreement and Northern Ireland Act 1998 illustrate this general point supremely. While the motives of the various governments may be variously high-minded, the result is a settlement that is politically brokered by higher forces and imposed through a constitutionalism made up of conventional ideas of state executives and legislatures, assemblies and, most of all, laws passed by a sovereign Parliament. As the Northern Ireland Act 2000, which temporarily

objects far flung in space and time—in the street, the schoolroom, the home, the operating theatre, the prison cell. Such 'action at a distance' inescapably depends on a whole variety of alliances and lash-ups between diverse and competing bodies of expertise, criteria of judgement and technical devices that are far removed from the 'political apparatus' as traditionally conceived . . . 'The state' is neither the only force engaged in the government of conduct nor the hidden hand orchestrating the strategies and techniques of . . . all those others seeking to act upon conduct . . . and to shape it to certain ends": "Government and Control" in (2000) 40 *British Journal of Criminology* 323.

[4] Foucault *supra* n. 2.

suspended the Assembly, illustrates all too well, such a settlement can be revoked with a stroke of the same sovereign power that instituted it.

In many ways the Northern Ireland Act 2000 marks out the boundaries of British constitutionalism generally and the limits of even the modified, consocial version developed in Northern Ireland. The whole idea of building an ever more elaborate institutional edifice to include all political elements, while at the same time drawing upon the language of rights to reserve certain matters from the agenda of short-term politics as a method of resolving the problems arising from a wider politics of disagreement, comes to a natural conclusion in the difficulties that have surrounded the Northern Ireland Act 1998 and the Belfast Agreement. The fact that even this most sophisticated structure remains highly problematic suggests that an idea of constitutional regeneration at a much more profound level is probably necessary, at least as an additional underpinning of what has been achieved so far. Whatever the eventual political fate of the Good Friday Agreement, arguably what is required is a genuine *constitutive* change; a transformation at the ethical level where it is the conduct of conduct that is being agreed. Such an idea of constitutional regeneration must be different from and subtler than simply adding on new layers of representation or holding together what is already there through impasse-favouring or consensus requiring arrangements in an Assembly. While the institutions of formal government may not provide a full solution to the political/constitutional issue of "who governs?" they can at least provide a place to "park" the problem. Meanwhile, however, there is still the more deep-seated issue about *how* to govern. It is here that ideas of governance and in particular the role of the voluntary sector may be of interest. It may be that new forms of civil society involvement developing beyond the formal state and its institutions provides an indication of how the values behind the settlement can be developed in all the realities of governance as it actually takes place now.

In seeking to make this argument this account will review briefly the settlement structures and suggest that the role of government may have changed since last there was devolution in Northern Ireland. Next the history of involvement of the voluntary sector in governance in the Northern Ireland context will be outlined to indicate its particular potential for development. Finally, the positive advantages of opening up a new democratic space through developing the role of the sector in the processes of governance will be reviewed and the value of a constitutional renewal project will be considered.

THE STRUCTURES OF SETTLEMENT

The constitutional solution to the historic problem of Northern Ireland is in many ways unique but at the same time parts of it remain thoroughly traditional. It is a British solution *mutatis mutandis,* where the chief aim is to restore institutions of representative democracy in order to provide a democratic space where accountable, self-government can take place against a background of what will be hopefully an economy and society growing in stability and self-confidence in the absence of chronic conflict. The achievement of brokering agreement is generally viewed as a triumph of (British, American and Irish) political will in securing a deal in the unpromising circumstances of recalcitrant local politics. Achieving agreement has required a number of interesting additional structures, such as those for North-South cooperation, intergovernmental relations between Britain and Ireland and relationships among all the people of these islands. It has also necessitated an important human rights and equality agenda (and this is where the uniqueness of the settlement largely lies). A reading of the Belfast Agreement alongside the Northern Ireland Act 1998 also shows the novel feature of a constitutional settlement that recognises that issues of devolved government in Northern Ireland have an associated (and troublesome) additional political agenda of weapons decommissioning, prisoner release, and the reform of policing and criminal justice. However important and unique these features are (and the equality and rights agenda which will be returned to later are of particular value), it is undeniable that the main tangible result of the settlement is a structure—an Assembly and an extensive and expensive administrative structure for devolved government.

More important even than the cost of the elaborate new structures of government for a population of only 1.6 millon is the fact that what is now possible and what is expected of governments generally has changed. One of the more resonant contemporary descriptions of the Belfast Agreement and the structures it promised was that it amounted to "Sunningdale for slow learners".[5] The reference here is of course to the not dissimilar package of legislative assembly, power-sharing executive and all-Ireland bodies that had a very short life in 1974. But while the structures are redolent of that previous exercise in devolution, it is important to remember that the whole project of government generally has changed radically since the 1970s. No longer do governments anywhere

[5] This remark is generally attributed to the Deputy First Minister, Seamus Mallon.

expect to fund, plan and deliver the whole range of social goods to its citizens within its defined territory. The general move from government to governance, with its emphasis on globalisation and the hollowing out of the nation state, is well documented.[6] There is now not only multilevel governance but also multi-form (or multi-format) governance too where the actions of the state are augmented by interventions from elsewhere in the market or the voluntary sector. Government in a 1970s model of parliamentary institutions and departments of state who tax widely, spend high and make big choices has been replaced everywhere by notions of governance where opportunities for making significant changes are more limited and the emphasis is on "steering" rather than "rowing". Northern Ireland can not be expected to be exempt from world-wide changes. Indeed it has been argued elsewhere that although the structures of settlement may provide a more or less satisfactory answer to a political problem, they do not offer a very satisfactory solution to a governmental problem.[7]

The project and aim of government has changed, and although it is necessary to acknowledge an important distinction between devolved government and mere local government, a paragraph from the 1998 Green Paper on modernising local government captures well the new dispensation for government in Northern Ireland where it says:

> "The days of the all-purpose authority that planned and delivered everything are gone. It is in a partnership with others—public agencies, private companies, community groups, and voluntary organisations that local government's future lies. Local authorities deliver important services but their distinctive leadership role will be to weave and knit together the contributions of the various local stakeholders".[8]

This realisation of the facilitating, brokering role of government in circumstances where most of the budget is already pre-spent and

[6] See, e.g., G Majone "The Rise of the Regulatory State in Europe", (1994) 17 *West European Politics* 77–110; D Held *Democracy and the Global Order: From the Modern State to Cosmopolitan Governance* (Cambridge, Polity Press, 1995) and R Rhodes *Understanding Governance: Policy Networks, Governance, Reflexivity and Accountability* (Buckingham, Open University Press, 1997) and A Gray and B Jenkins "From Public Administration to Public Management: Reassessing a Revolution?" (1995) 73 *Public Administration* 17.

[7] See further, J Morison "Constitutionalism and Change: Representation, Governance and Participation in the New Northern Ireland" (1999) 22 *Fordham International Law Journal* 1608.

[8] Department of the Environment, Transport and Regions Consultative Paper, *Modernising Local Government: Local Democracy and Community Leadership* (1998) para. 1.9.

choices are mainly about where exactly to enforce budgetary cuts, in turn brings us back to ideas of governmentality. We are returned to Foucault and his idea of the art of government not being about simply legislatures and executives but addressing questions of "How to govern oneself, how to be governed, how to govern others, by whom the people will accept to be governed, and how to become the best possible governor".[9] In particular this brings us to understanding how power is deployed through a whole series of networks and alliances, non-state and semi-state bodies, coalitions of influence and concentrations of power. Government is thus to be seen as, in the words of Rose and Millar, "a domain of strategies, techniques and procedures through which different forces seek to render programmes operable".[10]

This is the future of government—or rather governance as it might now be better termed to indicate how top-down, direct central government steering is being replaced by ideas of autonomous institutions and groupings at different levels and from different sites within the polity exercising self-responsibility and unforced cooperation within inter-organisational networks and concentrations of power that exist now within civil society and in the private sector as much as in state bureaucracies. Some parts of the new system in Northern Ireland seem ready for this new dispensation. Paradoxically it is in the area of policing, traditionally seen as one of the most basic functions of the classical state, that significant potential development can be seen. The Patten Report on policing is indicative of this new role for formal government. Although the Report is subtitled "Policing in Northern Ireland", it is not about operational policing, but rather a report on the structures that should be implemented in order to allow for policing. It is suffused with ideas of governmentality. The Report stresses that the police should be accountable within "a real partnership between the police and the community—government agencies, non-governmental organisations, families, citizens; a partnership based on openness and understanding; a partnership in which policing reflects and responds to the community's needs".[11] This involves, in the words of the Report, becoming "more decentralised; for the management style, which should become more open and delegated; and for the manner of policing down to beat level, which

[9] Foucault (1991) *supra* n. 1 at p. 87.

[10] Rose and Miller *supra* n. 1 at p. 183.

[11] *A New Beginning: Policing In Northern Ireland, The Report for the Independent Commission on Policing for Northern Ireland* (1999) (The Patten Report) 1:16. It should be noted that the Police (Northern Ireland) Act 2000 offers severely restricted version of the Report's recommendations.

should become more orientated towards active problem-solving and crime prevention, rather than more traditional, reactive enforcement".[12] At the same time the whole report is imbued with stressing the importance of human rights, community policing, partnership, neighbourhood and problem-solving. Management is to become more decentralised with an increased emphasis on ideas of efficiency, effectiveness and accountability. The use of information technology and techniques of audit are strongly endorsed. The Report not only offers a strategy for a radical transformation of policing but also a blueprint for how government generally might see its future role as a facilitator using a variety of forces from different sites in the public, private and third sectors to make things happen rather than seeking to power everything with purely public resources.

While arguing for a view of government that accommodates this more sophisticated understanding of the limits and potential of public power, this account will focus on the role of the voluntary sector in the governance of the new Northern Ireland. After considering the already highly developed nature of the sector in the context of pre-settlement mechanisms of rule, the capacity of the sector to provide an engine of renewal in addition to the Assembly and its various structures will be assessed.

THE EVOLUTION OF THE VOLUNTARY SECTOR IN NORTHERN IRELAND

There is a long history of voluntary activity in Ireland from the end of the eighteenth century.[13] Some have speculated that this degree of development was caused by a politics of exclusion from the formal state which had the result that alternative structures grew up among the nationalist community, and this in turn caused a response in the Protestant community.[14] Certainly there is now a belief that voluntary activity is highly developed in Northern Ireland with, for example, a former Secretary of State on record as estimating the sector to be 25 per cent larger than in Great Britain.[15] The Northern Ireland Council for

[12] *Ibid.*

[13] See further A Williamson "The Origins of the Voluntary Action in Belfast" in N Acheson and A Williamson (eds) *Voluntary Action and Social Policy in Northern Ireland* (Aldershot, Avebury, 1995) pp. 161–80.

[14] J Schense "Creating Space for Change: Can the Voluntary Sector Help End Northern Ireland's Troubles?" (1998) 11 *Harvard Human Rights Journal* 149–85.

[15] During a speech at the AGM and Conference of the Northern Ireland Council for Voluntary Action, Belfast, 20 June 1997, Dr Mowlam suggested that there are over 30,000 full time equivalent staff in Northern Ireland and more than 65,000 volunteers.

Voluntary Action (NICVA) maintains a list of some 5,000 organisations on their database.[16] Of course it is now customary everywhere for former opposition politicians in power to acknowledge links with the bodies who may have nurtured them at earlier stages and claim that the voluntary sector in their jurisdiction is particularly developed.[17] Nevertheless, it is probably true that in the recent past the voluntary and community sector performed a *different* and *wider* role in Northern Ireland than its counterparts in Great Britain and this ranges through service provision to a more engaged policy development role.

Voluntary sector as an adjunct to direct rule: "civil servants without ties"?

The system of "direct rule" which continued for more than twenty-five years, with only a limited interruption caused by the brief restoration of devolution in 1973–74, offered particular opportunities to the voluntary sector in Northern Ireland. Characterised as it was by an absence of a nexus between the local political process and mechanisms of government, direct rule in some ways allowed the sector to act as an alternative site of politics and as an unofficial opposition. In the absence of a local assembly and with only very limited local council involvement[18] direct

See <<http://www.nio.gov.uk/press/970620cc.htm>>. An economist estimates that the scale of government assistance exceeds £150 million per annum and that the services provided by the voluntary sector, in market terms, might be valued as over £200 million, nearly 2 per cent of GDP. See further J Simpson "Government Financial Support for Voluntary Sector Organisations: a Preliminary Analysis and Discussion" in N Acheson and A Williamson (eds) *supra* n. 13 at pp. 181–93.

[16] See also NICVA, *The State of the Sector II: Northern Ireland Voluntary Sector Almanac 1998* (1998) and Department for Social Development, *Consultation Document on Funding for the Voluntary and Community Sector* (April 2000) which suggests that the sector provides employment for 33,000 people and has a gross annual income of £500 million.

[17] In Wales the First Secretary has estimated that there are about 25,000 organisations directly employing some 13,000 people with 1.9 million volunteers contributing the equivalent of 15 per cent of gross domestic product in Wales: *The Official Report of the Welsh Assembly*, 21 July 1999, p. 59. In Scotland too there is a view that the voluntary sector has "a long and proud" tradition. The Deputy Minister for Communities suggests that there are more than 44,000 voluntary organisations and more than 50 per cent of the adult population volunteers on a regular basis with an overall contribution representing 3 per cent of Scotland's gross domestic product: *Scottish Parliament Official Report*, vol. 2, No. 8 (23 September 1999) col. 769.

[18] Local authorities, seen by civil rights campaigners of the 1960s as epitomising unionist domination and abuse of powers, were effectively stripped of all powers except for the most minor in areas of environmental health, tourism, recreation and refuse collection in the McCrory Review of 1972.

rule was characterised by a number of negative features. Legislation was made by orders in council with the effect that law that was primary in substance was made by means that were secondary in terms of the level of scrutiny and debate involved.[19] Much legislation was passed as a "read across" from Great Britain and arguably did not benefit from local input. Phrases like "helicopter rule" or "consular government" capture something of the flavour of a mechanism of government whereby a Secretary of State from the Westminster administration, holding a seat in an English Parliamentary constituency, was brought in to head up a governing structure that depended for local information on the Northern Ireland Civil Service rather than local politicians.

To counter these negative features, and the perceived democratic deficit that they brought, the voluntary sector was to some degree encouraged to become involved in government. In part the sector could bring a degree of local expertise and knowledge. To an extent also the sector may have brought a degree of legitimacy to state action, particularly in politically sensitive areas such as, for example, the work of the Northern Ireland Association for the Care and Resettlement of Offenders with regard to prisoners' families. From the sector's point of view, such a relationship with government had certain advantages. General efforts at depoliticising service delivery by removing it from both local government control and distancing it from direct rule mechanisms provided an opportunity for the sector to become involved in a service delivery role.[20] Some parts of the sector, particularly those staffed by individuals who might well elsewhere have entered political life but who were not attracted by the local political scene, may also have welcomed an opportunity to become more closely involved in a policy development role. Indeed there are suggestions that successive Secretaries of State, even those drawn from the Conservative Party, may have found voluntary sector personnel more familiar and easier to deal with than local party politicians.[21] Sweeney has argued that

[19] See further, B Hadfield "Legislating for Northern Ireland as Westminster" in M Connnolly and S Loughlin (eds) *Public Policy in Northern Ireland: Adoption or Adaptation* , (Belfast, Policy Research Institute, 1990) pp. 55–75.

[20] See, e.g., how provision for personal social services developed through statutory bodies, quangos and voluntary organisations (B Caul and S Herron, *A Service for the People: Origins and Development of the Personal Social Services of Northern Ireland* (1992). This was in line with more general trends (see, e.g., Central Personal Social Services Advisory Committee, *Report of the Sub-Committee on Support for Voluntary Organisation* (DHSS, 1979)) but arguably had particular application in Northern Ireland.

[21] Arguably the special circumstances of Northern Ireland meant that the although the Thatcher revolution did not by-pass the region, its impact there was reduced and uneven.

political fall-out from the Anglo-Irish Agreement in 1985 provided an important additional opportunity for strategists from the community and voluntary sector to influence government and, he maintains, "a cadre of senior civil servants were equally determined to experiment with bold new approaches to tackling community differences".[22] The role of the sector was enhanced further through the establishment of structures to deal with urban disadvantage such as the Belfast Action Teams and Making Belfast Work in 1987 and The Londonderry Initiative in 1989 which were subsequently re-cast with even further voluntary sector involvement. The Department of Agriculture developed a community-based rural development programme in the early 1990s leading to the establishment of the Rural Development Council and, subsequently, a series of local area-based rural regeneration strategy groups. The sector itself initiated the Community Development Review Group in 1989 to review community action and development and this led to government responding through the publication in 1993 of a *Strategy for the Support of the Voluntary Sector and for Community Development in Northern Ireland*[23] and the establishment of a Voluntary Activity Unit to facilitate interdepartmental consideration of issues affecting the voluntary sector and those involved in community development.[24]

Of course it is important not to view the sector as a monolith, and in particular there are significant (if somewhat blurred) distinctions between the voluntary sector and the community sector. The voluntary sector can be viewed generally as being larger, more established and better organised and with closer links to government, even if only in terms of funding. In contrast to the more professionalised voluntary sector, the community sector tends generally to be more locally organised and oriented and run by volunteers. There are also perhaps distinctions between the community relations approach, stressing

(See further, F Gaffikin and M Morrissey *Northern Ireland: The Thatcher Years* (1990).) This greater willingness to spend money on various social problems so as not to exacerbate the effects of the Troubles may have had the effect of ensuring relatively good relations between the voluntary sector and successive direct rule administrations.

[22] P Sweeney, "A View from the Voluntary Sector" in *People and Government: Questions for Northern Ireland* (Belfast, Joseph Rowntree Foundation and Chief Executives' Forum, 1998) p. 60

[23] Department of Health and Social Services, *Strategy for the Support of the Voluntary Sector and for Community Development in Northern Ireland* (HMSO, 1993).

[24] See further, J Kearney "The Development of Government Policy and its Strategy towards the Voluntary and Community Sectors" in N Acheson and A Williamson (eds) *supra* n. 21 at pp. 11–32.

tolerance of difference and improving understanding between two communities, as opposed to the community development approach which emphasises the revitalisation of economic and social infrastructure.[25] It is important also not to view the relationship with government as always being one of mutual cooperation and satisfaction. Very positive features, such as for example the Targeting Social Need (TSN) initiative introduced in 1991 or the earlier introduction of the Department of Economic Development's Action for Community Employment (ACE) scheme which provided a significant fillip to the sector, must be balanced against more negative aspects such as criticism of how TSN was organised and the notorious Hurd Criteria where government sought to control those bodies to whom financial assistance was given via ACE on the basis of political vetting.[26]

However, expressing it at its strongest, direct rule presented parts of the sector with an opportunity to engage in a relationship with the Northern Ireland departments and other structures of direct rule. Government engaged in dialogue with non-governmental organisations about policy and empowered them to deliver services and advance conflict resolution strategies. Of course there always remained an imbalance in the relationship in so far as government generally was the funder and the sector was in the role of applicant. However, there were alternative funders too and a range of other arenas where politics could develop. A variety of other, non-domestic government bodies played a very significant role in allowing another politics to grow up by underwriting the role and development of the voluntary sector. European institutions and structures in particular afforded the voluntary sector opportunities to by-pass domestic government institutions and engage in politics on different terms. The role of various human rights non-governmental organisations in lobbying various international bodies around state infringements of the European Convention on Human Rights provides an example of this. The role of NICVA's European Affairs Unit in making recommendations as to the

[25] See further the analysis provided by the sources cited at *supra* n. 16.

[26] The then Secretary of State Douglas Hurd stated in a parliamentary written answer (Hansard, 27 June 1985) that it was not in the public interest to give grant aid to organisations that had "sufficiently close links with paramilitary organisations to give rise to a grave risk that to give support . . . would have the effect of improving the standing, or furthering the aims of a paramilitary organisation, whether directly or indirectly". This had the effect of cutting off grant aid from about thirty organisations including Irish language classes and pre-school playgroups and creating a suspicion of inept government vetting.

Structural Funds plan for 1995–99 which itself contained two Measures of Community Infrastructure in its Physical and Social Environment Sub-Programme provides another example of how the sector can bypass domestic state structures and become involved in using the "external constitution" to progress its agenda. The role of the sector in the District Partnerships established by the European Special Support Programme for Peace and Reconciliation with its budget of 44.2 million ecus with matching government funding indicates too how successful the sector became in enmeshing itself in governance at all levels.[27] At its very highest this may have amounted to a sophisticated form of governmental dialogue or a "communicative constitutionalism" where the sector was one important element among the many involved in governance in Northern Ireland.[28]

THE VOLUNTARY SECTOR AND THE NEW DEMOCRATIC SPACE OF REPRESENTATIVE POLITICS

It seems clear that the voluntary sector had a distinctive role and a particular opportunity during the time of direct rule. British Government strategy centred around policing the crisis while awaiting the moment of grand scale political agreement which could then be rendered into institutional form in a devolved assembly that simultaneously guarantees the Union and recognises and legitimates an all-Ireland dimension, while ameliorating the excesses of majoritarian government. Tactics varied around the key elements of security, economic support and community relations. While the security agenda could always trump the other elements there were nevertheless conditions where the voluntary sector (along with others) could have a particular role both in getting the business of government done in difficult circumstances and in developing new forms of governance and conflict resolution. The Agreement secured in April 1998 represented the historic achievement of British aims to resolve the (Northern) Irish issue and of course changed the agenda completely.

In the post-Agreement situation the voluntary sector is in a different position. The exact nature of the role that the sector will play in the

[27] See further J Hughes *et al Partnership Governance in Northern Ireland: The Path to Peace* (Dublin, Oak Tree Press, 1998).

[28] See further J Morison and S Livingstone *Reshaping Public Power: Northern Ireland the British Constitutional Crisis* (London, Sweet and Maxwell, 1995) pp. 138–49.

future remains unclear. For some, particularly those presently holding elected office, it may appear that now there is little need for a voluntary sector operating in an enhanced role: with the political process restored normal service has been resumed and the sector can be dismissed to the background.

Of course such a view is unrealistic. As has been argued already, it is not a 1970s-style government that has been restored. The project of government everywhere has changed. Government is now more of a project of governance and it is not only multilevel but also multi-form. Today the role of government involves drawing upon a whole range of forces and operating through a variety of complex networks, pro grammes, techniques and devices. The commands of parliaments (sovereign or devolved) and the budgetary mandates of political administrations now make up only part of what is involved in present day governance. As ideas of governmentality demonstrate, just as the concept of "a unified solitary social domain . . . is displaced by images of multiple communities, plural identities, and cultural diversity", so too has there been a recasting of the role of national governments who "no longer aspire to be the guarantor and ultimate provider of security" but instead should "be a partner, animator and facilitator for a variety of independent agents and powers, and should exercise only limited powers of . . . [their] . . . own".[29] It is no longer possible, even if it were desirable, that "big government" look after its "citizens" (or "subjects") from cradle to grave. Today information comes from many sources. Issues are framed by many perspectives with citizens being "responsibilised" in relation to a wide range of schemes, modalities and rationalities ranging from individual morality and organisational rationality to more formal audit and legal controls. Individuals have multiple identities, the state has been rolled back and the complex consequences of globalisation and localisation, privatisation and marketisation, as well as consumerisation and a host of other re-designations have the effect of altering the whole project of government and what is to be governed.

This realisation of the limits of traditional representative democracy may not yet have quite come in the early excitement of restored devolved government. However, in some ways it should not be a surprise. Not only are the seemingly wide powers and resources of the new government constrained by the need (familiar to the old Stormont regime) to replicate standards elsewhere in the United Kingdom, but

[29] Rose *supra* n. 3 at pp. 323–4.

there are other, newer limits. If the newly installed politicians now occupying the offices of government were to look at developments in governance elsewhere they would notice a reduction in capacity that comes both from a rolling back of the state that now manifests itself in multilevel and multi-form governance, and from attempts to involve others in the operation of governance in an effort to, in the words of Anthony Giddens, "democratize democracy".[30] Furthermore, a proper understanding of the past and present role of the voluntary sector in Northern Ireland itself, combined with an accurate reading of the full implications of the Agreement and Northern Ireland Act 1998, would show the centrality of ideas of consultation and participation and suggest that the sector's role as an alternative democratic space beyond formal politics is more likely to increase than diminish. Each of these elements must be considered in turn.

New patterns of governance

The hollowing out of the state everywhere has been expressed in terms of reduced capacity and competence as well as in the involvement of a variety of other bodies from the private and voluntary sectors in dealing with a whole range of issues from social services, education, housing and homelessness, and leisure. Even local government in Great Britain has increasingly reoriented itself around the idea of the voluntary sector being significant in terms of service delivery.[31] Indeed, ideas about "best value" in providing services not only mean that the sector will be involved in competition to provide services but may be involved in the consultation to determine exactly what best value actually is in any given situation and in devising the performance indicators to ensure that it has been delivered.[32] With low turnouts for local elections in Britain diminishing arguments about legitimacy from formal

[30] A Giddens *The Third Way: The Renewal of Social Democracy* (London, Polity Press 1988) pp. 70–8.

[31] See I Leigh "The Legal Framework for Community Involvement" in A Dunn (ed.) *The Voluntary Sector, the State and the Law* (Oxford, Hart Publishing, 2000). Leigh details how, for example, one metropolitan borough council distributed grants to the voluntary sector amounting to £5.9 million from twenty-one different council budgets for a whole variety of services and initiatives (at p. 13).

[32] See the White Paper entitled *Modern Local Government: In Touch with the People* (Cm 4014, 1998), ch. 7 which sets out how best value differs from compulsory competitive tendering by widening the idea of quality means and involving consultation with local users and their representatives.

democratic mandates, there is an increasing role for voluntary sector organisations in local policy networks too.[33] Interest in developing "joined up" government has led to increased voluntary sector involvement in statutory plans and in community plans which seek to factor in all local stakeholders and providers from the public, private and voluntary sectors.[34] Similar interest in new formats of governance have led to not only a series of policy initiatives ranging from New Deal for Communities to Health Action Zones but also to the development of local authority companies, suggestions for Public Partnership Limited Companies which can formally take on council business' and the development of ideas of "beacon councils" which can be allowed greater freedom because of strong existing performance.[35] The role of the voluntary sector in partnership with government and the private sector is enhanced in many of these new formats. Partnerships are needed too to bid for certain additional moneys from, for example, the Single Regeneration Budget.

Beyond experimentation with new formats for governance to fill in the spaces left by the rolling back of the state, there is a genuine interest in using the voluntary sector as an important part of a "third way" between reliance on the state and wholehearted dependency on market forces.[36] The voluntary sector is thought especially appropriate to deal with some issues, such as particularly social exclusion, which are beyond the reach of government and outside the interest of the private sector. In recognition of the important potential of the sector, government is attempting to build a new relationship with the voluntary and community sector. This involves both recognition of its increasing role and the formalising of the association between parts of the sector and the state through the development of a series of compacts in all four constituent parts of the United Kingdom.[37]

[33] The average turnout in the 1999 local council election was 29 per cent. For discussion of local policy networks see further D Wilson, C Game, S Leach and G Stoker *Local Government in the United Kingdom* (Basingstoke, Macmillan, 1994) ch. 16; W Kickert, E Klijin and A Koopenhan (eds) *Network Management in the Public Sector* (London, Sage, 1997.)

[34] See the White Paper *Modern Local Government: In Touch with the People*, *supra* n. 32 at ch. 8.

[35] See Leigh *supra* n. 31.

[36] See Giddens *supra* n. 38 and T Blair *The Third Way: New Politics for the New Century* (London, The Fabian Society, 1998) especially chs 4 and 5.

[37] See further J Morison "The Government-Voluntary Sector Compacts: Governance, Governmentality, and Civil Society" (2000) 27 *Journal of Law and Society* 98–132.

Many of these things, like the compacts, have a direct equivalent in Northern Ireland. Although the state was never rolled back as far in Northern Ireland as elsewhere, the public sector was subject to similar changes within the general trend.[38] Of course, for the reasons mentioned earlier, the impact of these changes was more uneven. Nevertheless, new forms of governance do exist in Northern Ireland and do afford particular opportunities to the voluntary sector.

For example, there is the Northern Ireland version of the "best value" initiative that is to replace compulsory competitive tendering there sometime after 2001. It provides perhaps an important opportunity for the sector to develop a double role both as provider and a standard-setting body within the context of government spending programmes. As mentioned already, the idea of best value generally is about not only delivering services to clear standards, covering familiar principles relating to cost and efficiency, but also about meeting standards reflecting *quality*. While aspects of the idea of quality are nationally defined by central government, and underpinned through key national performance measures, there is an opportunity for elements of it to be discovered by local authorities in a participatory and consultative process at local level. District Councils in Northern Ireland (and perhaps other authorities spending public money) must, like local authorities in Great Britain, consult with local people as to what they want, and devise a corporate plan which sets out objectives and resources and contains mechanisms for measuring performance against stated objectives. 25 per cent of expenditure will be reviewed each year and emerging performance indicators will be subject to external scrutiny and audit. The duty to consult with users and user groups is itself of particular value and importance in creating a democratic space of real involvement beyond the formal halls of the Assembly. However, there is also clearly potential for "quality" to be defined to include more democratic elements requiring, for example, further consultation, fair employment practices, improved community relations, more transparent and accountable organisation and so on. The role of the sector here, and indeed in monitoring that quality is delivered, is potentially significant and may allow it to assist in mainstreaming a democratic agenda within elements of basic service delivery.

[38] See further J Morison "The Public Sector in a Divided Society" in *People and Government: Questions for Northern Ireland* (Belfast, Joseph Rowntree Foundation and Chief Executives' Forum 1998) pp. 109–26.

Another example of a heightened role for the sector within new governance structures relates to the local commissioning arrangements which are due to replace aspects of the internal market within the system of integrated health and personal social services in Northern Ireland. This perhaps illustrates the limited role of government as a direct provider (and also, the reduced scope for any new devolved administration to effect real change, independent of what is occurring in Great Britain) as well as new possibilities for the sector to become further involved in partnership and consultation. The original report announcing the end of the existing Fundholding Scheme offered a choice between only two variations: one involving the existing Health and Social Services Boards and one with new Local Care Agencies—but both involving the new local commissioning arrangements.[39] Although the final decision is to be left to the new Assembly, government has made it clear that the preferred option is a system of five new Health and Social Care Partnerships supported by community based Primary Care Co-operatives. These are to act in partnership with "local communities and other organisations which have a role to play in improving health and well-being" in order to commission health and social services.[40]

The idea of partnership government generally, which is gaining rapidly in popularity in Great Britain, is in fact more developed in Northern Ireland than elsewhere.[41] There is a version of partnership providing one of the basic tenets of the Belfast Agreement.[42] Other developments of the

[39] See *Fit for the Future: A Consultation Document on the Government's Proposals for the Future of the Health and Personal Social Services in Northern Ireland* (April 1998) at <http://www.dhssni.gov.uk/the_department/publications/fitforthefuture/fulldoc.html> which develops the general approach of the White Paper, The New NHS—Modern and Dependable (Cm 3807, 1997) setting out the future for the Health Service in England.

[40] See Department of Health and Social Services, *Fit for the Future—A New Approach* (March 1999) at <http://www.dhssni.gov.uk/hpss/publications/fftf/FFTF proposals.htm> at p. 9.

[41] See Morison *supra* n. 45 at pp. 103–8 for discussion of how partnership and civil society involvement fits in to third way political development in a general UK context. See Hughes *et al supra* n. 35 for a discussion of partnership in Northern Ireland.

[42] See the foundational "Declaration of Support" that appears at the beginning of The Belfast Agreement where it states "we are committed to partnership, equality and mutual respect as the basis for relationships" (para. 3). Indeed partnership ideas can be observed all over structures in Northern Ireland. North, South, West and Greater East Belfast all have partnerships for development and there are Rural Development Partnerships and numerous Local Action Partnerships. As has been noted already, even the Patten Report on policing *supra* n. 11, contains ideas about "a real partnership between the police and the community government agencies, non-governmental organisations, families and citizens" (at para. 1.16).

idea build upon the District Partnerships mentioned above. These structures were established as a result of structural funding (of some £1 billion) that came to Northern Ireland from the European Union. The basic structure for disbursing parts of this money was reproduced in twenty-six District Partnership Boards (and in the central Northern Ireland Partnership Board) whereby eight politicians, seven representatives from the voluntary sector and five from business, three trade unionists and one representative each from the Rural Development Council and the former equivalent to what is now the Department of Regional Development joined together. Partnerships operated with some basic concepts including customer focus, participation, collaboration, innovation, transparency and local resources (in terms of materials and expertise). Interestingly, partnerships were required to meet Policy Appraisal and Fair Treatment (PAFT) and Targeting Social Need (TSN) criteria which effectively ensured that their decisions were gender and community proofed. There is some debate about whether partnerships were simply at the end of a long funding chain or whether they were something extra and different. At their least they were perhaps a means of service delivery that reached parts that more formal government could not, although officious audit and financial checks reduced their scope. At their most, they were an embryonic form of participatory decision-making with considerable potential to enrich formal, representative democracy.

There is now further money (400 million euros) in a second round of the Special Support Programme for Peace and Reconciliation and significant Structural Fund money for Northern Ireland (given despite the fact that the region no longer has "Objective One" status). These monies (known colloquially as Peace II) will be available from 2000 to 2006 and they have significant potential to develop further the sorts of structures and practices of partnership governance that grew up with the earlier funding. However, there have been concerns expressed that the monies are targeted narrowly at ideas of economic regeneration rather than at also developing the social infrastructure and processes of reconciliation that might be thought necessary to underpin any such reconstruction.[43] Nevertheless, the potential for newer

[43] See, e.g., NICVA, *NICVA Response to the Government Consultation Paper on Northern Ireland Structural Funds Plan 2000–2006* (Belfast, NICVA, March 2000) and the Democratic Dialogue Briefing Paper, *Structurally Unsound: The Northern Ireland Bids for Further EU Monies* (Belfast, Democratic Dialogue, March 2000). There are also concerns that the partnership element whereby business, politics and the voluntary

participatory forms of governance to exist alongside restored representative government remains.

Governance and the Agreement: developing a new democratic space

Having characterised the Agreement and the Northern Ireland Act 1998 as being mainly about institutions and restoring traditional representative democracy, it must also be admitted that suffusing the whole structure are the elements of human rights, equality and indeed partnership. While some of this is about creating further institutions, such as the Human Rights Commission and the Equality Commission—and there may be an element of ritualistic obeisance to the new totem of rights and equality—there are significant features too. Important aspects of these may involve the voluntary sector in creating and operating within a new participatory space.

It has been mentioned already how partnership is described as being central to the Agreement[44] and, indeed, it is present in some form in the consocial aspects of the formal arrangements for government. Again, as already mentioned, partnership (in the more developed sense of working with civil society and others) is an important element in proposals on policing, and it is there too, up to a degree, in relation to criminal justice.[45] However, it is in pursuit of equality values and human rights that partnership and participation can be seen as having most application to the voluntary sector and its role in developing a forum beyond traditional politics where many voices can be heard in a participatory act of genuine *constitution*.

In addition to the other consultation requirements discussed above that relate to governance structures everywhere, there are clearly some that are unique to the Northern Ireland settlement. Building upon the earlier PAFT scheme, section 75 of the Northern Ireland Act 1998 imposes a statutory duty on public authorities to promote equality of opportunity.[46] In contrast to the Human Rights Act 1998 where the

sector were more or less equally represented will be replaced by ideas of District Councils "taking the lead" which may also undermine the participatory dimension.

44 See *supra* n. 42.

45 In a rather thin section on "Rights and Principles", the Criminal Justice Review Group identify as one of the "common values" of the criminal justice system an idea of "partnership between the criminal justice system, the community, and other external bodies", see *Review of the Criminal Justice System in Northern Ireland* (Belfast, Stationary Office, 2000) p. 30.

46 See C McCrudden, "Mainstreaming Equality in the Governance of Northern

definition of a public body is potentially problematic, public authorities are listed in Schedule 2 to the Northern Ireland Act 1998 and Schedule 9 to that Act requires bodies on the list to submit to the Equality Commission a scheme showing how it proposes to fulfil its obligations. Voluntary sector bodies are not included in these requirements. However, paragraph 5 of Schedule 9 to the 1998 Act requires public authorities to consult with persons likely to be affected by the scheme and with representatives of such persons. This brings the voluntary sector into the heart of the consultation process. Early evidence suggests that many public authorities will not stint in sending out draft equality schemes for comment.[47] It is likely that voluntary sector bodies will be involved too in consultation with regard to enforcement. This mainstreaming not only meets best practice guidelines from the Council of Europe[48] but, as the Equality Commission *Guide to the Statutory Duties* puts it, it "should enhance the crucial link between government and 'civil society', encouraging greater participation in government and leading to greater accountability in government decision making".[49]

Perhaps an indication even more illustrative of the potential of the Agreement to deliver democratic process (as opposed to only structure) lies with the debate about the bill of rights for Northern Ireland. It is here that the vital importance of the voluntary sector contribution in augmenting the traditional political institutions can be seen very clearly. Strand Three of the Agreement, where it relates to Rights, Safeguards and Equality of Opportunity, envisages in paragraph 4 that the Human Rights Commission will consult widely and advise on rights appropriate to Northern Ireland and supplementary to those in

Ireland" (1999) 22 *Fordham International Law Journal* 1696, for discussion of the origins and development of this initiative.

47 The Equality Commission guidelines detail what each element of an equality scheme should include and it refers to the need for consultation that is "timely, open and inclusive" (Equality Commission for Northern Ireland, *Guide to Statutory Duties: A Guide to the Implementation of the statutory duties on public authorities arising from Section 75 of the Northern Ireland Act 1998* (Belfast, Equality Commission, 2000) at para. 4.1(2)(c)). For many public authorities this has entailed simply asking everyone. For example, in April 2000 the Royal Hospitals placed advertisements in daily papers and circulated their draft equality scheme to a circulation list of some 335 bodies nearly all of whom are in the voluntary sector: see the Royal Hospitals, *Promoting Equality of Opportunity and Good Relations: Draft Equality Scheme* (2000) Appendix 1.

48 Council of Europe *Gender Mainstreaming: Conceptual Framework, Methodology and Presentation of Good Practices* (Strasbourg, Council of Europe, 1998).

49 Equality Commission for Northern Ireland *supra* n. 47 at para. 1.7.

the European Convention of Human Rights such as should be included in Westminster legislation. In one sense this involves simply adding to the burden of participation and overstretches the patience of those being continually consulted. In this way it may contribute to the danger of consultation becoming merely a matter of form rather than an aspect of a genuinely participatory process. However, it does in reality reveal and illustrate the vitally important role that the voluntary sector has in both establishing, and then operating within, an additional democratic space.

It is important that rights, and the use that they will be put to within the new Northern Ireland, is not something that is decided by the political parties alone (or even by elite groups of activists and scholars drawing upon international best practice). Rights must be fully constitutive of the new society. They must be the foundation for participation on an equal basis and by all groups and individuals within society with an equal voice that is worthy of respect. Rights must not be seen as belonging to one or other tradition or even as something of short-term value in smoothing over the aftermath of the conflict. Rights emphasising participation, as well as the usual safeguards against arbitrary power, are an important foundation for a proper civic dialogue about the way in which people in Northern Ireland wish to live together. If political discourse is to be widened beyond the debating chambers of the Assembly, where politicians register their tribal affiliation and must struggle to act in ways that transcend it,[50] then a widening of the political space is required. This can be achieved in part by defining a rights base that provides the foundation on which individual citizens can take part in dialogue within safe limits and where their voice will be guaranteed to be heard in a deliberative process.[51] The voluntary sector, who are largely outside the traditional limits of politics, are central in ensuring that the sort of rights considered appropriate are those that guarantee real and effective participation. They are also of course important in themselves in conducting this participatory

[50] Under the standing orders of the Assembly each member must formally declare him or herself as a unionist, nationalist or Other. This has particular importance with regard to certain "key decisions" where cross-community support is required.

[51] Of course, while there will be participation by many in the debate, and arguments for a rights base that enhances further participation, there will also be those who, tired of consultation and politics, seek simply what Unger characterises as an extended social democracy that ensures the "efficiencies, the equities and decencies making individual action effective". See R Unger, *What Should Legal Analysis Become?* (London, Verso, 1996) p. 138. Rights must protect their position too: their role is not solely to promote participation.

discourse through the means mentioned already and in other, as yet unknown, ways.

In general the creation of this new democratic space will mean consultation along the lines outlined already with regard to the Bill of Rights process. It will mean hearing voices from outside traditional politics, perhaps, as suggested, from the voluntary sector itself in its role as champion of various unheard interests in all the best value consultation exercises and partnership mechanisms. Indeed, there is evidence that the voluntary sector may well have a particular role to play on a number of issues such as gender equality, social exclusion and rural development where otherwise affected groups might have to struggle too hard to get their voice heard.[52] Participatory democracy will involve the voluntary sector more generally in more long-term projects of planning and peace-building. The short-termism and problem fixing of politicians, although necessary, is not particularly suited for addressing issues such as what the new society should look like in twenty years time. Civil society is particularly important in times of transition and there are jobs which, arguably, belong there most of all.[53] Even within the limits of more traditional forms of democracy, participatory rights must mean transparency in the operation of the Assembly so that those who represent us and spend our money are subject to a proper accountability based on an informed civic discourse between the governors and the governed.

More generally there is the main argument that democracy cannot work effectively without alternative spaces. The general project of reviving politics and democratising democracy factors in an important role for civil society and the voluntary sector. Ideas from a whole range of theorists relating to the way in which a revived participatory politics can be developed open up possibilities for an enhanced role for the

[52] See for example the role of the voluntary sector already in this regard in M Smyth "Women, Peace, Community Relations and Voluntary Action" in N Acheson and A Williamson (eds) *supra* n. 13 at pp. 145–60 and J Armstrong and A Kilmurray "Voluntary Action, Rural Policy and Social Development" in *ibid.* at pp. 115–34. The strategic framework for the next decade that is presently being devised for the Northern Ireland Voluntary Trust (NIVT) indicates a continuing, enhanced role for the sector in tackling the agenda for social inclusion and peace-building: see NIVT, *Driving Social Change: A Strategy for Inclusion* (Belfast, NIVT, April 2000).

[53] Walzer argues "no state can survive for long if it is wholly alienated from civil society . . . The production and reproduction of loyalty, civility, political competence, and trust in authority are never the work of the state alone, and the effort to go it alone—one meaning of totalitarianism—is doomed to failure": M Walzer, "The Civil Society Argument" in R Beiner (ed.) *Theorizing Citizenship* (New York, State University of New York Press, 1995) p. 153, at p. 168.

voluntary sector both within Blairite "third way" thinking and beyond it. Thus, for example, accounts range from those who require simply a more participatory form of politics—where the sector with its links to groups that are not perhaps adequately given voice in the sectarian politics of traditional representative politics can have a particular role—to those theories which identify wider ideas of community, sympathy and compassion as being necessary. In this way ideas of participatory politics[54] shade into more fundamental notions of a politics of association. Here, for example, Unger refers to ideas of "solidarity" and the deeper sense of belonging that people must develop in order to foster an idea of sympathy that is necessary to constitute community.[55] Richard Rorty argues that post-modern politics involves restoring an ethic of "human solidarity"[56] while theorists such as Nussbaum maintain that new ideas of citizenship will require an education in sensitivity and compassion for others in order to produce an idea of social justice founded on a "compassionate imagination".[57] Even Jacques Derrida has developed ideas about "friendship" as the non-foundational foundation for a post-modern politics, arguing that democracy is essentially "a community of friends".[58]

Whichever basis is adopted as underlying the new democratic space, or whatever route taken towards reviving politics and democratising democracy, the role of the voluntary sector and civil society generally is clearly a potentially important one beyond the placid limits of formal politics.[59] Indeed it would seem to involve much more even than the

[54] For an overview of the range of approaches in this context see Richard Blaug "New Theories of Discursive Democracy: A User's Guide" (1996) 22 *Philosophy and Social Criticism* 49; J Morison "The Case Against Constitutional Reform?"(1998) 25 *Journal of Law and Society* 528–34.

[55] See, in particular, *Knowledge and Politics* (New York, Free Press, 1975) pp. 3–5, 220–1 and *Passion: An Essay on Personality* (New York, Free Press, 1984).

[56] R Rorty, *Contingency, Irony and Solidarity* (Cambridge, Cambridge University Press, 1989).

[57] N Nussbaum, *Cultivating Humanity: A Classical Defense of Reform in Liberal Education* (Harvard, Havard University Press, 1997). See also Drucilla Cornell who argues that politics now requires us to identify "a common good" which is beyond the reach of a simple application of reason and can be found instead by individuals actively participating together in compassion and sympathy: see "Towards a Modern/ Postmodern Reconstruction of Ethics" (1985) 113 *University of Pennsylvania Law Review* 345; Drucilla Cornell *The Philosophy of the Limit* (New York, Routledge, 1992).

[58] J Derrida *The Politics of Friendship* (London, Verso, 1997).

[59] Although of course even if one is thinking only about how to revive traditional representative politics in Northern Ireland, there is clearly a need to establish a dynamic of rotation within the political structures. With almost every conceivable political figure already occupying one or more of the many offices available, the voluntary sector may appear as a potentially productive source of new political blood.

planned Civic Forum envisages, where the First Minister must make arrangements "for obtaining from the Forum the view of the community on social, economic and cultural matters".[60] Rather than being a formal institutional component, this element will be a space or process. Arguably the bill of rights debate will both usher in and be the foundation of this new participatory space where the voluntary sector will be involved with many others in giving democratic voice to a range of groups at the most appropriate level. With participation as a basic right it opens the door to a whole range of other rights from consultation and veto rights, through equality rights to information and minimum consensus rights. It also suggests a whole range of new mechanisms for giving voice, from preferenda and citizens' juries through to consensus conferences and new uses of information technology.[61] In short it opens up the potential for a re-invigoration of democracy.

CONCLUSIONS: GOVERNMENTALITY AND RENEWAL

It is perhaps odd that this account which began by suggesting that the elaborate structures of formal representative structure were perhaps a little overblown should end by arguing for the further development of another, albeit informal, democratic space. Of course part of the argument here is that although devolved government may have been restored, the project of government has changed such that the elected Assembly and new government departments can not hope to carry out all the functions involved in present day governance. Another part of the argument is that the voluntary sector in Northern Ireland, having played a particular and important role during the period of direct rule, is simply too important a resource to be lost, even if such disentanglement from multi-form governance were possible. It would seem wrong if the democratic deficit of the pre-devolution period where there was little representative democracy were to be replaced by a new democratic deficit where the quantity of traditional, institutional democracy crowds out the participatory democratic life that grew up in the shadow of direct rule. Fortunately, such a possibility seems remote no

[60] See the Belfast Agreement and the Northern Ireland Act 1998, s. 56.

[61] See further Morison *supra* n. 54 at pp. 528–35 for discussion of how participatory models challenge the conflation of the state apparatus with the public sphere of discourse and association, and suggest not only a different foundation for existing rights but also a variety of new rights and ways of achieving them.

matter what some of the new political representatives might wish for. Changes in governance everywhere, and in Northern Ireland in particular with regard to consultation requirements and the equality and human rights agenda, mean that the role and influence of the sector will persist, and may increase in the longer term.

But beyond arguments about the changing nature of the state and the multi-level and multiform nature of modern governance, there is a particular need for the voluntary sector in the new Northern Ireland. The voluntary sector have an already established role in giving voice to various groups and individuals who might otherwise be marginalised by the political process. This occurs through the various consultation roles with which they are increasingly involved. Such processes provide an embryonic version of a new democratic space, beyond formal politics, where real renewal can take place. The bill of rights debate, where the sector should play an important role in securing both traditional protections and mechanisms for further participation, provides an important illustration of the general renewal and reconstitutive debate that must now go on in the shadows of the big institutional "fix". A properly rooted settlement must involve people working out the details of how they want to live among themselves and establishing the basis of participation and democracy. It must be properly *constitutive* in fundamental and ongoing sense.

The governmentality approach which downgrades the importance of the state and looks instead at how power is actually constituted through society offers a better way to approach the issue than does a formal constitutionalism emphasising state and structure. It recognises that real constitutional agreement cannot be imposed or conferred. It must be facilitated in a maieutic process. It must be seen to be "natural" with notions of sharing power and acknowledging rights coming from the individual rather than from being merely legal rights. If one sees only a legal right then it is capable of being taken away by law. If, as Foucault says, we understand only "the political institution" and the "legal subject" rather than a concept of governmentality which allows us to "bring out the freedom of the subject and its relationship to others"[62] we have only a formal, and ultimately formally revocable, settlement. Instead we must follow through on Foucault's account and seek to analyse power "on the basis of freedom, strategies and governmentality".[63] It is only in this way that we can see and recognise the

[62] Foucault *supra* n. 2.
[63] *Ibid.*

basis of the way in which we live with each other, and are variously involved in the project of governmentality which is, after all, the "very stuff of ethics".

ANNEX

Agreement Reached in the Multi-Party Negotiations

CONTENTS

DECLARATION OF SUPPORT

1. We, the participants in the multi-party negotiations, believe that the agreement we have negotiated offers a truly historic opportunity for a new beginning.

2. The tragedies of the past have left a deep and profoundly regrettable legacy of suffering. We must never forget those who have died or been injured, and their families. But we can best honour them through a fresh start, in which we firmly dedicate ourselves to the achievement of reconciliation, tolerance, and mutual trust, and to the protection and vindication of the human rights of all.

3. We are committed to partnership, equality and mutual respect as the basis of relationships within Northern Ireland, between North and South, and between these islands.

4. We reaffirm our total and absolute commitment to exclusively democratic and peaceful means of resolving differences on political issues, and our opposition to any use or threat of force by others for any political purpose, whether in regard to this agreement or otherwise.

5. We acknowledge the substantial differences between our continuing, and equally legitimate, political aspirations. However, we will endeavour to strive in every practical way towards reconciliation and rapprochement within the framework of democratic and agreed arrangements. We pledge that we will, in good faith, work to ensure the success of each and every one of the arrangements to be established under this agreement. It is accepted that all of the institutional and constitutional arrangements—an Assembly in Northern Ireland, a North/South Ministerial Council, implementation bodies, a British-Irish Council and a British-Irish Intergovernmental Conference and any amendments to British Acts of Parliament and the Constitution of Ireland—are interlocking and interdependent and that in particular the functioning of the Assembly and the North/South Council are so closely inter-related that the success of each depends on that of the other.

6. Accordingly, in a spirit of concord, we strongly commend this agreement to the people, North and South, for their approval.

CONSTITUTIONAL ISSUES

1. The participants endorse the commitment made by the British and Irish Governments that, in a new British-Irish Agreement replacing the Anglo-Irish Agreement, they will:

(i) recognise the legitimacy of whatever choice is freely exercised by a majority of the people of Northern Ireland with regard to its status, whether they prefer to continue to support the Union with Great Britain or a sovereign united Ireland;

(ii) recognise that it is for the people of the island of Ireland alone, by agreement between the two parts respectively and without external impediment, to exercise their right of self-determination on the basis of consent, freely and concurrently given, North and South, to bring about a united Ireland, if that is their wish, accepting that this right must be achieved and exercised with and subject to the agreement and consent of a majority of the people of Northern Ireland;

(iii) acknowledge that while a substantial section of the people in Northern Ireland share the legitimate wish of a majority of the people of the island of Ireland for a united Ireland, the present wish of a majority of the people of Northern Ireland, freely exercised and legitimate, is to maintain the Union and, accordingly, that Northern Ireland's status as part of the United Kingdom reflects and relies upon that wish; and that it would be wrong to make any change in the status of Northern Ireland save with the consent of a majority of its people;

(iv) affirm that if, in the future, the people of the island of Ireland exercise their right of self-determination on the basis set out in sections (i) and (ii) above to bring about a united Ireland, it will be a binding obligation on both Governments to introduce and support in their respective Parliaments legislation to give effect to that wish;

(v) affirm that whatever choice is freely exercised by a majority of the people of Northern Ireland, the power of the sovereign government with jurisdiction there shall be exercised with rigorous impartiality on behalf of all the people in the diversity of their identities and traditions and shall be founded on the principles of full respect for, and equality of, civil, political, social and cultural rights, of freedom from discrimination for all citizens, and of parity of esteem and of just and equal treatment for the identity, ethos, and aspirations of both communities;

(vi) recognise the birthright of all the people of Northern Ireland to identify themselves and be accepted as Irish or British, or both, as they may so choose, and accordingly confirm that their right to hold both British and Irish citizenship is accepted by both Governments and would not be affected by any future change in the status of Northern Ireland.

2. The participants also note that the two Governments have accordingly undertaken in the context of this comprehensive political agreement, to propose and support changes in, respectively, the Constitution of Ireland and in British legislation relating to the constitutional status of Northern Ireland.

ANNEX A

Draft clauses/Schedules for incorporation in British legislation

1. (1) It is hereby declared that Northern Ireland in its entirety remains part of the United Kingdom and shall not cease to be so without the consent of a majority of the people of Northern Ireland voting in a poll held for the purposes of this section in accordance with Schedule 1.
(2) But if the wish expressed by a majority in such a poll is that Northern Ireland should cease to be part of the United Kingdom and form part of a united Ireland, the Secretary of State shall lay before Parliament such proposals to give effect to that wish as may be agreed between Her Majesty's Government in the United Kingdom and the Government of Ireland.

2. The Government of Ireland Act 1920 is repealed; and this Act shall have effect notwithstanding any other previous enactment.

SCHEDULE 1

Polls for the purpose of section 1

1. The Secretary of State may by order direct the holding of a poll for the purposes of section 1 on a date specified in the order.

2. Subject to paragraph 3, the Secretary of State shall exercise the power under paragraph 1 if at any time it appears likely to him that a majority of those voting would express a wish that Northern Ireland should cease to be part of the United Kingdom and form part of a united Ireland.

3. The Secretary of State shall not make an order under paragraph 1 earlier than seven years after the holding of a previous poll under this Schedule.

4. (Remaining paragraphs along the lines of paragraphs 2 and 3 of existing Schedule 1 to 1973 Act.)

ANNEX B

Irish Government draft legislation to amend the Constitution

Add to Article 29 the following sections:

7. 1. The State may consent to be bound by the British-Irish Agreement done at Belfast on the day of 1998, hereinafter called the Agreement.

   2. Any institution established by or under the Agreement may exercise the powers and functions thereby conferred on it in respect of all or any part of the island of Ireland notwithstanding any other provision of this Constitution conferring a like power or function on any person or any organ of State appointed under or created or established by or under this Constitution. Any power or function conferred on such an institution in relation to the settlement or resolution of disputes or controversies may be in addition to or in substitution for any like power or function conferred by this Constitution on any such person or organ of State as aforesaid.

   3. If the Government declare that the State has become obliged, pursuant to the Agreement, to give effect to the

amendment of this Constitution referred to therein, then, notwithstanding Article 46 hereof, this Constitution shall be amended as follows:

i. the following Articles shall be substituted for Articles 2 and 3 of the Irish text:

   "2. [Irish text to be inserted here]

   3. [Irish text to be inserted here]"

ii. the following Articles shall be substituted for Articles 2 and 3 of the English text:

   "Article 2

   It is the entitlement and birthright of every person born in the island of Ireland, which includes its islands and seas, to be part of the Irish nation. That is also the entitlement of all persons otherwise qualified in accordance with law to be citizens of Ireland. Furthermore, the Irish nation cherishes its special affinity with people of Irish ancestry living abroad who share its cultural identity and heritage.

   Article 3

   1. It is the firm will of the Irish nation, in harmony and friendship, to unite all the people who share the territory of the island of Ireland, in all the diversity of their identities and traditions, recognising that a united Ireland shall be brought about only by peaceful means with the consent of a majority of the people, democratically expressed, in both jurisdictions in the island. Until then, the laws enacted by the Parliament established by this Constitution shall have the like area and extent of application as the laws enacted by the Parliament that existed immediately before the coming into operation of this Constitution.

   2. Institutions with executive powers and functions that are shared between those jurisdictions may be established by their respective responsible authorities for stated purposes and may exercise powers and functions in respect of all or any part of the island."

iii. the following section shall be added to the Irish text of this Article:

   "8. [Irish text to be inserted here]"

and

iv. the following section shall be added to the English text of this Article:
"8. The State may exercise extra-territorial jurisdiction in accordance with the generally recognised principles of international law."
4. If a declaration under this section is made, this subsection and subsection 3, other than the amendment of this Constitution effected thereby, and subsection 5 of this section shall be omitted from every official text of this Constitution published thereafter, but notwithstanding such omission this section shall continue to have the force of law.
5. If such a declaration is not made within twelve months of this section being added to this Constitution or such longer period as may be provided for by law, this section shall cease to have effect and shall be omitted from every official text of this Constitution published thereafter.

STRAND ONE

Democratic Institutions in Northern Ireland

1. This agreement provides for a democratically elected Assembly in Northern Ireland which is inclusive in its membership, capable of exercising executive and legislative authority, and subject to safeguards to protect the rights and interests of all sides of the community.

The Assembly

2. A 108-member Assembly will be elected by PR(STV) from existing Westminster constituencies.

3. The Assembly will exercise full legislative and executive authority in respect of those matters currently within the responsibility of the six Northern Ireland Government Departments, with the possibility of taking on responsibility for other matters as detailed elsewhere in this agreement.

4. The Assembly—operating where appropriate on a cross-community basis—will be the prime source of authority in respect of all devolved responsibilities.

Safeguards

5. There will be safeguards to ensure that all sections of the community can participate and work together successfully in the operation of these institutions and that all sections of the community are protected, including:

(a) allocations of Committee Chairs, Ministers and Committee membership in proportion to party strengths;
(b) the European Convention on Human Rights (ECHR) and any Bill of Rights for Northern Ireland supplementing it, which neither the Assembly nor public bodies can infringe, together with a Human Rights Commission;
(c) arrangements to provide that key decisions and legislation are proofed to ensure that they do not infringe the ECHR and any Bill of Rights for Northern Ireland;
(d) arrangements to ensure key decisions are taken on a cross-community basis:
    (i) *either* parallel consent, i.e. a majority of those members present and voting, including a majority of the unionist and nationalist designations present and voting;
    (ii) *or* a weighted majority (60%) of members present and voting, including at least 40% of each of the nationalist and unionist designations present and voting.

    Key decisions requiring cross-community support will be designated in advance, including election of the Chair of the Assembly, the First Minister and Deputy First Minister, standing orders and budget allocations. In other cases such decisions could be triggered by a petition of concern brought by a significant minority of Assembly members (30/108).
(e) an Equality Commission to monitor a statutory obligation to promote equality of opportunity in specified areas and parity of esteem between the two main communities, and to investigate individual complaints against public bodies.

Operation of the Assembly

6. At their first meeting, members of the Assembly will register a designation of identity—nationalist, unionist or other—for the pur-

poses of measuring cross-community support in Assembly votes under the relevant provisions above.

7. The Chair and Deputy Chair of the Assembly will be elected on a cross-community basis, as set out in paragraph 5(d) above.

8. There will be a Committee for each of the main executive functions of the Northern Ireland Administration. The Chairs and Deputy Chairs of the Assembly Committees will be allocated proportionally, using the d'Hondt system. Membership of the Committees will be in broad proportion to party strengths in the Assembly to ensure that the opportunity of Committee places is available to all members.

9. The Committees will have a scrutiny, policy development and consultation role with respect to the Department with which each is associated, and will have a role in initiation of legislation. They will have the power to:

- consider and advise on Departmental budgets and Annual Plans in the context of the overall budget allocation;
- approve relevant secondary legislation and take the Committee stage of relevant primary legislation;
- call for persons and papers;
- initiate enquiries and make reports;
- consider and advise on matters brought to the Committee by its Minister.

10. Standing Committees other than Departmental Committees may be established as may be required from time to time.

11. The Assembly may appoint a special Committee to examine and report on whether a measure or proposal for legislation is in conformity with equality requirements, including the ECHR/Bill of Rights. The Committee shall have the power to call people and papers to assist in its consideration of the matter. The Assembly shall then consider the report of the Committee and can determine the matter in accordance with the cross-community consent procedure.

12. The above special procedure shall be followed when requested by the Executive Committee, or by the relevant Departmental Committee, voting on a cross-community basis.

13. When there is a petition of concern as in 5(d) above, the Assembly shall vote to determine whether the measure may proceed without reference to this special procedure. If this fails to achieve support on a cross-community basis, as in 5(d)(i) above, the special procedure shall be followed.

Executive authority

14. Executive authority to be discharged on behalf of the Assembly by a First Minister and Deputy First Minister and up to ten Ministers with Departmental responsibilities.

15. The First Minister and Deputy First Minister shall be jointly elected into office by the Assembly voting on a cross-community basis, according to 5(d)(i) above.

16. Following the election of the First Minister and Deputy First Minister, the posts of Ministers will be allocated to parties on the basis of the d'Hondt system by reference to the number of seats each party has in the Assembly.

17. The Ministers will constitute an Executive Committee, which will be convened, and presided over, by the First Minister and Deputy First Minister.

18. The duties of the First Minister and Deputy First Minister will include, *inter alia*, dealing with and coordinating the work of the Executive Committee and the response of the Northern Ireland administration to external relationships.

19. The Executive Committee will provide a forum for the discussion of, and agreement on, issues which cut across the responsibilities of two or more Ministers, for prioritising executive and legislative proposals and for recommending a common position where necessary (e.g. in dealing with external relationships).

20. The Executive Committee will seek to agree each year, and review as necessary, a programme incorporating an agreed budget linked to policies and programmes, subject to approval by the Assembly, after scrutiny in Assembly Committees, on a cross-community basis.

21. A party may decline the opportunity to nominate a person to serve as a Minister or may subsequently change its nominee.

22. All the Northern Ireland Departments will be headed by a Minister. All Ministers will liaise regularly with their respective Committee.

23. As a condition of appointment, Ministers, including the First Minister and Deputy First Minister, will affirm the terms of a Pledge of Office (Annex A) undertaking to discharge effectively and in good faith all the responsibilities attaching to their office.

24. Ministers will have full executive authority in their respective areas of responsibility, within any broad programme agreed by the Executive Committee and endorsed by the Assembly as a whole.

25. An individual may be removed from office following a decision of the Assembly taken on a cross-community basis, if (s)he loses the confidence of the Assembly, voting on a cross-community basis, for failure to meet his or her responsibilities including, inter alia, those set out in the Pledge of Office. Those who hold office should use only democratic, non-violent means, and those who do not should be excluded or removed from office under these provisions.

Legislation

26 The Assembly will have authority to pass primary legislation for Northern Ireland in devolved areas, subject to:

(a) the ECHR and any Bill of Rights for Northern Ireland supplementing it which, if the courts found to be breached, would render the relevant legislation null and void;
(b) decisions by simple majority of members voting, except when decision on a cross-community basis is required;
(c) detailed scrutiny and approval in the relevant Departmental Committee;
(d) mechanisms, based on arrangements proposed for the Scottish Parliament, to ensure suitable co-ordination, and avoid disputes, between the Assembly and the Westminster Parliament;
(e) option of the Assembly seeking to include Northern Ireland provisions in United Kingdom-wide legislation in the Westminster Parliament, especially on devolved issues where parity is normally maintained (e.g. social security, company law).

27. The Assembly will have authority to legislate in reserved areas with the approval of the Secretary of State and subject to Parliamentary control.

28. Disputes over legislative competence will be decided by the Courts.

29. Legislation could be initiated by an individual, a Committee or a Minister.

Relations with other institutions

30. Arrangements to represent the Assembly as a whole, at Summit level and in dealings with other institutions, will be in accordance with paragraph 18, and will be such as to ensure cross-community involvement.

31. Terms will be agreed between appropriate Assembly representatives and the Government of the United Kingdom to ensure effective co-ordination and input by Ministers to national policy-making, including on EU issues.

32. Role of Secretary of State:

(a) to remain responsible for NIO matters not devolved to the Assembly, subject to regular consultation with the Assembly and Ministers;
(b) to approve and lay before the Westminster Parliament any Assembly legislation on reserved matters;
(c) to represent Northern Ireland interests in the United Kingdom Cabinet;
(d) to have the right to attend the Assembly at their invitation.

33. The Westminster Parliament (whose power to make legislation for Northern Ireland would remain unaffected) will:

(a) legislate for non-devolved issues, other than where the Assembly legislates with the approval of the Secretary of State and subject to the control of Parliament;
(b) to legislate as necessary to ensure the United Kingdom's international obligations are met in respect of Northern Ireland;
(c) scrutinise, including through the Northern Ireland Grand and Select Committees, the responsibilities of the Secretary of State.

34. A consultative Civic Forum will be established. It will comprise representatives of the business, trade union and voluntary sectors, and such other sectors as agreed by the First Minister and the Deputy First Minister. It will act as a consultative mechanism on social, economic and cultural issues. The First Minister and the Deputy First Minister will by agreement provide administrative support for the Civic Forum and establish guidelines for the selection of representatives to the Civic Forum.

Transitional arrangements

35. The Assembly will meet first for the purpose of organisation, without legislative or executive powers, to resolve its standing orders and working practices and make preparations for the effective functioning of the Assembly, the British-Irish Council and the North/South Ministerial Council and associated implementation bodies. In this transitional period, those members of the Assembly serving as shadow Ministers shall affirm their commitment to non-violence and exclusively peaceful and democratic means and their opposition to any use or threat of force by others for any political purpose; to work in good faith to bring the new arrangements into being; and to observe the spirit of the Pledge of Office applying to appointed Ministers.

Review

36. After a specified period there will be a review of these arrangements, including the details of electoral arrangements and of the Assembly's procedures, with a view to agreeing any adjustments necessary in the interests of efficiency and fairness.

ANNEX A

Pledge of office

To pledge:

(a) to discharge in good faith all the duties of office;
(b) commitment to non-violence and exclusively peaceful and democratic means;
(c) to serve all the people of Northern Ireland equally, and to act in accordance with the general obligations on government to promote equality and prevent discrimination;
(d) to participate with colleagues in the preparation of a programme for government;
(e) to operate within the framework of that programme when agreed within the Executive Committee and endorsed by the Assembly;

(f) to support, and to act in accordance with, all decisions of the Executive Committee and Assembly;
(g) to comply with the Ministerial Code of Conduct.

Code of Conduct

Ministers must at all times:

- observe the highest standards of propriety and regularity involving impartiality, integrity and objectivity in relationship to the stewardship of public funds;
- be accountable to users of services, the community and, through the Assembly, for the activities within their responsibilities, their stewardship of public funds and the extent to which key performance targets and objectives have been met;
- ensure all reasonable requests for information from the Assembly, users of services and individual citizens are complied with; and that Departments and their staff conduct their dealings with the public in an open and responsible way;
- follow the seven principles of public life set out by the Committee on Standards in Public Life;
- comply with this code and with rules relating to the use of public funds;
- operate in a way conducive to promoting good community relations and equality of treatment;
- not use information gained in the course of their service for personal gain; nor seek to use the opportunity of public service to promote their private interests;
- ensure they comply with any rules on the acceptance of gifts and hospitality that might be offered;
- declare any personal or business interests which may conflict with their responsibilities. The Assembly will retain a Register of Interests. Individuals must ensure that any direct or indirect pecuniary interests which members of the public might reasonably think could influence their judgement are listed in the Register of Interests.

STRAND TWO

North/South Ministerial Council

1. Under a new British/Irish Agreement dealing with the totality of relationships, and related legislation at Westminster and in the Oireachtas, a North/South Ministerial Council to be established to bring together those with executive responsibilities in Northern Ireland and the Irish Government, to develop consultation, co-operation and action within the island of Ireland—including through implementation on an all-island and cross-border basis—on matters of mutual interest within the competence of the Administrations, North and South.

2. All Council decisions to be by agreement between the two sides. Northern Ireland to be represented by the First Minister, Deputy First Minister and any relevant Ministers, the Irish Government by the Taoiseach and relevant Ministers, all operating in accordance with the rules for democratic authority and accountability in force in the Northern Ireland Assembly and the Oireachtas respectively. Participation in the Council to be one of the essential responsibilities attaching to relevant posts in the two Administrations. If a holder of a relevant post will not participate normally in the Council, the Taoiseach in the case of the Irish Government and the First and Deputy First Minister in the case of the Northern Ireland Administration to be able to make alternative arrangements.

3. The Council to meet in different formats:

(i) in plenary format twice a year, with Northern Ireland representation led by the First Minister and Deputy First Minister and the Irish Government led by the Taoiseach;
(ii) in specific sectoral formats on a regular and frequent basis with each side represented by the appropriate Minister;
(iii) in an appropriate format to consider institutional or cross-sectoral matters (including in relation to the EU) and to resolve disagreement.

4. Agendas for all meetings to be settled by prior agreement between the two sides, but it will be open to either to propose any matter for consideration or action.

5. The Council:

(i) to exchange information, discuss and consult with a view to co-operating on matters of mutual interest within the competence of both Administrations, North and South;

(ii) to use best endeavours to reach agreement on the adoption of common policies, in areas where there is a mutual cross-border and all-island benefit, and which are within the competence of both Administrations, North and South, making determined efforts to overcome any disagreements;

(iii) to take decisions by agreement on policies for implementation separately in each jurisdiction, in relevant meaningful areas within the competence of both Administrations, North and South;

(iv) to take decisions by agreement on policies and action at an all-island and cross-border level to be implemented by the bodies to be established as set out in paragraphs 8 and 9 below.

6. Each side to be in a position to take decisions in the Council within the defined authority of those attending, through the arrangements in place for co-ordination of executive functions within each jurisdiction. Each side to remain accountable to the Assembly and Oireachtas respectively, whose approval, through the arrangements in place on either side, would be required for decisions beyond the defined authority of those attending.

7. As soon as practically possible after elections to the Northern Ireland Assembly, inaugural meetings will take place of the Assembly, the British/Irish Council and the North/South Ministerial Council in their transitional forms. All three institutions will meet regularly and frequently on this basis during the period between the elections to the Assembly, and the transfer of powers to the Assembly, in order to establish their modus operandi.

8. During the transitional period between the elections to the Northern Ireland Assembly and the transfer of power to it, representatives of the Northern Ireland transitional Administration and the Irish Government operating in the North/South Ministerial Council will undertake a work programme, in consultation with the British Government, covering at least 12 subject areas, with a view to identifying and agreeing by 31 October 1998 areas where co-operation and implementation for mutual benefit will take place. Such areas may include matters in the list set out in the Annex.

9. As part of the work programme, the Council will identify and

agree at least 6 matters for co-operation and implementation in each of the following categories:

(i) matters where existing bodies will be the appropriate mechanisms for co-operation in each separate jurisdiction;
(ii) matters where the co-operation will take place through agreed implementation bodies on a cross-border or all-island level.

10. The two Governments will make necessary legislative and other enabling preparations to ensure, as an absolute commitment, that these bodies, which have been agreed as a result of the work programme, function at the time of the inception of the British-Irish Agreement and the transfer of powers, with legislative authority for these bodies transferred to the Assembly as soon as possible thereafter. Other arrangements for the agreed co-operation will also commence contemporaneously with the transfer of powers to the Assembly.

11. The implementation bodies will have a clear operational remit. They will implement on an all-island and cross-border basis policies agreed in the Council.

12. Any further development of these arrangements to be by agreement in the Council and with the specific endorsement of the Northern Ireland Assembly and Oireachtas, subject to the extent of the competences and responsibility of the two Administrations.

13. It is understood that the North/South Ministerial Council and the Northern Ireland Assembly are mutually inter-dependent, and that one cannot successfully function without the other.

14. Disagreements within the Council to be addressed in the format described at paragraph 3(iii) above or in the plenary format. By agreement between the two sides, experts could be appointed to consider a particular matter and report.

15. Funding to be provided by the two Administrations on the basis that the Council and the implementation bodies constitute a necessary public function.

16. The Council to be supported by a standing joint Secretariat, staffed by members of the Northern Ireland Civil Service and the Irish Civil Service.

17. The Council to consider the European Union dimension of relevant matters, including the implementation of EU policies and programmes and proposals under consideration in the EU framework. Arrangements to be made to ensure that the views of the Council are

taken into account and represented appropriately at relevant EU meetings.

18. The Northern Ireland Assembly and the Oireachtas to consider developing a joint parliamentary forum, bringing together equal numbers from both institutions for discussion of matters of mutual interest and concern.

19. Consideration to be given to the establishment of an independent consultative forum appointed by the two Administrations, representative of civil society, comprising the social partners and other members with expertise in social, cultural, economic and other issues.

ANNEX

Areas for North-South co-operation and implementation may include the following:

1. Agriculture—animal and plant health.
2. Education—teacher qualifications and exchanges.
3. Transport—strategic transport planning.
4. Environment—environmental protection, pollution, water quality, and waste management.
5. Waterways—inland waterways.
6. Social Security/Social Welfare—entitlements of cross-border workers and fraud control.
7. Tourism—promotion, marketing, research, and product development.
8. Relevant EU Programmes such as SPPR, INTERREG, Leader II and their successors.
9. Inland Fisheries.
10. Aquaculture and marine matters
11. Health: accident and emergency services and other related cross-border issues.
12. Urban and rural development.

Others to be considered by the shadow North/ South Council.

STRAND THREE

British–Irish Council

1. A British–Irish Council (BIC) will be established under a new British–Irish Agreement to promote the harmonious and mutually beneficial development of the totality of relationships among the peoples of these islands.

2. Membership of the BIC will comprise representatives of the British and Irish Governments, devolved institutions in Northern Ireland, Scotland and Wales, when established, and, if appropriate, elsewhere in the United Kingdom, together with representatives of the Isle of Man and the Channel Islands.

3. The BIC will meet in different formats: at summit level, twice per year; in specific sectoral formats on a regular basis, with each side represented by the appropriate Minister; in an appropriate format to consider cross-sectoral matters.

4. Representatives of members will operate in accordance with whatever procedures for democratic authority and accountability are in force in their respective elected institutions.

5. The BIC will exchange information, discuss, consult and use best endeavours to reach agreement on co-operation on matters of mutual interest within the competence of the relevant Administrations. Suitable issues for early discussion in the BIC could include transport links, agricultural issues, environmental issues, cultural issues, health issues, education issues and approaches to EU issues. Suitable arrangements to be made for practical co-operation on agreed policies.

6. It will be open to the BIC to agree common policies or common actions. Individual members may opt not to participate in such common policies and common action.

7. The BIC normally will operate by consensus. In relation to decisions on common policies or common actions, including their means of implementation, it will operate by agreement of all members participating in such policies or actions.

8. The members of the BIC, on a basis to be agreed between them, will provide such financial support as it may require.

9. A secretariat for the BIC will be provided by the British and Irish Governments in co-ordination with officials of each of the other members.

10. In addition to the structures provided for under this agreement, it will be open to two or more members to develop bilateral or multilateral arrangements between them. Such arrangements could include, subject to the agreement of the members concerned, mechanisms to enable consultation, co-operation and joint decision-making on matters of mutual interest; and mechanisms to implement any joint decisions they may reach. These arrangements will not require the prior approval of the BIC as a whole and will operate independently of it.

11. The elected institutions of the members will be encouraged to develop interparliamentary links, perhaps building on the British–Irish Interparliamentary Body.

12. The full membership of the BIC will keep under review the workings of the Council, including a formal published review at an appropriate time after the Agreement comes into effect, and will contribute as appropriate to any review of the overall political agreement arising from the multi-party negotiations.

British-Irish Intergovernmental Conference

1. There will be a new British–Irish Agreement dealing with the totality of relationships. It will establish a standing British–Irish Intergovernmental Conference, which will subsume both the Anglo-Irish Intergovernmental Council and the Intergovernmental Conference established under the 1985 Agreement.

2. The Conference will bring together the British and Irish Governments to promote bilateral co-operation at all levels on all matters of mutual interest within the competence of both Governments.

3. The Conference will meet as required at Summit level (Prime Minister and Taoiseach). Otherwise, Governments will be represented by appropriate Ministers. Advisers, including police and security advisers, will attend as appropriate.

4. All decisions will be by agreement between both Governments. The Governments will make determined efforts to resolve disagreements between them. There will be no derogation from the sovereignty of either Government.

5. In recognition of the Irish Government's special interest in Northern Ireland and of the extent to which issues of mutual concern arise in relation to Northern Ireland, there will be regular and frequent meetings of the Conference concerned with non-devolved Northern

Ireland matters, on which the Irish Government may put forward views and proposals. These meetings, to be co-chaired by the Minister for Foreign Affairs and the Secretary of State for Northern Ireland, would also deal with all-island and cross-border co-operation on non-devolved issues.

6. Co-operation within the framework of the Conference will include facilitation of co-operation in security matters. The Conference also will address, in particular, the areas of rights, justice, prisons and policing in Northern Ireland (unless and until responsibility is devolved to a Northern Ireland administration) and will intensify co-operation between the two Governments on the all-island or cross-border aspects of these matters.

7. Relevant executive members of the Northern Ireland Administration will be involved in meetings of the Conference, and in the reviews referred to in paragraph 9 below to discuss non-devolved Northern Ireland matters.

8. The Conference will be supported by officials of the British and Irish Governments, including by a standing joint Secretariat of officials dealing with non-devolved Northern Ireland matters.

9. The Conference will keep under review the workings of the new British-Irish Agreement and the machinery and institutions established under it, including a formal published review three years after the Agreement comes into effect. Representatives of the Northern Ireland Administration will be invited to express views to the Conference in this context. The Conference will contribute as appropriate to any review of the overall political agreement arising from the multi-party negotiations but will have no power to override the democratic arrangements set up by this Agreement.

Rights, Safeguards and Equality of Opportunity

Human rights

1. The parties affirm their commitment to the mutual respect, the civil rights and the religious liberties of everyone in the community. Against the background of the recent history of communal conflict, the parties affirm in particular:

- the right of free political thought;
- the right to freedom and expression of religion;

- the right to pursue democratically national and political aspirations;
- the right to seek constitutional change by peaceful and legitimate means;
- the right to freely choose one's place of residence;
- the right to equal opportunity in all social and economic activity, regardless of class, creed, disability, gender or ethnicity;
- the right to freedom from sectarian harassment; and
- the right of women to full and equal political participation.

United Kingdom legislation

2. The British Government will complete incorporation into Northern Ireland law of the European Convention on Human Rights (ECHR), with direct access to the courts, and remedies for breach of the Convention, including power for the courts to overrule Assembly legislation on grounds of inconsistency.

3. Subject to the outcome of public consultation underway, the British Government intends, as a particular priority, to create a statutory obligation on public authorities in Northern Ireland to carry out all their functions with due regard to the need to promote equality of opportunity in relation to religion and political opinion; gender; race; disability; age; marital status; dependants; and sexual orientation. Public bodies would be required to draw up statutory schemes showing how they would implement this obligation. Such schemes would cover arrangements for policy appraisal, including an assessment of impact on relevant categories, public consultation, public access to information and services, monitoring and timetables.

4. The new Northern Ireland Human Rights Commission (see paragraph 5 below) will be invited to consult and to advise on the scope for defining, in Westminster legislation, rights supplementary to those in the European Convention on Human Rights, to reflect the particular circumstances of Northern Ireland, drawing as appropriate on international instruments and experience. These additional rights to reflect the principles of mutual respect for the identity and ethos of both communities and parity of esteem, and—taken together with the ECHR—to constitute a Bill of Rights for Northern Ireland. Among the issues for consideration by the Commission will be:

- the formulation of a general obligation on government and public bodies fully to respect, on the basis of equality of treatment,

the identity and ethos of both communities in Northern Ireland; and
- a clear formulation of the rights not to be discriminated against and to equality of opportunity in both the public and private sectors.

New institutions in Northern Ireland

5. A new Northern Ireland Human Rights Commission, with membership from Northern Ireland reflecting the community balance, will be established by Westminster legislation, independent of Government, with an extended and enhanced role beyond that currently exercised by the Standing Advisory Commission on Human Rights, to include keeping under review the adequacy and effectiveness of laws and practices, making recommendations to Government as necessary; providing information and promoting awareness of human rights; considering draft legislation referred to them by the new Assembly; and, in appropriate cases, bringing court proceedings or providing assistance to individuals doing so.

6. Subject to the outcome of public consultation currently underway, the British Government intends a new statutory Equality Commission to replace the Fair Employment Commission, the Equal Opportunities Commission (NI), the Commission for Racial Equality (NI) and the Disability Council. Such a unified Commission will advise on, validate and monitor the statutory obligation and will investigate complaints of default.

7. It would be open to a new Northern Ireland Assembly to consider bringing together its responsibilities for these matters into a dedicated Department of Equality.

8. These improvements will build on existing protections in Westminster legislation in respect of the judiciary, the system of justice and policing.

Comparable steps by the Irish Government

9. The Irish Government will also take steps to further strengthen the protection of human rights in its jurisdiction. The Government will, taking account of the work of the All-Party Oireachtas Committee on the Constitution and the Report of the Constitution Review Group, bring forward measures to strengthen and underpin the constitutional

protection of human rights. These proposals will draw on the European Convention on Human Rights and other international legal instruments in the field of human rights and the question of the incorporation of the ECHR will be further examined in this context. The measures brought forward would ensure at least an equivalent level of protection of human rights as will pertain in Northern Ireland. In addition, the Irish Government will:

- establish a Human Rights Commission with a mandate and remit equivalent to that within Northern Ireland;
- proceed with arrangements as quickly as possible to ratify the Council of Europe Framework Convention on National Minorities (already ratified by the UK);
- implement enhanced employment equality legislation;
- introduce equal status legislation; and
- continue to take further active steps to demonstrate its respect for the different traditions in the island of Ireland.

A Joint Committee

10. It is envisaged that there would be a joint committee of representatives of the two Human Rights Commissions, North and South, as a forum for consideration of human rights issues in the island of Ireland. The joint committee will consider, among other matters, the possibility of establishing a charter, open to signature by all democratic political parties, reflecting and endorsing agreed measures for the protection of the fundamental rights of everyone living in the island of Ireland.

Reconciliation and victims of violence

11. The participants believe that it is essential to acknowledge and address the suffering of the victims of violence as a necessary element of reconciliation. They look forward to the results of the work of the Northern Ireland Victims Commission.

12. It is recognised that victims have a right to remember as well as to contribute to a changed society. The achievement of a peaceful and just society would be the true memorial to the victims of violence. The participants particularly recognise that young people from areas affected by the troubles face particular difficulties and will support the

development of special community-based initiatives based on international best practice. The provision of services that are supportive and sensitive to the needs of victims will also be a critical element and that support will need to be channelled through both statutory and community-based voluntary organisations facilitating locally-based self-help and support networks. This will require the allocation of sufficient resources, including statutory funding as necessary, to meet the needs of victims and to provide for community-based support programmes.

13. The participants recognise and value the work being done by many organisations to develop reconciliation and mutual understanding and respect between and within communities and traditions, in Northern Ireland and between North and South, and they see such work as having a vital role in consolidating peace and political agreement. Accordingly, they pledge their continuing support to such organisations and will positively examine the case for enhanced financial assistance for the work of reconciliation. An essential aspect of the reconciliation process is the promotion of a culture of tolerance at every level of society, including initiatives to facilitate and encourage integrated education and mixed housing.

RIGHTS, SAFEGUARDS AND EQUALITY OF OPPORTUNITY

Economic, Social and Cultural Issues

1. Pending the devolution of powers to a new Northern Ireland Assembly, the British Government will pursue broad policies for sustained economic growth and stability in Northern Ireland and for promoting social inclusion, including in particular community development and the advancement of women in public life.

2. Subject to the public consultation currently under way, the British Government will make rapid progress with:

(i) a new regional development strategy for Northern Ireland, for consideration in due course by the Assembly, tackling the problems of a divided society and social cohesion in urban, rural and border areas, protecting and enhancing the environment, producing new approaches to transport issues, strengthening the physical infrastructure of the region, developing the advantages and resources of rural areas and rejuvenating major urban centres;

(ii) a new economic development strategy for Northern Ireland, for consideration in due course by the Assembly, which would provide for short and medium term economic planning linked as appropriate to the regional development strategy; and

(iii) measures on employment equality included in the recent White Paper ("Partnership for Equality") and covering the extension and strengthening of anti-discrimination legislation, a review of the national security aspects of the present fair employment legislation at the earliest possible time, a new more focused Targeting Social Need initiative and a range of measures aimed at combating unemployment and progressively eliminating the differential in unemployment rates between the two communities by targeting objective need.

3. All participants recognise the importance of respect, understanding and tolerance in relation to linguistic diversity, including in Northern Ireland, the Irish language, Ulster-Scots and the languages of the various ethnic communities, all of which are part of the cultural wealth of the island of Ireland.

4. In the context of active consideration currently being given to the UK signing the Council of Europe Charter for Regional or Minority Languages, the British Government will in particular in relation to the Irish language, where appropriate and where people so desire it:

- take resolute action to promote the language;
- facilitate and encourage the use of the language in speech and writing in public and private life where there is appropriate demand;
- seek to remove, where possible, restrictions which would discourage or work against the maintenance or development of the language;
- make provision for liaising with the Irish language community, representing their views to public authorities and investigating complaints;
- place a statutory duty on the Department of Education to encourage and facilitate Irish medium education in line with current provision for integrated education;
- explore urgently with the relevant British authorities, and in co-operation with the Irish broadcasting authorities, the scope for achieving more widespread availability of Teilifis na Gaeilige in Northern Ireland;

- seek more effective ways to encourage and provide financial support for Irish language film and television production in Northern Ireland; and
- encourage the parties to secure agreement that this commitment will be sustained by a new Assembly in a way which takes account of the desires and sensitivities of the community.

5. All participants acknowledge the sensitivity of the use of symbols and emblems for public purposes, and the need in particular in creating the new institutions to ensure that such symbols and emblems are used in a manner which promotes mutual respect rather than division. Arrangements will be made to monitor this issue and consider what action might be required.

Decommissioning

1. Participants recall their agreement in the Procedural Motion adopted on 24 September 1997 "that the resolution of the decommissioning issue is an indispensable part of the process of negotiation", and also recall the provisions of paragraph 25 of Strand 1 above.

2. They note the progress made by the Independent International Commission on Decommissioning and the Governments in developing schemes which can represent a workable basis for achieving the decommissioning of illegally-held arms in the possession of paramilitary groups.

3. All participants accordingly reaffirm their commitment to the total disarmament of all paramilitary organisations. They also confirm their intention to continue to work constructively and in good faith with the Independent Commission, and to use any influence they may have, to achieve the decommissioning of all paramilitary arms within two years following endorsement in referendums North and South of the agreement and in the context of the implementation of the overall settlement.

4. The Independent Commission will monitor, review and verify progress on decommissioning of illegal arms, and will report to both Governments at regular intervals.

[5.]

6. Both Governments will take all necessary steps to facilitate the decommissioning process to include bringing the relevant schemes into force by the end of June.

Security

1. The participants note that the development of a peaceful environment on the basis of this agreement can and should mean a normalisation of security arrangements and practices.

2. The British Government will make progress towards the objective of as early a return as possible to normal security arrangements in Northern Ireland, consistent with the level of threat and with a published overall strategy, dealing with:

(i) the reduction of the numbers and role of the Armed Forces deployed in Northern Ireland to levels compatible with a normal peaceful society;
(ii) the removal of security installations;
(iii) the removal of emergency powers in Northern Ireland; and
(iv) other measures appropriate to and compatible with a normal peaceful society.

3. The Secretary of State will consult regularly on progress, and the response to any continuing paramilitary activity, with the Irish Government and the political parties, as appropriate.

4. The British Government will continue its consultation on firearms regulation and control on the basis of the document published on 2 April 1998.

5. The Irish Government will initiate a wide-ranging review of the Offences Against the State Acts 1939–85 with a view to both reform and dispensing with those elements no longer required as circumstances permit.

Policing and Justice

1. The participants recognise that policing is a central issue in any society. They equally recognise that Northern Ireland's history of deep divisions has made it highly emotive, with great hurt suffered and sacrifices made by many individuals and their families, including those in the RUC and other public servants. They believe that the agreement provides the opportunity for a new beginning to policing in Northern Ireland with a police service capable of attracting and sustaining support from the community as a whole. They also believe that this agree-

ment offers a unique opportunity to bring about a new political dispensation which will recognise the full and equal legitimacy and worth of the identities, senses of allegiance and ethos of all sections of the community in Northern Ireland. They consider that this opportunity should inform and underpin the development of a police service representative in terms of the make-up of the community as a whole and which, in a peaceful environment, should be routinely unarmed.

2. The participants believe it essential that policing structures and arrangements are such that the police service is professional, effective and efficient, fair and impartial, free from partisan political control; accountable, both under the law for its actions and to the community it serves; representative of the society it polices, and operates within a coherent and co-operative criminal justice system, which conforms with human rights norms. The participants also believe that those structures and arrangements must be capable of maintaining law and order including responding effectively to crime and to any terrorist threat and to public order problems. A police service which cannot do so will fail to win public confidence and acceptance. They believe that any such structures and arrangements should be capable of delivering a policing service, in constructive and inclusive partnerships with the community at all levels, and with the maximum delegation of authority and responsibility, consistent with the foregoing principles. These arrangements should be based on principles of protection of human rights and professional integrity and should be unambiguously accepted and actively supported by the entire community.

3. An independent Commission will be established to make recommendations for future policing arrangements in Northern Ireland including means of encouraging widespread community support for these arrangements within the agreed framework of principles reflected in the paragraphs above and in accordance with the terms of reference at Annex A. The Commission will be broadly representative with expert and international representation among its membership and will be asked to consult widely and to report no later than Summer 1999.

4. The participants believe that the aims of the criminal justice system are to:

- deliver a fair and impartial system of justice to the community;
- be responsive to the community's concerns, and encouraging community involvement where appropriate;

- have the confidence of all parts of the community; and
- deliver justice efficiently and effectively.

5. There will be a parallel wide-ranging review of criminal justice (other than policing and those aspects of the system relating to the emergency legislation) to be carried out by the British Government through a mechanism with an independent element, in consultation with the political parties and others. The review will commence as soon as possible, will include wide consultation, and a report will be made to the Secretary of State no later than Autumn 1999. Terms of Reference are attached at Annex B.

6. Implementation of the recommendations arising from both reviews will be discussed with the political parties and with the Irish Government.

7. The participants also note that the British Government remains ready in principle, with the broad support of the political parties, and after consultation, as appropriate, with the Irish Government, in the context of ongoing implementation of the relevant recommendations, to devolve responsibility for policing and justice issues.

ANNEX A

Commission on Policing for Northern Ireland

Terms of reference

Taking account of the principles on policing as set out in the agreement, the Commission will inquire into policing in Northern Ireland and, on the basis of its findings, bring forward proposals for future policing structures and arrangements, including means of encouraging widespread community support for those arrangements.

Its proposals on policing should be designed to ensure that policing arrangements, including composition, recruitment, training, culture, ethos and symbols, are such that in a new approach Northern Ireland has a police service that can enjoy widespread support from, and is seen as an integral part of, the community as a whole.

Its proposals should include recommendations covering any issues such as re-training, job placement and educational and professional development required in the transition to policing in a peaceful society.

Its proposals should also be designed to ensure that:

- the police service is structured, managed and resourced so that it can be effective in discharging its full range of functions (including proposals on any necessary arrangements for the transition to policing in a normal peaceful society);
- the police service is delivered in constructive and inclusive partnerships with the community at all levels with the maximum delegation of authority and responsibility;
- the legislative and constitutional framework requires the impartial discharge of policing functions and conforms with internationally accepted norms in relation to policing standards;
- the police operate within a clear framework of accountability to the law and the community they serve, so:
  - they are constrained by, accountable to and act only within the law;
  - their powers and procedures, like the law they enforce, are clearly established and publicly available;
  - there are open, accessible and independent means of investigating and adjudicating upon complaints against the police;
  - there are clearly established arrangements enabling local people, and their political representatives, to articulate their views and concerns about policing and to establish publicly policing priorities and influence policing policies, subject to safeguards to ensure police impartiality and freedom from partisan political control;
  - there are arrangements for accountability and for the effective, efficient and economic use of resources in achieving policing objectives;
  - there are means to ensure independent professional scrutiny and inspection of the police service to ensure that proper professional standards are maintained;
- the scope for structured co-operation with the Garda Siochana and other police forces is addressed; and
- the management of public order events which can impose exceptional demands on policing resources is also addressed.

The Commission should focus on policing issues, but if it identifies other aspects of the criminal justice system relevant to its work on policing, including the role of the police in prosecution, then it should draw the attention of the Government to those matters.

The Commission should consult widely, including with non-governmental expert organisations, and through such focus groups as they consider it appropriate to establish.

The Government proposes to establish the Commission as soon as possible, with the aim of it starting work as soon as possible and publishing its final report by Summer 1999.

ANNEX B

Review of the Criminal Justice System

Terms of reference

Taking account of the aims of the criminal justice system as set out in the Agreement, the review will address the structure, management and resourcing of publicly funded elements of the criminal justice system and will bring forward proposals for future criminal justice arrangements (other than policing and those aspects of the system relating to emergency legislation, which the Government is considering separately) covering such issues as:

- the arrangements for making appointments to the judiciary and magistracy, and safeguards for protecting their independence;
- the arrangements for the organisation and supervision of the prosecution process, and for safeguarding its independence;
- measures to improve the responsiveness and accountability of, and any lay participation in the criminal justice system;
- mechanisms for addressing law reform;
- the scope for structured co-operation between the criminal justice agencies on both parts of the island; and
- the structure and organisation of criminal justice functions that might be devolved to an Assembly, including the possibility of establishing a Department of Justice, while safeguarding the essential independence of many of the key functions in this area.

The Government proposes to commence the review as soon as possible, consulting with the political parties and others, including non-governmental expert organisations. The review will be completed by Autumn 1999.

Prisoners

1. Both Governments will put in place mechanisms to provide for an accelerated programme for the release of prisoners, including transferred prisoners, convicted of scheduled offences in Northern Ireland or, in the case of those sentenced outside Northern Ireland, similar offences (referred to hereafter as qualifying prisoners). Any such arrangements will protect the rights of individual prisoners under national and international law.

2. Prisoners affiliated to organisations which have not established or are not maintaining a complete and unequivocal ceasefire will not benefit from the arrangements. The situation in this regard will be kept under review.

3. Both Governments will complete a review process within a fixed time frame and set prospective release dates for all qualifying prisoners. The review process would provide for the advance of the release dates of qualifying prisoners while allowing account to be taken of the seriousness of the offences for which the person was convicted and the need to protect the community. In addition, the intention would be that should the circumstances allow it, any qualifying prisoners who remained in custody two years after the commencement of the scheme would be released at that point.

4. The Governments will seek to enact the appropriate legislation to give effect to these arrangements by the end of June 1998.

5. The Governments continue to recognise the importance of measures to facilitate the reintegration of prisoners into the community by providing support both prior to and after release, including assistance directed towards availing of employment opportunities, re-training and/or re-skilling, and further education.

VALIDATION, IMPLEMENTATION AND REVIEW

Validation and implementation

1. The two Governments will as soon as possible sign a new British–Irish Agreement replacing the 1985 Anglo-Irish Agreement, embodying understandings on constitutional issues and affirming their solemn commitment to support and, where appropriate, implement the

agreement reached by the participants in the negotiations which shall be annexed to the British–Irish Agreement.

2. Each Government will organise a referendum on 22 May 1998. Subject to Parliamentary approval, a consultative referendum in Northern Ireland, organised under the terms of the Northern Ireland (Entry to Negotiations, etc.) Act 1996, will address the question: "Do you support the agreement reached in the multi-party talks on Northern Ireland and set out in Command Paper 3883?". The Irish Government will introduce and support in the Oireachtas a Bill to amend the Constitution as described in paragraph 2 of the section "Constitutional Issues" and in Annex B, as follows: (a) to amend Articles 2 and 3 as described in paragraph 8.1 in Annex B above and (b) to amend Article 29 to permit the Government to ratify the new British–Irish Agreement. On passage by the Oireachtas, the Bill will be put to referendum.

3. If majorities of those voting in each of the referendums support this agreement, the Governments will then introduce and support, in their respective Parliaments, such legislation as may be necessary to give effect to all aspects of this agreement, and will take whatever ancillary steps as may be required including the holding of elections on 25 June, subject to parliamentary approval, to the Assembly, which would meet initially in a "shadow" mode. The establishment of the North–South Ministerial Council, implementation bodies, the British–Irish Council and the British–Irish Intergovernmental Conference and the assumption by the Assembly of its legislative and executive powers will take place at the same time on the entry into force of the British–Irish Agreement.

4. In the interim, aspects of the implementation of the multi-party agreement will be reviewed at meetings of those parties relevant in the particular case (taking into account, once Assembly elections have been held, the results of those elections), under the chairmanship of the British Government or the two Governments, as may be appropriate; and representatives of the two Governments and all relevant parties may meet under independent chairmanship to review implementation of the agreement as a whole.

Review procedures following implementation

5. Each institution may, at any time, review any problems that may arise in its operation and, where no other institution is affected, take

remedial action in consultation as necessary with the relevant Government or Governments. It will be for each institution to determine its own procedures for review.

6. If there are difficulties in the operation of a particular institution, which have implications for another institution, they may review their operations separately and jointly and agree on remedial action to be taken under their respective authorities.

7. If difficulties arise which require remedial action across the range of institutions, or otherwise require amendment of the British–Irish Agreement or relevant legislation, the process of review will fall to the two Governments in consultation with the parties in the Assembly. Each Government will be responsible for action in its own jurisdiction.

8. Notwithstanding the above, each institution will publish an annual report on its operations. In addition, the two Governments and the parties in the Assembly will convene a conference 4 years after the agreement comes into effect, to review and report on its operation.

AGREEMENT BETWEEN THE GOVERNMENT OF THE UNITED KINGDOM OF GREAT BRITAIN AND NORTHERN IRELAND AND THE GOVERNMENT OF IRELAND

The British and Irish Governments:

Welcoming the strong commitment to the Agreement reached on 10th April 1998 by themselves and other participants in the multi-party talks and set out in Annex 1 to this Agreement (hereinafter "the Multi-Party Agreement");

Considering that the Multi-Party Agreement offers an opportunity for a new beginning in relationships within Northern Ireland, within the island of Ireland and between the peoples of these islands;

Wishing to develop still further the unique relationship between their peoples and the close co-operation between their countries as friendly neighbours and as partners in the European Union;

Reaffirming their total commitment to the principles of democracy and non-violence which have been fundamental to the multi-party talks;

Reaffirming their commitment to the principles of partnership, equality and mutual respect and to the protection of civil, political, social, economic and cultural rights in their respective jurisdictions;

Have agreed as follows:

Article 1

The two Governments:

(i) recognise the legitimacy of whatever choice is freely exercised by a majority of the people of Northern Ireland with regard to its status, whether they prefer to continue to support the Union with Great Britain or a sovereign united Ireland;

(ii) recognise that it is for the people of the island of Ireland alone, by agreement between the two parts respectively and without external impediment, to exercise their right of self-determination on the basis of consent, freely and concurrently given, North and South, to bring about a united Ireland, if that is their wish, accepting that this right must be achieved and exercised with and subject to the agreement and consent of a majority of the people of Northern Ireland;

(iii) acknowledge that while a substantial section of the people in Northern Ireland share the legitimate wish of a majority of the people of the island of Ireland for a united Ireland, the present wish of a majority of the people of Northern Ireland, freely exercised and legitimate, is to maintain the Union and accordingly, that Northern Ireland's status as part of the United Kingdom reflects and relies upon that wish; and that it would be wrong to make any change in the status of Northern Ireland save with the consent of a majority of its people;

(iv) affirm that, if in the future, the people of the island of Ireland exercise their right of self-determination on the basis set out in sections (i) and (ii) above to bring about a united Ireland, it will be a binding obligation on both Governments to introduce and support in their respective Parliaments legislation to give effect to that wish;

(v) affirm that whatever choice is freely exercised by a majority of the people of Northern Ireland, the power of the sovereign government with jurisdiction there shall be exercised with rigorous impartiality on behalf of all the people in the diversity of their identities and traditions and shall be founded on the principles of full respect for, and equality of, civil, political, social and cultural rights, of freedom from discrimination for all citizens, and of parity of esteem and of just and equal treatment for the identity, ethos and aspirations of both communities;

(vi) recognise the birthright of all the people of Northern Ireland to identify themselves and be accepted as Irish or British, or both, as they may so choose, and accordingly confirm that their right to hold both British and Irish citizenship is accepted by both Governments and would not be affected by any future change in the status of Northern Ireland.

Article 2

The two Governments affirm their solemn commitment to support, and where appropriate implement, the provisions of the Multi-Party Agreement. In particular there shall be established in accordance with the provisions of the Multi-Party Agreement immediately on the entry into force of this Agreement, the following institutions:

(i) a North/South Ministerial Council;
(ii) the implementation bodies referred to in paragraph 9(ii) of the section entitled "Strand Two" of the Multi-Party Agreement;
(iii) a British–Irish Council;
(iv) a British–Irish Intergovernmental Conference.

Article 3

(1) This Agreement shall replace the Agreement between the British and Irish Governments done at Hillsborough on 15th November 1985 which shall cease to have effect on entry into force of this Agreement.

(2) The Intergovernmental Conference established by Article 2 of the aforementioned Agreement done on 15th November 1985 shall cease to exist on entry into force of this Agreement.

Article 4

(1) It shall be a requirement for entry into force of this Agreement that:

(a) British legislation shall have been enacted for the purpose of implementing the provisions of Annex A to the section entitled "Constitutional Issues" of the Multi-Party Agreement;
(b) the amendments to the Constitution of Ireland set out in Annex B to the section entitled "Constitutional Issues" of the

Multi-Party Agreement shall have been approved by Referendum;

(c) such legislation shall have been enacted as may be required to establish the institutions referred to in Article 2 of this Agreement.

(2) Each Government shall notify the other in writing of the completion, so far as it is concerned, of the requirements for entry into force of this Agreement. This Agreement shall enter into force on the date of the receipt of the later of the two notifications.

(3) Immediately on entry into force of this Agreement, the Irish Government shall ensure that the amendments to the Constitution of Ireland set out in Annex B to the section entitled "Constitutional Issues" of the Multi-Party Agreement take effect.

In witness thereof the undersigned, being duly authorised thereto by the respective Governments, have signed this Agreement.

Done in two originals at Belfast on the 10th day of April 1998.

For the Government of the United Kingdom of Great Britain and NorthernIreland

For the Government of Ireland

ANNEX 1

The Agreement Reached in the Multi-Party Talks

ANNEX 2

Declaration on the Provisions of Paragraph (vi) of Article 1 in Relationship to Citizenship

The British and Irish Governments declare that it is their joint understanding that the term "the people of Northern Ireland" in paragraph (vi) of Article 1 of this Agreement means, for the purposes of giving effect to this provision, all persons born in Northern Ireland and having, at the time of their birth, at least one parent who is a British citizen, an Irish citizen or is otherwise entitled to reside in Northern Ireland without any restriction on their period of residence.

Index